Automated Vehicles Symposium 2018

San Francisco, California, USA
9 - 12 July 2018

Volume 1 of 2

ISBN: 978-1-5108-6840-3

Printed from e-media with permission by:

Curran Associates, Inc.
57 Morehouse Lane
Red Hook, NY 12571

Some format issues inherent in the e-media version may also appear in this print version.

Printed by Curran Associates, Inc. (2018)

For permission requests, please contact Association for Unmanned Vehicle Systems International at the address below.

Association for Unmanned Vehicle Systems International
2700 South Quincy Street
Suite 400
Arlington, VA 22206
USA

Phone: (703) 845-9671
Fax: (703) 845-9679

www.auvsi.org

Additional copies of this publication are available from:

Curran Associates, Inc.
57 Morehouse Lane
Red Hook, NY 12571 USA
Phone: 845-758-0400
Fax: 845-758-2633
Email: curran@proceedings.com
Web: www.proceedings.com

TABLE OF CONTENTS

VOLUME 1

BREAKOUT SESSIONS

VOLUME 2

ENTERPRISE SOLUTION SERIES

PLENARY SESSIONS

AUTOMATED VEHICLES SYMPOSIUM 2018
USERS. VEHICLES. INFRASTRUCTURE.
Driven by AUVSI and TRB

Breakout Session 1

Objectives

- Understand machine vision system interactions with traffic control devices
- Learn about machine vision system failure modes and countermeasures
- **Identify state of readiness assessment initiatives worldwide**
- **Identify paths forward for readiness assessment framework in North America**

Summary of Key Findings and Lessons Learned

- Consistency and logical application may be more important than compliance
- **Sensor fusion is critical, excepting GNSS, which is considered unreliable for the trajectory task**
- Large shortfalls exist in MV capabilities relative to more complex environments
- **Readiness framework development goes beyond Level of Automation or Operating Domain**
- Readiness asesssement will require multi-criteria topology and uniform rating practices

AUTOMATED VEHICLES SYMPOSIUM 2018
USERS. VEHICLES. INFRASTRUCTURE.
Driven by AUVSI and TRB

Breakout Session 1

Research Needs and Action Items

- Connect MV experts with TSMO engineers, traffic design engineers and TCD manufacturers
  - Need to identify shortcomings on both sides and collaborative gap-filling
  - Structuring a national dialog in cooperation with the FHWA, AASHTO, NCUTCD, ITE, TRB, SAE, and others
- Understand desired outcomes from advancements in sensor systems
  - Practices and policy development to facilitate readiness
  - Identify key problems/scenarios and work the problems to resolution
- Convene independent consortium to develop readiness criteria, assessment methodologies, and rating/needs identification databases; key in determining minimum investment required
  - Determine standards/development body for these rating systems and criteria
  - Identify risk of "stamp of approval" in tort liability
  - Key component of interim need to identify handoff need/locations for various domains and regimes

AVS 2018 Proceedings Summary

Breakout Session 1
Reading the Road Ahead: Traffic Control Devices and Machine Vision Interactions

Organizers

Scott O. Kuznicki, P.E., Director of Safety and Mobility Technology Solutions, Transpo Group
Brian Watson, Director of New Programs, American Traffic Safety Services Association

Reporters

Scott O. Kuznicki and Ken Smith

Session Focus

This session is the second in the Reading the Road Ahead series, a continuation of the original 2016 breakout session on infrastructure readiness. Building on the 2017 session success in identifying key issues, the AVS 2018 Reading the Road Ahead workshop was designed to address gaps from previous sessions.

The key objective of the session was to aid attendees in developing an understanding machine vision system interactions with traffic control devices and a working knowledge of machine vision system failure modes and countermeasures. Understanding the current and near-term capabilities of MV systems with regard to sensing, identification, and use of TCD information continued to lay the groundwork for identifying state of readiness assessment initiatives worldwide and identifying paths forward for development and deployment of a road assessment framework in North America.

Session Summary

Opening Remarks, Safety and Comfort Briefing
Averaged 60 attendees for the entire session

Scott Kuznicki presented a summary of the ongoing work to development a road assessment system for self-driving transport.

Robert Seidl and Leo McCloskey hosted an interactive session titled "How do Machines See?" which described the capabilities of various MV systems, including LiDAR, radar, camera vision (visible light), ultrasonic, and GNSS and inertia-based systems.

- **Consistency and logical application may be more important than compliance**
- **Sensor fusion is critical, excepting GNSS, which is considered unreliable for the trajectory task**
- **Large shortfalls exist in MV capabilities relative to more complex environments**

- **Readiness framework development goes beyond Level of Automation or Operating Domain**
- **Readiness asesssement will require multi-criteria topology and uniform rating practices**

<u>Suggested Action Items</u>

Develop a National Dialogue

Connect MV experts with TSMO engineers, traffic design engineers and TCD manufacturers. Need to identify shortcomings on both sides and collaborative gap-filling. Structuring a national dialog in cooperation with the FHWA, AASHTO, NCUTCD, ITE, TRB, SAE, and others

Define Desired Outcomes in Machine Vision/TCD Interactions

Identify key problems/scenarios and work the problems to resolution. Practices and policy development to facilitate readiness

Convene Consortium to Develop Readiness Criteria

Convene independent consortium to develop readiness criteria, assessment methodologies, and rating/needs identification databases; key in determining minimum investment required

Determine standards/development body for these rating systems and criteria

Identify risk of "stamp of approval" in tort liability

Key component of interim need to identify handoff need/locations for various domains and regimes

Rijkswaterstaat
Ministerie van Infrastructuur en Waterstaat

EU Road Readiness, Discussing the ODD framework

AVS2018 Breakout session #1
"Reading the road ahead"
9 July 2018, San Francisco

Tom Alkim
Senior Advisor
Connected & Automated Driving

Regeerakoord (coalition agreement)

Vertrouwen in de toekomst

Regeerakkoord 2017 – 2021

VVD, CDA, D66 en ChristenUnie

10 oktober 2017

3.2 Mobiliteit

Een slim en duurzaam vervoerssysteem waarvan de delen naadloos op elkaar aansluiten. Zo willen we Nederland mobiel en bereikbaar houden. Nu de economie weer goed draait, is een extra investering in infrastructuur nodig en mogelijk om toenemende drukte op de weg, het spoor, het water en in de lucht te verminderen. Tegelijkertijd nemen we maatregelen om de belasting voor het klimaat, de luchtkwaliteit en de leefomgeving te beperken. Innovatie biedt daarbij enorme kansen. De technologische ontwikkeling biedt de mogelijkheid om uiteindelijk tot een meer geïntegreerd vervoerssysteem te komen dat steeds schoner wordt.

Personenvervoer

- **Bij ontwerp, aanleg en onderhoud van infrastructuur houden we rekening met zelfrijdende voertuigen en benodigde systemen in of langs de weg. Overheidsinformatie over verkeer wordt zoveel mogelijk via open data beschikbaar gesteld voor voertuigen, apps en reisplanners. Om ieders privacy te waarborgen leggen we spelregels vast over de eigendom en het gebruik van reisdata.**

When designing, building or maintaining infrastructure, take into account self driving vehicles and the required systems in or along the road.

Intertraffic 2018

1. "I want to ensure our **infrastructure is ready for connected and automated driving**. I want to sit down with other road operators and the automotive and telecom sector. To discuss what infrastructure we'll need in order for the new generation of vehicles to deliver the greatest gains to society."

2. "I want to **enable the next generation of vehicles to actually take the road**. So I'm creating a legal framework for automated driving. Laying down requirements for reliability and safety that cars must meet before they can hit the road. **A driving licence for self-driving cars**, if you like. Not for the driver – but for the car itself!"

3. "I want to **create room for new transport concepts and make sure people can use them easily**. For instance, 'mobility as a service' – transport that's available when you need it, to wherever you want to go, using your choice of transport modes"

https://www.government.nl/documents/speeches/2018/03/26/speech-by-cora-van-nieuwenhuizen-minister-of-infrastructure-and-water-management-at-the-opening-of-intertraffic-amsterdam-20-march-2018

Studies

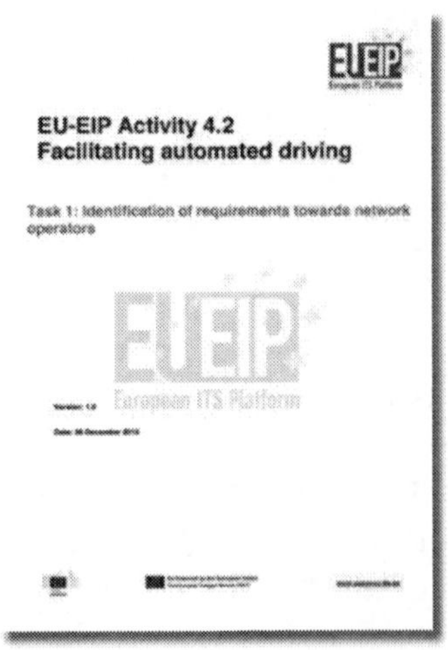

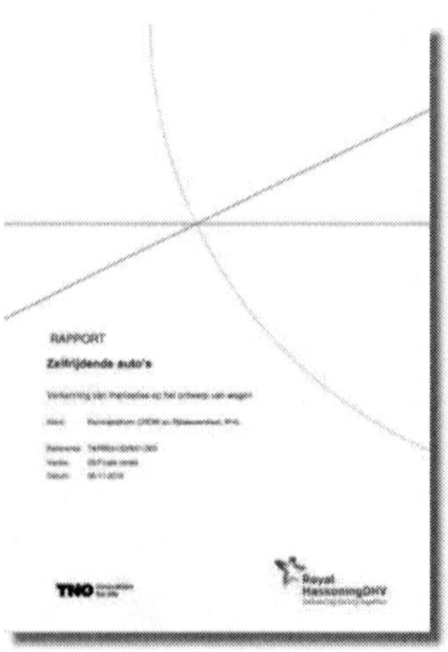

http://knowledgeagenda.connekt.nl/engels/

EU EIP sa4.2

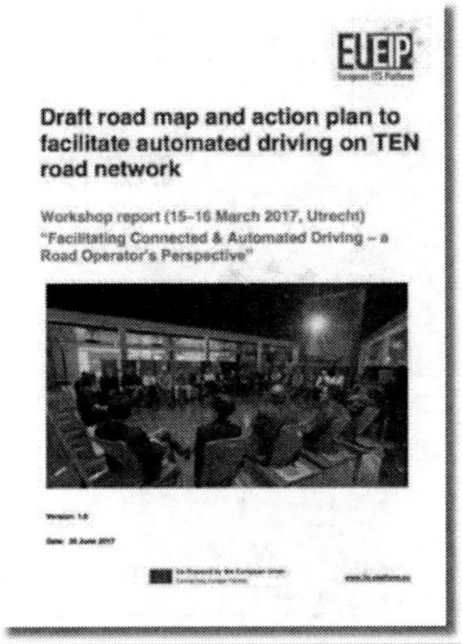

- Still a lot of uncertainty
- Need to have a dialogue with automotive industry
- SAE levels are too general for detailed dialogue about requirements, functional description is needed
- Mixed traffic vs dedicated infrastructure, different requirements
- What about (physical) transition zones?
- Adding elements vs leaving out elements

Deliverable is published in June 2017:
https://www.its-platform.eu/filedepot_download/1950/6112

Past the hype?

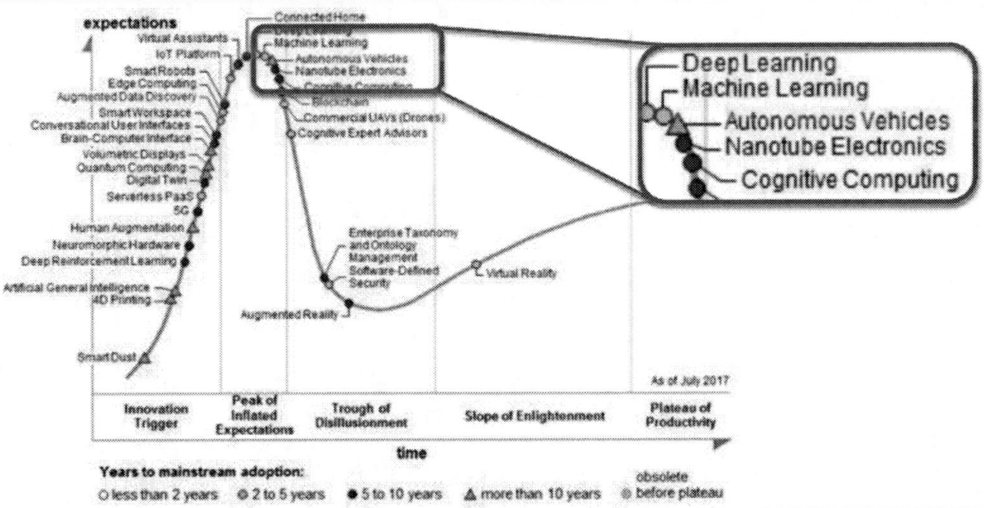

ODD framework

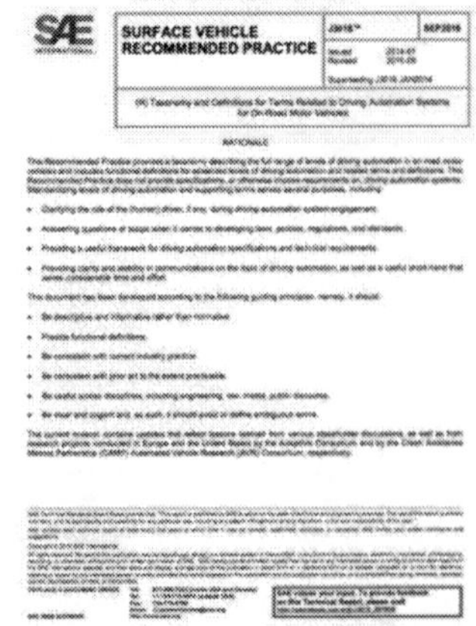
SURFACE VEHICLE
RECOMMENDED PRACTICE

Level	Name	Narrative definition	DDT: Sustained lateral and longitudinal vehicle motion control	DDT: OEDR	DDT fallback	ODD
Driver performs part or all of the DDT						
0	No Driving Automation	The performance by the driver of the entire *DDT*, even when enhanced by active safety systems.	*Driver*	*Driver*	*Driver*	n/a
1	Driver Assistance	The sustained and ODD-specific execution by a driving automation system of either the lateral or the longitudinal vehicle motion control subtask of the DDT (but not both simultaneously) with the expectation that the driver performs the remainder of the DDT.	*Driver and System*	*Driver*	*Driver*	Limited
2	Partial Driving Automation	The sustained and ODD-specific execution by a driving automation system of both the lateral and longitudinal vehicle motion control subtasks of the DDT with the expectation that the driver completes the OEDR subtask and supervises the driving automation system.	**System**	*Driver*	*Driver*	Limited
ADS ("System") performs the entire DDT (while engaged)						
3	Conditional Driving Automation	The sustained and ODD-specific performance by an ADS of the entire DDT with the expectation that the DDT fallback-ready user is receptive to ADS-issued requests to intervene, as well as to DDT performance-relevant system failures in other vehicle systems, and will respond appropriately.	*System*	**System**	*Fallback-ready user (becomes the driver during fallback)*	Limited
4	High Driving Automation	The sustained and ODD-specific performance by an ADS of the entire DDT and DDT fallback without any expectation that a user will respond to a request to intervene.	*System*	*System*	**System**	Limited
5	Full Driving Automation	The sustained and unconditional (i.e., not ODD-specific) performance by an ADS of the entire DDT and DDT fallback without any expectation that a user will respond to a request to intervene.	*System*	*System*	*System*	**Unlimited**

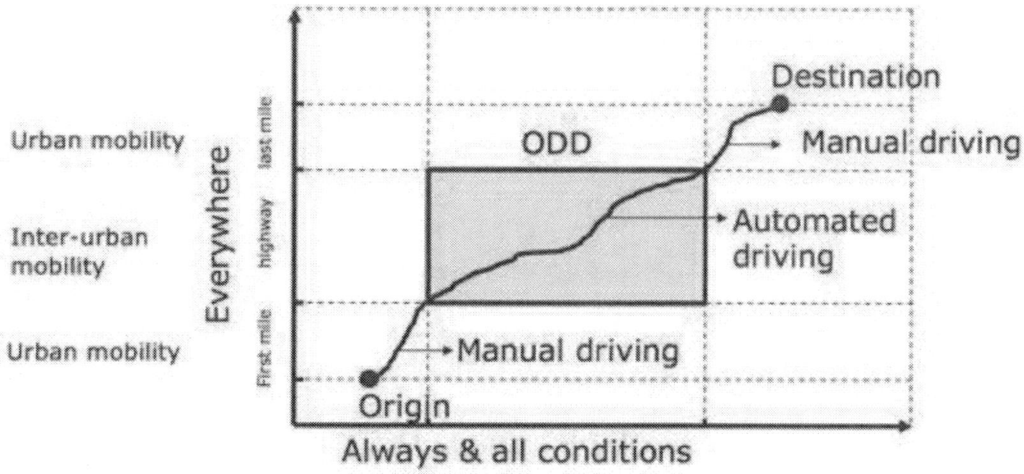
Destination
Urban mobility
ODD
Manual driving
last mile
Everywhere
highway
Inter-urban
mobility
Automated
driving
First mile
Urban mobility
Manual driving
Origin
Always & all conditions

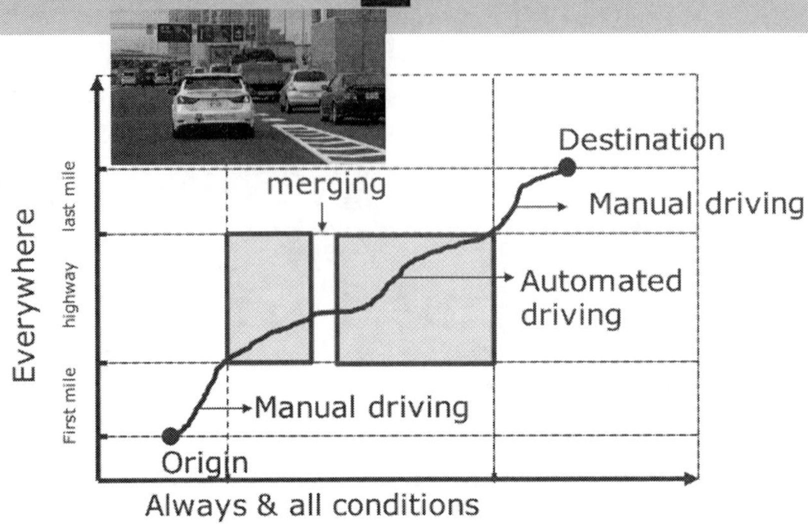

Destination
merging
Manual driving
Automated driving
Everywhere
last mile
highway
First mile
Manual driving
Origin
Always & all conditions

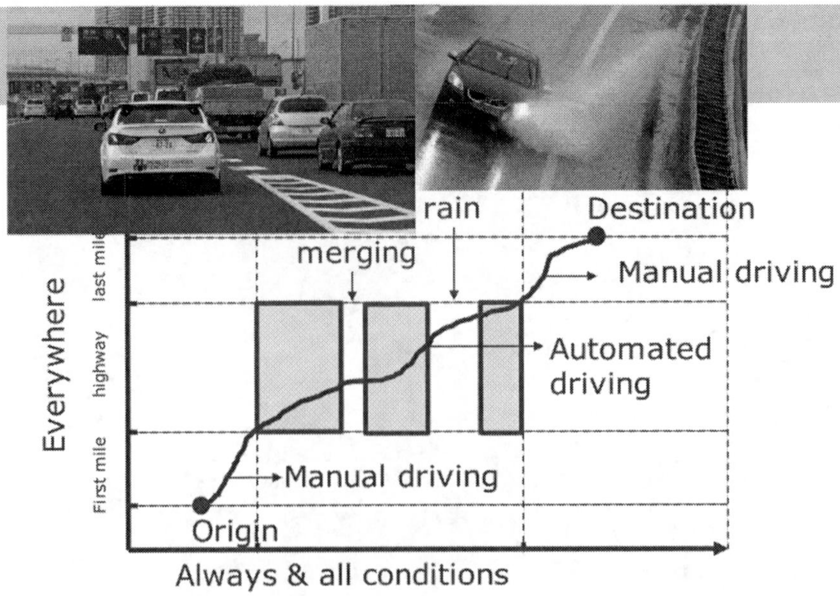

rain
Destination
merging
Manual driving
Automated driving
Everywhere
last mile
highway
First mile
Manual driving
Origin
Always & all conditions

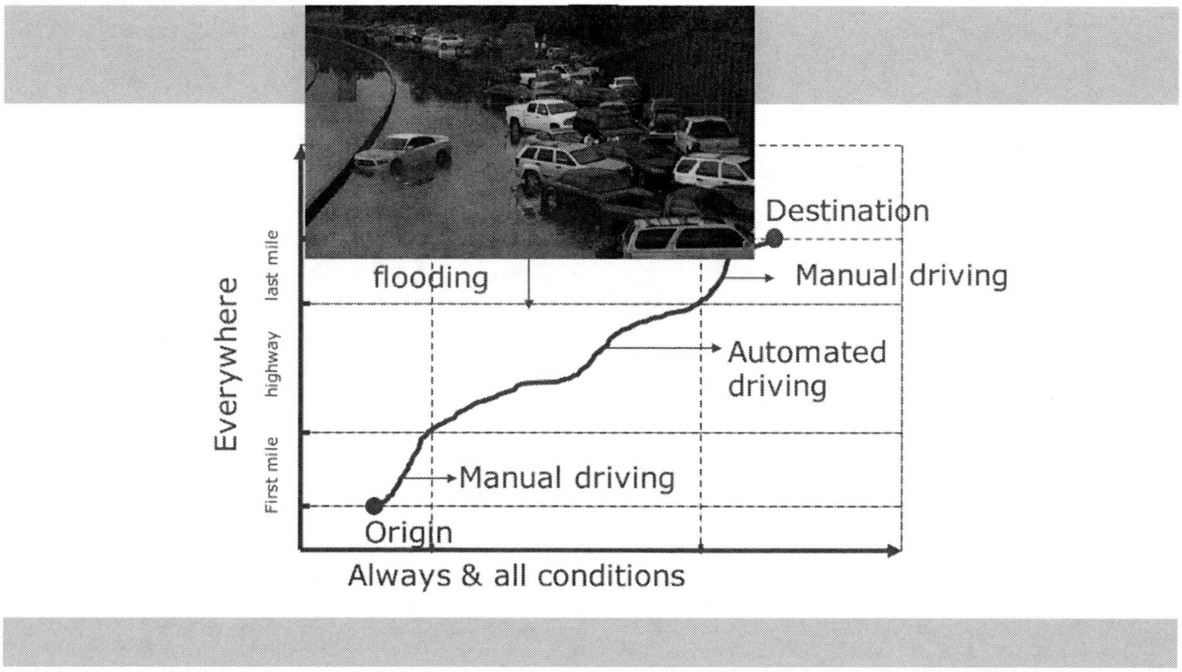

Destination
flooding
Manual driving
Automated driving
Manual driving
Origin
Everywhere
First mile
highway
last mile
Always & all conditions

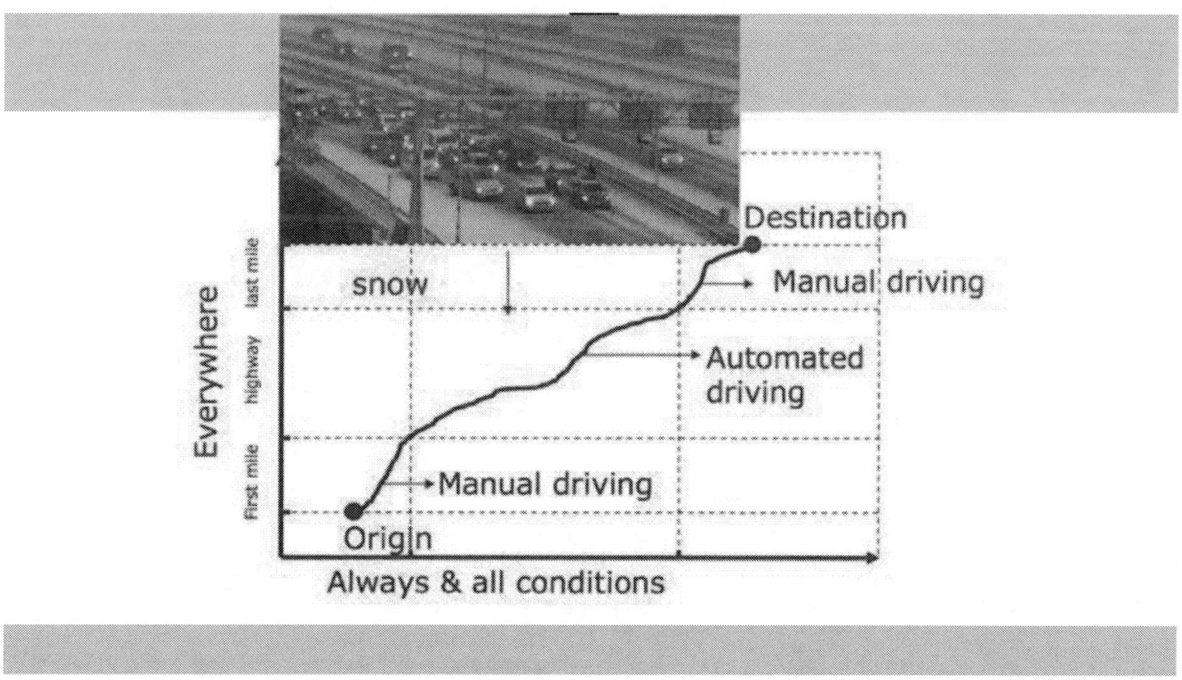

Destination
snow
Manual driving
Automated driving
Manual driving
Origin
Everywhere
First mile
highway
last mile
Always & all conditions

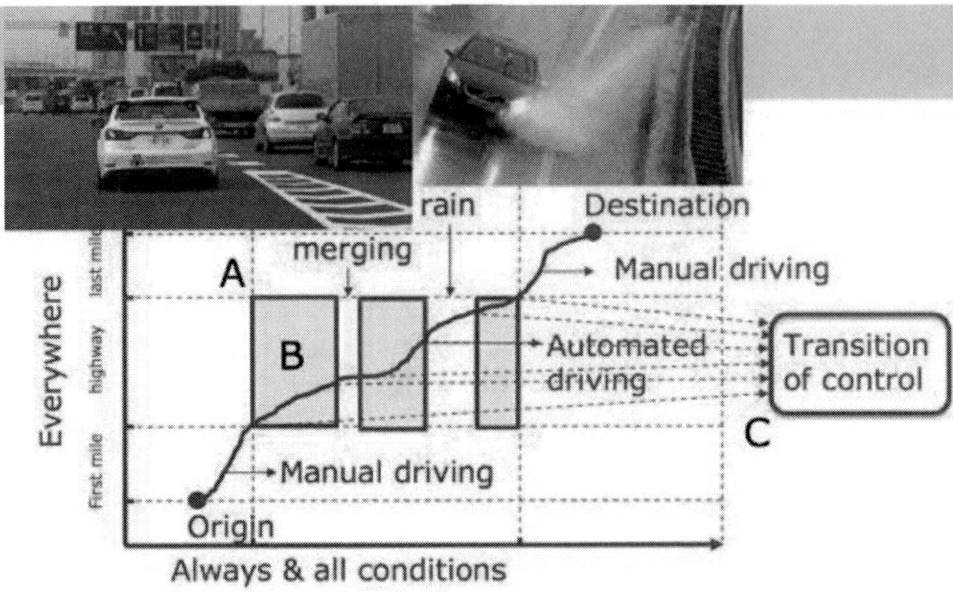

Important research areas

A - How to optimize the ODD
- Various stakeholders involved, what roles and responsibilities?
- How to organise a constructive dialogue?

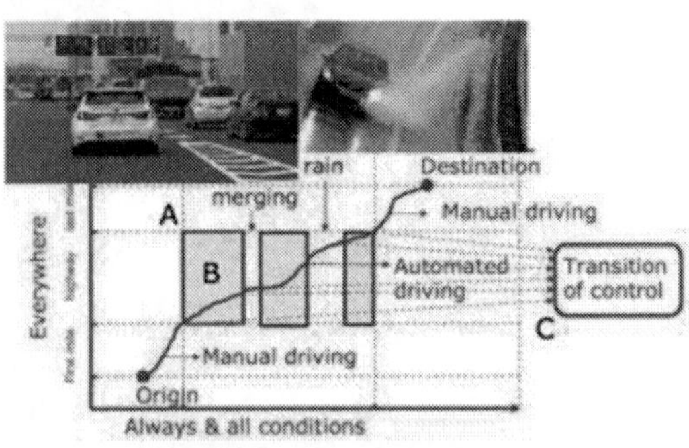

1. "I want to ensure our **infrastructure is ready for connected and automated driving**. I want to sit down with other road operators and the automotive and telecom sector. To discuss what infrastructure we'll need in order for the new generation of vehicles to deliver the greatest gains to society."

Important research areas

C - Transition of Control (ToC)

- Will happen a lot
- Needs to be safe and swift
- Difference between expected and un-expected ToCs

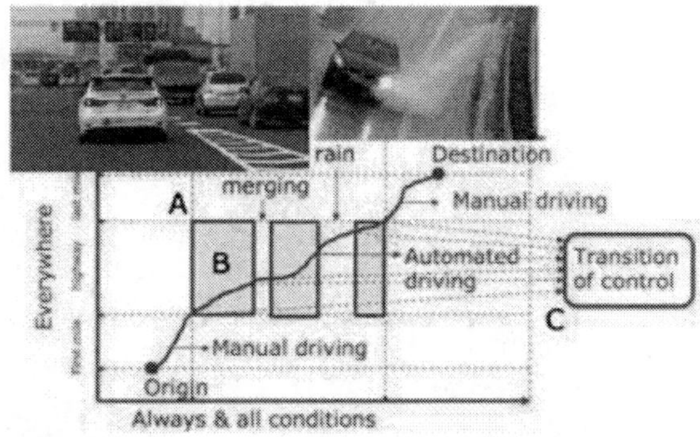

Important research areas

B – Appropriate behaviour within ODD

- detection – context – appropriate behaviour
- Who determines? Multiple perspective (OEM – road operator – driver/passenger)
- How to assess vehicle behaviour? ("Driver's licence"?)
- Mixed traffic challenge

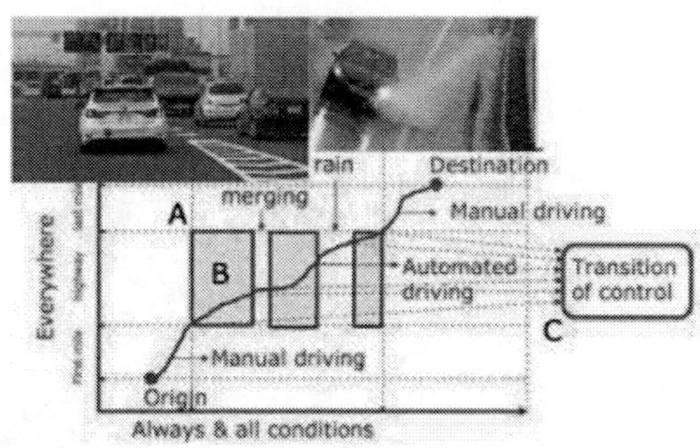

2. "I want to **enable the next generation of vehicles to actually take the road**. So I'm creating a legal framework for automated driving. Laying down requirements for reliability and safety that cars must meet before they can hit the road. **A driving licence for self-driving cars**, if you like. Not for the driver – but for the car itself!"

Context

Context

Context

Context

Concrete examples as input for constructive dialogue

- Infrastructure design / layout

- Detection errors / mistakes

- Unexpected situations / edge cases

- Adverse weather conditions

Disengagement reports

https://www.dmv.ca.gov/portal/dmv/detail/vr/autonomous/disengagement_report_2017

Disengagement reports

Company	Autonomous miles	Disengagements	Rate per 1000 miles
Google	635868	124	0.20
Cruise	10015	284	28.36
Nissan	4099	28	6.83
Delphi	3125	178	56.95
Bosch	983	1442	1466.94
Mercedes	673	336	498.95
BMW	638	1	1.57
Ford	590	3	5.08
Tesla	550	182	330.91
Honda	0	0	0.00
VW	0	0	0.00

Image: Mark Harris; Source: California DMV

https://spectrum.ieee.org/cars-that-think/transportation/self-driving/the-2578-problems-with-self-driving-cars

Disengagement reports; more factors impact ODD

- System fault
- Control discrepancy
- Perception discrepancy
- Unwanted maneuver
- Planning discrepancy
- Map discrepancy
- Localization discrepancy
- Weather conditions
- Road surface conditions
- Construction
- Emergencies
- Accidents or collisions
- Planned test of technology

- (Heavy) pedestrian traffic
- Cyclist
- Complete lane change
- Traffic light detection
- Construction
- Localization divergence
- Poor lane markings
- Vehicle cut in
- Car blocking lane
- Other driver, unexpected behaviour
- Undesired motion planning behavior
- Precautionary for nearby cyclist
- Error in perception system detection

- Cyclist riding from sidewalk into crosswalk
- Precautionary disengagement for long yellow light
- Precautionary disengagement for jay walker
- Precautionary disengagement for late cut-in
- Precautionary disengagement for pedestrian near curb
- Precautionary for vehicle exiting parking lot
- Vehicle in cross traffic ran red light
- Localization error
- Sensor fusion error

Disengagement causes (San Jose State University)

Taxonomy for Disengagement Causes

1. Human Factors: the human driver decides to initiate the disengagement of the AV technology (e.g., discomfort, lack of trust)
2. System Failure: hardware and/or software failure (e.g., incorrect prediction of behavior, hardware discrepancy)
3. External Condition/Visibility: triggers related to the external environment (e.g., debris on road, excessive pedestrian traffic)
4. Other: not directly traceable to previous macro-categories

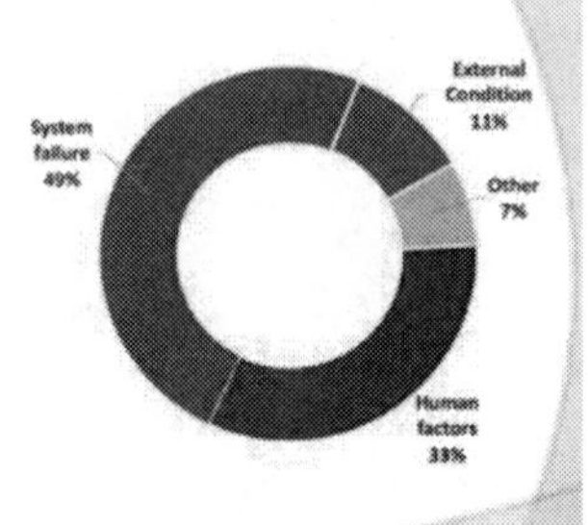

T3e Webinar – Talking Technology and Transportation in Education 30

https://www.pcb.its.dot.gov/t3/s180206_Are_Autonomous_Vehicles_Safe.asp

Joint stakeholder workshop

discuss in a multi-stakeholder setting:

- Optimization of the ODD
- "Appropriate" behavior of the automatec vehicle within the ODD, specifically for mixed traffic conditions.
- Explore how to set up a constructive dialogue with the developers of automated vehicles and other stakeholders to see what each can contribute from it's own role and responsibility.

Thank you for your attention!
tom.alkim@rws.nl

How Do Machines See?

Leo McCloskey, Echodyne
Robert Seidl, Motus Ventures

ECHODYNE motus ventures

Machine Perception is the Goal

Machine perception is the capability of a computer system to interpret data in a manner that is similar to the way humans use their senses to relate to the world around them. The basic method that the computers take in and respond to their environment is through the attached hardware.

Machine perception - Wikipedia
https://en.wikipedia.org/wiki/Machine_perception

https://www.google.com/search?q=what+is+machine+perception&oq=what+is+machine+perce&aqs=chrome.0.0j69i57j0l2.2870j1j7&sourceid=chrome&ie=UTF-8

ECHODYNE motus ventures 2

Typical AV Software Stack

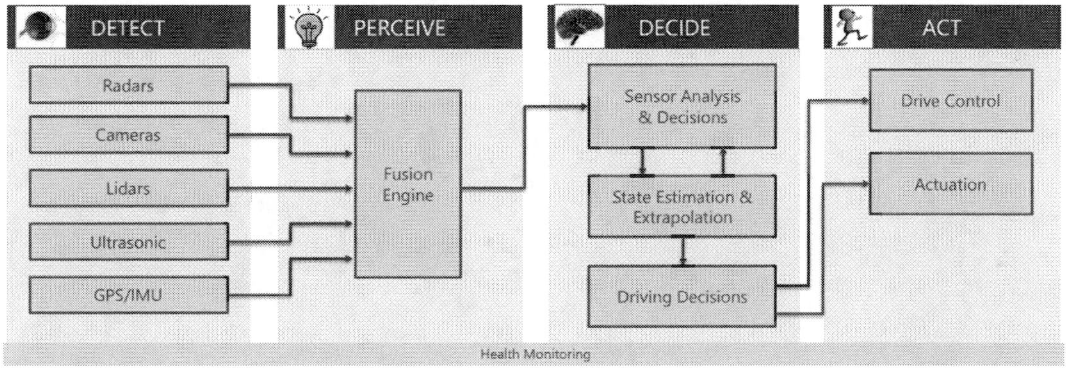

ECHODYNE motus ventures

3

AI Needs Contextual Data...

Sensor information needs to be transformed into a domain that a machine can interpret, where algorithms that perform decision making can extract the useful semantic information they need to operate.

Sensors. Data. Probabilities. Confidence.

ECHODYNE motus ventures

4

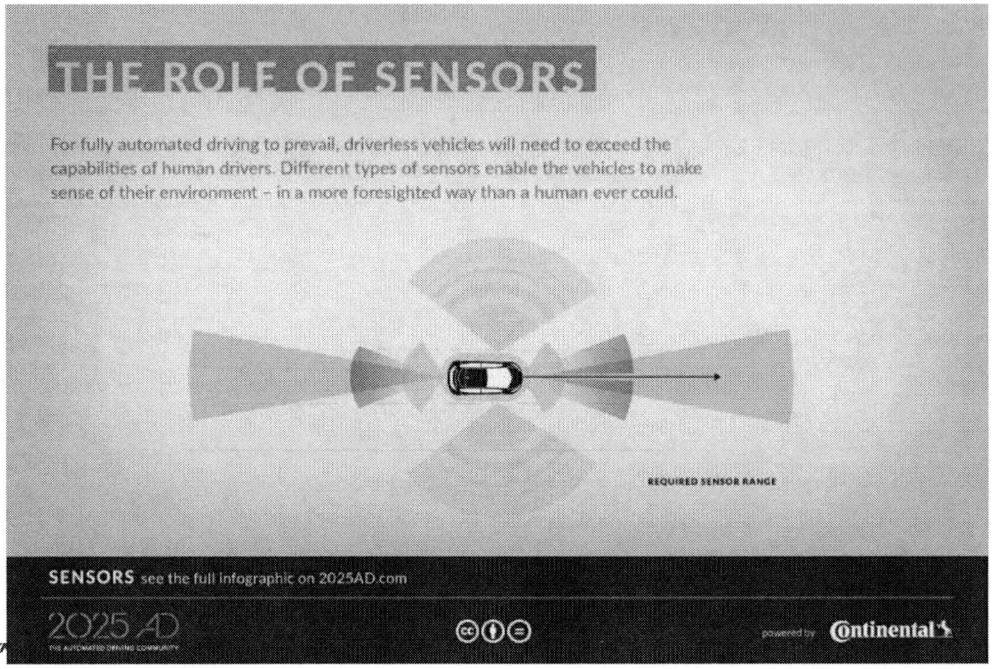
THE ROLE OF SENSORS
For fully automated driving to prevail, driverless vehicles will need to exceed the capabilities of human drivers. Different types of sensors enable the vehicles to make sense of their environment – in a more foresighted way than a human ever could.
REQUIRED SENSOR RANGE
SENSORS see the full infographic on 2025AD.com
2025 AD
powered by Continental

5

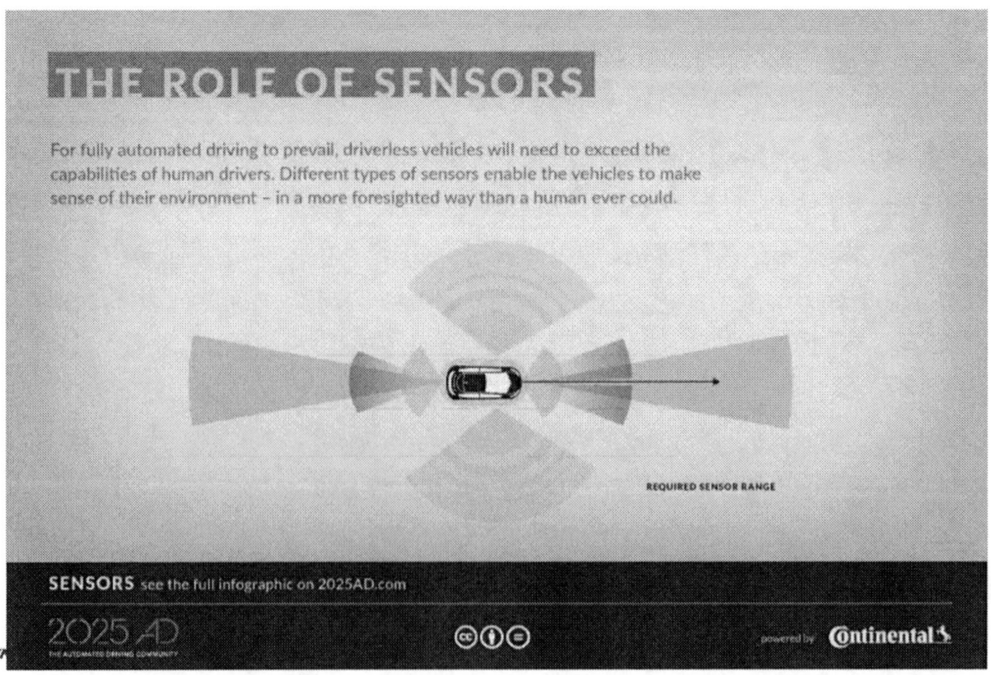
THE ROLE OF SENSORS
For fully automated driving to prevail, driverless vehicles will need to exceed the capabilities of human drivers. Different types of sensors enable the vehicles to make sense of their environment – in a more foresighted way than a human ever could.
REQUIRED SENSOR RANGE
SENSORS see the full infographic on 2025AD.com
2025 AD
powered by Continental

6

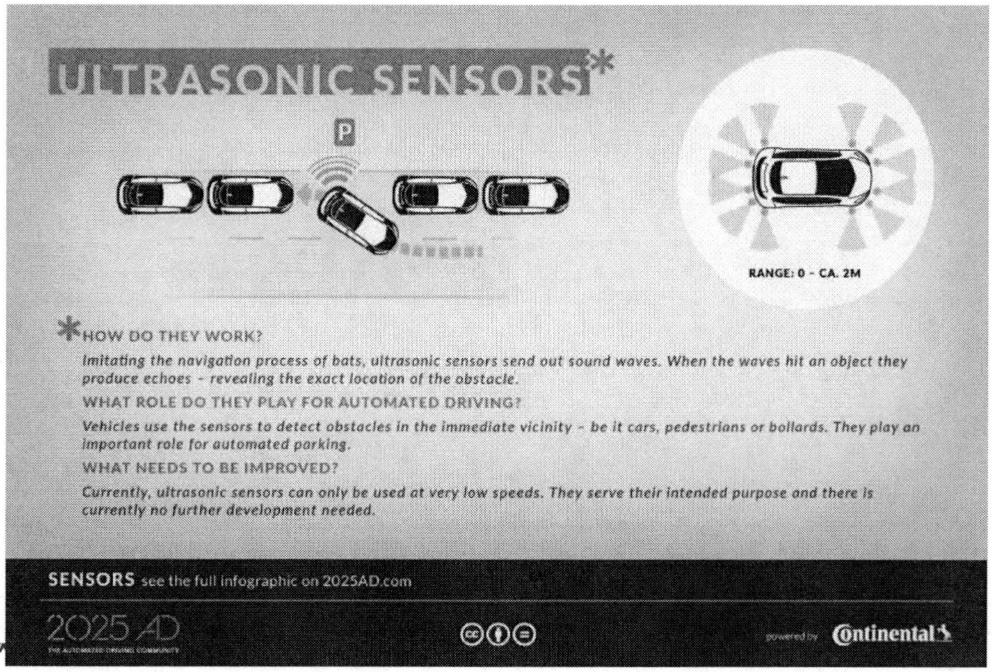
ULTRASONIC SENSORS*
P
RANGE: 0 - CA. 2M
*HOW DO THEY WORK?
Imitating the navigation process of bats, ultrasonic sensors send out sound waves. When the waves hit an object they produce echoes - revealing the exact location of the obstacle.
WHAT ROLE DO THEY PLAY FOR AUTOMATED DRIVING?
Vehicles use the sensors to detect obstacles in the immediate vicinity - be it cars, pedestrians or bollards. They play an important role for automated parking.
WHAT NEEDS TO BE IMPROVED?
Currently, ultrasonic sensors can only be used at very low speeds. They serve their intended purpose and there is currently no further development needed.
SENSORS see the full infographic on 2025AD.com
2025 AD
powered by Continental

7

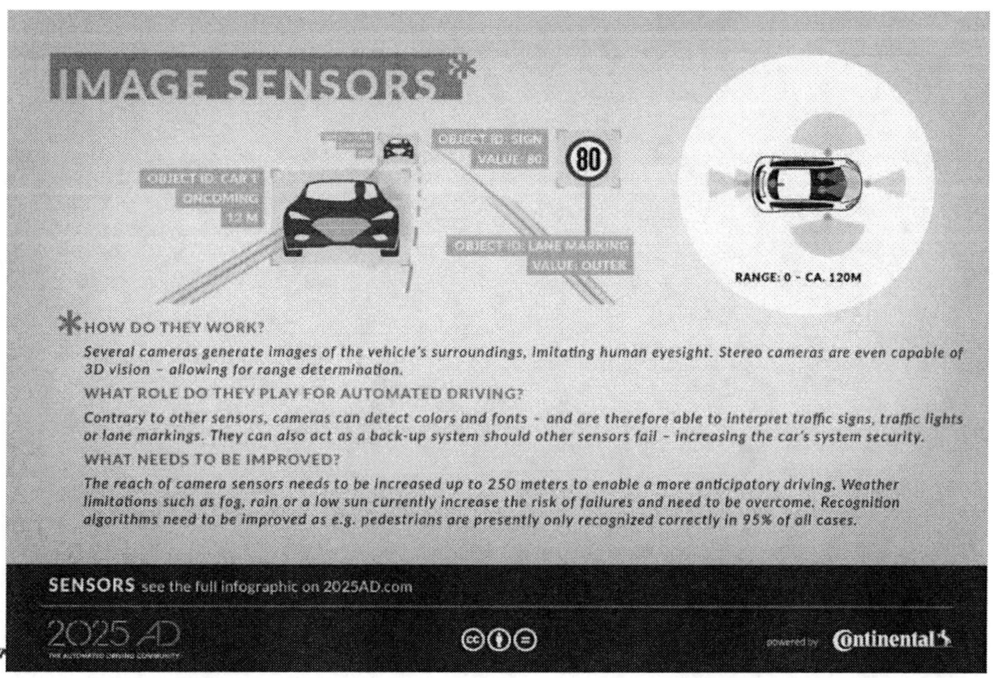
IMAGE SENSORS*
OBJECT ID: CAR 1
ONCOMING
12 M
OBJECT ID: SIGN
VALUE: 80
80
OBJECT ID: LANE MARKING
VALUE: OUTER
RANGE: 0 - CA. 120M
*HOW DO THEY WORK?
Several cameras generate images of the vehicle's surroundings, imitating human eyesight. Stereo cameras are even capable of 3D vision - allowing for range determination.
WHAT ROLE DO THEY PLAY FOR AUTOMATED DRIVING?
Contrary to other sensors, cameras can detect colors and fonts - and are therefore able to interpret traffic signs, traffic lights or lane markings. They can also act as a back-up system should other sensors fail - increasing the car's system security.
WHAT NEEDS TO BE IMPROVED?
The reach of camera sensors needs to be increased up to 250 meters to enable a more anticipatory driving. Weather limitations such as fog, rain or a low sun currently increase the risk of failures and need to be overcome. Recognition algorithms need to be improved as e.g. pedestrians are presently only recognized correctly in 95% of all cases.
SENSORS see the full infographic on 2025AD.com
2025 AD
powered by Continental

8

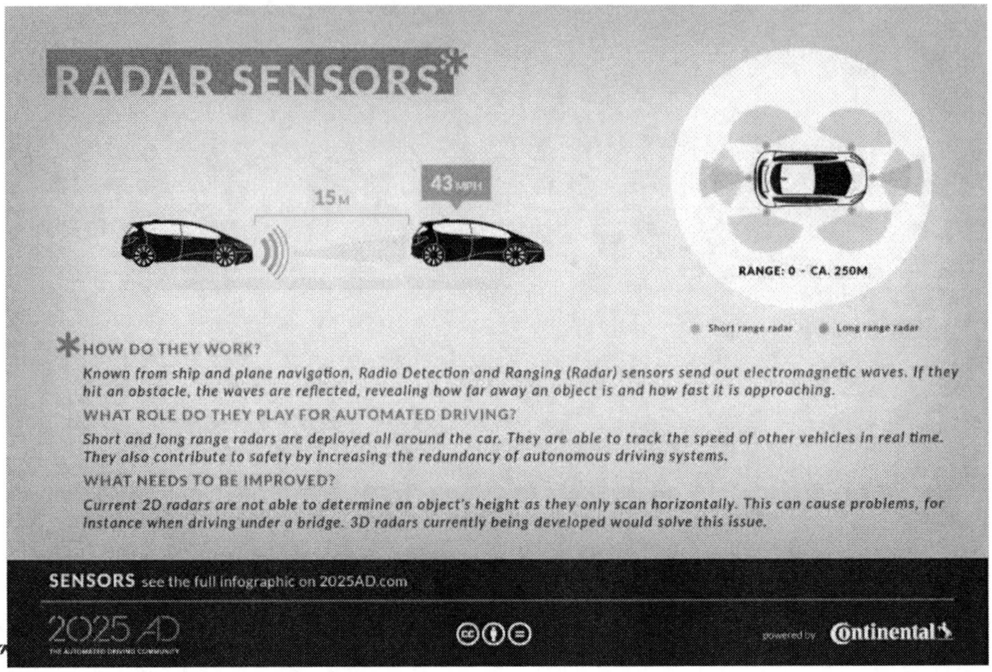
RADAR SENSORS
15M
43 MPH
RANGE: 0 - CA. 250M
Short range radar
Long range radar
HOW DO THEY WORK?
Known from ship and plane navigation, Radio Detection and Ranging (Radar) sensors send out electromagnetic waves. If they hit an obstacle, the waves are reflected, revealing how far away an object is and how fast it is approaching.
WHAT ROLE DO THEY PLAY FOR AUTOMATED DRIVING?
Short and long range radars are deployed all around the car. They are able to track the speed of other vehicles in real time. They also contribute to safety by increasing the redundancy of autonomous driving systems.
WHAT NEEDS TO BE IMPROVED?
Current 2D radars are not able to determine an object's height as they only scan horizontally. This can cause problems, for instance when driving under a bridge. 3D radars currently being developed would solve this issue.
SENSORS see the full infographic on 2025AD.com
2025 AD
powered by Continental

9

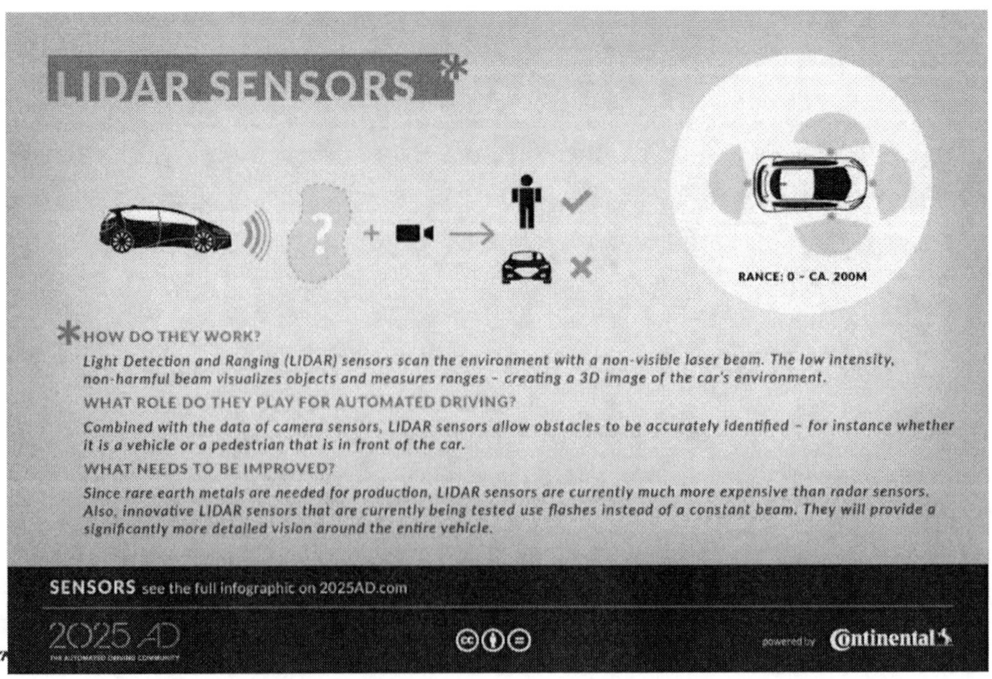
LIDAR SENSORS
?
RANCE: 0 - CA. 200M
HOW DO THEY WORK?
Light Detection and Ranging (LIDAR) sensors scan the environment with a non-visible laser beam. The low intensity, non-harmful beam visualizes objects and measures ranges - creating a 3D image of the car's environment.
WHAT ROLE DO THEY PLAY FOR AUTOMATED DRIVING?
Combined with the data of camera sensors, LIDAR sensors allow obstacles to be accurately identified - for instance whether it is a vehicle or a pedestrian that is in front of the car.
WHAT NEEDS TO BE IMPROVED?
Since rare earth metals are needed for production, LIDAR sensors are currently much more expensive than radar sensors. Also, innovative LIDAR sensors that are currently being tested use flashes instead of a constant beam. They will provide a significantly more detailed vision around the entire vehicle.
SENSORS see the full infographic on 2025AD.com
2025 AD
powered by Continental

10

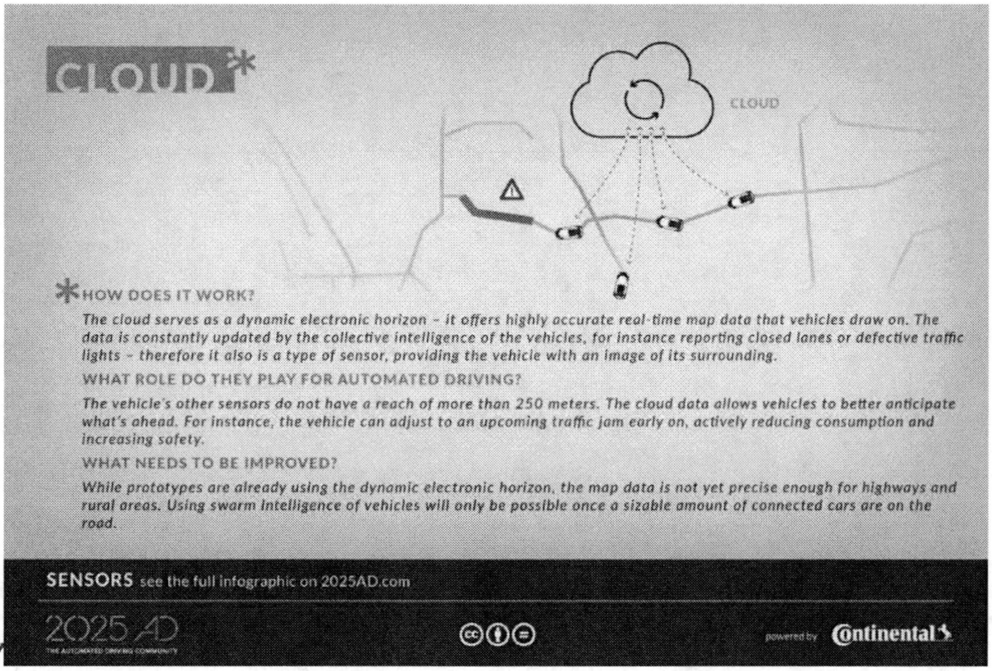
CLOUD*
CLOUD
*HOW DOES IT WORK?
The cloud serves as a dynamic electronic horizon – it offers highly accurate real-time map data that vehicles draw on. The data is constantly updated by the collective intelligence of the vehicles, for instance reporting closed lanes or defective traffic lights – therefore it also is a type of sensor, providing the vehicle with an image of its surrounding.
WHAT ROLE DO THEY PLAY FOR AUTOMATED DRIVING?
The vehicle's other sensors do not have a reach of more than 250 meters. The cloud data allows vehicles to better anticipate what's ahead. For instance, the vehicle can adjust to an upcoming traffic jam early on, actively reducing consumption and increasing safety.
WHAT NEEDS TO BE IMPROVED?
While prototypes are already using the dynamic electronic horizon, the map data is not yet precise enough for highways and rural areas. Using swarm intelligence of vehicles will only be possible once a sizable amount of connected cars are on the road.
SENSORS see the full infographic on 2025AD.com
2025 AD
powered by Continental

ECHODY

11

That was easy – we're done right?

ECHODYNE motus ventures

17

ECHODYNE motus ventures

18

ECHODYNE motus ventures 19

Phew, now we're done?

ECHODYNE motus ventures 20

ECHODYNE motus ventures

Single Sensor Information Alone is *Necessary but Insufficient*

ECHODYNE motus ventures

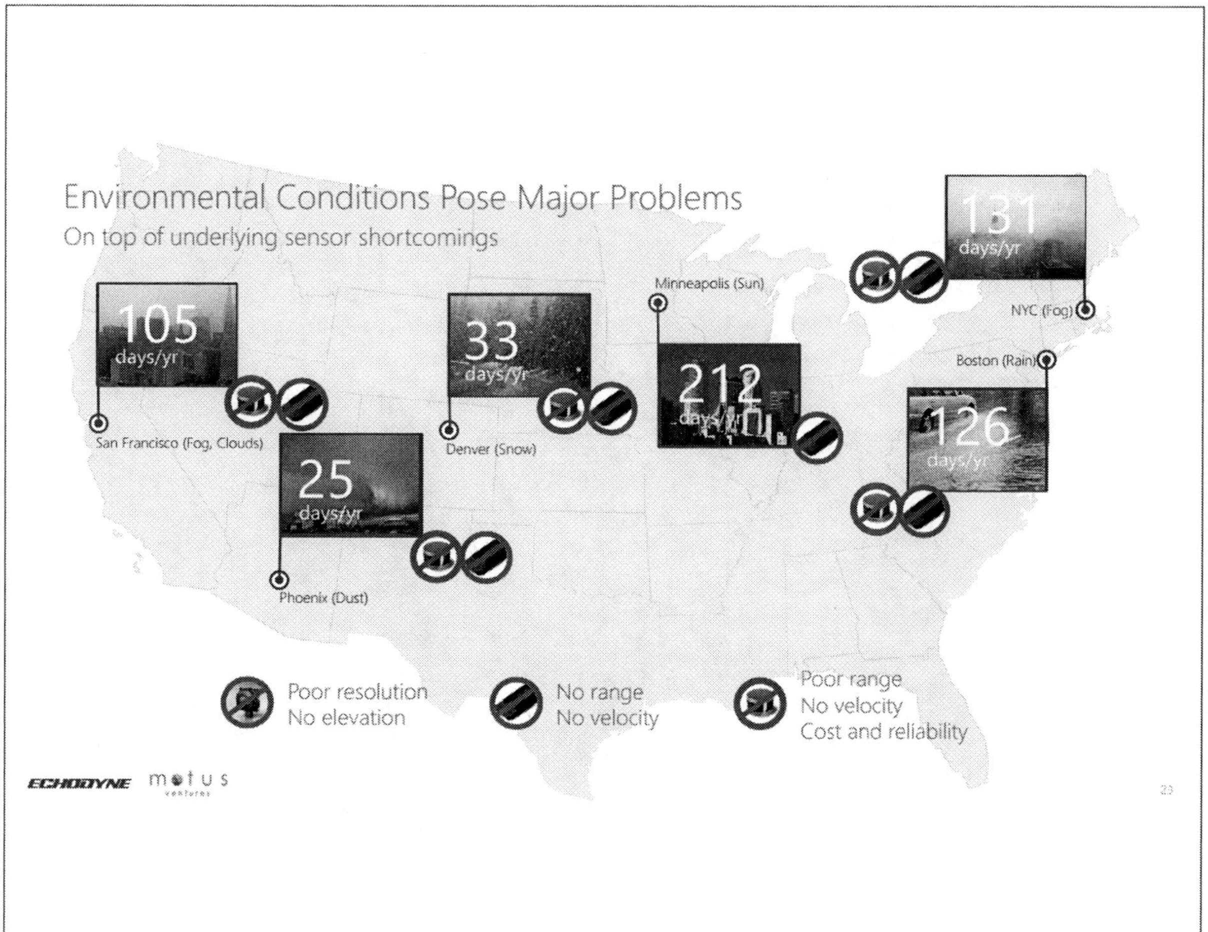
Environmental Conditions Pose Major Problems
On top of underlying sensor shortcomings
105
days/yr
San Francisco (Fog, Clouds)
25
days/yr
Phoenix (Dust)
33
days/yr
Denver (Snow)
Minneapolis (Sun)
212
days/yr
131
days/yr
NYC (Fog)
Boston (Rain)
126
days/yr
Poor resolution
No elevation
No range
No velocity
Poor range
No velocity
Cost and reliability
ECHODYNE
motus
ventures
23

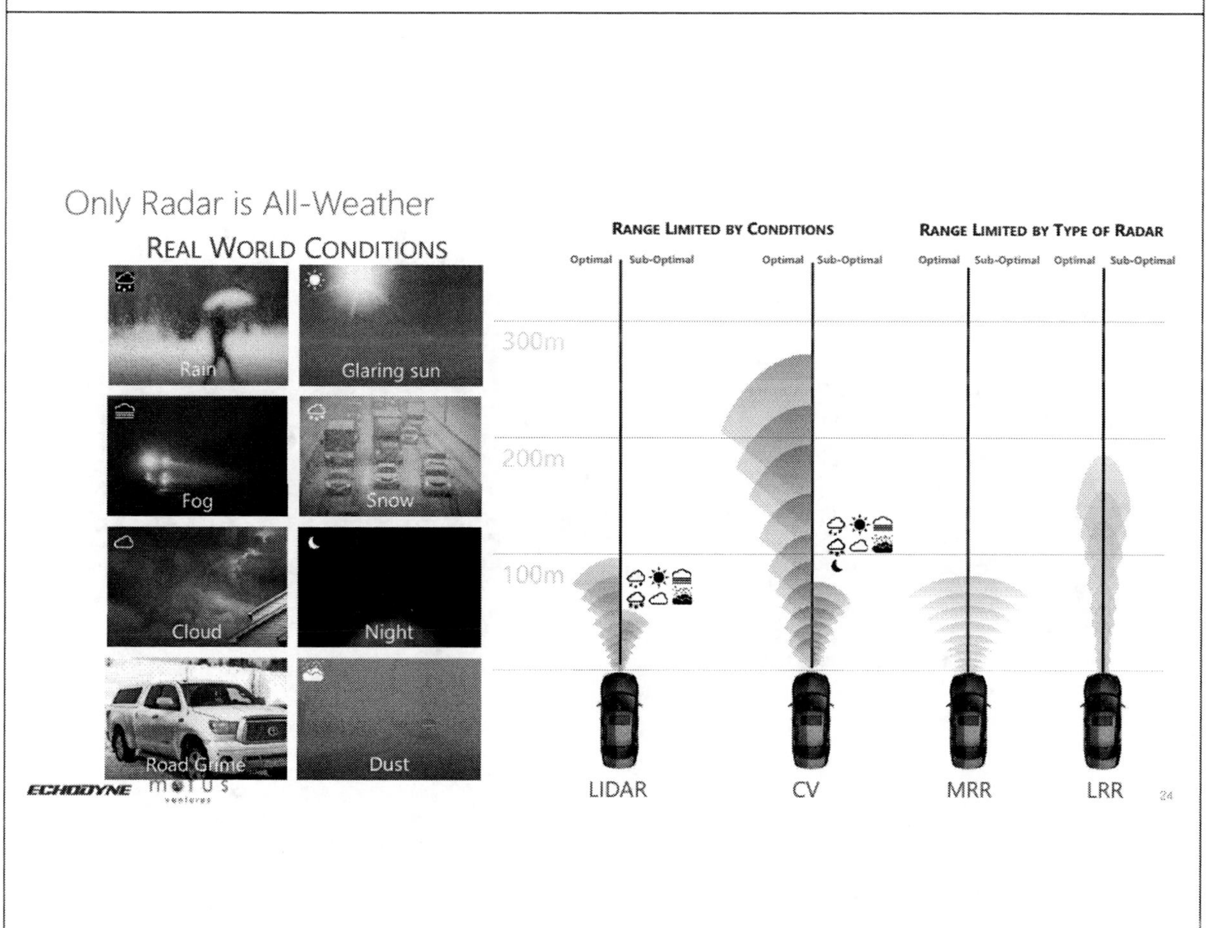
Only Radar is All-Weather
Real World Conditions
Rain
Glaring sun
Fog
Snow
Cloud
Night
Road Grime
Dust
Range Limited by Conditions
Range Limited by Type of Radar
Optimal
Sub-Optimal
300m
200m
100m
LIDAR
CV
MRR
LRR
ECHODYNE
motus
ventures
24

MESA Radar Vision for Autonomous Cars and Trucks

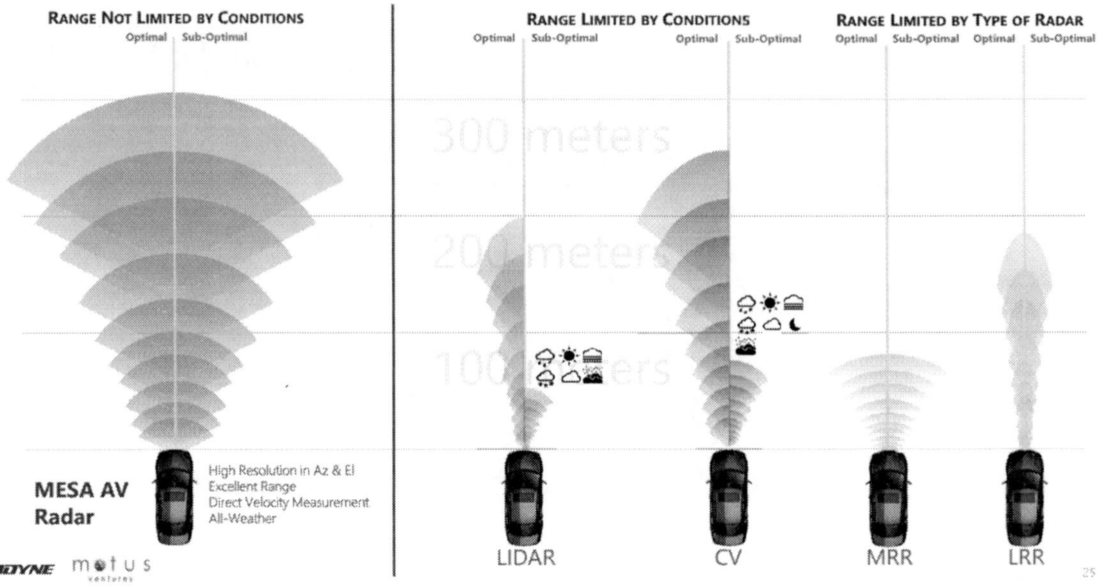

ECHODYNE motus ventures

Vision Sensors Are NOT Foolproof

TRAPPING A SELF-DRIVING CAR IS SURPRISINGLY EASY

POSTED BY DERRICK ROSSIGNOL ON MARCH 21, 2017 SHARE

https://nerdist.com/trap-a-self-driving-car/

Researchers hack a self-driving car by putting stickers on street signs

https://autoblog.com/2017/08/04/self-driving-car-sign-hack-stickers/

ECHODYNE motus ventures

Beyond AV Sight: I2V Can Bolster AI Confidence

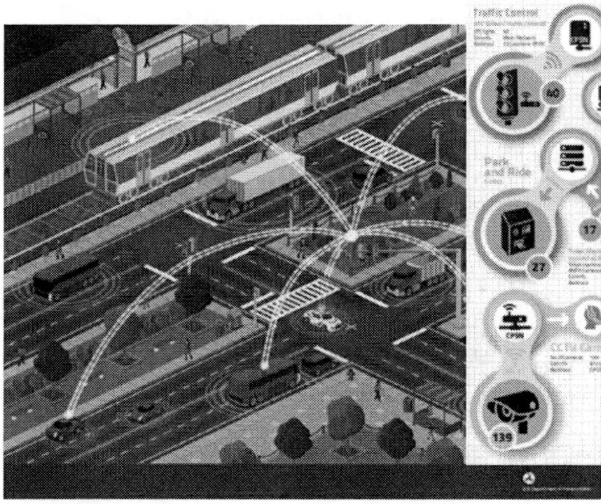

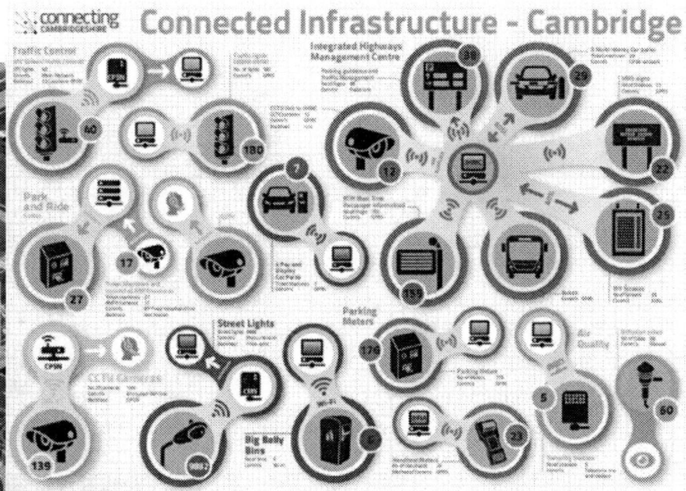

Do AVs Need Traffic Control Devices?

Road Signs
Does an AV need to read a Sign? Would HD Map + I2V be as effective?
Do we need Exit signs? Stop signs?

Traffic Signals
Does an AV need to read a Signal? Would SPaT + V2x be as effective?
Would a traffic signal aware of vehicular needs be necessary for governance?

Road Markings
Does an AV need road markings? Would HD GPS + I2V be as effective?

Islands
Probably no reason to remove, but would we build more?

The Next 2-15yrs are Much Different Than Following Years

Maturing Autonomy

- Vehicles improving capabilities
  - Primarily existing vehicle types adapted for AV
- Maturity likely to be use case driven (eg, urban, highways)
- Infrastructure needs swamped by break/fix
- Vehicles will learn to make due with whatever is present
  - Will *all* Agencies use special roadway striping paint?

Full Autonomy

- Vehicles likely to be remade
  - Designed for Autonomy
- Vehicles need:
  - Safe reliable roadways
  - Years of sensor fusion & AV stack development
  - Reliable and proper connectivity (DSRC or 5G?)
- How will the Agency role change?

The next 15 years may decide the Agency role in an Autonomous Vehicle Age

ECHODYNE motus ventures

29

Thank You

Leo McCloskey, Echodyne
Robert Seidl, Motus Ventures

ECHODYNE motus ventures

AUTOMATED VEHICLES SYMPOSIUM 2018

Driven by: AUVSI & TRB

SESSION 1

Reading the Road Ahead: Traffic Control Devices and Machine Vision Interactions

Automated Vehicles Symposium
TRB AND AUVSI

Monday, July 9th, 2018
1:30 PM TO 5:30 PM

AUTOMATED VEHICLES SYMPOSIUM 2018

Driven by: AUVSI & TRB

SESSION 1

Reading the Road Ahead: Traffic Control Devices and Machine Vision Interactions

Scott O. Kuznicki, P.E.
Director, Safety and Mobility Technology Solutions
Transpo Group

Brian Watson
Director of New Programs
American Traffic Safety Services Association

transpogroup

ATSSA
SAFER ROADS SAVE LIVES

AUTOMATED VEHICLES SYMPOSIUM 2018
Driven by: AUVSI & TRB
SESSION 1
SELF-DRIVING CARS ARE NOT ANTI-SOCIAL!
#AVS18
@scottkuznicki
@auvsi
@trb

AUTOMATED VEHICLES SYMPOSIUM 2018
Driven by: AUVSI & TRB
SESSION 1
ELECTRONIC SIGN-IN SHEET
scott.kuznicki@transpogroup.com

AUTOMATED VEHICLES SYMPOSIUM 2018

Driven by: AUVSI & TRB

SESSION 1

Reading the Road Ahead

INTRODUCTIONS & Setting the Stage	1:30 PM - 2:00 PM
How do Machines See? An INTERACTIVE Overview	2:00 PM - 3:00 PM
AFTERNOON BREAK	3:00 PM - 3:15 PM
European Road Readiness	3:15 PM - 3:35 PM
TCDs for Machine Vision	3:35 PM - 4:00 PM
Machine Vision System Failure Modes	4:00 PM - 4:30 PM
Many Sensors, One Picture	4:30 PM - 5:00 PM
Developing a Level of Readiness Framework	5:00 PM - 5:30 PM

AUTOMATED VEHICLES SYMPOSIUM 2018

Driven by: AUVSI & TRB

SESSION 1

Reading the Road Ahead: Traffic Control Devices and Machine Vision Interactions

SETTING THE STAGE
Correlating Machine Vision Performance and Operating Design Domain

MODERATED BY

Scott O. Kuznicki, P.E.

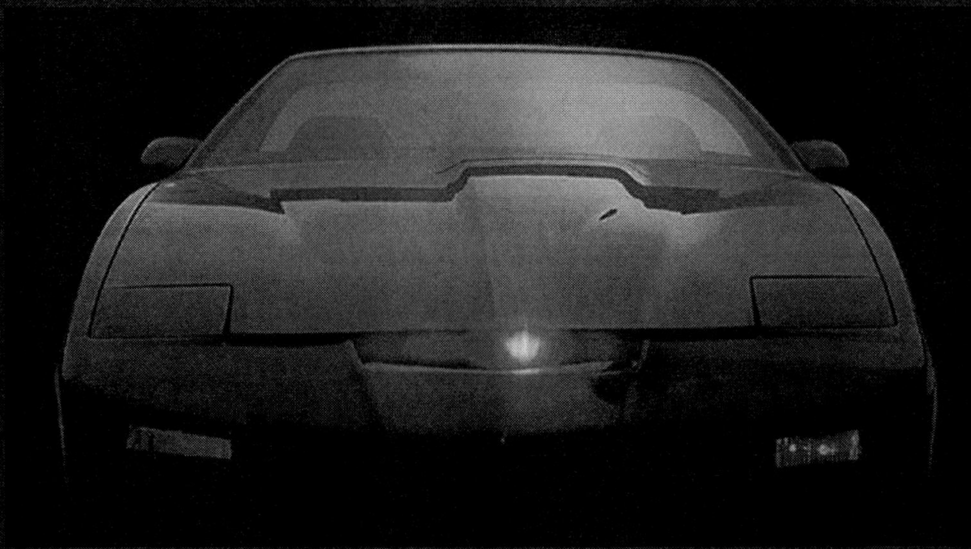

AVS 2018 - Session 1

AUTOMATED VEHICLES SYMPOSIUM 2018

Driven by: AUVSI & TRB

SESSION 1

INFRASTRUCTURE TYPES

VISIBLE	INVISIBLE
ACTIVE	PASSIVE
SECURED	OPEN
PROVEN	NASCENT
EXISTING	PROPOSED
PERVASIVE	RARE

AVS 2018 - Session 1 9

WSDOT

"Achieve superior competence in your strategic areas of focus and obtain confidence from manufacturers for deployment here in Washington State."

10

WSDOT's profiled markings are LIDAR-compliant and benefit human drivers, too

Pavement markings are key information for visual-light machine vision systems

11

EXIT 26
821
NORTH TO
823
Selah

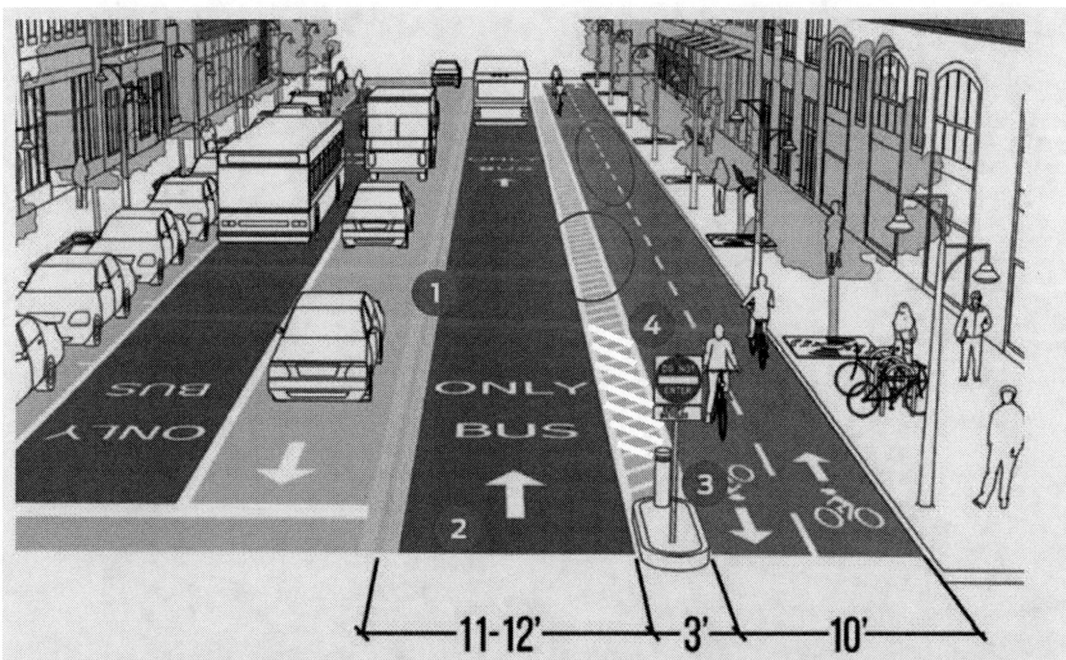

A contraflow transit lane is designed to prevent or discourage incursions, while reducing friction with adjacent traffic. Dimensions shown are recommended unless otherwise noted.

AVS 2018 - Session 1 17

AUTOMATED VEHICLES SYMPOSIUM 2018

Driven by: AUVSI & TRB

SESSION 1

ISSUES

- MAINTAINING INTEROPERABILITY WITH HUMANS
- OPERATIONAL DESIGN DOMAIN DEFINITION
- LIMITED CORRELATION WITH SAE LEVELS OF AUTOMATION
- MULTI-FACETED CRITERIA WITH NO SINGLE GRADE
- FUNDING AND PRIORITIES
- CHANGING CAPABILITIES
-

AVS 2018 - Session 1 18

AVS 2018 - Session 1

AUTOMATED VEHICLES SYMPOSIUM 2018
Driven by: AUVSI & TRB
SESSION 1
Reading the Road Ahead:
Traffic Control Devices and
Machine Vision Interactions
DISCUSSION
Correlating Machine Vision Performance
and Operating Design Domain
MODERATED BY
Scott O. Kuznicki, P.E.
Brian Watson

What is the goal of transportation?
Why do we have cars & trucks?

Safe efficient transportation of humans and goods

ECHODYNE

2

Crash data & Causes

- Speeding
- Drunk driving
- Distracted driving
- Weather
- Poor driving (red lights, stop signs, etc)
- Driver fatigue

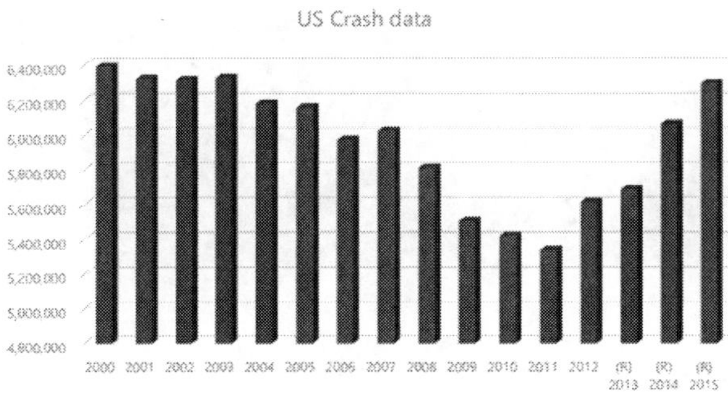

https://bts.gov/content/transportation-accidents-mode

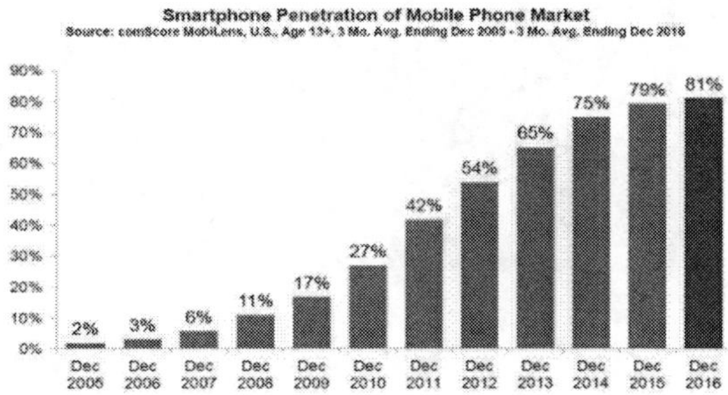

https://comscore.com/Insights/Blog/US-Smartphone-Penetration-Surpassed-80-Percent-in-2016

ECHODYNE

3

"As drivers, humans are incredibly inefficient at driving and navigating due to poor decision making, distractions and our, often, emotional responses when sitting behind the wheel and interacting with our surroundings. It is not my intention to offend anyone, even if you possess a driver skill equal to Lewis Hamilton."

Rethinking The Consequences Of Technology Disruption
Disruption, 18 June 2018 , Jonathan Carrier
https://disruptionhub.com/rethinking-technology-disruption-jonathan-carrier-car-industry-2196/

ECHODYNE

4

To Achieve The Goal, Humans Should Give Up Driving

But When Can We Trust The Vehicle?

ECHODYNE

5

Clearly, Autonomy Hasn't Arrived Yet

BUSINESS NEWS JUNE 21, 2018 / 11:48 PM / 14 DAYS AGO

Uber car's 'safety' driver streamed TV show before fatal crash: police

Heather Somerville, David Shepardson

FORT LAUDERDALE

2 teens dead after Tesla crashes into wall in Fort Lauderdale

Third teen seriously hurt after being thrown from car, police say

By Tim Swift · Local10.com Digital Editor, Janine Stanwood · Anchor/Reporter

Posted: 7:15 PM, May 08, 2018
Updated: 8:46 AM, May 09, 2018

Bloomberg

Hyperdrive

Tesla Model X in California Crash Sped Up Prior to Impact

By Alan Levin and Ryan Beene
June 7, 2018, 7:09 AM PDT *Updated on June 7, 2018, 4:21 PM PDT*

ECHODYNE

6

Machine Perception is the Goal

Machine perception is the capability of a computer system to interpret data in a manner that is similar to the way humans use their senses to relate to the world around them. The basic method that the computers take in and respond to their environment is through the attached hardware.

Machine perception - Wikipedia
https://en.wikipedia.org/wiki/Machine_perception

https://www.google.com/search?q=what+is+machine+perception&oq=what+is+machine+perce&aqs=chrome.0.0j69i57j0l2.2870j1j7&sourceid=chrome&ie=UTF-8

ECHODYNE 7

Typical AV Software Stack .. Is Woefully Incomplete

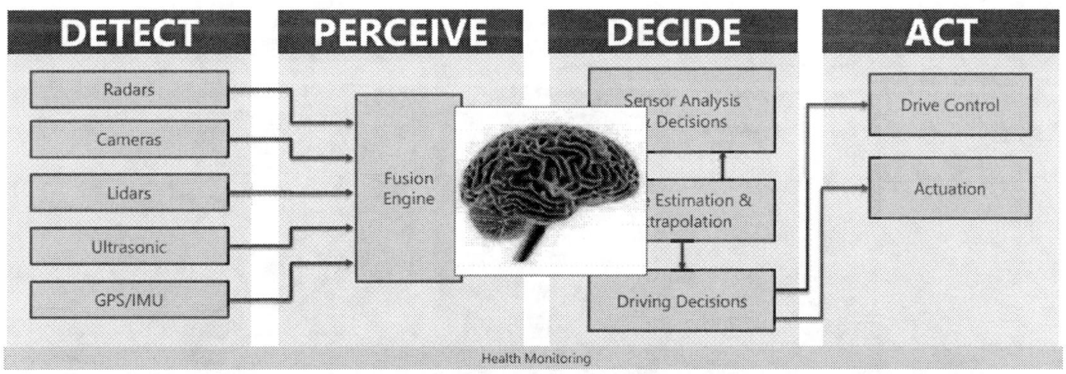

How does this compare to perfected human driving activity?

ECHODYNE 8

What happens at the scene of an accident?

- Investigators don't just "**see**" the accident scene, they…
  - Examine
  - Investigate
  - Probe
  - Interrogate
  - Inspect

- Accident scene details
- Weather
- Time of Day
- Math & Science

- Photos and Photos
- Witnesses
- Infrastructure data (video)
- Vehicle data
- Phone data

ECHODYNE

How do humans perceive their environment?

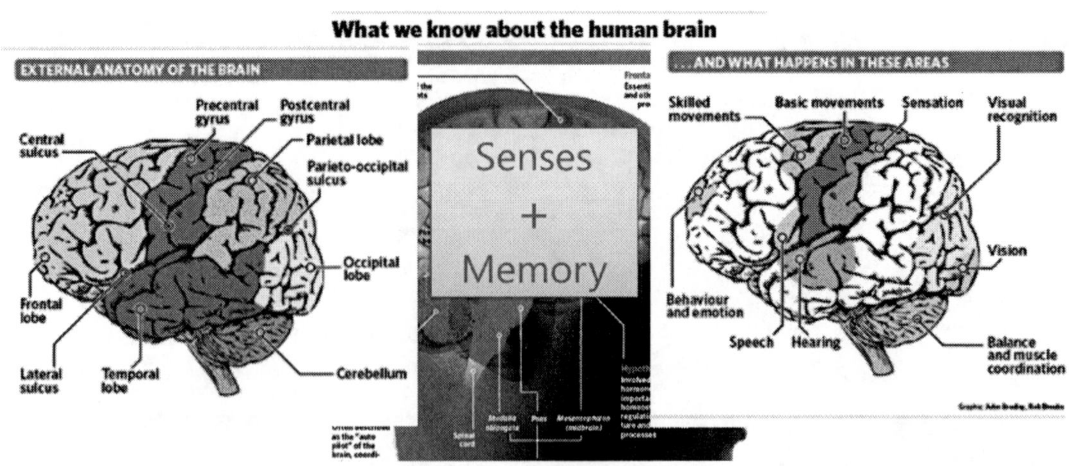

ECHODYNE

How should a vehicle perceive its environment?

<u>Senses</u>

Vision:
- Radar
- Lidar
- Camera
- Ultrasonic

Hearing:
- V2V
- V2I
- I2V
- Offboard Services

https://designboom.com/art/olaf-mooij-braincar/

<u>Memory</u>

Maps:
- High res 3D

Repetition:
- Pattern behavior

Updateable from Offboard Services

ECHODYNE

Connected Vehicles Improve Autonomy

Fully Connected Vehicle

U.S. Department of Transportation 2

ECHODYNE

Sensors flow large streams of data to Fusion and Analysis in ONE direction

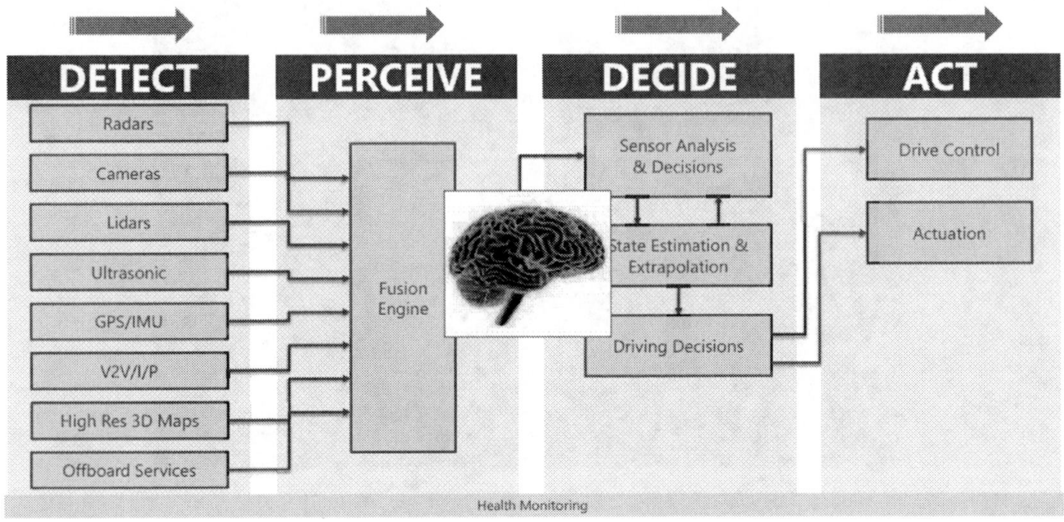

ECHODYNE 13

What happens at the scene of an accident?

- Investigators don't just "**see**" the accident scene, they…
  - Examine
  - Investigate
  - Probe
  - Interrogate
  - Inspect

ECHODYNE 14

When a person hears an unknown sound or sees something in peripheral vision, what do they do? What would you do?

ECHODYNE 15

AV software stack, like a human, needs to interrogate its environment
The sensor stack must be task-able, invokable, interrogative

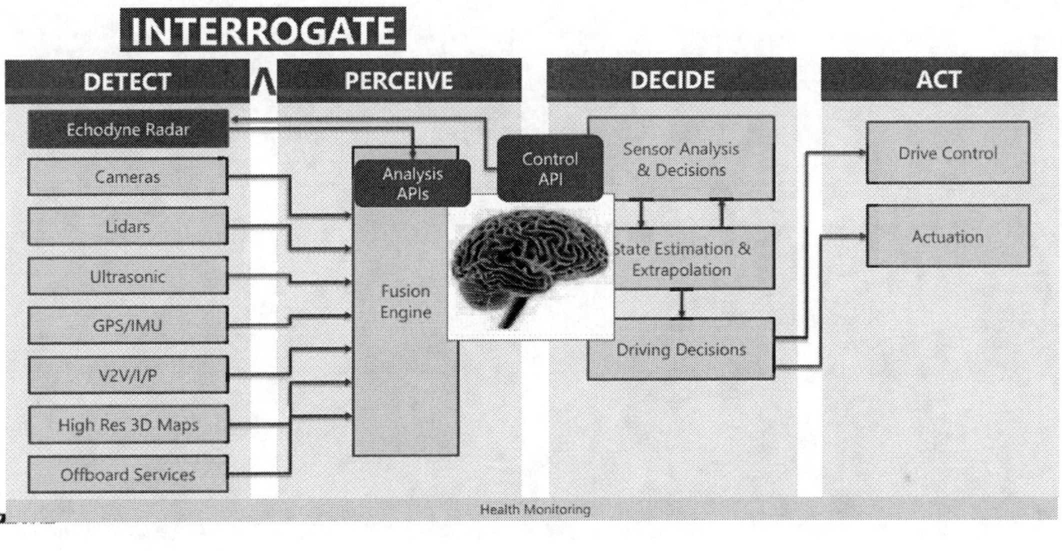

ECHO... 16

Solutions to Edge Cases Requires More Than Just Sensors

Detection and Awareness

- Can a sensor detect a tire, box, or other debris in the roadway?
- It all depends on…
- But, what if vehicles, infrastructure, and offboard services each reported the debris?
- Faster response from local agency to clear debris

Individual vehicle detection is laudable goal.
Collective data leading to greater awareness is a better goal.

ECHODYNE

17

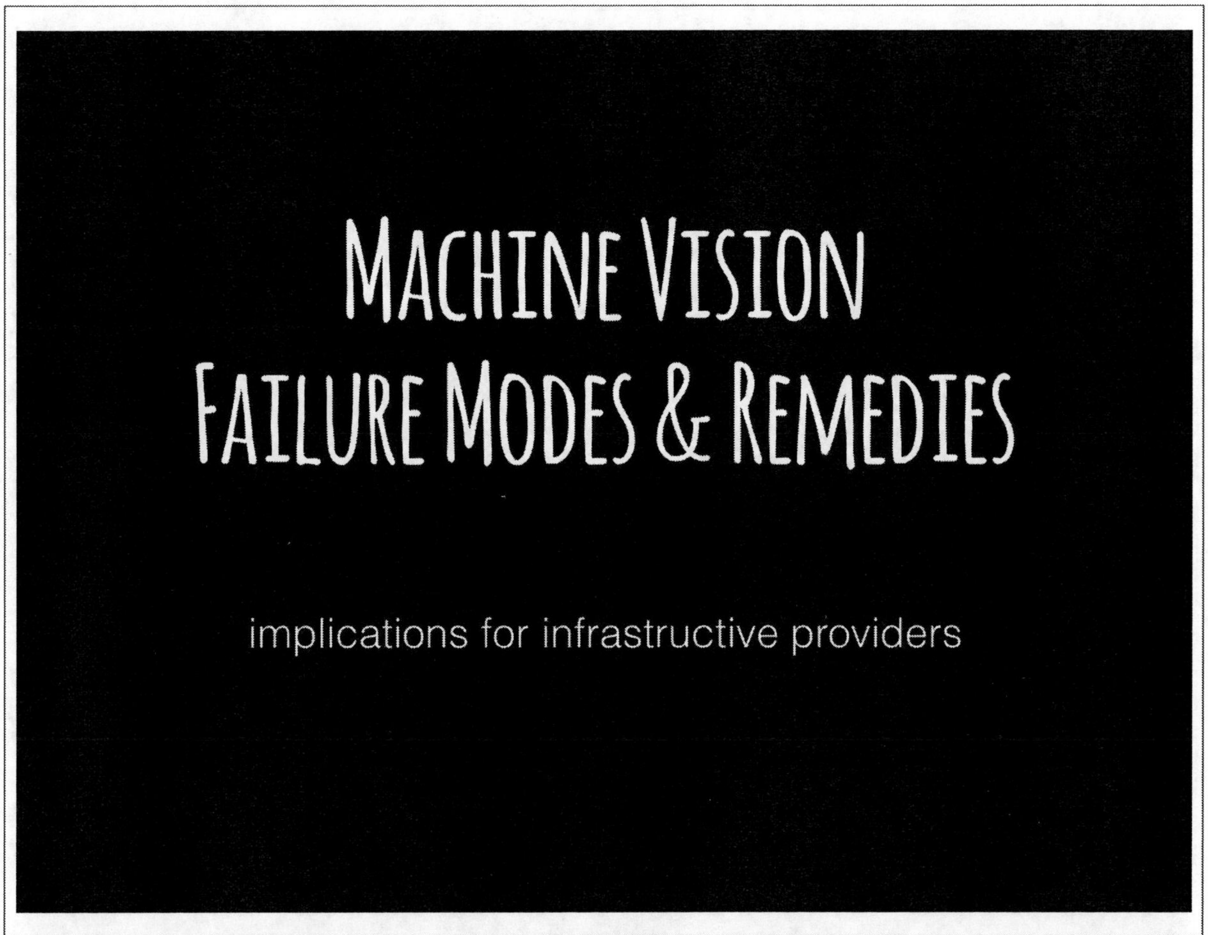
MACHINE VISION
FAILURE MODES & REMEDIES
implications for infrastructive providers

motus
ventures
Early stage investments
Autonomous Systems
Sensors • Software • Services
QUANERGY
metamoto
Civil Maps
evconnect
METAWAVE
ARGUS
CYBER SECURITY
AURO
LOCOMATION

Potential Failure Areas

- Environment
- Sensor
- Perception stack

AV Sensors for Traffic & Safety

Environment

Dispersive medium #1: fog, dust, snow, rain, smoke

Dispersive medium #2: exhaust plumes

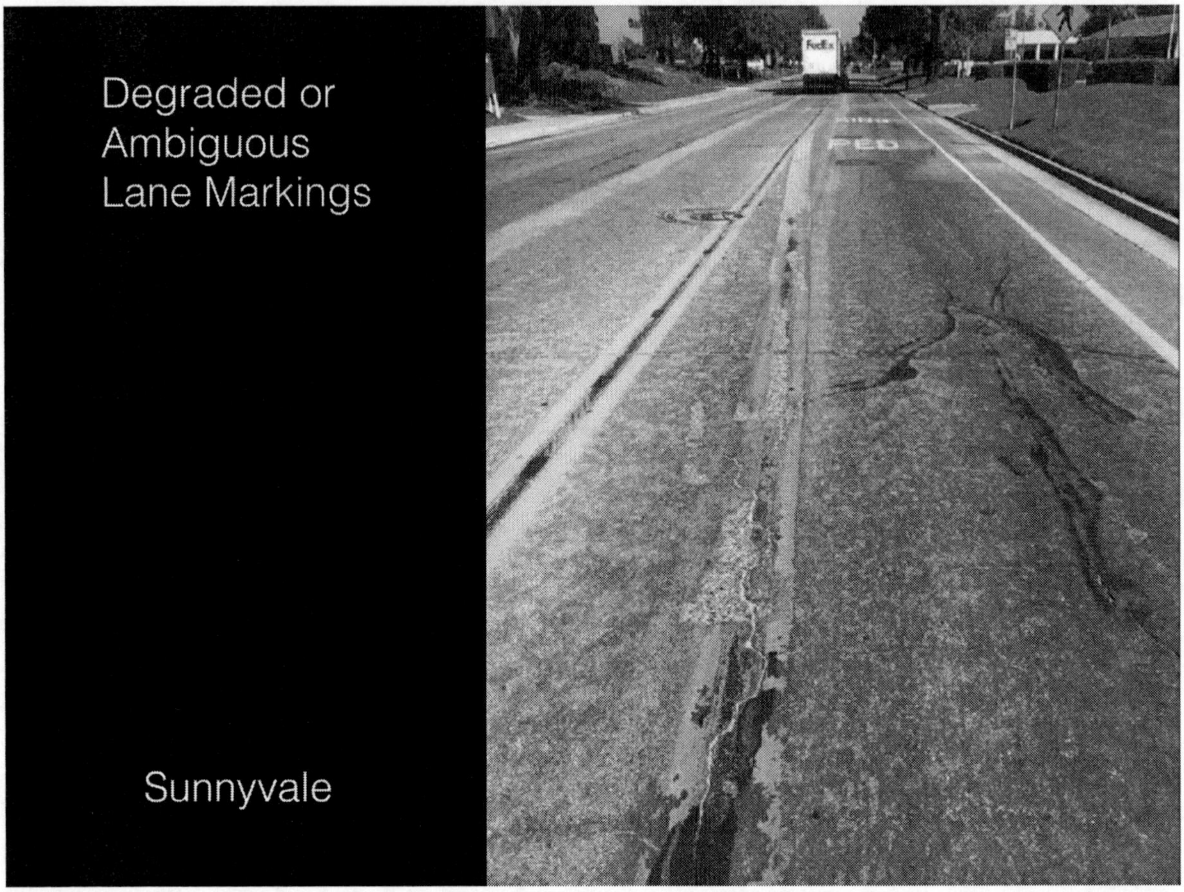
Degraded or
Ambiguous
Lane Markings
Sunnyvale

Recommendation: at least remove the most ambiguous

No lane markings visible…
DriversEdGuru.com

↑ Detroit
Flint →

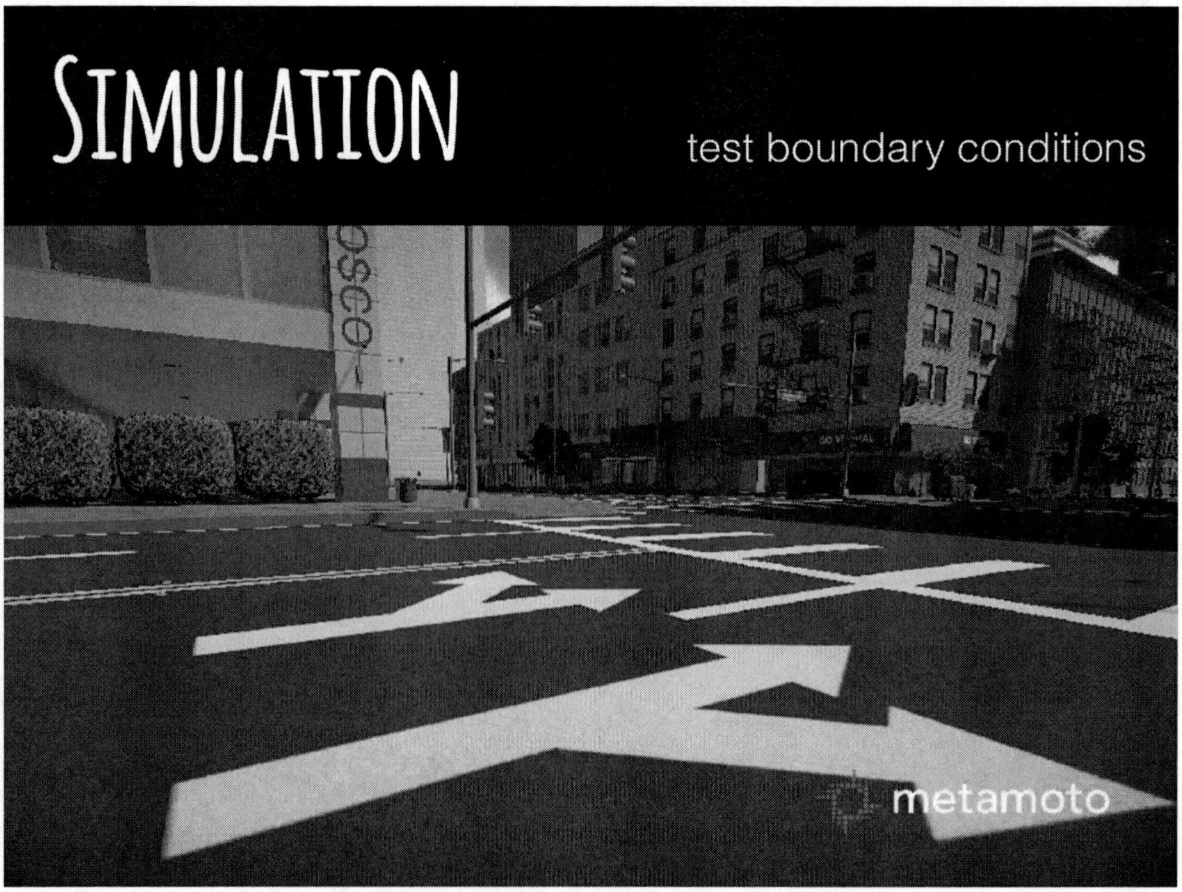
SIMULATION
test boundary conditions
metamoto

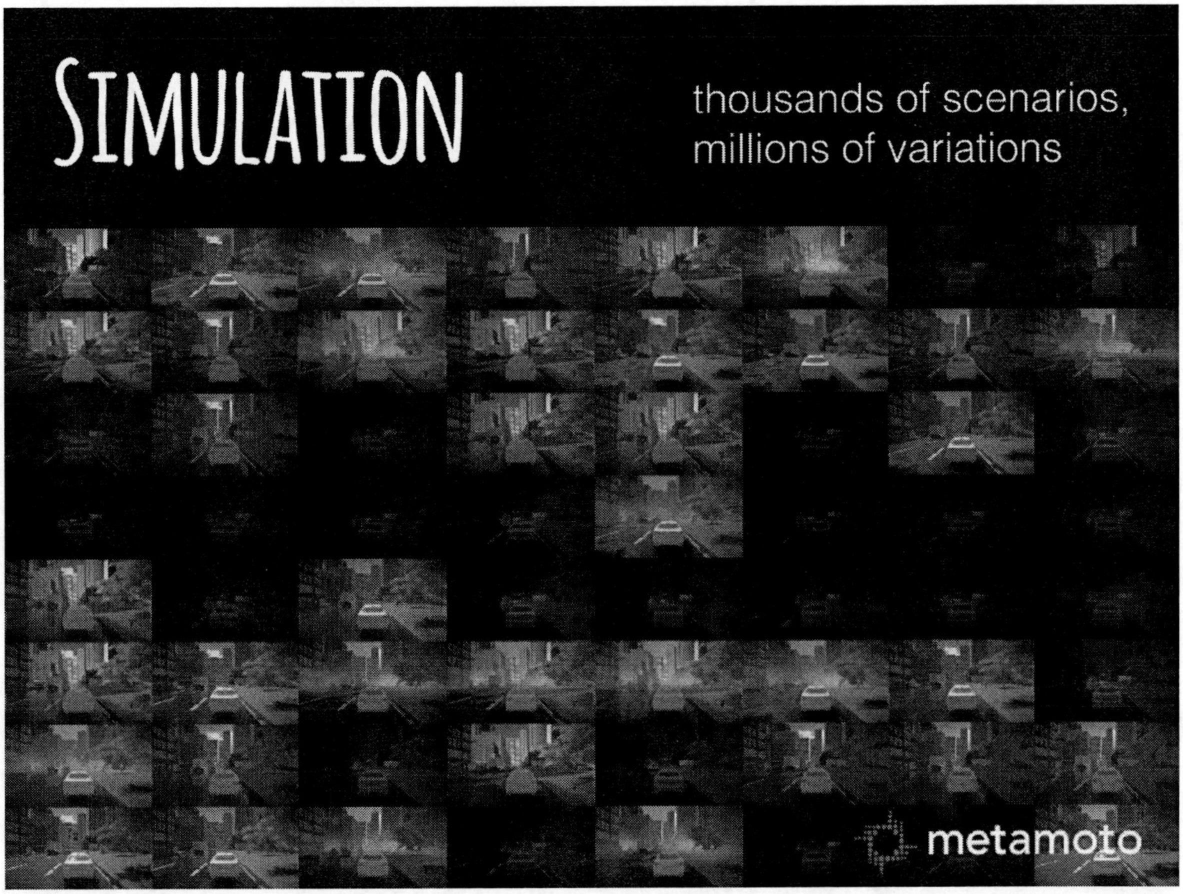
SIMULATION
thousands of scenarios,
millions of variations
metamoto

Degraded or
Invisible Signage

Bouncing, swaying traffic lights (rural US special)

HD Maps
lanes, signs, signals
Civil Maps

HD Maps
use while driving:
know lane boundaries and semantics
focus camera processing on signals
Civil Maps

HD Maps

Sensor

Dirty lens #1: seagull vs Google Streetview
Las Palmas Ave
N June
Seward St
N Hudson Ave

Dirty lens #2: Water, dust, mud
Scratches, chips, cracks
Graceful degradation...

JAMMING: BLINDING CAMERA, DAZZLING

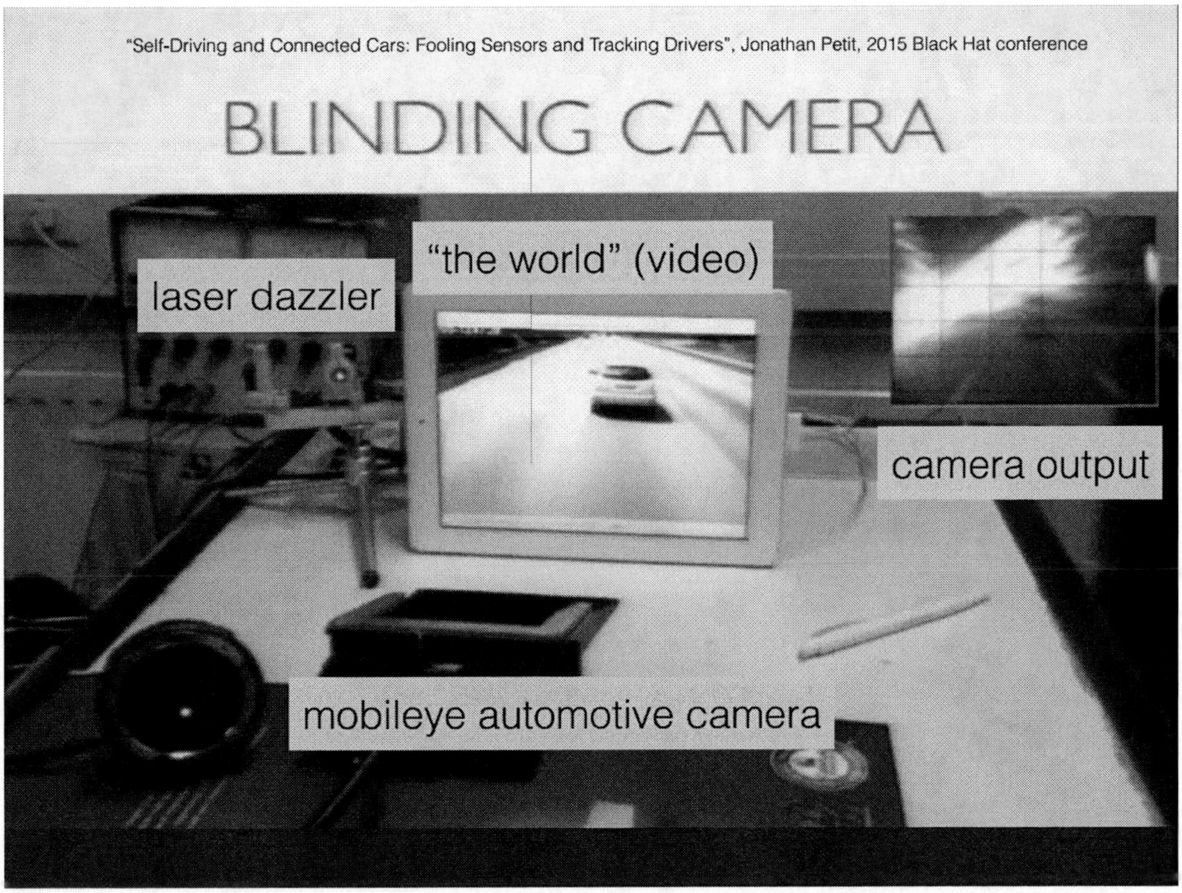
"Self-Driving and Connected Cars: Fooling Sensors and Tracking Drivers", Jonathan Petit, 2015 Black Hat conference
BLINDING CAMERA
laser dazzler
"the world" (video)
camera output
mobileye automotive camera

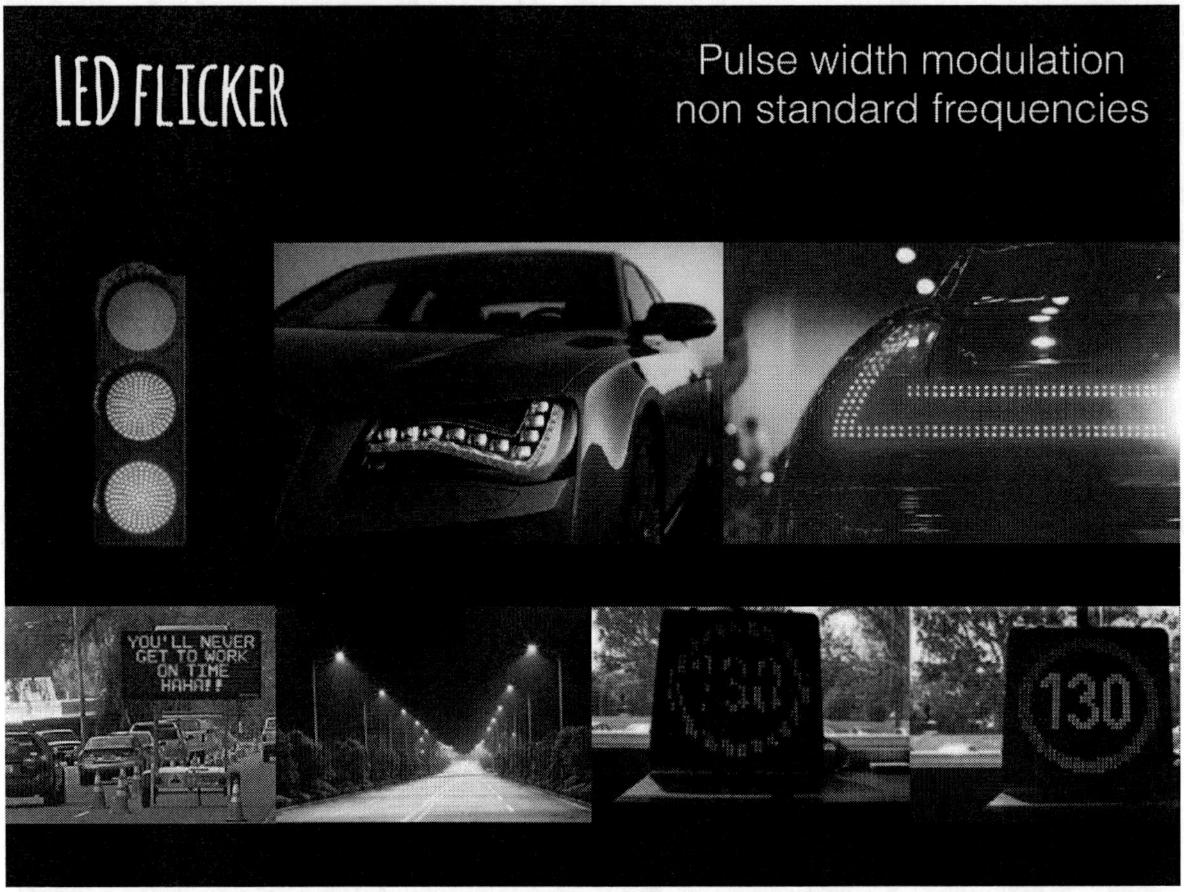
LED FLICKER
Pulse width modulation
non standard frequencies
YOU'LL NEVER
GET TO WORK
ON TIME
HAHA!!
130
130

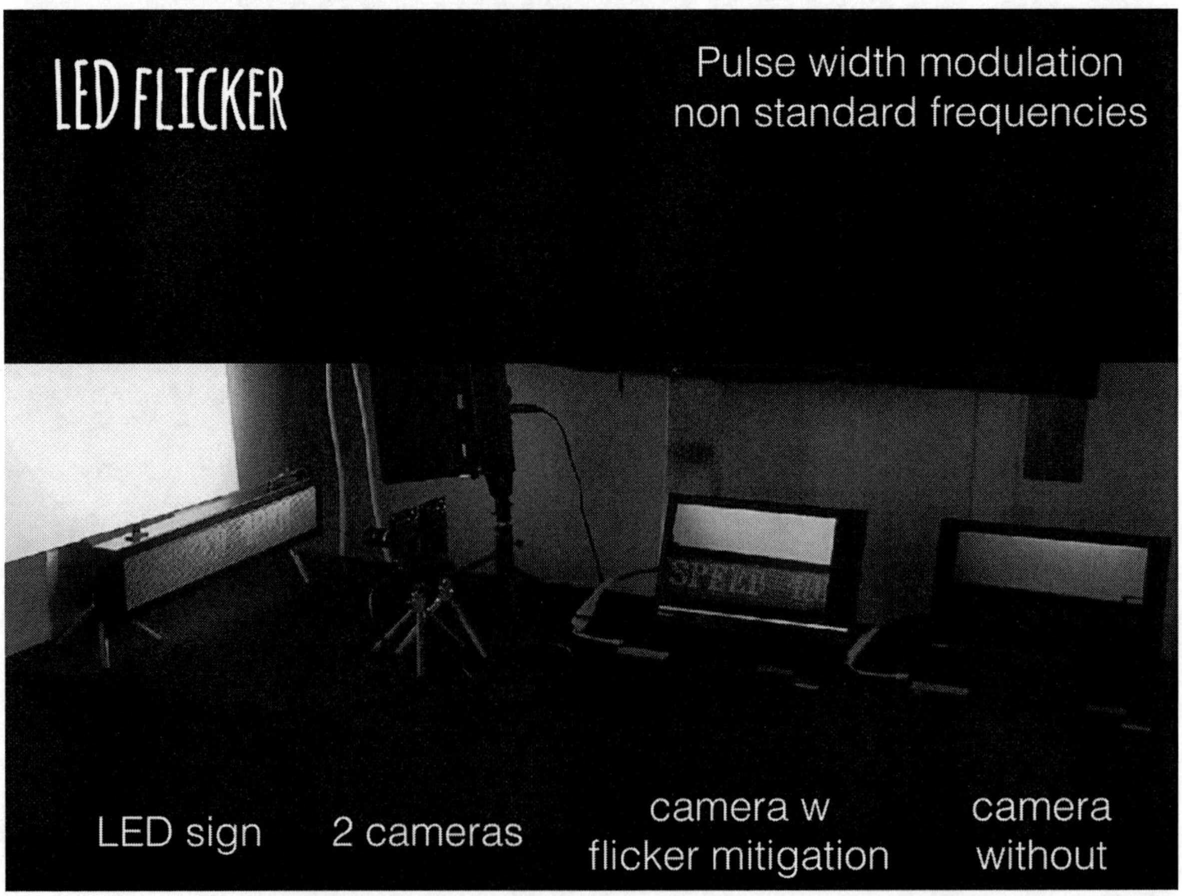
LED FLICKER
Pulse width modulation
non standard frequencies
LED sign
2 cameras
camera w
flicker mitigation
camera
without

Perception stack

BCAA Traffic Safety Foundation and the public awareness group preventable.ca

car
car
person
bicycle
bicycle
bicycle
Bosch eBike
Systems
BOSCH
Invented for life

Cars
Signs
Buildings
Humans
Vegetation
Electric pole
sign
sky
building
Pink = “Car”

May 7, 2016 Florida

Eye contact
Head pose, gaze estimation
THE EASIEST WAY TO AVOID A COLLISION IS TO ALWAYS MAKE EYE CONTACT WITH DRIVERS.
PEDESTRIAN SAFETY

BEHAVIOR PREDICTION

Intent-Aware Long-Term Prediction of Pedestrian Motion
http://vision.ucla.edu/papers/karasevAHS16.pdf

BEHAVIOR PREDICTION
Chris Urmson 2015 TED talk

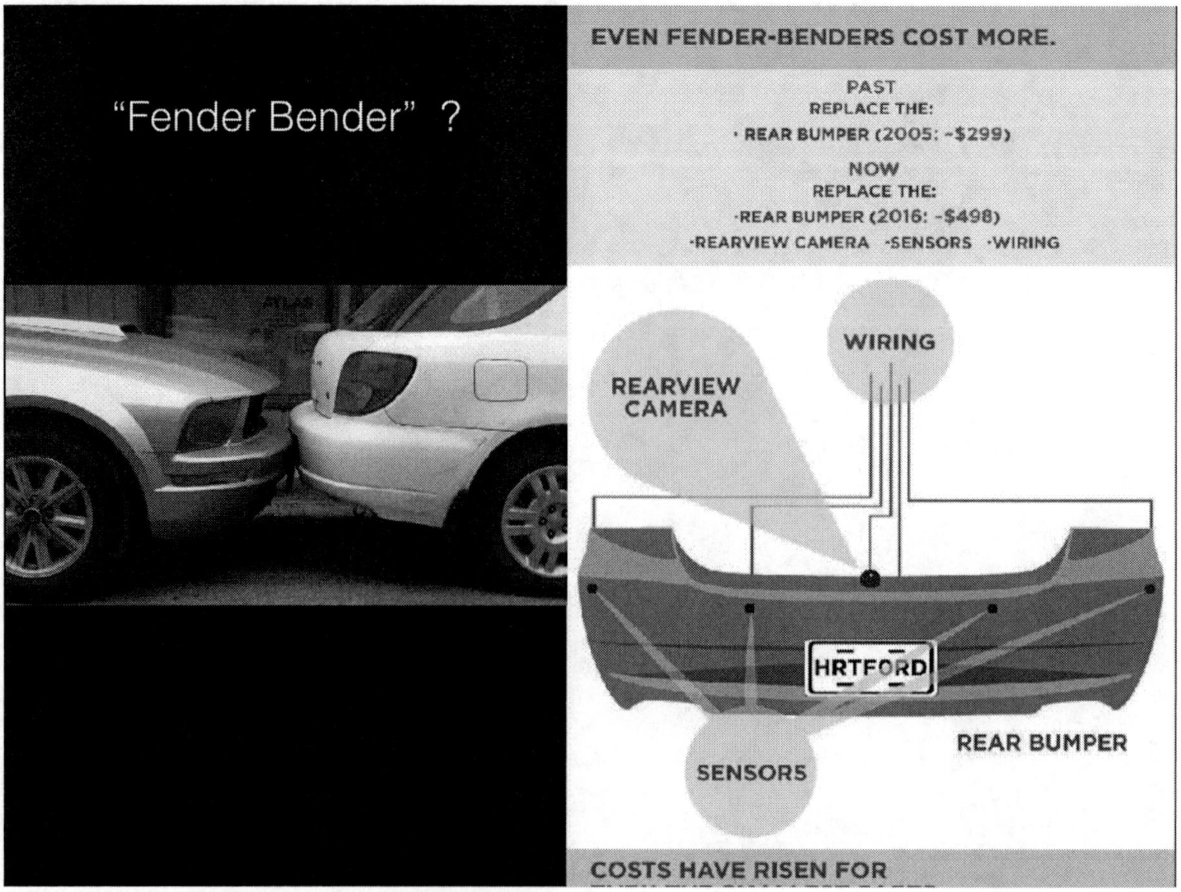
"Fender Bender" ?
EVEN FENDER-BENDERS COST MORE.
PAST
REPLACE THE:
· REAR BUMPER (2005: ~$299)
NOW
REPLACE THE:
·REAR BUMPER (2016: ~$498)
·REARVIEW CAMERA ·SENSORS ·WIRING
WIRING
REARVIEW CAMERA
HRTFORD
REAR BUMPER
SENSORS
COSTS HAVE RISEN FOR

Windscreen repair: sensor recalibration
• Drive around for ~30 mins
• In good light
• On roads with good markings
KM67 JFK

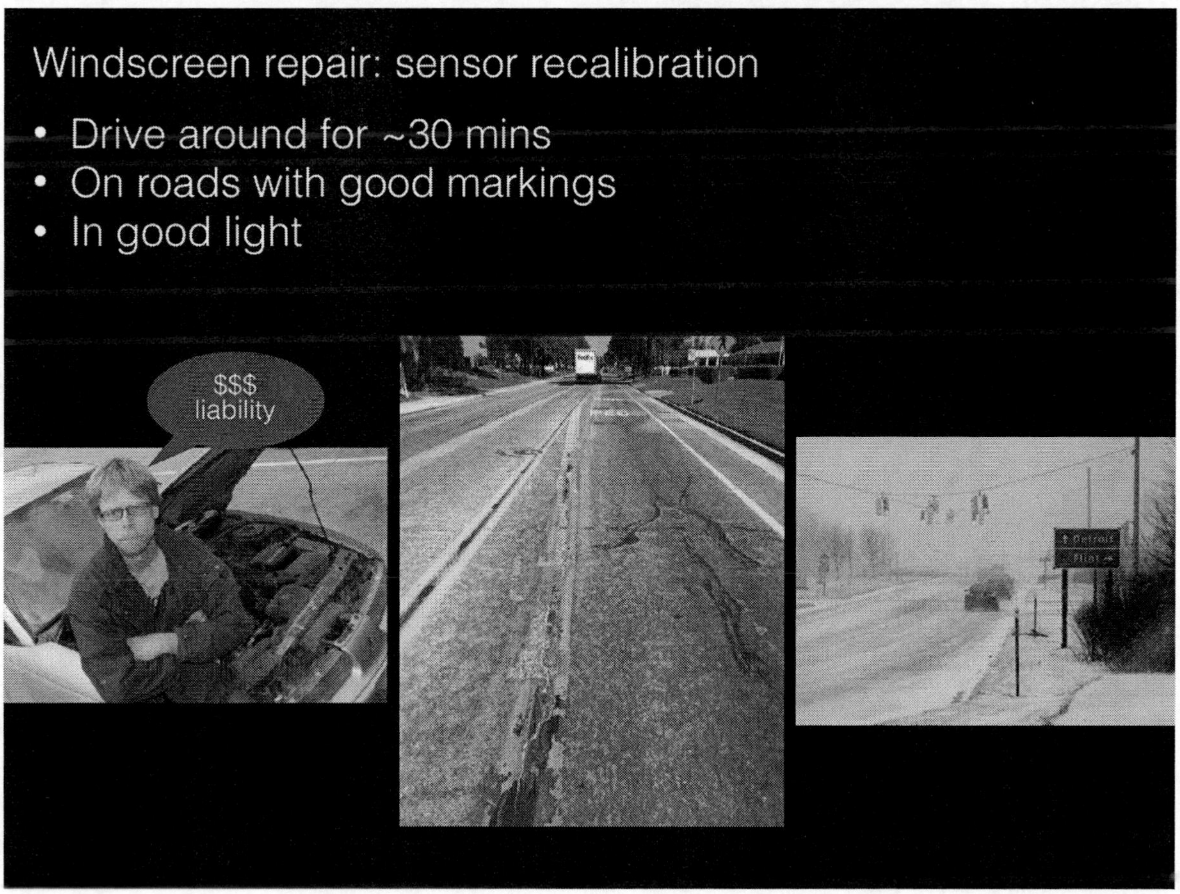
Windscreen repair: sensor recalibration
• Drive around for ~30 mins
• On roads with good markings
• In good light
$$$ liability

VEHICLE SENSORS APPLIED TO TRAFFIC, SAFETY ?

INTERSECTIONS

Overcoming Camera limitations
1280 Lawrence (NB View)
1280 Traffic moving toward the camera is traveling NB on Lawrence
1281 Lawrence (SB View)
1282 Duane (EB View)
1283 Oakmead (WB View)
CENTRAL EXPY
Available Cameras
Future: Capitol
NOTICE: These video images are not recorded or archived
metrotech

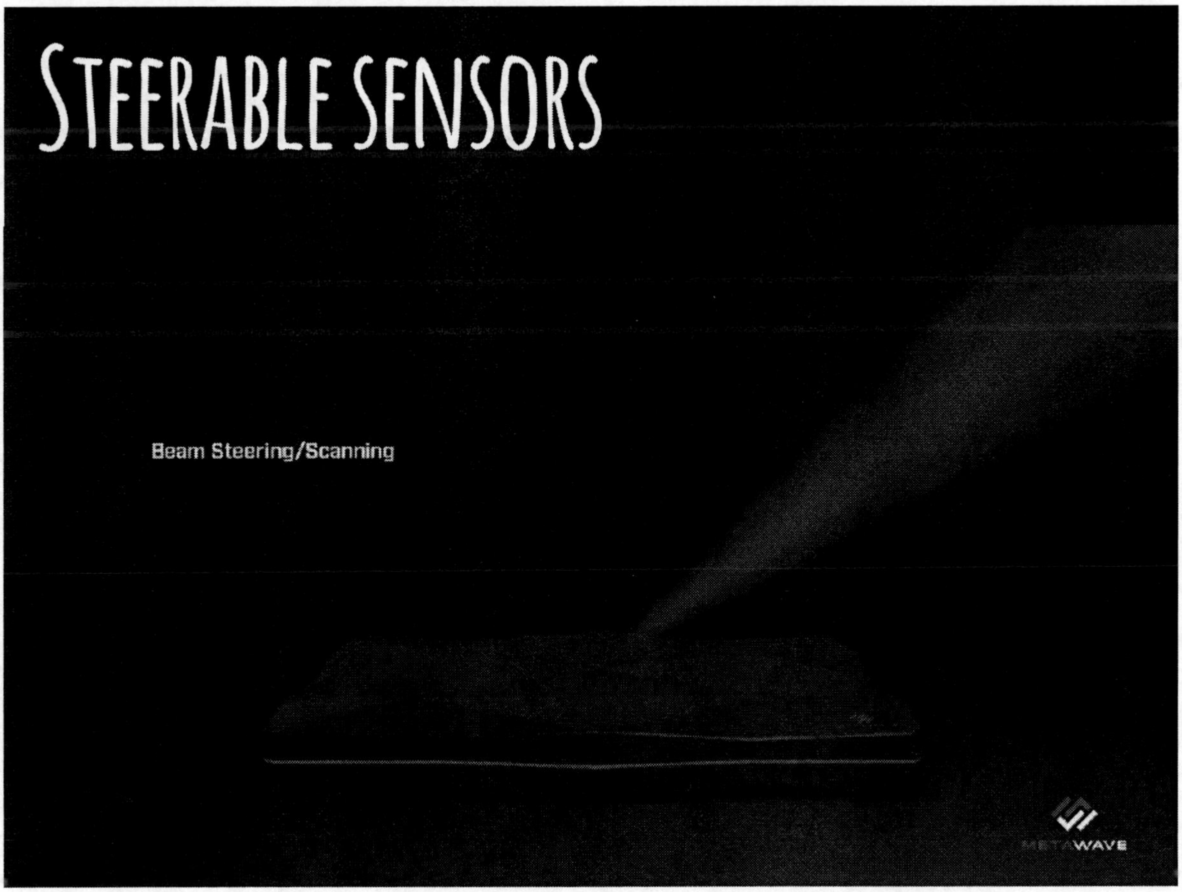
STEERABLE SENSORS
Beam Steering/Scanning
METAWAVE

WARLORD simulation by Metamoto
congested highway scene tracking multiple objects
WAVE
metamoto

STEERABLE RADAR
any time of day
all weather
long range into block
cars, bicycles, pedestrians

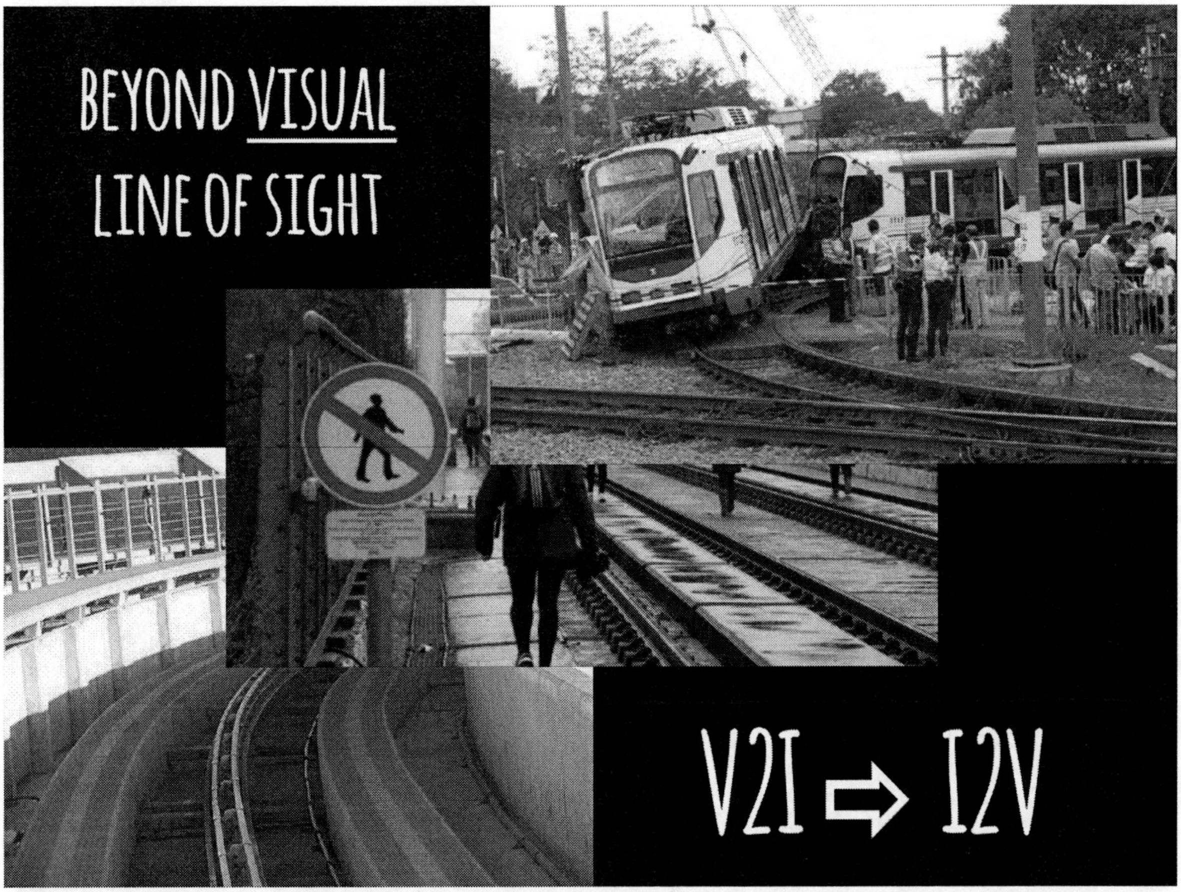

ROADS

TRAFFIC
ROAD QUALITY
ACCIDENT RISK

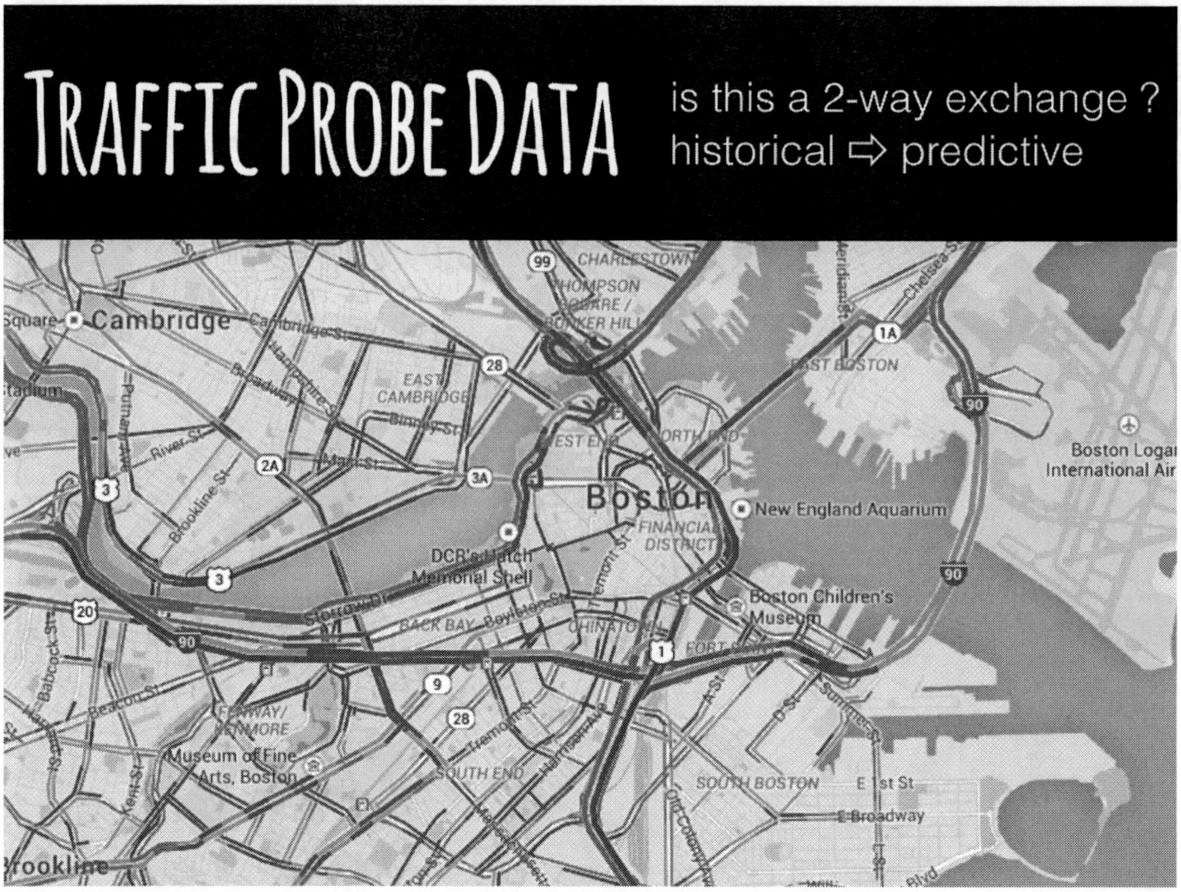
TRAFFIC PROBE DATA
is this a 2-way exchange ?
historical ⇨ predictive
Cambridge
EAST CAMBRIDGE
Boston
FINANCIAL DISTRICT
New England Aquarium
DCR's Hatch Memorial Shell
Boston Children's Museum
BACK BAY
CHINATOWN
EAST BOSTON
SOUTH END
SOUTH BOSTON
Museum of Fine Arts, Boston
Brookline
E Broadway
E 1st St

OTHER PROBE DATA
road quality index / map ?
I-280

OTHER PROBE DATA
road gators, debris

OTHER PROBE DATA
near misses / accidents
predictive heatmap

Robert Seidl
robert@motusventures.com
@motusventures.com
motus
ventures

3M has a long history of innovation in infrastructure materials

1939
1st reflective Stop Sign Installed

1939
1st attempt at reflective pavement markings

3M

2

Infrastructure has a key role to play in the successful migration to Connected and Automated Vehicles (CAVs)

WE BELIEVE THAT...

...humans and assisted/automated vehicles will coexist for many years

...simple solutions can enhance safety and redundancy

....as in aerospace, redundancy is critical for robust and safe CAV navigation

...there may be many solutions for "average" conditions, but safe navigation of "corner case conditions" (adverse weather & work zones) will require advanced solutions

...Infrastructure materials can and should be optimized to be more apparent to machine sensing and vision systems

...Infrastructure will directly impact of where CAVs can be driven

3M Transportation Safety Division

Nearly 80 years of innovation toward creating a safer and more secure world

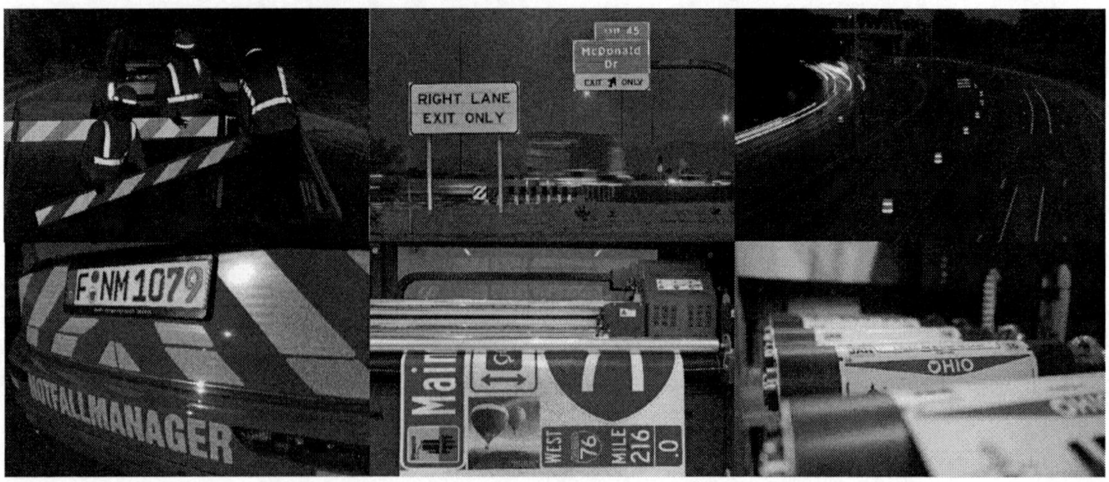

 3M 4

Modified High Visibility of Safety Vest

Studying the effects of additional retroreflective features

- Detection
- Wearer position and activity
- Distance from sensor

Employs convolutional neural network

Exploring both day and night conspicuity

Challenges

- Constantly changing orientation
- Occlusion
- Relatively small area versus viewing distance
- Regulatory constraints (*e.g.* ANSI/ISEA 107-2015) for high visibility safety applications

Modified High Visibility of Safety Vest

Night Activity Classification

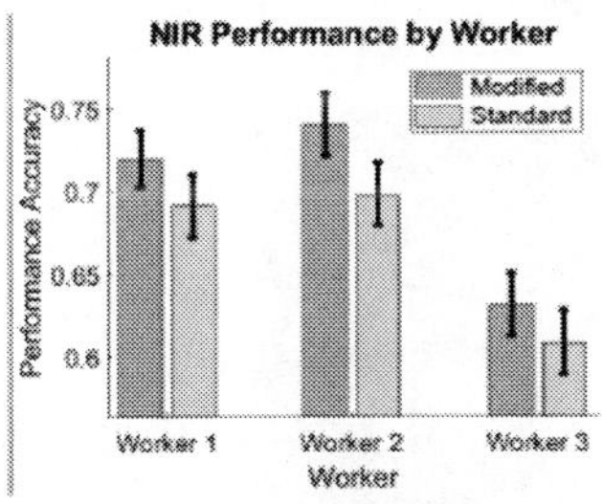

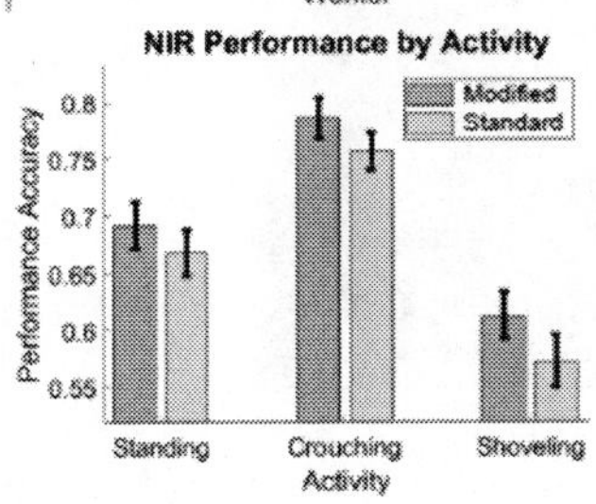

Initial results

- Day-time detection achieved at long distances for fluorescent yellow high-visibility vests using CNN algorithm
- Night-time detection improvement using near-IR retroreflective signal
  - Strong effect of work zone lighting and retroreflective clutter
  - Potential for improved robustness for position, activity classification at different orientations

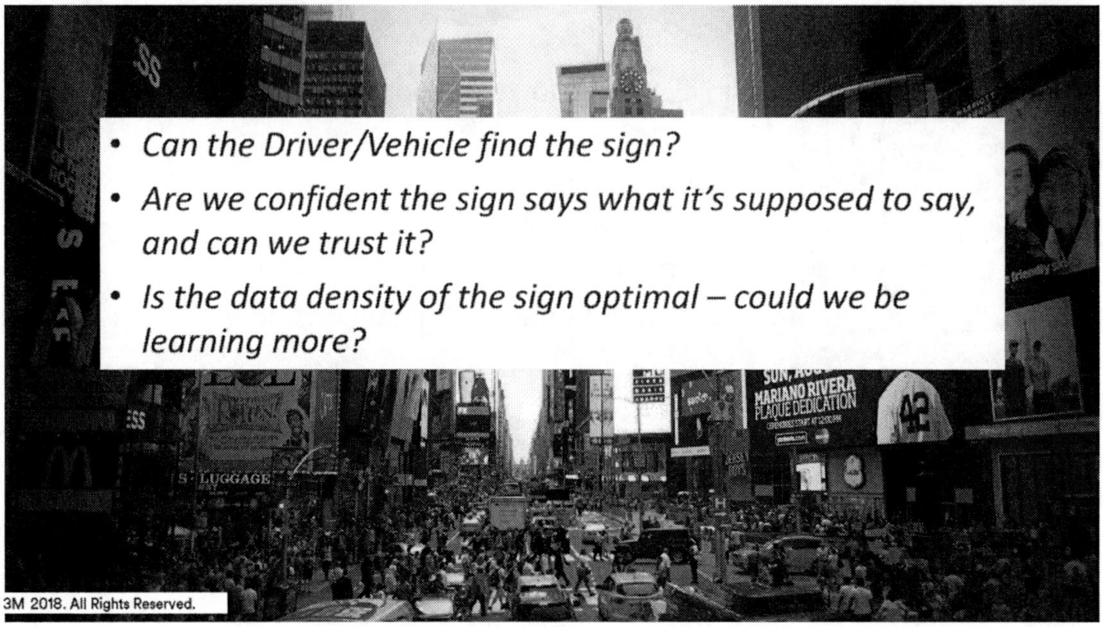

- *Can the Driver/Vehicle find the sign?*
- *Are we confident the sign says what it's supposed to say, and can we trust it?*
- *Is the data density of the sign optimal – could we be learning more?*

3M™ "Smart Code Signs"
Human and Machine Readable Signage

Visible Light

Visible Spectrum Image

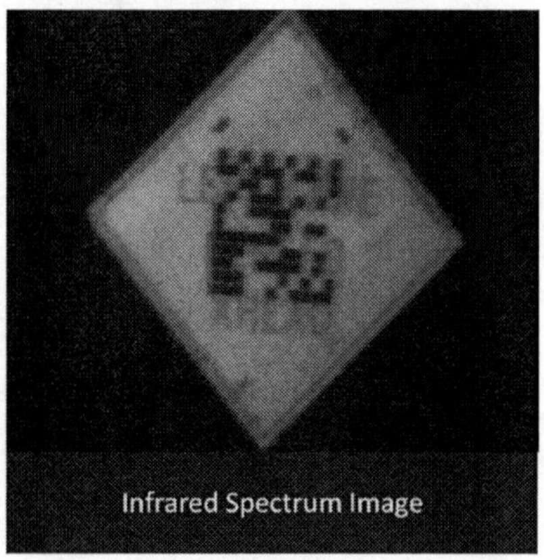

Infrared Spectrum Image

Near-IR Light

3M™ "Smart Code Signs" - Layout

Reduced overhead, designed for small payloads

	3M™ Smart Code	Micro QR	Data Matrix
Min Size	9x7	13x13	12x12
Data Payload	24 bits	24 bits	24 bits
Error Correction	12 bits	8 bits	~16 bits
Number Finder Modules	15	129	88

"Invisible" 2D Bar Code to Enable Machine Readability of Road Signs – Material and Software Solutions
ITSA June 4-7, 2018, Snyder, Dunn, Howard, Potts and Hansen, 3M Co

3M™ Smart Code Sign Layout

Error correction for occlusions typical with signs

✓ Valid (3 blocks corrected)

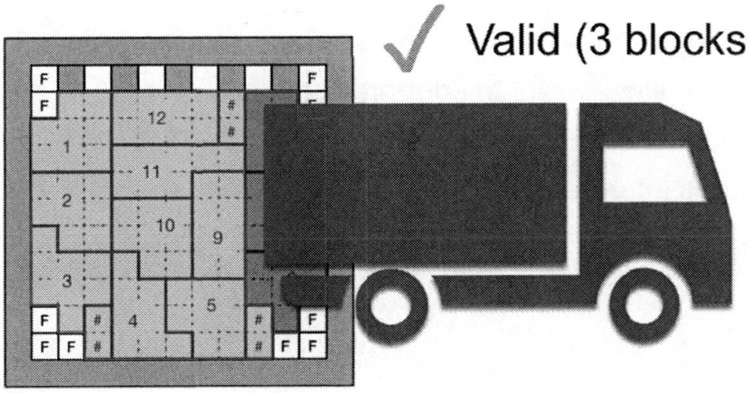

3M™ Smart Code Field Test Results

- Testing between May & Sept 2017 at speeds of 35-90 kph.
- Avg. max. read distance was 95 m daylight, 86 m at night.
- Outdoor durability acceptable
- Accelerated weathering tests suggest lifetime > 5 yrs (evaluation is ongoing)

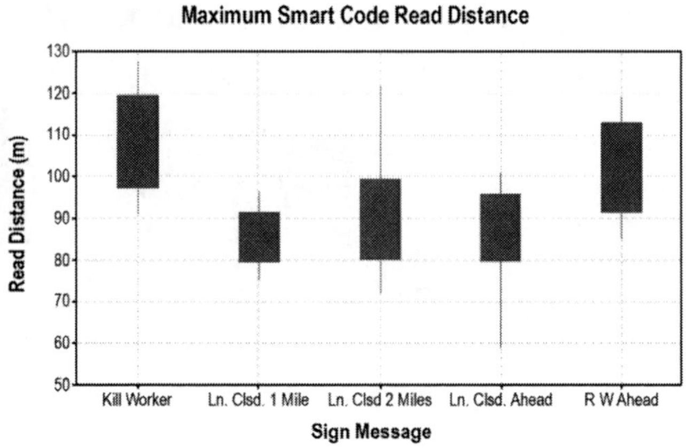

3M™ Smart Code Sign Features

- Maintains human and vision system performance
- Each sign has a unique ID
- Static and/or dynamic metadata (cached or requested)
  - GPS –high quality fiducial to support localization
  - Local/variable conditions & rules
  - Unambiguous interpretation
- Trustable system for metadata & verification

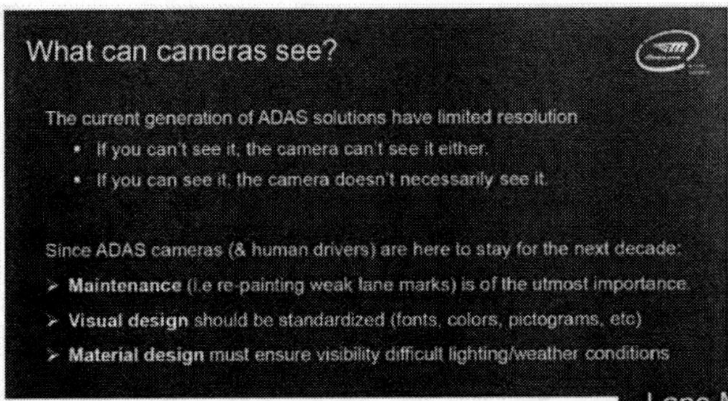

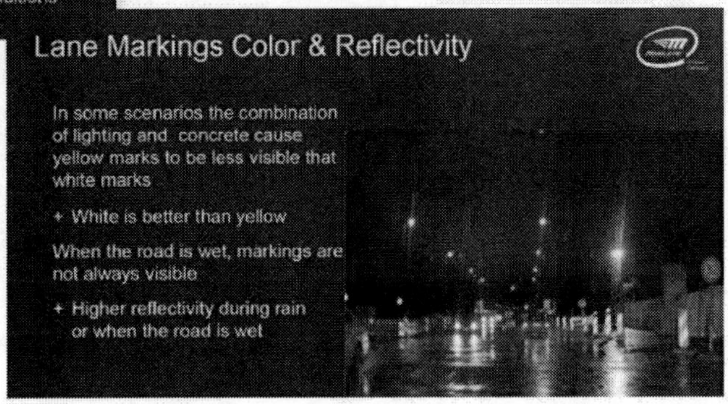

Presented by Mobileye (Epstein)
CEN TC226; April 19, 2018

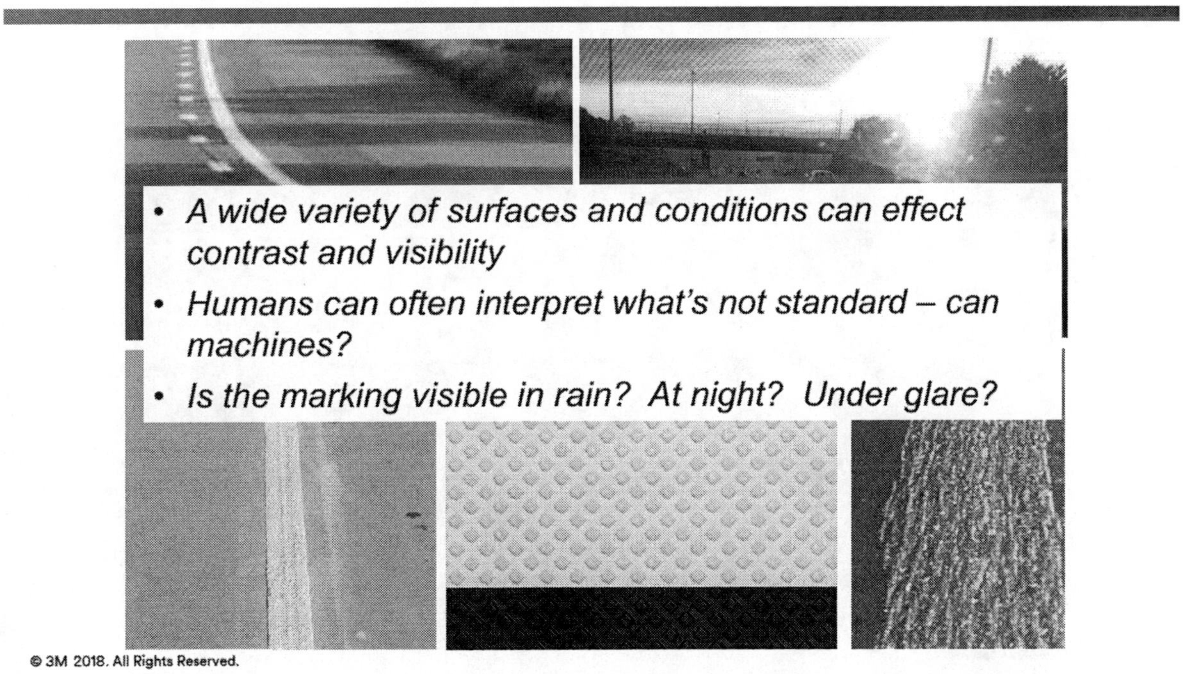

Wet-retroreflection In Action

Yellow Line = Optimized for wet reflectivity

White Line & Arrow = Not optimized for wet reflectivity

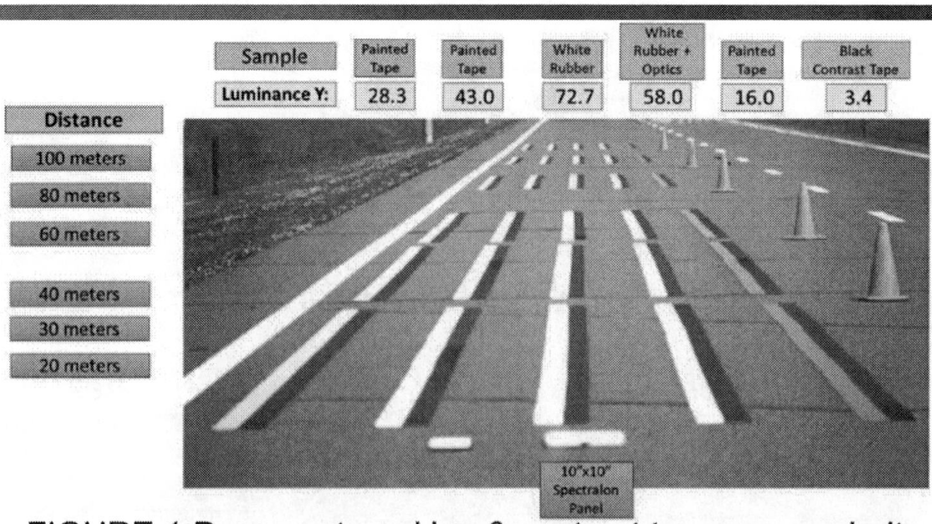

FIGURE 1 Pavement marking & contrast tapes on asphalt

IMPROVED DAYTIME DETECTION OF PAVEMENT MARKINGS WITH MACHINE VISION CAMERAS, ITSA June 4-7, 2018, Whitney, Hedblom and Clear, 3M Co

Edge detection (Sobel contrast gradient magnitude) as a function of distance and width (tape on asphalt)

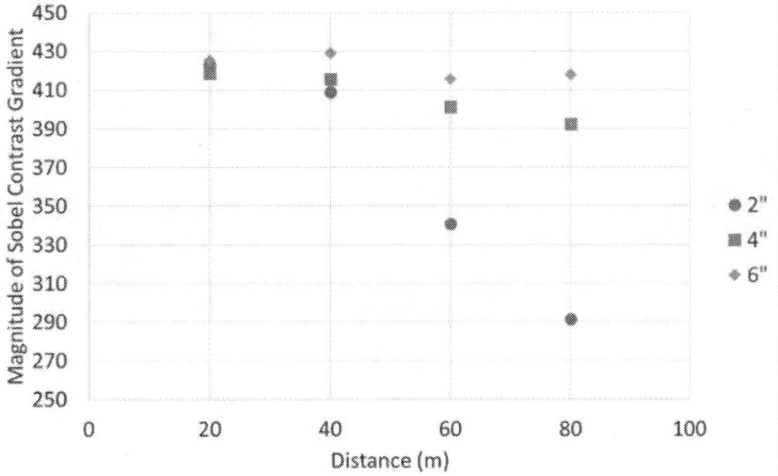

Characteristics of PM for optimal detection by ADAS

- Higher luminance over all lighting conditions means more light is available to each pixel on visible camera to enable detection
- Higher contrast over all lighting conditions improves differentiation between marking and pavement substrate and improves detection
- Edge detection is typically improved by higher cap Y and especially by the use of contrast tape
- Wet retroreflective markings improve detection in nighttime and low-light wet conditions for both humans and cameras

Partnering with Global Industry Leaders

It is about an integrated ecosystem to keep drivers and non-drivers safe

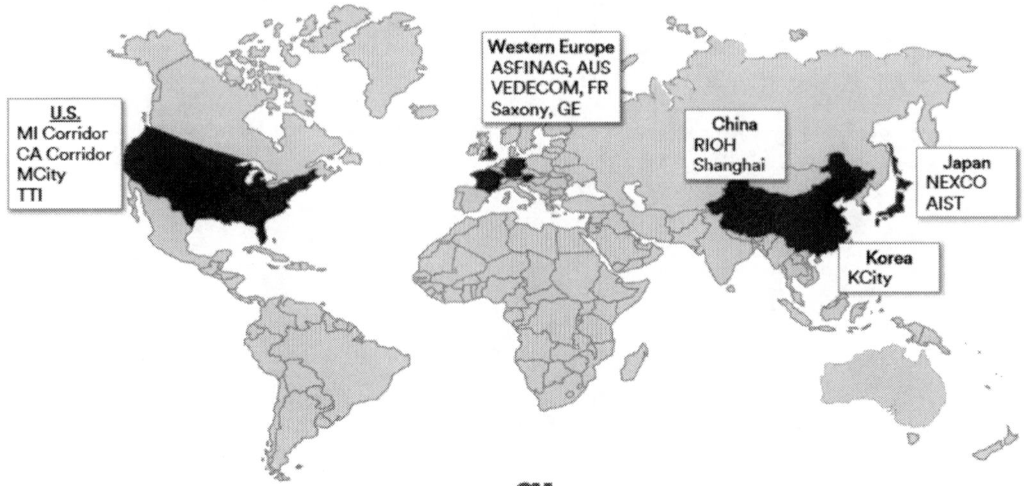

3M

3M Science. Applied to Life.™

Texas A&M Transportation Institute

3M – TTI Research Partnership at RELLIS Campus

- What: 3M Investment of $1.6 MM over 3 years
  - Support two PhD students working on technologies to enable redundancy and safety for CAVs
  - Build a 1000 foot controlled rain test range
  - Provide additional equipment supporting research projects (e.g. sensors and instruments)
- Why:
  - Develop infrastructure materials which leverage the capabilities of evolving vehicle sensors while continuing to serve human drivers
  - Identify infrastructure materials which enable CAVs to function reliably during inclement weather
  - Develop standards which support safety for CAVs
  - Build a diverse, multidisciplinary workforce needed to advance and deploy CAV technology

3M

Current Test Corridor Activities: USA

Objectives:

1) Integrate 3M™ Smart Code Signs and lane markings on active highway corridors
2) Share 3M™ technology and clarify value proposition with partners
   - AOEMs/Tiers - gain insights into vehicle utility
   - DOTs - to understand needs and public benefit
3) To share source code with selected partners to optimize integration

3M

Current Test Corridor Activities: USA

1) Minnesota – MnDOT
   a) Albertville
   b) 3M™ Cottage Grove wet reflective lane markings (with TTI)
2) Michigan – MDOT
   a) I75 northbound from Troy
   b) M59 westbound toward Pontiac
3) California - Caltrans
   a) ~6 miles on Highway 85 in Cupertino, CA
   b) Stevens Creek Blvd to Winchester Blvd

3M

Albertville Corridor

Smart code sign reading over extended time periods

3M

ity":10.0,"l
ity":"None"
ity":"None"
ity":10.0,"l
ity":"None"
ity":"None"
ity":"None"
ity":10.0,"l
ity":10.0,"l
ity":10.0,"l
ity":10.0,"l
ity":10.0,"l
ity":"None"
ity":"None"
ity":"None"
ity":"None"
ity":"None","lon":-93.705661667,"timestamp":"2018-04-09T12:59:49.411199+00:00","velocity":15.793,"lat":45.260156667,"message"
ity":"None","lon":-93.705661667,"timestamp":"2018-04-09T12:59:49.594949+00:00","velocity":15.793,"lat":45.260156667,"message"

3M

Michigan Corridor Ready to Go Live✓

I-75 (Chrysler Hwy) north from Troy

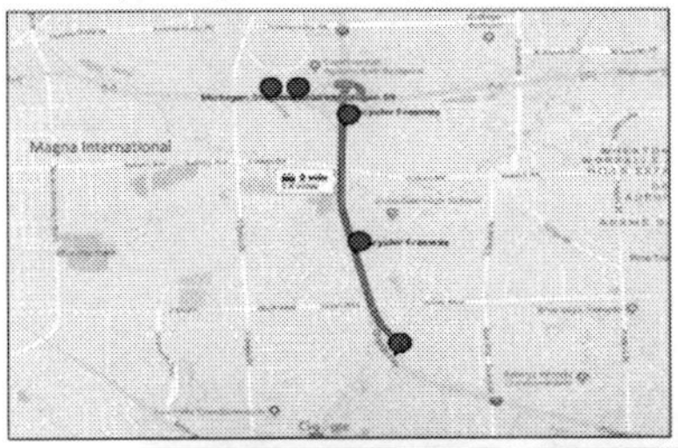

- **Corridor is convenient to partners near Detroit**
- **Smart code signs are in place**
- **Plans for installation of lane markings are on-going with MDOT; installation by early Q3.**

3M

CC85: California Test Corridor

- Pavement markings
  - Skips: 380 AW
  - Edge lines: Thermo with All Weather Elements Series 50
- Smart Code Signs
  - Embedded NIR codes
- Installation Summer 2018

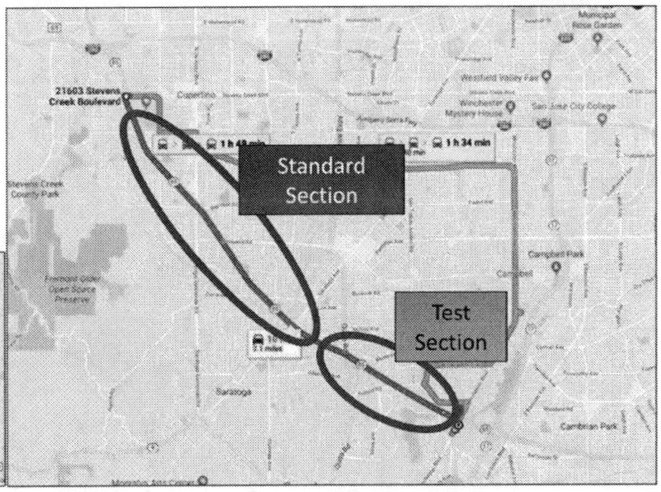

3M

Conclusions

- Infrastructure materials can and should be optimized to be more apparent to machine sensing and vision systems
- Corridors are real world activities to engage partners (DOT's, AOEMs and tiers) and to clarify value proposition of concepts
- Standardization of next generation infrastructure and sensors is critical
- Redundancy is critical in non-optimal (real-world) driving scenarios
- Infrastructure will directly impact of where CAVs can be driven

Questions?

Thank you!

AUTOMATED VEHICLES SYMPOSIUM 2018

BREAKOUT SESSION

Breakout Session #2:

Preparing for Automated Vehicles and Shared Mobility: What Are the Priorities?

The National Academies of
SCIENCES · ENGINEERING · MEDICINE

TRB
TRANSPORTATION RESEARCH BOARD

The National Academies of
SCIENCES · ENGINEERING · MEDICINE

FORUM MEETING #2: JULY 8-9, 2018; SAN FRANCISCO, CA

Forum on Preparing for Automated Vehicles & Shared Mobility

The National Academies of
SCIENCES · ENGINEERING · MEDICINE
TRB
TRANSPORTATION RESEARCH BOARD

Forum on Preparing for Automated Vehicles & Shared Mobility

The National Academies of
SCIENCES · ENGINEERING · MEDICINE
TRB
TRANSPORTATION RESEARCH BOARD

- Bring together public, private, and research organizations to share perspectives on critical issues surrounding the deployment of automated vehicles and shared mobility. Focus on the discussion, identification, and facilitation of evidence-based research needed to deploy these technologies in a manner and timeframe that informs policy to meet long-term goals.
  - The long-term goals include increasing safety, reducing congestion, enhancing accessibility, increasing environmental and energy sustainability, and encouraging economic development and equity.

The National Academies of
SCIENCES · ENGINEERING · MEDICINE

TRB
TRANSPORTATION RESEARCH BOARD

Forum on Preparing for Automated Vehicles & Shared Mobility

The National Academies of
SCIENCES · ENGINEERING · MEDICINE
TRB
TRANSPORTATION RESEARCH BOARD

- Federal/State/Local Agencies (11)
- Private Sector (8)
- Academic/Research Institutions (7)
- National Academies' Boards (7)
- TRB Committees (7)
- Partner Associations (6)

The National Academies of
SCIENCES · ENGINEERING · MEDICINE

TRB
TRANSPORTATION RESEARCH BOARD

Products of the Forum

- Research roadmap
  - NCHRP Project 20-102 on Connected & Automated Vehicles
- White papers
- Authored perspectives
- Workshops and sessions
- *TR News* articles
- Forum website
- Forum annual report
- Research partnerships

AUTOMATED VEHICLES SYMPOSIUM 2018

BREAKOUT SESSION

Breakout Session #2:
Preparing for Automated Vehicles and Shared Mobility:
What Are the Priorities?

1. Hear what key leaders feel are the highest priority research needs
2. Review the catalogue of research needs for AVs
3. Rank the research needs by priority
   - You get to vote!
4. Identify next steps to generate and share answers

AUTOMATED VEHICLES SYMPOSIUM 2018

BREAKOUT SESSION

Breakout Session #2:
Preparing for Automated Vehicles and Shared Mobility:
What Are the Priorities?

DEPARTMENT OF TRANSPORTATION
UNITED STATES OF AMERICA

The National Academies of
SCIENCES · ENGINEERING · MEDICINE

TRB
TRANSPORTATION RESEARCH BOARD

AUTOMATED VEHICLES SYMPOSIUM 2018

BREAKOUT SESSION

Breakout Session #2:
Preparing for Automated Vehicles and Shared Mobility:
What Are the Priorities?

NCHRP

NATIONAL
COOPERATIVE
HIGHWAY
RESEARCH
PROGRAM

The National Academies of
SCIENCES · ENGINEERING · MEDICINE

TRB
TRANSPORTATION RESEARCH BOARD

Forum on Preparing for Automated Vehicles & Shared Mobility

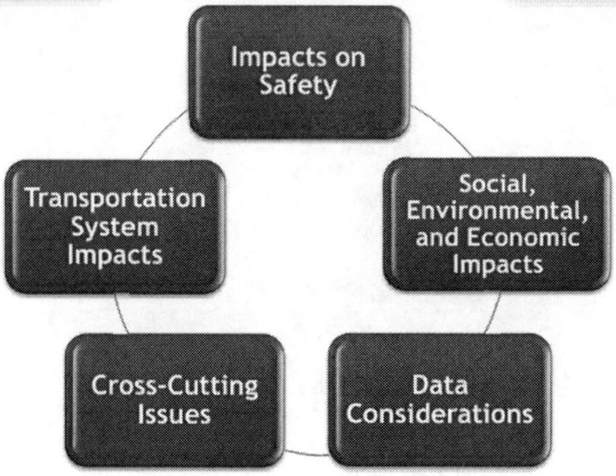

The National Academies of
SCIENCES · ENGINEERING · MEDICINE

TRB
TRANSPORTATION RESEARCH BOARD

IMPACTS ON SAFETY

- Potential adverse impacts on safety due to AVs
- Determining how safe is safe enough?
- State and local policies to ensure safety prior to deployment
- Potential safety scenarios during the transition to highly automated vehicles
- Liability in a world of AVs
- Impacts on law enforcement and first responders
- Minimum set of safety data needed for AV operations and crashes
- Impacts of shared mobility on safety
- Safe operations of commercial vehicles

The National Academies of
SCIENCES · ENGINEERING · MEDICINE

TRB
TRANSPORTATION RESEARCH BOARD

TRANSPORTATION SYSTEM IMPACTS

- Infrastructure Enablers for AVs & Shared Mobility
- Critical paths to level 4/5 automation for light and heavy-duty vehicles (use cases)
- Synergy within the transportation ecosystem
- Potential impacts of higher level automated vehicles and shared mobility on traveler behavior and freight movement
- Impacts of shared mobility on transit
- Impacts on infrastructure funding
- AVs impact on asset management practices
- Implications for work zones

The National Academies of
SCIENCES · ENGINEERING · MEDICINE

TRB
TRANSPORTATION RESEARCH BOARD

SOCIAL, ENVIRONMENTAL, ENERGY, & ECONOMIC IMPACTS

- Net positive and negative social impacts of AV deployment and shared mobility
- Addressing social inclusion and equity
- Impacts on land use, and how land use impacts AVs and shared mobility
- Integrating shared mobility for mega-regions
- Evaluation of pilot deployments to determine contributions to various societal goals
- Preparing the future workforce

The National Academies of
SCIENCES · ENGINEERING · MEDICINE

TRB
TRANSPORTATION RESEARCH BOARD

DATA ANALYSIS & CONSIDERATONS

- Models for sharing of data
- Getting the most out of "Big Data"
- Meeting cybersecurity and privacy challenges
- Framework for automated/connected vehicle pilot and smart cities data analytics for policy guidance

The National Academies of
SCIENCES · ENGINEERING · MEDICINE

TRB
TRANSPORTATION RESEARCH BOARD

CROSS-CUTTING TOPICS

- Alternative scenarios for synergy among automated vehicles, shared mobility, & alternative fuels
- Systems approach – how will this all work from beginning to end with all players in ecosystem?
- Implications for transportation planning and planning models
- What constitutes success/failure of pilots and deployments?
- Education
- Precursory policy analysis for these technologies and services
- Cooperative national research plan for automated vehicles and shared mobility systems

The National Academies of
SCIENCES · ENGINEERING · MEDICINE

TRB
TRANSPORTATION RESEARCH BOARD

Research Needs National Academies/TRB Forum on Preparing for Automated Vehicles & Shared Mobility

The National Academies of
SCIENCES · ENGINEERING · MEDICINE

TRB
TRANSPORTATION RESEARCH BOARD

Forum Research Needs

#10: Impacts on land use, and how land use impacts AVs and shared mobility

- Facilitating active/livable communities
- Impacts on the built environment
- What's in it for rural areas?

The National Academies of
SCIENCES · ENGINEERING · MEDICINE

TRB
TRANSPORTATION RESEARCH BOARD

The National Academies of
SCIENCES · ENGINEERING · MEDICINE
TRB
TRANSPORTATION RESEARCH BOARD

Forum Research Needs

#9: Implications for transportation planning and planning models

- Modeling the impacts of increased penetration of AVs & shared mobility
- New planning tools - revisiting the traditional 4-step planning process
- Moving to objective-based planning
- Evolution to near-term or real-time planning
- Planning for rural areas

The National Academies of
SCIENCES · ENGINEERING · MEDICINE
TRB
TRANSPORTATION RESEARCH BOARD

Forum Research Needs

#8: Impacts of shared mobility on transit, and vice-versa

- Helping transit agencies solve first mile/last mile issues
- Models for integration of AVs & shared mobility with transit/micro-transit

The National Academies of
SCIENCES · ENGINEERING · MEDICINE
TRB
TRANSPORTATION RESEARCH BOARD

Forum Research Needs

#7: Potential impacts of higher level automated vehicles and shared mobility on traveler behavior and freight movement

- Impacts of shared mobility on VMT & system capacity
- Behavior of other road users around highly automated vehicles
- Should zero occupancy vehicles be regulated?

The National Academies of
SCIENCES · ENGINEERING · MEDICINE

TRB
TRANSPORTATION RESEARCH BOARD

The National Academies of
SCIENCES · ENGINEERING · MEDICINE
TRB
TRANSPORTATION RESEARCH BOARD

Forum Research Needs

#6: Addressing social inclusion and equity

- Helping to ensure equity of access to AVs and shared mobility
- How to best serve those with special needs

The National Academies of
SCIENCES · ENGINEERING · MEDICINE

TRB
TRANSPORTATION RESEARCH BOARD

The National Academies of
SCIENCES · ENGINEERING · MEDICINE
TRB
TRANSPORTATION RESEARCH BOARD

Forum Research Needs

#5: Net positive (and opportunities to mitigate negative) social impacts of AV deployment and shared mobility

- Managing unintended consequences (e.g., security, privacy, labor impacts, insurance)
- What are the right metrics and measurements that should be used to improve social outcomes?
- Diverse planning tools to address key social and environmental transportation issues

The National Academies of
SCIENCES · ENGINEERING · MEDICINE

TRB
TRANSPORTATION RESEARCH BOARD

The National Academies of
SCIENCES · ENGINEERING · MEDICINE
TRB
TRANSPORTATION RESEARCH BOARD

Forum Research Needs

#4: Infrastructure Enablers for AVs & Shared Mobility

- Future designs of highways, streets, intersections, etc.
- At what point should we dedicate lanes to AVs?
- Infrastructure needs for V2I
- Establishing a stable technology environment to enable public agencies to support connectivity
- Impacts on public agencies' procurement policies
- Impacts on existing standards and standards-development processes

The National Academies of
SCIENCES · ENGINEERING · MEDICINE

TRB
TRANSPORTATION RESEARCH BOARD

The National Academies of
SCIENCES · ENGINEERING · MEDICINE
TRB
TRANSPORTATION RESEARCH BOARD

Forum Research Needs

#3: State and local policies to ensure safety prior to deployment

- Clarifying federal and state responsibilities
- Roles of simulation, modeling, and off-road testing vs. on-road testing
- Risk management strategies during on-road testing
- Standards for vehicle responses in emergency situations

The National Academies of
SCIENCES · ENGINEERING · MEDICINE

TRB
TRANSPORTATION RESEARCH BOARD

The National Academies of
SCIENCES · ENGINEERING · MEDICINE
TRB
TRANSPORTATION RESEARCH BOARD

Forum Research Needs

#2: Potential safety scenarios during the transition to highly automated vehicles

- Educating drivers on AV capabilities
- Re-engaging drivers' attention when human intervention needed
- Examining the role of human behavior in maintaining safe operation
- Implications of long term mixed vehicle fleet
- Interacting/communicating intent with manually operated vehicles and vulnerable road users

The National Academies of
SCIENCES · ENGINEERING · MEDICINE

TRB
TRANSPORTATION RESEARCH BOARD

The National Academies of
SCIENCES · ENGINEERING · MEDICINE
TRB
TRANSPORTATION RESEARCH BOARD

Forum Research Needs

#1: Models for Sharing of Data

- Public sector use of private sector data
- Sharing of safety & crash data
- Making data available for research and planning models
- Development of V2X communication protocols and procedures
- Protocols for data sharing and management for real-time operations & freight supply chains
- Use of transportation data to support Smart Cities
- Access to data by law enforcement agencies

The National Academies of
SCIENCES · ENGINEERING · MEDICINE

TRB
TRANSPORTATION RESEARCH BOARD

The National Academies of
SCIENCES · ENGINEERING · MEDICINE
TRB
TRANSPORTATION RESEARCH BOARD

Response to Forum Top 10 Research Needs

- Greg Winfree: Texas A&M Transportation Institute
- Jane Lappin: Toyota Research Institute
- Catherine McGhee: Virginia Transportation Research Council
- Reema Griffith: Washington State Transportation Commission
- Audience

The National Academies of
SCIENCES · ENGINEERING · MEDICINE

TRB
TRANSPORTATION RESEARCH BOARD

AUTOMATED VEHICLES SYMPOSIUM 2018

BREAKOUT SESSION

Breakout Session #2:
Preparing for Automated Vehicles and Shared Mobility:
What Are the Priorities?

YOUR TURN TO VOTE!

The National Academies of
SCIENCES · ENGINEERING · MEDICINE

TRB
TRANSPORTATION RESEARCH BOARD

AUTOMATED VEHICLES SYMPOSIUM 2018

BREAKOUT SESSION

Breakout Session #2:
Preparing for Automated Vehicles and Shared Mobility:
What Are the Priorities?

What is Being Done to Address These Needs?

- Peter Sweatman: TRB Forum and CAVita
- King Gee: AASHTO
- Brian Cronin: U.S. DOT

The National Academies of
SCIENCES · ENGINEERING · MEDICINE

TRB
TRANSPORTATION RESEARCH BOARD

TRB Forum Top 10 Priority Research Areas: Next Steps

- Identify & pursue action item for each:
  - Prepare Research Problem Statement?
  - Prepare white paper (including research already completed, underway, or planned)?
  - Assign a Forum Working Group?
- Presentations/dissemination to research organizations, other audiences/groups
  - Pursue needed research by an organization or partnership of organizations
- Other?

The National Academies of
SCIENCES · ENGINEERING · MEDICINE

TRB
TRANSPORTATION RESEARCH BOARD

Forum on Preparing for Automated Vehicles & Shared Mobility

The National Academies of
SCIENCES · ENGINEERING · MEDICINE

TRB
TRANSPORTATION RESEARCH BOARD

Forum Website: https://trb.org/AVSMForum

Staff Contacts:

- Mark Norman (mnorman@nas.edu)
- Katherine Kortum (kkortum@nas.edu)
- Patrice Davenport (pdavenport@nas.edu)
- Adrienne Archer (aarcher@nas.edu)
- Ray Derr (rderr@nas.edu)
- Rich Cunard (rcunard@nas.edu)
- Neil Pedersen (npedersen@nas.edu)

The National Academies of
SCIENCES · ENGINEERING · MEDICINE

TRB
TRANSPORTATION RESEARCH BOARD

Breakout Session 02: Preparing for Automated Vehicles and Shared Mobility: What Are the Priorities?

Organizers

Katherine Kortum, Transportation Research Board
Mark Norman, Transportation Research Board
Richard Cunard, Transportation Research Board
Ray Derr, Transportation Research Board

Reporter

Katherine Kortum, Transportation Research Board

Session Focus

This session reviewed the top 10 research priorities as defined by TRB's Forum on Preparing for Automated Vehicles and Shared Mobility. The Forum was launched in early 2018 to facilitate the evidence-based research that is needed to deploy automated vehicles and shared mobility in a manner and timeframe that informs policy to meet long-term goals. The Forum co-chairs summarized the group's research priorities, a panel of respondents gave their thoughts on the priorities, and the audience members voted on their own set of priorities.

Session Summary

Neil Pedersen, TRB Executive Director, moderated the session.

Brian Cronin of FHWA provided an overview of USDOT work on automated vehicle research. He described the Automated Driving Systems 2.0 policy, which clarified voluntary safety self-assessment processes. The update coming later in 2018, to be called 3.0, will broaden the policy's considerations to reflect multimodal operations. The Accessible Transportation Technologies Research Initiatives (ATTRI) program is designed to enable a complete automated trip and has several applications in development. The USDOT has also convened a working group of several early deployers of low-speed automated shuttles around the country to document their emerging findings. FHWA is engaged in cooperative automated vehicle testing with the Virginia DOT and Transurban, studying speed harmonization, vehicle platooning, and cooperative merging on a managed facility where access is limited to cooperative automated vehicles.

FRA is working on rail grade crossing issues, and FTA has a Strategic Transit Automation Research Plan with seven demonstrations planned through 2022. FMCSA is developing baseline safety performance measures for evaluation of commercial motor vehicles that are operated autonomously. MARAD is considering automated low-speed truck operations at ports and warehouses. NHTSA has Federal Motor Vehicle Safety Standards research projects that have an overall goal of identifying potential barriers for self-certification and compliance verification of innovative new vehicle designs precipitated by automated driving systems. PHMSA is also trying to determine when it is safe enough to use an automated vehicle to assist with the movement of hazardous materials.

Ram Kandarpa of Booz Allen Hamilton summarized their work as contractors on NCHRP 20-102, a project that is updating AASHTO's connected and automated vehicle research roadmap. The first task is commenting on the current research plan, written four years ago, and the 19 problem statements arose from it. Booz Allen reviewed which of these were complete or underway, and which were still valid four years later. The second task was to develop a new research catalog, which has 62 potential projects, and develop one-page problem summaries for each. The final task is to develop white papers on selected topics three times per year for two years.

A Dutch participant shared that they have undergone a similar process and have developed approximately 450 research needs.

Greg Winfree and **Peter Sweatman**, co-chairs of TRB's Forum, described the activities that the group has undertaken to date. Peter Sweatman led a discussion of the top ten priorities as chosen by the Forum members. They are as follows:

1. Models for sharing of data
2. Potential safety scenarios during the transition to highly automated vehicles
3. State and local policies to ensure safety prior to deployment
4. Infrastructure enablers for AVs and shared mobility
5. Net positive and opportunities to mitigate negative social impacts of AV deployment and shared mobility
6. Addressing social inclusion and equity
7. Potential impacts of higher level automated vehicles and shared mobility on traveler behavior and freight movement
8. Impacts of shared mobility on transit, and vice-versa
9. Implications for transportation planning and planning models
10. Impacts on land use, and how land use impacts AVs and shared mobility

Each of these topic areas has several questions within it, and Forum members will be refining these research needs in coming months, as well as outlining which organizations are already considering some of these questions.

A national panel provided their responses to the Forum's top ten priorities. **Reema Griffith** of the Washington State Transportation Commission generally agreed with the priorities, but felt that sustainable finance questions were missing from the list. She also feels that the public needs to be brought along with the experts and the research. **Catherine McGhee**, representing both Virginia DOT and the Virginia Transportation Research Council, emphasized the need to accommodate different environments in urban and rural settings. She also urged researchers to influence the development of shared mobility and AVs rather than merely accommodating it. **Ed Bradley** of Toyota Research Institute described Toyota's two overarching priorities in their research and development efforts: safety and user education. **Chandra Bhat**, of The University of Texas at Austin, began by expressing his appreciation that so many sectors are working together. The number of deaths that still occur on the roadway is "humbling," and provides an opportunity for collective action. He suggested the use of time and considerations of quality of life to be the center of focus for future actions. **King Gee** of AASHTO described the states as fifty living laboratories for automated vehicles and called for a third Strategic Highway Research Program (SHRP3) to address these issues.

When Neil Pedersen opened the discussion to the audience, people brought up a variety of topics. Audience members had differing viewpoints on the future ability of highly automated vehicles to operate seamlessly in mixed traffic. Some believed that insurance companies, pricing, and market forces could drive the adoption of automated vehicles; others believed that consumers will drive the adoption, citing the fact that alternate fuel vehicles still make up less than 4% of new vehicle sales despite high-level promotions and incentives.

Education was a subject of much discussion, and there was general audience consensus that education of policy makers and the public should have been part of the Forum's top ten list. Providing direct exposure to and the chance to ride in AVs or shared rides is likely to help address many people's concerns about fear of change.

Booz Allen Hamilton polled the audience in order to determine the group's sense of priorities and how they compared with the Forum's. As a starter question, approximately half the audience believed that France would win the world cup.

Breakout session members voted on the following research priorities:

1. **Safety**: state and local policies to ensure safety prior to deployment
2. **Safety**: potential safety scenarios during the transition to highly automated vehicles
3. **Safety**: determining how safe is safe enough
4. **Safety**: minimum set of safety data needed for AV operations and crashes
5. **Systems**: infrastructure enablers for AVs and shared mobility
6. **Systems**: potential impacts of high level AVs and shared mobility on travel behavior and freight
7. **Social/Environmental**: net positive and negative social impacts of AVs and shared mobility
8. **Social/Environmental**: addressing social inclusion and equity
9. **Data**: Models for sharing of data
10. **Cross-cutting**: Education

Clearly there were both similarities and differences between the voting of the breakout session attendees and Forum members. Safety and data sharing models were a top priority of both, but research questions about travel behavior, societal impacts, and education had a stronger showing among breakout group members. Forum members placed an overall higher emphasis on applied research.

Suggested Action Items

The breakout session members suggested that the Forum should consider how to transition from research to education to deployment of automated vehicles. There were also calls for additional international collaboration, both on research and on standards.

AUTOMATED VEHICLES SYMPOSIUM 2018
USERS. VEHICLES. INFRASTRUCTURE.
Driven by AUVSI and TRB

Breakout Session #2

Summary of Key Findings and Lessons Learned

- USDOT CAV activities span all modes of transportation, from PHMSA to highways
- Booz Allen Hamilton is in the midst of updating AASHTO's connected/automated vehicle research roadmap
  - Status of projects from last research roadmap
  - Gaps in current understanding
  - Current research needs
- TRB Forum members and breakout participants have both similarities and differences in their research priorities
  - Both prioritize safety and data sharing
  - Breakout attendees gave a higher priority to travel behavior and societal impacts
  - Forum members gave a higher priority to applied research needs

AUTOMATED VEHICLES SYMPOSIUM 2018
USERS. VEHICLES. INFRASTRUCTURE.
Driven by AUVSI and TRB

Breakout Session #2

Research Needs

Top research priorities of the breakout session attendees:

- **Safety**: state and local policies to ensure safety prior to deployment
- **Safety**: potential safety scenarios during the transition to highly automated vehicles
- **Safety**: determining how safe is safe enough
- **Safety**: minimum set of safety data needed for AV operations and crashes
- **Systems**: infrastructure enablers for AVs and shared mobility
- **Systems**: potential impacts of high level AVs and shared mobility on travel behavior and freight
- **Social/Environmental**: net positive and negative social impacts of AVs and shared mobility
- **Social/Environmental**: addressing social inclusion and equity
- **Data**: Models for sharing of data
- **Cross-cutting**: Education

U.S. Department of Transportation

U.S. DOT Automated Vehicle Research Activities

Brian Cronin, FHWA

July 2018

Automated Vehicles Symposium

U.S. DOT Automation Policy

- **Automated Driving Systems (ADS) 2.0**: A Vision for Safety (September 2017)
  - Clarifies Voluntary Safety Self-Assessment process
  - Emphasis on motor vehicle safety
- **2018 Update**
  - Multimodal, surface transportation
  - Broadens considerations to reflect multimodal responsibilities (e.g. operations)

More information on ADS 2.0 is available on the NHTSA website: https://www.nhtsa.gov/technology-innovation/automated-vehicles

2

Research Approach and Highlights

Exploring how to ensure safe, accessible, and efficient integration of automation

Accessible Transportation Technologies Research Initiative

Enabling the Complete Trip

Current ATTRI Applications in Development

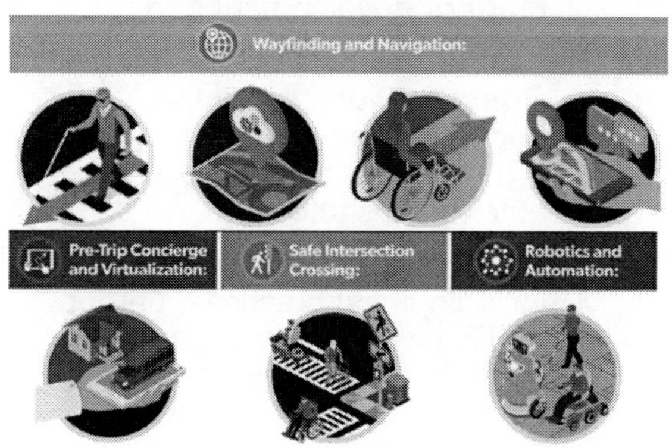

http://www.its.dot.gov/research_areas/attri/index.htm

Data For Automated Vehicle Integration

Voluntary data exchanges to accelerate the safe integration of AVs

WORK ZONE DATA EXCHANGE

- **Purpose**
  - Voluntary adoption of a basic work zone data spec
- **Outcomes within 6 months**
  - Active work zone data feed
  - Feed is used in a meaningful way
- **Big Picture Outcome**
  - Repeatable approach to harmonize local data

Components of a TTC Zone

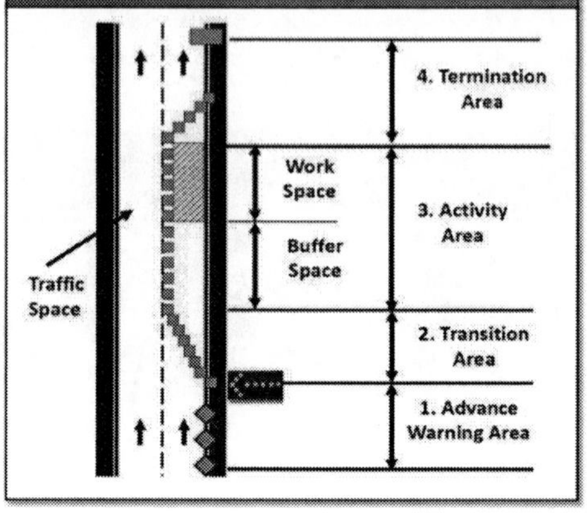

Automated Low-Speed Shuttles

- **Monitor** development and deployment activity
- **Convene** working group of early deployer communities and other organizations interested in small, automated shuttles.
- **Document Emerging Findings** – including best practices, barriers, and lessons learned from early deployers and technology suppliers

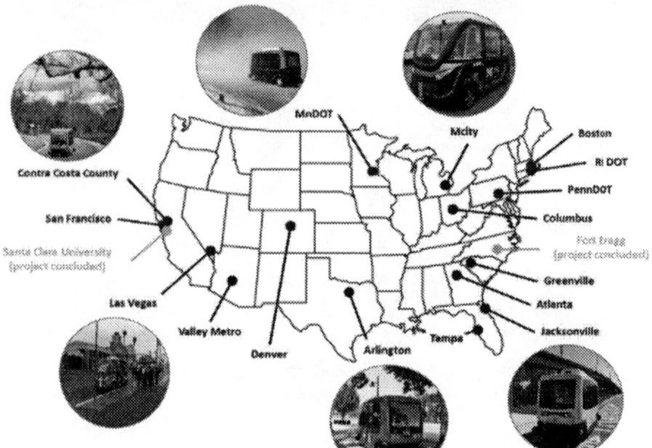

Working Group Member Communities

Impact Assessment

- Framework for AV Impacts
  - Mobility/energy/emissions modeling on a freeway
  - Safety baseline development
  - System dynamics of broader impacts
  - International collaboration on Key Performance Indicators
- Economic and Workforce Impacts from Automation
  - New project to analyze potential impacts from vehicle automation the macro economy including: GDP growth, employment/jobs, wages, consumption, exports, etc.

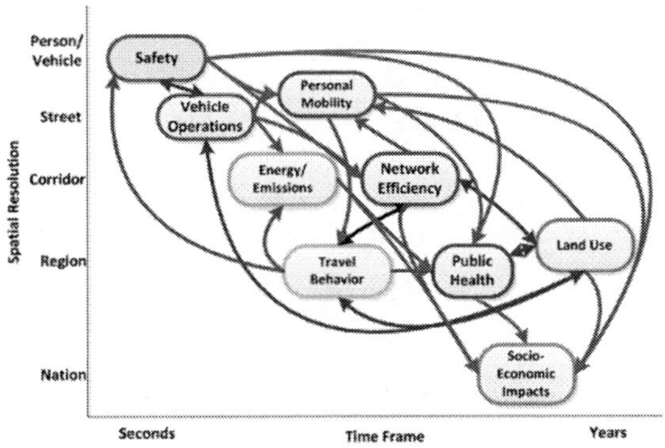

7

FHWA: Cooperative Automation Testing

Cooperative Automated Vehicle Testing with VDOT and Transurban

Objective: Investigate the combination of speed harmonization, vehicle platooning, and cooperative merging at an entrance ramp for a single-lane, managed facility with access limited to cooperative automated vehicles.

Findings: Completed 21 runs with 5 FHWA test vehicles that confirm earlier simulation results that throughput can realistically be increased by more than 50 percent with greater stability and reliability

.

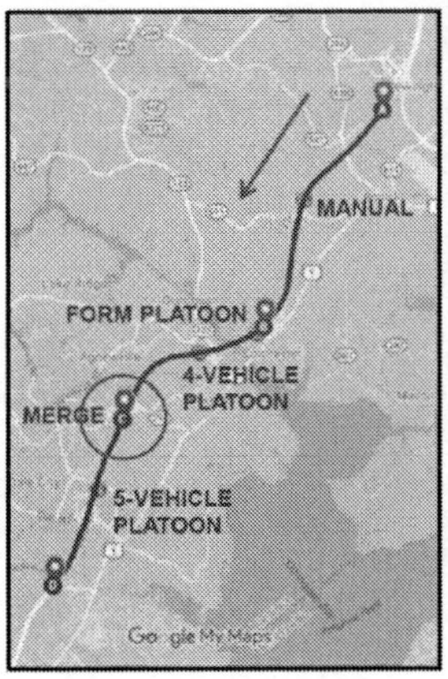

Google Maps Image of the Sites for the Cooperative Automated Vehicle Testing

FHWA: Scenario Planning for Connected and Automated Vehicles

Purpose: To equip agencies with information and tools to consider the uncertainties of CV/AV deployment

- Created scenarios of potential CV/AV deployment
- Conducted workshops to refine test scenarios
- Assessed overarching scenario impacts and implications

Outputs: Practitioner guidance/guidebook, qualitative assessment of impacts and policy implications arising from future described in scenarios

Scenario Building Blocks

FMCSA: Development of Baseline Safety Performance Measures for Automated CMVs

- **Overview: Create baseline safety performance measures** to evaluate automated commercial motor vehicle (CMV) safety systems through analysis of data from CMV naturalistic driving and driver/fleet management systems (100m hours, 33,000 collision events).

- **Goal:** Establish a baseline against which the safety performance of automated CMV functions can be evaluated.

FRA: Automated Vehicles at Highway-Rail Grade Crossings

- The Federal Railroad Administration is researching the interaction between Automated Vehicles (AV) and rail crossings.
- The objective of this project is to explore what information AVs will need to negotiate Highway-Rail Intersections (HRI)
- Deliverables
  - Literature Review
  - Stakeholder Engagement
  - System Requirements

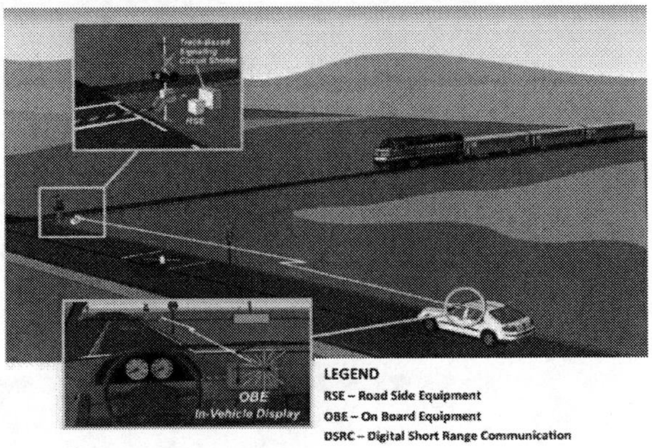

11

FTA: Strategic Transit Automation Research Plan

7 demonstrations planned through 2022

Research in progress:

- Transit Bus Applications of Light and Commercial Vehicle Automation Technology
- Automation Policy Review
- Test Facility Requirements for Automated Transit Buses
- User Acceptance and Human Factors for Automated Transit Buses

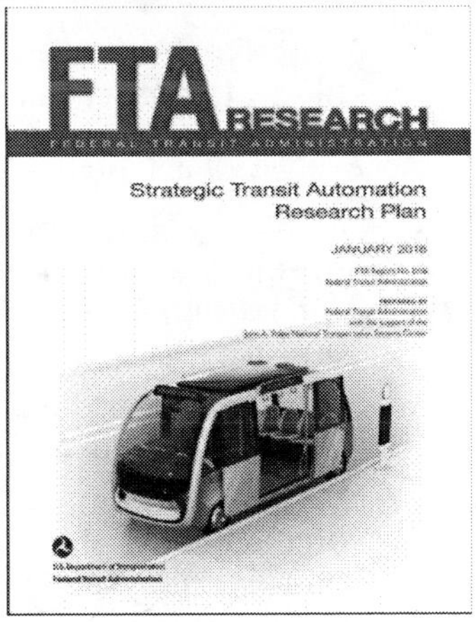

https://www.transit.dot.gov/automation-research

12

MARAD/FMCSA: Feasibility Study: Low-Speed Automated Truck Queue at Ports and Warehouses

- Exploring application of automation to low-speed commercial vehicle operations at port terminals and warehouses
- Review of related studies and papers
- Surveys of industry and technology stakeholders (currently underway)
- Technology scan of existing or near-term enabling technologies
- Regulatory changes, cost estimation and safety benefits
- Report expected in Fall 2018

NHTSA: FMVSS Considerations for Vehicles with Automated Driving Systems

- **Goal**: Research and identify potential barriers for self-certification and compliance verification of innovative new vehicle designs precipitated by Automated Driving Systems (ADSs)
- Developed four conceptual vehicle designs using generic features and characteristics expected in ADS-equipped vehicles with no human-operated controls
- Significant progress in translation of regulatory text has been made
- **Next step** is testing FMVSS test procedures for vehicles without traditional human-operated controls to assess objectivity and repeatability
- **Contractor**: VTTI

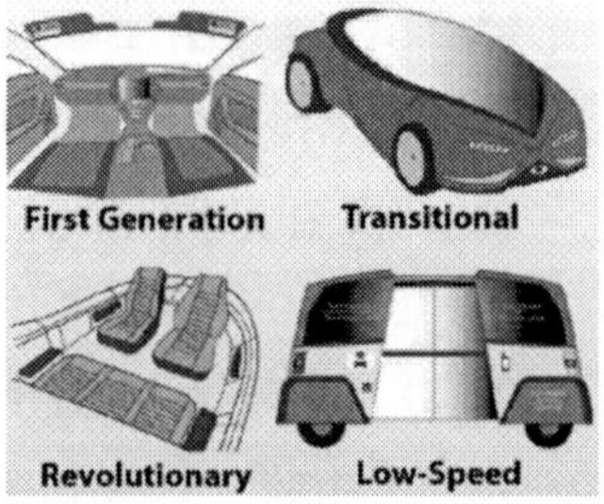

Source: VTTI

14

PHMSA: Carriage of Hazardous Materials by Automated Vehicles

- **Overview:**
  - The Pipeline and Hazardous Materials Safety Administration is researching the potential for applying automation for hazardous materials transportation.
- **Objective:**
  - To identify, characterize, and quantify risks associated with pairing automation technologies and hazardous materials transport.

transportation.gov/av

Booz | Allen | Hamilton

NCHRP 20-102(19) Update AASHTO's Connected Vehicle/Automated Vehicle Research Roadmap

TRB Forum on Preparing for AV & Shared Mobility Meeting #2

JULY 2018

CONSULTING | ANALYTICS | DIGITAL SOLUTIONS | ENGINEERING | CYBER

Booz | Allen | Hamilton

NCHRP 20-102(19) PROJECT

PROJECT TITLE:

- Update AASHTO's Connected Vehicle/Automated Vehicle Research Roadmap

PROJECT OBJECTIVES:

- Update and maintain the research roadmap to reflect the current landscape.

PROJECT REQUIREMENTS

- Literature review and review of relevant research underway or planned by other research institutions,
- Information gathering from
  - Relevant AASHTO events,
  - Annual Automated Vehicle Symposium sponsored by TRB and AUVSI, and
  - TRB Forum on Preparing for Automated Vehicles and Shared Mobility.
- Desk scans of particular topics.

Booz | Allen | Hamilton

TASK 1 – COMMENTARY ON CURRENT RESEARCH PLAN

- Submitted a Task 1 Report summarizing:
  - Previous Roadmap Topics
  - Commentary on Completed Projects
  - Review of Completed Projects
  - Gaps in Research Implementation

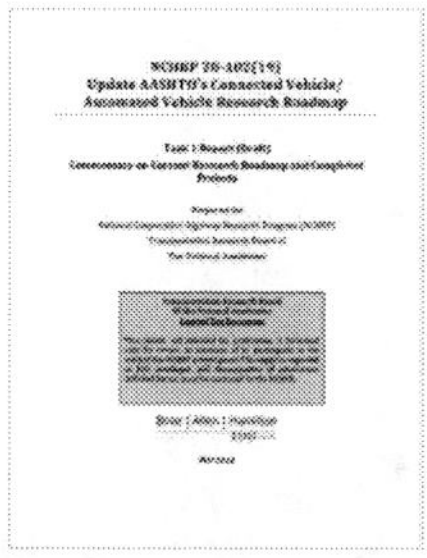

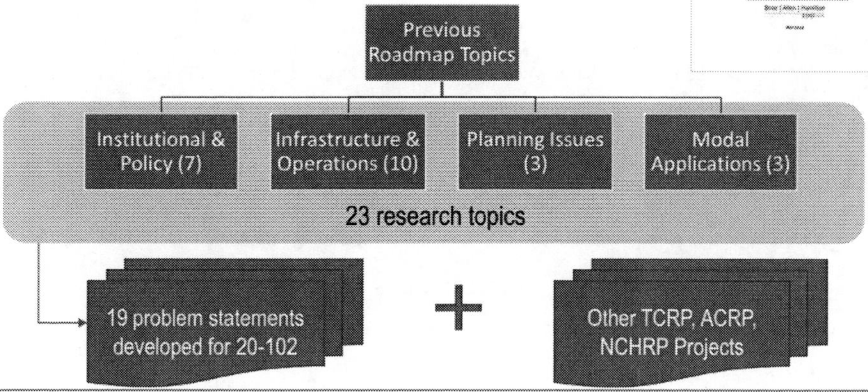

Booz | Allen | Hamilton

NCHRP 20-102 PROJECTS STATUS (1/2)

Project No.	Title	Status	
20-102(1)	Policy and Planning Actions to Internalize Societal Impacts of CV and AV Systems into Market Decisions	Completed in May 2017 by TTI	✓
20-102(2)	Impacts of Regulations and Policies on CV and AV Technology Introduction in Transit Operations	Completed in May 2017 by Kimley Horn	✓
20-102(3)	Challenges to CV and AV Application in Truck Freight Operations	Completed in January 2017 by Booz Allen	✓
20-102(4) + 08-116	Framework for Managing Data from Emerging Transportation Technologies to Support Decision-Making	Not started	✕
20-102(5)	Strategic Communications Plan for Project 20-102	Not started	✕
20-102(6)	Road Markings for Machine Vision	On-going work conducted by TTI	▶
20-102(7)	Implications of Automation for Motor Vehicle Codes	Completed in February 2018 by VTTI	✓
20-102(8)	Dedicating Lanes for Priority or Exclusive Use by CVs and AVs	Completed in April 2018 by Booz Allen	✓
20-102(9)	Providing Support to the Introduction of CV/AV Impacts into Regional Transportation Planning and Modeling Tools	On-going work conducted by TTI	▶
20-102(10) + 03-127	Cybersecurity Implications of CV/AV Technologies on State and Local Transportation Agencies	On-going work conducted by SWRI	▶

NCHRP 20-102 PROJECTS STATUS (2/2)

Project No.	Title	Status
20-102(11)	Mobility-on-Demand and Automated Driving Systems: A Framework for Public-Sector Assessment	On-going conducted by Booz Allen ▶
20-102(12)	Business Models to Facilitate Deployment of CV Infrastructure to Support AV Operations	On-going conducted by WSP ▶
20-102(13)	Planning Data Needs and Collection Techniques for CV/AV Applications	Not started ×
20-102(14)	Data Management Strategies for CV/AV Applications for Operations	Not started ×
20-102(15)	Impacts of Connected and Automated Vehicle Technologies on the Highway Infrastructure	On-going conducted by Booz Allen ▶
20-102(16)	Preparing Traffic Incident Management (TIM) Responders for Connected Vehicles and Automated Vehicles	Not started ×
20-102(17)	Deployment Guidance for CV Applications in the Open Source Application Development Portal	Not started ×
20-102(18)	Minimum Safety Data Needed for Automated Vehicle Operations and Crash Analysis	Not started ×
20-102(19)	Update AASHTO's Connected Vehicle/Automated Vehicle Research Roadmap	On-going conducted by Booz Allen ▶

Booz | Allen | Hamilton

NON 20-102 PROJECTS FROM TRB

Project No.	Title	Status
ACRP 11-03 (S03-11)	Transportation Network Companies: Challenges and Opportunities for Airport Operators	Completed in August 2017 ✓
ACRP 03-48	Advanced Ground Vehicle Technologies (AGVT) for Airside Operations	Not started ×
TCRP J-05 (18-03)	Workplace Implications of Autonomous Vehicles on the Transit Workforce	Not started ×
TCRP J-11 (21)	The Impact of New Technology-Enabled Mobility Services on Public Transportation	Completed in May 2016 ✓
TCRP J-11 (26)	Collaborations and Partnerships between Public Transportation and Transportation Network Companies	Not started ×
TCRP J-11 (27)	Analysis of Low-Speed Automated Vehicle (LSAV): Pilots and Deployments	Not started ×
20-24(111)	State CEO Leadership Forum on Connected & Autonomous Vehicles and Transportation Infrastructure Readiness	Completed in August 2017 ✓
20-24(112)	Connected Road Classification System (CRCS) Development	On-going work ▶
08-117	Impact of Transformational Technologies on Land Use and Transportation	Not started ×

Booz | Allen | Hamilton

This document is confidential and intended solely for the client to whom it is addressed.

Booz | Allen | Hamilton

GAPS ANALYSIS CONSIDERATIONS (1/2)

- FAVP revisions and iterations.
- USDOT and other projects that were funded in the past four years.
- Greater emphasis on data-related themes with the expansion of data availability, collection, sharing, aggregation, and re-dissemination through crowd-sourced, private, and public-sector sources facilitated through Application Program Interfaces (APIs) and other third-party tools.
- Shared economy research areas, including shared mobility and mobility-on-demand are increasingly impacting transportation systems operations and planning, travel demand modeling, and even revenue-driving business models.
- Taxonomy adoption and changes such as SAE J3016 for ADS and SAE J3613 for shared mobility and enabling technologies.
- The state-of-the-industry with respect to automation is more advanced in all modes, including automated truck platooning, automated trains, low-speed automated shuttles, automated personal vehicles (higher-levels of automation becoming market ready), etc.
- Growing recognition throughout the industry on the importance of issues, such as ethics, personal privacy protection, and cybersecurity for the implementation of road vehicle automation systems.

Booz | Allen | Hamilton

GAPS ANALYSIS CONSIDERATIONS (2/2)

- Implemented regulatory frameworks by states to govern the testing and public operation of Highly Automated Vehicles (HAV).
- Better understanding of road vehicle automation operational concepts.
- Active role by IOOs in supporting CAV testing on facilities, engaging in CV pilot deployments, and considering CAV impacts on both short- and long-term planning exercises and modeling, as well as on operations and maintenance of new devices.
- Realization of issues such as adverse weather, human factors, and under-developed infrastructure on the operation of highly automated vehicles.

GAPS IN RESEARCH FROM PREVIOUS ROADMAP (1/3)

Title	Comments
Harmonization of state regulations	While some of the aspects of this research was undertaken as part of NCHRP 20-102(7) on Implications of Motor Vehicle Codes, this research requires ongoing updates based on changes in the state and federal regulatory and legislative policies. As the USDOT releases updates to its original guidance, and federal legislation possibly takes shape, the state regulatory processes need to be reviewed to determine the extent to which harmonization is achievable.
Federal-state-local boundaries of responsibility	This research need is still valid, and it could be tied to another research need that assesses the CV/AV deployment scenarios.
Lessons learned from other transportation technology roll outs	There are lessons to be learned from CV pilot implementations, smart cities, AV pilot programs conducted by state and local agencies, Mobility-on-Demand (MOD) Sandbox demonstrations, etc.
Lessons learned from CV Pilot Deployments	There are multiple projects aiming at documenting the lessons learned from CV pilot deployments. For example, NCHRP 20-102(17) aims at looking into developing deployment guidance for CV applications on OSADP. In addition, a review of lessons learned from the "data management" piece of CV Pilots is included in other projects, such as NCHRP 08-116. There are also numerous non-USDOT sponsored projects aiming at deployment of Connected Vehicle applications in response to the National SPaT Challenge. Consequently, there may be value to consolidating lessons learned from all these deployments.
Tools for predicting AV/CV impacts	CV/AV impacts is a widely researched topic, but most of these studies rely on assumptions that may be outdated. Developing tools or a toolkit to predict the CV/AV impacts would serve as a common platform for agencies to justify investments. This was one of the topics that required such substantial resources that it could not be funded within the resources that were already available, indicating the need for a higher level of investment.

This document is confidential and intended solely for the client to whom it is addressed. 8

Booz | Allen | Hamilton

GAPS IN RESEARCH FROM PREVIOUS ROADMAP (2/3)

Title	Comments
CV/AV applications for maintenance fleets	CV/AVs have great potential in maintenance fleets, such as snow removal, work-zone dampeners, etc. Some of these would be even near-term owing to the controlled environments they operate in.
Relationships of Connected Vehicle and Automated Vehicle systems.	Given the importance of taking a holistic view that treats the infrastructure and the vehicles as portions of an integrated system rather than as distinct systems, there needs to be some basic research identifying the opportunities and challenges of both scenarios.
Traffic control strategies with consideration of AV	Given that legacy vehicles will still dominate the traffic mix for the foreseeable near- and medium-term, this research could wait. However, there is merit to researching the impacts of CAVs on existing traffic control strategies and how those strategies can be enhanced to support mixed traffic including CAVs and conventional vehicles.
Geometric design concepts for AV systems	Given that legacy vehicles will still dominate the traffic mix for the foreseeable near- and medium-term, this research could wait. However, there is merit to researching alternative design concepts for AV-only facilities, such as dedicated AV lanes.
Workforce capability strategies for state and local agencies	In most cases, the agency workforce that is affected by CAVs is broad. For example, it could impact traffic engineers, planners, TIM-responders, emergency maintenance personnel, work-zone contractors etc. Detailed study of implications on each of these categories would entail a significant effort. The current funded roadmap deals with implications on TIM-responders, and is covered under NCHRP 20-102(16).

This document is confidential and intended solely for the client to whom it is addressed. 9

Booz | Allen | Hamilton

GAPS IN RESEARCH FROM PREVIOUS ROADMAP (3/3)

Title	Comments
Including consideration of AV systems in the regional planning process	The team also sees that expanding the scope beyond AVs, to include implications of MOD and shared mobility on regional planning process is also important. NCHRP 20-102(09) formed a good first step in this series, but it was limited in scope. This is a resource-intensive activity that needs additional resources invested.
Assessing transportation system impacts of CV/AV Effects of AV/CV on land use, travel demand, and traffic impact models	This is still one of the high-priority research needs that requires on-going and in-depth research. While topic 2.2 focused on developing the tools, this topic focuses on specific impacts summarized for different scenarios of CV/AV deployment. Research projects, such as NCHRP 08-117 (pending) and USDOT-funded "Benefits estimation framework for Automated Vehicles," are conducting some of this research, but are only taking the first small steps of many that will be needed, and the first small investments out of a larger total that will be needed.
Benefit-Cost analysis of AV transit systems	This research need is still very relevant to public transit agencies to plan their investments. Although some of this is partially covered under certain TCRP J-11 Task Orders, a dedicated study assessing the benefits and costs of automated transit systems, including LSAVs, is important.

TASK 2 – CATALOG OF WORTHY RESEARCH TOPICS

TASK OBJECTIVE:

- To develop a catalog of research projects that should be pursued to assess the open issues and unanswered questions.
- Catalog developed in 2018 will be updated again in 2019.

OUTPUT:

- Developed a DRAFT Catalog with 62 topics under five different areas:

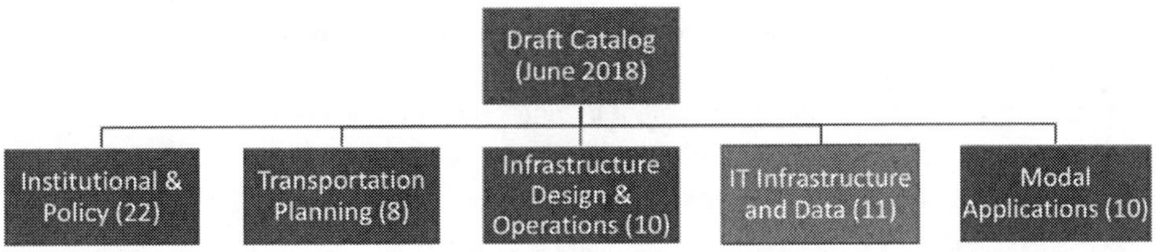

Booz | Allen | Hamilton

TASK 2 – METHODOLOGY

- Developed a DRAFT Catalog with 62 topics under five different areas:

Develop High-level Topic Areas for Classification → Scan of Potential Sources of Ideas → Consolidation of Ideas to a Complete List → Prioritization of Ideas → Develop & Present Catalog of Worthy Projects

Develop High-level Topic Areas for Classification
- Institutional and Policy
- Transportation Planning
- Infrastructure Design and Operations
- IT Infrastructure and Data
- Modal Applications

Scan of Potential Sources of Ideas
- Sourced from TRB Forum.
- SMEs within the team.
- SMEs outside the team/industry needs search.
- Proposed by the team.

Consolidation of Ideas to a Complete List
- Draft List submitted to the panel with 62 project needs.
- Presented at AVS 2018.

Prioritization of Ideas
- Panel survey and comments
- TRB Forum Priority List

Develop & Present Catalog of Worthy Projects
- 1-page problem statement on each.
- List will be updated in 2019

2019 Updates

Booz | Allen | Hamilton

INSTITUTIONAL AND POLICY TOPICS (1/2)

1. Clarification of federal-state relationship in the Federal Automated Vehicle Policy 3.0; State-Local and Federal boundaries of responsibilities.
2. Guidance standards for CV/AV pilot testing.
3. Research alternative non-governmental mechanisms for vetting automated vehicle safety.
4. Effective interactions between AVs and law enforcement and first responders.
5. Research focusing on guidance for CV/AV implication in public agencies and public policy.
6. Development of public education and outreach materials regarding the perception of vehicle automation.
7. Development of common system guidelines for reporting vehicle testing and operations for public awareness.
8. Review of regulatory approaches and development of innovating approaches.
9. Impacts of AVs and Shared AVs on public transportation.
10. Harmonization of state regulations.
11. Lessons learned from other transportation technology roll outs.
12. Workforce capability strategies for state and local agencies.

This document is confidential and intended solely for the client to whom it is addressed.

Booz | Allen | Hamilton

INSTITUTIONAL AND POLICY TOPICS (2/2)

13. Potential Safety Impacts of AVs.
14. Determining how safe is safe enough.
15. Liability in a world of AVs.
16. Impacts of shared mobility during evacuations, disaster relief etc.
17. Critical paths to Highly Automated Vehicles.
18. Synergy within the transportation ecosystem.
19. Evaluation of pilot deployments to determine contributions to various societal goals.
20. Framework for connected vehicle pilot and smart cities data analytics for policy guidance.
21. Alternative scenarios for synergy among automated vehicles, shared mobility, & alternative fuels.
22. Certification of AV performance and safety.

Booz | Allen | Hamilton

TRANSPORTATION PLANNING TOPICS

1. Infrastructure and land use impacts of shared automated vehicles.
2. Equity impacts of SAVs on communities.
3. Research focusing on guidance for socio-economic impacts of transformational technologies.
4. Research focused on urban planning and infrastructure development.
5. Tools for predicting AV/CV impacts.
6. Including consideration of AV systems in the regional planning process.
7. Potential impacts of higher level automated vehicles and shared mobility on traveler behavior and freight movement.
8. Implications for transportation planning and planning models.

INFRASTRUCTURE DESIGN AND OPERATIONS

1. Infrastructure modifications to improve the Operational Domain of Automated Vehicles.
2. Infrastructure analysis of automation systems and work zone interactions.
3. Safety Implications in a mixed vehicle environment.
4. Traffic management strategies with consideration of AV.
5. Geometric design concepts for AV systems.
6. Infrastructure enablers for connected and automated vehicles and shared mobility - Near term and Mid-term.
7. Long-term infrastructure enablers for CAVs/SAVs.
8. Impacts on infrastructure funding.
9. AVs impact on asset management practices.
10. Implications for work zones.

Booz | Allen | Hamilton

IT INFRASTRUCTURE AND DATA TOPICS

1. Recommendations for privacy and open data standards with respect to AVs/CVs.
2. Developing frameworks and best practices for managing data from emerging technologies.
3. Minimum set of data sharing between public agencies and private deployers.
4. Data sharing between AVs and law enforcement, first responders.
5. Minimum set of safety data needed for AV operations and crash-investigations.
6. Getting the most out of "Big Data".
7. Models for exchange of data.
8. Meeting cybersecurity and privacy challenges.
9. Framework for connected vehicle pilot and smart cities data analytics for policy guidance.
10. Data formatting standards for AVs.
11. Data analytics mechanisms to improve infrastructure-AV integration.

Booz | Allen | Hamilton

Booz | Allen | Hamilton

MODAL APPLICATIONS TOPICS

1. Analysis on the impacts of Advanced Automated Transit.
2. State and local impacts of automated freight transportation systems.
3. CV/AV applications for maintenance fleets.
4. Benefit-Cost analysis of AV transit systems.
5. Critical paths to highly automated transit and freight vehicles.
6. Impacts of shared mobility on transit.
7. CAVs and shared mobility impacts on travelers with accessibility restrictions.
8. Alternative scenarios for synergy among automated vehicles, shared mobility, & alternative fuels.
9. Preparing drivers and travelers for automated transit/freight vehicles.
10. Consumer willingness to share "driverless" AV taxis with strangers.

Booz | Allen | Hamilton

NEXT STEPS

- Task 2:
  - Prioritization of Research Ideas
  - Development of 1-page problem summaries.

- Task 3:
  - White-papers on selected topics 3x each year.

This document is confidential and intended solely for the client to whom it is addressed.

AUTOMATED VEHICLES SYMPOSIUM 2018
USERS. VEHICLES. INFRASTRUCTURE.
Driven by AUVSI and TRB

Breakout Session 3 : FHWA Road Weather Management Program: Investigating Automation Technologies and Adverse Weather

Summary of Key Findings and Lessons Learned

- How does the vehicle know when the weather puts the vehicle outside of its ODD? (Open Question)
- Should there be a standardized weather component of ODD? (how is it defined, who is responsible to measure, who should convey, etc.)
  - Visibility
  - Friction
  - Pavement condition (e.g., flooded road)
  - Wind

AUTOMATED VEHICLES SYMPOSIUM 2018
USERS. VEHICLES. INFRASTRUCTURE.
Driven by AUVSI and TRB

Breakout Session 3: FHWA Road Weather Management Program: Investigating Automation Technologies and Adverse Weather

Research Needs

- Technology solutions
  - Vehicles need multiple, integrated sensors (cameras, radar, LiDAR, etc.) for confident localization in all conditions
  - Multiple, external sources (e.g., via V2V and V2I) allow the AV to "see" conditions farther ahead, extend decision horizon (e.g., spatial temporal high-accuracy geo-referenced details)
  - What road weather data needs to be collected, quality checked, assimilated and disseminated and by whom?
- Should there be a published standard of performance for AVs and methods of assessment for OEMs to use?

AUTOMATED VEHICLES SYMPOSIUM 2018 — USERS. VEHICLES. INFRASTRUCTURE. — Driven by AUVSI and TRB

Breakout Session # 4
Testbed Evolution and Collaboration

Summary of Key Findings and Lessons Learned

- Need for minimum level of performance, operational capabilities, certification requirements, etc. for automated vehicles
- Legislative requirements and policies will provide clarity on rules of operations
- Public agencies and industry partners will need to work together to progress through stages to open road testing and deployment, including planning, programming, funding, demonstrating, and conduction leadership and public outreach
- Sharing lessons learned, porting, and scaling proven demos will accelerate their adoption and replication elsewhere
- More to come in white paper

AUTOMATED VEHICLES SYMPOSIUM 2018 — USERS. VEHICLES. INFRASTRUCTURE. — Driven by AUVSI and TRB

Breakout Session # 4
Testbed Evolution and Collaboration

Summary of Key Findings and Lessons Learned

- Consider licensing of vehicles and driving system
- Define a set if limited Operational Design Domains
- Provide means for coordination and collaboration among testbeds
- Repeat tests across multiple sites and compare
- Look for best practices from aviation

AUTOMATED VEHICLES SYMPOSIUM 2018
USERS. VEHICLES. INFRASTRUCTURE.
Driven by AUVSI and TRB

Breakout Session #5

Summary of Key Findings and Lessons Learned

- **Vehicle Security**: Authentication and data trustworthiness are key to protect against external and internal attackers. A lot of research topics were proposed and put together to provide a comprehensive security system.
- **Infrastructure Security**: Authentication and data trustworthiness are key to protect against external and internal attackers. Traffic Management Centers should start thinking about cybersecurity incident response plan as field devices lack protection.
- **Human Factors** in Connected and Automated Vehicle (CAV) cybersecurity has been overlooked. We should consider humans (e.g. passenger, remote operator) in the decision process.
- **Resilience**: The CAV system will fail. Be prepared. Humans can provide a built-in resilience into the system (e.g. absorb and recover from failed system quickly).
- **Panel**: A lot of synergy between vehicle / infrastructure cybersecurity, human factors, and resilience.

AUTOMATED VEHICLES SYMPOSIUM 2018
USERS. VEHICLES. INFRASTRUCTURE.
Driven by AUVSI and TRB

Breakout Session #5

Research Needs

- Design of a **misbehavior detection** system for connected, automated, and electric vehicle. Build attacks in simulation to foster research.
- **Investigate human reactions to cyber-incidents**, how they recover from errors, and design appropriate training and feedback. Include them in reaction after detecting misbehavior.
- **Apply resilience approach to vehicle architecture** and compare to the NIST cybersecurity framework for example.
- Identify **compositional properties** and apply them to each subsystem (infrastructure, vehicle, human) to build a resilient CAV ecosystem.

AUTOMATED VEHICLES SYMPOSIUM 2018

Cybersecurity Breakout session

July 9, 2018, 13:30 – 16:30

Room: Continental 3

OnBoardSecurity

Rules

- No PRESS access
- Chatham House rules:
  - no one participating in the discussion should be identified by name or employer;
  - no tweeting or other social media from other session participants is permitted that violates that commitment.

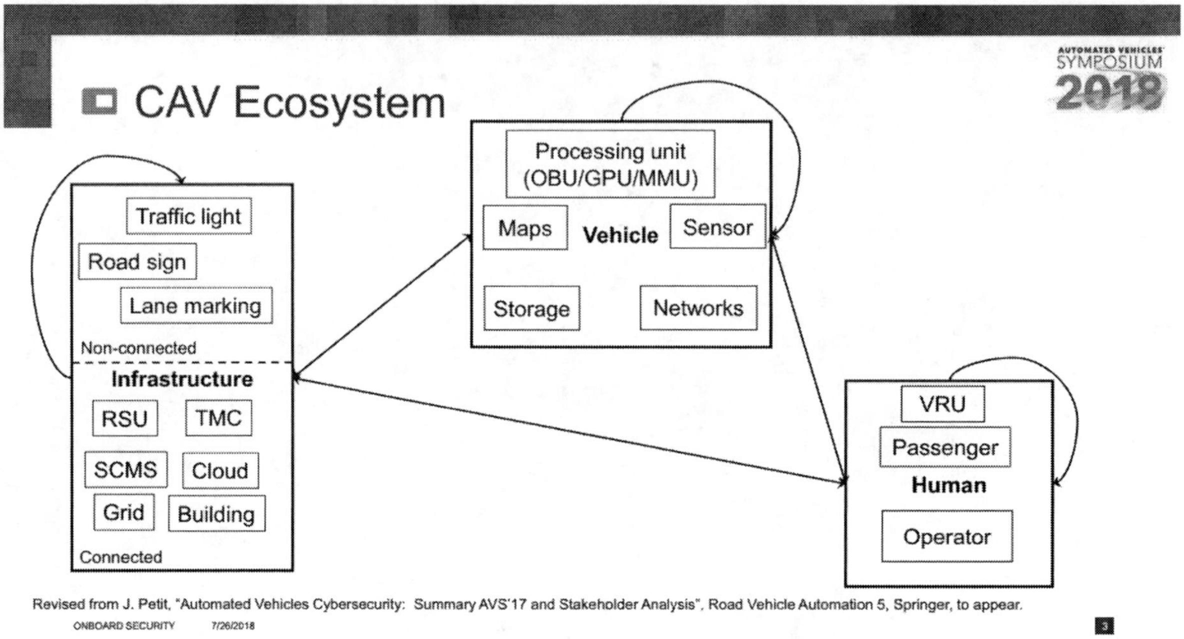
CAV Ecosystem
AUTOMATED VEHICLES SYMPOSIUM 2018
Processing unit (OBU/GPU/MMU)
Maps
Vehicle
Sensor
Storage
Networks
Traffic light
Road sign
Lane marking
Non-connected
Infrastructure
RSU
TMC
SCMS
Cloud
Grid
Building
Connected
VRU
Passenger
Human
Operator
Revised from J. Petit, "Automated Vehicles Cybersecurity: Summary AVS'17 and Stakeholder Analysis", Road Vehicle Automation 5, Springer, to appear.
ONBOARD SECURITY 7/26/2018
3

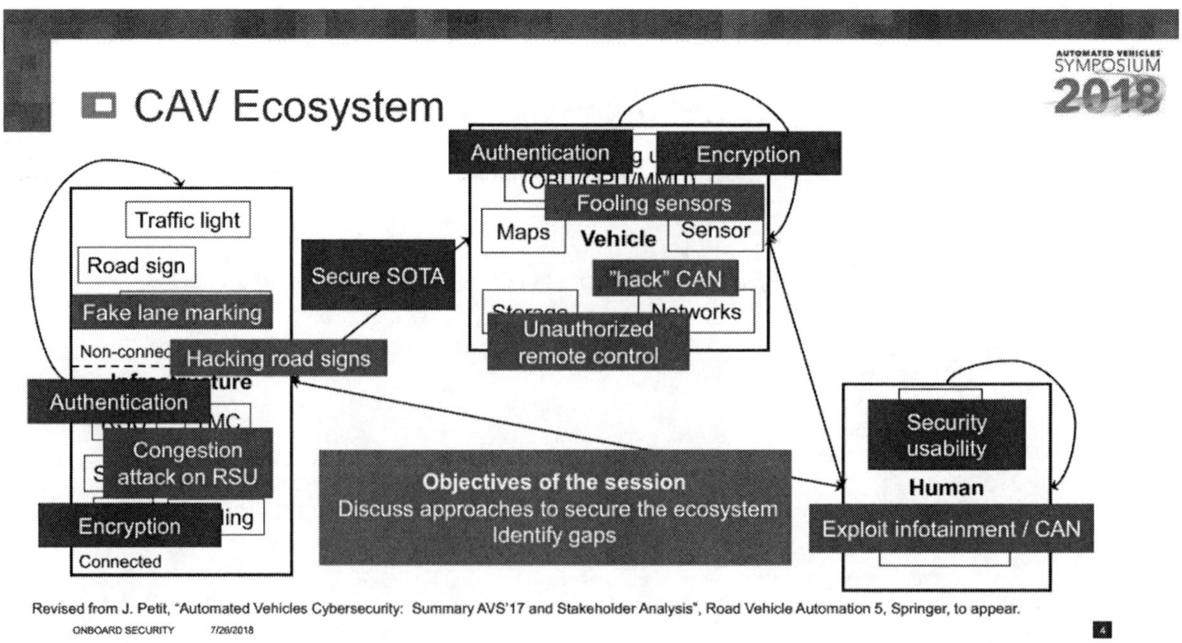
CAV Ecosystem
AUTOMATED VEHICLES SYMPOSIUM 2018
Authentication
Encryption
Fooling sensors
Maps
Vehicle
Sensor
"hack" CAN
Secure SOTA
Unauthorized remote control
Traffic light
Road sign
Fake lane marking
Hacking road signs
Authentication
Congestion attack on RSU
Encryption
Connected
Objectives of the session
Discuss approaches to secure the ecosystem
Identify gaps
Security usability
Human
Exploit infotainment / CAN
Revised from J. Petit, "Automated Vehicles Cybersecurity: Summary AVS'17 and Stakeholder Analysis", Road Vehicle Automation 5, Springer, to appear.
ONBOARD SECURITY 7/26/2018
4

Agenda

AUTOMATED VEHICLES SYMPOSIUM 2018

Time	Topic	Speaker
13:30 – 13:40	Welcome and Introductory Remarks	Jonathan Petit *OnBoard Security*
13:40 – 14:10	Connected and Automated Vehicle Security	Jonathan Petit *OnBoard Security*
14:10 – 14:40	Systemic Cybersecurity: Improving TMS Field Network Resilience	Robert Heller *Southwest Research Institute*
14:40 – 15:10	Cybersecurity and Human Factors	Shannon C. Roberts *UMass Amherst*
15:10 – 15:30	Break	
15:30 – 16:00	Intelligent Transportation System Resilience	Avi Mersky *U.S. Army Engineer Research and Development Center*
16:00 – 16:30	Panel Discussion	All speakers

ONBOARD SECURITY 7/26/2018 5

Breakout Session 05: Cybersecurity of Automated Vehicles Ecosystem

Organizers

Dr. Jonathan Petit, Senior Director of Research, OnBoard Security

Reporters

Dr. Jonathan Petit, Senior Director of Research, OnBoard Security

Session Focus

Presenting the status of research in securing connected and automated vehicles CAV). Covering the three main components in the CAV ecosystem: vehicle, infrastructure, humans and discussing how to build-in resilience.

Session Summary

PowerPoints of most of the presentations are available on the AUVSI website at: [add link]

Suggested Action Items

NOTES:
#attendees: ~50

Jonathan
Overview of the session
Vehicle security: overview of challenges and solutions
Questions about simulator and should we rely on it
Question about score being static or dynamic

Robert
Traffic Management System (TMS) composed of field devices being networked and not secure (default password/username)
Present NCHRP 03-127
26 different state DOT are surveyed the field devices being deployed and looked at common threats (human harm, loss of confidence in TMS, traffic disruption) following STRIDE.
Conflict resolution between traffic controller (SPaT, TIM - traffic message sign)
Task 3 is to look at classes of attacks and provide recommendations
Question: any findings from Task 2? The conflict resolution.

Camera captures opening of traffic cabinet and can check if this is a technician. Removing the card from the cabinet put the traffic in flash and it would have to resolve this before having a green.

STRIDE done at infrastructure and vehicle level. Are they compatible?

Shannon
Training, transfer of control
Change the culture around cybersecurity because humans are not aware of automotive cybersecurity
Training and feedback systems improve driver's effectiveness.
Driving simulator study will be performed at UMass/Amherst.
Monitor facial expression, action, heart rate -> interview (what information would you have liked to have to help you understand the situation?)
Human-centered design: see how people can recover from errors (→ similarity with resilience)

Test cases: screen freeze, vehicle stops, unresponsive brakes, etc. What about GPS spoofing (see differences between what they live through)

Question: response of driver in seconds, there is a lot of variability in how fast people can take over.
Starting in Sept. 2018

Avi
Explained what is resilience: ability to absorb and recover from events. To get resilience, one should start by identifying the critical functionalities of the system (e.g. data security, availability).
Two approaches to measure resilience: metrics (by interviewing expert), model (of network / process) in order to find the loss function.

By simulating events (and knowing the normal traffic) you can measure the effects of different infection rate. This tool could also be used offensively to identify the most critical roads to attack.

Question: what differences exist between resilience (network science) and the NIST cybersecurity framework?
Resilience seems to rely on good data. How much quality does it require?

Often when a city is affected by disruption, it is mainly due to its constrained network. So, we can't simply change the road network (on the fly). One way to approach is to focus on recovery.

Panel
Questions for resilience

- What properties of attack affect the most the absorb / recover phases?
- How to add resilience during design? What does that mean?

- Does recovery require involvement of human and therefore is linked to HF?

Questions for human factor

- What types of attack would you test on users in the driving simulator?
- Should users be involved to mitigate cyberattacks?
- In the reaction of an attack should the handover consider the level of attention and knowledge of the user?
- How to notify users that attack mitigation is going on and consider all possible type of user (e.g. visually impaired, kids)?
- What data should be sent to remote operator in order to facilitate response?
- If Level 5 AV, do we still need training and feedback of user?

HF clearly depends on the development in HMI, applications, activities performed within the vehicle.

Questions for infrastructure

- What is the top priority to secure infrastructure?
- What is considered outside of the security domain?
- Traffic Management System field devices are often unsecure, is that due to lack of training of human? (relation to HF)

Questions for all

- How sensitive to regulation is each system?

Resilience / HF: Humans provide a built-in absorption of events. For example, at a traffic light (all red), the drivers know how to solve the situation and not get stuck forever.

Resilience: there is a difference between nature and attacks. Nature mostly attacks capacity while attack targets control. However, some more catastrophic events could affect both capacity and control (loss of power, snowstorm blocking roads). Failures (from nature or attack) will have different behavior. So, could that help detection of attack? How much different are they? And is the reaction the same no matter what the failure is?
HF: do users response differently to incident due to nature or to attack?

Infrastructure / resilience: capacity is key
TMCs start slowly to think about response plan to cyberattack. Until now, they were relying on air-gap.
TMS field devices are not supposed to push information to TMC (only pull from TMC). So, if you see a TMS field device traffic while not requested it means something wrong is going on.

HF: Humans are pretty good at handling incident (they know how to delegate to others passenger when overloaded). But what about level 4 automation where passengers will have less awareness?

By building resilience, can it "corner" the attacker to a limited set of attack possibilities?

Comment from audience: always assume components are untrustworthy.
If each component has its own security methodology, we should aim at harmonization. Is SAE J3061 a good start?
It feels like we are waiting for a catastrophic event to wake up and integrate security. This sounds counter-intuitive as attacks are safety issues!
Insurance companies will have an important role (esp. in terms of liability). Insurance companies would require audit of cybersecurity protection, which would force adoption.
Regulation (NHTSA) could also influence adoption, but this wasn't considered in the revisit of FMVSS.

Cybersecurity of Traffic Management Systems

SOUTHWEST RESEARCH INSTITUTE®

NCHRP 03-127

TRB
TRANSPORTATION RESEARCH BOARD

SwRI PRAETORIAN

Agenda

- Program Goals
- Program Stakeholders
- Program Research Team
- Program Process
- Program Tasks
- Schedule
- Upcoming Items

SwRI PRAETORIAN

Program Goals

- Improve cybersecurity posture of Traffic Management Systems (TMSs)
- How?
  - Review State of the art across multiple disciplines
  - Assess representative TMS systems and equipment
  - "Red Team" high risk equipment
  - Develop Guidance for state and local agencies that aids in identifying:
    - Risks to their current field networks
    - Recommended changes they may implement to reduce those risks
    - Implications of CV and AV technologies on the field networks
    - Best practices for wide deployment of CV and AV technologies
  - Promote adoption and industry participation

Program Stakeholders

- NCHRP
  - State, county, and city transportation agencies
  - Metropolitan Planning Organizations (MPOs)
  - FHWA Division Office
  - Transit agencies and operators
  - Highway service patrol/contractors
  - State and local law enforcement
  - Fire departments and emergency medical services
  - Transportation Systems Cybersecurity Framework (TSCF)

Program Research Team

- Southwest Research Institute
  - San Antonio, TX
  - 70 years R&D
  - 20+ years CS efforts, commercial & government
  - 25+ years ATMS development, 40+ ATMS deployments
  - Industry leaders in CAV (V2X)
- Praetorian Cybersecurity
  - Austin, TX
  - Experts in embedded devices, cloud, cryptanalysis, and IoT
  - CS incident response and forensics
  - Secure development lifecycles and red team methodologies

Program Process

- Research and Model Development
  - Literature Search
  - Model V2X and TMS equipment
  - Identify Cyber Security Goals

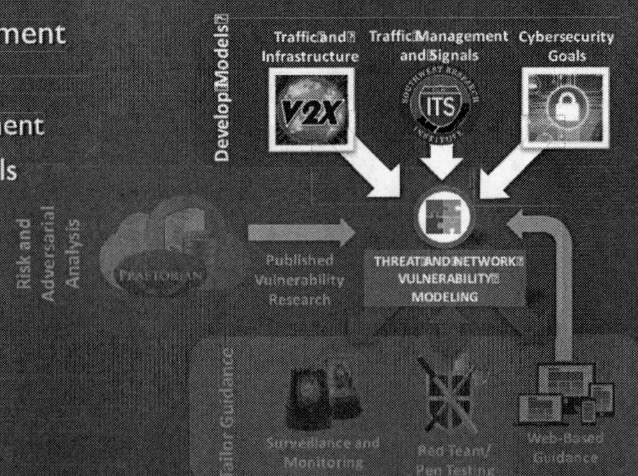

Program Process

- Risk and Adversarial Analysis
  - Identify high risk equipment
  - Evaluate equipment security
  - Refine models
  - Responsible disclosure
  - Red Team Review

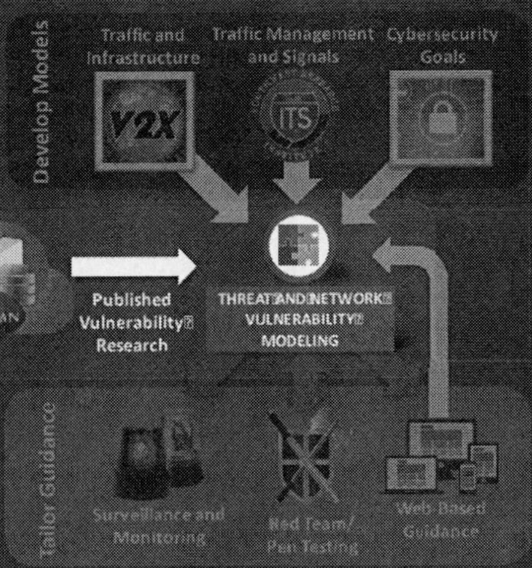

SwRI PRAETORIAN Cybersecurity of Traffic Management Systems - NCHRP 03-127 INTELLIGENT SYSTEMS swri.org 7

Program Process

- Guidance Development
  - Use models to develop guidance
  - Help TMSs mature System CS
  - Tailor to TMSs needs
  - Feedback improvements into models/framework
  - Host workshops
  - Promote adoption into industry

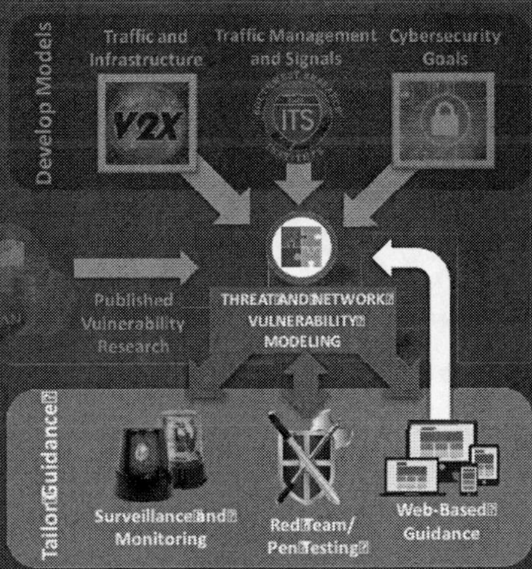

SwRI PRAETORIAN Cybersecurity of Traffic Management Systems - NCHRP 03-127 INTELLIGENT SYSTEMS swri.org 8

Program Process

- Broken down into following tasks:
  - T1 – Literature and Ongoing Efforts Review
  - T2 – Risk Assessment of Typical TMS Designs
  - T3 – Adversarial Assessment of High-Priority Systems
  - T4 – Cyber-Attack Mitigation and Response Guidance Tool Development and Workshop
  - T5 – Cybersecurity Primer Development for Connected Vehicles (CV) and Automated Vehicle (AV) Technology Deployment

SwRI PRAETORIAN Cybersecurity of Traffic Management Systems - NCHRP 03-127 INTELLIGENT SYSTEMS swri.org 9

Task 1 – Literature Review

- Reviewed state of the art across multiple disciplines
- Identified works that can be leveraged for TMS
- Investigated existing CS maturity models and guidance

- Delivered (NCHRP03-127_Cybersecurity_Literature_Review.pdf)
- Ongoing throughout project period of performance

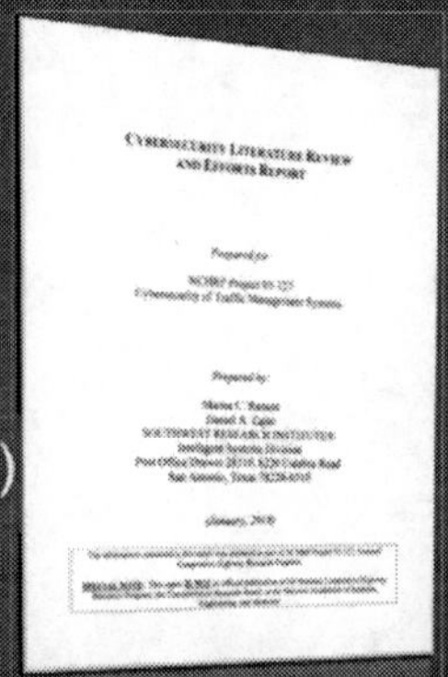

SwRI PRAETORIAN Cybersecurity of Traffic Management Systems - NCHRP 03-127 INTELLIGENT SYSTEMS swri.org 10

Task 2 – Risk Assessment of TMS

- Identify high priority equipment
- Recommend mitigating strategies
- Recommend best practices
- CS Modeling of TMS
  - Models of TMS equipment and networks
  - Integrate into framework for estimating risk
  - Prioritize CS evaluation of equipment

- Delivered, Under review

RISK ASSESSMENT PROCESS

Identify Assets of Typical TMS Design
Create Models: TMS, Field Devices, V2X
Decompose the TMS Design
Identify the Threats
Document The Threats
Rate The Threats

SwRI PRAETORIAN Cybersecurity of Traffic Management Systems - NCHRP 03-127 INTELLIGENT SYSTEMS swri.org 11

Task 3 – Adversarial Assessment

- "Red Teaming" or Penetration Testing
- Provide security baseline of high-priority devices
- Recommend mitigations
- Responsibly disclose identified vulnerabilities
- Update/refine models

SwRI PRAETORIAN Cybersecurity of Traffic Management Systems - NCHRP 03-127 INTELLIGENT SYSTEMS swri.org 12

Task 4 – CS Guidance Development

- Web-based guidance
  - Adapt to TMS budgets and risk exposure
  - Configurable for equipment and complexity
  - Produce guided self assessment
- Recommend countermeasures
- Identify standards for AV/CV integration
- Recommend best practices

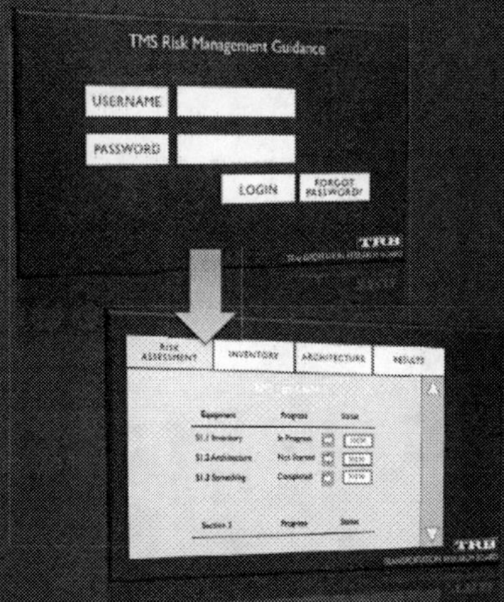

Task 5 – Attack Mitigation and Response

- Provide CS guidance for vehicle-to-infrastructure (V2I) systems
- Promote adoption by state and local transportation communities
- Evaluate applicability of Transportation Systems Cybersecurity Framework (TSCF)

Current Status

- Literature search published to project page (Link)
- Risk Assessment under NAS review for publishing
- Seeking partners for equipment
- Seeking input from TMS and those involved in CAV technologies
- Developing web guidance tool
  - Ensuring data sensitivity requirements are met
  - Testing usability of tool for different users

SwRI PRAETORIAN Cybersecurity of Traffic Management Systems - NCHRP 03-127 INTELLIGENT SYSTEMS swri.org 15

Schedule

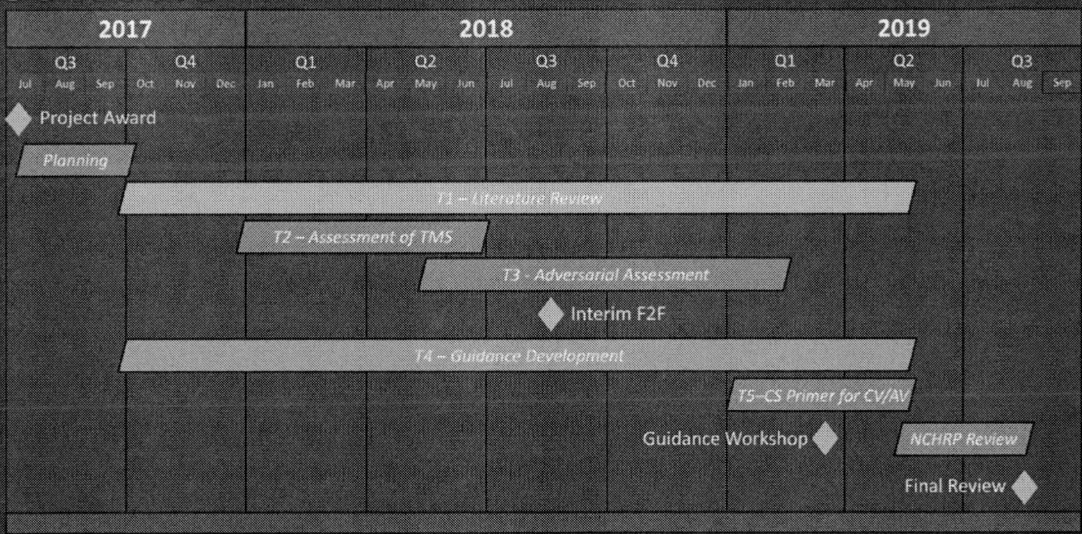

SwRI PRAETORIAN Cybersecurity of Traffic Management Systems - NCHRP 03-127 INTELLIGENT SYSTEMS swri.org 16

Questions?

- For more information please contact:

Project Manager:
Marisa Ramon
Marisa.Ramon@SwRI.org
(210)-522-3869

Principle Investigator:
Ben Abbott, Ph.D.
Ben.Abbott@SwRI.org
(210)-522-2802

Domain Manager
Robert Heller, Ph.D.
Robert.Heller@SwRI.org
(210)-522-3824

SwRI PRAETORIAN Cybersecurity of Traffic Management Systems - NCHRP 03-127 INTELLIGENT SYSTEMS 17
swri.org

Quantifying Cyber-Resilience

Igor Linkov

Risk and Decision Sciences Team Lead,

US Army Corps of Engineers;

Adjunct Professor, Carnegie Mellon University

Igor.Linkov@USACE.Army.mil, 16172330969

Avi Chaim Mersky

ORISE Post-Doc, US Army Corps of Engineers;

Avi.C.Mersky@USACE.Army.mil

AVS, July 2018

US Army Corps of Engineers.

US Army Engineer Research and Development Center

2500 Employees

Over 1000 engineers and scientists 28% PhDs; 43 % MS degrees, $1B Budget Annually

Cold Regions Research Engineering Laboratory (Hanover, NH)

Risk and Decision Science Team (15 employees, Boston)

Topographic Engineering Center (Alexandria, VA)

Construction Engineering Research Laboratory (Champaign, IL)

Environmental Laboratory
Coastal & Hydraulics Laboratory
Geotechnical & Structures Laboratory
Information Technology Laboratory
Headquarters (Vicksburg, MS)

ERDC

BUILDING STRONG®

Innovative solutions for a safer, better world

2

What Does Cyber/Physical Resilience Really Means?

https://www.airforcetimes.com/news/your-air-force/2016/01/15/northrop-lays-out-vision-for-cyber-resilient-next-gen-fighter/

3

Risk -- "a situation involving exposure to danger [threat]."

Security -- "the state of being free from danger or threat."

Resilience -- "the capacity to recover quickly from difficulties."

Definitions by Oxford Dictionary

30 | NATURE | VOL 555 | 1 MARCH 2018

Don't conflate risk and resilience

'Risk' and 'resilience' are fundamentally different concepts that are often conflated. Yet maintaining the distinction is a policy necessity. Applying a risk-based approach to a problem that requires a resilience-based solution, or vice versa, can lead to investment in systems that do not produce the changes that

Igor Linkov, Benjamin D. Trump *US Army Corps of Engineers, Concord, Massachusetts, USA.*
Jeffrey Keisler *University of Massachusetts Boston, USA.*
igor.linkov@usace.army.mil

BUILDING STRONG®

ERDC
Innovative solutions for a safer, better world

4

Calls for Increased Resilience

The White House
Office of the Press Secretary
For Immediate Release October 31, 2013

Presidential Proclamation -- Critical Infrastructure Security and Resilience Month, 2013

CRITICAL INFRASTRUCTURE SECURITY AND RESILIENCE MONTH, 2013

BY THE PRESIDENT OF THE UNITED STATES OF AMERICA

A PROCLAMATION

Over the last few decades, our Nation has grown increasingly dependent on critical infrastructure, the b
our national and economic security. America's critical infrastructure is complex and diver
both cyberspace and the physical world -- from power plants, bridges, and interstates to
massive electrical grids that power our Nation. During Critical Infrastructure Security anc
resolve to remain vigilant against foreign and domestic threats, and work together to furt
systems, and networks.

"resilience" means the ability to anticipate, prepare for, and adapt to changing conditions and withstand, respond to, and recover rapidly from disruptions.

The White House
Office of the Press Secretary
For Immediate Release May 11, 2017

Presidential Executive Order on Strengthening the Cybersecurity of Federal Networks and Critical Infrastructure

EXECUTIVE ORDER

(vi) Effective immediately, it is the policy of the executive branch to build and maintain a modern, secure, and more **resilient executive branch IT architecture.**

BUILDING STRONG®

5

System Risk/Security and Resilience

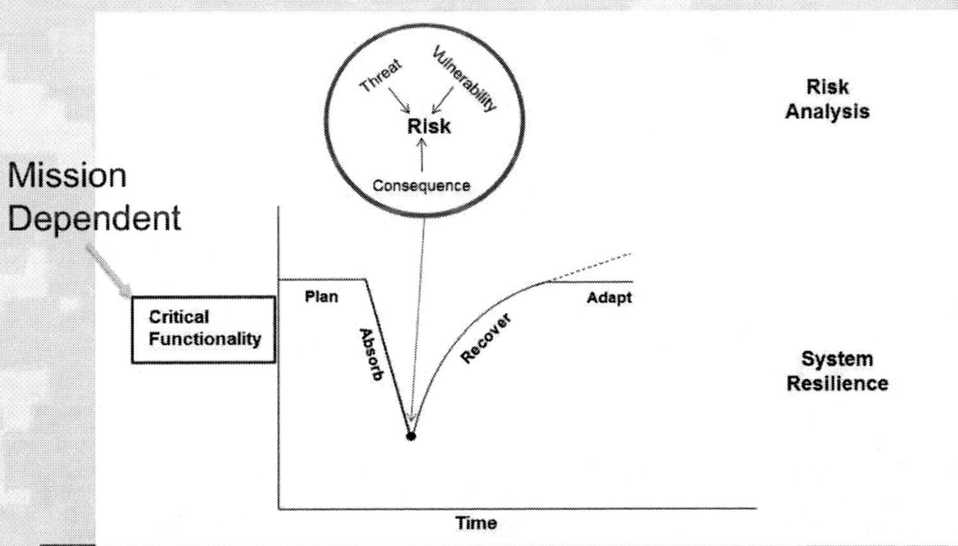

After Linkov et al, Nature Climate Change 2014

ERDC
Innovative solutions for a safer, better world

BUILDING STRONG®

6

What is the Critical Functionality of a Cyber-Connected System?

Military and Weapon Systems
- Data Security
- Accessibility
- Operational effectiveness
- Continuous functionality

Civilian Systems
- **These are the same as in civilian systems**

BUILDING STRONG®

ERDC

Innovative solutions for a safer, better world

7

opinion & comment

NATURE CLIMATE CHANGE | VOL 4 | JUNE 2014 | www.nature.com/natureclimatechange

COMMENTARY:

Changing the resilience paradigm

Igor Linkov, Todd Bridges, Felix Creutzig, Jennifer Decker, Cate Fox-Lent, Wolfgang Kröger,

The human body is resilient in its ability to persevere through infections or trauma. Even through severe disease, critical life functions are sustained and the body recovers, often adapting by developing immunity to further attacks of the same type. Our society's critical infrastructure — cyber, energy, water, transportation and communication — lacks the same degree of resilience, typically losing essential functionality following adverse events.

"Your body has an incredible system called white blood cells that attack and try to manage that virus in such a way that prevents it from harming the body. The systems in 2030 will have something very similar."

***Tom Vice**, president of Northrop's aerospace sector, on 6th Gen Fighter*

8

Video Presentation in Original e-media

BUILDING STRONG®

ERDC

Innovative solutions for a safer, better world

9

How to Measure Resilience?

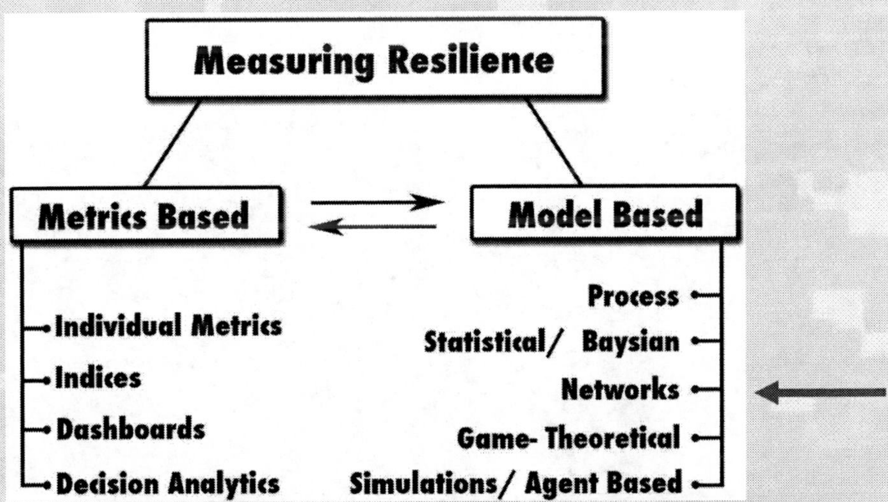

After Linkov and Kott, 2018

BUILDING STRONG®

ERDC

Innovative solutions for a safer, better world

Vision for Cyber/Physical Resilience

Physical Resilience Domains

Real world

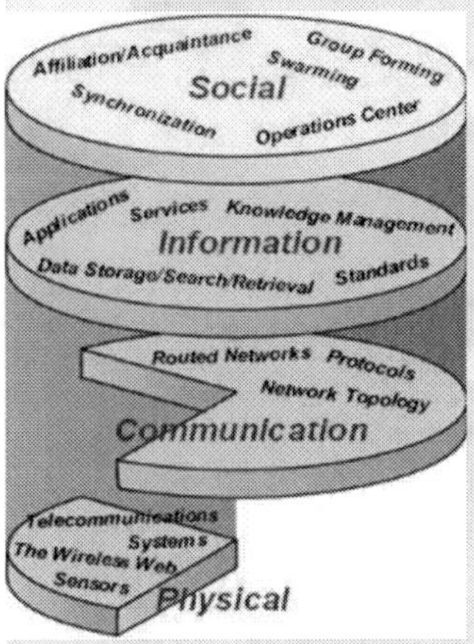

Model

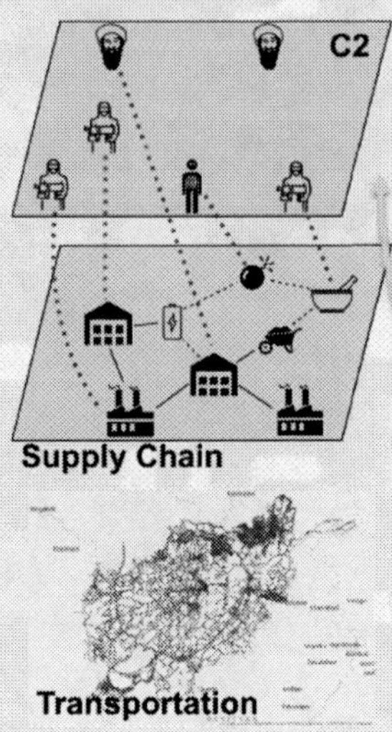

Cyber Resilience Domains

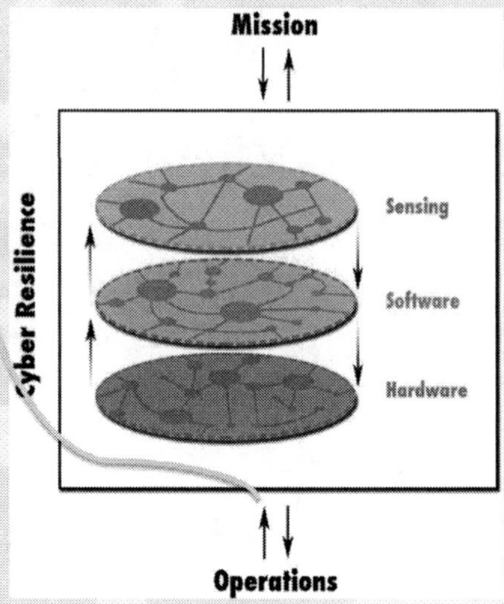

Case Study 1: Transportation Networks

BUILDING STRONG®

Case 2: Cyber Insider Risk/Resilience

Constrained Individual Latitude

Organizational Structure

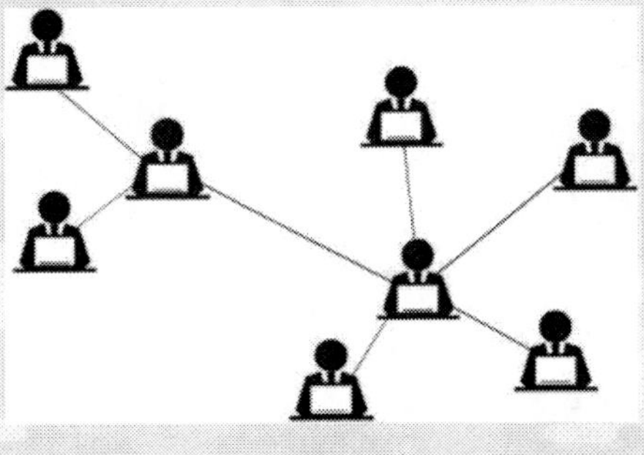

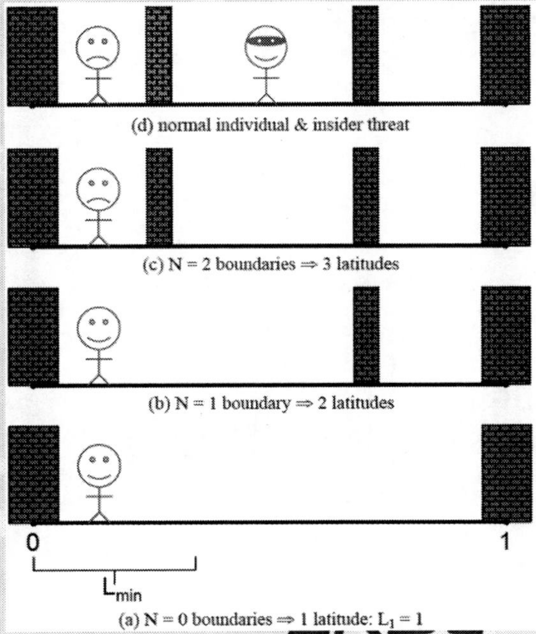

Adapted from Kepner et al (ESD, 2015)

BUILDING STRONG®

Innovative solutions for a safer, better world

13

Missions

Transportation

Cyber

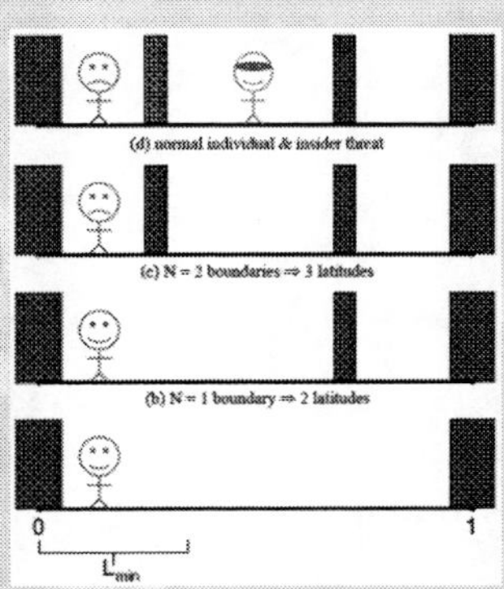

Ganin et al. *Science Advances*, 2017, V. 3(12), p. e1701079.

14

Kepner et al. *Environment Systems and Decisions*, 2015, V. 35(4), p. 504.

Resilience

Transportation System

(Road networks for 40 US Cities)

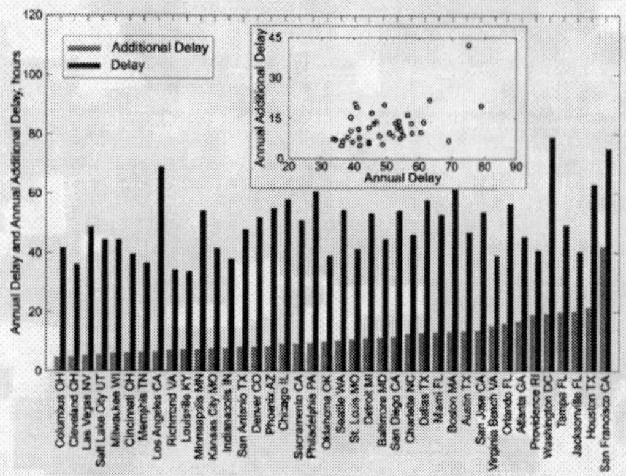

Cyber System

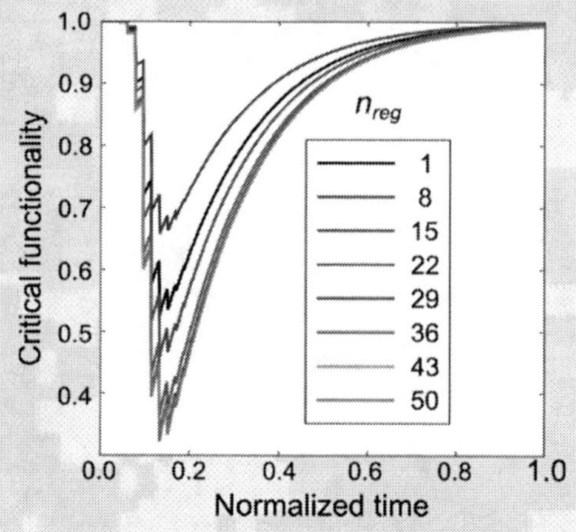

Ganin et al. *Science Advances*, 2017, V. 3(12), p. e1701079.

Gisladottir et al. *Risk Analysis*, 2017, V. 37(9), p. 1644.

Trade Space

Transportation System

(Road networks for 40 US Cities)

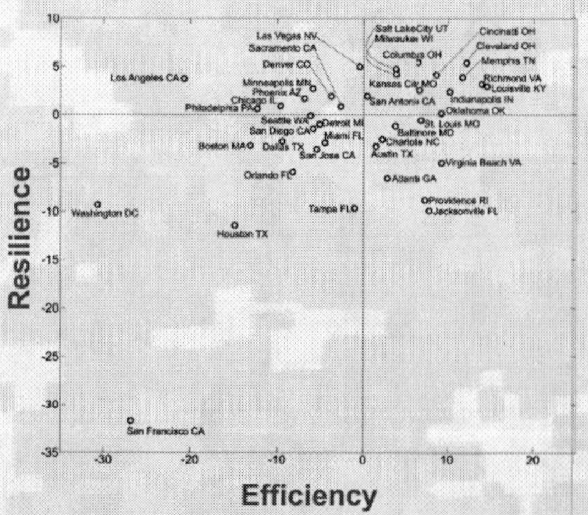

Cyber System

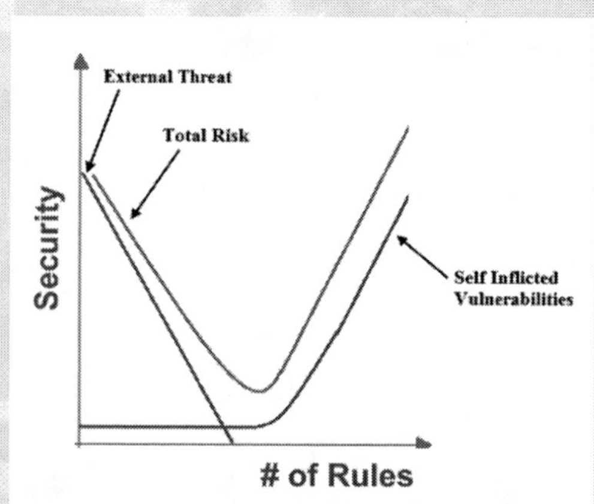

Ganin et al. *Science Advances*, 2017, V. 3(12), p. e1701079.

Gisladottir et al. *Risk Analysis*, 2017, V. 37(9), p. 1644.

Trade Space

Transportation System

(Road networks for 40 US Cities)

Poor Efficiency:

System cannot not accommodate a large volume of commuters driving at the same time.

Traffic congestions are predictable and are typically of moderate level.

Lack of Resilience:

System cannot recover from adverse events (car accidents, natural disasters)

Traffic congestions are **not predictable** and of variable scale.

Cyber System

Poor Performance:

System cannot not accommodate a large volume of data requests or handle complex computations.

Employees and clients are stressed to follow burdensome procedures.

Lack of Resilience:

System is susceptible to attacks degrading data confidentiality, integrity, and availability (lack of robustness/security).

Recovery from attacks or adaptation to regulatory changes is lengthy (lack of recoverability/flexibility).

ERDC

BUILDING STRONG®

Innovative solutions for a safer, better world

17

Cyber Attacks on Transportation

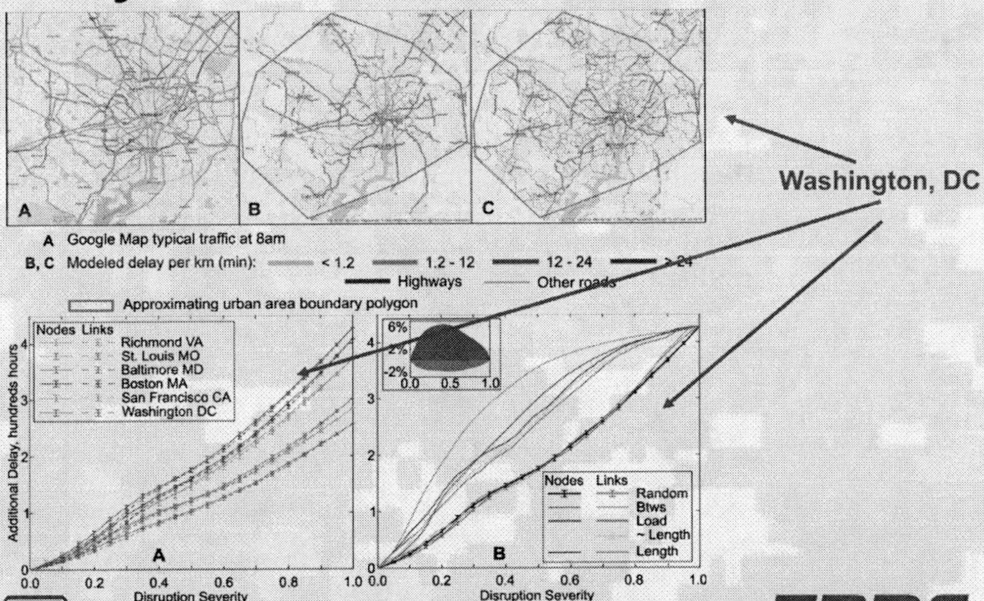

After Ganin et al., 2018 (under review)

ERDC

BUILDING STRONG®

Innovative solutions for a safer, better world

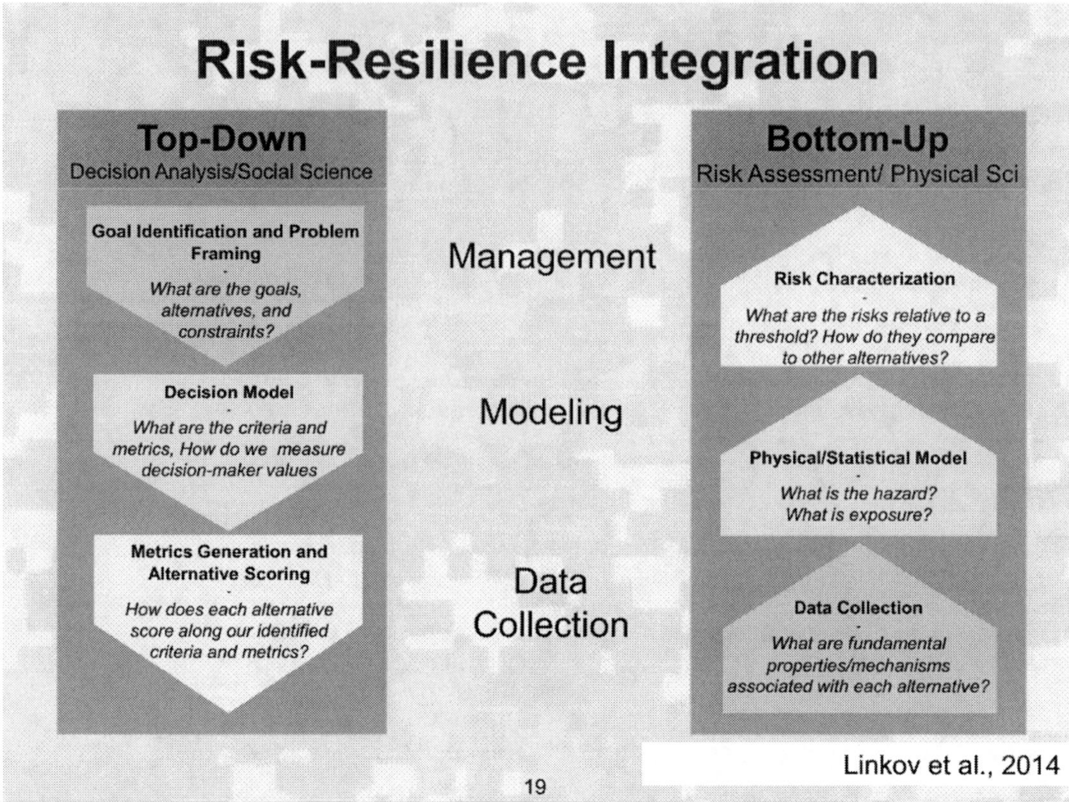
Risk-Resilience Integration
Top-Down
Decision Analysis/Social Science
Goal Identification and Problem Framing
What are the goals, alternatives, and constraints?
Decision Model
What are the criteria and metrics, How do we measure decision-maker values
Metrics Generation and Alternative Scoring
How does each alternative score along our identified criteria and metrics?
Management
Modeling
Data Collection
Bottom-Up
Risk Assessment/ Physical Sci
Risk Characterization
What are the risks relative to a threshold? How do they compare to other alternatives?
Physical/Statistical Model
What is the hazard? What is exposure?
Data Collection
What are fundamental properties/mechanisms associated with each alternative?
Linkov et al., 2014
19

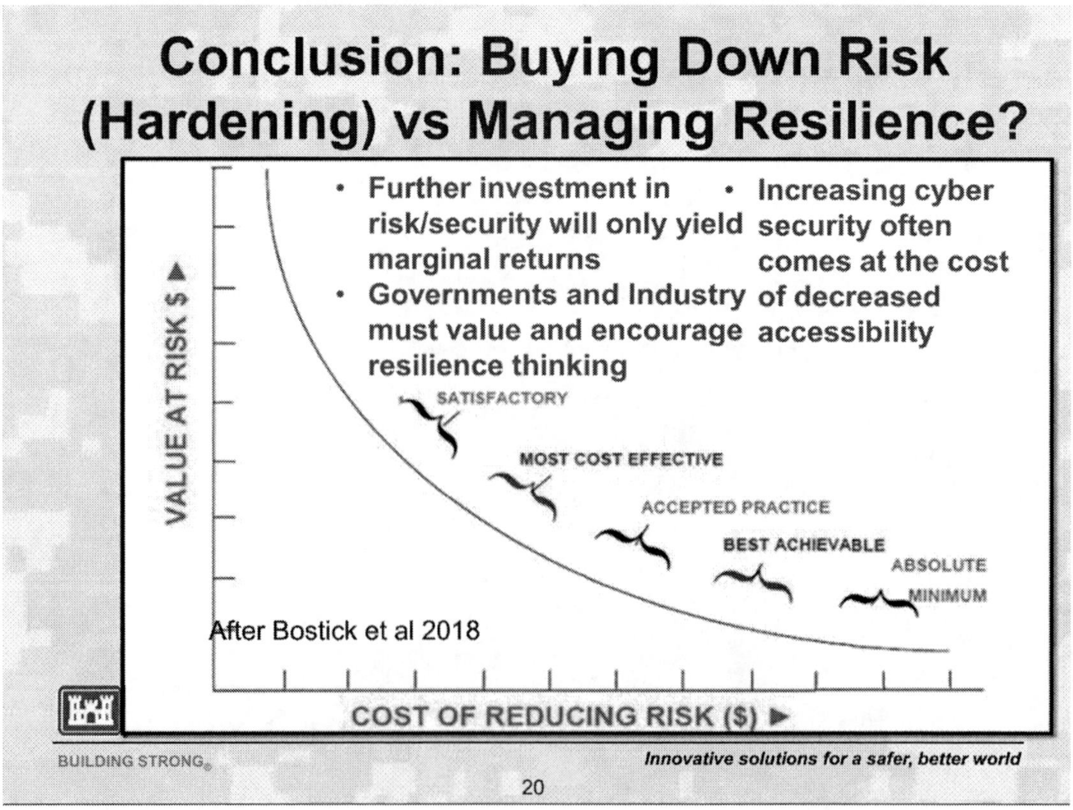
Conclusion: Buying Down Risk (Hardening) vs Managing Resilience?
Further investment in risk/security will only yield marginal returns
Governments and Industry must value and encourage resilience thinking
Increasing cyber security often comes at the cost of decreased accessibility
VALUE AT RISK $ ▶
SATISFACTORY
MOST COST EFFECTIVE
ACCEPTED PRACTICE
BEST ACHIEVABLE
ABSOLUTE MINIMUM
After Bostick et al 2018
COST OF REDUCING RISK ($) ▶
BUILDING STRONG®
Innovative solutions for a safer, better world
20

Publications

1) Kott, A., Linkov, I. eds. (2018). **Cyber Resilience in Systems and Networks.** Springer, Amsterdam.
2) Linkov, I., Palma-Oliveira, J.M., eds (2017). **Risk and Resilience.** Springer, Amsterdam.
3) Florin, M.V., Linkov, I., eds. (2017). **International Risk Governance Council (IRGC) Resource Guide on Resilience.** International Risk Governance Center, Switzerland
4) Linkov, I., Trump, B.D., Keisler, J.M. (2018). **Risk and resilience must be independently managed.** Nature 555:30
5) Kurth, M., Keenan, J.M., Sasani, M., Linkov, I. (2018, in press). **Defining resilience for the US building industry.** Building Research and Innovation.
6) Bostick, T.P., Lambert, J.H., Linkov, I. (2018). **Resilience Science, Policy and Investment for Civil Infrastructure.** Reliability Engineering & System Safety **175**:19-23.
7) Massaro, E., Ganin, A., Linkov, I., Vespignani, A. (2018). **Resilience management of networks during large-scale epidemic outbreaks.** Science Reports 8:1859.
8) Ganin, A., Kitsak, M., Keisler, J., Seager, T., Linkov, I., (2017). **Resilience and efficiency in transportation networks.** Science Advances 3:e1701079.
9) Marchese, D., Reynolds, E., Bates, M.E., Clark, S.S., Linkov, I. (2018). **Resilience and sustainability: similarities and differences.** Sci Total Environ. 613-614:1275-83.
10) Marchese, D., & Linkov, I. (2017). **Can You Be Smart and Resilient at the Same Time?** *Environ. Sci. Technol.* 2017, 51, 5867−5868
11) Connelly, E. B., Allen, C. R., Hatfield, K., Palma-Oliveira, J. M., Woods, D. D., & Linkov, I. (2017). **Features of resilience.** *Environ Systems and Decisions*, *37*(1), 46-50.
12) Allen, C.R., Bartlett-Hunt, S., Bevans, R.A., Linkov, I. (2016). **Avoiding decline: fostering resilience and sustainability in midsize cities.** Sustainability 8:844
13) DiMase D, Collier ZA, Linkov I (2016, on-line) **Traceability and Risk Analysis Strategies for Addressing Counterfeit Electronics in Supply Chains.** Risk Analysis.
14) Thorisson, H., Lambert, J.H., Cardenas, J.J., Linkov, I., (2017). **Resilience Analytics with Application to Power Grid of a Developing Region.** Risk Analysis 37:1268
15) Gisladottir, V., Ganin, A., Keisler, J.M., Kepner, J., Linkov, I., (2017). **Resilience of Cyber Systems with Over- and Under-regulation** Risk Analysis 37:1644
16) Bakkensen, L., Fox-Lent, C., Read, L., and Linkov, I. (2016). **Validating Resilience and Vulnerability Indices in the Context of Natural Disasters.** Risk Analysis 37:982
17) Linkov, I., Larkin, S., Lambert, J.H. (2015). **Concepts and approaches to resilience in governance.** Environment, Systems, and Decisions 35:219-228.
18) Ganin, A., Massaro, E., Keisler, J., Kott, A., Linkov, I. (2016). **Resilient Complex Systems and Networks.** Nature Scientific Reports 6,19540.
19) Fox-Lent, C., Bates, M. E., Linkov, I. (2015). **A Matrix Approach to Community Resilience Assessment.** Environment, Systems, and Decisions 35(2):205-219.
20) Larkin, S., Fox-Lent C., Linkov, I. (2015). **Benchmarking Agency and Organizational Practices in Resilience Decision Making.** Environ., Syst., & Dec. 35(2):185-195.
21) DiMase D, Collier ZA, Linkov I (2015). **Systems Engineering Framework for Cyber Physical Security and Resilience.** Environment, Systems, and Decisions 35:291.
22) Sikula, N.R., Linkov, I., (2015). **Risk Management Isn't Enough: A Conceptual Model for Resilience.** Environ., Syst., & Dec. 35:219-228.
23) Linkov, I., Fox-Lent, C., Keisler, J., Della-Sala, S., Siweke, J. (2014). **Plagued by Problems: Resilience Lessons from Venice** .Environment, Systems, Decision 34:378
24) Collier, Z.A., Linkov, I., DiMase, D., Walters, S., Lambert, J.(2014). **Risk-Based Cybersecurity Standards: Policy Challenges and Opportunities.** *Computer* 47:70
25) Linkov, I, Kröger, W., Levermann, A., Renn, O. et al. (2014). **Changing the Resilience Paradigm.** *Nature Climate Change* 4:407
26) Roege, P., Collier, Z.A., Mancillas, J., McDonagh, J., Linkov, I. (2014). **Metrics for Energy Resilience.** *Energy Policy* Energy Policy 72:249
27) Eisenberg, D.A., Linkov, I., Park, J., Chang, D., Bates, M.E., Seager, T., (2014). **Resilience Metrics: Lessons from Military Doctrines.** *Solutions* 5:76
28) Linkov, I., Eisenberg, D. A., Plourde, K., Seager, T. P., Allen, J., Kott, A (2014). **Resilience Metrics for Cyber Systems.** *Environment, Systems and Decisions* 33:471
29) Park, J., Seager, T, Linkov, I., (2013). **"Integrating risk and resilience approaches to catastrophe management in engineering systems,"** *Risk Analy.*, 33(3), pp. 356.
30) Linkov, I., Eisenberg, D. A., Bates, M. E., Flynn, S. E., Seager, T. P. (2013). **Measurable Resilience for Actionable Policy.** *Environ. Science & Technology* 47:10108

21

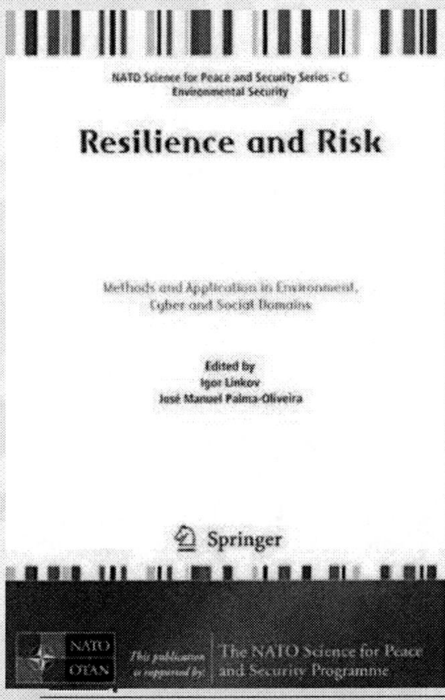

BUILDING STRONG®

SRA

Coming to South Africa in 2019

FIFTH WORLD CONGRESS ON RISK
DEVELOPMENT AND RESILIENCE

Cape Town International Convention Centre
Cape Town, South Africa
May 6-8, 2019

THE WORLD CONGRESS ON RISK is organised by the **Society for Risk Analysis (SRA)** to grow innovation and knowledge across risk analysis and management communities, researchers, practitioners, policymakers and related stakeholders. The event seeks to stimulate ideas and solutions for regional and global risk challenges. The past World Congresses in Singapore (2015), Sydney (2012), Guadalajara (2008) and Brussels (2003) engaged thousands of scholars and professionals from more than forty countries. In 2019, the SRA brings the World Congress to Cape Town, South Africa, where organisations, companies, academia and individuals will gather with a theme of "**Development and Resilience**," across topics of importance to both developed and developing countries, including:

- Emerging technologies and innovation
- Environment, ecology, climate
- Agriculture, food and water supply
- Human health and safety
- Law, policy and governance
- Business processes and standards
- Population and workforce behaviors
- Disaster preparedness and resilience
- Energy, transportation, logistics
- Poverty in rural and urban areas
- Infrastructure systems
- Economics, finance and fraud-related issues in enterprise and government
- Ethnic and socio-economic risks

22

Security and Resilience of Information Systems Affected by Hybrid Threats

NATO Workshop

Estonia, 26-29 Aug 2018

NATO OTAN *This workshop is supported by:* The NATO Science for Peace and Security Programme

BUILDING STRONG®

ERDC

Innovative solutions for a safer, better world

23

ADDITIONAL SLIDES

BUILDING STRONG®

ERDC

Innovative solutions for a safer, better world

24

international risk
governance center

Environ Syst Decis (2017) 37:46–50
DOI 10.1007/s10669-017-9634-9

Features of resilience

Elizabeth B. Connelly[1] · Craig R. Allen[2] ·
David D. Woods[5] · Igor Linkov[6]

RESOURCE GUIDE

Resilience

An edited collection of authored pieces comparing, contrasting, and integrating risk and resilience with an emphasis on ways to measure resilience

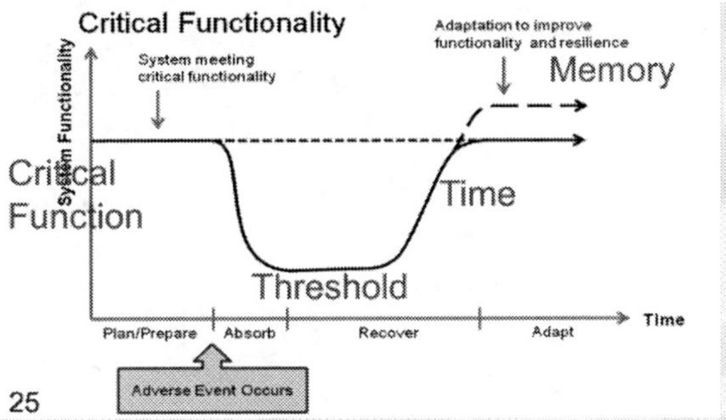

25

Resilient System and Hierarchal Systems

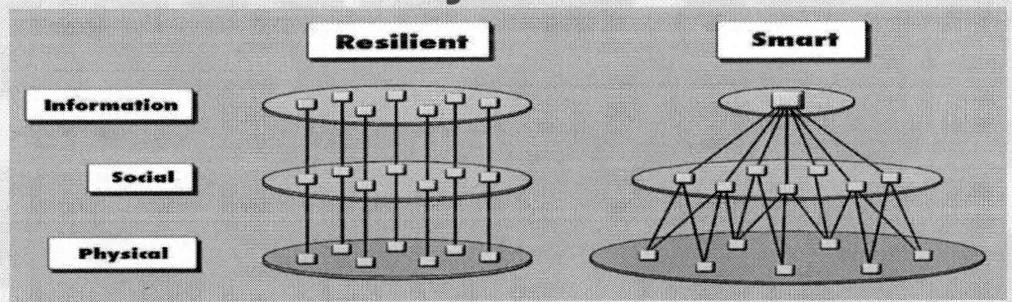

- Fully Redundant
- Greater maintenance requirements
- Functional during disruption
- **Less efficient during random attacks**

- Observe emergent patterns
- Centralized decision making
- No redundancy
- **Prone to targeted attacks**

Resilience Needs to be Quantified

Marchese, D., and Linkov, I. (2017). "Can You Be Smart and Resilient at the Same Time?" *Environmental Science & Technology*, 51(11), 5867–5868.

26

AUTOMATED VEHICLES SYMPOSIUM 2018
Connected and Automated Vehicle Security
Jonathan Petit
July 9, 2018
OnBoardSecurity

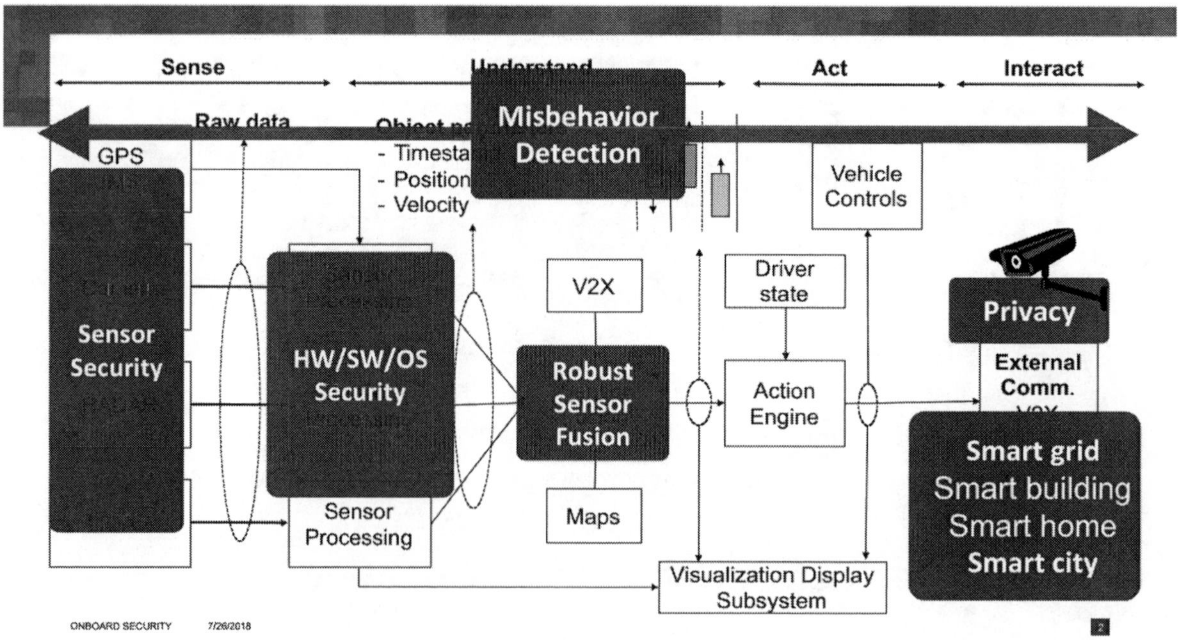
Sense
Act
Interact
Raw data
GPS
- Timesta
- Position
- Velocity
Misbehavior Detection
Vehicle Controls
Sensor Security
HW/SW/OS Security
V2X
Driver state
Robust Sensor Fusion
Action Engine
Privacy
External Comm.
Smart grid
Smart building
Smart home
Smart city
Sensor Processing
Maps
Visualization Display Subsystem
ONBOARD SECURITY
7/26/2018
2

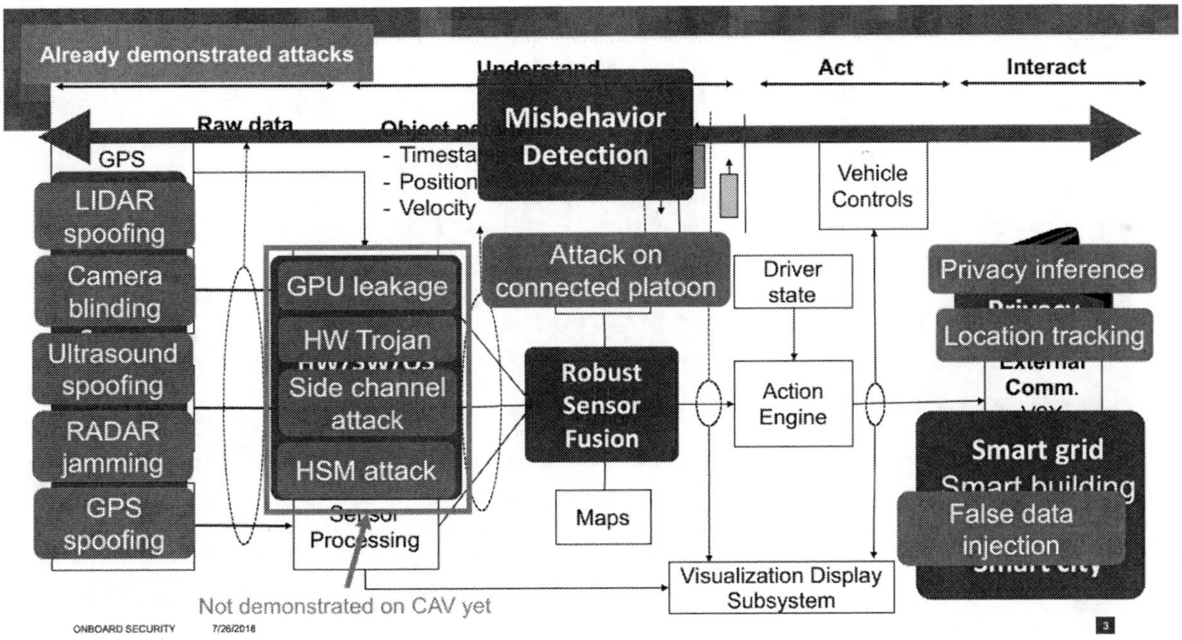
Already demonstrated attacks
Act
Interact
Misbehavior Detection
Raw data
GPS
- Position
- Velocity
Vehicle Controls
LIDAR spoofing
Camera blinding
Ultrasound spoofing
RADAR jamming
GPS spoofing
GPU leakage
HW Trojan
Side channel attack
HSM attack
Attack on connected platoon
Driver state
Privacy inference
Location tracking
Robust Sensor Fusion
Action Engine
Comm.
Smart grid
False data injection
Maps
Processing
Visualization Display Subsystem
Not demonstrated on CAV yet
ONBOARD SECURITY
7/26/2018
3

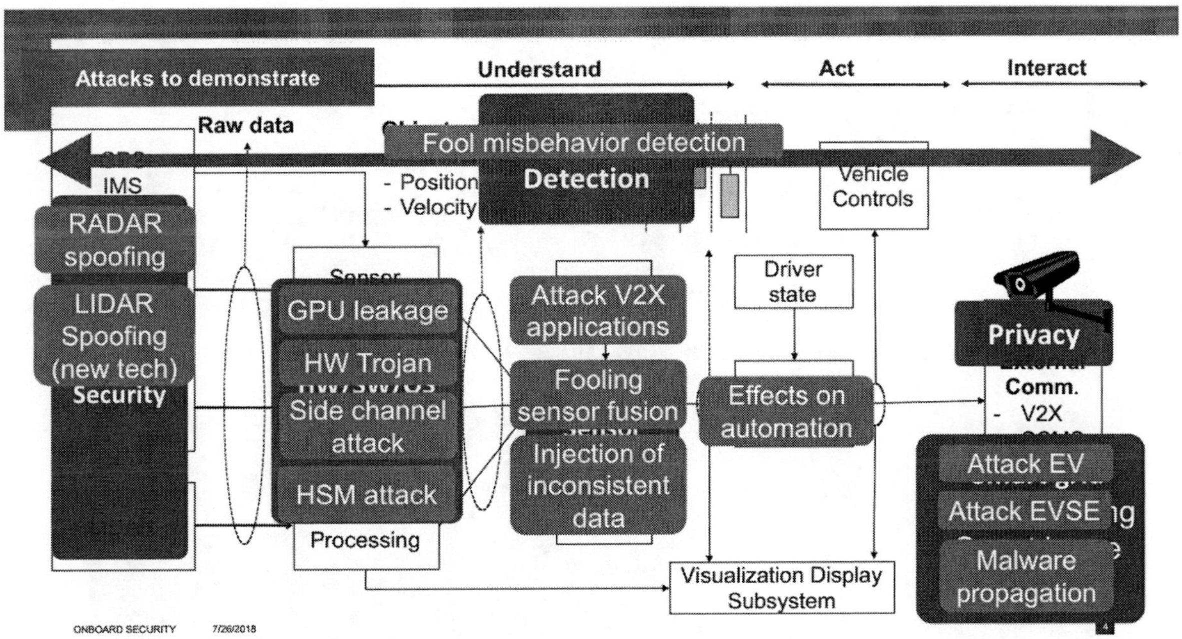
Attacks to demonstrate
Understand
Act
Interact
Raw data
Fool misbehavior detection
Detection
IMS
- Position
- Velocity
Vehicle Controls
RADAR spoofing
LIDAR Spoofing (new tech)
Security
GPU leakage
HW Trojan
Side channel attack
HSM attack
Processing
Attack V2X applications
Fooling sensor fusion
Injection of inconsistent data
Driver state
Effects on automation
Privacy
Comm.
- V2X
Attack EV
Attack EVSE
Malware propagation
Visualization Display Subsystem
ONBOARD SECURITY
7/26/2018

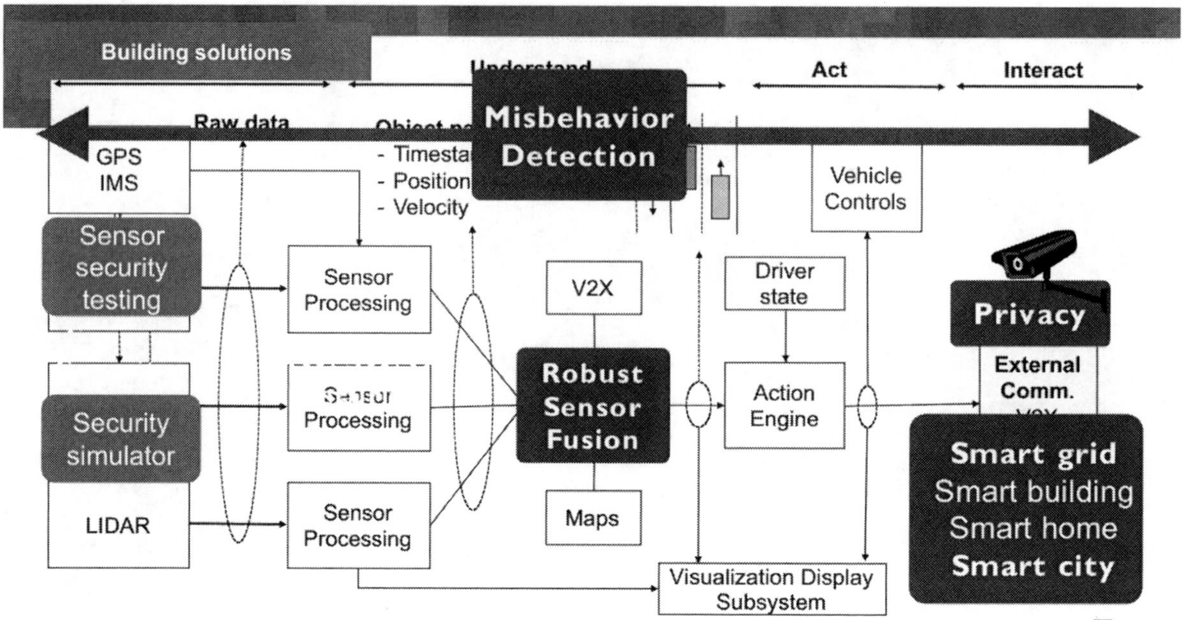

ONBOARD SECURITY 7/26/2018 5

AV Security Simulator

- Partnership with Oxford University
- Using CARLA simulator
- Test camera, LIDAR attacks
- **Outcomes**: add attacks as "environmental perturbations", foster research

ONBOARD SECURITY 7/26/2018 6

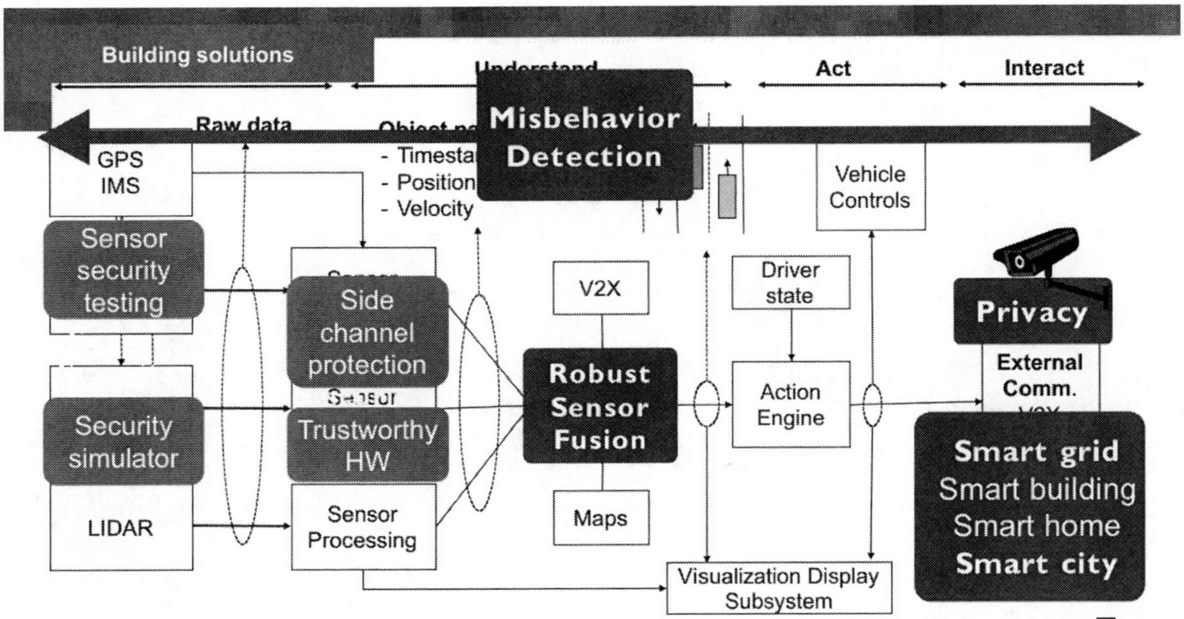

Hardware Trojan: Processing + network

- Results from UT/Dallas
- **Todos:**
  - Investigate HW Trojan on cellular and DSRC chipsets (used in OEM or aftermarket devices)
  - Can HW Trojan be installed in HSM?
  - Check backdoortolerance.org

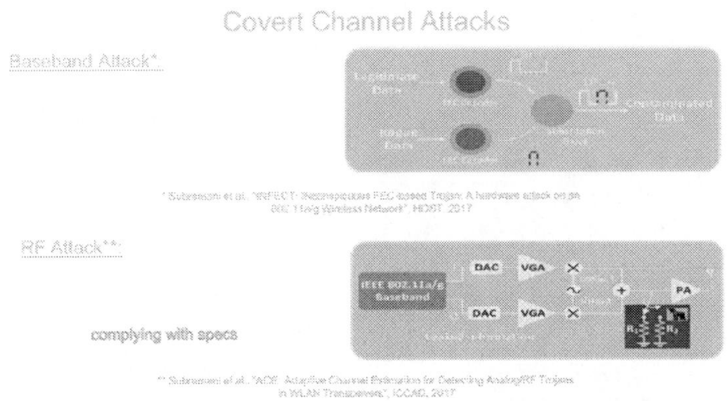

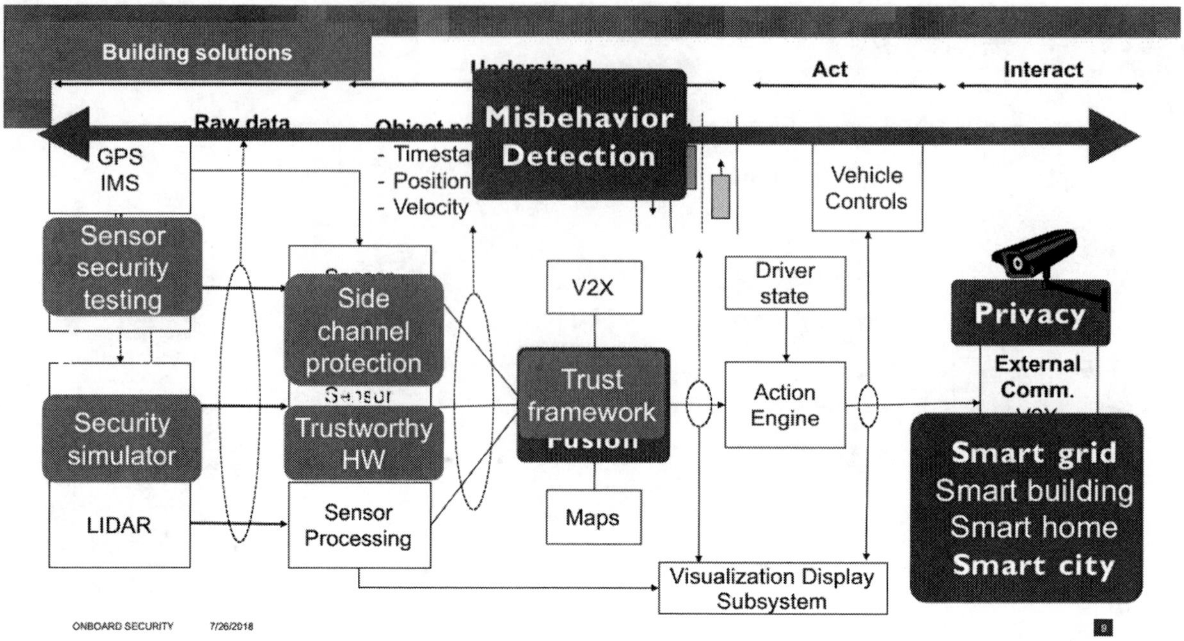
Building solutions
Act
Interact
Raw data
Misbehavior Detection
GPS
IMS
- Timesta
- Position
- Velocity
Vehicle Controls
Sensor security testing
Side channel protection
V2X
Driver state
Privacy
Trust framework
Action Engine
External Comm.
Security simulator
Trustworthy HW
Smart grid
Smart building
Smart home
Smart city
LIDAR
Sensor Processing
Maps
Visualization Display Subsystem
ONBOARD SECURITY 7/26/2018
9

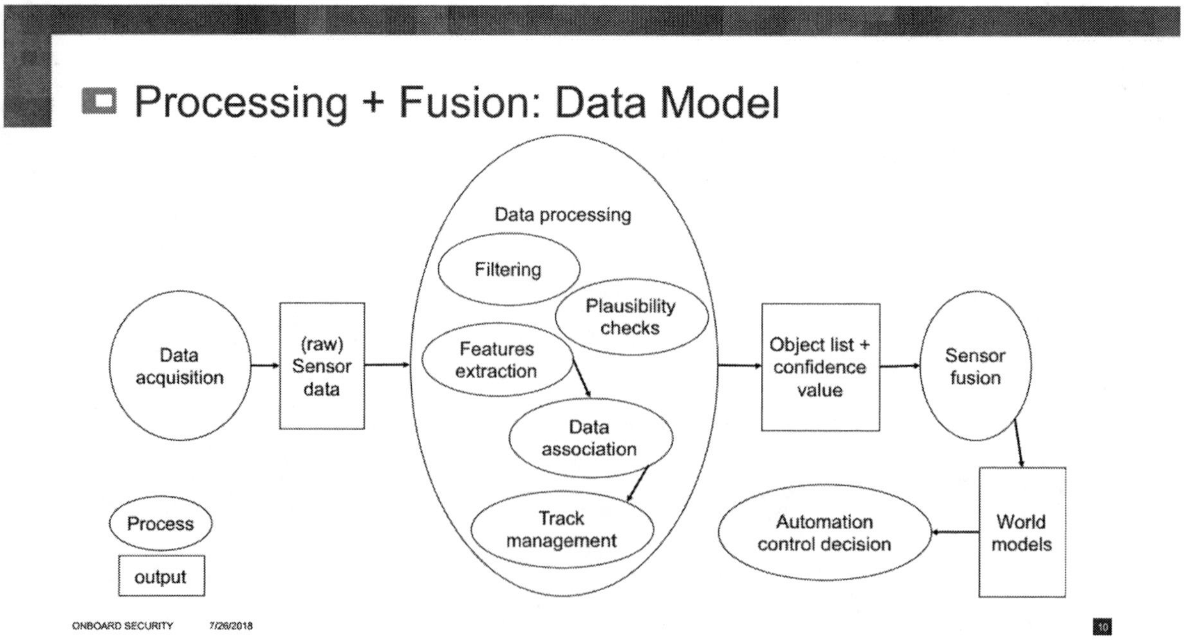
Processing + Fusion: Data Model
Data processing
Filtering
Plausibility checks
Data acquisition
(raw) Sensor data
Features extraction
Object list + confidence value
Sensor fusion
Data association
Track management
Automation control decision
World models
Process
output
ONBOARD SECURITY 7/26/2018
10

V2X + AV sensor fusion

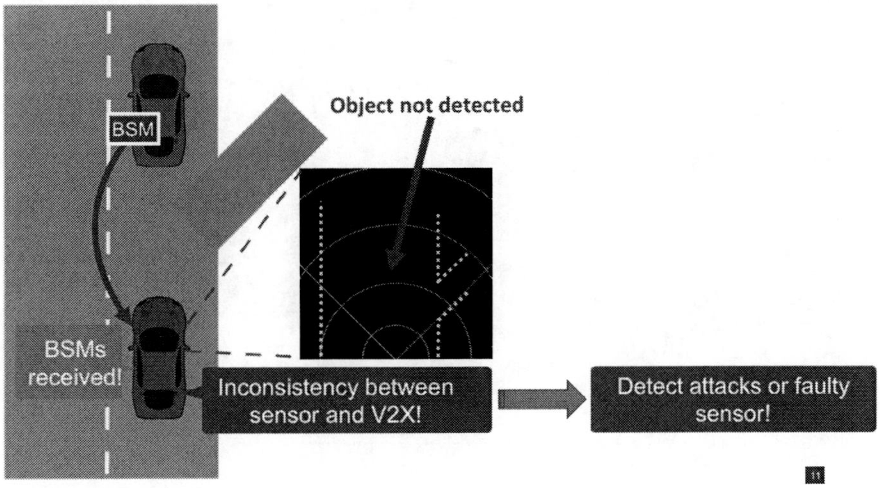

V2X + AV sensor fusion

- Yuan *et al.*, "Object Matching for Inter-Vehicle Communication Systems—An IMM-Based Track Association Approach With Sequential Multiple Hypothesis Test," in *IEEE Transactions on Intelligent Transportation Systems*, Dec. 2017. → **shows benefits of V2X for data association**.
- **Todo**: full-fledged trust framework for CAV data fusion

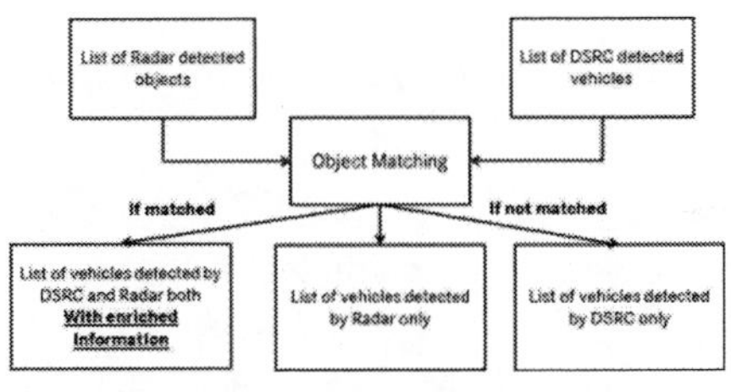

Fig. 1. Overview of the information association/fusion concept.

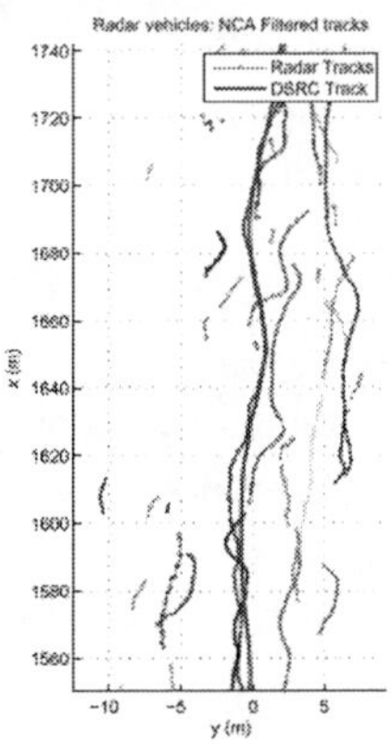

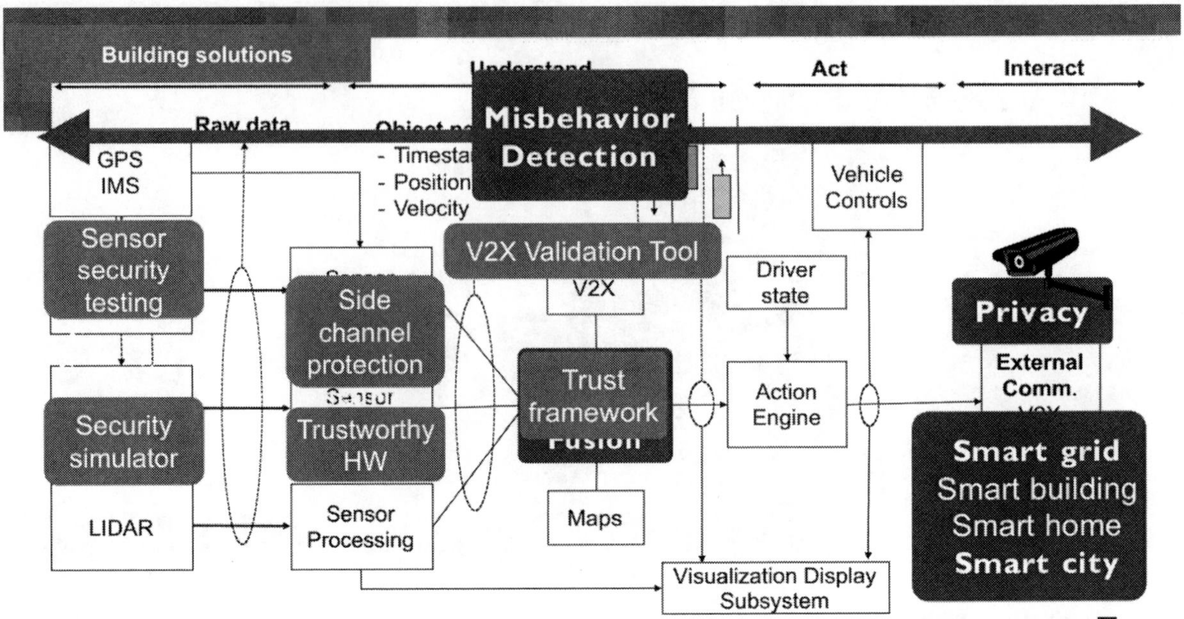

V2X Validation Tool

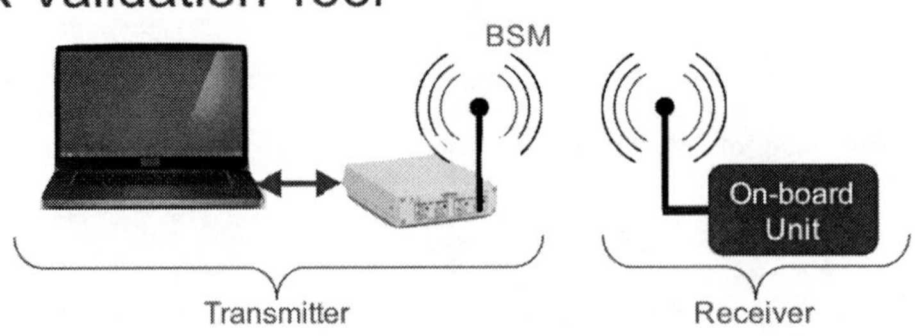

- **Goal**: Stress-test OBU implementation and detect vulnerabilities
- **Features**:
  - Works on a Raspberry Pi 3
  - Send BSMs: valid, invalid, signed, unsigned, etc.
  - Fuzzer
  - Monitor impact on receiver (computation / communication overhead, memory, energy)

V2X Validation Tool

Demo at BlackHat US 2018

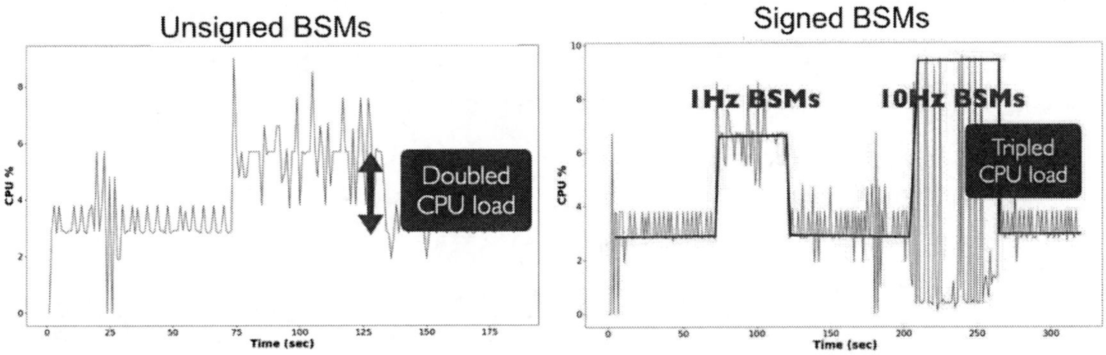

ONBOARD SECURITY 7/26/2018 15

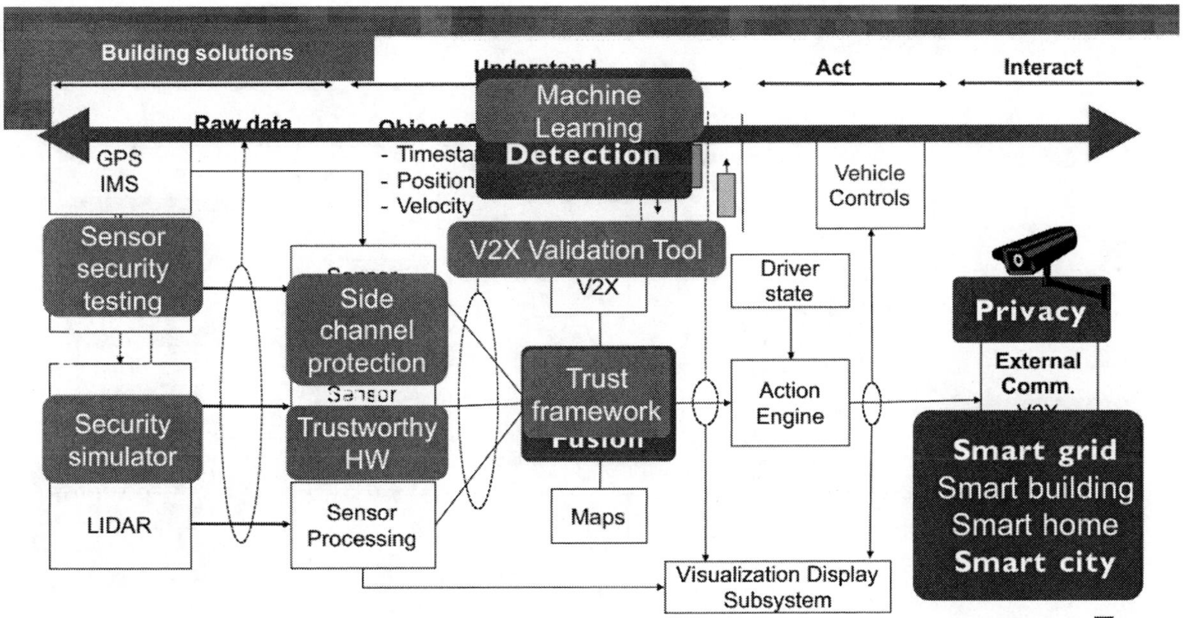

ONBOARD SECURITY 7/26/2018 16

Machine Learning for Misbehavior detection

- Statistics: Pearson correlation matrix
- Supervised learning: k-NN, SVM, Random Forest
- Unsupervised learning: (variational) autoencoders
- Deep learning: Generative Adversarial Network
- Think Counter-ML

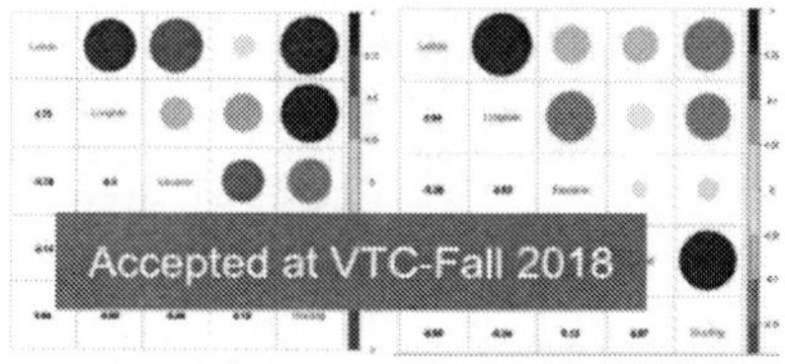

Fig. 5. Attack 1: Victim car (left) and Ghost car (right)

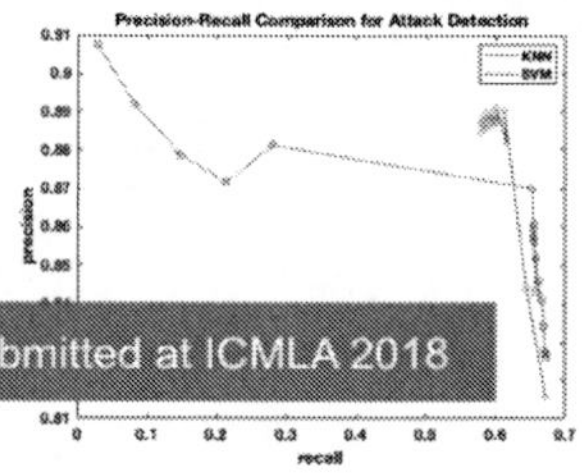

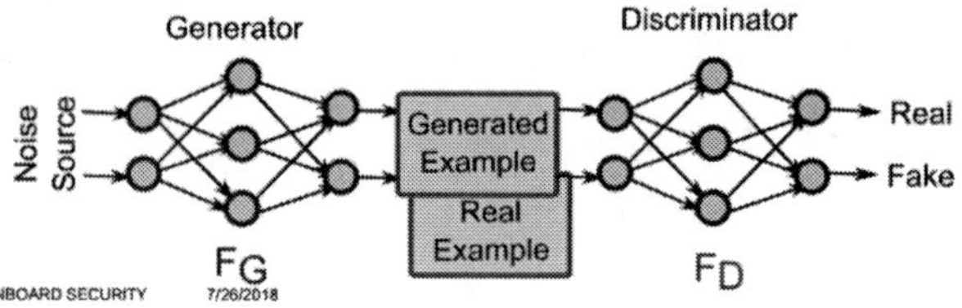

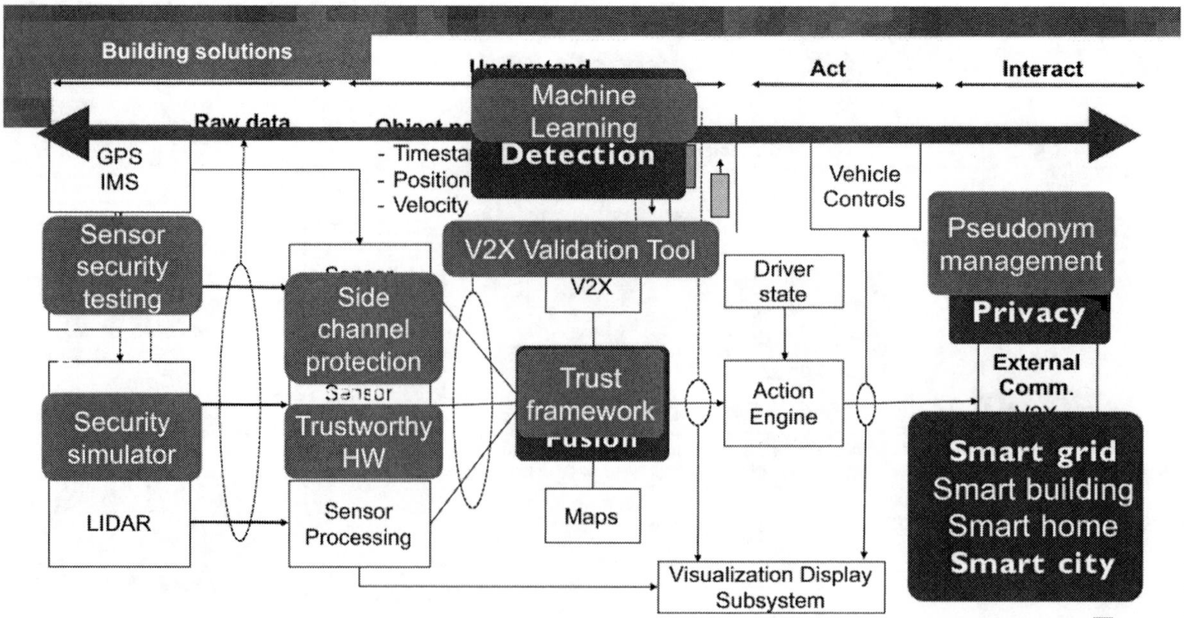

V2X Privacy

- Offensive: Simulation of Neural Network for better tracking
- Intrinsic privacy: Simulation of different road topologies
- Simulation of silent period strategies
- Effects on safety (below is an example)
  - Entry Condition: Speed below 25 mph
  - Exit Condition: Speed above 30 mph

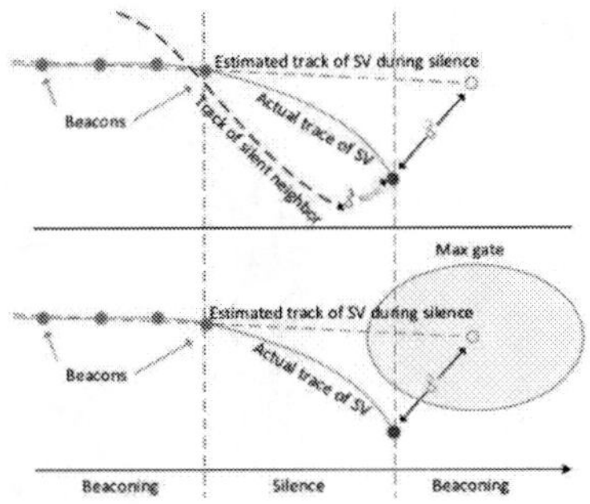

Scenario Name	Conflict Type				
	Total	*unclassified*	*crossing*	*rear end*	*lane change*
No Silent Period	7107	0	258	6800	49
With Silent Period	9737	0	311	9359	67

Building solutions

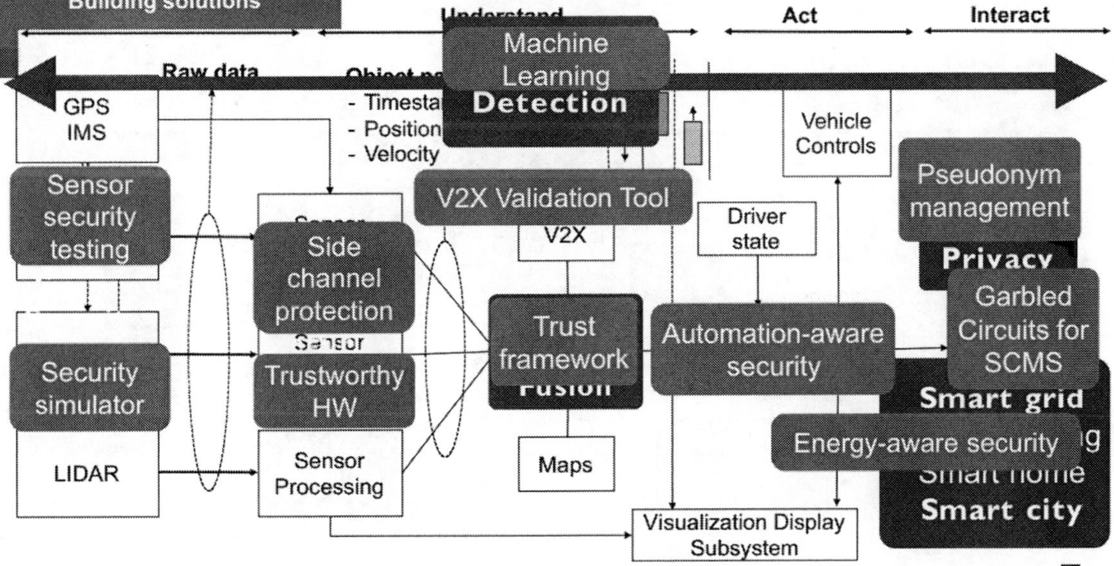

Fitting solutions together

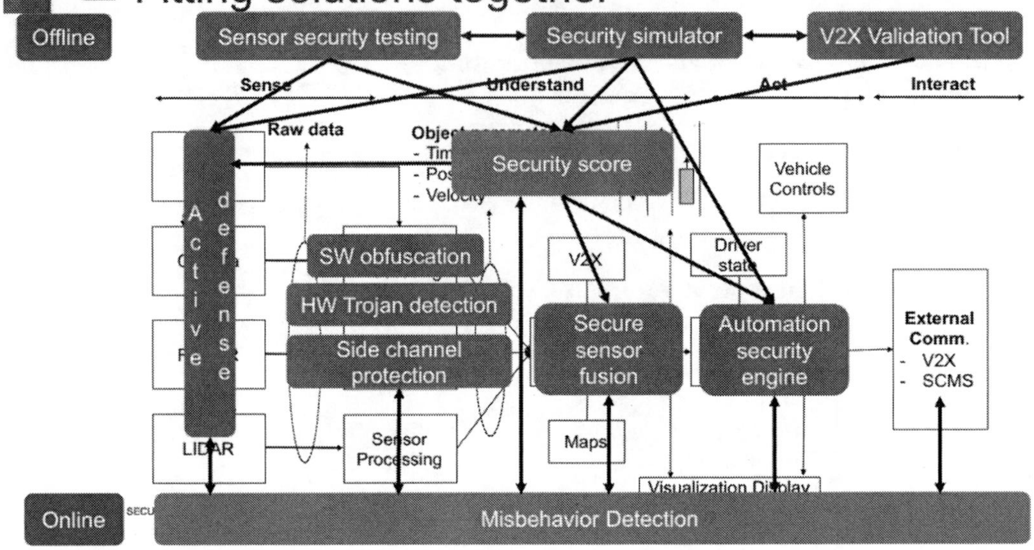

21

Conclusion

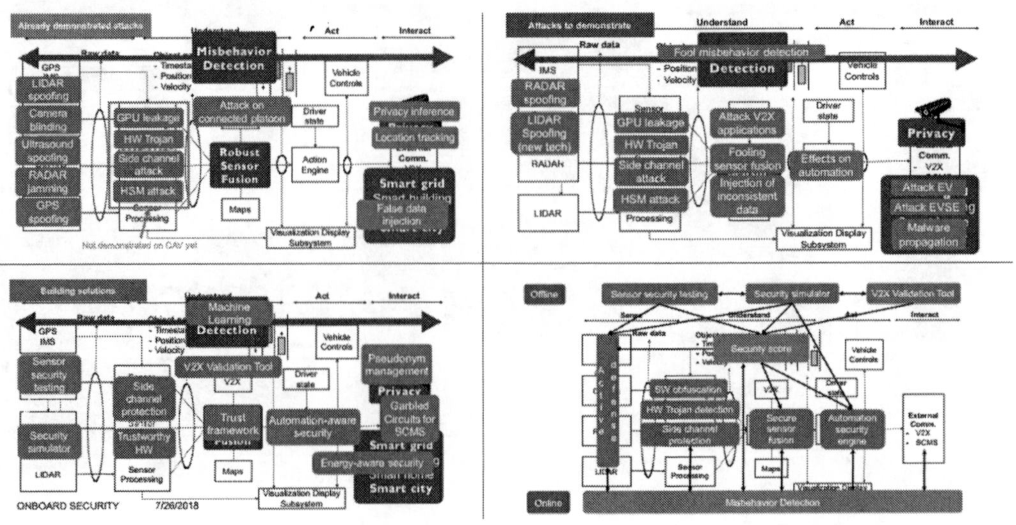

22

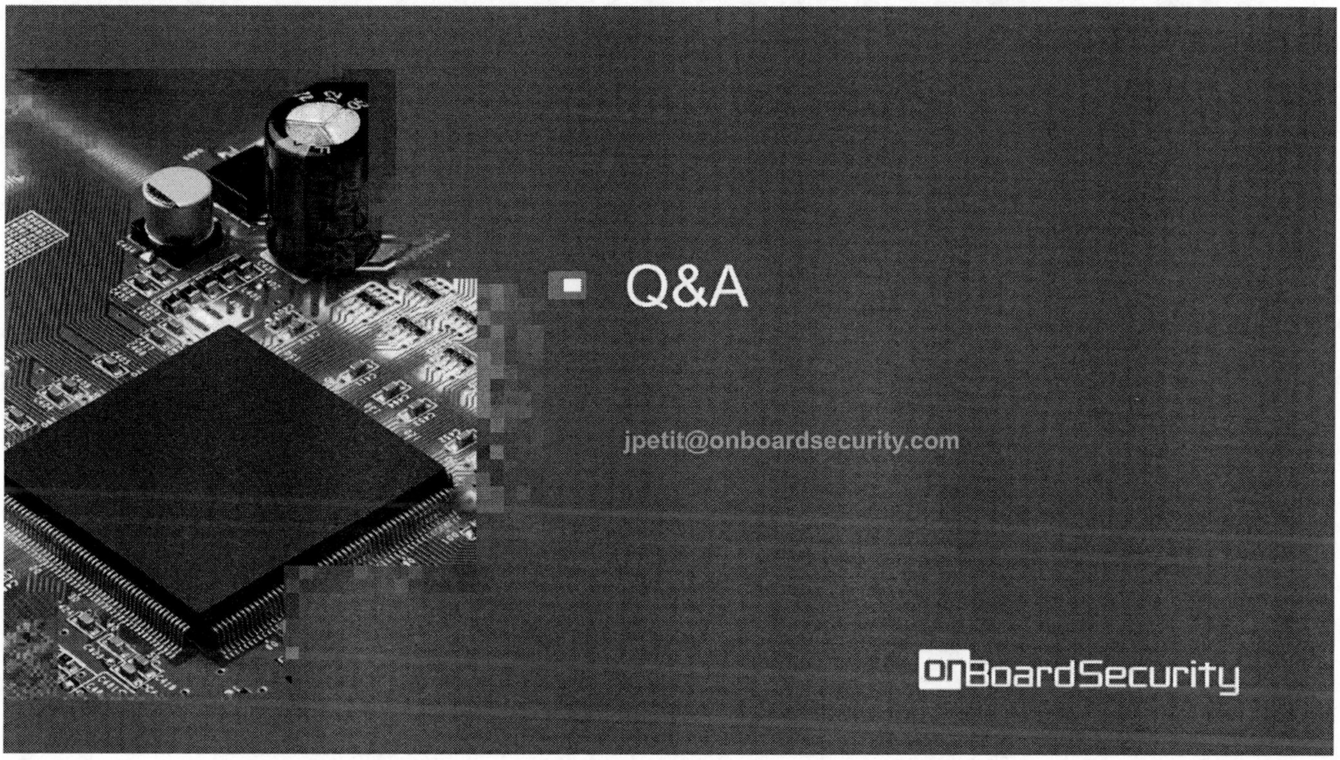
Q&A
jpetit@onboardsecurity.com
OnBoardSecurity

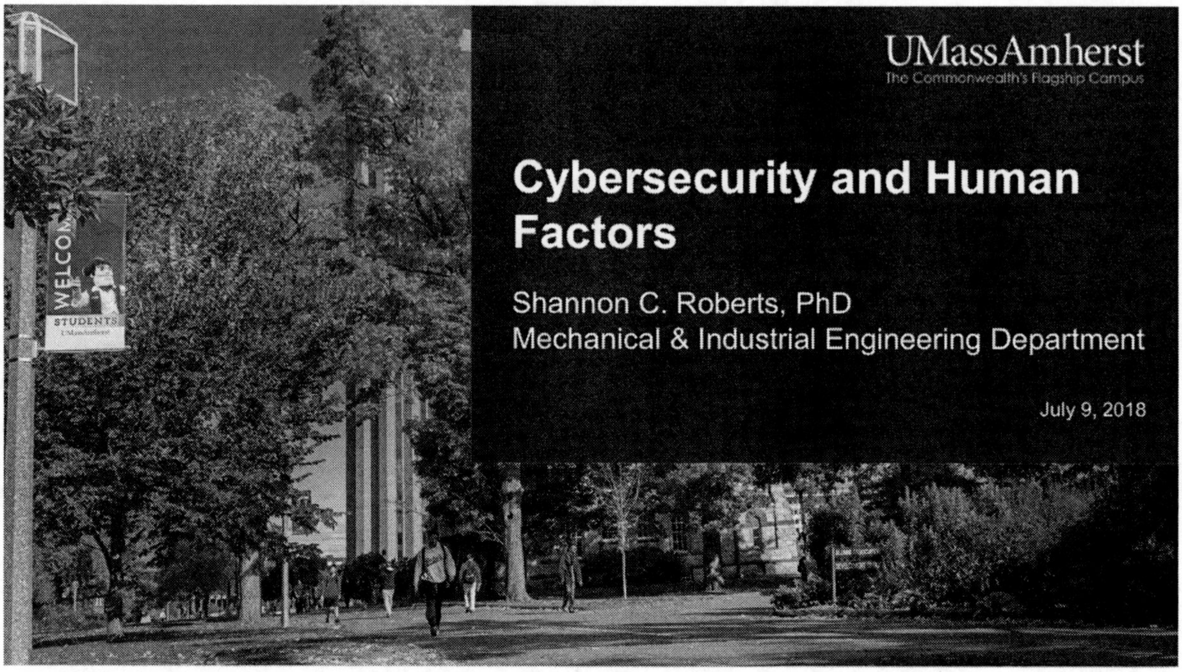

UMassAmherst
The Commonwealth's Flagship Campus

Cybersecurity and Human Factors

Shannon C. Roberts, PhD
Mechanical & Industrial Engineering Department

July 9, 2018

UMassAmherst

HUMAN PERFORMANCE LAB

- Research Expertise
  - Training and feedback systems
  - Automation and transfer of control
  - Roadway design
  - Vulnerable populations
- Facilities
  - Driving Simulator
  - Virtual Reality Headsets
  - Eye Trackers
  - Physiological Monitoring

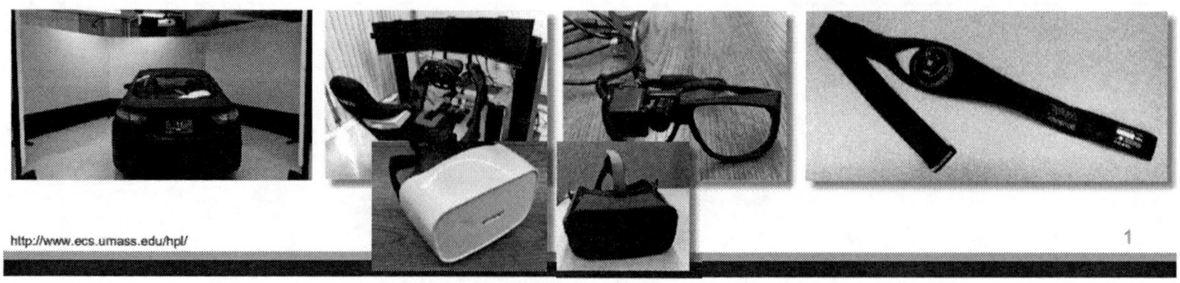

http://www.ecs.umass.edu/hpl/

1

HUMAN
PERFORMANCE LAB

UMassAmherst

Current Efforts to Improve Cybersecurity

Develop technologies and mechanisms to identify, protect, detect, respond, and recover from malicious attacks, e.g.,

- Firewalls
- Anti-virus programs
- Software updates
- User best practices (don't plug in strange devices)

2

HUMAN
PERFORMANCE LAB

UMassAmherst

What are we Missing?

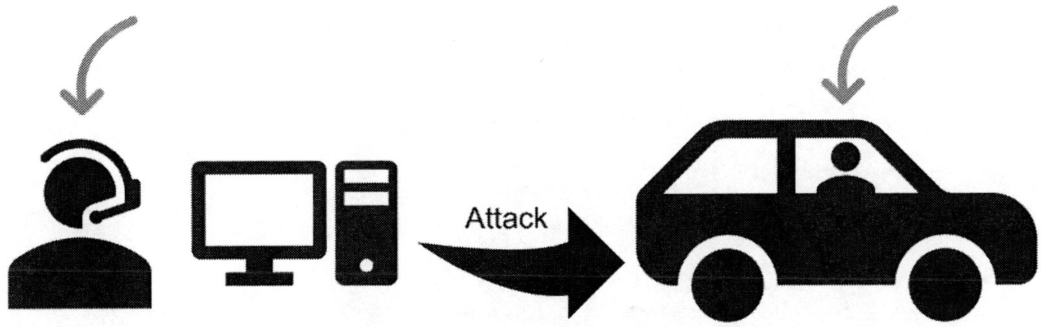

3

HUMAN
PERFORMANCE LAB

UMassAmherst

Humans and Cybersecurity—the downside

Most people are careless and clueless with respect to cybersecurity of their personal devices, e.g.,

- Use weak passwords (Florencio & Herley, 2007)
- Succumb to phishing attempts (Schupak, 2015)
- Do not use the principle of least privilege (Motiee, Hawkey, & Beznosov, 2010)

4

HUMAN
PERFORMANCE LAB

UMassAmherst

Humans and Cybersecurity—the upside

- Most people don't know about ***vehicle*** cybersecurity →
- We still have time to shape the culture!

(Eiza, & Ni, 2017)

5

HUMAN
PERFORMANCE LAB

UMassAmherst

Humans in Vehicles—the downside

- Drivers are responsible for maintaining vehicle control
- **BUT**, over 90% of crashes are due to human error (Lum & Reagan, 1995)

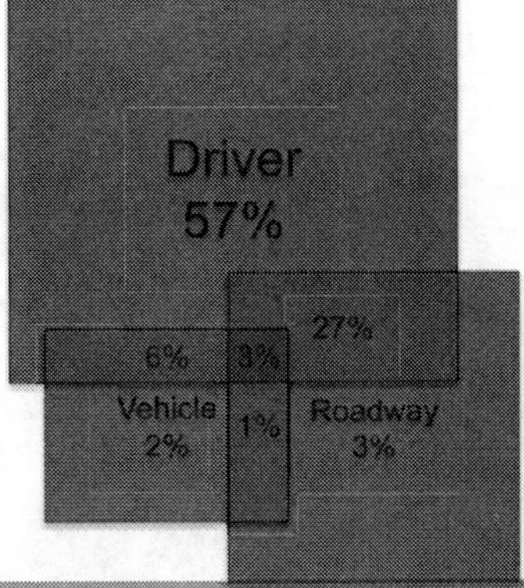

6

HUMAN
PERFORMANCE LAB

UMassAmherst

Humans in Vehicles—the upside

- Example: unintended acceleration—drivers press the brake pedal to no avail, do not use the emergency brake, do not shift gears, etc. (Pollard et al., 2017)
- **BUT**, when drivers know what to do, they can respond within seconds (Duncan, Williams, & Brown, 1991; Pollard et al., 2017)

7

HUMAN
PERFORMANCE LAB

UMassAmherst

Humans in Automated Vehicles—mixed

- Drivers are not aware of vehicle functionality and may place too much trust in the technology (Abraham et al., 2017; Banks et al., 2018)
- Drivers can take back control (e.g., when automation fails) within seconds (Eriksson and Stanton, 2017)

8

HUMAN
PERFORMANCE LAB

UMassAmherst

How can we Improve the Human?

- Training—teach the driver what to do before they encounter a situation

- Feedback systems—give information to the driver in real time to encourage safe behavior

9

HUMAN PERFORMANCE LAB

UMassAmherst

Examples of Training and Feedback Systems

- Computer security
  - Training can improve people's ability to recognize fraudulent emails (Kumaraguru et al., 2010)
  - Feedback can help avoid malware attempts (Akhawe & Felt, 2013)
- Transportation safety
  - Trained drivers can properly respond to unintended acceleration (Pollard et al., 2017)
  - Feedback can encourage fuel efficiency via better speed management (Birrell & Young, 2013)

10

HUMAN PERFORMANCE LAB

UMassAmherst

Study: Purpose

1. Develop training and in-vehicle message systems to improve drivers' awareness of and response to cybersecurity incidents, and
2. Test the effectiveness of the training and in-vehicle message systems with a driving simulator study

11

HUMAN PERFORMANCE LAB

UMassAmherst

Study: Phase 1

Iterative development of the training and feedback system

Observation and Interview → 1st design iteration

12

HUMAN PERFORMANCE LAB

UMassAmherst

Study: Phase 1

Iterative development of the training and feedback system

Observation and Interview → 1st design iteration

Co-design Session → 2nd design iteration

13

HUMAN
PERFORMANCE LAB

UMassAmherst

Study: Phase 1

Iterative development of the training and feedback system

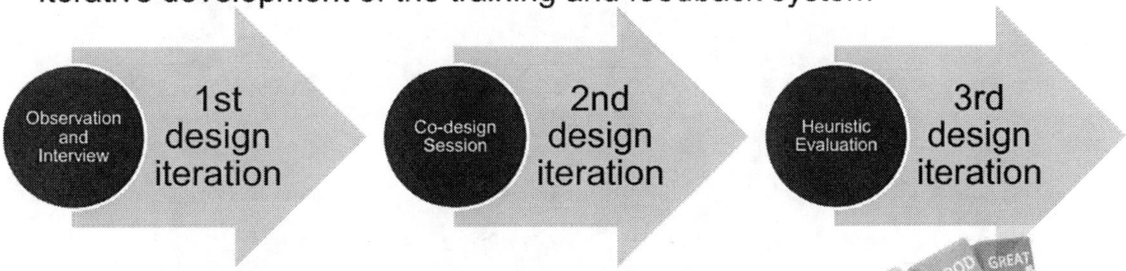

14

HUMAN
PERFORMANCE LAB

UMassAmherst

Study: Example Training System

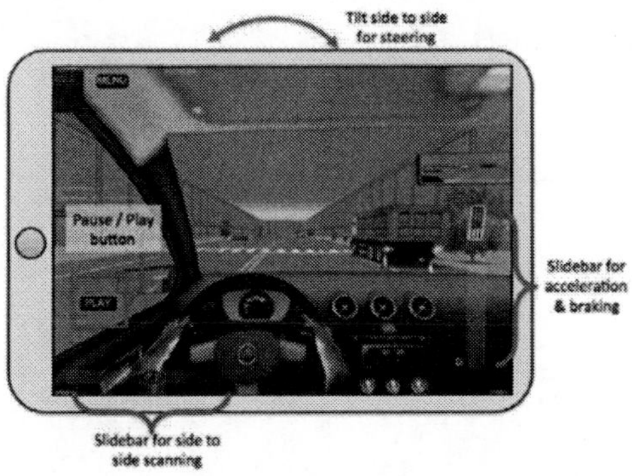

(Zafian et al., 2016)

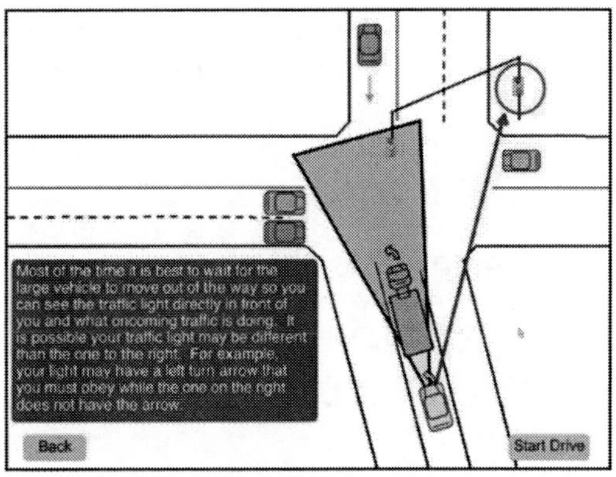

(Thomas et al., 2017)

15

HUMAN
PERFORMANCE LAB

UMassAmherst

Study: Example Feedback Systems

16

HUMAN
PERFORMANCE LAB

UMassAmherst

Study: Phase 2

Evaluate the training and feedback systems in a driving simulator study where the vehicle will be hacked, e.g.,

- Speedometer becomes nonresponsive
- Screen freezes
- Warning lights illuminate
- Brakes stop functioning
- Vehicle turns off

17

HUMAN
PERFORMANCE LAB

UMassAmherst

Study: Phase 2

We will measure:

- Eye glances
- Vehicle behavior
- Response times
- Subjective assessment of systems

18

HUMAN
PERFORMANCE LAB

UMassAmherst

Shannon C. Roberts
Asst. Professor of Industrial Engineering
scroberts@umass.edu
https://people.umass.edu/scroberts/

19

UMassAmherst

The Commonwealth's Flagship Campus

20

HUMAN
PERFORMANCE LAB

UMassAmherst

References

- Abraham, H., Seppelt, B., Mehler, B., & Reimer, B. (2017, September). What's in a Name: Vehicle Technology Branding & Consumer Expectations for Automation. In *Proceedings of the 9th International Conference on Automotive User Interfaces and Interactive Vehicular Applications* (pp. 226-234). ACM.
- Akhawe, D., & Felt, A. P. (2013). Alice in warningland: a large-scale field study of browser security warning effectiveness. *Proceedings of the 22nd USENIX Security Symposium*, 257–272.
- Banks, V. A., Eriksson, A., O'Donoghue, J., & Stanton, N. A. (2018). Is partially automated driving a bad idea? Observations from an on-road study. *Applied Ergonomics*, *68*(October 2017), 138–145. http://doi.org/10.1016/j.apergo.2017.11.010
- Birrell, S. A., & Young, M. S. (2013). Smart Driving Assistance Systems: Designing and Evaluating Ecological and Conventional Displays. In M. A. Regan, J. D. Lee, & T. W. Victor (Eds.), *Driver Distraction and Inattention: Advances in Research and Countermeasures2* (pp. 373–388). Burlington, VT: Ashgate.
- Duncan, J., Williams, P., & Brown, I. (1991). Components of driving skill: experience does not mean expertise. Ergonomics, 34(7), 919–937. Retrieved from http://dx.doi.org/10.1080/00140139108964835
- Eiza, M. H., & Ni, Q. (2017). Driving with sharks: Rethinking connected vehicles with vehicle cybersecurity. IEEE Vehicular Technology Magazine, 12(2), 45-51.
- Eriksson, A., & Stanton, N. A. (2017). Takeover Time in Highly Automated Vehicles: Noncritical Transitions to and from Manual Control. *Human Factors*, *59*(4), 689–705. http://doi.org/10.1177/0018720816685832
- Florencio, D., & Herley, C. (2007). A large-scale study of web password habits. In *Proceedings of the 16th international conference on World Wide Web - WWW '07* (pp. 657–665). http://doi.org/10.1145/1242572.1242661 .
- Kumaraguru, P., Sheng, S., Acquisti, A., Cranor, L. F., & Hong, J. (2010). Teaching Johnny not to fall for phish. *ACM Transactions on Internet Technology*, *10*(2), 1–31. http://doi.org/10.1145/1754393.1754396 .
- Motiee, S., Hawkey, K., & Beznosov, K. (2010). Do windows users follow the principle of least privilege? *Proceedings of the Sixth Symposium on Usable Privacy and Security - SOUPS '10*. http://doi.org/10.1145/1837110.1837112 .
- Lum, H. & Reagan, J. A. (1995). Interactive Highway Safety Design Model: Accident Prevention Module. *Public Safety*, *59* (2). http://www.fhwa.dot.gov/publications/publicroads/95winter/p95wi14.cfm
- Pollard, J., Fisher, D., Guglielmi, J., Mattson, A., Young, J., & Lucibello, G. (2017). Driver Ability to Cope with Unintended Acceleration by Shifting Gears in a Simulator Experiment. In *Transportation Research Board 96th Annual Meeting*. Washington DC.
- Schupak, A. (2015, May 12). Majority of Americans fall for email phishing scams. *CBS News*. Retrieved from http://www.cbsnews.com/news/majority-of-americans-fall-for-email-phishing-scams-cbs-intel-security-quiz/ .
- Thomas, F. D., Rilea, S. L., Blomberg, R. D., Peck. R. C., & Korbelak, K. T. (2016, January). *Evaluation of the safety benefits of the risk awareness and perception training program for novice teen drivers* (Report No. DOT HS 812 235). Washington, DC: National Highway Traffic Safety Administration.
- Zafian, T. M., Samuel, S., Coppola, J., Neill, E. G. O., Romoser, M. R. E., & Fisher, D. L. (2016). On-Road Effectiveness of a Tablet-Based Teen Driver Training Intervention. In *Proceedings of the Human Factors and Ergonomics Society Annual Meeting* (pp. 1926–1930). Washington DC.

21

AUTOMATED VEHICLES SYMPOSIUM 2018 USERS. VEHICLES. INFRASTRUCTURE. Driven by AUVSI and TRB

Breakout Session #6 From Automated Vehicles to Automated Transportation Systems

Summary of Key Findings and Lessons Learned

- Take a System View! Not separate vehicle, infrastructure, and driver/occupants
- Learn lessons from other modes: rail, air, marine transport
- Seek opportunities for early deployments on dedicated facilities, such as Managed Lanes
- Cooperation between vehicles, and between vehicles and the infrastructure, is critical!
- Take a structured approach (crawl, walk, run) including:
  - Simulation and Modelling
  - Field Testing in a controlled environment (AV Proving Grounds?)
  - Pilot Deployments

AUTOMATED VEHICLES SYMPOSIUM 2018 USERS. VEHICLES. INFRASTRUCTURE. Driven by AUVSI and TRB

Breakout Session #6 From Automated Vehicles to Automated Transportation Systems

Research Needs

- Greater collaboration/interaction between Infrastructure Owner-Operators and Vehicle Manufacturers
- Determine if the results of small-scale field tests can be used to extrapolate the benefits of large-scale deployments
- Create (standardize?) better models for more accurate simulation results
- Collect more field test data to calibrate and validate the accuracy of models (iterative process)

AUTOMATED VEHICLES SYMPOSIUM 2018
USERS. VEHICLES. INFRASTRUCTURE.
Driven by AUVSI and TRB

Breakout Session # 7

Summary of Key Findings and Lessons Learned

Session: AVs & Vulnerable Road Users: Envisioning a Healthy, Safe & Equitable Future

- Session Goal: Expand the discussion surrounding the interaction between VRUs of all types and AV systems
  - Particularly exploring considerations related to mobility and equity
- Research Talks:
  - Azra Habibovic, Pravin Varaiya, Fred Feng, Tabitha Combs
  - Presented recent research on ways that AVS and/or drivers currently interact with VRUs, as well as an overview of the equity-related challenges that will be faced as we transition to AVs
- Panel Discussion: The role of Equity Considerations in AV/VRU Interactions
  - Moderator: Mohammed Yousuf
  - Panelists: Bernard Soriano, Ryan Snyder, Robert Wall Emerson, Sarah McCullough
  - Examined the many ways that equity considerations come into play with AVS deployment, ways to engage communities to fairly distribute benefits & risks

AUTOMATED VEHICLES SYMPOSIUM 2018
USERS. VEHICLES. INFRASTRUCTURE.
Driven by AUVSI and TRB

Breakout Session # 7

Recommended Action Items

- Research needed to evaluate on-road as well as trip-level behavior
- Recognize the tradeoff & interdependence between safety & access to transportation
- Explore ways to shift the culture to prioritize pedestrians, maximize access and safety
- Identify needs & risks for people with disabilities in particular, within & outside vehicles
- Find ways to engage at-risk communities including low-SES areas
- Recognize & learn from the history of how transportation has affected communities, particularly low SES
- Work with stakeholders at all levels from community to OEMs, bring in government at all levels – local, state & federal to identify & solve problems before they

AUTOMATED VEHICLES SYMPOSIUM 2018 — Driven by AUVSI and TRB
USERS. VEHICLES. INFRASTRUCTURE.

Breakout Session # 8: Autonomous Vehicles: Driving Employment for People with Disabilities

Summary of Key Findings and Lessons Learned

- OEMs have created an aftermarket to address the physical accessibility of automobiles, including access for wheelchair users, those who use other mobility aids such as walkers, and those of short stature.
  - Discussed strategies that could work to address physical accessibility of AVs.
- AV accessibility will need to include people with sensory disabilities, including those who are blind or who have low vision, and those who are deaf or hard of hearing.
  - Discussed strategies that could work to address accessibility for those with sensory disabilities.
- One segment of the market for AVs is individuals with cognitive disabilities, such as people on the autism spectrum, traumatic brain injury, and people with intellectual disabilities such as Down Syndrome.
  - Discussed strategies that could work to address accessibility for people with cognitive disabilities.
- Although AV deployment initially focuses on urban areas, this technology shows strong potential to increase transportation access in rural areas.
  - Discussed strategies to ensure that people in urban, suburban, and rural areas can all benefit equitably from AV deployment.
- Federal and State Governments have begun to create policy regarding the development and deployment of AVs.
  - Discussed strategies that federal and state policymakers could employ to ensure AV accessibility that will result in improved access to employment for Americans with disabilities.

AUTOMATED VEHICLES SYMPOSIUM 2018 — Driven by AUVSI and TRB
USERS. VEHICLES. INFRASTRUCTURE.

Breakout Session # 8: Autonomous Vehicles: Driving Employment for People with Disabilities

Research Needs

- Just as the rapid increase in the availability of ICT has greatly increased the employment potential of PWD, AVs will allow millions to get to workplaces, and thereby greatly expand their employment potential.
- If AVs are not accessible, however, an inequitable situation becomes worse. Just as inaccessible job application software tilts the playing field toward those applicants without disabilities, inaccessible AVs will do the same, and PWD will fall even further behind in the job market.
- How do we leverage the excitement around AVs to meet the transportation needs of all Americans?

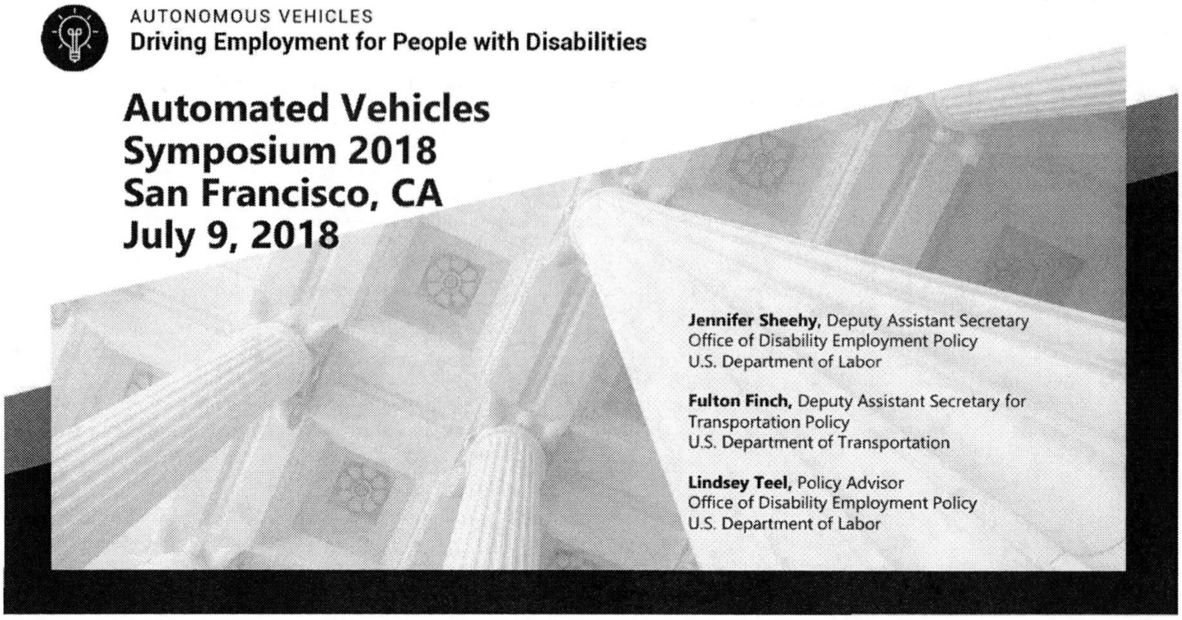

AVs and Disability Employment

- DOL/ODEP and DoT have long history of collaborating together on initiatives to increase transportation access for people with disabilities.
- DOL/ODEP also has a history of working with the private sector on ensuring accessibility of information and communication technology (ICT).
  - ✓Just as the rapid increase in the availability of ICT has greatly increased the employment potential of PWD, AVs will allow millions to get to workplaces, and thereby greatly expand their employment potential.
  - ✓If AVs are *not* accessible, however, an inequitable situation becomes worse. Just as inaccessible job application software tilts the playing field toward those applicants without disabilities, inaccessible AVs will do the same, and PWD will fall even further behind in the job market.

Idea Generation – Accessibility & Mobility Disabilities

OEMs have created an aftermarket to address the physical accessibility of automobiles, including access for wheelchair users, those who use other mobility aids such as walkers, and those of short stature.

Group discussion on strategies that could work to address physical accessibility of AVs.

Idea Generation – Accessibility & Sensory Disabilities

AV accessibility will need to include people with sensory disabilities, including those who are blind or who have low vision, and those who are deaf or hard of hearing.

Group discussion on strategies that could work to address accessibility for those with sensory disabilities.

Idea Generation– Accessibility & Cognitive Disabilities

One segment of the market for AVs is individuals with cognitive disabilities, such as people on the autism spectrum, traumatic brain injury, and people with intellectual disabilities such as Down Syndrome.

Group discussion on strategies that could work to address accessibility for people with cognitive disabilities.

Idea Generation – Equitable AV Deployment

Although AV deployment initially focuses on urban areas, this technology shows strong potential to increase transportation access in rural areas.

Group discussion on strategies to ensure that people in urban, suburban, and rural areas can all benefit equitably from AV deployment.

Idea Generation – The Government's Role

Federal and State Governments have begun to create policy regarding the development and deployment of AVs.

Group discussion on strategies that federal and state policymakers could employ to ensure AV accessibility that will result in improved access to employment for Americans with disabilities.

Thank You for Sharing Your Feedback & Ideas!

Check out our accessible transportation online crowdsourcing community to continue the discussion!

TransportationInnovation.ePolicyWorks.org

To see more about what the Department of Transportation is doing around automation, visit:

www.dot.gov/av

AUTOMATED VEHICLES SYMPOSIUM 2017

USERS. VEHICLES. INFRASTRUCTURE.

Breakout Session # 9
An AV Crash Occurs: What Happens Next?

Summary of Key Findings and Lessons Learned:

- Consensus that standards are required.
- Plaintiff and Defense Lawyers continue to debate about appropriate liability regimes.
- Importance of data collection, storage, verification, and security methods.
- The insurance industry is working on new products to address risks stemming from AVs.
- Mixed fleets introduce additional complexities for a foreseeable future.
- There is a mismatch between consumer expectations and current AV technologies' performance.

Session Speakers:

Todd Benoff, Alston & Bird; Ryan Gammelgard, State Farm; Dorothy Glancy, Santa Clara University School of Law; Carl Schultheisz, National Transportation Safety Board; Don Slavik, Slavik Law Firm; Susan Clare, King & Spalding; Ryan Harrington, Exponent; Kristin Kolodge, JD Power

AUVSI TRB

AUTOMATED VEHICLES SYMPOSIUM 2017

USERS. VEHICLES. INFRASTRUCTURE.

Breakout Session # 9
An AV Crash Occurs: What Happens Next?

Recommended Action Items

- Develop standards and speed timeframes for revisions to FMVSS.
- Close the gap between consumer expectations and reality.
- Product liability laws and lemon laws need to be revised to account for the complexities of AVs, including over-the-air updates.

AUVSI TRB

AUTOMATED VEHICLES SYMPOSIUM 2018
USERS. VEHICLES. INFRASTRUCTURE.
Driven by AUVSI and TRB

Breakout Session # 10 – Building Automation into **Urban and Metropolitan Mobility Planning**

Summary of Key Findings and Lessons Learned

- Modellers, planners and policy makers are **actively pursuing automation** as a mobility option. But **little actual automation content** in Long-range plans (US)/ Sustainable Urban Mobility Plans (EU)
- There is a **growing consensus** about the direction of opportunities and challenges.
- **Shared mobility** and **last mile transport** are priority areas. Common question: what will be **the new mix** of shared/ on demand, mainstream collective and personal ownership-based transport (and cycling and walking)
- **Uncertainties** about deployment, user response and impacts are major barriers to more concrete planning. **Missing data** and **methodologies** are a major professional challenge.
- **Pilots** are providing a basis to verify assumptions and collect data.
- Is there **a new playbook** metropolitan mobility planning being written?

AUTOMATED VEHICLES SYMPOSIUM 2018
USERS. VEHICLES. INFRASTRUCTURE.
Driven by AUVSI and TRB

Breakout Session # 10 – Building Automation into **Urban and Metropolitan Mobility Planning**

Recommendations and Research Needs

- Planning for automation should be **data-based** and **policy-guided**.
- **Scenario planning** and other exploratory analysis can help to handle uncertainty. More work on adapted **methodologies** and use of **new data sources** is needed.
- Wide **collaboration with stakeholders and new actors** is a key requirement for planning.
- Links to **energy, land use, environment, social inclusion** should be considered.
- Should there be an international dialogue about **"automation policy principles"**?

AUTOMATED VEHICLES SYMPOSIUM 2018 Driven by AUVSI and TRB

USERS. VEHICLES. INFRASTRUCTURE.

Breakout Session # 10 – Building Automation into **Urban and Metropolitan Mobility Planning**

Thanks!

Organisers/ speakers:
•**Siegfried Rupprecht**, Rupprecht Consult – Forschung & Beratung GmbH
•**Scott Smith**, US DOT, Volpe Center
•**Elizabeth Machek**, US DOT, Volpe Center
•**Bart van Arem**, Delft University of Technology
Speakers:
•**John Corbin**, FHWA
•**William Riggs**, University of San Francisco
•**Aernout Aki Ackerman**, MRDH Metropolitan Region Rotterdam The Hague
•**Augustin Wegscheider**, Boston Consulting Group
•**Melissa Ruhl, William Baumgardner,** Arup
•**Tilly Chang**, SFCTA

Building Automation into Urban and Metropolitan Mobility Planning

Monday, July, 9, 1:30 PM – 5:30 PM Golden Gate 4

Agenda

The session will examine the following areas:

- building automation into long-range transport plans (US) sustainable urban mobility plans (EU) and other policy documents (and the processes that help produce them),
- methodologies and tools that help planners to reduce uncertainties in the planning process (e.g. scenario and use case development, models, other techniques),
- lessons learned and results emerging from pilots in cities and regions,

1330	**Introduction,** Siegfried Rupprecht and Scott Smith
1340	**Planning for automation** • Siegfried Rupprecht – *Automation in the European guidelines for mobility planning (SUMP)* • John Corbin, FHWA – *National Dialogue on Highway Automation*
1410	**Methodologies and tools** • Scott Smith, Volpe / US DOT – *Modeling automation: report from the TRB Innovations in Travel Modeling Conference, June 2018* • Bart van Arem, TU Delft - *Scenarios, methods and tools to explore the urban and regional mobility impacts of automated driving* • William Riggs, University of San Francisco - *Hacking the future of streets: Participatory design of streets for autonomous vehicles*
1510	**Break**
1520	**Lessons learnt in cities** • Aernout Aki Ackerman, MRDH Metropolitan Region Rotterdam The Hague, *The automated vehicles for the last mile innovation network* • Augustin Wegscheider, Boston Consulting Group, *Using advanced analytics to inform urban mobility planning* • Melissa Ruhl, Arup, *Regional Strategies for an Autonomous Future* • Tilly Chang, SFCTA, *New mobility in San Francisco*

1620	**Round Tables – What would you like to see in an automation action plan?**
	• Discussed from different perspectives. For example:
	  o **Long-range planning**
	  o **Parking, curb space allocation**
	  o **Infrastructure (road) owners**
	  o City services (snow removal, trash collection)
	  o **New mobility services, public transit**
	  o Freight and delivery services
	• Questions
	  o What would be the main actions (now, 5+, 15+ years)
	  o Which "golden rules"/ guiding principles would be included?
	  o Which (emerging) good practice of planning for automation is available?

1715	**Summary** Reports from Round Tables (by moderators)

AUVSI TRB

What would you like to see in an automation action plan?

General Questions

- **Actions to take now? In 5 years? Later?**
- **What guiding principles would be included?**
- **What (emerging) planning practices are available?**

Long range planning

- What types of decisions are we most concerned about? Major infrastructure? Policy?
- What are the largest areas of uncertainty?
- Do we have the right tools, data and processes? What do we need?
- What have we learned from recent unplanned changes (e.g., the rise of TNCs) that can inform our future planning?
- How do policies and regulations need to change to better support appropriate long range planning, given these uncertainties?

Parking, curb space allocation

- How will land use regulations (e.g., zoning) affect the deployment of automated mobility services in cities?
- Where do we expect to see initial impacts?
- Automation may free up urban space by reducing the need for parking. How can we identify and repurpose space gained through automation?
- What are our ideas for land use in general (e.g., freight hubs vs. recreation areas) and at the curb? Which designs support automated services that support urban policy goals?
- Are there experiences that we can learn from today?

New mobility services, public transport

- How can automated transport improve last mile transport?
- How can automated transport improve bus service operation?
- Will Level 4 automated on demand transport replace public transport or regular taxi services?
- How important is automation for Mobility as a Service?
- What degree of coordination is necessary with existing transit operators and other mobility providers (including new service models, stakeholders such as real estate developers)?

Infrastructure (road) owners

- How and when to make roads (urban, rural, freeway) automation ready? What are the key elements: road marking, traffic signs, geometry
- Do we need high tech or low tech roads? (C-ITS, positioning, maps, V2X). What minimum regular maintenance levels are needed to support automation effectively?
- Do we allow automation and manual driving on the same road?
- Do we need to regulate the use of automated vehicles? Which conditions would apply?
- Are urban automation corridors/ lanes (exclusively for C-ITS) a desirable concept, in terms of traffic efficiency, safety, access, vulnerable road users? Can we mix automated road use and pedestrian/ bicycle traffic in low speed urban environments?

City services (snow removal, road cleaning, maintenance, trash collection)

- More than an interesting "niche"? What are interesting use cases of highly automated service vehicles in transport and infrastructure management?
- What are the possible impacts of autonomous service vehicles (e.g., in traffic management, cost reduction, data quality)?
- Are there other synergies with automated city service vehicles (electric charging, data collection, confidence in automation)?

Freight and delivery services

- What disruptions are you seeing today, with the rise of e-commerce?
- How will automation change long haul freight? Local delivery?
- Many forms of automated freight have been envisioned: truck platoons, local delivery trucks, small sidewalk vehicles, drones. What do you see as most likely? Most in need of a response from the city?
- What opportunities do you see for automated freight to improve the quality of life in cities?
- What can transport authorities/ MPOs do to support automated freight delivery that is in line with city goals?

METROPOOLREGIO
ROTTERDAM DEN HAAG

Innovation Network AVLM

AV Symposium, San Francisco
9 juli 2018

METROPOOLREGIO
ROTTERDAM DEN HAAG

Metropolitan region Rotterdam The Hague (MRDH)

- Regional cooperation
- 23 municipalities

Roles:

- Transportation authority
- Public transport management
- Economic development

METROPOOLREGIO
ROTTERDAM DEN HAAG

Roadmap Next Economy

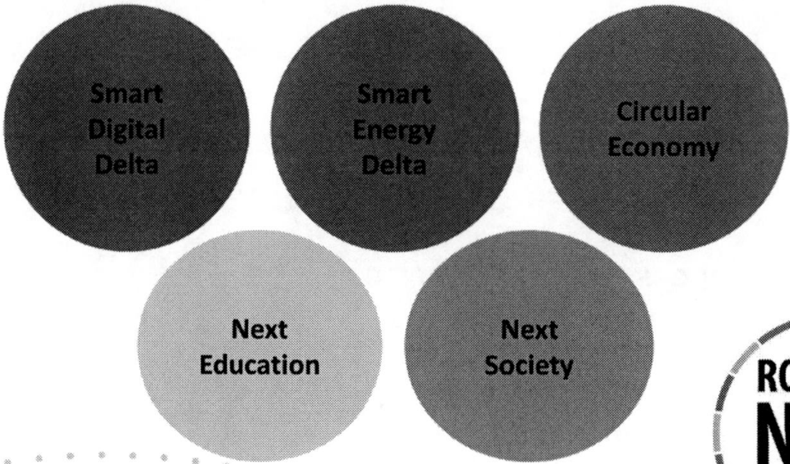

ROADMAP NEXT ECONOMY

METROPOOLREGIO
ROTTERDAM DEN HAAG

Innovation network AVLM

- Better public transport
- Cost reduction
- More revenue
- Safer mode of transport
- Environmental benefits
- Economic development

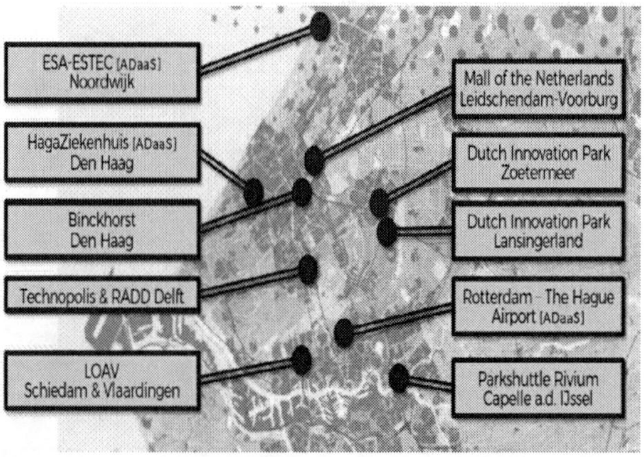

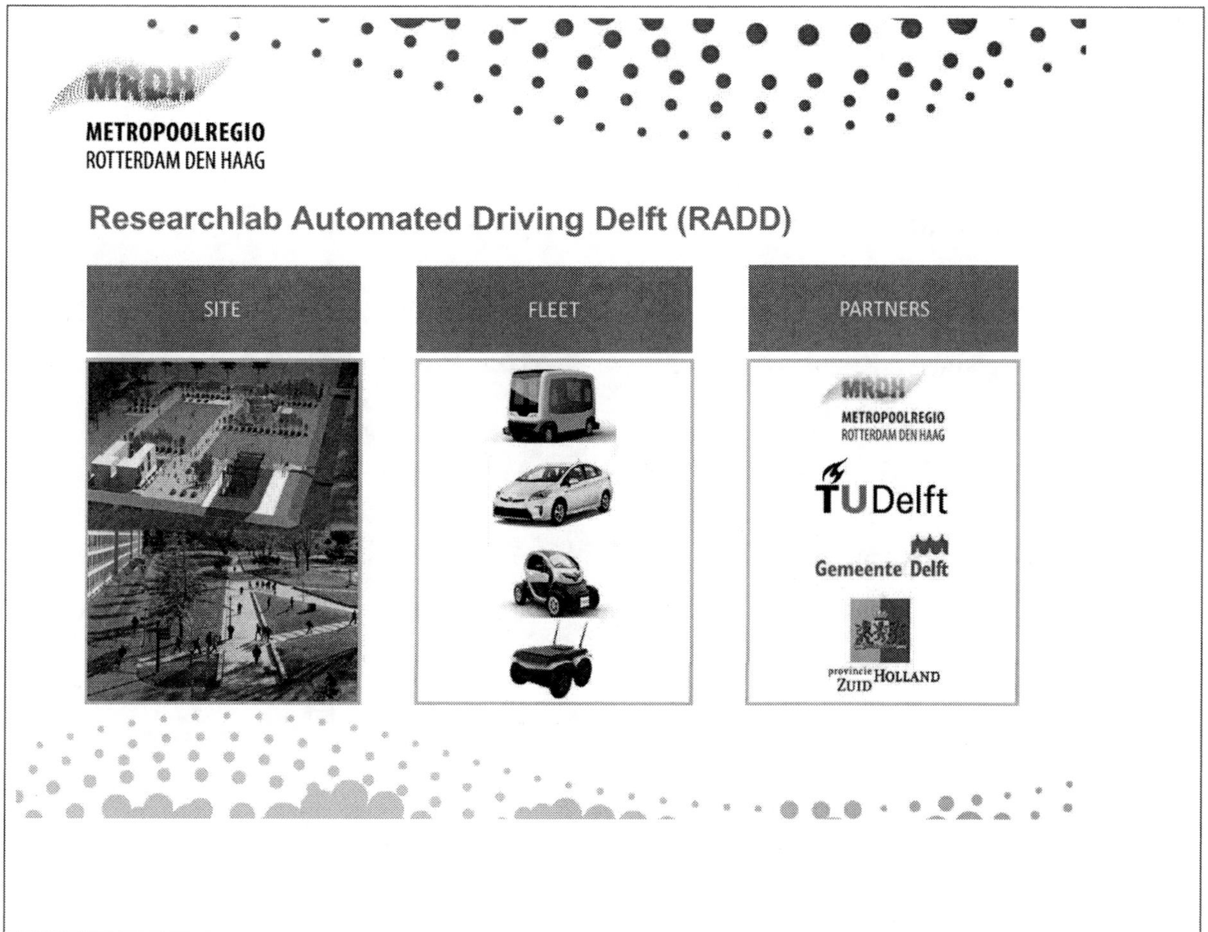
MRDH
METROPOOLREGIO
ROTTERDAM DEN HAAG
Researchlab Automated Driving Delft (RADD)
SITE
FLEET
PARTNERS
MRDH
METROPOOLREGIO
ROTTERDAM DEN HAAG
TUDelft
Gemeente Delft
provincie ZUID HOLLAND

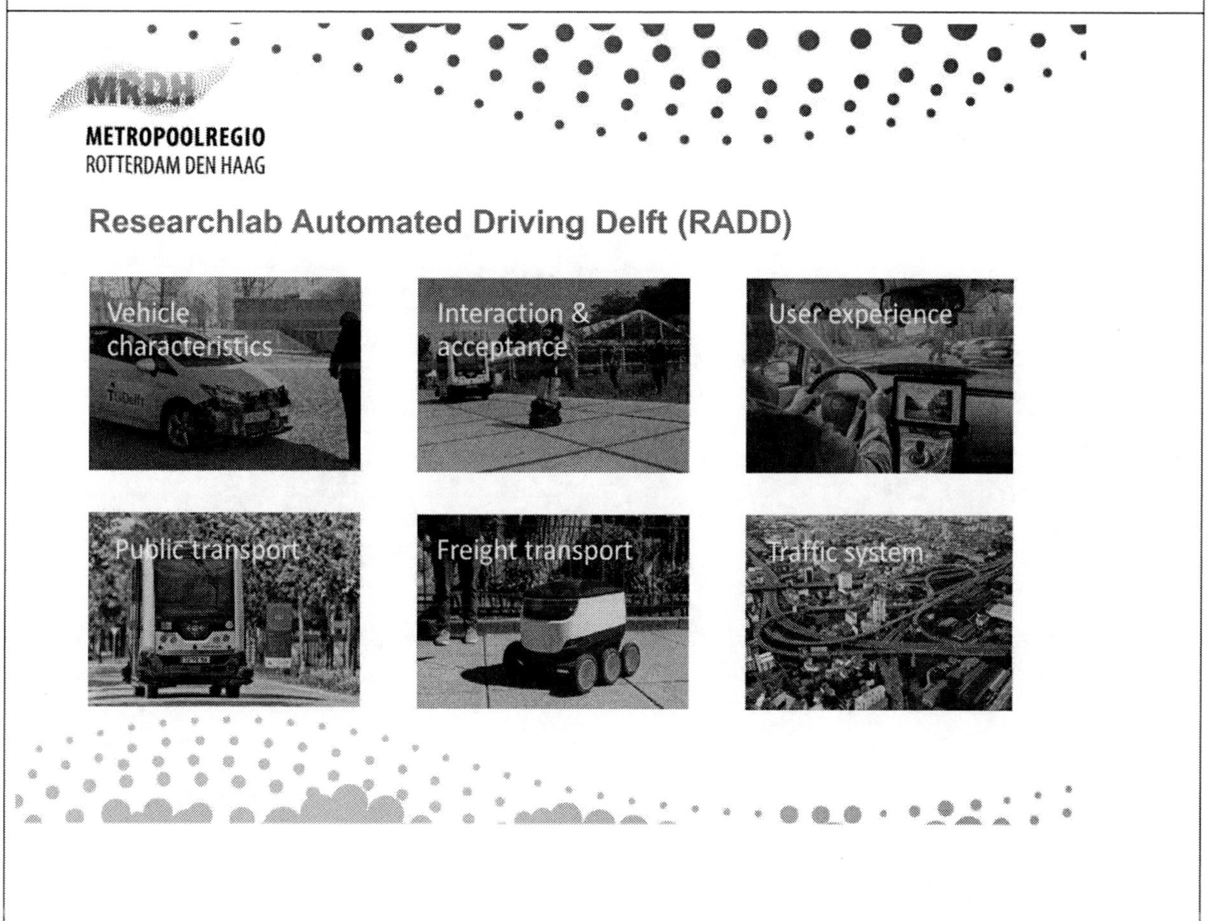
MRDH
METROPOOLREGIO
ROTTERDAM DEN HAAG
Researchlab Automated Driving Delft (RADD)
Vehicle characteristics
Interaction & acceptance
User experience
Public transport
Freight transport
Traffic system

METROPOOLREGIO
ROTTERDAM DEN HAAG

Leidschendam-Voorburg – Mall of the Netherlands

- Aligned with city objectives
- Private partner
- Transportation value

+ 50% usage public transport

+ 68% usage elderly

+ 40% commuters

METROPOOLREGIO
ROTTERDAM DEN HAAG

Parkshuttle Rivium 1.0 and 2.0 – anno 2001

2getthere

METROPOOLREGIO
ROTTERDAM DEN HAAG

Parkshuttle Rivium 3.0

- Innovative mobility hub
- 3rd generation AV-shuttle

Phased implementation

- Fase 1: upgrade current track
- Fase 2a: mixed traffic (2018-2019)
- Fase 2b: fully operational (2019)

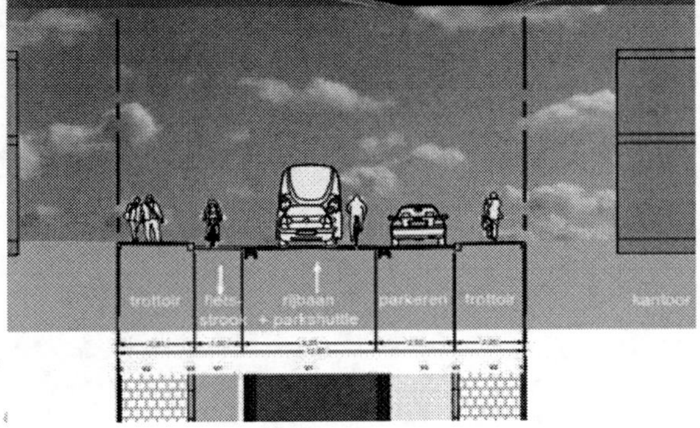

METROPOOLREGIO
ROTTERDAM DEN HAAG

Focus and role MRDH

1. Local business cases and exploitation
2. Cooperate to accelerate learning
3. Co-finance
4. Facilitate creation legal framework
5. European cooperation

METROPOOLREGIO
ROTTERDAM DEN HAAG

Lessons learned

- Multiple locations and cooperation needed to accelerate the learning
- Insight in exploitation and resources needed for large scale deployment
- Economies of scale

METROPOOLREGIO
ROTTERDAM DEN HAAG

Thank you!

Metropolitan region of Rotterdam The Hague

- Christine Baukema: c.baukema@mrdh.nl
- Patrick van Norden: p.vannorden@mrdh.nl
- Aernout Aki Ackerman: aki@mabsconsultancy.nl

NATIONAL DIALOGUE ON
HIGHWAY AUTOMATION

U.S. Department of Transportation
Federal Highway Administration

Briefing to the 2018 TRB Automated Vehicle Symposium,

July 9-12, 2018

John Corbin, Connected Automated Vehicle Program Manager
FHWA Office of Operations
John.corbin@dot.gov

Aspects of Highway Automation Readiness

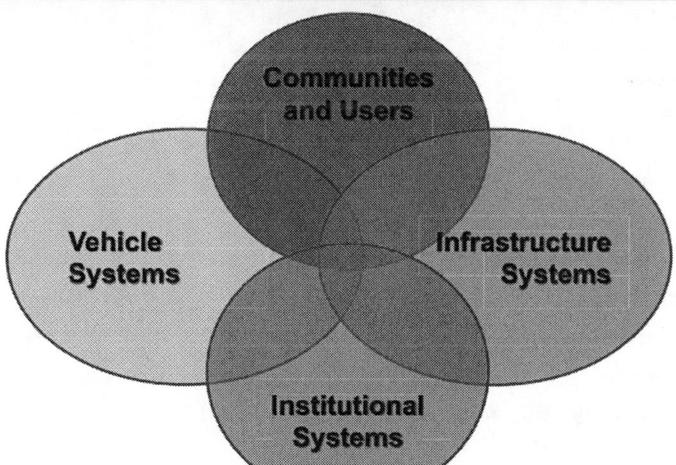

NATIONAL DIALOGUE ON
HIGHWAY AUTOMATION

U.S. Department of Transportation
Federal Highway Administration

USDOT Activities in Automation

Event	Date	Summary / Outcomes
USDOT releases *Automated Driving Systems (ADS) 2.0: A Vision for Safety*	September 12, 2017	• Replaces the 2016 Federal Automated Vehicles Policy. • More information on ADS 2.0 is available on the NHTSA website: https://www.nhtsa.gov/technology-innovation/automated-vehicles.
Roundtable on Data for Automated Vehicle Safety	December 7, 2017	• Demonstrated multimodal alignment around "One DOT" approach to Federal automated vehicle policy. • Brought together over 60 participants from government, private sector, nonprofit organizations, universities, and research centers. • Gathered feedback on USDOT's Guiding Principles and Draft Framework.
Automated Vehicles 3.0 announced	January 10, 2018	• Secretary Chao announced work on a follow-up Automated Vehicles 3.0 document, with a possible release date in 2018.
Public Listening Summit on Automated Vehicle Policy	March 1, 2018	• Stakeholder engagement summit with senior leadership at USDOT. • Focused on key cross-modal issues important to successful integration of automated vehicles.
National Dialogue Introductory Webinar	May 8, 2018	• Introductory webinar introducing the National Dialogue. • 360+ attendees. • Recording: https://connectdot.connectsolutions.com/p52h2c59wp92/.

NATIONAL DIALOGUE ON
HIGHWAY AUTOMATION

U.S. Department of Transportation
Federal Highway Administration

4

USDOT Activities in Automation (continued)

FHWA Request for Information Key Themes:

- Greater **Uniformity and Quality** in road markings and traffic control devices would enable automation.
- **Uncertainty** in infrastructure investment and allocation of limited resources are key concerns for State and local agencies.
- Certain **Data Elements** around the roadway environment are useful for industry, State, and local DOTs to share and could improve automation operations.
- Conducting **Pilots** and supporting pilot testing are important for facilitating learning, collaboration, and information sharing.
- FHWA should take a **Leadership** role in convening stakeholders to encourage collaboration.

NATIONAL DIALOGUE ON
HIGHWAY AUTOMATION

U.S. Department of Transportation
Federal Highway Administration

6

Purpose of the National Dialogue

1. **LISTEN:** Gather input from a diverse group of stakeholders.
2. **ENGAGE:** Facilitate information sharing among industry, public agencies, associations and others.
3. **INFORM:** Raise awareness of FHWA and USDOT activities in automation and emerging technologies.
4. **EVOLVE:** Adapt existing programs, policies and research to address automation.

NATIONAL DIALOGUE ON HIGHWAY AUTOMATION

U.S. Department of Transportation Federal Highway Administration

7

Desired Outcomes of the National Dialogue

FHWA will use inputs to:

1. Assess national issues and priorities
2. Develop technical guidance, best practices and standards
3. Support and coordinate necessary research
4. Adapt programs and policies
5. Create a national community or coalition

NATIONAL DIALOGUE ON HIGHWAY AUTOMATION

U.S. Department of Transportation Federal Highway Administration

8

National Dialogue Schedule

Month	Event	Location
June 7	**National Dialogue Launch Workshop**	Cobo Center, Detroit, MI
June 26-27	**National Workshop 1** Planning and Policy Considerations for Highway Automation	Science History Institute Philadelphia, PA
July 12	**Automated Vehicle Symposium** FMCSA-FHWA Truck Automation Listening Session	San Francisco, CA
August 1-2	**National Workshop 2** Digital Infrastructure and Data Considerations for Highway Automation	DoubleTree Hilton Airport Seattle, WA
Early September	**National Workshop 3** Freight Considerations for Highway Automation	Chicago, IL
October 24-25	**National Workshop 4** Operations Considerations for Highway Automation	Phoenix, AZ
Week of Nov. 12	**National Workshop 5** Infrastructure Design and Safety Considerations for Highway Automation	Austin, TX

NATIONAL DIALOGUE ON HIGHWAY AUTOMATION

U.S. Department of Transportation
Federal Highway Administration

20

Contact:
HighwayAutomation@dot.gov

https://ops.fhwa.dot.gov/automationdialogue/index.htm

(Search on "FHWA National Dialogue")

NATIONAL DIALOGUE ON HIGHWAY AUTOMATION

U.S. Department of Transportation
Federal Highway Administration

15

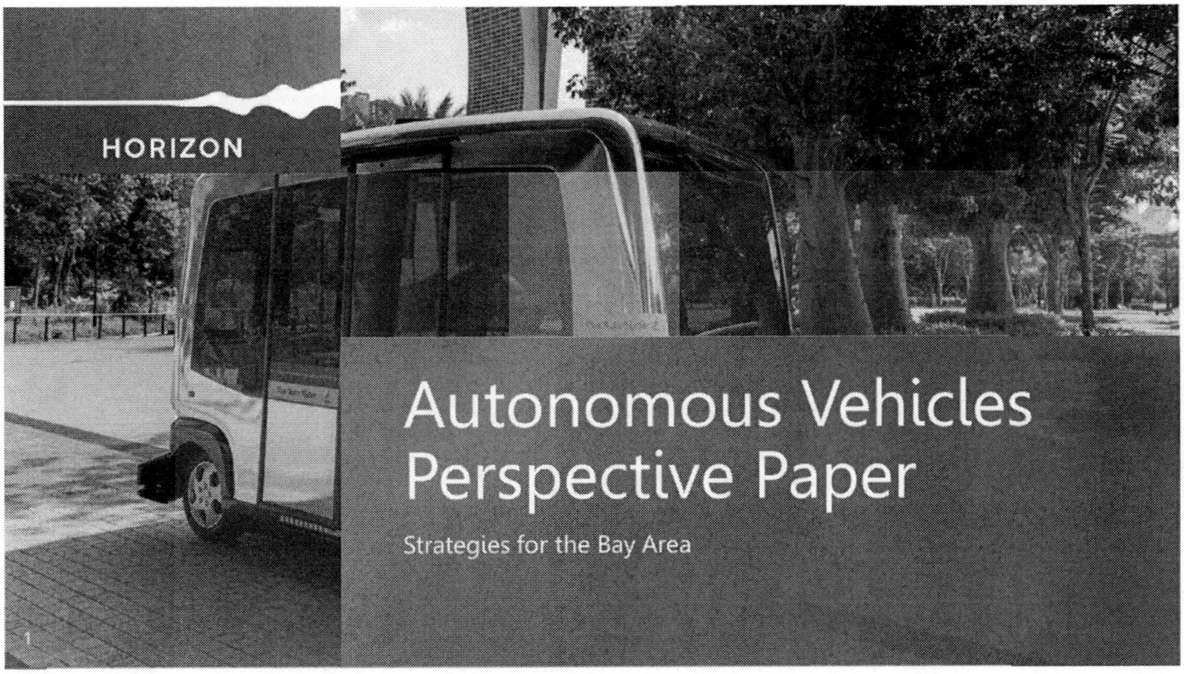
HORIZON
Autonomous Vehicles Perspective Paper
Strategies for the Bay Area
1

Horizon is exploring how **economic, environmental, technological, and political uncertainties** may create new challenges – or exacerbate existing ones – for the Bay Area over the coming decades.
HORIZON
Futures Planning
Perspective Papers
Project Performance
PLAN BAY AREA 2050
For more information, go to:
mtc.ca.gov/horizon

HORIZON
Overview
• Autonomous Vehicles 101
• Implications and Strategies
• Horizon Guiding Principles
• Opportunities and Risks
• "Big Ideas" and Applications for the Bay Area
• Next Steps
HORIZON

HORIZON
Autonomous Vehicles 101
4

Autonomous Vehicles Components

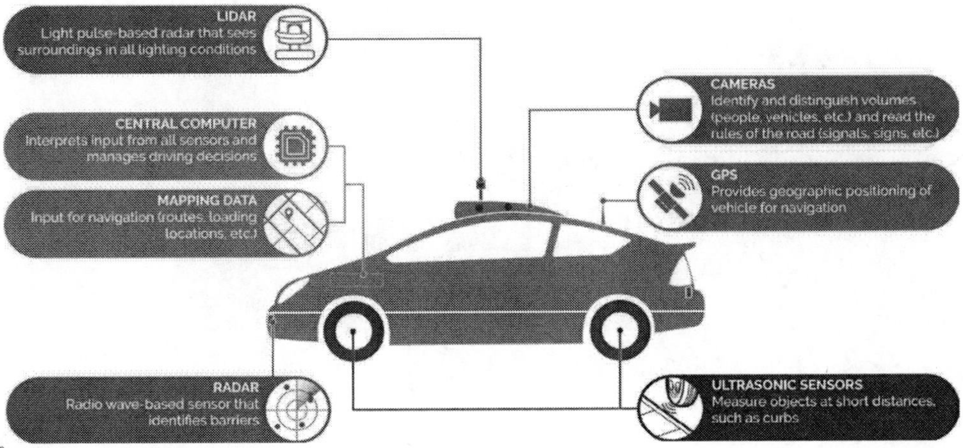

5 Introduction

HORIZON

Levels of Automation

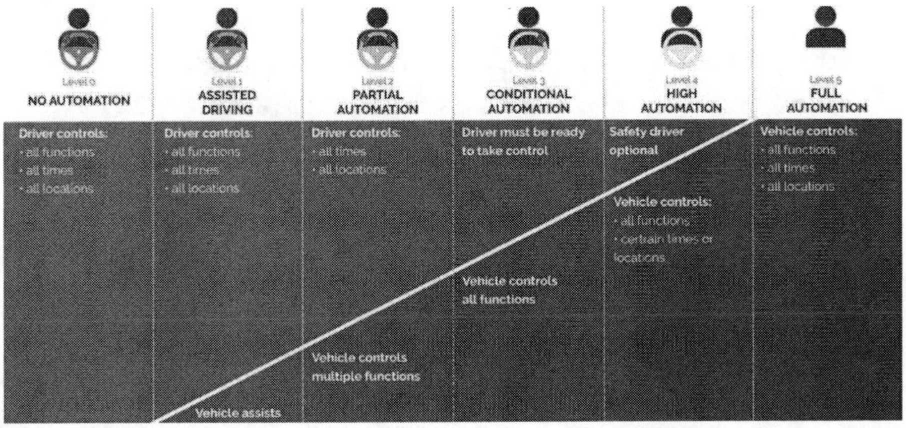

6 Introduction

HORIZON

When do AVs become commonplace?

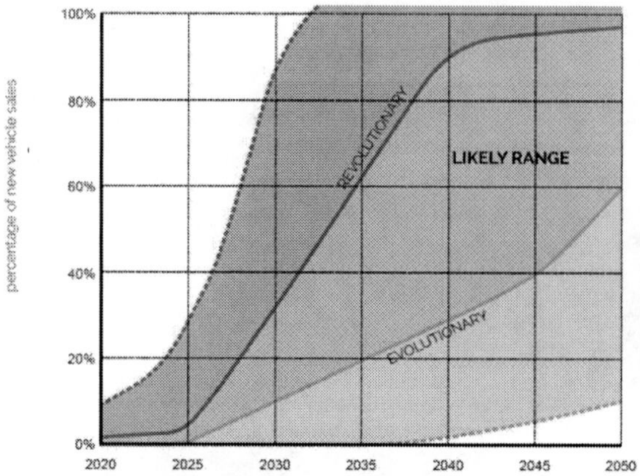

Fully Autonomous Vehicle (L4/5) uptake predictions based on high disruption scenarios, indicates possible percentage of new car sales 2016 to 2050.

Revolutionary

- Technology breakthroughs
- Regulatory resolutions
- Shared model, at much lower cost than ownership
- Rapid adoption

Evolutionary

- Slower technology development and rollout
- Owned AV model with cost premium
- Slower adoption

HORIZON

The future is highly uncertain

TIMING **3 to 13** years until L5 AVs available for purchase

SAFETY **+40% to +90%** increase in safety

CAPACITY **0% to +45%** increase in roadway capacity

DEMAND **+5% to +40%** increase in VMT

ENERGY/EMISSIONS **-50% to + 100%** change in GHGs

HORIZON

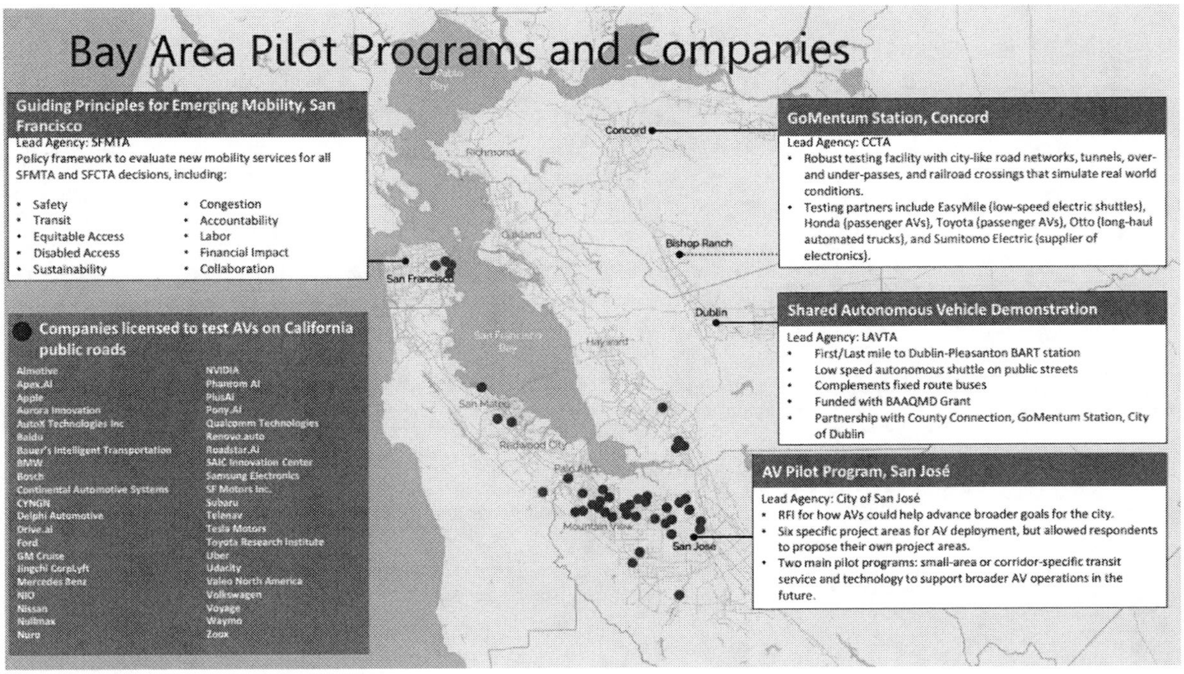

Bay Area Pilot Programs and Companies
Guiding Principles for Emerging Mobility, San Francisco
Lead Agency: SFMTA
Policy framework to evaluate new mobility services for all SFMTA and SFCTA decisions, including:
• Safety
• Transit
• Equitable Access
• Disabled Access
• Sustainability
• Congestion
• Accountability
• Labor
• Financial Impact
• Collaboration
Companies licensed to test AVs on California public roads
Almotive
Apex.AI
Apple
Aurora Innovation
AutoX Technologies Inc
Baidu
Bauer's Intelligent Transportation
BMW
Bosch
Continental Automotive Systems
CYNGN
Delphi Automotive
Drive.ai
Ford
GM Cruise
Jingchi CorpLyft
Mercedes Benz
NIO
Nissan
Nullmax
Nuro
NVIDIA
Phantom AI
PlusAI
Pony.AI
Qualcomm Technologies
Renovo.auto
Roadstar.AI
SAIC Innovation Center
Samsung Electronics
SF Motors Inc.
Subaru
Telenav
Tesla Motors
Toyota Research Institute
Uber
Udacity
Valeo North America
Volkswagen
Voyage
Waymo
Zoox
GoMentum Station, Concord
Lead Agency: CCTA
• Robust testing facility with city-like road networks, tunnels, over- and under-passes, and railroad crossings that simulate real world conditions.
• Testing partners include EasyMile (low-speed electric shuttles), Honda (passenger AVs), Toyota (passenger AVs), Otto (long-haul automated trucks), and Sumitomo Electric (supplier of electronics).
Shared Autonomous Vehicle Demonstration
Lead Agency: LAVTA
• First/Last mile to Dublin-Pleasanton BART station
• Low speed autonomous shuttle on public streets
• Complements fixed route buses
• Funded with BAAQMD Grant
• Partnership with County Connection, GoMentum Station, City of Dublin
AV Pilot Program, San José
Lead Agency: City of San José
• RFI for how AVs could help advance broader goals for the city.
• Six specific project areas for AV deployment, but allowed respondents to propose their own project areas.
• Two main pilot programs: small-area or corridor-specific transit service and technology to support broader AV operations in the future.
Concord
Richmond
Bishop Ranch
San Francisco
Oakland
Dublin
Hayward
San Mateo
Redwood City
Palo Alto
Mountain View
San Jose

HORIZON
Implications and Strategies

FINAL GUIDING PRINCIPLES

The San Francisco Bay Area Aspires To Be:

AFFORDABLE — All Bay Area residents and workers have sufficient housing options they can afford – households are economically secure.

CONNECTED — An expanded, well-functioning transportation system connects the Bay Area – fast, frequent and efficient intercity trips are complemented by a suite of local transportation options, connecting communities and creating a cohesive region.

DIVERSE — The Bay Area is an inclusive region where people from all backgrounds, abilities, and ages can remain in place – with access to the region's assets and resources.

HEALTHY — The region's natural resources, open space, clean water and clean air are conserved – the region actively reduces its environmental footprint and protects residents from environmental impacts.

VIBRANT — The Bay Area region is an innovation leader, creating quality job opportunities for all and ample fiscal resources for communities.

Icons Credit: The Noun Project

HORIZON

AFFORDABLE

Horizon Guiding Principle – All Bay Area residents and workers have sufficient housing options they can afford – households are economically secure.

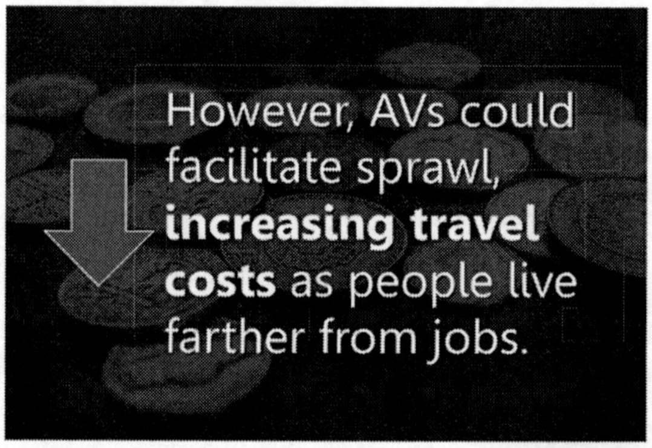

HORIZON

Housing Opportunity Sites in an Autonomous Future

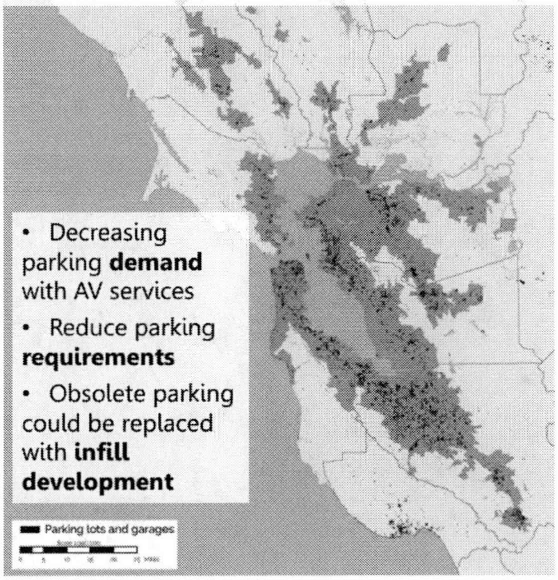

- Decreasing parking **demand** with AV services
- Reduce parking **requirements**
- Obsolete parking could be replaced with **infill development**

AFFORDABLE

Priority Strategies

- Repurpose off-street parking for **infill development**
- Institute **parking maximums** for both on- and off-street parking supply
- Retain or strengthen urban growth boundaries to **control greenfield development**

HORIZON

CONNECTED

Horizon Guiding Principle – An expanded, well-functioning transportation system connects the Bay Area – fast, frequent and efficient intercity trips are complemented by a suite of local transportation options, connecting communities and creating a cohesive region.

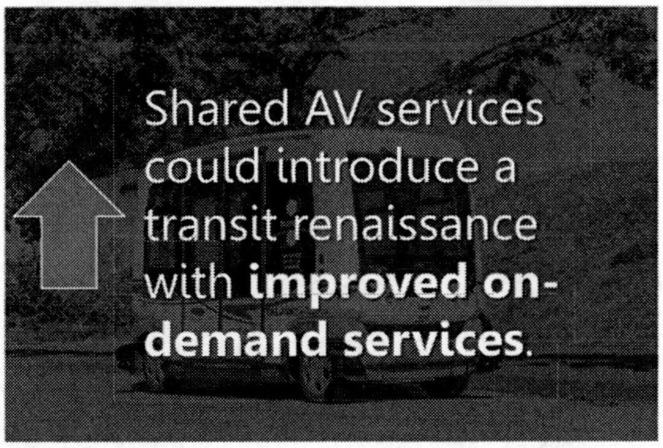

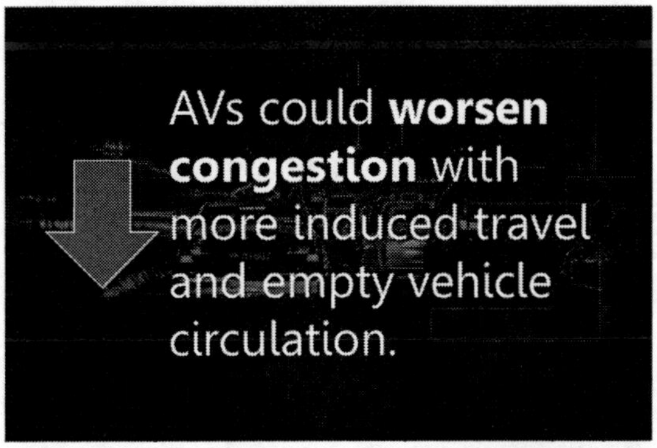

HORIZON

Regional Autonomous Demand-Responsive Transit

CONNECTED

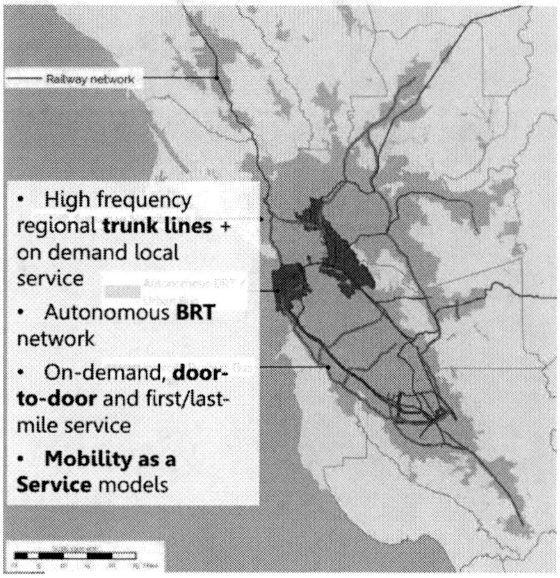

- High frequency regional **trunk lines** + on demand local service
- Autonomous **BRT** network
- On-demand, **door-to-door** and first/last-mile service
- **Mobility as a Service** models

Priority Strategies

Double down on **high-capacity bus and rail corridors**

Innovate suburban transit with autonomous, **demand-responsive microtransit**

Develop a **mobility as a service** platform to provide a unified and equitable gateway to services and information

HORIZON

Dynamic Pricing Opportunities in an AV Future

CONNECTED

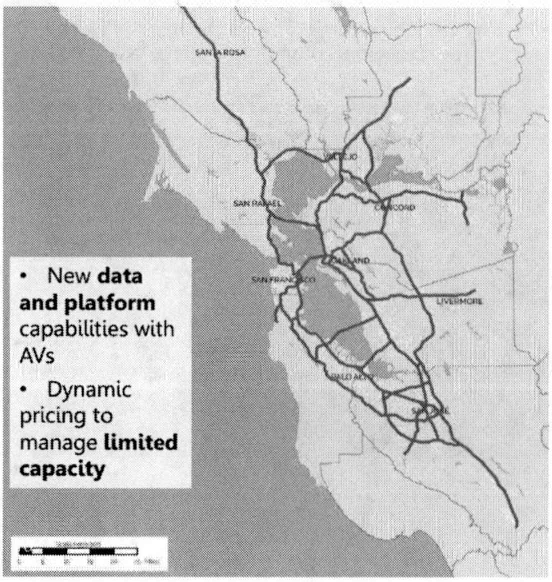

- New **data and platform** capabilities with AVs
- Dynamic pricing to manage **limited capacity**

Priority Strategies

Price mobility fairly through **dynamic road pricing**

Design **smart streets** with dynamic allocation of street and curb space

Develop industry-wide **data sharing protocols** to provide real-time information to connected AVs

HORIZON

DIVERSE

Horizon Guiding Principle – The Bay Area is an inclusive region where people from all backgrounds, abilities, and ages can remain in place – with access to the region's assets and resources.

Mobility options could proliferate with new business models, benefitting people from all backgrounds, abilities and ages.

AVs could widen the **equity gap** with declining public transit, service disparities, job loss, digital divide.

17 Implications and Strategies

HORIZON

Equitable AV Services

DIVERSE

- **Require accountability**: targets, metrics, monitoring, improvement
- Target strategies for **specific equitable outcomes**.
- Focus all strategies on **inclusive prosperity**.

Priority Strategies

Mandate **equitable provision** of mobility services with transparent reporting

Subsidize public **transit innovations,** replacing fixed route transit in Communities of Concern

Prioritize **AV mobility services or programs** that serve Communities of Concern

HORIZON

HEALTHY

Horizon Guiding Principle – The region's natural resources, open space, clean water and clean air are conserved – the region actively reduces its environmental footprint and protects residents from environmental impacts.

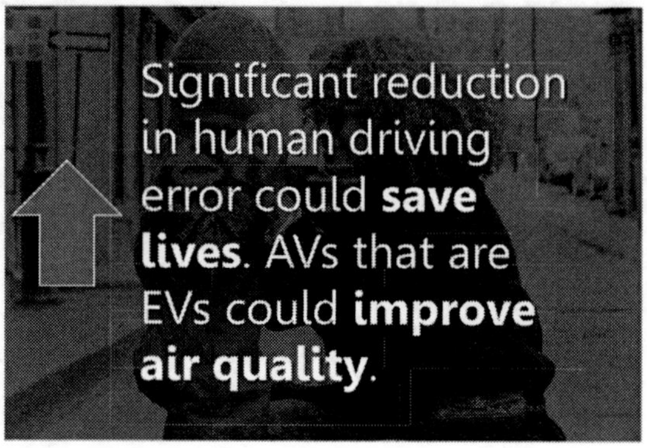

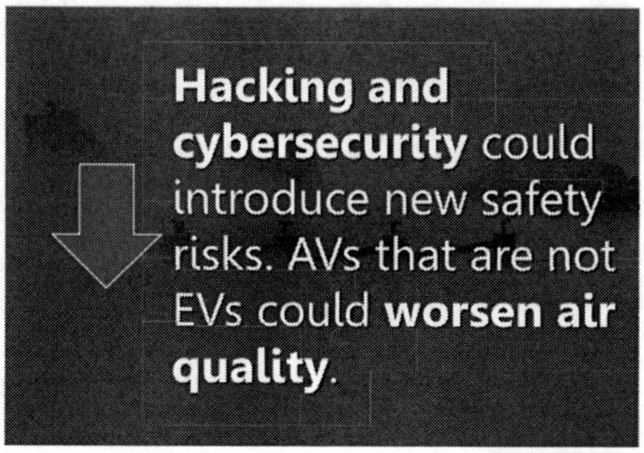

HORIZON

HEALTHY

Priority Strategies

Cap **speed limits** in downtowns and neighborhoods

Mandate that **all AVs are EVs** and invest in the necessary infrastructure

Develop "bounty program" to reduce **hacking vulnerability**

HORIZON

VIBRANT

Horizon Guiding Principle – The Bay Area region is an innovation leader, creating quality job opportunities for all and ample fiscal resources for communities.

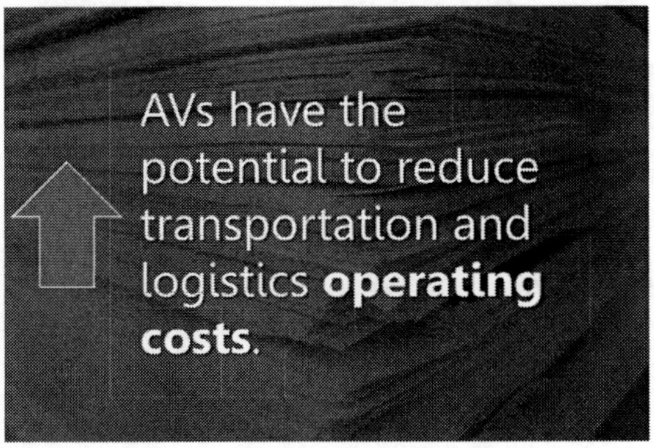

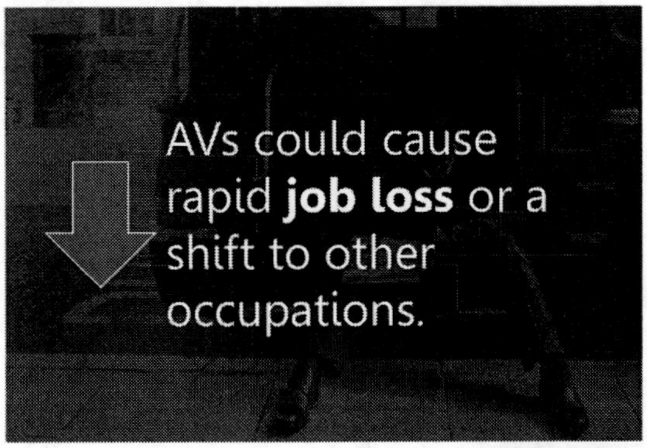

HORIZON

VIBRANT

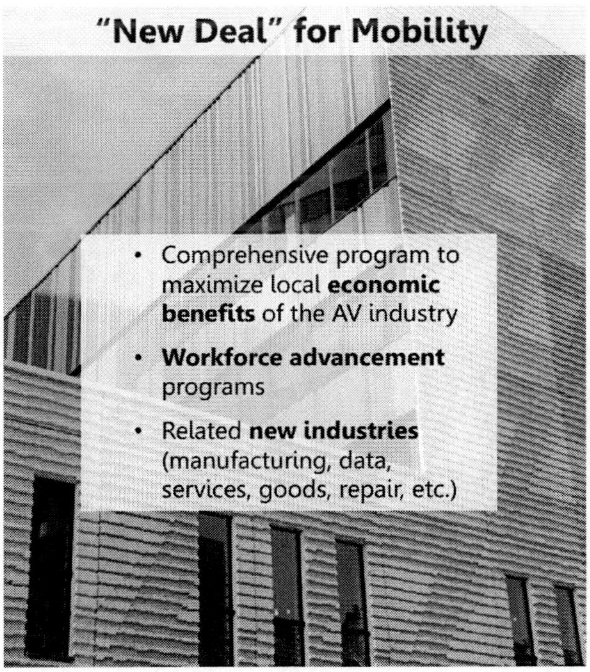

Priority Strategies

Strengthen the capacity of **training programs** to expand opportunities for workers in the AV industry

Target job clusters on **industrially-zoned land** for production, distribution, and repair

Pilot **innovative AV applications** that could spur new job opportunities

HORIZON

Affordable	Connected	Diverse	Healthy	Vibrant
Housing Opportunity Sites	**Fair Pricing & Autonomous Transit**	**Equitable Outcomes**	**Vision Zero 2.0**	**New Deal for Mobility**
Increase affordable housing supply	Expand access to high quality transportation	Ensure universal access to AV services	Save lives and improve air quality	Expand prosperity and access to jobs

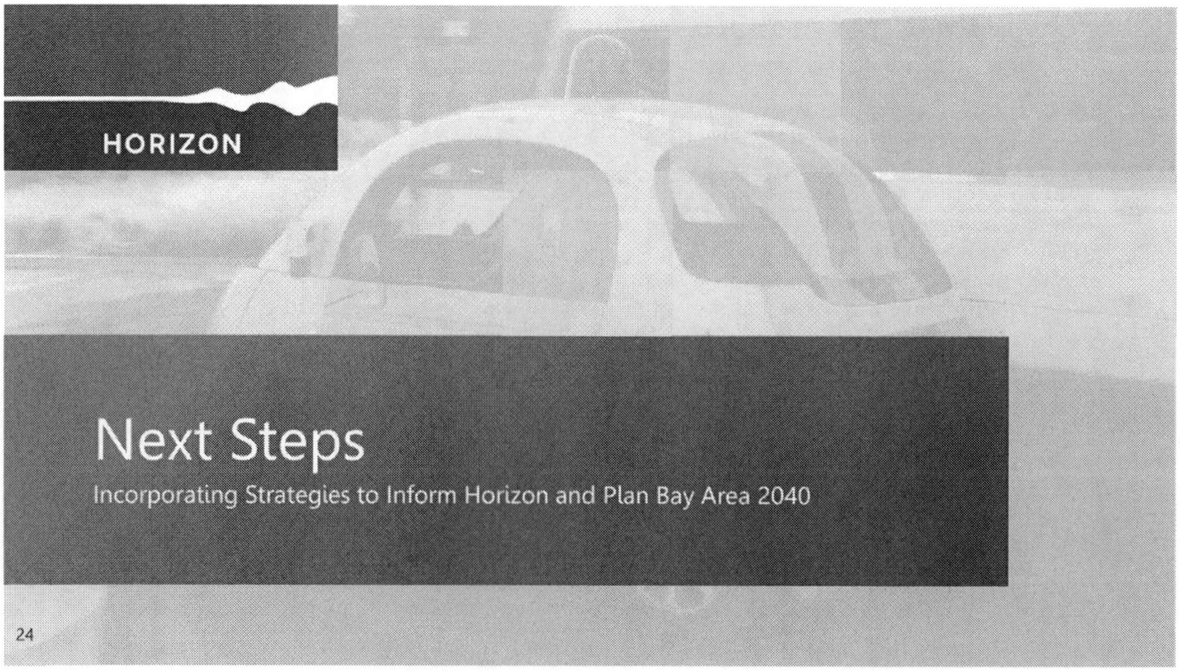

Next Steps

Incorporating Strategies to Inform Horizon and Plan Bay Area 2040

Planning for an Uncertain Future

HORIZON

HORIZON
Thank you!
27

European Platform on Sustainable Urban Mobility Plans

CoEXist

RUPPRECHT CONSULT Forschung & Beratung GmbH

2020 CIVITAS SUMPS-UP

Automation in the European Guidelines for Mobility Planning (SUMP)

Siegfried Rupprecht, Cologne/ Germany

AUTOMATED VEHICLES SYMPOSIUM 2018

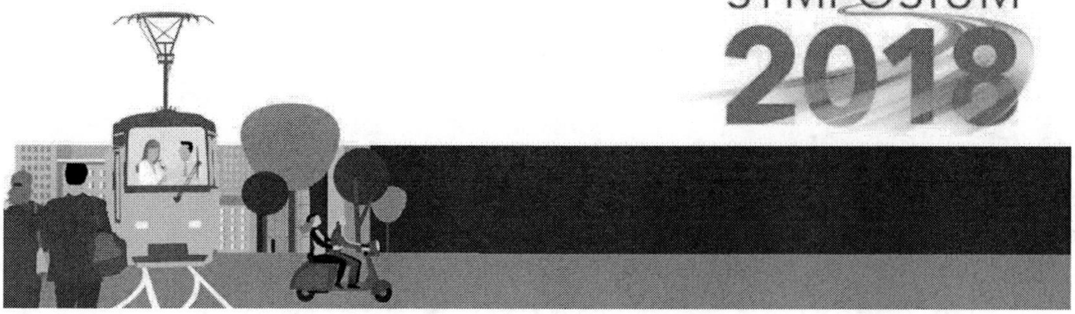

European Platform on Sustainable Urban Mobility Plans

AUTOMATED VEHICLES SYMPOSIUM 2018

Sustainable Urban Mobility Plans

Concept of "Sustainable Urban Mobility Plan" (SUMP)

- part of Urban Mobility Package, Dec 2013
- concept definition (tool to implement policies)
- "recommendation" (but precondition for some funding programmes)

SUMP Guidelines

- based on a systematic stakeholder consultation process, published in Dec 2013
- strongly promoted, good support available
- widely used in Europe (and internationally)

A European success story!

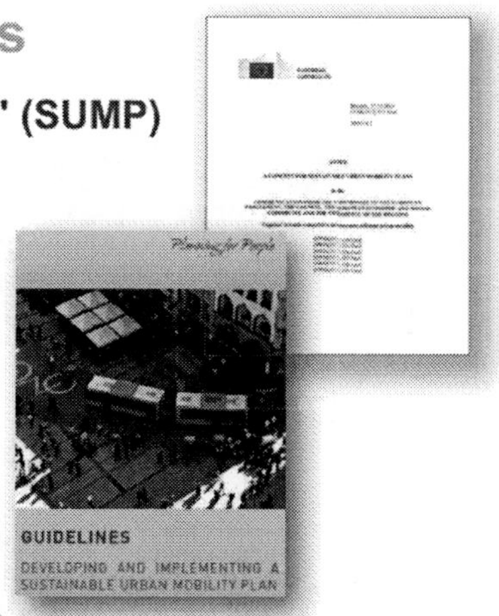

Eltis

www.eltis.org/mobility-plans

European Platform
on Sustainable Urban
Mobility Plans

AUTOMATED VEHICLES SYMPOSIUM 2018

Why do we need an update of the SUMP Guidelines?

- Wealth of **practical implementation experience** for inclusion in new guidance.
- **Specialised guiding material** from various projects.
- **New mobility developments** (emerging technologies, new policy challenges and opportunities) require a rethinking and extension of current guidelines, including:
  - Shared mobility concepts/ Mobility as a Service
  - Urban air mobility/ drones
  - **Road automation**

www.eltis.org/mobility-plans

European Platform
on Sustainable Urban
Mobility Plans

The SUMP Cycle

AUTOMATED VEHICLES SYMPOSIUM 2018

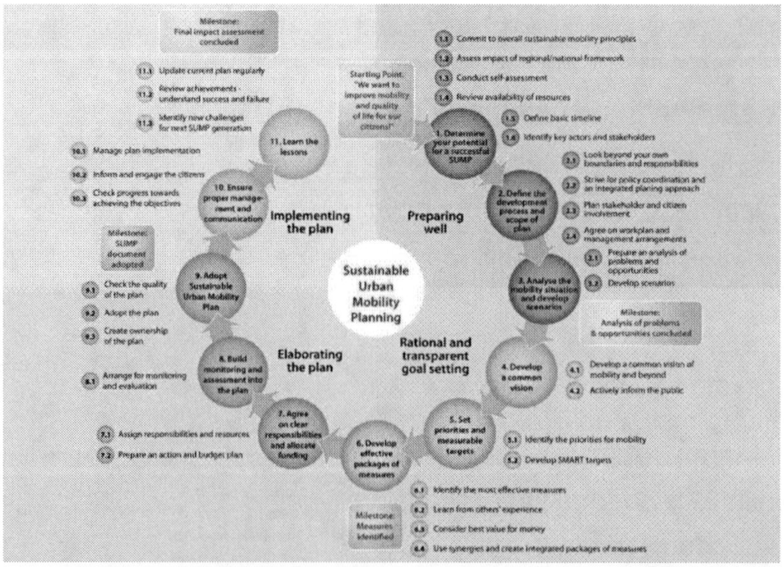

www.eltis.org/mobility-plans

European Platform on Sustainable Urban Mobility Plans

AUTOMATED VEHICLES SYMPOSIUM 2018

The SUMP Principles

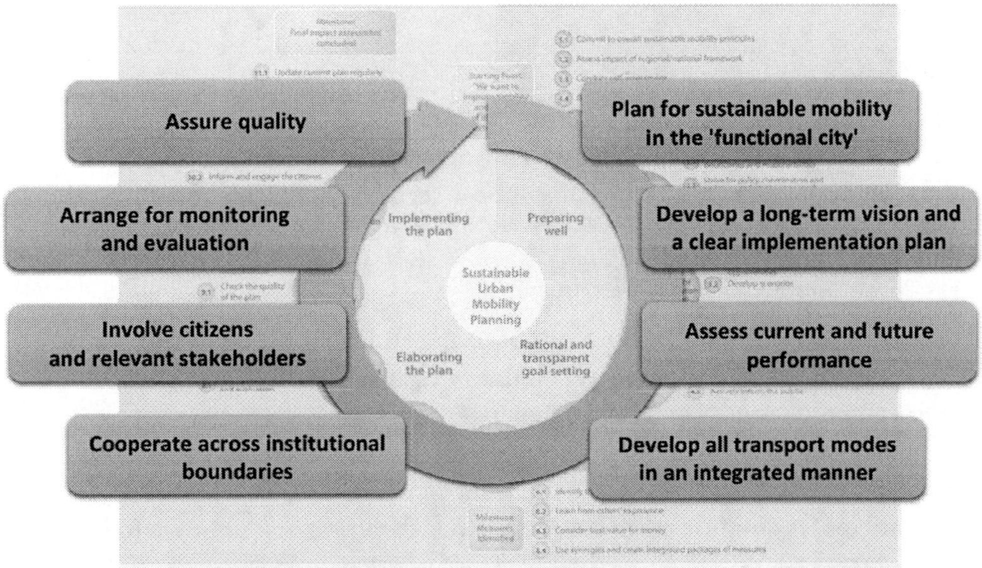

www.eltis.org/mobility-plans

European Platform on Sustainable Urban Mobility Plans

AUTOMATED VEHICLES SYMPOSIUM 2018

Principles of SUMP – and automation

- Aim for **sustainable (automated) mobility**!
  - How will automation contribute to solving major **urban challenges**?
    - Improved **quality of life** and contribution to **economic**, **environmental**, **social** development?
    - Better management of urban **space**?
    - Improved **safety** (also for vulnerable road users); (cyber-)**security** and privacy concerns?
    - Better performance of <u>**overall**</u> **transport system**?
  - Which opportunities/ threats provides automation, what are our city's (region's) strengths/ weaknesses in dealing with it? ("**automation SWOT**")
- Plan for the 'functional city' **– automation in the metro region**!
  - Plan across municipal **boundaries**?
  - Balance **impacts** throughout the **(metropolitan) region**?

www.eltis.org/mobility-plans

European Platform on Sustainable Urban Mobility Plans

AUTOMATED VEHICLES SYMPOSIUM 2018

Principles of SUMP – and automation

- Develop a **long-term vision** of automation!
  - **Scenario techniques** to deal better with uncertainty?
  - Understand external factors in **'big disruption'** (acceptance of business models, legal framework, vehicle deployment)? What are **interactions/ dynamics**?
  - What can we do **now**, in five years, later?
- **Assess expected performance** of an automated transport system!
  - Are adequate **tools** (e.g. modelling) and **data** available?
  - What are **key challenges** that automation needs to solve?
  - Which **skills** are we missing? Which **institutional** structures need updating?
  - Which **revenue streams** are we loosing, what do we need to compensate?
  - Which **policies/ regulations** make automation less desirable (**"policy audit"**)

www.eltis.org/mobility-plans

European Platform on Sustainable Urban Mobility Plans

AUTOMATED VEHICLES SYMPOSIUM 2018

Principles of SUMP – and automation

- **Cooperate** widely: across **institutional** boundaries, involve **citizens** and **stakeholders**!
  - Who are relevant old and new **actors**, what is the new **playbook** (e.g. energy suppliers, OEMs, IT startups, real estate)? Who benefits from what?
  - How can we contribute to a realistic **perception and image** of automated vehicles (public/ decision makers)?
  - What do we need **state/ national government** to regulate?
- Arrange for **monitoring** and **evaluation**!
  - What are automation **SMART targets**? How can we measure?
  - How would a more **dynamic system** of monitoring and decision making increase responsiveness to rapid change?
  - Who **collects data**, who owns/ creates **revenues** from it?

www.eltis.org/mobility-plans

European Platform
on Sustainable Urban
Mobility Plans

AUTOMATED VEHICLES SYMPOSIUM 2018

Principles of SUMP – and automation

- Develop an action plan for **all (automated) transport modes**!
  - A new **collective mobility** paradigm: shared use/ ownership, last/ first mile, mainstream transit? Future of public transit **operators**? New actors, new synergies?
  - What can we do about **freight** (e.g. pods on sidewalks, convert parking houses to freight hubs, curb management?)
  - **Traffic management** policy (automated corridors, priorities, signage)
  - **Infrastructure** development (C-ITS equipment, intersections, parking spaces, drop-off/ pick-up zones, electric charging, hubs and interchanges, maintenance)
  - Interactions with **walking, cycling**, placemaking? Priorities?
  - Effective combinations of **policy tools**? (taxes, parking fees, zoning laws, persuasion)?
  - **Service delivery** (private – public)? Which **business models**?

www.eltis.org/mobility-plans

European Platform
on Sustainable Urban
Mobility Plans

AUTOMATED VEHICLES SYMPOSIUM 2018

Getting ready for automation (example)

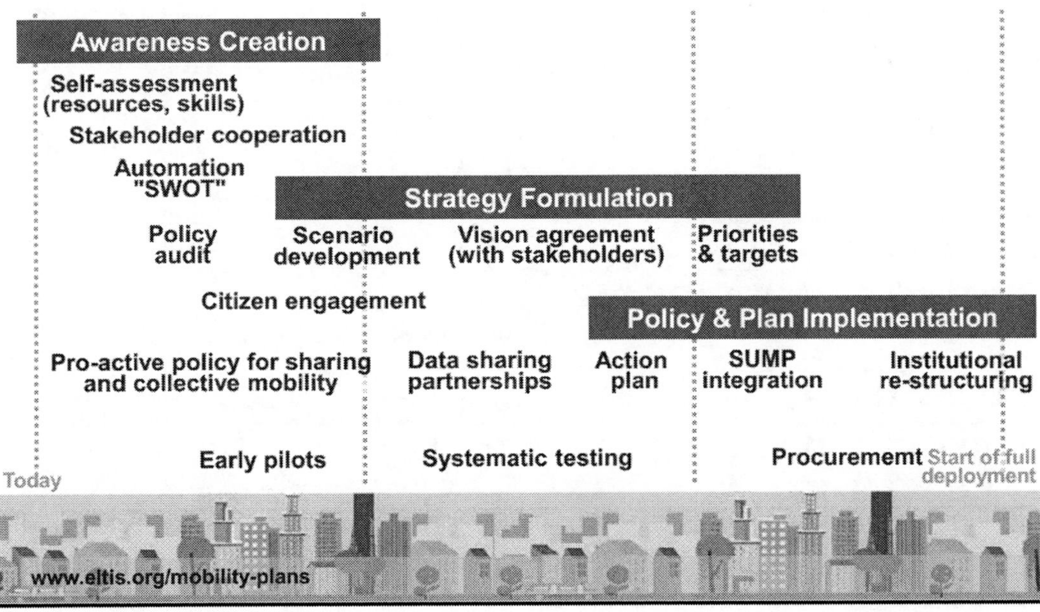

www.eltis.org/mobility-plans

European Platform on Sustainable Urban Mobility Plans

AUTOMATED VEHICLES SYMPOSIUM 2018

What will the new Guidelines include?

- Improved quality & relevance of **core guidance**
  - more **good practice** examples
  - **action plan** preparation
  - references to expansions & new contexts (**knowledge map**)
- Expansions to **new areas** of urban mobility
  - Guidelines: freight, suburban areas, shared mobility & MaaS, financing & procurement
  - Practitioner briefs: alternative fuel strategies, **automation**, air mobility
- **Contextualisation** for new planning situations
  - Guidelines: small & medium cities

www.eltis.org/mobility-plans

European Platform on Sustainable Urban Mobility Plans

AUTOMATED VEHICLES SYMPOSIUM 2018

Tentative timeline

SUMP 2.0 Tasks	2018											2019						
	Feb	Mar	Apr	May	Jun	Jul	Aug	Sep	Oct	Nov	Dec	Jan	Feb	Mar	Apr	May	Jun	Jul
Update/input during **SUMP Platform Coordination Group Meeting**; endorsement of focal topics for guidelines by EC SUMP projects																		
Guidelines work by EC SUMP projects and additional coordination as required																		
Related survey published on SUMP conference registration site (tbc)																		
Whole strand (quarter) of SUMP2018 conference dedicated to topics relevant to the revision of the SUMP guidelines (tbc)																		
Public launch of process at SUMP2018 conference and input from relevant sessions and coordination with EC SUMP projects																		
Consolidation of input from SUMP2018 and coordination with EC SUMP projects																		
Guidelines work by EC SUMP projects and additional coordination as required																		
Update/input during SUMP Platform Coordination Group Meeting																		
Guidelines work by EC SUMP projects and additional coordination as required																		
Endorsement in principle at SUMP Coordination Group Meeting Consolidation by Editorial Board (tbc)																		
Approval of draft by MOVE/INEA																		
Presentation of advanced draft at SUMP2019																		
Consolidation of input from SUMP2019 and coordination with EC SUMP projects																		
Final approval (via email) by MOVE/INEA/SUMP Platform Coordination Group; Publication on Eltis; Promotion and dissemination (by EC and others), (SUMP Platform Group Meeting)																		

public launch at SUMP Conference, 14 May 2018

interaction with SUMP stakeholders

drafting and review of new guidelines

endorsement and presentation at SUMP Conference May 2019

www.eltis.org/mobility-plans

European Platform on Sustainable Urban Mobility Plans

AUTOMATED VEHICLES SYMPOSIUM 2018

For discussion: 'automation planning fundamentals'?

- **Policy and planning**
  - integration into **planning frameworks** (guided by policy goals)
  - active policy & incentives for **shared and collective services**
  - regulation, taxation for **non-desirable automation** (e.g. parking)
  - **financing** of service/ infrastructure development
  - wider **context of digitalisation** ('disruption') and new **mindsets**
- **Infrastructure**
  - **usability of roads** for all (non-instrumented) road users
  - **design** standards for roads and IT equipment
    - priority for good signage and maintenance?
    - segregation of automated vehicles only as exception?

www.eltis.org/mobility-plans

European Platform on Sustainable Urban Mobility Plans

AUTOMATED VEHICLES SYMPOSIUM 2018

For discussion: 'automation planning fundamentals'?

- Management of **space**
  - **redevelopment** of space, placemaking
  - **curb** design/ **parking** management
  - **land use** policies?
- **Planning** methods & process
  - differentiate between **time horizons**
  - availability of appropriate **data** and **tools**
  - ability to deal with **uncertainty**: scenarios, monitoring
- First **references**
  - APA Policy Principles
  - Policy papers from POLIS, NACTO, UITP

www.eltis.org/mobility-plans

European Platform on Sustainable Urban Mobility Plans

AUTOMATED VEHICLES' SYMPOSIUM 2018

Siegfried Rupprecht

s.rupprecht@rupprecht-consult.eu

RUPPRECHT CONSULT
Forschung & Beratung GmbH

CoEXist

2020 CIVITAS
SUMPS-UP

CoEXist and CIVITAS SUMPs-Up projects have received funding from the European Union's Horizon 2020 research and innovation programme.

www.eltis.org/mobility-plans

Innovations in Travel Modeling: summary of a conference

Scott Smith, Volpe Center, US DOT

Automated Vehicles Symposium

Breakout session on Urban and Metropolitan Planning

12 July 2018

7th Innovations in Travel Modeling Conference

- Conference occurs in even-numbered years
- ~300 attendees
- Sponsored by committees that are in the TRB Travel Analysis Methods section

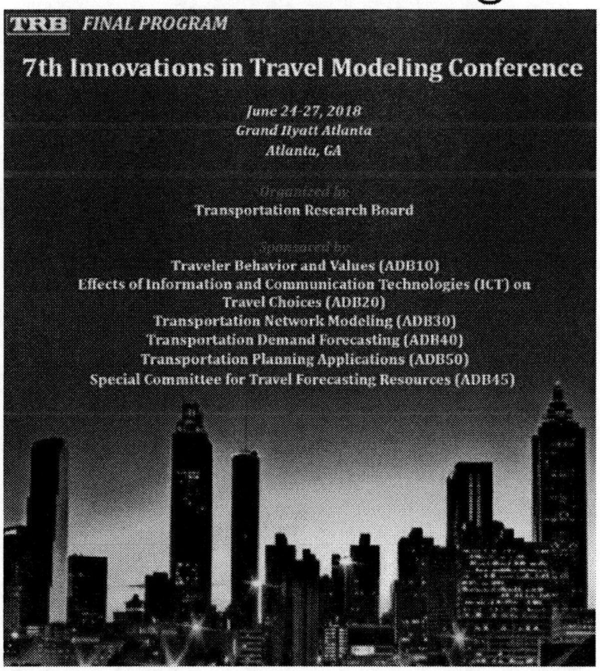

Travel Modeling and Automation

From the conference program:

With the dawn of a new era of transformative disruptions in transportation – characterized by automation and connectivity, autonomy and sharing, and on-demand mobility and delivery – the need for innovations in travel modeling has never been greater. Transport modelers are being challenged every day to explore new frontiers in big data analytics and harness the power of machine and deep learning algorithms to simulate and forecast mobility demand and travel patterns under a wide range of future scenarios of automation and connectivity.

Themes

- It is an unsettling time to be a planner/modeler
  - Some of our fundamental assumptions on modeling are being challenged
    - We now have new modes (TNCs, shared vehicles, automation)
  - Given the uncertainties, a greater interest in scenario modeling and exploratory analysis
- Need to consider the social and policy context
  - Land use, urban form
  - Traveler attitudes
- Consider new data sources and modeling approaches
  - "Big data" – mobile phone, GPS
  - Machine learning, agent-based models

Broad themes related to automation

- In 2016, just starting to talk AVs; in 2018, they were a major part of the conference
- Automation increases the uncertainty
  - New modes
  - Technological progress
  - User response
  - Policy response
- Several presentations on incorporating automation into planning models

Relevant sessions

- TMIP Exploratory Modeling and Analysis Workshop
- Understanding, Modeling and Steering Mobility Futures
- Soft and Powerful: On Incorporating Values and Perceptions into Travel Forecasting
- Emerging Travel Modeling Needs under Autonomous and Connected Technologies
- Papers
  - A Time-space-speed Network Model for Trajectory Optimization of Connected Automated Vehicles
  - Modeling Willingness-to-Share Trips in an Autonomous Vehicle Future: A Stochastic Psychological Latent Construct Approach

Modeling AVs in metro. areas

- Incorporating Connected and Autonomous Vehicles and Ride-Hailing Services in the Traditional Four Step Model (Dallas/Ft. Worth, Texas)
- Using an Activity-Based Model with Dynamic Traffic Simulation to Explore Scenarios for Private and Shared Autonomous Vehicle Use in Jacksonville (Florida)
- Modeling Connected and Automated Vehicles in a Four-Step Travel Demand Model for the Chittenden County Region (Vermont)
- A Framework for Modeling Connected and Autonomous Vehicles in the New Michigan Statewide Model
- Development and Application of a Model to Estimate Driverless Autonomous Vehicle Trips (Minneapolis/St. Paul, Minnesota)
- Modeling the New Mobility: Integrating Autonomous Vehicles, the Sharing Economy and the Impacts of E-Commerce into a Model Framework (Vancouver, British Columbia)
- Shared Autonomous Vehicles as a Replacement for Buses: A Simulation Study in Berlin
- Estimating the Potential Impacts of AVs and TNCs using an Activity-Based Travel Demand Model in MTP/SCS Scenario Development (Sacramento, California)
- Impact Assessment of Autonomous DRT Systems (Berlin)
- Autonomous Vehicle (AV) Scenarios using a VisionEval model in Oregon

Key uncertainties – demand side

- Auto-ownership
  - Privately owned AVs
  - Shared services
- Level of car-sharing and ride-sharing
  - What is the disutility from sharing a journey?
- Change in parking location and behavior
- Empty vehicle trips
  - Fleet repositioning
  - Intra-family repositioning
  - Avoidance of expensive parking
- Value of time
- Induced trips

Key uncertainties – supply side

- Different capacity consumption by CAVs
- CAV infrastructure (signals, lanes, etc.)
- Incidents
- Shared fleet sizes and operations

For more information

Conference presentations/papers to be posted on Travel Forecasting Resource, at

http://www.tfresource.org

Look for 7th ITM Conference, Atlanta

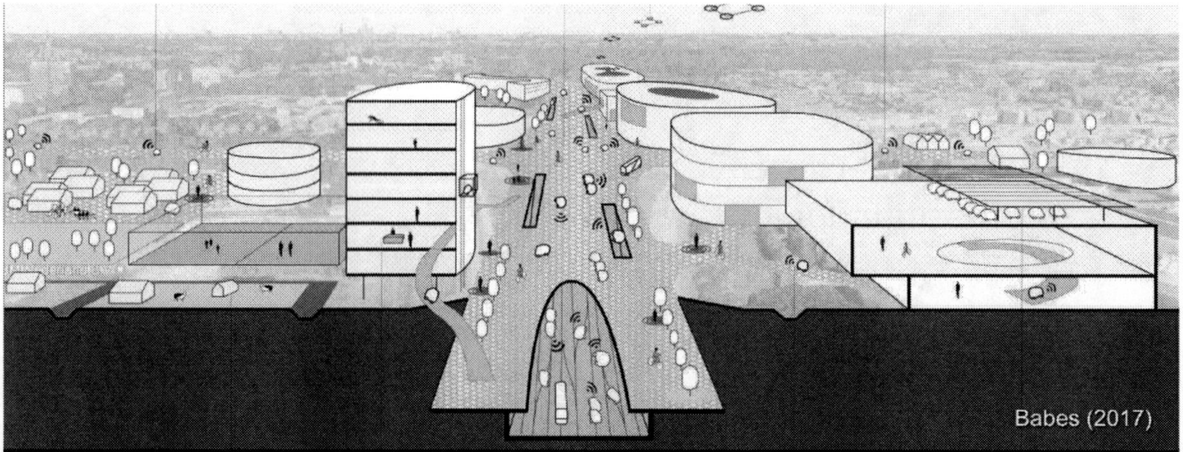

Scenarios, methods and tools to explore the urban and regional mobility impacts of automated driving

Bart van Arem, Delft University of Technology, The Netherlands

2018 Automated Vehicle Symposium, San Francisco, 2-5 July 2018

Spatial and Transport Impacts of Automated Driving

2016-2020, 2,5 M€ , www.stad.tudelft.nl

Automated driving

Driver assistance/ Partial automation

Driver needs to be able to intervene at all times

Automated parking, autocruise

Conditional/ High automation

Vehicle in control in special conditions

Taxibots, platooning, automated highways

Comfort, efficiency, safety, costs →

→ Mode choice, location choice, urban and transport planning

TU Delft

Many questions …

When fully automated vehicles will hit the market?

Will we travel safer?

Are we going to own or share cars?

Will we need more or less road infrastructures?

Will we still need buses?

Will there be more or less congestion?

Will we drive longer or shorter distances?

How much on-street and off-street parking spaces will still be needed?

How will cities evolve?

Will we consume more or less energy to travel?

TU Delft

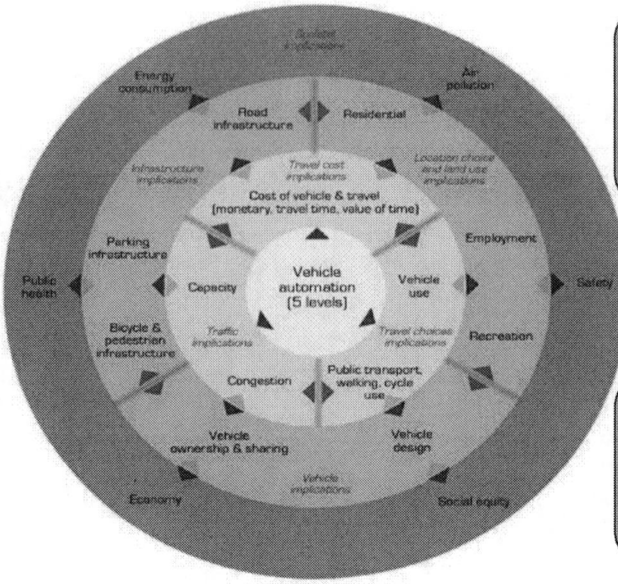

Much progress short term and small scale impacts on driver behaviour and traffic flow.

Research on longer term, indirect, wider scale impacts on mobility, logistics, residential patterns and spatial-economic structure in its infancy.

TU Delft

Milakis et al (2017), Policy and society related implications of automated driving, Journal of ITS.

Spatial and Transport Impacts of Automated Driving

- Reclaim infrastructure and parking space
- Improve environmental quality
- Improve social quality
- Improve economic conditions
- AVs: shared and electric

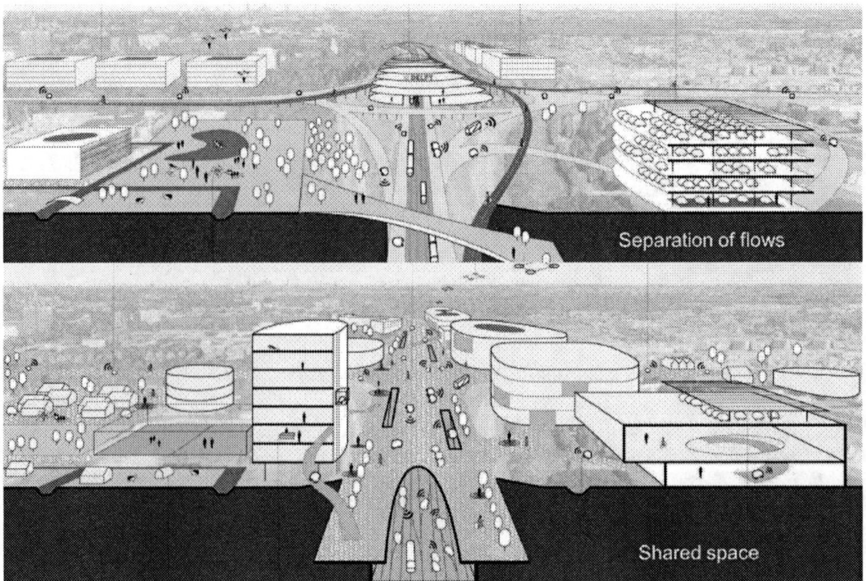

TU Delft

Babes (2017), A Method to Assess the Impact of Automated Vehicles on Urban Liveability in the Rotterdam The Hague Metropolitan Region, European Post-Master in Urbanism (EMU), P5 report, repository.tudelft.nl

Spatial and Transport Impacts of Automated Driving

AV ...in standby | **High** technological development | **AV ...in bloom**

- Fully automated & cooperative vehicles (V2V) in 2030.
- Legislation inflexibility for AV. Transport policies restraining use of AV. High regulation of AV trials.
- Modest economic growth.
- "Wait and see..." customers' attitude, mid-low demand for AV.
- No major environmental problems, but still low penetration of electric vehicles.

- Fully automated & cooperative vehicles (V2V & V2I) in 2025.
- Laws allowing AV traffic. Limited regulation of AV trials. Public investments on AV research and on smart infrastructure.
- High economic growth.
- Positive customers' attitudes, strong demand for AVs.
- Limited environmental problems. Clean technologies prevail.

Restrictive AV policies ⟷ **Supportive** AV policies

- Fully automated vehicles in 2045.
- Limited legislation for AV integration. No AV trials allowed.
- Recessive economy, high unemployment.
- Negative customers' attitude, almost no demand for AV.
- Important environmental problems. Very slow transition to low-carbon economy.

- Fully automated & cooperative (V2I) vehicles in 2040.
- Progressive legislation for AV integration. No regulation of AV trials. Promotional campaigns.
- Slow economic growth.
- "Not really interested..." customers' attitude, low demand for AV.
- Increased environmental problems. Transport sector still among major polluters.

AV ...in doubt | **Low** technological development | **AV ...in demand**

Milakis et al (2017), Development and transport implications of automated vehicles in the Netherlands: scenarios for 2030 and 2050, European Journal for Transport and Infrastructure Research 17(1), 2017, pp. 63-85.

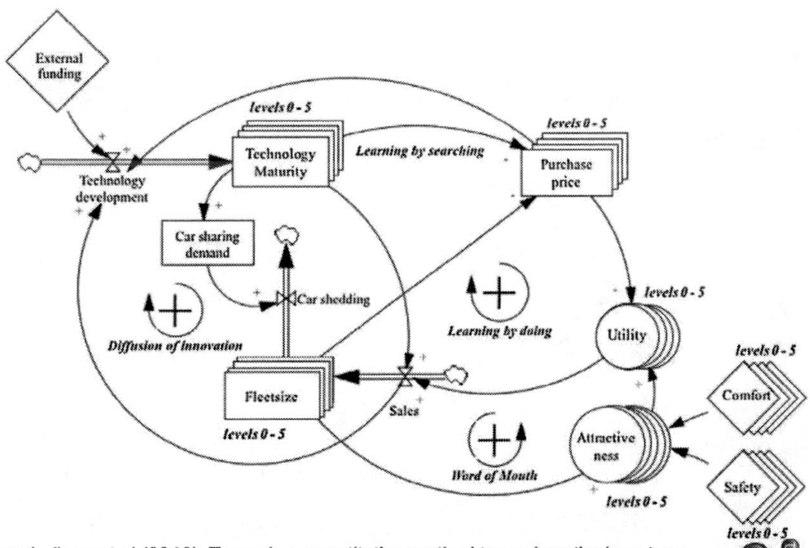

Nieuwenhuijsen et al (2018), Towards a quantitative method to analyze the long-term innovation diffusion of automated vehicles technology using system dynamics, Transportation Research Part C 86 (2018) 300–327

Reference scenario

AV in bloom conservative

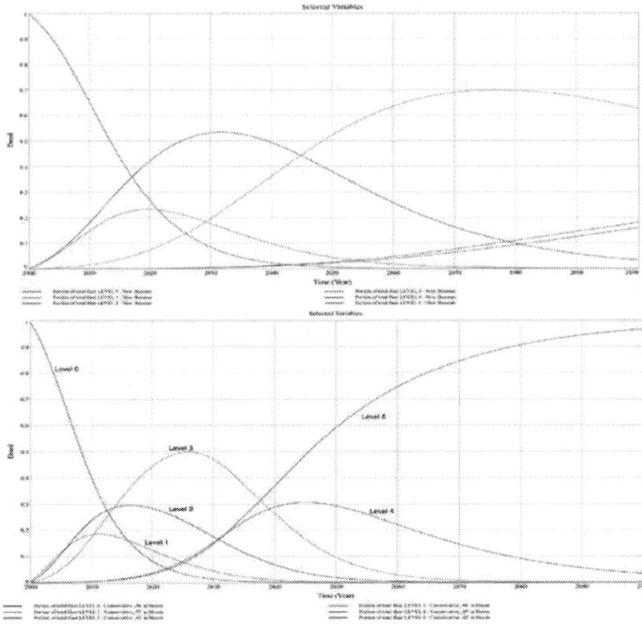

TUDelft

Spatial and Transport Impacts of Automated Driving

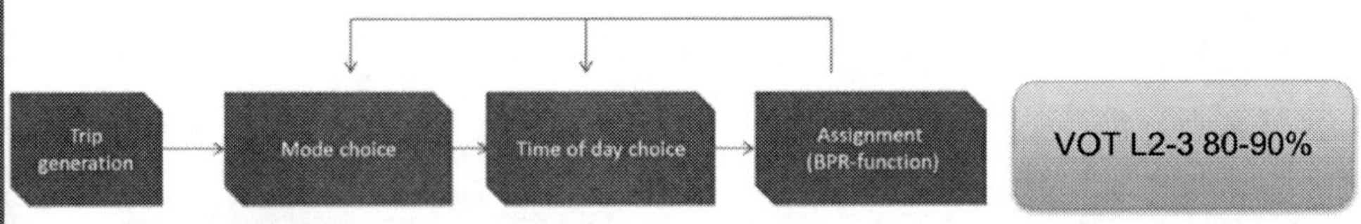

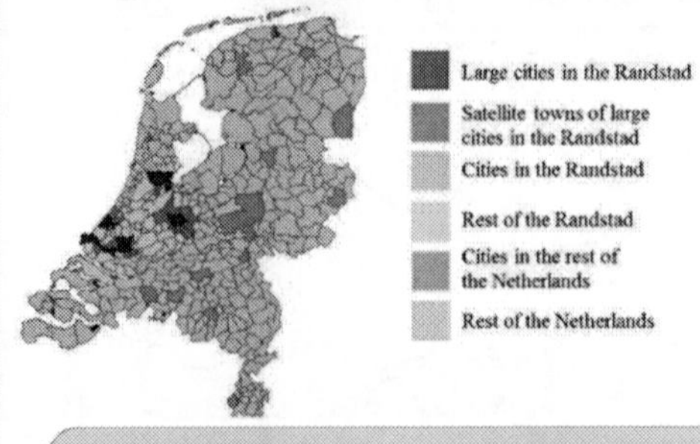

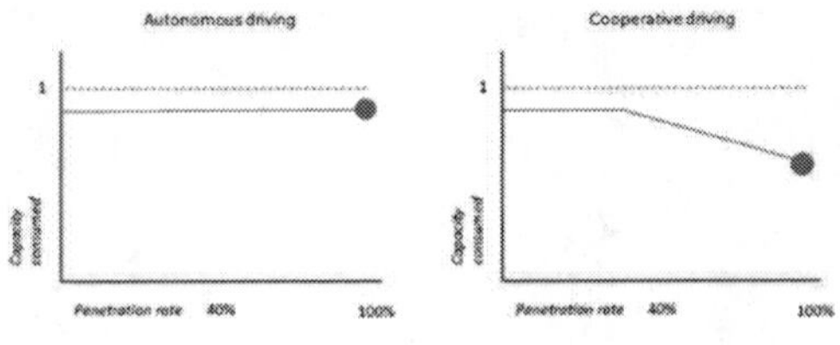

Hybrid system dynamics/transport model with half year increments

TUDelft

Puylaert et al (2017), Mobility impacts of early forms of automated driving – A system dynamic approach, Transport Policy, in press (2017) 1–9

Spatial and Transport Impacts of Automated Driving

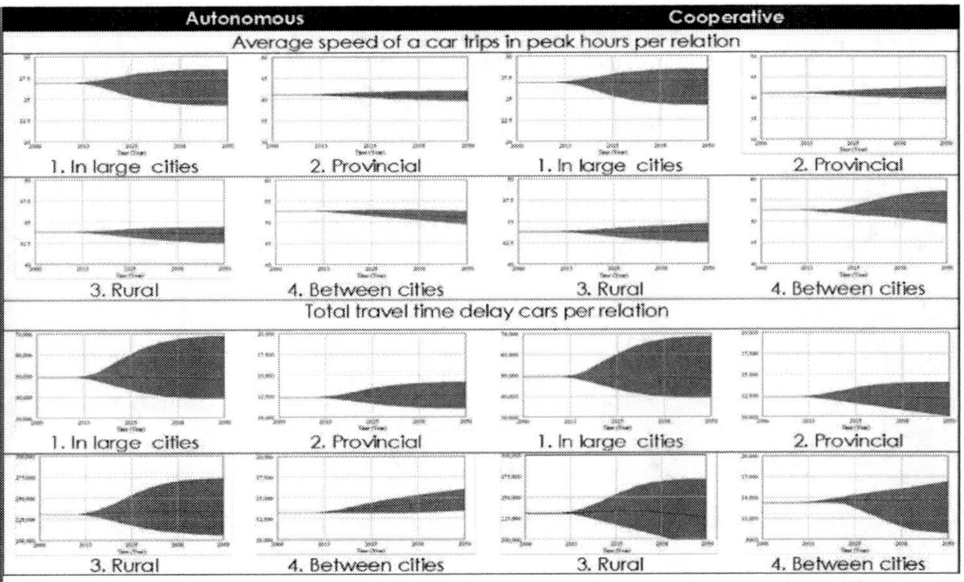

TUDelft

Spatial and Transport Impacts of Automated Driving

Network design and impacts of Automated Driving

High automation on designated roads

AV has 95% VOTT of regular vehicles

Passenger Car Equivalent AV:
95% penetration rate <= 40%
90% penetration >40%

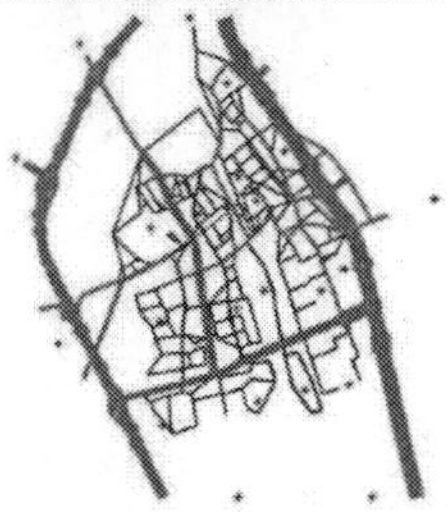

Level 4 enabled network of Delft

Automated Vehicles may travel further to be on L4 roads

Automated Vehicles lead to shorter travel time on L4 roads

TUDelft

Spatial and Transport Impacts of Automated Driving

User equilibrium static assignment

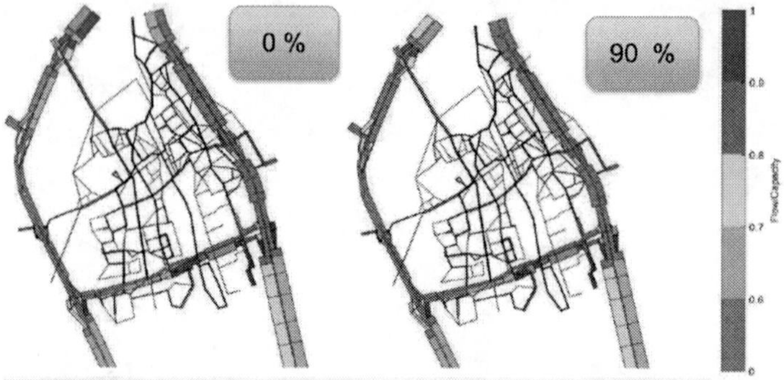

Penetration Rate	Total Travel Cost (€)	Total Travel Time (h)	Total Travel Distance (km)
0	71265	3451	211580
10%	70897	3448	211686
50%	67574	3438	212911
90%	64634	3429	213971
Max improvement	- 9%	- 1%	+1%

Next steps
Dynamic assignment
Multi-user class
Optimal Network Design

TU Delft

Madadi et al (2018) Automated Driving and infrastructure network design (Proc TRB 2018)

Automated Vehicles in National Market and Capacity Analysis (NMCA)

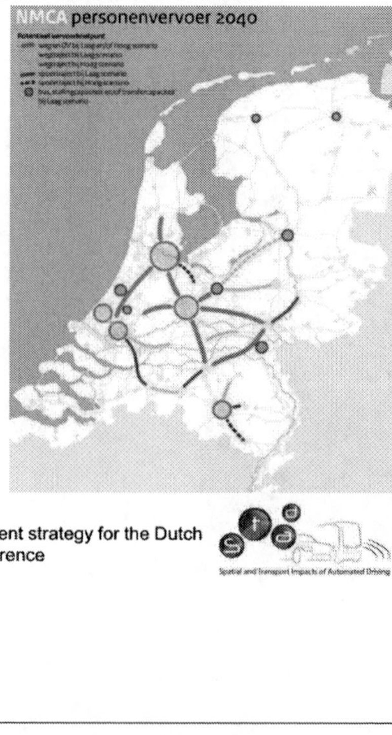

NMCA

Updated every 4 year to identify main transport problems

Used to support major transport infrastructure decisions

Typical horizon 20 years

Uses Dutch National Transport Model (LMS)

What if AVs could deliver substantial capacity improvement in 20 years?

TU Delft

Smit et al (2017), Will Self-Driving cars impact the long term investment strategy for the Dutch national trunk road system? Proceedings European Transport Conference

Results* motorways

	AV Penetration rate cars	AV Penetration rate trucks	PCU car HWN	PCU truck HWN*	ΔVOT car	ΔVOT truck
Truck platooning	0%	40%	1	0,75	0%	-20%
Autonomous	30%	40%	1,15	0,75	-5%	-20%
Cooperative	30%	40%	0,7	0,75	-5%	-20%
Cooperative VOT	30%	40%	0,7	0,75	-20%	-20%

Capacity –4,5%

Capacity + 9%

KM driven	Morning peak	Evening peak	Other	Total
Truck platooning	100.9	100.8	100.9	**100.8** ↑
Autonomous	99.1	100.2	99.0	**99.8** ↓
Cooperative	105.3	103.2	105.4	**103.9** ↑
Cooperative VOT	106.4	105.0	106.7	**105.5** ↑

Vehicle loss hours	Morning peak	Evening peak	Other	Total
Truck platooning	97.6	95.9	99.6	**97.8** ↓
Autonomous	103.6	107.9	104.7	**105.3** ↑
Cooperative	91.0	80.0	91.9	**87.9** ↓
Cooperative VOT	94.0	83.9	95.1	**91.3** ↓

TU Delft

* Results are indications. Functionality to assess impacts of AVs is still experimental.

Spatial and Transport Impacts of Automated Driving

TAKE AWAYS

AV Scenarios, methods, tools for transport planners

Scenarios developing world-wide – choose one or make your own

System dynamics and basic transport models provide first order impacts

Ranges available for changes roadway capacity and Value of Time

Consider AV in relation to C-ITS EVs, sharing, parking (etc)

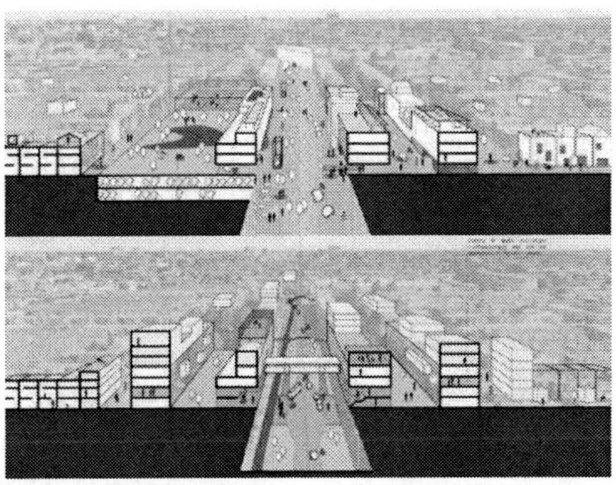

Open research issues

Improve models using real-world experience
Extend to land use and urban design

TU Delft

Spatial and Transport Impacts of Automated Driving

BCG

THE BOSTON CONSULTING GROUP

Using Advanced Analytics to inform Urban Mobility Planning

Augustin K. Wegscheider

9 JULY 2018

WORLD ECONOMIC FORUM

CITY of BOSTON

Convergence of three trends will drive massive change

Sharing

+

Autonomy

+

Electrification

1

SAE Vehicle economics are very compelling

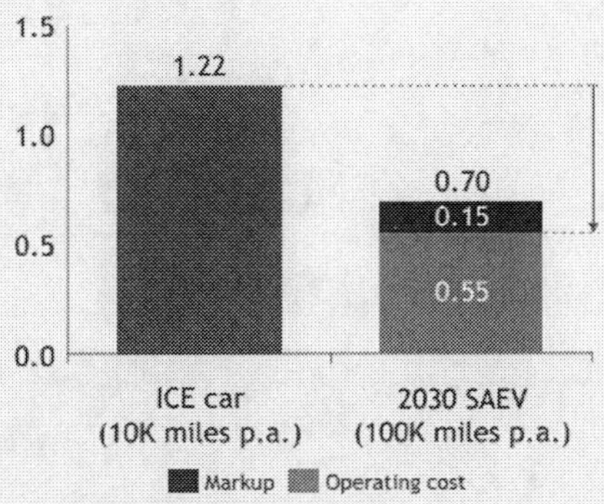

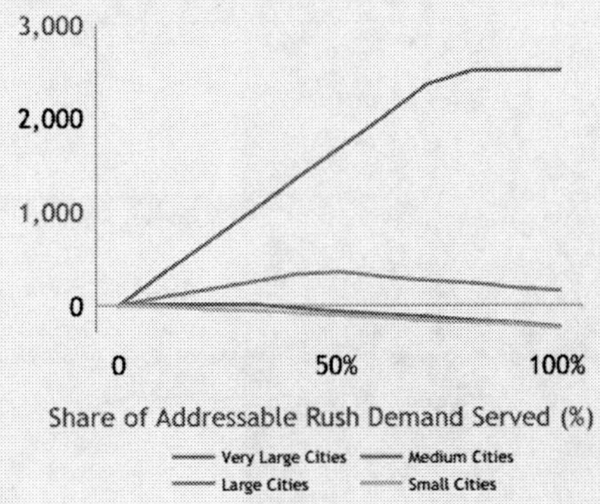

Source: BCG analysis

2

About ¼ of miles traveled could be SAEV by 2030

Geography	Total Pop. (M)	Total miles driven (%)	Profit max share of total miles served	Revenue max share of total miles served
Very large cities	95	26%	47%	51%
Large cities	80	25%	39%	49%
Medium cities	65	18%	0%	0%
Small cities	33	10%	0%	0%
Rural areas	50	21%	0%	0%
US Overall	~320	100%	22%	26%

Revenue maximizing while still profitable

Source: Texas A&M Traffic Institute, US Census Data, BCG Analysis

3

City of Boston: Collaboration with WEF and BCG to get AVs on road and study AV impact
CITY of BOSTON
WORLD ECONOMIC FORUM
BCG
ENGIE
door2door
Swiss Re
Baidu 百度
DHL
Keolis
VOLKSWAGEN
Audi
BRIDGESTONE
aramco
MIT
AIRBUS
GM
MINISTRY OF TRANSPORT
GAC MOTOR
massDOT
Land Transport Authority
DENSO
HYUNDAI
Drive your way
VOLVO
BOSCH
SIEMENS
lyft
UNITED ARAB EMIRATES MINISTRY OF CABINET AFFAIRS
RENAULT NISSAN
RTA
ROADS & TRANSPORT AUTHORITY
QUALCOMM
Sompo Japan Nipponkoa
ups
chargepoint
StateFarm
nuTonomy
INRIX
LAND ROVER
JAGUAR
DRIVE SWEDEN
transdev
ERICSSON
innogy
ZURICH
TOYOTA
OPTIMUS RIDE
4

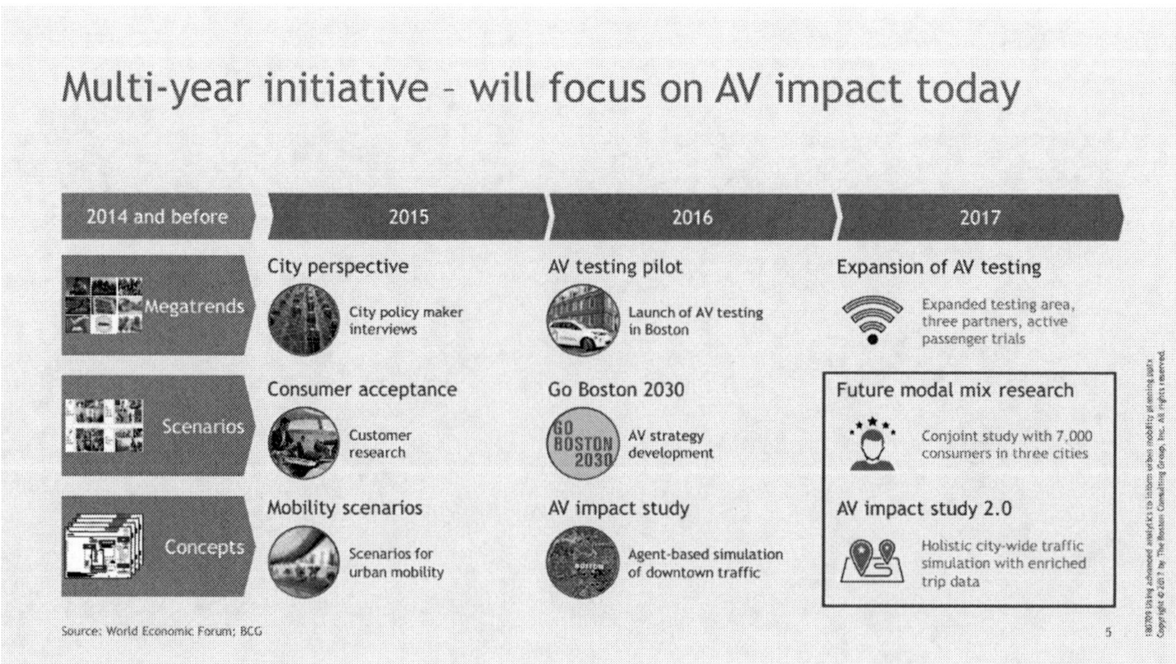
Multi-year initiative – will focus on AV impact today
2014 and before
2015
2016
2017
Megatrends
Scenarios
Concepts
City perspective
City policy maker interviews
Consumer acceptance
Customer research
Mobility scenarios
Scenarios for urban mobility
AV testing pilot
Launch of AV testing in Boston
Go Boston 2030
GO BOSTON 2030
AV strategy development
AV impact study
Agent-based simulation of downtown traffic
Expansion of AV testing
Expanded testing area, three partners, active passenger trials
Future modal mix research
Conjoint study with 7,000 consumers in three cities
AV impact study 2.0
Holistic city-wide traffic simulation with enriched trip data
Source: World Economic Forum; BCG
5

Conducted large-scale conjoint study in three cities

Representative sample

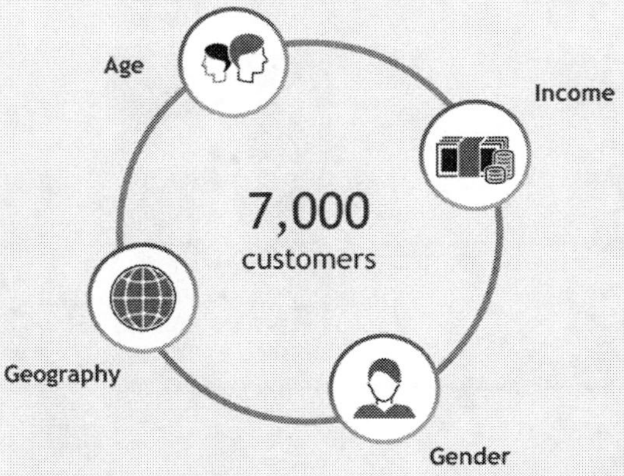

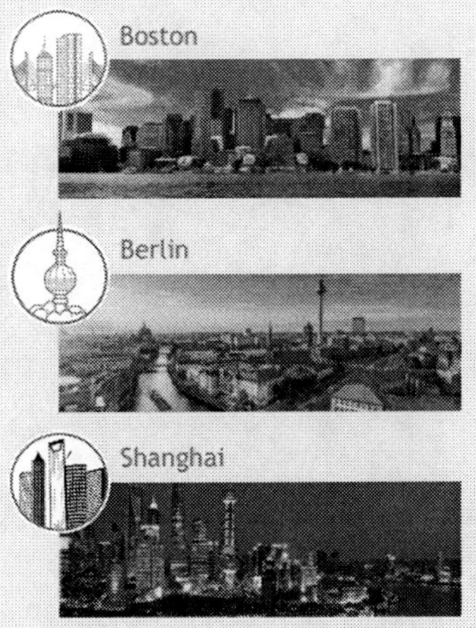

Source: World Economic Forum; BCG Analysis

6

Top 5 findings from our conjoint study across 3 cities

1. Cities globally move to 30-40% mobility on demand
2. Mass transit ridership drops in urban areas
3. AV adoption varies across city—correlated to income levels
4. The shorter the trip, the higher the AV adoption
5. 20-25% will use a personal car no matter what

Note: 7,000 respondents across three cities (Berlin, Boston, Shanghai)
Source: World Economic Forum, BCG Analysis

7

Cities globally move to 30-40% mobility on demand

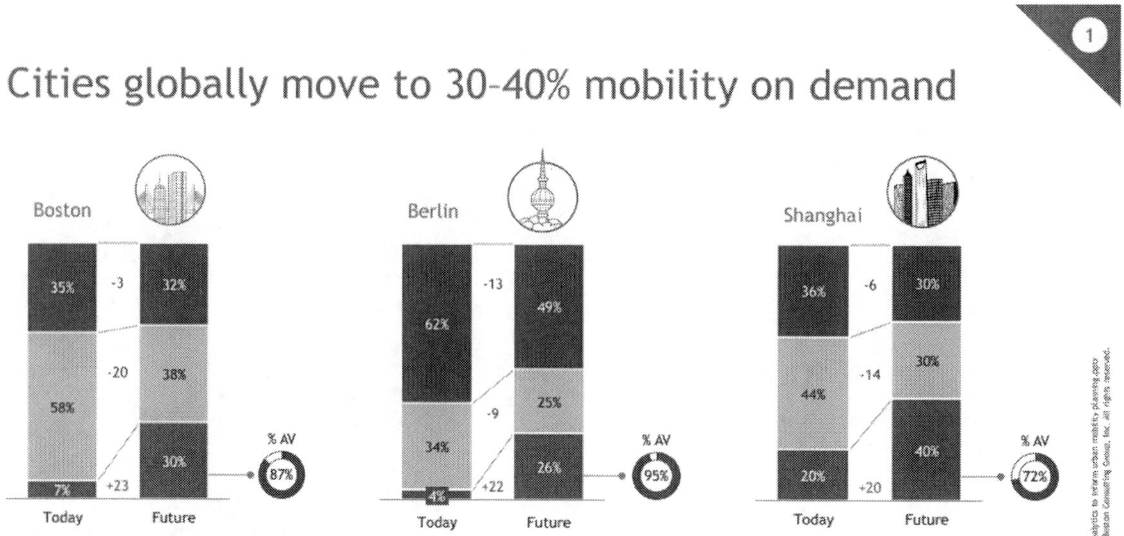

Mass transit = Bus/subway + Commuter rail
Personal car = Personal car + Autonomous personal car
Mobility on demand = Taxi/ride sharing + Autonomous taxi + Autonomous shared taxi + Autonomous minibus

Source: World Economic Forum; BCG Analysis

8

Mass transit ridership drops in urban areas

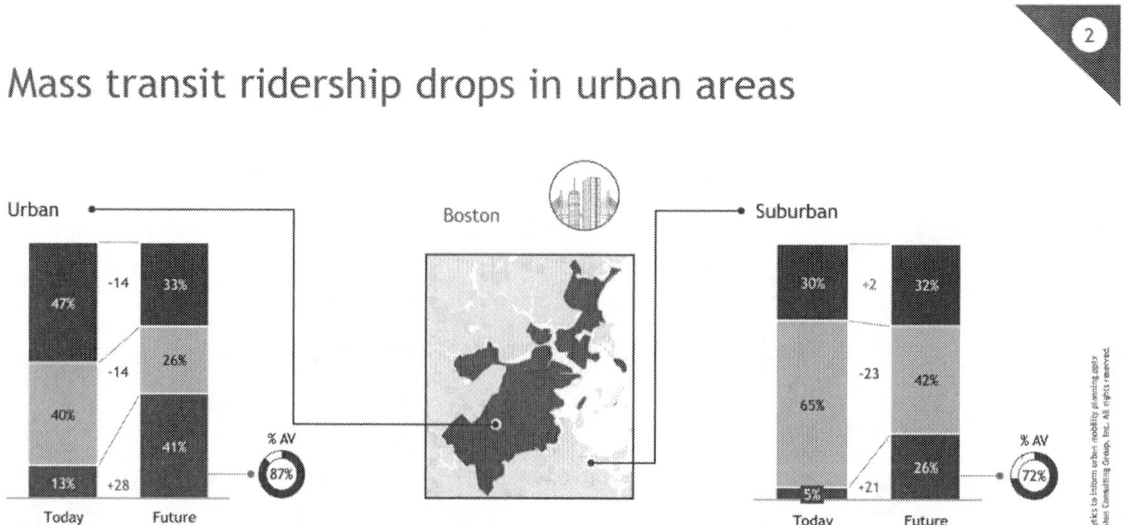

Mass transit = Bus/subway + Commuter rail
Personal car = Personal car + Autonomous personal car
Mobility on demand = Taxi/ride sharing + Autonomous taxi + Autonomous shared taxi + Autonomous minibus

Source: World Economic Forum; BCG Analysis

9

The AV impact study for Boston unfolded over two years

We started with an agent-based simulation in downtown Boston...

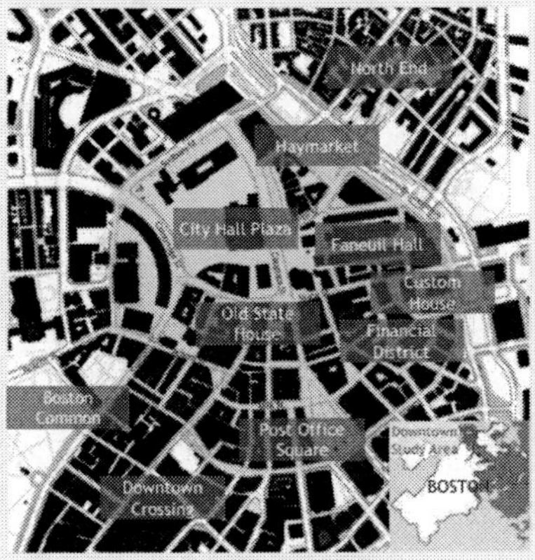

...and expanded to a holistic city-wide simulation with a research-based future scenario

1. Expanded study to cover entire City of Boston

2. Added enriched trip data and commercial vehicles
   CITY of BOSTON | ups | INRIX

3. Determined future modal mix through conjoint study with 2,400 consumers

4. Quantified traffic efficiency gains from AV technology

10

Simulation shows traffic level on each street

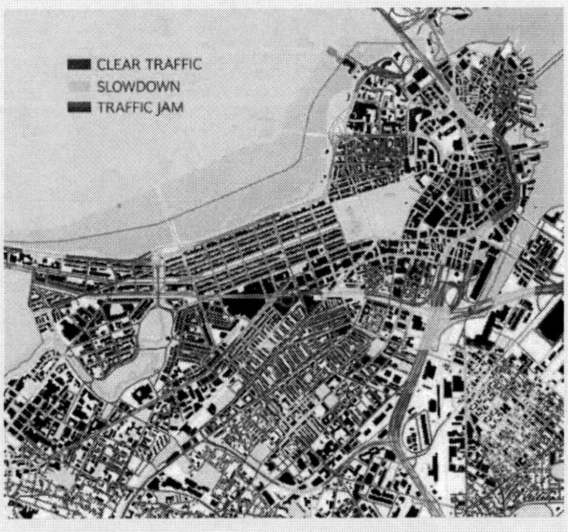

11

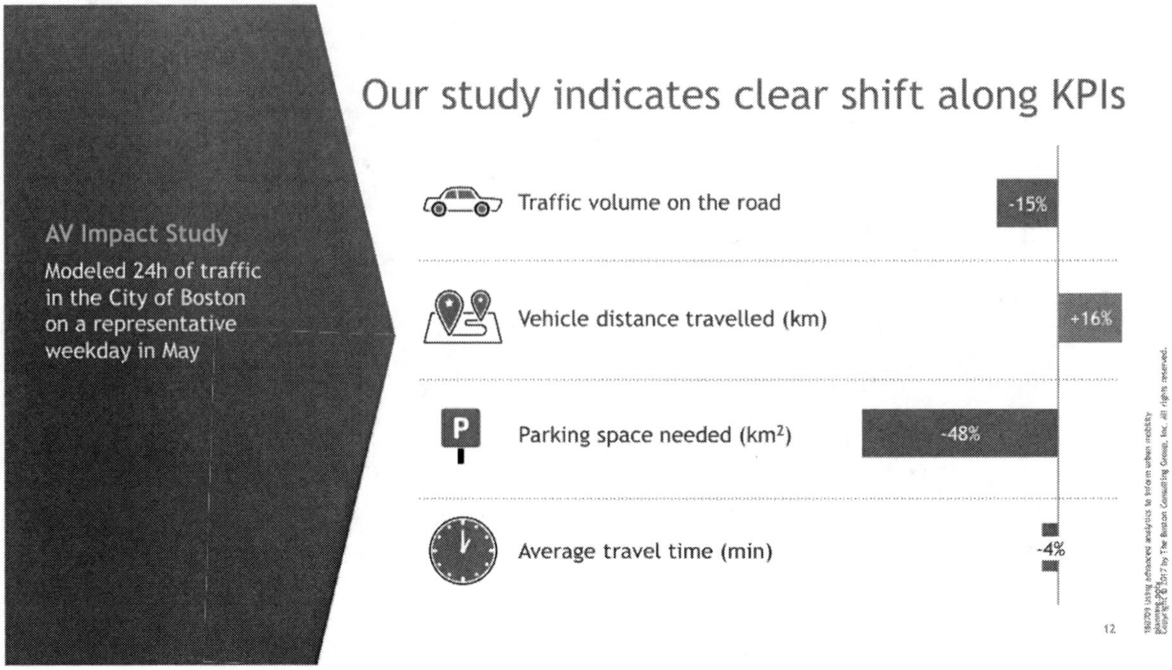
AV Impact Study
Modeled 24h of traffic in the City of Boston on a representative weekday in May
Our study indicates clear shift along KPIs
Traffic volume on the road
-15%
Vehicle distance travelled (km)
+16%
Parking space needed (km²)
-48%
Average travel time (min)
-4%
12
Copyright © 2017 by The Boston Consulting Group, Inc. All rights reserved.

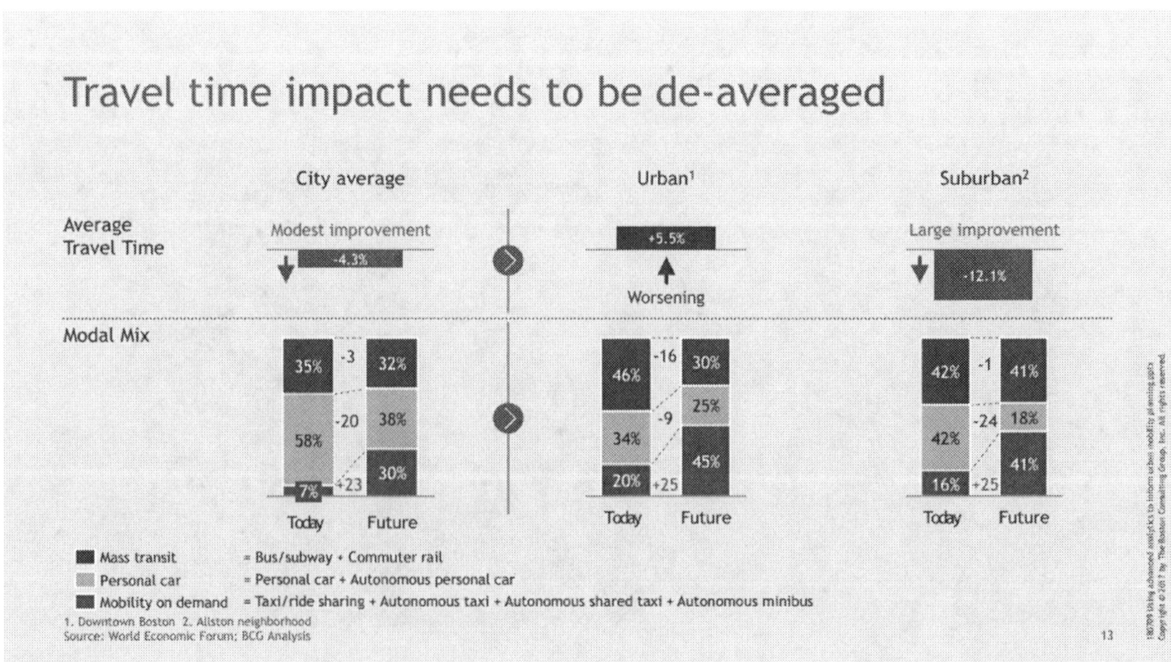
Travel time impact needs to be de-averaged
City average
Urban¹
Suburban²
Average Travel Time
Modest improvement
-4.3%
+5.5%
Worsening
Large improvement
-12.1%
Modal Mix
35%
-3
32%
58%
-20
38%
7%
+23
30%
46%
-16
30%
34%
-9
25%
20%
+25
45%
42%
-1
41%
42%
-24
18%
16%
+25
41%
Today
Future
Mass transit = Bus/subway + Commuter rail
Personal car = Personal car + Autonomous personal car
Mobility on demand = Taxi/ride sharing + Autonomous taxi + Autonomous shared taxi + Autonomous minibus
1. Downtown Boston 2. Allston neighborhood
Source: World Economic Forum; BCG Analysis
13

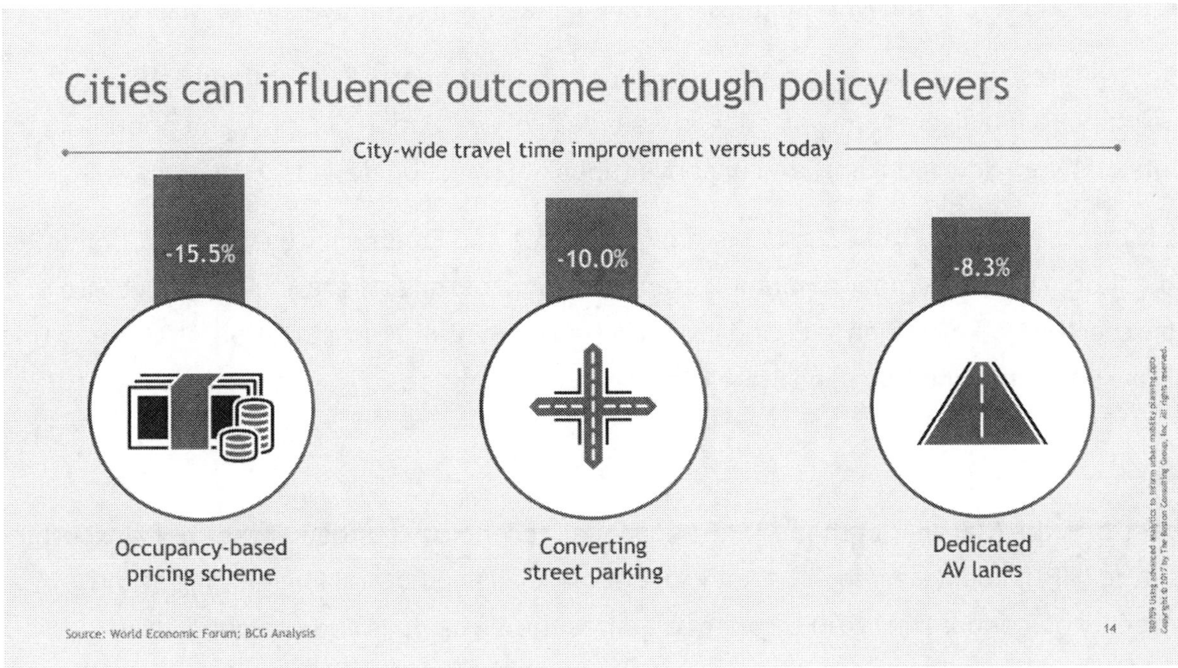

Cities can influence outcome through policy levers
City-wide travel time improvement versus today
-15.5%
-10.0%
-8.3%
Occupancy-based pricing scheme
Converting street parking
Dedicated AV lanes
Source: World Economic Forum; BCG Analysis
14
Copyright © 2017 by The Boston Consulting Group, Inc. All rights reserved.

BCG
THE BOSTON CONSULTING GROUP
bcg.com

Dr. Terry Yang is an Assistant Professor in Transportation Engineering at the University of Utah. He received his Ph.D. and M.S. degrees in Civil Engineering from University of Maryland.

His current research focuses on traffic operations under connected automated vehicles environment, intelligent transportation system design, evacuation planning and operation, traffic safety, network flow modeling, and unconventional intersection design.

He is member of TRB Traffic Signal System Committee and Emergency Evacuation Committee, and editorial board member of multiple transportation journals. His research projects are sponsored by National Science Foundation, US Department of Transportation, Utah Department of Transportation, and Maryland State Highway Administrations.

Dr. Letter is the Manager of the I-STREET advanced technology testbed at the UF Transportation Institute (UFTI) and Research Assistant Professor of Civil and Coastal Engineering at the University of Florida. The I-STREET testbed utilizes the UF campus and surrounding neighborhoods to test connected and autonomous vehicles and devices that support them. Letter coordinates all efforts related to planning and implementation of the advanced technologies testbed. This includes project management and coordination with UF's testbed partners, the City of Gainesville and the Florida Department of Transportation. Letter works with the project's principal investigators and industry partners to coordinate software, hardware, and application efforts to ensure consistency and interoperability of the various components. He is also develops proposals and works on grants related to the testbed, and leads the effort to publicize testbed activities to the wider transportation community.

Dale Thompson is a Lead Research Engineer in the FHWA Office of Operations Research and Development at the Turner-Fairbank Highway Research Center. He serves as the Transportation Enabling Technologies Team Leader where he coordinates automated vehicle research in the Saxton Transportation Operations Lab and Data Testbed. Dale supports a broad portfolio of cooperative automation research including truck and light vehicle platooning, speed harmonization, and signalized intersection approach and departure.

Dale has prior experience in the ITS Joint Program Office supporting connected vehicle policy direction, data acquisition and open data access. He also served in local government with

Maricopa County DOT leading the AZTech ITS Partnership and on active duty in the United States Air Force as a Systems and Data Integration Officer.

Dale has a Bachelor's of Science in Aerospace Engineering from the University Of Notre Dame, and a Master's of Science in Aerospace Sciences from Embry-Riddle Aeronautical University.

Blaine Leonard is employed by the Utah Department of Transportation (UDOT) in Salt Lake City, where he is the Technology & Innovation Engineer. In this role, he is responsible for traffic management technologies, and leads the planning for connected and automated vehicles. He co-chairs the American Association of State Highway and Transportation Officials (AASHTO) Connected and Automated Vehicles Working Group and leads the SPaT Challenge Tactical Working Group within the Vehicle to Infrastructure Deployment Coalition (V2I DC). Prior to joining UDOT, Blaine spent 20 years in the consulting engineering business.

Mr. Leonard served as the President of the American Society of Civil Engineers in 2010. He is a licensed engineer in six western states and is the Vice Chair of the Utah Professional Engineers and Professional Land Surveyors Licensing Board.

Mike Mercer, CSEP is a senior staff analyst in the Federal Civilian Solutions Division of Noblis. He provides systems engineering support to the U.S. DOT in the areas of ITS standards, program assessment, project management, and transportation research.

AVS2018

Integrating Freeway and Arterial Traffic Control under Vehicle-to-Infrastructure Communications Environment

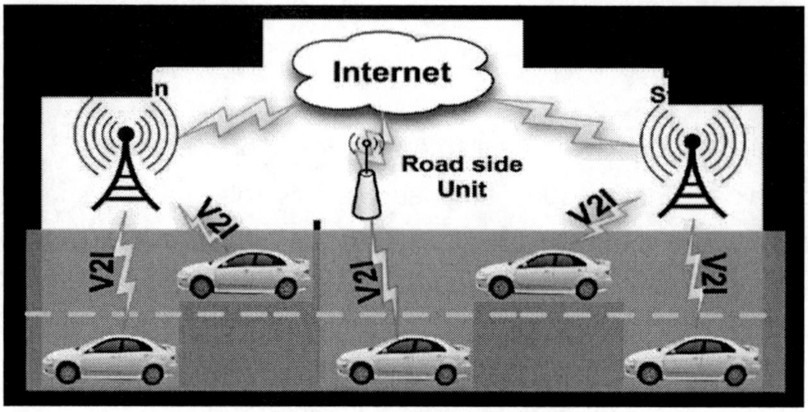

Alexander Skabardonis, David Kan
PATH UC Berkeley
July 9, 2018, San Francisco, CA

Problem Statement

- Ramp metering: common strategy to manage freeway congestion and prevent "capacity drop" [1]
- Independent signal control: cause of queue spillback

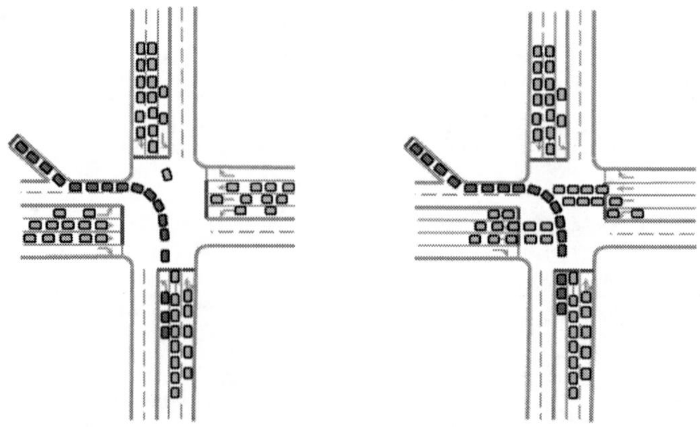

[1] Cassidy and Rudjanakanoknad, 2005; Zhang and Levinson, 2010; Kim and Cassidy, 2012

1

Queue Override

- **Prevents on-ramp queue spillback**
  - Suspends or adjusts ramp metering
  - Freeway congestion/capacity drop ~ 10%

2

Vehicle to Infrastructure Communication

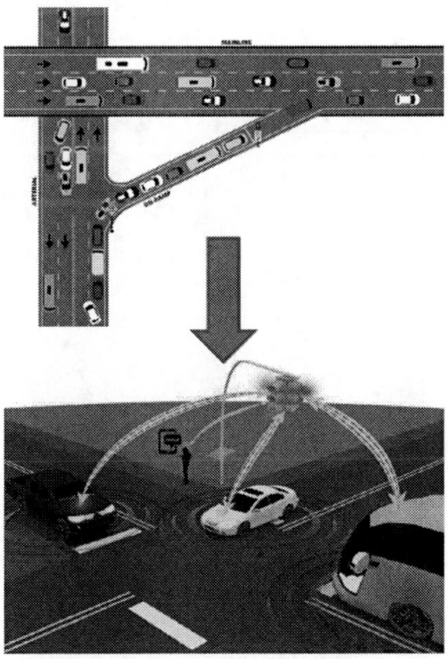

- **Vehicles can send real-time on-ramp queue length data to the traffic signal**
- **Traffic signals can adjust the cycle length in real-time**
  - Avoid long platoons entering the freeway on-ramp from the nearby arterials
  - Prevent queue spillback

3

Proposed Signal Control Strategy (1)

- **Ramp metering control unchanged**
- **Integrate ramp metering controllers**
- **Reduced and variable cycle length**
- **Mitigate both on-ramp and arterial spillback**
- **Example signalized intersection near freeway on-ramp**

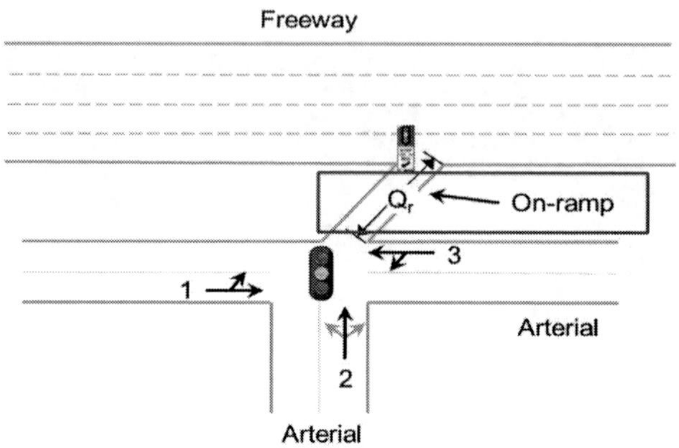

4

Proposed Signal Control Strategy (2)

On-ramp excess accumulation

- $Q(t) + G_1 \cdot s_1 \cdot \beta_1 + G_2 \cdot s_2 \cdot \beta_2 - G_1 \cdot r(t) - G_2 \cdot r(t) \leq Q_r$

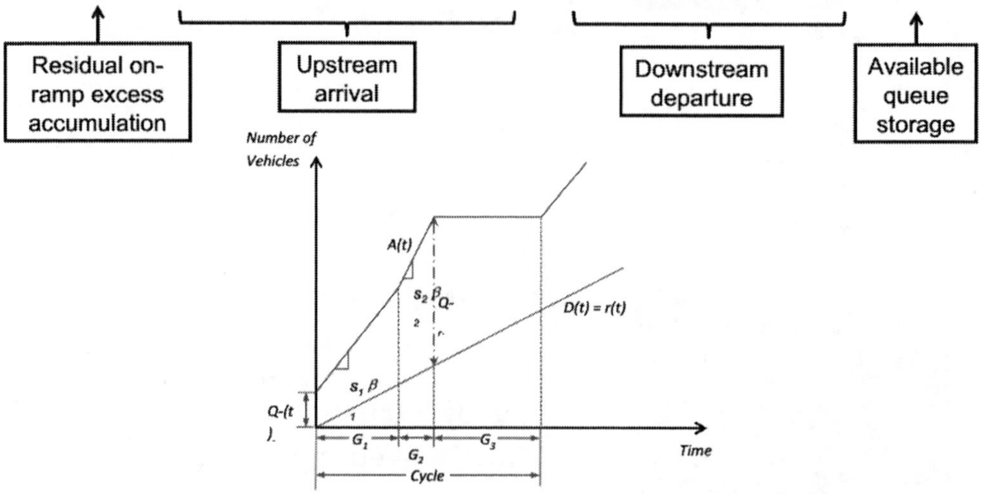

5

Proposed Signal Control Strategy (3)

- Phases 1 and 2: shorter green times
- Clear on-ramp queues in phase 3

Before

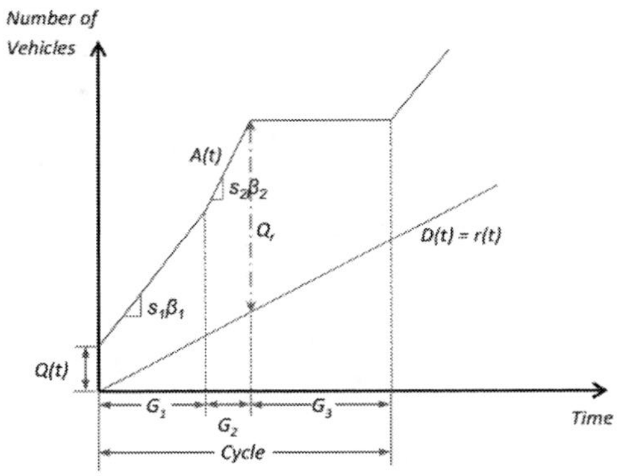

After

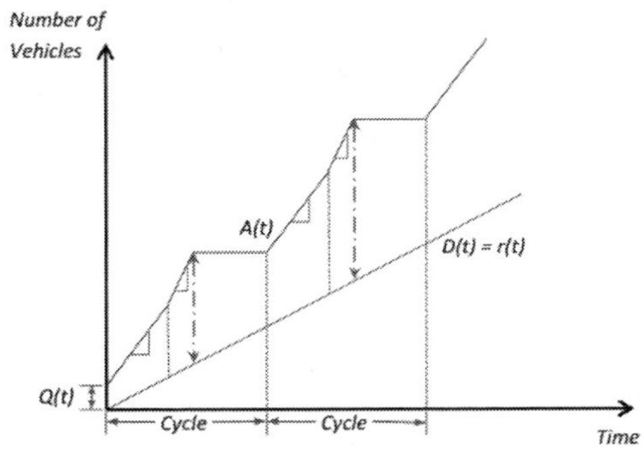

6

Proposed Signal Control Strategy (4)

- On-ramp residual excess accumulation can be determined by tracking vehicles entering and leaving the on-ramp:

$$Q(0) = 0$$

$$Q(1) = Q(0) + A(1) - D(1)$$

$$\vdots$$

$$Q(t) = Q(t-1) + A(t) - D(t)$$

- Maintain the same green time distribution:

$$g_1 = \frac{y_1}{Y} \cdot (C - 3l)$$

$$g_2 = \frac{y_2}{Y} \cdot (C - 3l)$$

7

Proposed Signal Control Strategy (5)

- Cycle length upper limit:

$$C \le \frac{[Q_r - Q(t) + r(t) \cdot 2l] \cdot Y + 4l \cdot \left[\sum_{i=1,2} s_i \beta_i y_i - \sum_{i=1,2} r(t) y_i\right]}{\left[\sum_{i=1,2} s_i \beta_i y_i - \sum_{i=1,2} r(t) y_i\right]}$$

- Updated every cycle
- Does not provide progression/maximum bandwidth

8

Simulation Test

- **Test site: NB I-680/Capitol Ave Corridor, San Jose, CA**

  - **Recurrent bottleneck –AM peak (7:00AM – 9:30AM)**
  - **Aimsun Model**
    - **Enhanced driving behavior model [1]**
  - **Calibrated to replicate field data**
  - **Before: metering <u>with</u> queue override, <u>long</u> cycle lengths**
  - **After: metering <u>without</u> queue override, <u>short</u> cycle lengths**
  - **20 replications**

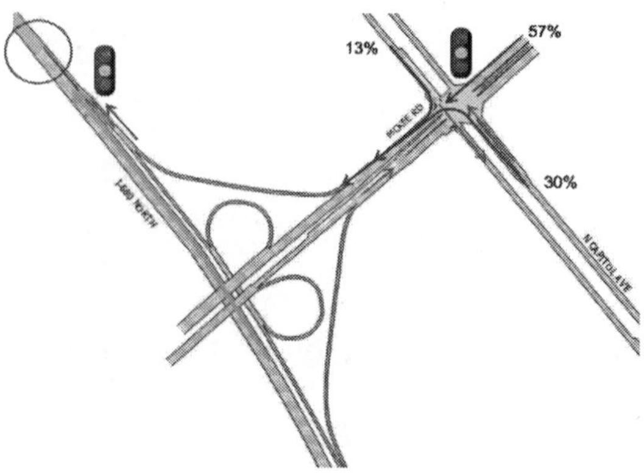

[1] *Lu, Kan, Shladover, 2017*

9

Simulation Results: Test Corridor (1)

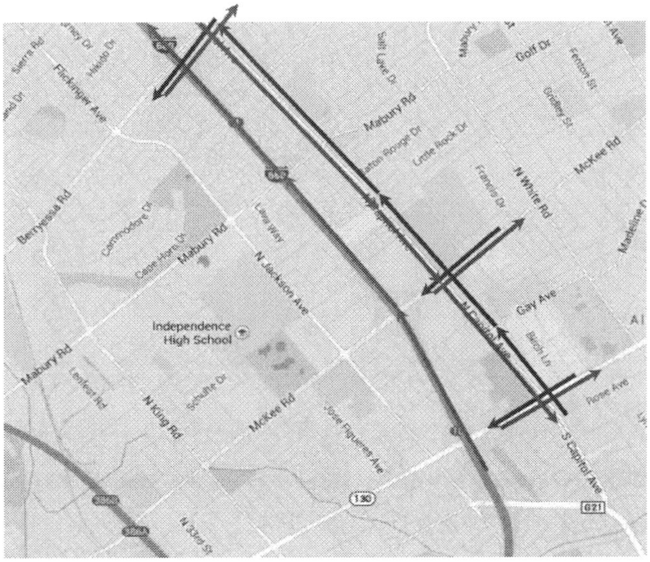

10

Simulation Results: Test Corridor (2)

I. Freeway

	Before		After		% Difference	
	Freeway Mainline					
	Total Delay (veh-hr)	Total Distance Traveled (veh-mile)	Total Delay (veh-hr)	Total Distance Traveled (veh-mile)	Change in Total Delay	Change in Total Distance Traveled
I-680 NB	833.41	43104.13	740.64	44792.95	-11.13%	3.92%

II. Arterial

Average Delay on Main Parallel Arterial (min/veh)			
Capitol Ave NB	8.63	10.51	21.84%
Capitol Ave SB	5.72	5.91	3.33%
Average Delay of Cross Street (sec/veh)			
Alum Rock WB	48.05	47.33	-1.43%
Alum Rock EB	37.27	37.82	1.47%
McKee WB	56.76	52.34	-7.79%
McKee EB	28.92	16.51	-42.91%
Berryessa WB	47.27	39.26	-16.73%
Berryessa EB	50.50	37.55	-34.48%

III. Total System

	Total Delay (veh-hr)	Total Delay (veh-hr)	Change in Total Delay
Freeway & Arterial	2881.37	2727.19	-5.65%

11

Summary

- Vehicle-to-infrastructure communication allows for traffic signals to adjust cycle lengths based on on-ramp queue length
- Improved arterial signal timing can reduce arterial and freeway delay
- Sensitivity analysis: similar improvement when peak (7:30 AM – 8:30 AM) demand increases by 5-10%.

12

DATE 7/26/2018

LEARNING MOMENTS IN A CONNECTED WORLD

EXPERIENCES FROM A LARGE-SCALE TRANSIT SIGNAL PRIORITY IMPLEMENTATION IN WASHINGTON DC

AUTOMATED VEHICLES SYMPOSIUM 2018
SAN FRANCISCO, CALIFORNIA

KITTELSON & ASSOCIATES

Project Background

- DC Area Transit Signal Priority (TSP) deployment
  - Over 250 signalized intersections
  - Up to 1600 WMATA buses
  - 13+ priority bus routes
- Mix of agency and area types

metro

Metropolitan Washington Council of Governments

d.

VDOT

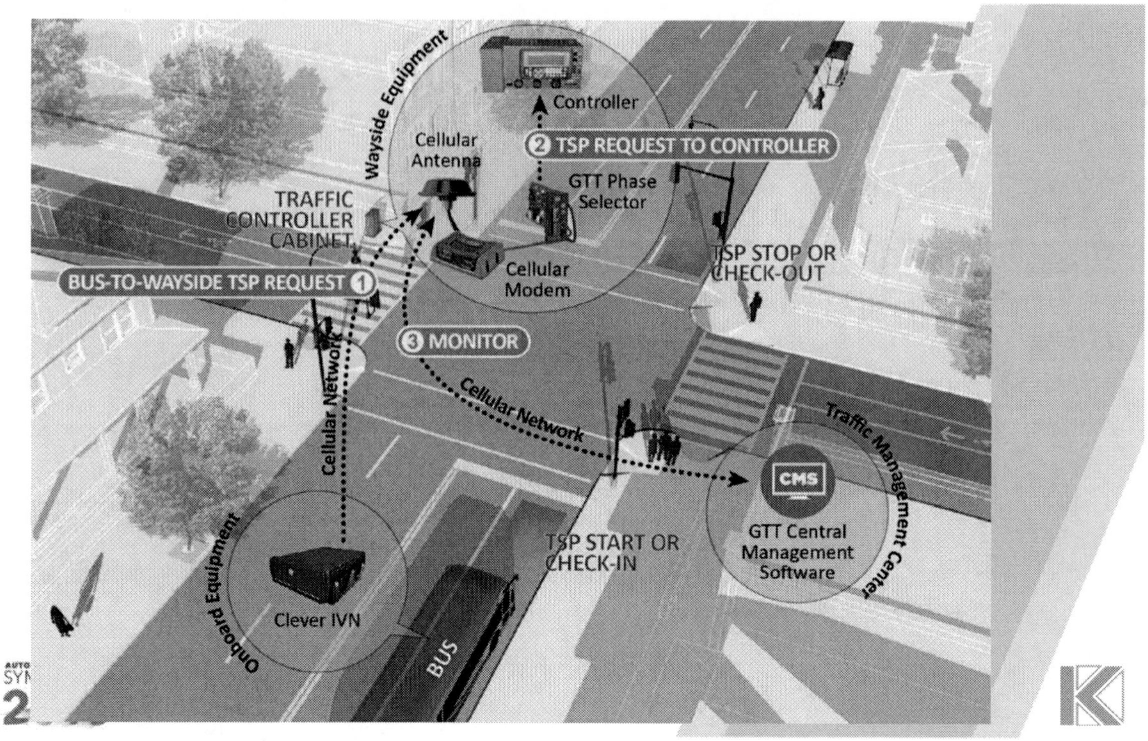

Transit Signal Priority

- Decentralized system
- "Smart Bus" upgrades
- Cellular communications
- Conditional priority
  - Lateness + others
- Signal control
  - Time of day directional

TSP ARCHITECTURE
NTCIP 1211 Example 4

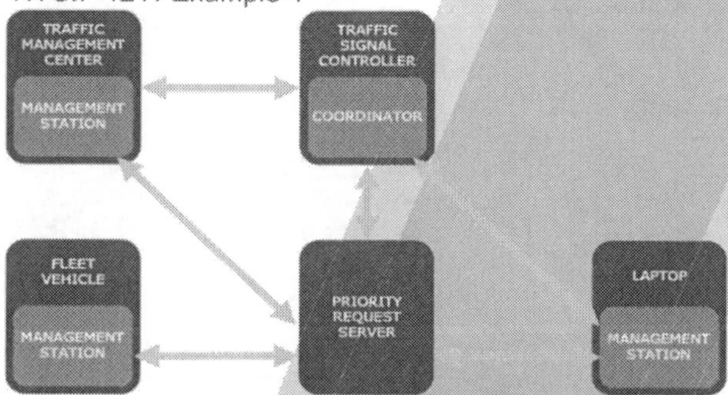

SYMPOSIUM 2018

Overview

- Successful scalable deployment
- Inter- and Intra- agency coordination
- Strides in capabilities
- Time and persistence

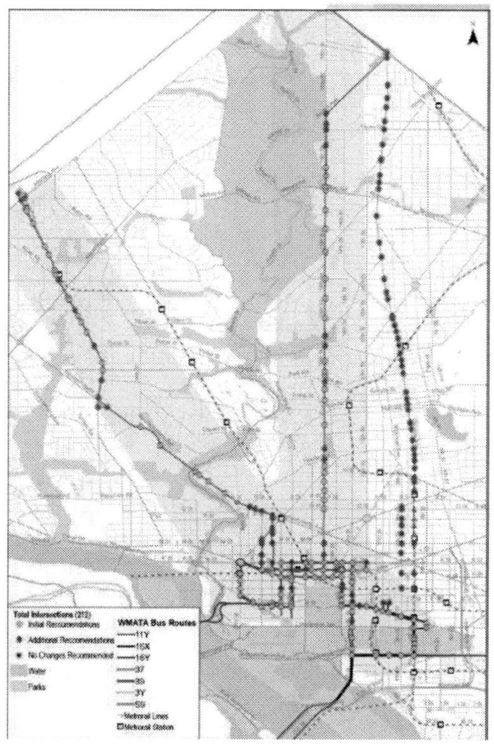

AUTOMATED VEHICLES SYMPOSIUM 2018

Perspectives

- Validation and Verification
- Precision and Accuracy

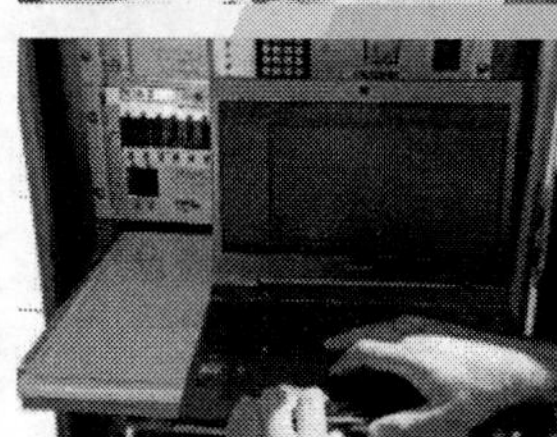

AUTOMATED VEHICLES SYMPOSIUM 2018

Communication Plan

- Check-in
- Updates
- Check-out

TSP Communication Zones at Closely Spaced Intersections

AUTOMATED VEHICLES SYMPOSIUM 2018

Mapped Communication Activity

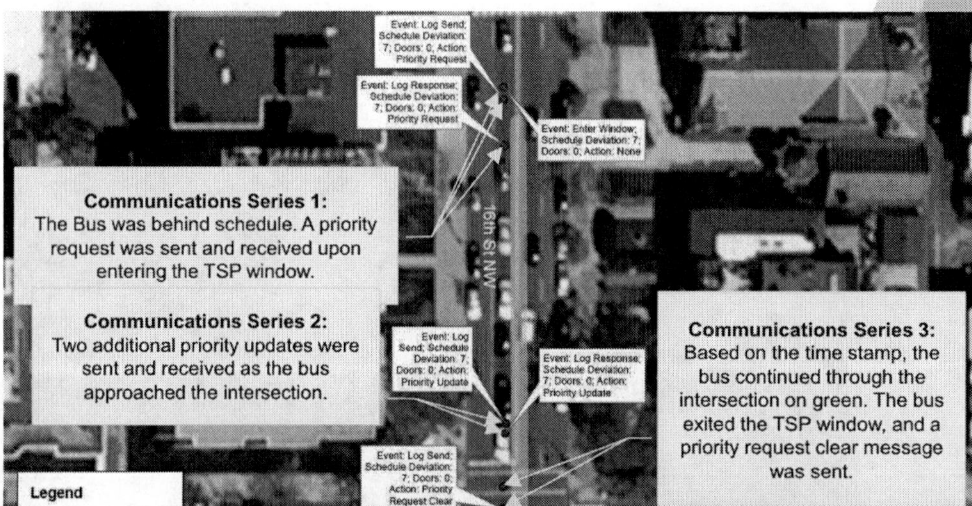

Activity Assessment

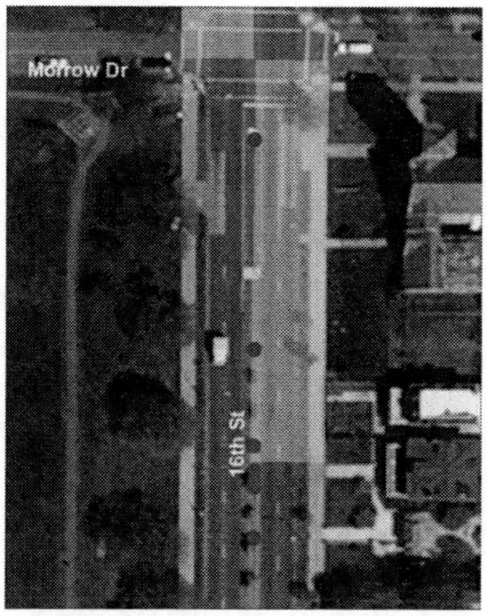

AUTOMATED VEHICLES SYMPOSIUM 2018

Assessing Variability

- Generate request
- Check-in
- Message packages sent
- Message received
- Process request

Time

AUTOMATED VEHICLES SYMPOSIUM 2018

Assessing Variability

- Sources of potential delay/latency
  - Processing
  - Communication
- Dynamic environment
  - Urban area
  - Vehicle movement

AUTOMATED VEHICLES SYMPOSIUM 2018

Outcomes

- Successful large-scale deployment
  - TSP installed,
  - Messages sent/received,
  - Data logging and analysis,
  - Refining parameters,
  - Performance measures
- Measurable activity and performance

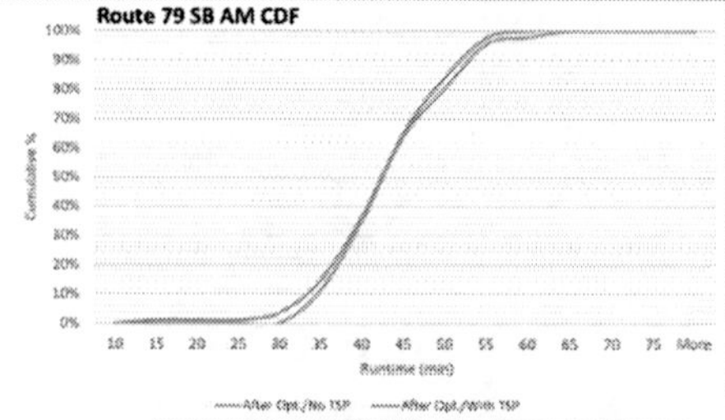

AUTOMATED VEHICLES SYMPOSIUM 2018

Project Assessment

	Validated	Verified
• Conditional TSP (PRG)	✓	✓
• Cellular Communication	✓	✓
• Signal System (PRS)	✓	✓
• End-to-End TSP Operations	✓	?

• TSP Implemented	✓
• Inter-agency Collaboration	✓

AUTOMATED VEHICLES SYMPOSIUM 2018

Lessons Learned

- Collaboration takes time and persistence
- Assess precision vs accuracy
- Scaling expectations and outcomes
- Looking forward to improve

AUTOMATED VEHICLES SYMPOSIUM 2018

Thank You
Kevin Lee
Kittelson & Associates
klee@kittelson.com
Bus Stop#
1000753
For the
Next Bus
Call
(202) 637-7000
or visit
wmata.com

V2I Deployment in Utah: The Redwood Road MMITSS Operation

Blaine D Leonard, P.E., F.ASCE
Technology & Innovation Engineer
Utah Department of Transportation

UDOT Keeping Utah Moving

Overview

- Deployment - Redwood Road MMITSS Deployment
  - Deployment Goals
  - Deployment Location – Redwood Road
  - Installation Details
  - Utah Version of MMITSS
- Operational Results
  - Data Used for Performance Assessment
  - Assessment Results

UDOT Keeping Utah Moving

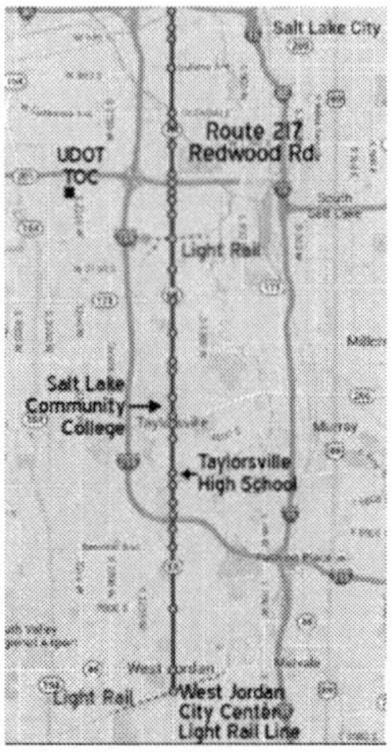

- Full DSRC Corridor
  - Facilitate testing / deployment
  - Prepare for equipped vehicles

- Transit Signal Priority for Improved Schedule Reliability
  - MMITSS Software (Utah Version)
  - Goal: increase from 86% to 94%
  - Minimal impact to other traffic

- Meet the SPaT Challenge

Utah Deployment Site – Redwood Road

Redwood Road
(1700 West)
(400 S to 8020 S)

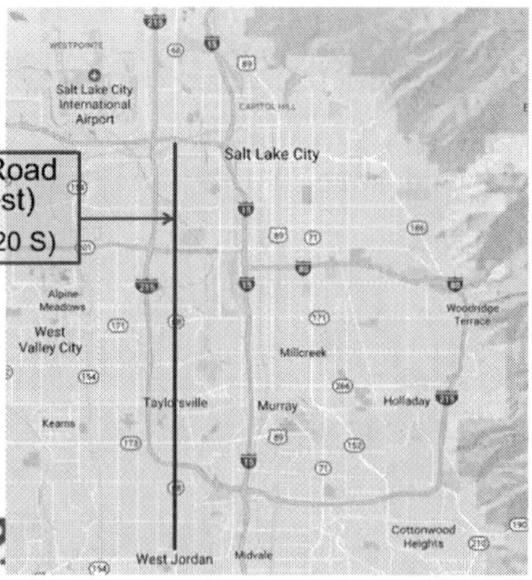

11-mile urban arterial corridor
- UDOT-owned corridor
- 30 signalized intersections
- Varies from 5 to 7 lanes
- ADT: 18,000 to 40,000
  - 60,000 peak at I-215
  - Truck Traffic: 24%
- Two light-rail crossings
- Demographic variety
  - Commercial / Retail
  - Residential
  - College / High School

Redwood Road DSRC Corridor

- Full Fiber Optic Connectivity
  - All signals connected to central system
    - Intelight MaxView
    - Signal Performance Measures (ATSPM)
- Two Brands of Signal Controller
  - Econolite (4) – Cobalt / ASC3
  - Intelight (26)
- Single Board Linux Computer (Beagle Bone)
- Four brands of DSRC RSUs/OBUs
    - Arada / Lear
    - Cohda
    - Savari

Signalized Intersections are the "low hanging fruit" for early V2I

DSRC Hardware Test Lab

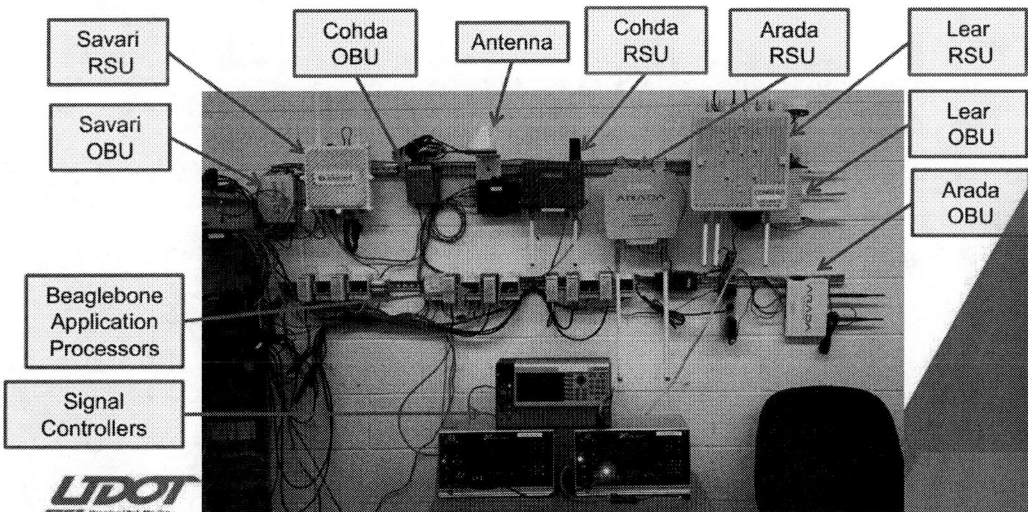

UDOT
Keeping Utah Moving

Redwood Road Installation

- RSU mounted on signal pole, mast arm, luminaire pole
  - Omni-directional antenna, but obstructions can impair signal
  - 300 meter range (nominal minimum – usually longer)
- Ethernet cable to cabinet

UDOT Keeping Utah Moving

On-Board Installation

- DSRC OBU
  - "Beaglebone" On-board Processor (OBP)
  - Mounted on pin-rail
- Power Supply
- Antenna (roof of bus)
  - DSRC and GPS

UDOT Keeping Utah Moving

User Interface

- Monitor & troubleshoot
- Observe data flow
- Demonstrate the system

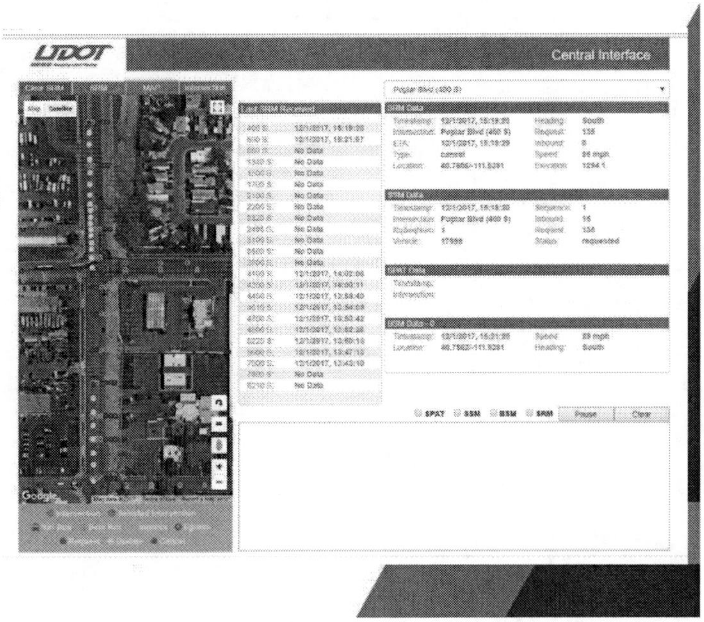

Utah MMITSS Schematic

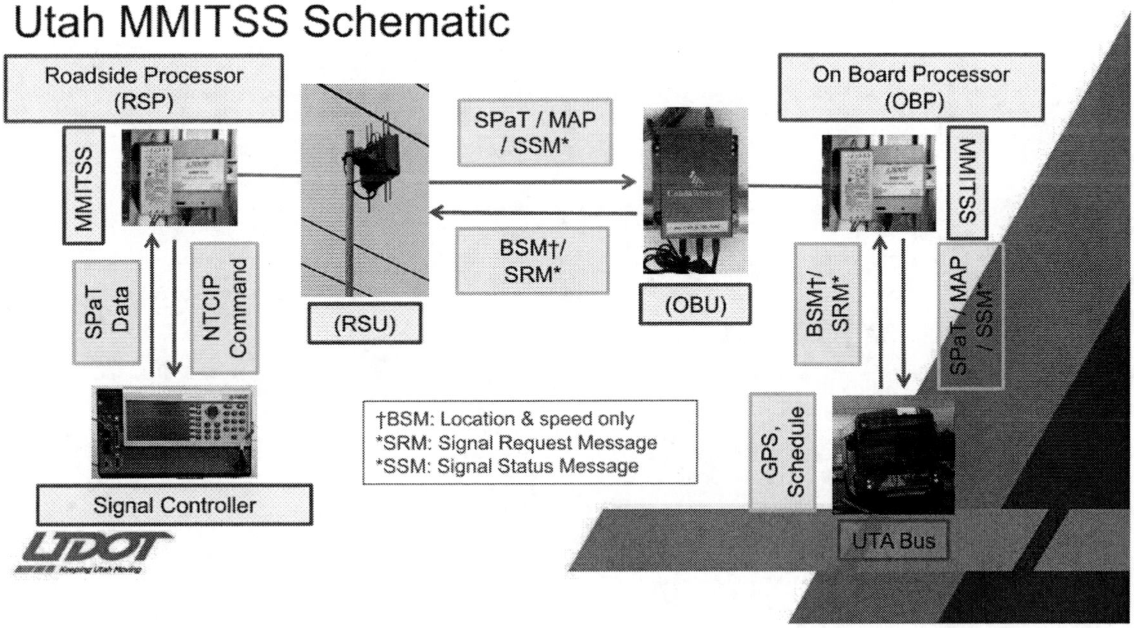

Redwood Road DSRC Database Diagram

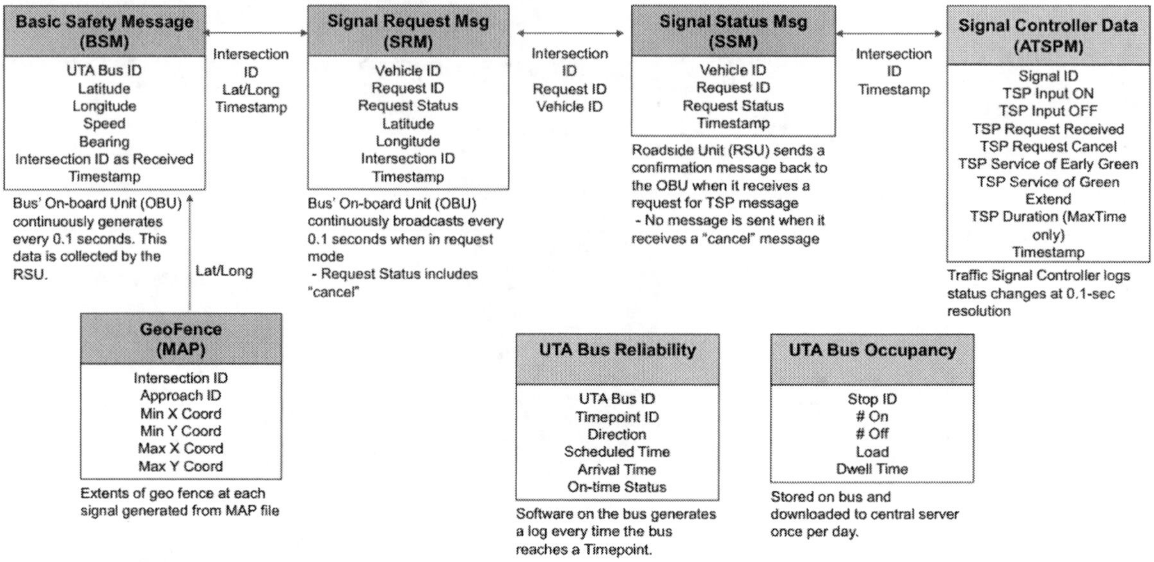

Basic Safety Message (BSM)

	Pk_Id	Timestamp	ReceivedAtIntersectionId	VehicleId	Latitude	Longitude	Elevation	Heading	Speed	DateCreated
1	23994	2018-01-12 14:09:53.0000000	7105	4	40.6849775	-111.9389926	1301.6	179.8875	50	2018-01-12 14:09:57.0000000
2	23995	2018-01-12 14:09:54.0000000	7105	4	40.684952	-111.9389923	1301.6	179.875	50	2018-01-12 14:09:57.0000000
3	23996	2018-01-12 14:09:54.0000000	7105	4	40.6849393	-111.938992	1301.6	179.5375	50	2018-01-12 14:09:57.0000000
4	23997	2018-01-12 14:09:54.0000000	7105	4	40.6849266	-111.938992	1301.6	180.0125	50	2018-01-12 14:09:57.0000000
5	23998	2018-01-12 14:09:54.0000000	7105	4	40.6849013	-111.9389915	1301.6	179.8	50	2018-01-12 14:09:57.0000000

Signal Request Message (SRM)

	Pk_Id	Timestamp	Seq...	VehicleId	Vehicle...	Vehicl...	Inbou...	Inters...	RequestId	Reque...	VehicleLatitude	VehicleLongitude	VehicleElev...	VehicleHe...	VehicleSpeed
1	1	2018-01-17 18:36:35.0...	1	4873	OBE	transit	15	7105	3	request	40.6806408	-111.9389637	1299.7	179.85	58
2	2	2018-01-17 18:36:35.0...	1	4873	OBE	transit	15	7105	3	request	40.6806408	-111.9389637	1299.7	179.85	58
3	3	2018-01-17 18:36:35.0...	1	4873	OBE	transit	15	7105	3	request	40.6806408	-111.9389637	1299.7	179.85	58
4	4	2018-01-17 18:36:35.0...	1	4873	OBE	transit	15	7105	3	request	40.6806408	-111.9389637	1299.7	179.85	58
5	5	2018-01-17 18:36:36.0...	1	4873	OBE	transit	15	7105	3	request	40.6806408	-111.9389637	1299.7	179.85	58

Signal Safety Message (SSM)

	Pk_Id	IntersectionId	Timestamp	RequestId	VehicleId	SequenceNumber	InboundLaneId	Status	DateCreated
1	1	7605	2018-01-17 18:37:40.0000000	5	4873	1	16	requested	2018-01-17 18:37:41.0000000
2	2	7605	2018-01-17 18:37:40.0000000	5	4873	1	16	requested	2018-01-17 18:37:41.0000000
3	3	7107	2018-01-17 18:38:05.0000000	6	4873	1	28	requested	2018-01-17 18:38:06.0000000
4	4	7107	2018-01-17 18:38:05.0000000	6	4873	1	28	requested	2018-01-17 18:38:06.0000000
5	5	7107	2018-01-17 18:38:42.0000000	6	4873	2	29	requested	2018-01-17 18:38:46.0000000

GeoFence (MAP)

	IntersectionID	Intersection	GeoCoordSys	Datum	MinOfX_COORD	MinOfY_COORD	MaxOfX_COORD	MaxOfY_COORD
1	7080	2770 South and Redwood Road	GCS_WGS_1984	D_WGS_1984	40.708410822000	-111.940361023000	40.713889558300	-111.938830214000
2	7090	400 South and Redwood Road	GCS_WGS_1984	D_WGS_1984	40.759560489990	-111.940649399630	40.762192548310	-111.937034413700
3	7091	500 South and Redwood Road	GCS_WGS_1984	D_WGS_1984	40.756297687700	-111.941018783000	40.759513640400	-111.937372501000
4	7092	Indiana Avenue and Redwood Road Intersection	GCS_WGS_1984	D_WGS_1984	40.750304576600	-111.940459416000	40.752359443000	-111.937508171000
5	7083	California Ave and Redwood Road	GCS_WGS_1984	D_WGS_1984	40.739575557320	-111.940012758390	40.741563004760	-111.938133642780

Signal Controller Data (ATSPM)

	id	tstamp	code	param
1	7103	2018-02-14 14:14:56.1000000	176	6
2	7103	2018-02-14 14:14:56.2000000	604	4
3	7103	2018-02-14 14:14:56.4000000	603	3
4	7103	2018-02-14 14:14:56.4000000	112	6
5	7103	2018-02-14 14:14:56.5000000	114	6
6	7103	2018-02-14 14:15:12.3000000	177	6
7	7103	2018-02-14 14:15:12.4000000	604	3
8	7103	2018-02-14 14:15:12.6000000	603	4
9	7103	2018-02-14 14:15:12.6000000	115	6

Intelight MaxTime

	id	tstamp	code	param
3	7116	2018-02-14 18:59:13.5000000	171	2
4	7116	2018-02-14 18:59:13.5000000	112	1
5	7116	2018-02-14 18:59:40.4000000	113	1
6	7116	2018-02-14 18:59:53.7000000	172	2
7	7116	2018-02-14 18:59:54.6000000	115	1

Econolite ASC/3 & Cobalt

Operational Results (Preliminary)

- Operational Since Nov 2017
- Transit Schedule Reliability
  - Improvement: 88% to 94% (Southbound)
  - Similar Results for:
    - Northbound Peak
    - Southbound Average
    - Northbound Average

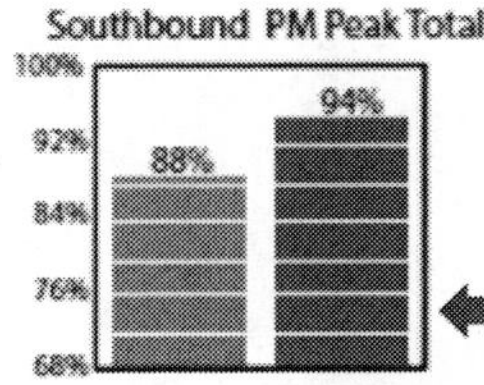

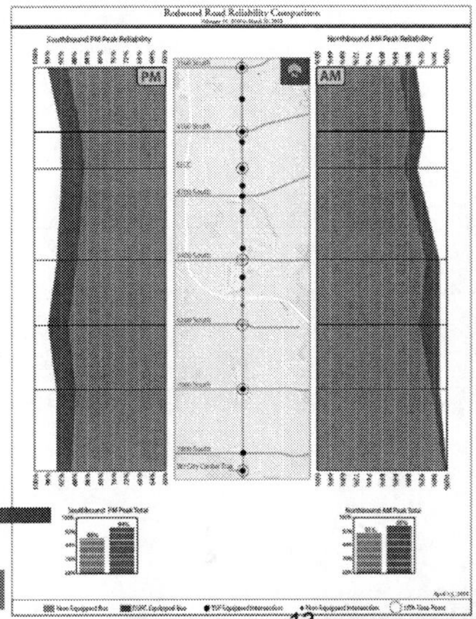

Additional Deployments

Provo-Orem BRT Project
- 47 Intersections
- Intersection Variety – skews / curves
- MMITSS-Utah
- Two DSRC Brands

Snow Plow Pre-emption Project
- Salt Lake Valley – 4 to 5 corridors
- Up to 120 Intersections

AUTOMATED VEHICLES SYMPOSIUM 2018
USERS. VEHICLES. INFRASTRUCTURE.
Driven by AUVSI and TRB

Breakout Session #12

Summary of Key Findings and Lessons Learned

- Onboard attendants – *Can't live with them, can't live without them*
- User acceptance – *Try it, you'll like it*
- Most difficult moment – *Between a truck and a hard place*
- Scalability of low speed – *Need for speed?*

AUTOMATED VEHICLES SYMPOSIUM 2018
USERS. VEHICLES. INFRASTRUCTURE.
Driven by AUVSI and TRB

Breakout Session #12

Research Needs

- Reliability in varying environments
- Safety validation
- Integration into wider transport system
- Role of infrastructure

Life in the Slow Lane: Automated Low-Speed Shuttles

Automated Vehicles Symposium 2018

Breakout Session 12

July 9, 2018

Icebreaker: Agree or Disagree?

Low-Speed Automated Shuttles

Automated shuttles should always have an attendant/operator on-board (now or future)

Agree!

Disagree!

Using an automated shuttle will be more cost-effective than traditional transit

Agree!

Disagree!

People with disabilities will be able to ride without assistance from an onboard attendant

Agree! Disagree!

Automated shuttles can serve transportation needs in my community without exceeding a speed of 25 miles per hour (40 kph)

Agree! Disagree!

Context: State of the Practice

Low-Speed Automated Shuttles in the United States

Low-Speed Automated Shuttle Characteristics

There is considerable interest and deployment activity in automated vehicles that follow a relatively new format:

- **Fully Automated Driving (Level 4):** Intended for use without a driver or operator.
- **Restricted Operational Design Domain (ODD):** Limited operating area, times, enviromnets.
- **Low Speeds:** Generally limited to 25 mph.
- **Shared Service:** Designed to carry the weight of 6-15 passengers.

Representative models

EasyMile EZ10

Navya Arma

Local Motors Olli

8

U.S. DOT/ITS JPO Project Overview

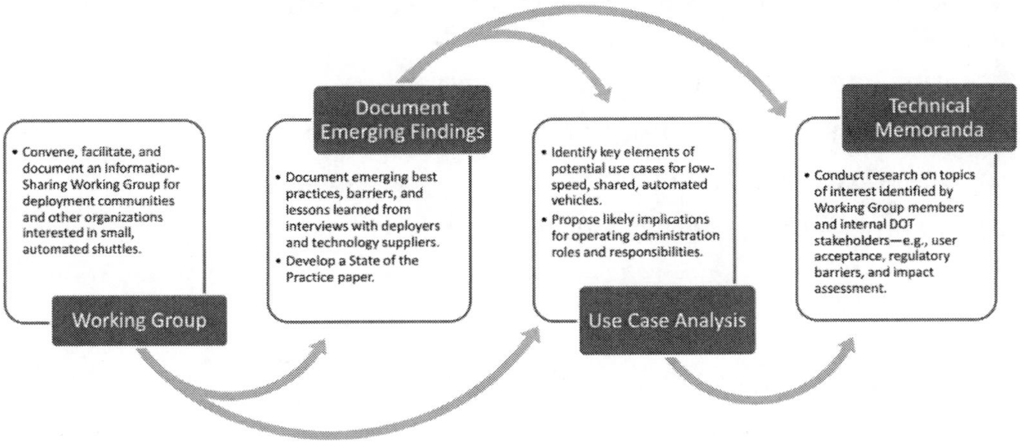

9

Working Group Member Projects

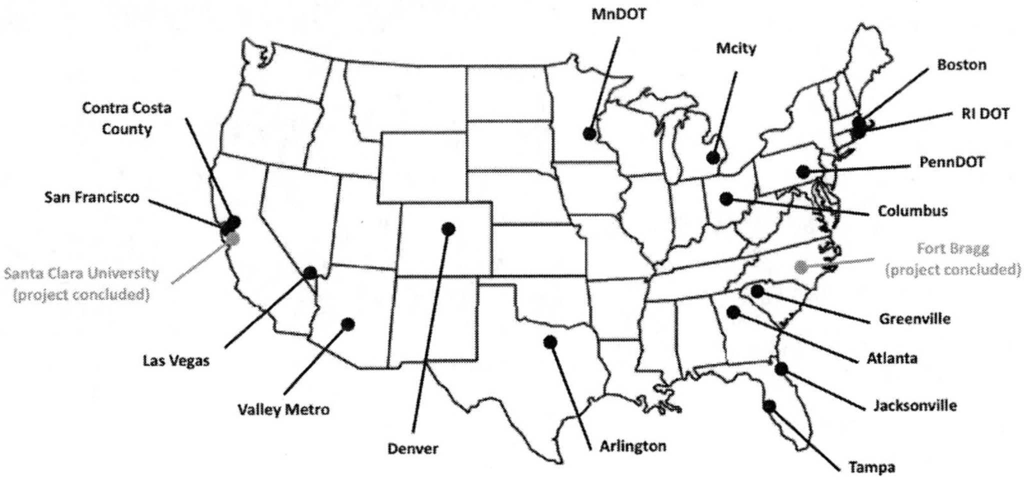

10

Who's doing what, where, and why?

In general...

- Increasing interest during period 2016-present
- Relatively high number of nontraditional project sponsors
- Short demonstrations and pilots
- Lots of enthusiasm; uncertain business models

11

Jacksonville Transportation Authority (JTA)

Ultimate Urban Circulator (U2C) Concept

- Convert the existing monorail to track for automated shuttles
- Expand the system to include at-grade routes

"Test and Learn" Track

- 1/3-mile linear access road
- Two stops with bus shelters
- Near Sports Complex, brewery, outdoor amphitheater and water taxi

EasyMile EZ10 Shuttle Testing

- Working to build public trust and confidence
- Creating focus group with diverse members
- Collecting feedback from rider surveys
- Coordinating with the city, sheriff's office, fire and rescue during events

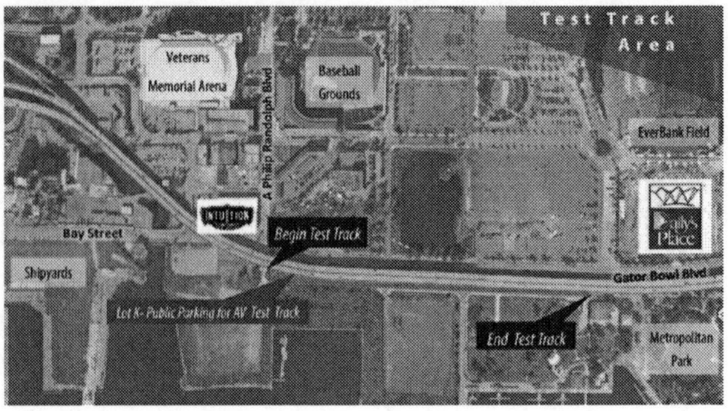

Identified Test Topics

- Acceleration/Deceleration
- Connected Vehicle Communications
- Cybersecurity
- Human Interaction
- Obstacle Navigation
- Safety
- Universal Access

12

San Francisco CTA/MTA Treasure Island Shuttle

Treasure Island Shuttle

- Joint project between San Francisco County Transportation Authority (SFCTA) and San Francisco Municipal Transportation Agency (SFMTA)
- In addition to SFCTA and SFMTA, partners include Treasure Island Development Authority and UC Berkeley
- Automated shuttle service to carry passengers around Treasure Island and Yerba Buena Island
- Planning work has suggested a system with three 12-passenger shuttles, though the number of shuttles and their specifications have not been set

Source: SFCTA

ATCMTD Grant

- $10.9 million award from the Advanced Transportation and Congestion Management Technologies Deployment (ATCMTD) grant program
- Portion of funds will support creation of an automated shuttle service

13

City of Arlington, Texas

Demonstration

- The Arlington Convention Center was the third stop of the "2017 Autonomous Vehicle Road Trip" event in February 2017.
- Community and media were invited to attend, and Arlington City Council members were among some of the first riders.

Milo Pilot Project

- In March 2017, the Arlington City Council approved a one-year lease agreement for two EasyMile EZ10 shuttles.
- Beginning in August 2017, the shuttles operated along select off-street trails in the Entertainment District during Stadium and Ballpark events.
- Due to construction in October 2017, the shuttles moved to a different trail within the Entertainment District.

Source: City of Arlington, TX

14

University of Michigan Shuttle

Organization

- Mobility Transformation Center, University of Michigan

Mcity Testing

- In December 2016, Mcity began working with a Navya Arma shuttle.
- The shuttle has been used to conduct research and provide tours.

North Campus Pilot

- Beginning in 2018, the shuttles will be used in a new campus mobility service for students, faculty, and staff. Currently the shuttles are being tested without passengers.
- The shuttles will drive along a two-mile route between the Lurie Engineering Center and the North Campus Research Complex (NCRC), picking up passengers every 10 minutes.
- The study will examine passenger reactions and user acceptance of the technology.

Source: University of Michigan

15

GoMentum Station / Bishop Ranch Shuttle

Testing and Deployment

- Initial tests began in fall 2016 at GoMentum Station (former military base)
- Phase 2 demonstrations began in March 2017 on a 250-yard loop in a parking lot at Bishop Ranch business park in San Ramon
- Shuttles received CA license plates in December 2017
- Awaiting permit from CA DMV to move testing to public roads
- Plans to expand service to the entire Bishop Ranch campus (including public streets) by 2019

Source: EasyMile

Automated Shuttles

- Two 12-passenger EasyMile EZ10 shuttles
- 12 mph top speed

16

Greenville Automated Taxi Shuttles

Automated Shuttle Pilot

- The deployment will initially use ARIBO shuttles from Fort Bragg, NC.
- The team has mapped roads near Clemson University International Center for Automotive Research (CU-ICAR) and are scheduling demonstration rides.
- Identified locations for deployment include:
- **Phase 0:** CU-ICAR campus, with service to the nearby Center for Manufacturing Innovation (CMI) at Greenville Technical College
- **Phase 1:** Verdae, an upscale residential, retirement, office, and retail area
- **Phase 2:** Parker, a low-income area with a need for non-emergency medical transportation and other fist/last-mile transportation solutions

Source: Google Maps

Source: FHWA Public Roads 80(2)

17

Panel Discussion

Low-Speed Automated Shuttles

Panelists

- Elizabeth Machek, U.S. DOT, *moderator*
- Jürg Michel, PostAuto
- Harri Santamala, Sensible 4
- Kristen Fernandez-Medina, TRL
- Jay Hietpas, MnDOT
- Sharad Agarwal, EasyMile
- Sarah Swigart, AAA

To scale up commercial operations we really have only two problems; open traffic and environment

sensible 4

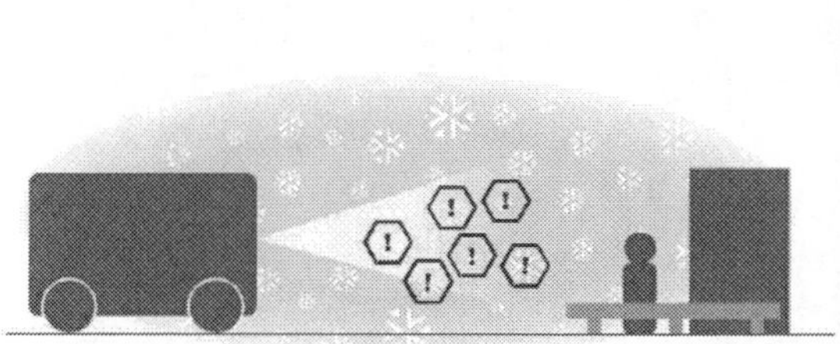

If it works here, it works anywhere

- Open-road pilot for customers
- 3.5 km
- Vantaa, Finland
- Operational during Nordic winter

*https://www.teslarati.com/ces-2018-rains-shuts-down-self-driving-demos/

www.sensible4.fi

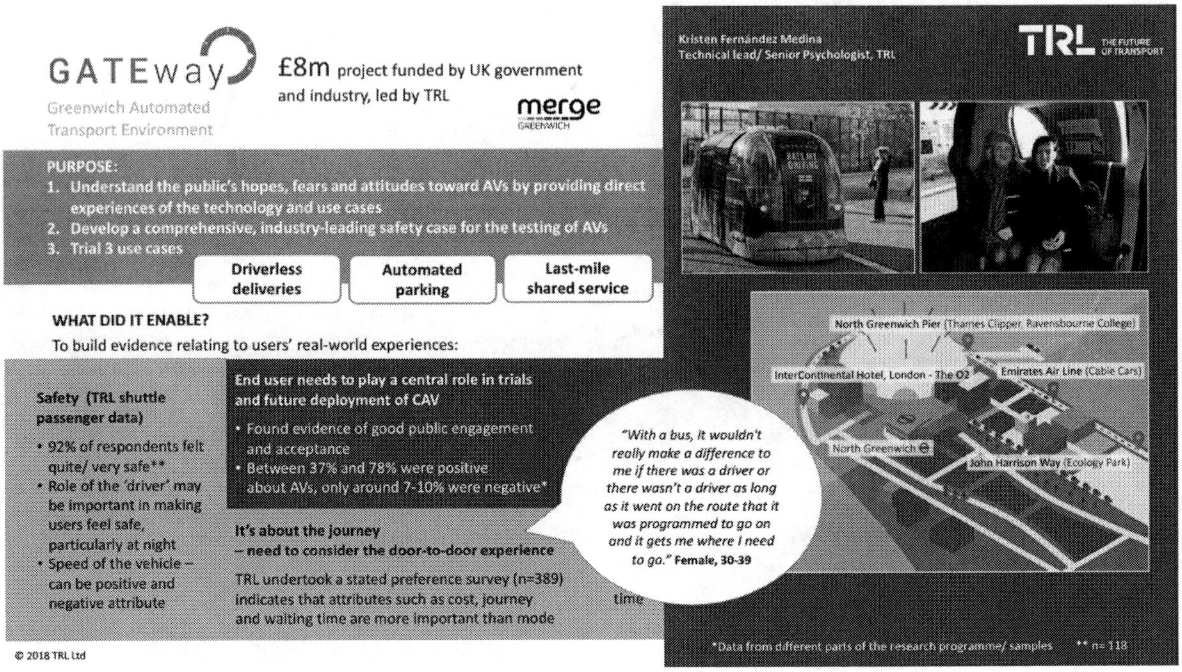
GATEway
Greenwich Automated Transport Environment
£8m project funded by UK government and industry, led by TRL
merge GREENWICH
Kristen Fernández Medina
Technical lead/ Senior Psychologist, TRL
TRL THE FUTURE OF TRANSPORT
PURPOSE:
1. Understand the public's hopes, fears and attitudes toward AVs by providing direct experiences of the technology and use cases
2. Develop a comprehensive, industry-leading safety case for the testing of AVs
3. Trial 3 use cases
Driverless deliveries
Automated parking
Last-mile shared service
WHAT DID IT ENABLE?
To build evidence relating to users' real-world experiences:
Safety (TRL shuttle passenger data)
• 92% of respondents felt quite/ very safe**
• Role of the 'driver' may be important in making users feel safe, particularly at night
• Speed of the vehicle – can be positive and negative attribute
End user needs to play a central role in trials and future deployment of CAV
• Found evidence of good public engagement and acceptance
• Between 37% and 78% were positive about AVs, only around 7-10% were negative*
It's about the journey – need to consider the door-to-door experience
TRL undertook a stated preference survey (n=389) indicates that attributes such as cost, journey and waiting time are more important than mode time
"With a bus, it wouldn't really make a difference to me if there was a driver or there wasn't a driver as long as it went on the route that it was programmed to go on and it gets me where I need to go." Female, 30-39
North Greenwich Pier (Thames Clipper, Ravensbourne College)
InterContinental Hotel, London - The O2
Emirates Air Line (Cable Cars)
North Greenwich
John Harrison Way (Ecology Park)
*Data from different parts of the research programme/ samples ** n= 118
© 2018 TRL Ltd

DEPARTMENT OF TRANSPORTATION
Minnesota Winter Weather Testing and Demonstrations
EASY MILE

AAA Free Self-Driving Shuttle Pilot Program in Las Vegas

- Launched November 2017 as the first self-driving shuttle with V2I on US public roads
  - 27,100 passengers and counting
  - Passengers "Rate the Driver"....gave Shuttle a 4.9 of 5 star rating
- Pilot demonstrated that consumer acceptance increases after experiencing the technology in a safe environment
  - AAA involvement influences acceptance
  - AAA staff on-site for public interaction, including curb management
  - The Navya shuttle is owned & operated by Keolis North America, with Safety Attendant with onboard

AAA Northern California, Nevada & Utah is a not for profit, member services organization that has advocated for safe mobility for our 5.8 mi regional and 60 million Members North America-wide since its founding over a century ago. ***Future is Here. Hop On!***

KEOLIS RTC

Passenger Feedback

"Can't wait for AVs and to never hear the phrase road rage again"

"My whole family felt like it was so futuristic and safe"

"I do trust it because its with AAA"

"I am very proud to be a AAA Member, learning about the sponsorship of this shuttle and their support of AVs"

EASY MILE

- 15 PAX
- 100% electric with no infrastructure changes
- Unlimited Routes Programmable
- Multiple redundancies built into the Safety Chain
- Ideal Vehicle Today for
  - Parking Lots
  - Hospitals
  - Corporate Campuses
  - Planned Communities
  - Industrial Parks
  - Hotels / Resorts

2
6

Small Group Discussion

Agree? Disagree?

Small Group - Instructions

- Introduce yourselves
- Select a reporter
- Discuss the statement on the worksheet at your table
  - Identify three arguments in its favor, and three against
- Record arguments on the worksheet
- If time – report out!
- Hand in your worksheet before you leave

AUTOMATED VEHICLES SYMPOSIUM 2017

USERS. VEHICLES. INFRASTRUCTURE.

Breakout Session # 13
Training Needs for Automated Driving

Summary of Key Findings and Lessons Learned: Invited Speakers

1. Do drivers understand their advanced driver assistance systems?
   - Ashley McDonald, National Advanced Driving Simulator
2. What is industry doing to train drivers?
   - Charles Green, General Motors
3. What have academics developed as possible training programs?
   - Charlie Klauer, VTTI
   - Mike Manser, TTI
4. What can we learn from other modes?
   - Steve Casner, NASA

AUVSI

TRB

AUTOMATED VEHICLES SYMPOSIUM 2017

USERS. VEHICLES. INFRASTRUCTURE.

Breakout Session # 13
Training Needs for Automated Driving

Recommended Action Items: Breakout Sessions and Invited Panelists

1. Who should be responsible for providing training?
2. What lessons from other transportation modes can be applied to driver training and automated vehicles?
3. How do we determine if training is successful?
4. Should training be integrated into a vehicle's technology suite or should training be delivered by traditional means
5. What existing and novel types of training should be provided?

- Panelists: David Kidd (Insurance Institute for Highway Safety), John Lenneman (Toyota Collaborative Research Institute), Elizabeth Mazzae (NHTSA), Cathy Curtis (AAMVA)

AUVSI

TRB

AUTOMATED VEHICLES SYMPOSIUM 2017 — USERS. VEHICLES. INFRASTRUCTURE.

Breakout Session # 14: Speed Dating in the Legal Coliseum

Summary of Key Findings and Lessons Learned:

1. More collaboration and partnership with cities is needed to shape AV development and policy.
2. Consistent, reliable, and mandatory collection of relevant data is needed to evaluate the safety of AV technology.
3. Gaining consumer trust is central to the deployment and success of AVs.

Session Moderator:
Bryant Walker Smith, University of South Carolina

Session Speakers:
Pete Gould, Shared Mobility Strategies
David Kidd, Insurance Institute for Highway Safety
Kristin Kolodge, J.D. Power
Sam Loesche, Teamsters
Brian Soublet, California DMV Legal Affairs
Daniel McGehee, University of Iowa
Sheldon Russell, Virginia Tech Transportation Institute
Stephen Wu, ABA Section on Science and Technology
Damon Porter, Global Automakers
Jack Weast, Intel/Mobileye
Clay Walker, Uniform Law Commission
Robert C. Passmore, PCI

AUVSI TRB

AUTOMATED VEHICLES SYMPOSIUM 2017 — USERS. VEHICLES. INFRASTRUCTURE.

Breakout Session # 14: Speed Dating in the Legal Coliseum

Recommended Action Items

1. Understand the voice of consumers in the product development process.
2. Develop metrics for AVs beyond crashes per VMT, including metrics for AV decision-making.
3. Ensure consumers have sufficient knowledge about how to safely operate automated vehicles.

AUVSI TRB

AUTOMATED VEHICLES SYMPOSIUM 2017

USERS. VEHICLES. INFRASTRUCTURE.

Breakout Session # 15: Trucking automation: Deployment Challenges and Opportunities

Key take-aways:

- Platooning
  - Very strong interest among many stakeholders – moving forward rapidly
  - Near-term: L1-L2 platooning, single brand, large fleets
  - Longer-term: Multi-fleet/multi-brand (ENSEMBLE)
  - Major field trials (both private and public funded) both in Europe and US, close collaboration between key stakeholders (trucking industry, OEMs, tech companies, state governments, enforcement)
- Stand-alone automation
  - Today mainly driven by tech companies
  - Near-term: Low-speed L4 automation (e.g., trailer switching and drayage)
  - Longer-term: L4 highway operations – strongest business case
  - Now: Operational (revenue-producing) highway operations with test drivers (L2) to validate L4
- **Start to see more clearly were we are heading**

AUVSI

TRB

AUTOMATED VEHICLES SYMPOSIUM 2017

USERS. VEHICLES. INFRASTRUCTURE.

Breakout Session # 2.2

What's next (what will we talk about next year)?

- Field trials demonstrating business models in revenue-producing operations
- Discussion on regulatory issues continues: Interpretation of FMCSR's, NHTSA policy doc AV3.0 (will address trucks), harmonization of state laws
- Safety assurance
- Liability and insurance
- Inspection and enforcement
- **How will the trucking ecosystem change with the deployment of automated trucking?**

AUVSI

TRB

VOLVO

Opportunities and challenges for the deployment of truck platooning

Dr. Aravind Kailas
Volvo Group North America

AVS, July 10th, 2018

Leading global manufacturer of trucks, buses, construction EQ, and marine and industrial engines

Build in **18** Countries

Sell in **190+** Markets

Engines: VOLVO PENTA

Buses: VOLVO, NOVABUS, PREVOST, SUNWIN

Construction Equipment: VOLVO, SDLG, TEREX

Trucks: VOLVO, MACK, UD, RENAULT TRUCKS, DONGFENG Trucks, EICHER

Volvo believes platooning offers major advantages for its customers and the society

SARTRE EU TPC SE4P ...

2009 PATP SERET On-road trials Today

Volvo Group North America

3

VOLVO

Texas A&M Transportation Institute

Development of Level 2 Truck Platooning in Texas

TTI Commercial Truck Platooning Research

Mike Lukuc
Program Manager/Research Scientist
Texas A&M Transportation Institute

Texas A&M Transportation Institute

Texas Freight Issues

- Projected changes between 2014 and 2040 in TX:
  - Freight tons moved is expected to increase 46% from 2.6 billion to nearly 3.8 billion tons
  - Freight tonnage moved by trucks is projected to double from over 1 billion to over 2.2 billion (translates to 89% VMT growth)
  - Truck trips on TX roadways are estimated to increase from 589,000 to over 1 million
- In 2013, Texas had over $1 billion in congestion cost to the trucking industry, ranking only behind California.
  - Dallas ranked 4th with over $406 million and Houston ranked 6th with over $373 million
  - I-35 through Austin is the most congested roadway segment in Texas based on trucking delays with over 116,000 hours of delay in 2013.

Texas A&M Transportation Institute

What is Truck Platooning?

- Extension of cooperative adaptive cruise control
- *Automated and precise lateral and longitudinal vehicle control (L2 automation)*
  - Maintain tight formation
  - Safely follow at short distances
- Lead truck: manually driven
- Following truck(s): driver engaged and monitoring environment and driving task
- Leverages ACC and collision mitigation system
  - Bendix® Wingman® Fusion™

Texas A&M Transportation Institute

Why Truck Platooning?

- Fuels savings
- Emission reductions
- Safety benefits
- Traffic network benefits
- Driver benefits
- Other

Texas A&M Transportation Institute

TTI Research Focus

- *Collaboration*: Bring together major partners who have committed resources in terms of in-kind matching of equipment, engineering services, and intellectual property.
- *Feasibility Assessment*: Assess feasibility of deploying 2-vehicle truck platoons on specific corridors in Texas in 5 to 10 years.
- *Implementation*: Performing systems engineering to develop the system, an implementation plan, and deployment guidance necessary for Phase III deployment
- Outreach, training and knowledge transfer

Texas A&M Transportation Institute

PHASE 1 RESEARCH ACTIVITIES

Texas A&M Transportation Institute

Phase 1 Activities

Foundational Studies	• Literature Review • Legislative Impacts • Liability Issues
Platooning	• Develop alternative concepts • Define corridor characteristics • Assess and identify corridors • Validate scenarios
System Development	• Operational requirements • Safety analyses • Specifications • System design
Proof of Concept Demonstration	• Vehicle Build • Functional integration • Integration • Demonstration

Texas A&M Transportation Institute

Legislative and Regulatory Analysis

- Analyzed potentially-relevant regulations, recommendations, and standards from FMCSA, FMVSS, NHTSA, and the Texas Transportation and Administrative Code
- Many areas *could* apply, but most do not seem insurmountable
- Potential shift between Federal and State authority

Texas A&M Transportation Institute

Liability Review and Interviews

- Liability from crashes will likely shift to AV OEMs and lead driver, but **uncertainty abounds**.
- Industry concerns about liability & uncertainty; cautious but curious about platooning and automation
- Some ideas to address industry liability concerns:
  - Specialized insurance, like purchasing groups and risk-retention groups
  - Contract-based risk management, like in construction

Texas A&M Transportation Institute

Identified Organizational Issues

- Held stakeholder interviews with truckers, owners, govt. agencies, trade associations
  - Findings of interest
    - Fuel savings most common cited benefit
    - Safety concerns about differences in vehicle power, torque, and loads
    - Construction zones are a concern
    - Rural, multilane roadways better suited
    - Identifiers on platooning trucks and signage on corridors are recommended
    - Training and public outreach are important

Deployment Site Characteristics

Initial Phase 3 Site Requirements

- ≥ 4-lane, rural interstate highways
- Low AADT
- High truck percentage → at least 15% of AADT.
- Relatively long stretches of highway between urban centers
- Posted speed limit ≥ 65 mph

Traffic Simulation: Fuel Consumption Results

- Average fuel savings in the range up to 6% - 8%
  - High volume and high MPR produce more savings, but *only in non-congested traffic conditions*
  - Consistent with industry and other research results under similar conditions
    - Lead vehicle: 4%-6% savings
    - Trailing vehicle: 8%-10% savings

Texas A&M Transportation Institute

Simulation: % Change in Vehicle Throughput

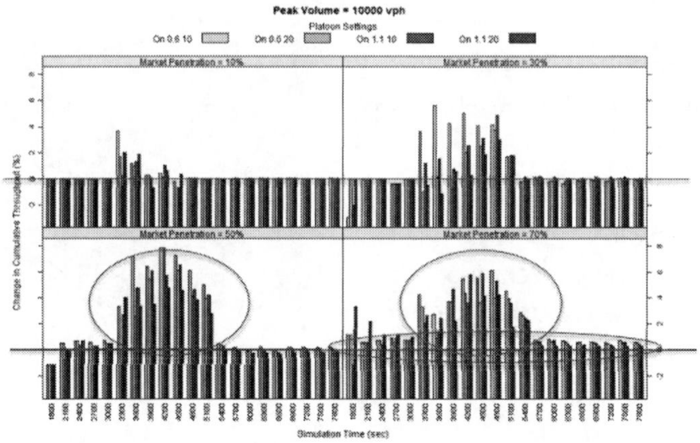

With platooning, there is a noticeable increase in throughput observed in high volume condition at MPR > 30%. *The maximum increase in throughput is in the range of 6-8% at 50% MPR*, tight following gap, and quick formation time.

Texas Department of Transportation | TEXAS A&M UNIVERSITY | Great Dane | Bendix | ZF TRW | DENSO | RICARDO | lytx | Argonne | TARDEC | INTERNATIONAL

Texas A&M Transportation Institute

PHASE 2 RESEARCH

Texas Department of Transportation | TEXAS A&M UNIVERSITY | Great Dane | Bendix | ZF TRW | DENSO | RICARDO | lytx | Argonne | TARDEC | INTERNATIONAL

Texas A&M Transportation Institute

Phase 2 Research

- Understand the global impact of automated truck platooning on TxDOT operations and develop guidance for TxDOT and other stakeholders
- Raise awareness of the value of advanced technologies and the project's contribution

Texas A&M Transportation Institute

Implementation Planning

- Purpose of task to identify and address "roadblocks" that might limit Commercial Truck Platooning in Texas
  - Governance
  - Regulatory
  - Enforcement
  - Operations
  - Legal
  - Education
  - Outreach
- Impact of Recent AV Legislation
  - Review AV governance models for states
  - Public Sector Stakeholder Workshop

Functional Safety for Commercial Truck Platooning

- A Highly Complex Cyber-Physical-System
- Safety achieved through:
  - Adoption of "Industry Best" development methodologies (Model Based Systems Engineering)
    - Requirements and Tests Driven, Architecture and Model Based Development, ISO 26262
  - Extensive testing for verification / some validation
    - Extensive verification (especially corner cases) using Virtual Simulations
    - Limited verification through controlled vehicle tests
    - Cautious and objective transition to naturalistic testing

Integrated Safety and Performance

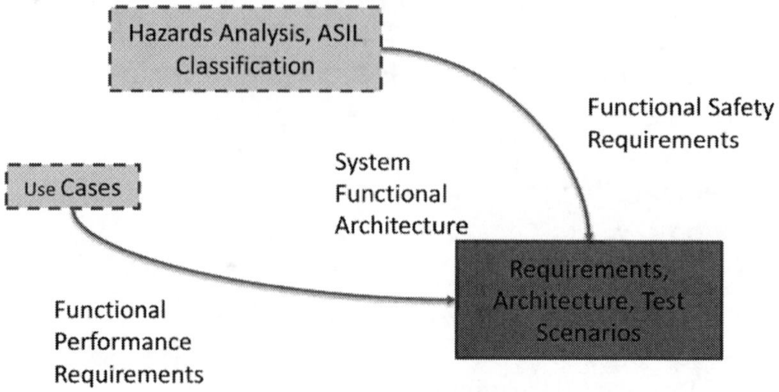

Texas A&M Transportation Institute

MBSE Process for Commercial Truck Platooning

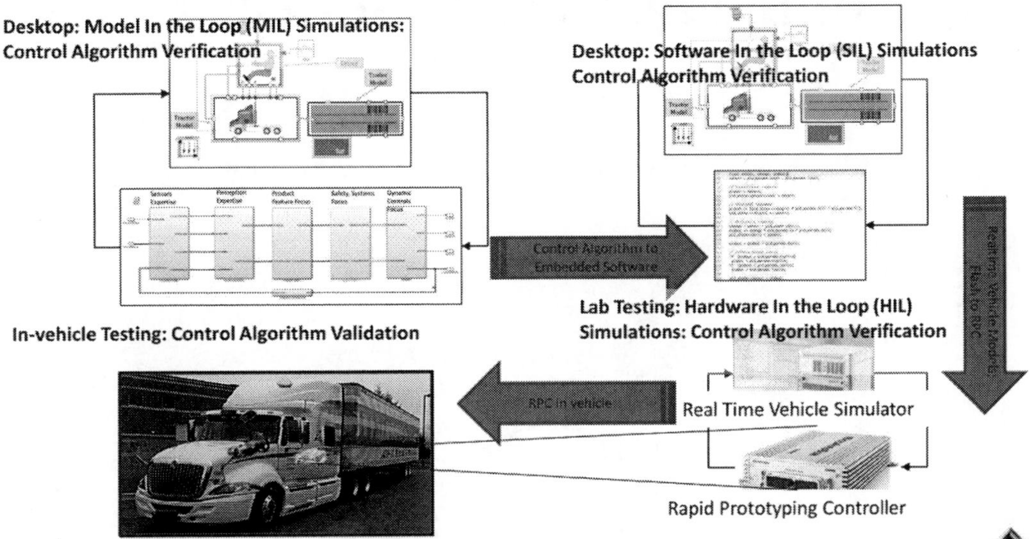

Texas A&M Transportation Institute

Systems Engineering - Functional Safety for Commercial Truck Platooning

- A Highly Complex Cyber-Physical-System
- Safety achieved through:
  - Adoption of "Industry Best" development methodologies (Model Based Systems Engineering)
    - Requirements and Tests Driven, Architecture and Model Based Development, ISO 26262
  - Extensive testing for verification / some validation
    - Extensive verification (especially corner cases) using Virtual Simulations
    - Limited verification through controlled vehicle tests
    - Cautious and objective transition to naturalistic testing

Texas A&M Transportation Institute

Implementation Guidance

Texas A&M Transportation Institute

Final Questions

SH 47

Texas A&M Transportation Institute

Contact Information

Mike Lukuc
Manager/Research Scientist
Connected and Automated Transportation Program

Texas A&M Transportation Institute
2929 Research Parkway
3135 TAMU
College Station, TX 77843-3135
Phone: 979-845-5239
Email: m-lukuc@tti.tamu.edu

ENSEMBLE

ENabling SafE Multi-Brand Platooning for Europe

Maurice Kwakkernaat

TNO innovation for life

AUTOMATED VEHICLES SYMPOSIUM 2018

Driven by: AUVSI

July 9–12, 2018

AUTOMATED VEHICLES SYMPOSIUM 2018

TNO innovation for life

ENABLING SAFE MULTI-BRAND PLATOONING FOR EUROPE

- **20 partners**, including **six EU truck manufacturers** and CLEPA representing **automotive suppliers**
- June 1st 2018 - 2021, **20 Million** Euro EC funding from H2020 ART Call

	Projects
ICT infrastructure for automated road transport	
• Design, implement and test ICT infrastructure for higher levels of automation (up to level 4)	**ICT4ART**
Multi-Brand platooning in real traffic conditions	
• Testing multi-brand platooning on European roads across national borders	**ENSEMBLE**
Urban road transport automation	
• Full scale demonstrations of fleets of autonomous mini-buses in low to medium demand areas	**AVENUE**
• Pre-commercial procurement of future autonomous bus urban level Operation Systems	**FABULOS**

2 | ENSEMBLE

10 July 2018

AUTOMATED VEHICLES SYMPOSIUM 2018

ENSEMBLE TEAM

TNO innovation for life

Participant No	Participant organisation name	Short name	Country
1 (Coordinator)	TNO	TNO	NL
2	DAF Trucks NV	DAF	NL
3	DAIMLER AG	DAIM	DE
4	IVECO	IVE	IT
5	MAN Truck & Bus AG	MAN	DE
6	SCANIA CV AB	SCA	SE
7	VOLVO GROUP	VOL	SE
8	CLEPA, the European Association of Automotive Suppliers	CLEPA	BE
9	ERTICO, European Road Transport Telematics Implementation Coordination Organisation	ERT	BE
10	IDIADA AUTOMOTIVE TECHNOLOGY, S.A.	IDIA	ES
11	IFSTTAR	IFS	FR
12	NXP	NXP	NL
13	ZF FRIEDRICHSHAFEN AG	ZF	DE
14	WABCO GMBH	WAB	DE
15	Kungliga Tekniska högskolan	KTH	SE
16	Vrije Universiteit Brussel	VUB	BE
17	Daimler Fleetboard GmbH	FLE	DE
18	FRENI BREMBO Spa	BREM	IT
19	Robert Bosch GMBH	BOSCH	DE
20	Continental Automotive GmbH	CONTI	DE

3 | ENSEMBLE 10 July 2018

AUTOMATED VEHICLES SYMPOSIUM 2018

TNO innovation for life

ENSEMBLE MAIN OBJECTIVES

- Pave the way for the **adoption of multi-brand truck platooning** in Europe
- Align and work on **standardization**
- Demonstrate differently branded trucks in one (or more) platoon(s)
  - Under **real-world traffic** conditions
  - Across national **borders**

4 | ENSEMBLE 10 July 2018

EXPECTED IMPACT

- Assess impacts on
  - Traffic safety
  - Throughput
  - Fuel economy

and ensure **acceptance** of truck platooning in Europe

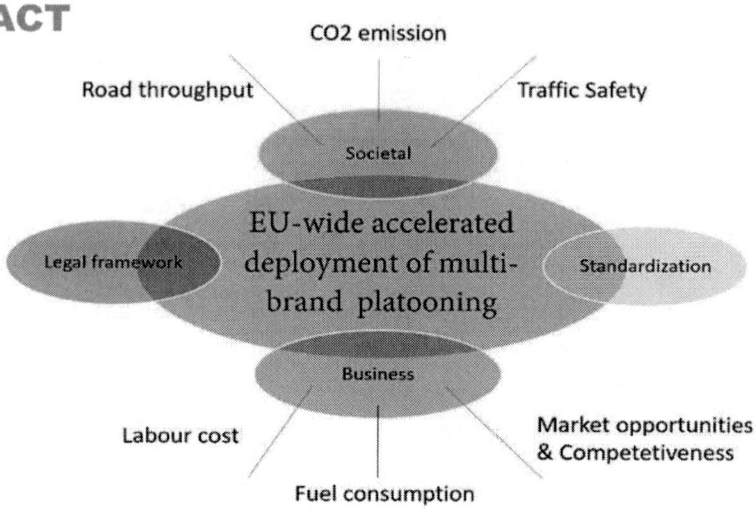

PROJECT STRUCTURE

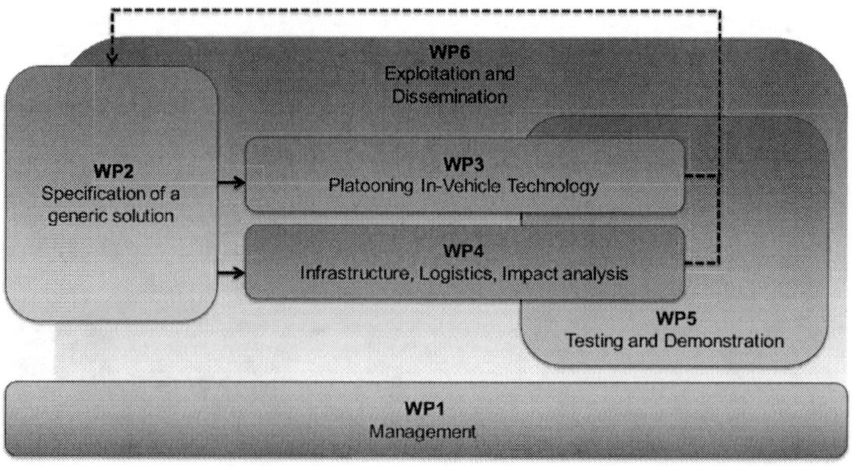

AUTOMATED VEHICLES SYMPOSIUM 2018

TNO innovation for life

WP2 SPECIFICATION OF A GENERIC SOLUTION

The objective of this work package is the definition of the **specifications** of the whole **multi-brand truck platooning concept** to be implemented within the testing and demonstration trucks of the 6 OEMs.

Starting from the current state of the art, common specification of the **Strategic**, **Tactical** and **Operational** Layers and their interfaces will be provided.

The **iteration** process to validate and modify the specification during the whole project life-cycle is an essential part of the work

AUTOMATED VEHICLES SYMPOSIUM 2018

TNO innovation for life

PLATOONING LAYERS

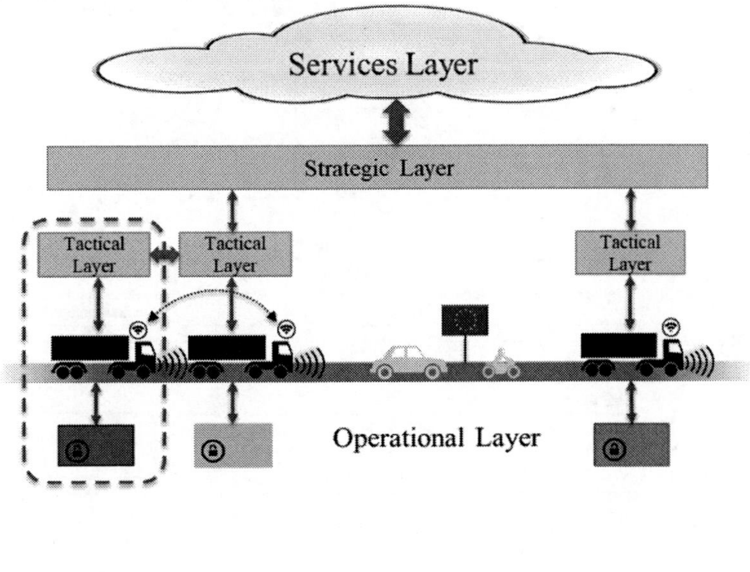

WP3 PLATOONING TECHNOLOGY

- **Design** and **implementation** of a platooning system according to the specifications of WP2
- Common functionality required for multi-brand platooning:
  - Platoon coordinator functionality
  - Mechanism to check consistency of messages
  - Functionality to guarantee safe behaviour of the platoon

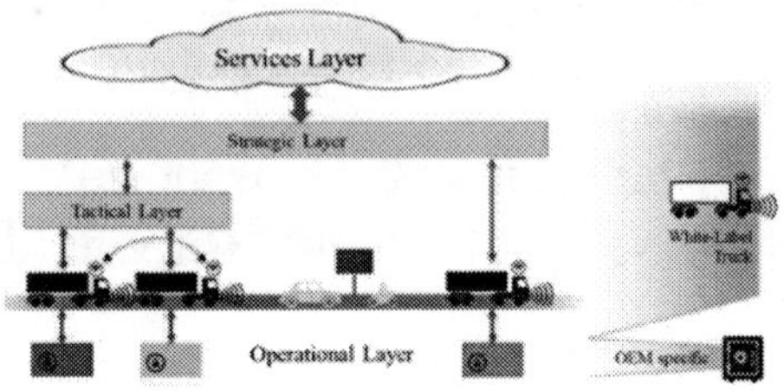

- At least Platooning Level A is implemented: **longitudinal automation**
- Implementations will be verified in WP5 against the specifications and requirements given in WP2

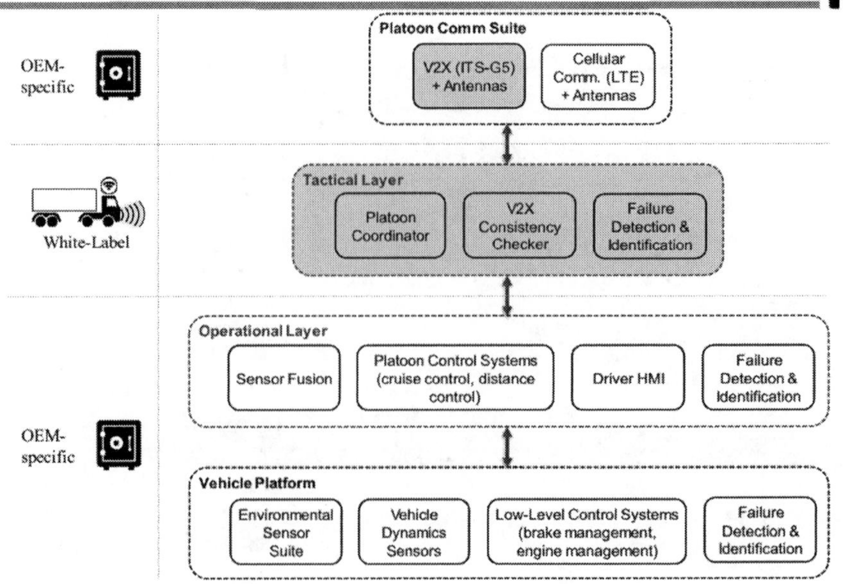

AUTOMATED VEHICLES SYMPOSIUM 2018 AUVSI TNO innovation for life

OTHER WPS

- WP4 Infrastructure and Logistics
  - Assessment of the impact of multi-brand platooning on road, services, economics, environment, truck drivers and traffic
- WP5 Testing and Demonstration
  - Methodology and test plan for Verification, Validation, Evaluation and Demonstration
  - Validation of the generic solution via physical tests in test tracks
  - Multi-brand platooning testing in **public roads**
  - Technical evaluation of the generic multi-brand platooning solution
  - Demonstration of the multi-brand platoon solution in public roads
- WP6 Communication and exploitation

11 | ENSEMBLE 10 July 2018

AUTOMATED VEHICLES SYMPOSIUM 2018 AUVSI TNO innovation for life

IMPORTANT DATES

- **December 2018**
  - Definition of specifications (use cases, scenarios, platooning levels, functional specs for operational, tactical layer and intelligent infrastructure, cloud API).
- **August 2019**
  - Communication functions and software available for implementation
- **February 2020**
  - Trucks implemented with hardware and software and ready for mono-brand platooning
- **April 2020**
  - Multi-brand platoon tests on test tracks
- **June – August 2020**
  - Multi-brand platoon tests on public roads (incl license exemption)
- **November 2020**
  - Service layer design for location based services
- **May 2021**
  - Final event!

12 | ENSEMBLE 10 July 2018

EU – US TWINNING

- Encouraged by EC and US DOT
- WHY?
  - Exchange knowledge and experience
  - Exploit synergies
- WHO? Research projects…
  - … of similar scope, objectives and timeline
  - …aligned to run in parallel & integrated fashion, for the mutual benefit of both sides
- WHAT to exchange? Examples:
  - information, data, methodologies,
  - Researchers, results, publications
  - joint workshops, visits, etc
- HOW?
  - Sponsoring agencies on either side identify project areas suitable for twinning
  - Voluntary basis, Full flexibility for defining activities
  - Regular interaction between project teams including (yearly) face-to-face meetings,
  - No transfer of project funds

European Commission

Thank you for your attention!

Making multi-brand platooning a reality in Europe

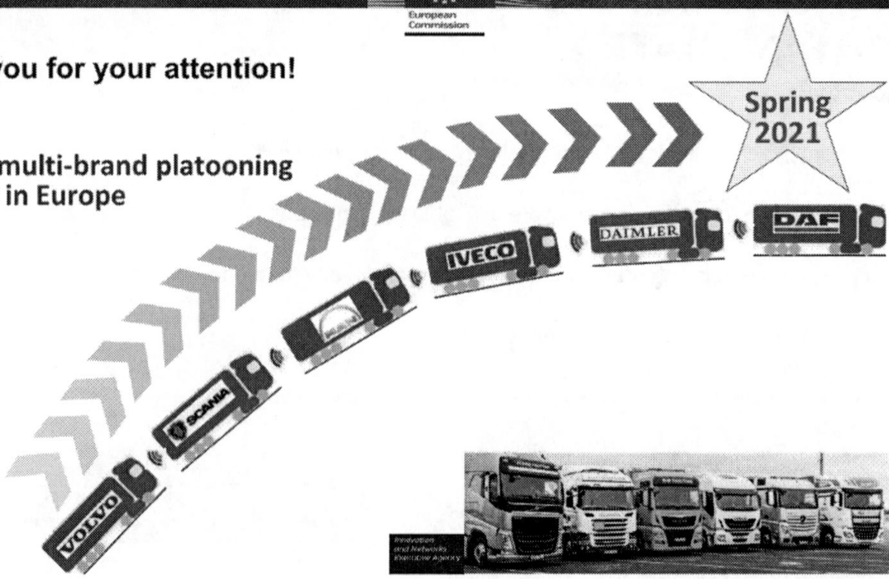

TRL THE FUTURE OF TRANSPORT

Richard Cuerden
Director, TRL Academy

AVS 2018

HelmUK

ADVANCED PLATOONING TRIALS

HelmUK

We are embarking on the world's first independent platooning trials to take place in a live operating environment to quantify real-world benefits.

This is not a demonstration.

2

ADVANCED PLATOONING TRIALS

Platooning technology on three trucks

ADVANCED PLATOONING TRIALS

HelmUK

Setting the Pace. Driving the Distance.

ADVANCED PLATOONING TRIALS

The Profile

This is an ambitious programme to collaborate with industry, our partners and the general public in evaluating how we can positively influence the UK's transport operations.

Government Green Lights Platooning Trial £8.1m Investment

"We are investing in technology that will improve people's lives. Advances such as lorry platooning could benefit businesses through cheaper fuel bills and other road users thanks to lower emissions and less congestion. But first we must make sure the technology is safe and works well on our roads, and that's why we are investing in these trials."

Paul Maynard, Transport Minister

Department for Transport

highways england

Project Objectives: Independent Impact Assessment of truck platooning in real world operating conditions

Deliver Safety & Cyber Security Evidence

- For platooning drivers
- For other road users
- Cyber security V2V control systems
- Platooning risk rated Strategic Road Network map

Quantify Environmental Benefits

- Fuel consumption
- Emissions

Determine Commercial Viability

- Effects on logistics schedules
- Vehicle maintenance
- Driver workload

Evaluate & Assess Impact

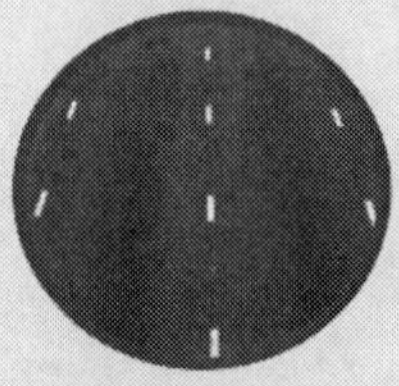

- Infrastructure
- Traffic management
- Platooning operators
- Human factors & behaviour

Acceptance of Technology & Standards

- Engage and educate general public
- Inform industry bodies
- Influence standards & regulations

ADVANCED PLATOONING TRIALS

HelmUK

The Profile

The Partners

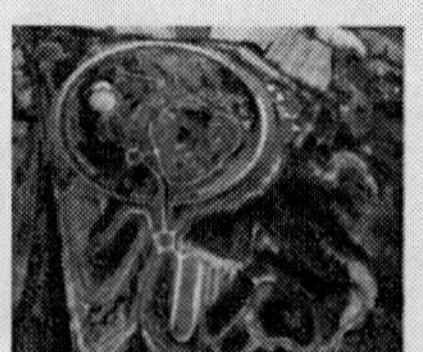

The Phases

The Planning

The Possibilities

Setting the Pace. Driving the Distance.

ADVANCED PLATOONING TRIALS

Strength through Partnership

Project Sponsors

Department for Transport | highways england

Consortium Lead

TRL

Consortium Partners

DAF A PACCAR COMPANY | DHL | CATAPULT | COSTAIN | RICARDO | MILLBROOK | TNO innovation for life | Apollo Vehicle Safety | TransportPR

Logistics, Operator, Behavioural Scientists, Technology, Data Analytics, Proving Ground, Engagement

Advisory Board

- International Experts
- Road Freight Community
- Road Operators
- OEMs
- Road User Groups
- Infrastructure Providers
- Road Safety Groups
- Emergency Services
- Communications & Electronic Security

HelmUK

Setting the Pace. Driving the Distance.

Timeline: Phased Approach

Phase 1	Phase 2	Phase 3	Phase 4
Detailed Planning, Risk Mapping & Track Testing	Initial Road Trial	Operator Trial	Analysis & Reporting

12 Months — 4 Months — 8 Months

Methodology: Phase 1

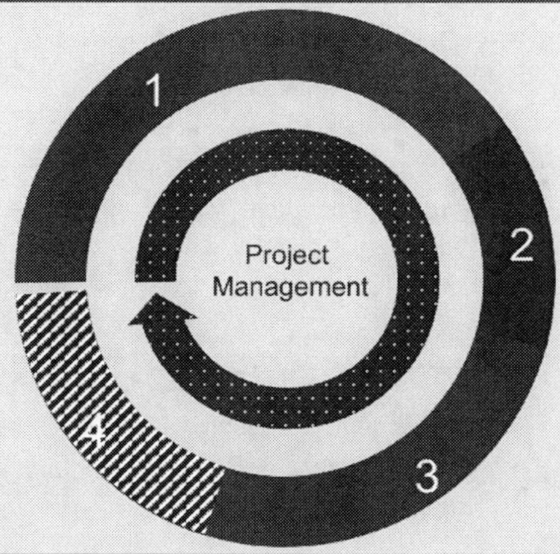

Detailed Planning, Risk Mapping & Track Testing

- Install platooning technology - twinned with the ENSEMBLE multi-brand EU platooning project
- Fit and test other monitoring equipment
- Verify system safety and functionality
- Measure fuel economy baseline
- Train drivers on track
- Select representative routes
  - Risk assessment of Strategic Road Network (SRN)
  - Assessment of commercial haulage routes

Methodology: Phase 1 – Simulators for testing and training

Example MiniDigiTruck Simulator

TRL's driving simulators will be used to:

- Train drivers
- Simulate scenarios which may be potentially dangerous to investigate in on-road conditions
- Measure the effects on other road users of marked or unmarked platoons
- Present lengthy platoons (>3 trucks) and measure how other road users respond
- Investigate obscuration of road signs
- Plus more …

New DigiCar Simulator

Methodology: Phase 2

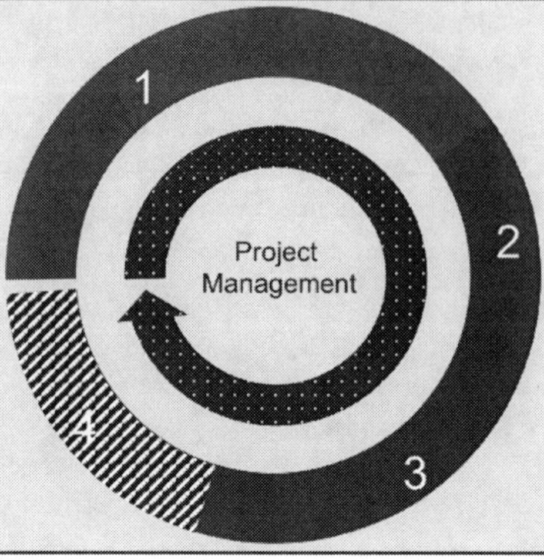

Initial Road Trial

- Verify platooning procedures
- Iteratively increase complexity of the on road operations in line with the safety case
    - Target: Three trucks with 0.5 sec gap
- Train drivers on road
- Assess ...
    - Driver workload
    - Other road user and vehicle behaviours
    - Fuel economy measurement on the road
    - Trial instrumentation and data capture
    - Effect on infrastructure

ADVANCED PLATOONING TRIALS

Methodology: Phase 3

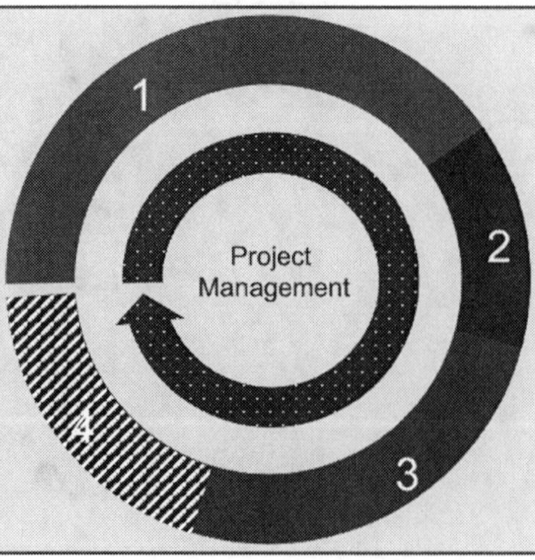

Operator Trial

- A commercial operator (DHL) will transport goods using three trucks fitted with platooning technology
- TRL will capture, store and manage data ...
    - Fuel economy 'in operation'
    - Safety proxies
    - Other road user and vehicle behaviours
    - Driver workload
    - Driver acceptance
    - Logistics operational performance
- Continuous safety monitoring

ADVANCED PLATOONING TRIALS

Methodology: Phase 3

140 journeys in platooning configurations

140 journeys in non-platooning configurations

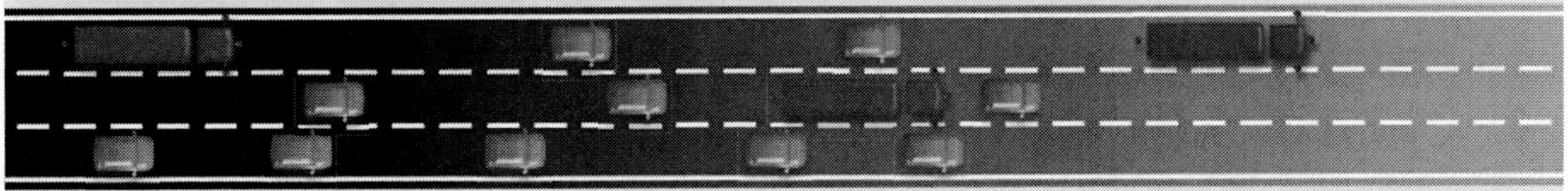

Only by running a trial of 280 journeys will enough data be captured to robustly provide the required input measures for the Impact Assessment

Methodology: Phase 4

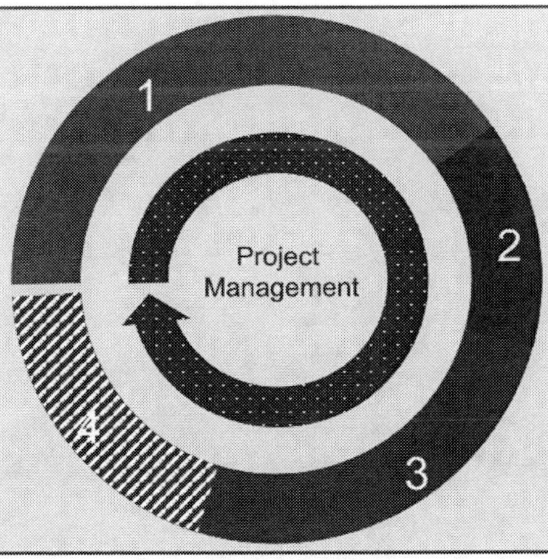

Analysis & Reporting

- Data analysis
- Safety case for platooning
- Business case for Platooning Service Provider
- Future Road Map
  - Technology companies
  - Haulage operators
  - Road authorities and regulators
  - + other stakeholders
- Active and full engagement of stakeholders
- Independent and robust findings
- **Real World Impact Assessment**

HelmUK

The Profile

The Partners

The Phases

The Planning

The Possibilities

Setting the Pace. Driving the Distance.

The Planning

Safety and risk management are paramount and are at the centre of every aspect of the platooning trials.

HelmUK

Setting the Pace. Driving the Distance.

ADVANCED PLATOONING TRIALS

The Possibilities

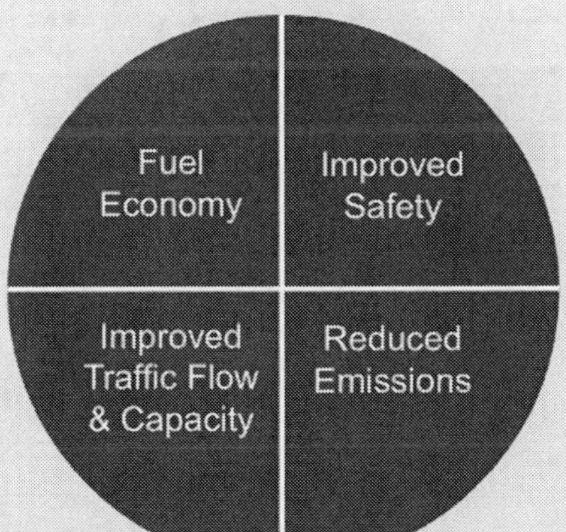

HelmUK is designed to evaluate potential benefits and dis-benefits of truck platooning, based on haulage operations that are as far as practicable representative of real world conditions:

- Drivers
- Technology, trucks and trailers
- Roads
- Commercial operations

ADVANCED PLATOONING TRIALS

The Possibilities

System is always 'on'

- No distraction
- Doesn't get tired
- Always alert

Safety & reliability

- Robust components
- ISO 26262

HelmUK

TRL THE FUTURE OF TRANSPORT

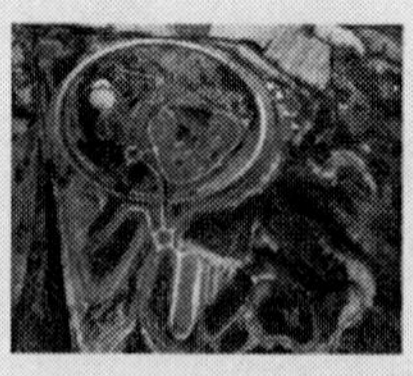

Richard Cuerden
Director, TRL Academy
Email: rcuerden@trl.co.uk
Twitter: @rcuerden_trl

Thank You

U.S. Department of Transportation
Federal Highway Administration

Office of Safety Research and Development

Human Factors Issues related to Truck Platooning Operations

Michelle Arnold, Ph.D., FHWA

Automated Vehicle Symposium
July 10, 2018

Truck Platoon Benefits

- Faster response time of following vehicles

↳ Shorter headway between trucks
  - Increases road capacity
  - Improves traffic flow
  - Reduces fuel consumption
  - Reduces CO^2 emissions

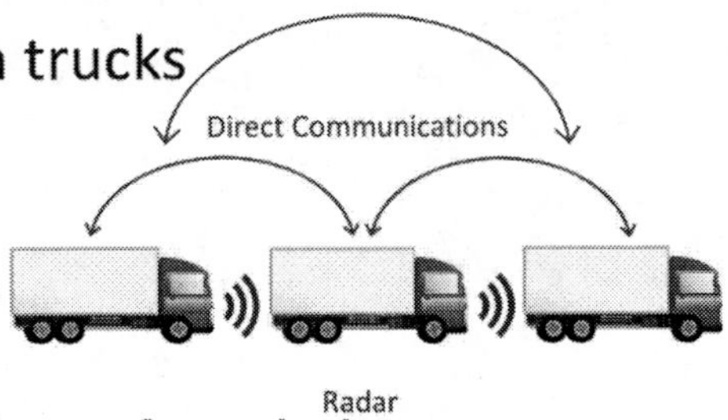

- Potentially could improve safety of truck drivers and road users
  - Prevents accidents caused by sleepiness/inattention and slow reaction speed

U.S. Department of Transportation
Federal Highway Administration

Research Goal

- Goal: investigate human factors issues associated with traveling near an automated truck platoon

- Main issues researched:
  - Freeway entry/exit
  - Visual indicator/other road user recognition of platooning
  - Prior knowledge of other road users

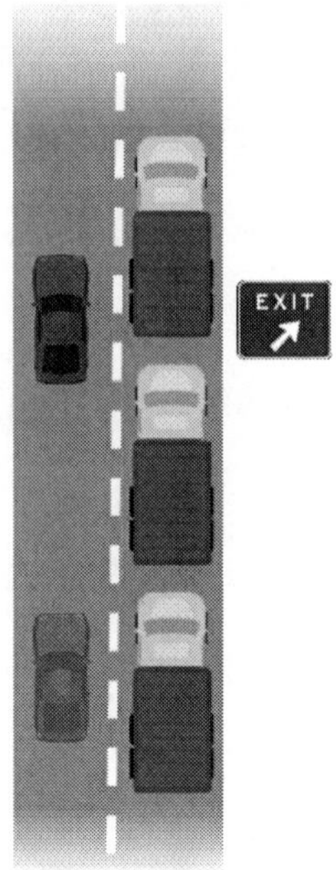

U.S. Department of Transportation
Federal Highway Administration

Study Progression

Preliminary → **Simulator** → **Test Track Letter Report**

Preliminary
- 1-2 Sign Lab Studies
- Test visual indicators
- Narrow down IV's & stimuli for MiniSim

Simulator
- 2 MiniSim studies
- Simulate merging, exiting scenarios

Test Track Letter Report
- High-level recommendations for feasible test track studies

U.S. Department of Transportation
Federal Highway Administration

Research Questions

1. Do automated truck platoons affect other road users' driving behavior?
   - In particular, do automated truck platoons affect other road users' ability to safely enter or exit the freeway?
2. Do automated truck platoons affect other road users' ability to reach an assigned destination (e.g., via sign occlusion, increased travel time, physical obstruction)?
3. Do changes in the characteristics of the automated truck platoon (e.g., gap distance, number of trucks) have a significant effect on driver behavior?
4. Does knowledge of automated platooning (in the form of provided explanations or signs identifying trucks) affect other road users' driving behavior?
   - If so, which forms of information and what level of detail are most useful to and/or easily comprehended by drivers?

U.S. Department of Transportation
Federal Highway Administration

Current Status

- Research began in 10/2017
- 30 month research project; complete in 2020

U.S. Department of Transportation
Federal Highway Administration

U.S. Department of Transportation
Federal Highway Administration
Office of Safety Research and Development
THANK YOU

Peloton

Peloton Technology
Overview on Driver-Assistive Truck Platooning
AVS18, July 2018

PELOTON TECHNOLOGY

Presentation Roadmap

- **Market Overview:** Global movement to market of Driver-Assistive Truck Platooning
- **Details of the Peloton PlatoonPro System:** Enhancing Driver Teamwork, Safety & Efficiency
- **Regulatory Picture:** Growing Allowance for Commercial Deployment of Truck Platooning

Peloton 2

PELOTON TECHNOLOGY

Accelerating Global Activity in Truck Platooning

EU Platooning Challenge 2016 & ENSEMBLE Multi-Brand Platooning 2019+

Canada - PIT 2009

Examples of Global Activity that span North America, Europe & Asia

EU (Sweden) - SARTRE 2009+

US – PATH, NREL, etc. '90s and ongoing

Germany – KONVOI 2005-09

US – Peloton, Navistar, Freightliner, other NA OEM Platooning activity 2013 -present

Peloton 3

PELOTON TECHNOLOGY

Driver-Assistive Truck Platooning Market Overview

Many Companies in US, Europe, and Asia Involved with Bringing Truck Platooning to Market

Peloton
Continental
DELPHI
DAF A PACCAR COMPANY
VOLVO
NAVISTAR
DAIMLER
SCANIA
NXP
MAN
IVECO
DENSO
HINO
Bendix
WABCO
Peterbilt A PACCAR COMPANY
KENWORTH A PACCAR COMPANY
Cummins

Peloton 4

PELOTON TECHNOLOGY

Peloton System: Driver Teamwork Not "Self-Driving" Trucks

Driver-Assistive Safety and Efficiency System:

- **All Drivers remain engaged at all times.**
- **Only pairs of trucks:** 2 drivers teaming up.
- **Trucks require best available Safety Systems:** Collision Avoidance, Disc brakes on all tractor axles.
- **Enhanced driver teamwork:** dedicated driver-to-driver radio, shared video, linked safety systems.
- **Limited to suitable conditions:** multi-lane divided, limited access highways; suitable weather & traffic.
- **Integrated with 3 of 4 Major NA Truck OEMs**

Peloton PLATOONPRO

Peloton 5

Peloton PlatoonPro: Driver Assistance, Not High Automation

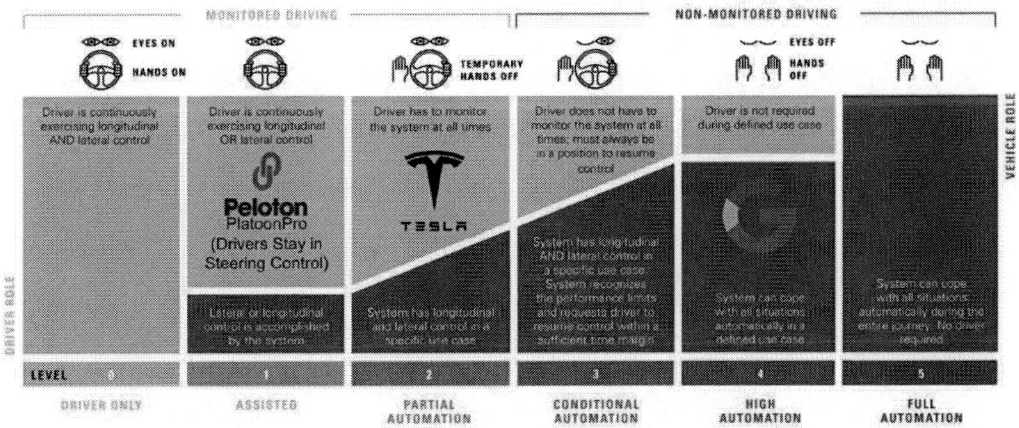

Graphic courtesy: ZF TRW

PELOTON TECHNOLOGY

We Start by Making Individual Trucks Safer at All Times

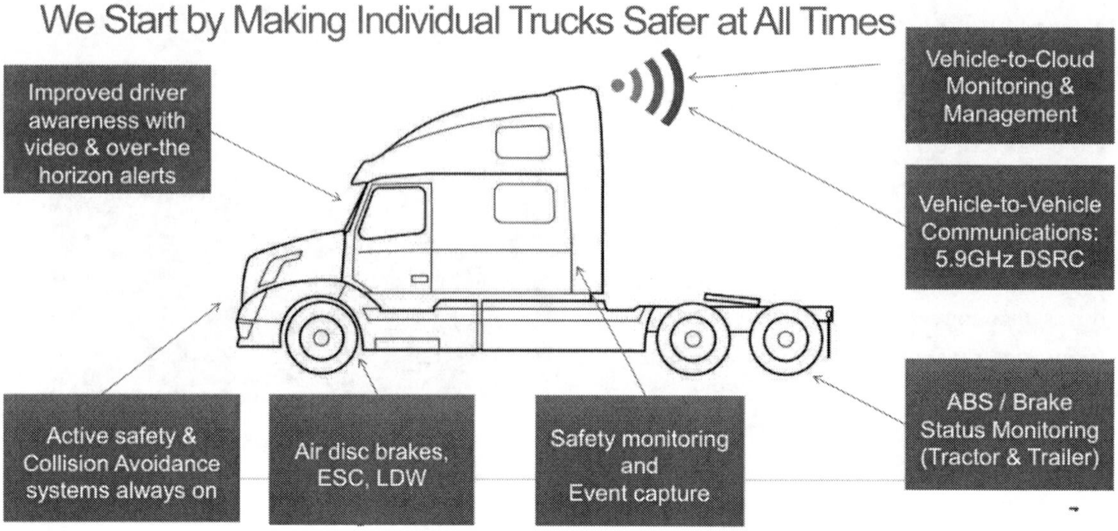

PELOTON TECHNOLOGY

Pairs of Trucks, Both Drivers Steering At All Times

Driver Teamwork via Linked Adaptive Cruise Control

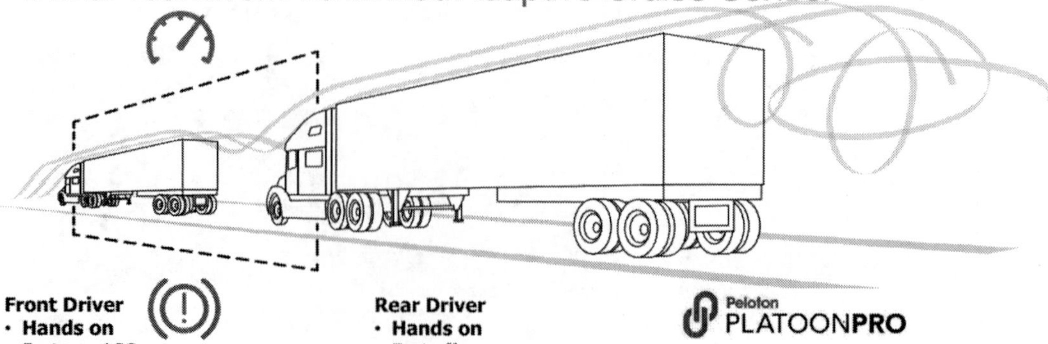

Front Driver
- **Hands on**
- Feet on+ACC
- Eyes/Mind on

Rear Driver
- **Hands on**
- Feet off
- Eyes/Mind on

Peloton PLATOONPRO

Peloton 8

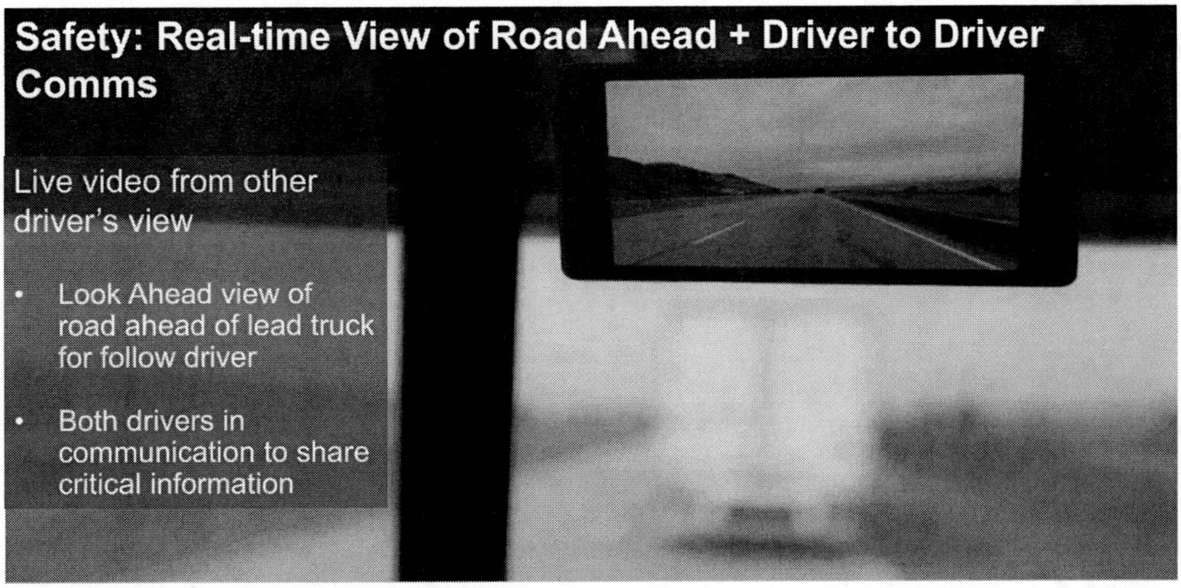

PELOTON TECHNOLOGY

Making Aerodynamic Following Safe: V2V linked Safety

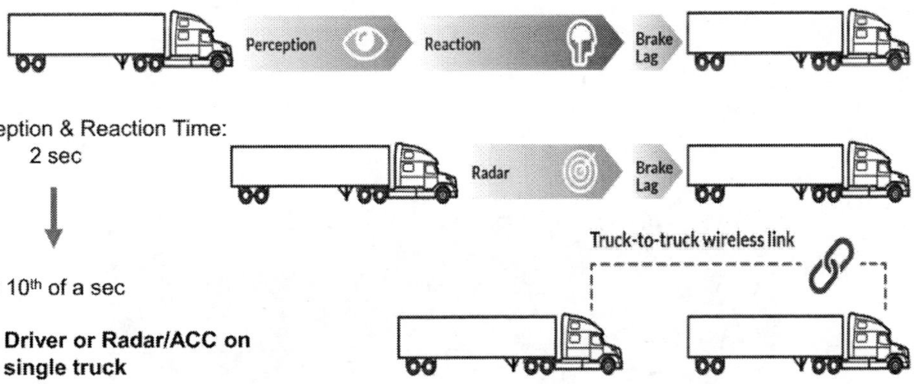

Driver Perception & Reaction Time:
2 sec

↓

< 10th of a sec

Faster than a Driver or Radar/ACC on a single truck

Peloton 10

PELOTON TECHNOLOGY

Safety: Handling Vehicle Cut-ins

Driver sees car cutting in and backs off
OR

If driver does not respond, system radar detects cut-in vehicle and automatically begins to back off follow truck

Follow truck will continue to back off to safe manual following distance (100+ ft) and then give full manual control back to follow driver

Peloton

11

PELOTON TECHNOLOGY

Safety: Only Enabled for Suitable Roads & Conditions

Peloton Network Operations Cloud (NOC) limits platooning to:

- Multi-lane, divided, limited access highways
- Moderate or low traffic conditions
- Good traction conditions (no heavy rain, sleet, ice or snow)
- Appropriate topography (good line of sight; no steep grades)

NOC provides over-the-horizon alerts to drivers on roadway conditions

Peloton

12

US Freight Arteries
Platooning focus: Multi-lane, divided, limited access highways

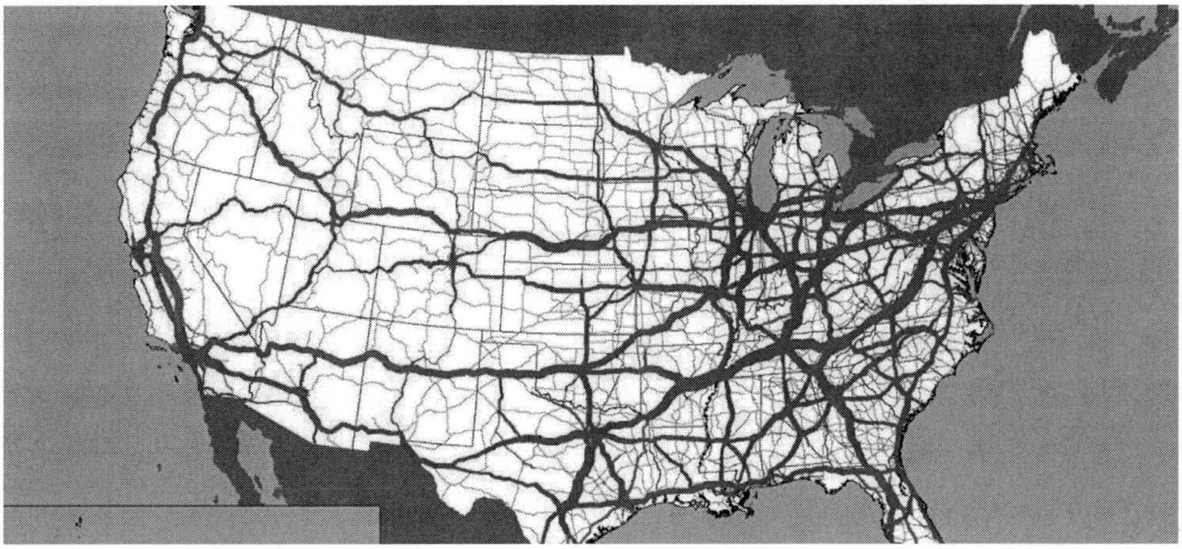

PELOTON TECHNOLOGY

Joint Integration & Safety Validation with Partners

Bendix

Peloton

Cummins

PACCAR

VOLVO

NAVISTAR®

Peloton

14

PELOTON TECHNOLOGY

Solid Demand From Top Fleets (including these and others)

FedEx Freight | Hub Group | SCHNEIDER

UPS | MB MARTIN BROWER | PTI PAPER TRANSPORT, INC.

J.B. HUNT | SWIFT | DOT

Peloton 15

PELOTON TECHNOLOGY

Progress on Allowance of Truck Platooning in the US

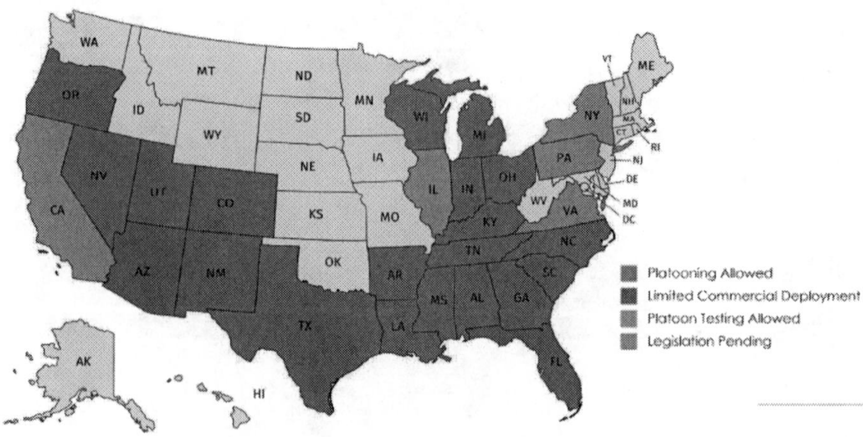

16

PELOTON TECHNOLOGY

Outlook

General Activity

- US: Peloton with several OEMs and Freightliner each bringing platooning into commercial ops
- International activity including:
  - EU Multi-brand platooning project: Platooning by all 6 European Truck Manufacturers, 2019+
  - Other European OEMs gearing up for commercial activity
  - Platooning activity being explored by Asian OEMs in Singapore, South Korea and China
  - Platooning projects have launched in the UK and Australia

Key Activity Ahead - Peloton

- Robust activity continues in CA as we complete integrations and final safety validation with OEMs
- Major fleet customer truck integration and trial preparations underway
- Commercial Freight activity scheduled for 2nd half of 2018
- Working with additional states on legislative and administrative allowance for platooning

Peloton 17

PELOTON TECHNOLOGY

Backup Slides

Peloton 18

PELOTON TECHNOLOGY

PlatoonPro: View from Follow Driver's Perspective

Video Presentation in
Original e-media

Peloton

19

PELOTON TECHNOLOGY

Detailed Video on Linked Safety & Braking

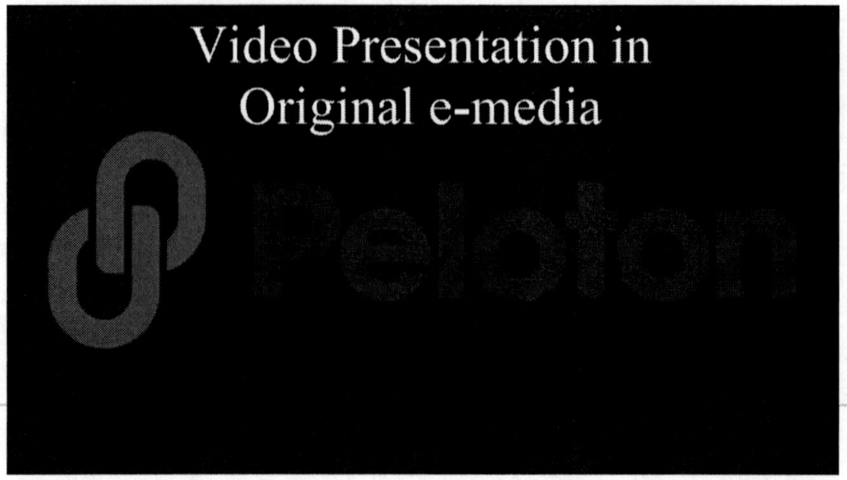

20

Auburn's Driver Assisted Truck Platooning

AUBURN
UNIVERSITY

David M. Bevly
Professor, Auburn University
GPS and Vehicle Dynamics Lab
dmbevly@eng.auburn.edu
(334) 844-3446

GAVLAB
GPS and Vehicle Dynamics Lab

Driver Assisted Truck Platooning

http://eng.auburn.edu/~dmbevly/FHWA_AU_TRUCK_EAR/

AUBURN
UNIVERSITY

- Sponsored by FHWA ($1.5M)
  - Demonstrate and evaluate prototype CACC System
    - Reduce fuel and emissions
    - Improve safety
    - Improve highway throughput
  - Show that CACC technology is ready for industry use and is valuable (~$1B-$5B in fuel savings)
- Investigating & report on multiple aspects
  - ME (lead) – automation and performance evaluation
  - ISE – logistics, HMMI, business analysis
  - CS – communications and cyber
  - AE – aerodynamic modeling
  - CE – traffic flow

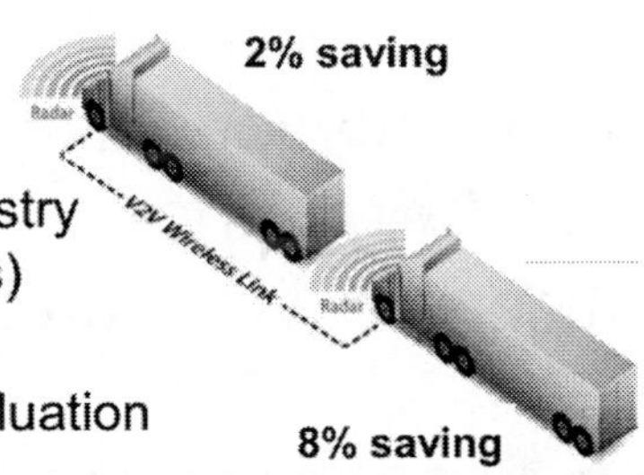

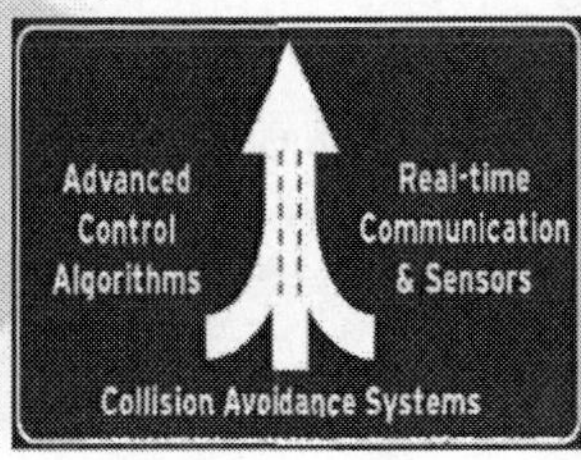

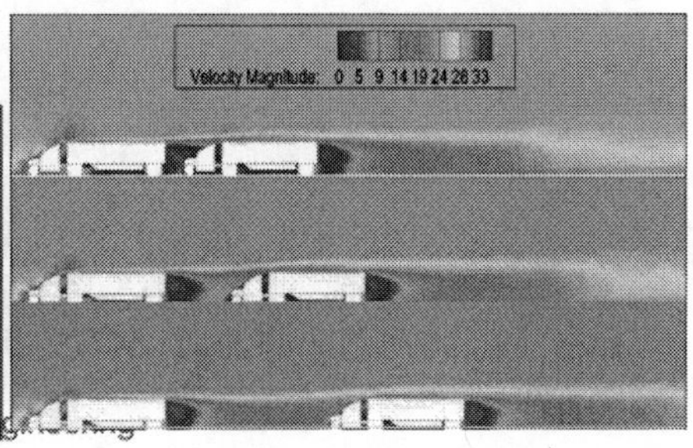

Military/Commercial Truck Convoy

- Sponsored by TARDEC
  - Demonstrate 4 vehicle convoy comprised of military and commercial heavy trucks
  - Show interoperability of DSRC V2V communication between Bosch and Denso radios
  - Investigate optimal following distances and convoy ordering

DATP Research Approach

- Utilize CFD analysis, simulation and experimentation for development and verification of research

Two Truck Platoon Fuel Savings

AUBURN UNIVERSITY

- Average of 5-7% fuel savings per truck

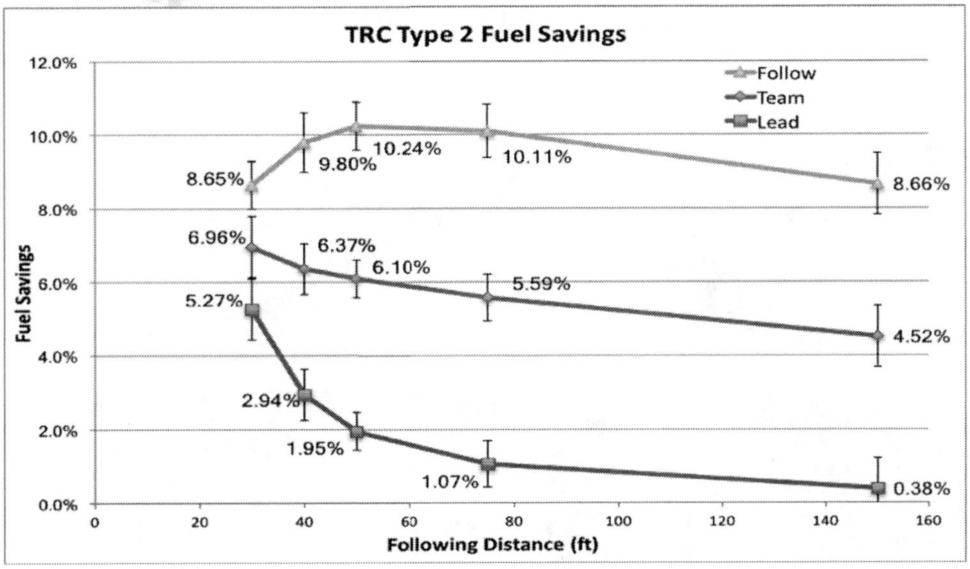

Samuel Ginn College of Engineering

Lateral Offset CFD Simulation Results

AUBURN UNIVERSITY

- CFD shows similar trends seen in decreased drag reduction on following truck at closer spacing
- Need to perform experiments to verify (we have been wrong before)

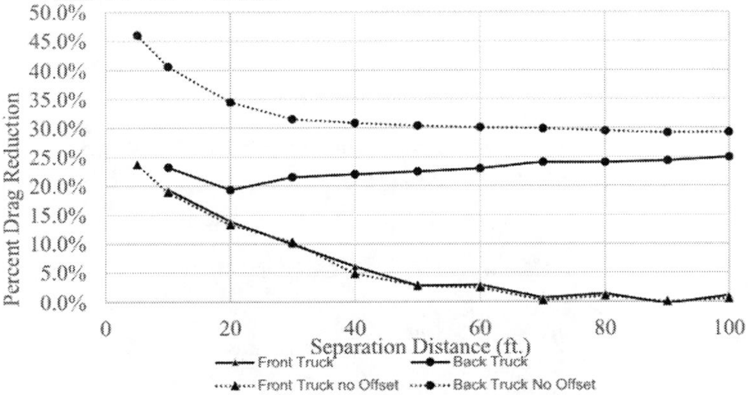

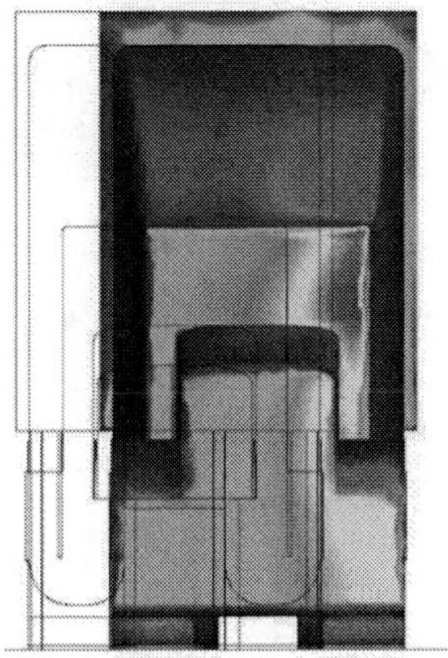

Samuel Ginn College of Engineering

Truck Setup

AUBURN UNIVERSITY

- Two Peterbilt 579 trucks
- Two GPS Receivers(Leader and Follower) 
- DSRC V2V Communication (Leader and Follower)
- Delphi ESR (Follower) 

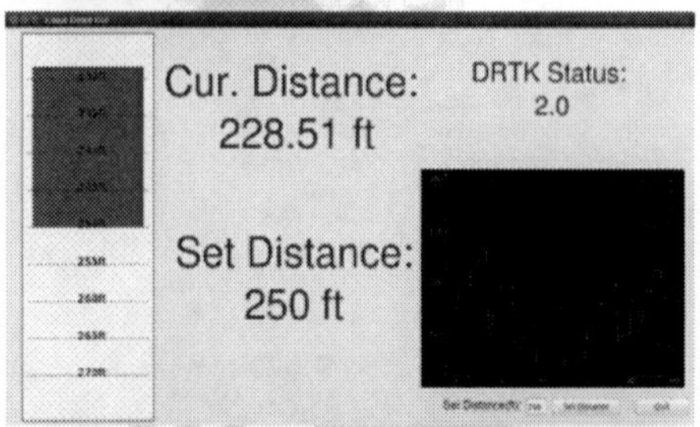

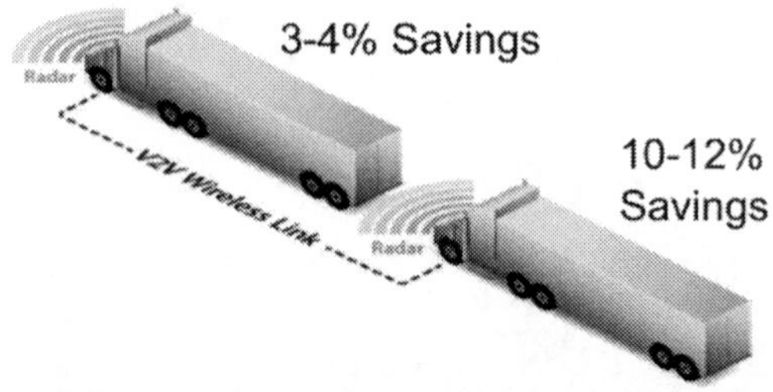

DRTK Aiding of Ranging Radar

AUBURN UNIVERSITY

Delphi ESR Reported Range Measurements

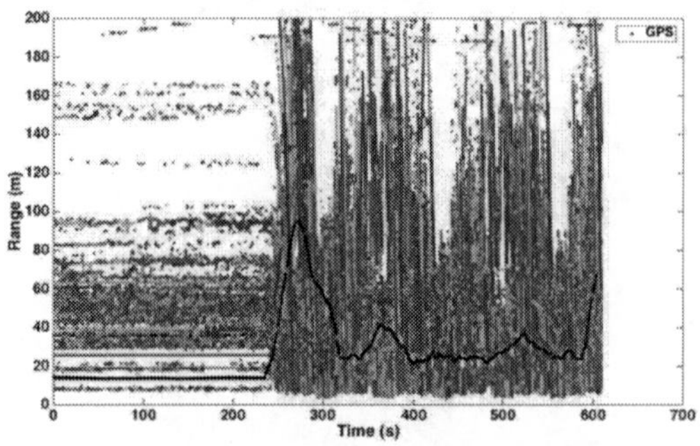

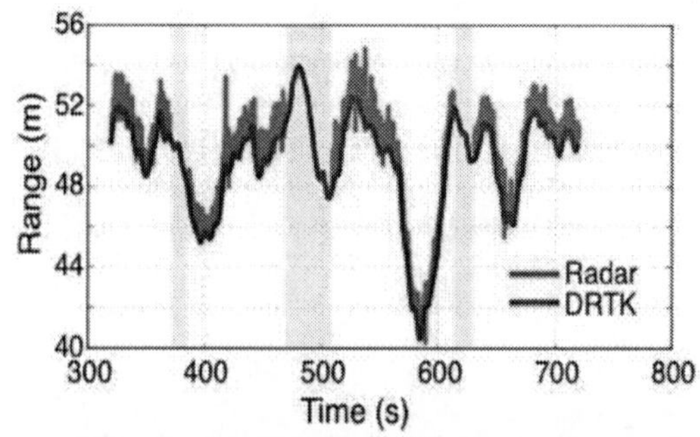

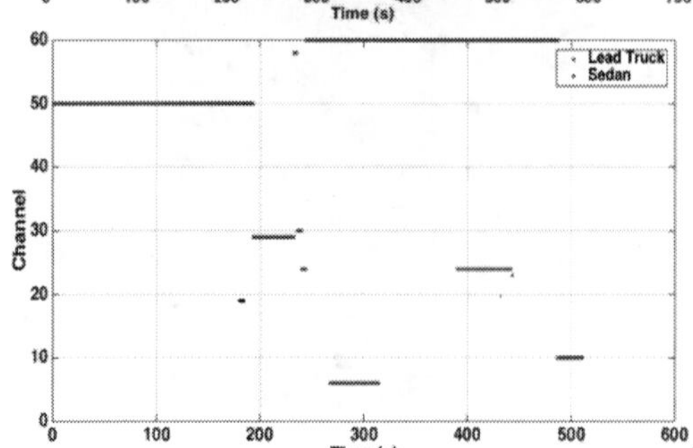

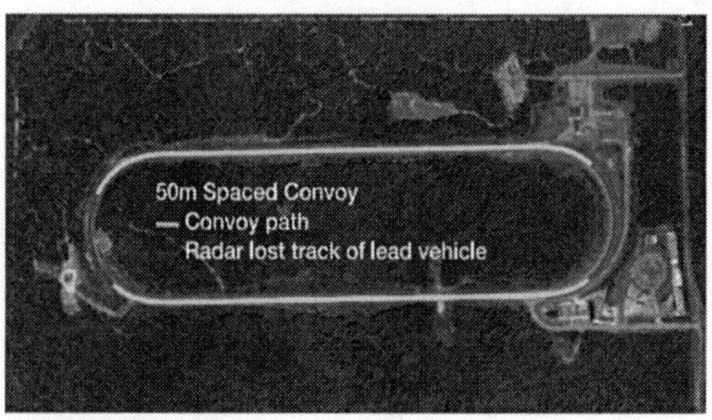

Experimental Runs

AUBURN UNIVERSITY

Experiments run on tests tracks, I-69 in Michigan, and Blue Water Bridge into Canada

Samuel Ginn College of Engineering 9

http://gavlab.auburn.edu

AUBURN UNIVERSITY

David M. Bevly
Professor
Department of Mechanical Engineering
Auburn University, AL 36849-5341

Director of Auburn University's
GPS and Vehicle Dynamics Lab

GAVLAB

DAIMLER

Daimler Truck Platooning

Ritchie W. Huang
Manager, Engineering & Safety
Daimler Trucks North America

The Daimler Family

Daimler is the global leader in mobility which includes the manufacturing of heavy trucks, buses, and luxury vehicles

DAIMLER

WESTERN STAR TRUCKS · FREIGHTLINER® · Mercedes-Benz · FUSO · smart · SETRA · DETROIT · Thomas Built Buses · AMG · moovel · RIDESCOUT™ · CAR2GO

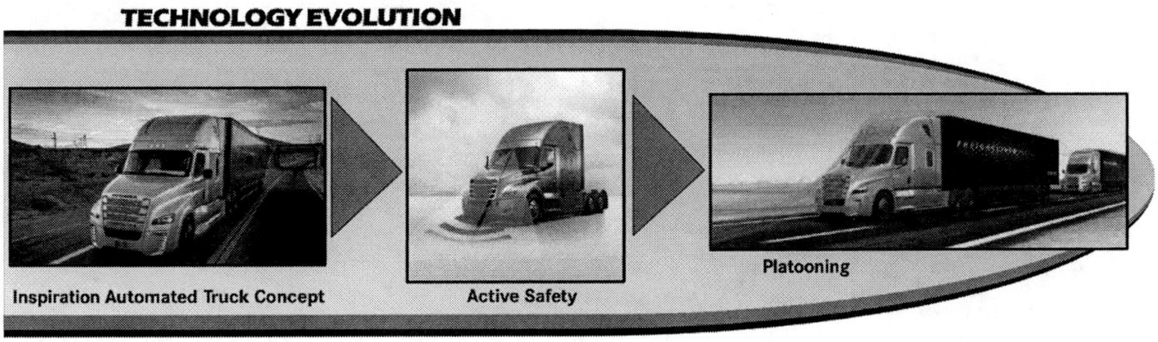

DAIMLER

Truck Platooning/Pairing

Safety

- Increased sensing capability
- Coordinated Braking
- Reduce Driver fatigue
- Active lane assist

Connectivity

- V2V Broadcast & Integration
- Infrastructure Integration
- Additional safety features

Efficiency

- Lower operating cost
- Lower emissions
- Greater range
- Congestion reduction

DAIMLER

V2V Enables Close Following By Reducing Reaction Times

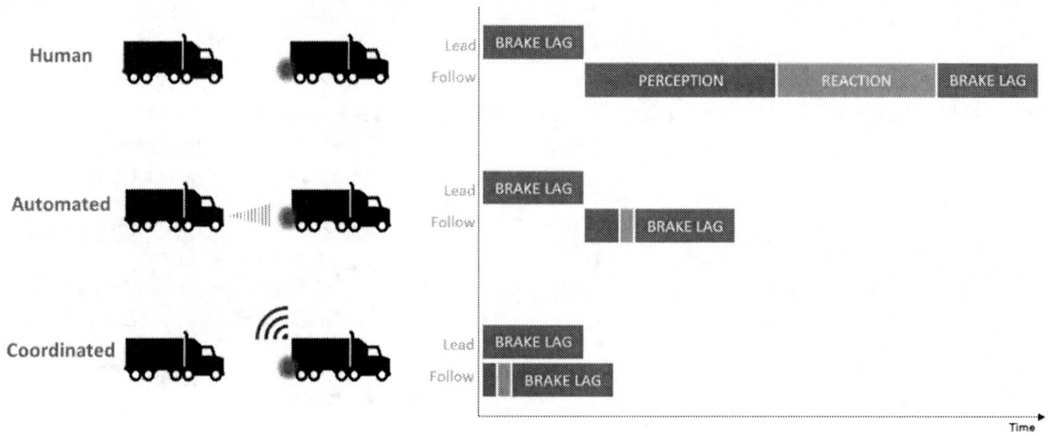

- 45% of braking distance depends on ***human perception and reaction***
- 38% of braking distance reduced by coordinated braking using DSRC

DAIMLER

Technical Design Challenges for Safe Implementation

- Vehicle positioning & identification
- Vehicle cut ins - allowing space for merging or overtaking vehicles
- Emergency braking scenarios
- Evasive maneuvers
- Environmental interference
- Single vs Multi-Lane use
- Coordinated Braking Strategy
- Functional safety

Platooning system safety tests being conducted at Daimler's closed course facility in Madras, OR

DAIMLER

Alignment and Collaboration with Regulatory Organizations

Regulatory Considerations for testing and deployment:

- Location restrictions (e.g. outside business district, city limits)
- Following distance and caravan laws
- Implications for bridge loading
- Liability
- External Visual indication (e.g. lights, signage)

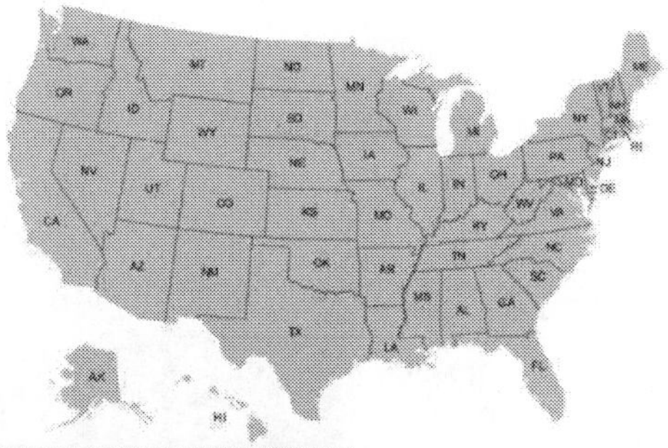

DAIMLER

Daimler Truck Platooning / 2018/ Page 7

Conclusions

- Vehicle Connectivity enables opportunities for safety and efficiency improvements
- Requires detailed system knowledge & deep system integration
- Platooning function leads to highly stringent safety goals and system requirements

... and don't forget ... Testing, Testing, Testing

Safety is our top priority!

DAIMLER

Daimler Truck Platooning / 2018/ Page 8

DAIMLER

AUTOMATED VEHICLES SYMPOSIUM 2018

BREAKOUT SESSION

BREAKOUT SESSION 15
Trucking Automation: Deployment Challenges and Opportunities

AUTOMATED VEHICLES SYMPOSIUM 2018

Predicting the future of automated trucking

Last year's session

- Key takeaways
  - Several strong business cases for automated trucking
  - Key deployment challenges: Safety assurance, public acceptance, human factors...
  - Trucking industry operates on small margins and requires fast and reliable return-on-investment
  - The return on investment from automation depends critically on the specific type of operations to be automated
  - Trucking industry is diverse and trucking operations are heterogeneous
  - Business landscape is changing with technology

Hard to predict how things will play out!

Engström, J., Bishop, R., Shladover, S., Murphy, M., O'Rourke, L., Voege, T., Denaro, B., Demato, R. and Demato, D. in press. Deployment of automated trucking: Challenges and opportunities. To appear in G. Meyer and S. Beiker (Eds.): *Road Vehicle Automation 5*. London: Springer

AUTOMATED VEHICLES SYMPOSIUM 2018

Agenda: Day 1: Truck Platooning

AUTOMATED VEHICLES SYMPOSIUM 2018

1:30 PM – 1:45PM Introduction (Johan Engström, VTTI)

1:45 PM – 3:00 PM Presentations: Current state-of-the-art and key deployment issues
Moderator: Johan Engstrom, VTTI

Speakers:
- *Overview of the current state-of-the-art on truck platooning* - **Steven Shladover, California PATH**
- *Real world truck platooning trials – an impact assessment for the UK* - **Richard Cuerden, TRL**
- *The Texas commercial truck platooning project* - **Mike Lukuc, Texas Transportation Institute**
- *Evaluation and testing of driver-assitive truck platooning* - **David Bevly, Auburn University**
- *Multi-brand truck platooning in Europe: The ENSEMBLE project* - **Maurice Kwakkernaat, TNO**

3.00 PM – 3:20 PM Break

3:20 PM – 5.00 PM Panel: Opportunities and challenges for the deployment of truck platooning applications

Moderator: Gene McHale, FHWA

Panelists:
- Steve Boyd, Peloton
- Michelle Arnold, FHWA
- Aravind Kailas, Volvo Group
- Tom Byron, Florida DOT
- Jim Epperson, CA Highway Patrol
- Ritchie Huang, Daimler

5.00 PM – 5:30 PM Wrap up Johan Engström (VTTI)

Agenda: Day 2: Stand-alone trucking automation applications

1:30 PM – 1:45PM Introduction (Andrew Krum, VTTI)

1:45 PM – 3.00 PM Presentations: Current state-of-the-art and key deployment issues

Moderator: Andrew Krum

Speakers:
- *Overview of the current state-of-the-art on stand-alone automated trucking applications* - **Richard Bishop, Bishop Consluting**
- *The last mile and yard operations* – **Steve Dellenback, SwRI**
- *Safety assurance for truck automation: Learning from ongoing passenger cars initiatives* - **Adrian Zlocki, FKA**
- *Effect of automated trucks on the US economy: Summary from an NSF workshop* - **Jeff Hickman, VTTI**
- *Building autonomous trucks* - **Vijaysai Patnaik, Waymo**

3:00 PM – 3:20 Break

3:20 PM – 5:00 PM Panel: Opportunities and challenges for the deployment of stand-alone trucking automation applications

Moderator: Richard Bishop, Bishop Consulting

Panelists:
- Kelly Regal, FMCSA
- Johnny Morris, Embark
- Mike Cammisa, ATA
- Stefan Seltz-Axmacher, Starsky Robotics
- Tommy Ruke, Motor Carrier Insurance Education Foundation
- Chuck Price, TU-Simple
- Paul Konasewich, PACCAR

5:00-5.30 PM Wrap-up & Conclusions (Richard Bishop, Bishop Consulting)

Assistant Chief
Jim Epperson
California Highway Patrol

AUVSI

Background

- American Association of Motor Vehicle Administrators (AAMVA) Autonomous Vehicle Working Group
- Commercial Vehicle Safety Alliance (CVSA) Enforcement and Industry Modernization Committee
- California Transportation Agency Working Group
- California Highway Patrol Working Group
- Oversee CHP Commercial Program

AUVSI

Short Term AV Issues

- Statutory Issues
  - Allowed if not Prohibited
  - Driver  Vs Driver 
- Response Issues
  - Enforcement
  - Crash investigation

AUVSI

Long Term AV Issues

- Systems management
  - Technology becomes the job
- Cybersecurity

- Sensor security
- Infrastructure (connected vehicle) security

AUVSI

Federal Oversight

- NHTSA
  - FMVSS
  - OEM
  - Remanufactures
- FMCSA
  - Commercial Vehicles
  - Equipment
  - HOS application
- FCC

AUVSI

Non-Federal Jurisdictions

- Aftermarket equipment of non-manufacturers
- Rules of the road
- Non-FMVSS covered items
- Infrastructure
- End user requirements
  - Maintenance
  - Updates
  - Operator Licensing
  - Registration

VSI

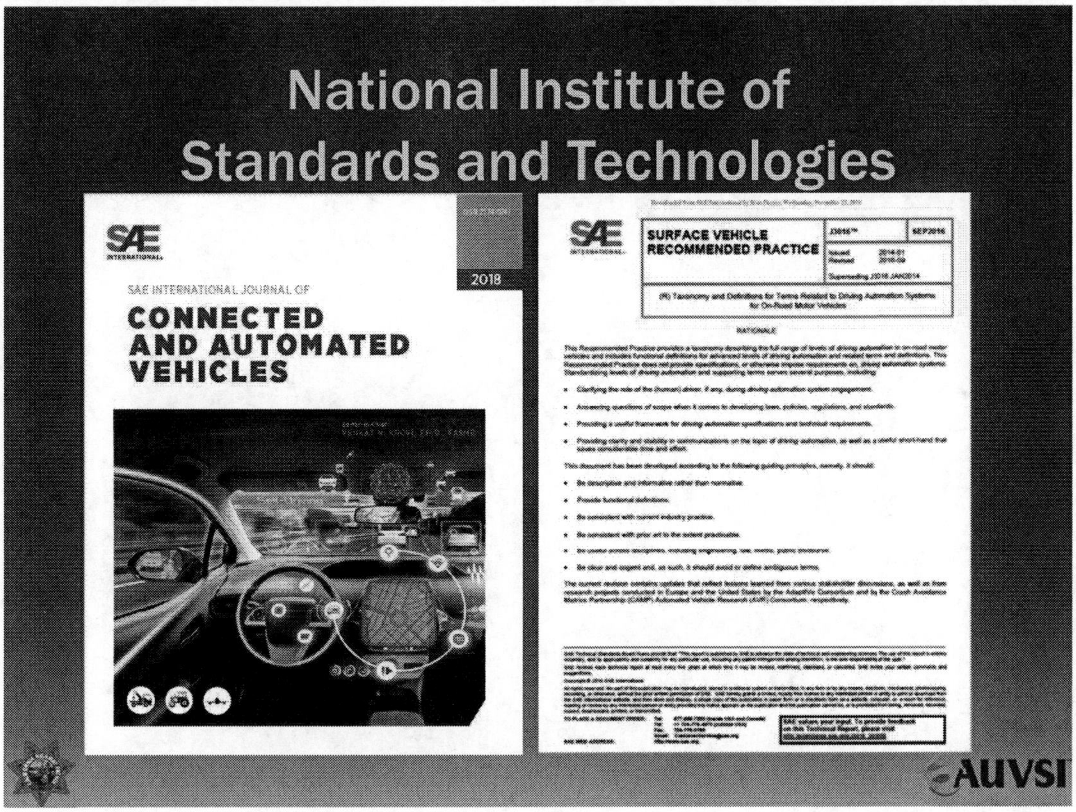
National Institute of
Standards and Technologies
SAE INTERNATIONAL
2018
SAE INTERNATIONAL JOURNAL OF
CONNECTED
AND AUTOMATED
VEHICLES
SAE INTERNATIONAL
SURFACE VEHICLE
RECOMMENDED PRACTICE
RATIONALE
AUVSI

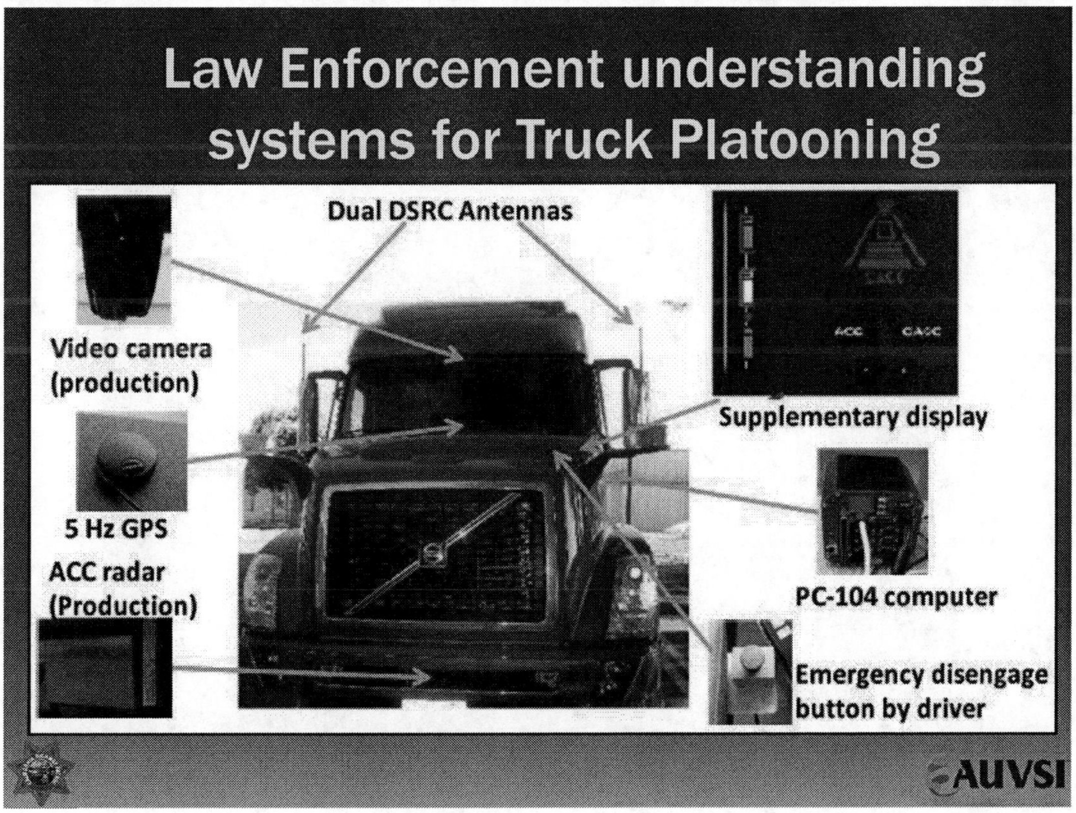
Law Enforcement understanding
systems for Truck Platooning
Dual DSRC Antennas
Video camera
(production)
5 Hz GPS
ACC radar
(Production)
Supplementary display
ACC
CACC
PC-104 computer
Emergency disengage
button by driver
AUVSI

Law Enforcement Considerations with Platooning

- Vehicle identification
- System engagement
- Crash investigation
- Inspection/Operational ability
- Cross-vendor standards
- Load Differences
- Driver engagement

AUVSI

Study of Truck Platooning

- Project work plan 2014-2017
- Develop & test Cooperative Adaptive Cruise Control
- Partnership between CalTrans, UC Berkeley PATH, Volvo & Cambridge Systematics, CHP
- Funded through FHWA Exploratory Advanced Research Program
- V2V and V2I for traffic management

AUVSI

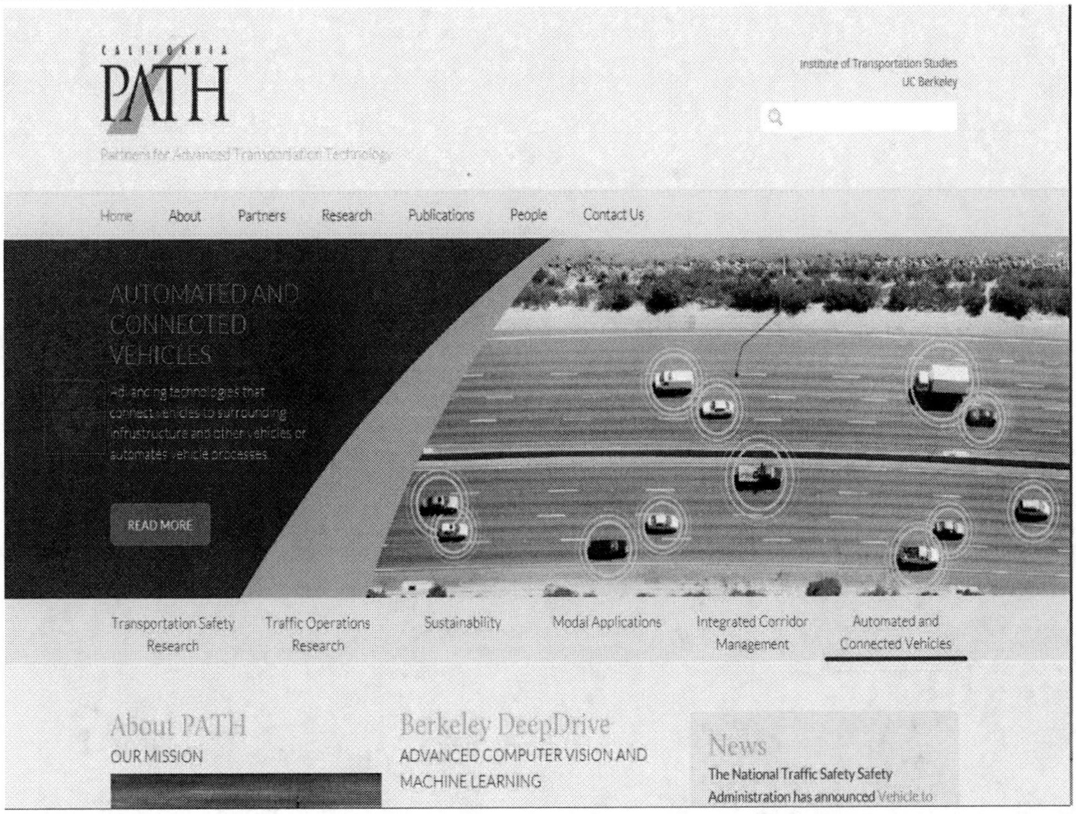
CALIFORNIA
PATH
Partners for Advanced Transportation Technology
Institute of Transportation Studies
UC Berkeley
Home
About
Partners
Research
Publications
People
Contact Us
AUTOMATED AND CONNECTED VEHICLES
Advancing technologies that connect vehicles to surrounding infrastructure and other vehicles or automates vehicle processes.
READ MORE
Transportation Safety Research
Traffic Operations Research
Sustainability
Modal Applications
Integrated Corridor Management
Automated and Connected Vehicles
About PATH
OUR MISSION
Berkeley DeepDrive
ADVANCED COMPUTER VISION AND MACHINE LEARNING
News
The National Traffic Safety Safety Administration has announced Vehicle to

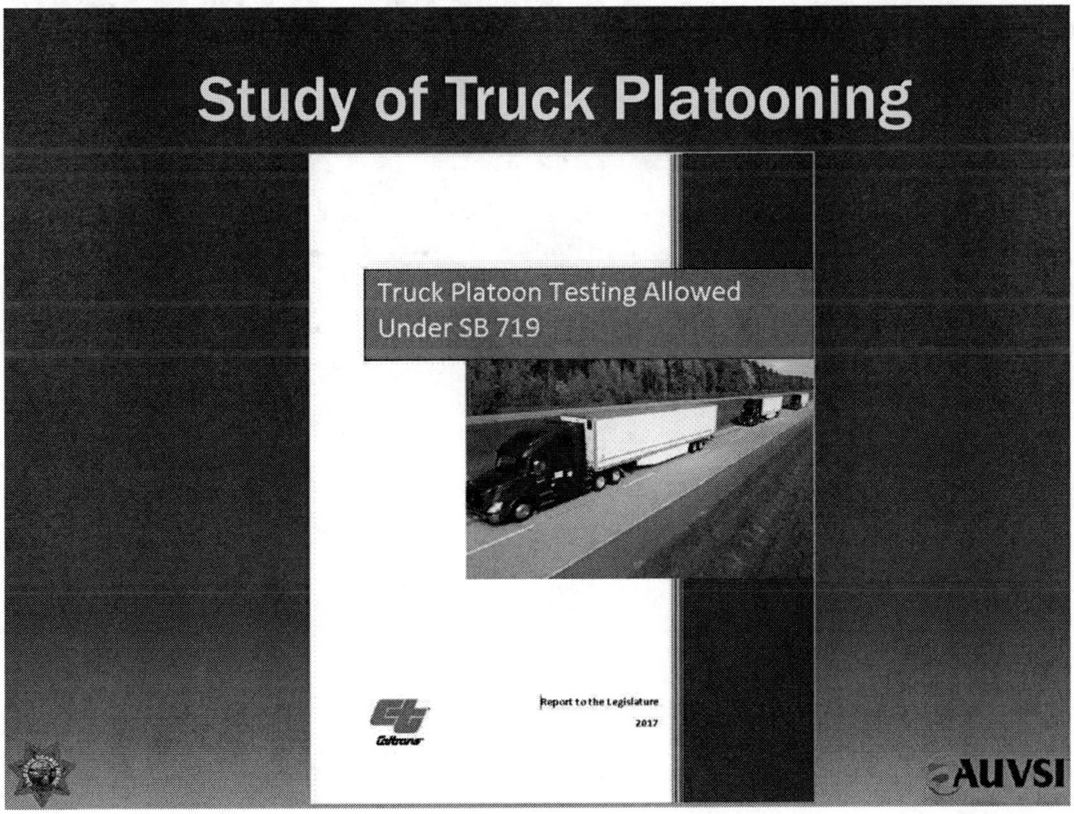
Study of Truck Platooning
Truck Platoon Testing Allowed Under SB 719
Report to the Legislature
2017
Caltrans
AUVSI

Jim Epperson
Assistant Chief

Enforcement and Planning
Division

California Highway Patrol
601 N. 7th Street
Sacramento, CA 95811
(916) 843-3330
JEpperson@chp.ca.gov

Truck Platooning:
State of the Art Review

Steven E. Shladover, Sc.D.
California PATH Program (Retired)
University of California, Berkeley
AVS18 Truck Automation Breakout Session #15
July 10, 2018

CALIFORNIA PATH

1

What are the benefits of truck platooning?

- Significant energy savings from aerodynamic drafting
- More stable vehicle following dynamics, reducing traffic flow disturbances and thereby saving additional energy and emissions
- Increased lane capacity and reduced congestion from improved traffic dynamics and shorter gaps
- *(Potential)* safety improvements
- *(When Level 3 automation of followers becomes feasible)* Improvement in truck driving working conditions, with more diverse assignments for drivers
- *(When Level 4 automation of followers becomes feasible)* Reduced need for truck drivers in followers

CALIFORNIA PATH

2

Technological Enablers

- Adaptive cruise control (forward ranging sensor, plus engine, braking and transmission control)
- Fast, highly reliable V2V communication
- Informative driver-vehicle interface
- Reliable, early detection of and response to cut-in vehicles
- (For L2+) Lane position detection and automatic steering control
- (For L3+) Central supervision, I2V comm.
- (For L4) Extensive safety assurance + possibly dedicated, segregated truck lanes

CALIFORNIA PATH

3

L1 Truck Platooning State of the Art

Automated longitudinal control only

- Cooperative ACC as first step (pre-platoon)
  - V2V peer-peer communication/coordination
  - Ad-hoc joining and leaving
  - Constant time-gap following
- L1 Platooning
  - Add coordination/supervision (central or by leader)
  - May extend to constant clearance distance gap and shorter distances
- Many R&D projects with homogeneous fleets, and ENSEMBLE project extending to multiple brands
- Active product development by several companies

CALIFORNIA PATH

4

Fuel Savings per Truck by Position and Average of 3 Trucks in L1 Platoon

Tests under U.S. DOE SMART Mobility Program and Transport Canada sponsorship by PATH, Volvo Group, National Research Council of Canada and NREL (2017)

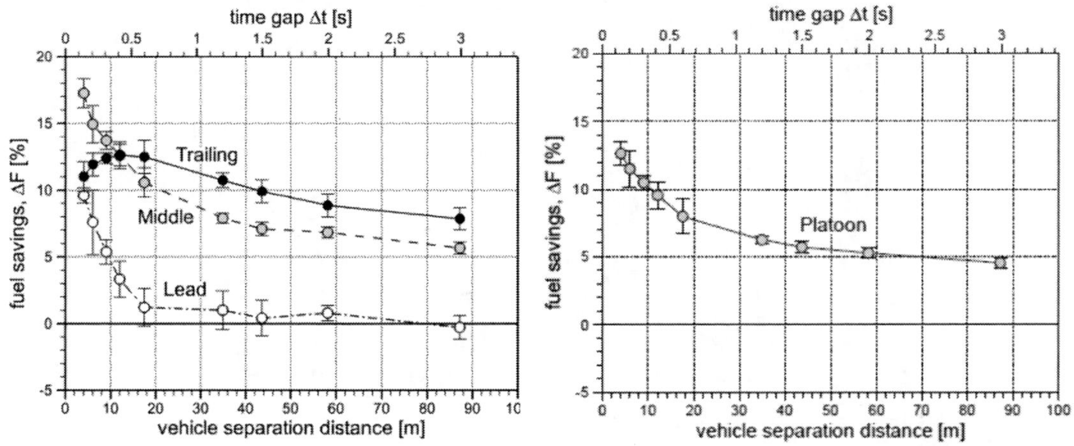

PATH

5

Summary of Main Fuel Consumption Test Results for 3-Truck Platoon

- Average savings of 3 trucks ranged from 4.5% at 87 m to 13% at 4 m, significantly better than 2 trucks at comparable gaps
- Even at 87 m, the second and third trucks saved 6% and 8% compared to solo driving, so trucks are already getting some drafting benefits now
- Cut-ins every 2 miles reduced fuel savings by only 1% - 1.5%
- Following light-duty vehicles can produce some fuel savings (1% - 2%)
- Long combination vehicle (two 53-ft. trailers) carrying same trailer load behind one tractor saved 28% energy compared to two solo tractor-trailer rigs

CALIFORNIA PATH

6

Driver Preferences for Platoon Gaps

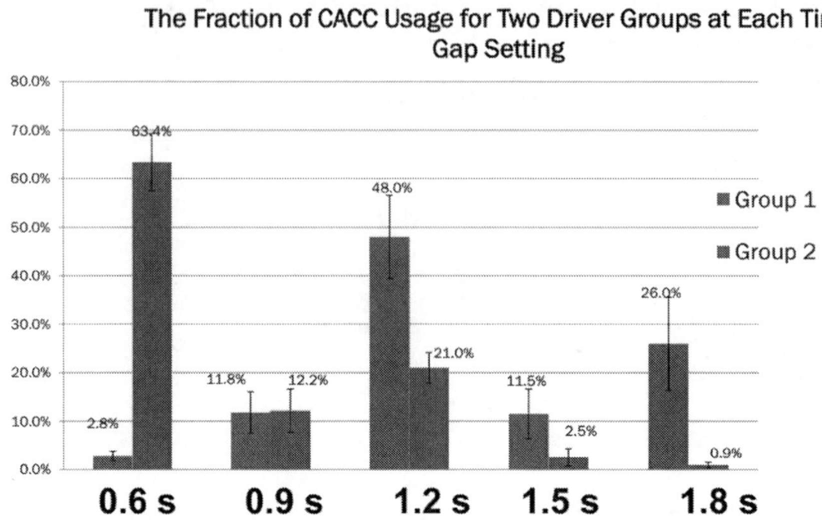

Group 1: More cautious, less experienced
Group 2: More aggressive and experienced

CALIFORNIA PATH

7

L2 Truck Platooning State of the Art

L1 platooning + automatic steering control

- **Automatic steering likely necessary for shorter longitudinal gaps (visibility limitations) – but at how short a gap?**
- **Risk of reduction of driver vigilance**

- **Multiple research projects have tested it, from CHAUFFEUR (1996-2004) to Konvoi, SARTRE, Energy ITS, etc.**
- **Several companies doing R&D on it**
- **No product releases yet**

CALIFORNIA PATH

8

L3 Truck Platooning State of the Art

L2 + driver can divert attention temporarily to other tasks, while remaining available to intervene when needed

- **Follower truck driver could work as sales person or logistics manager *en route***

- **Research needed on driver-vehicle interface designs to try to ensure driver availability when needed**
- **Remote supervision (by lead driver over V2V or central supervisor over I2V communication link) may be needed**
- **Passenger car applications within limited ODD likely to precede heavy trucks**

9

L4 Truck Platooning State of the Art

L3 + ability to ensure minimal risk condition without any human intervention (while operating within its specified Operational Design Domain – ODD)

- **L4 platoon followers likely to be coupled behind a leader driven at L0, L1 or L2**
- **Singapore seeking this for a 10 km route connecting two container terminals**
- **Safety assurance state of the art not sufficient to support this level of automation for mixed traffic and highway-speed operations**
  - **Likely to need segregated truck-only lanes to simplify the ODD.**

10

Highly Automated Road Freight Movement

AVS Automated Truck Breakout Session
July 11, 2018
Richard Bishop

BISHOP CONSULTING

Automated Road Freight: Where/How?

- Level 3 operations: driver role remains critical, numerous safety concerns
- Weak business case!
- If HOS regulations unchanged: no ROI

BISHOP CONSULTING

PELOTON TECHNOLOGY

Driver-Assistive Truck Platooning Market Overview

Many Companies in US, Europe, and Asia Involved with Bringing Truck Platooning to Market

Peloton 3

Automated Road Freight: Level 3 Operations

- Driver role remains critical, numerous safety concerns
- Weak business case!
- If HOS regulations unchanged: no ROI
- HOS regulations adapt to Level 3: could become highly useful

BISHOP CONSULTING

Level 3 Traffic Jam Pilot Opportunity: Port of Palm Beach Gate Congestion

Automated Road Freight: Key Use Cases

Near Term

- Level 1 Driver Assistive Truck Platooning
- Level 4 Low Speed Driverless

Longer Term

- Level 4 High Speed Driverless

Automated Road Freight: Key Use Cases

Near Term

- Level 1 Driver Assistive Truck Platooning
- Level 4 Low Speed Driverless
  - private property
  - highly limited public road operations

Longer Term

- Level 4 High Speed Driverless

Automated Trailer Switching: Stepping Stone to High Automation On-Road

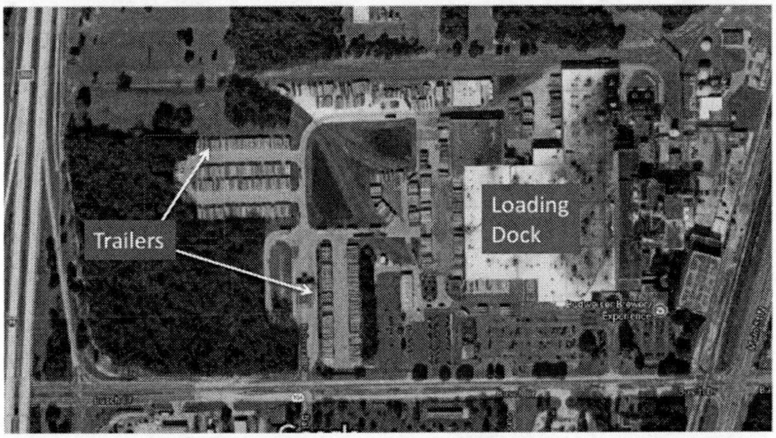

Anheuser Busch Brewery, Jacksonville, Florida

BISHOP CONSULTING

Drayage: Jacksonville Port

Level 4 Automation Within Immediate Port Area

- very short drays between sea terminal and rail intermodal facility
- ¼ mile
- one left turn
- one stop light

Automated Road Freight: Key Use Cases

Near Term

- Level 1 Driver Assistive Truck Platooning
- Level 4 Low Speed Driverless

Longer Term

- Level 4 High Speed Driverless:
- substantial business value at Level 4
- forget Level 5!

Highway Driverless Trucking

- Tech developers
  - Waymo
  - Uber
  - Embark
  - TU-Simple
- General concept:
  - "exit-to-exit" driverless, tranfer hubs at freeway interchanges
  - some deliver to end point
- Freight being hauled
  - Level 4 function operating as Level 2 (safety driver in seat)
- Challenges:
  - regulatory uncertainty in U.S.

BISHOP CONSULTING

Regulatory Complexity Over Time

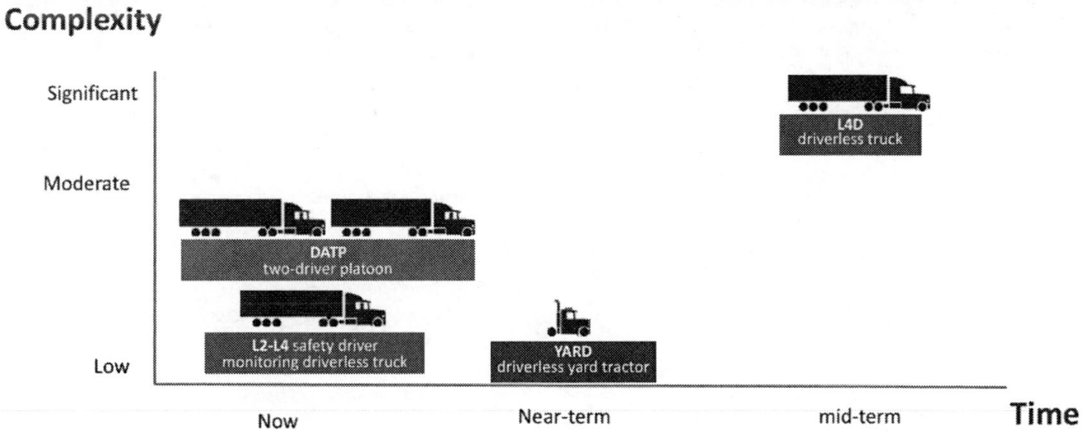

FMCSA Request for Comments on AD Testing and Deployment

- Request for Comments Concerning Federal Motor Carrier Safety Regulations (FMCSRs) Which May Be a Barrier to the Safe Testing and Deployment of Automated Driving Systems-Equipped Commercial Motor Vehicles on Public Roads
- Docket No. FMCSA-2018-0037
- Comments closed May 10, 2018
- 82 comments submitted
  - 6 companies
  - 56 individuals
  - 20 organizations

BISHOP CONSULTING

FMCSA Request for Comment: Notable Commenters

- AV System Developers
  - Embark
  - Tesla
  - Uber
- Motor Carriers
  - Amazon
  - American Trucking Associations
  - National Motor Freight Traffic Association
  - National Tank Truck Carriers
  - The Trucking Alliance
- Driver Groups
  - AFL-CIO
  - Owner Operator Independent Drivers Association
  - Commercial Vehicle Training Association
- Insurance
  - Insurance Institute for Highway Safety
  - Property Casualty Insurers
  - Travelers
- Government / Public Safety
  - American Association of Motor Vehicle Administrators
  - Commercial Vehicle Safety Alliance
- Other
  - Competitive Enterprise Institute
  - MITRE Corporation
  - Motor and Equipment Manufacturers Association
  - Securing America's Energy Future
  - Truck and Engine Manufacturers Association

BISHOP CONSULTING 14

L4 Highway Driverless: Quick Fix?

- Embark, Waymo, and Uber argue that L4 Driverless can be considered legal via an interpretation of the existing regulations.
- Proposing that FMCSA issue such a clarifying interpretation.
- Would result in rapid deployment of L4 Highway Driverless.
- FMCSA considering this and other comments.

Remote Truck Driving

- Recent Starsky Robotics demo in Florida.
- Remote driver 80 miles away viewed video feeds from the truck's onboard cameras.

BISHOP CONSULTING

Driverless Firms Addressing Workforce Issues

- Embark
  - "our goal isn't to replace the driver entirely"
  - seek for long-haul trips to be managed by fewer drivers
  - drivers are still expected to help with the parts of the route that don't involve freeway and driving
- Uber
  - recent blog post forecasts "an increase in better truck driving jobs" due to large increases in e-commerce and other factors

BISHOP CONSULTING

What about Platooning?

- Platooning benefit is additive to Driverless benefit.
- Platooning is an Operational Mode for Close Following
  - independent of Level of Automation
- Driverless trucks can Platoon

BISHOP CONSULTING

Bottom Lines

- Business Case strong
- Miles being accumulated rapidly with safety drivers
- Regulatory pathway not clear, being worked
- Workforce issues becoming better understood

BISHOP CONSULTING

Advancing Transportation through Innovation

Effect of Automated Trucks on the US Economy: Summary from an NSF workshop

Jeff Hickman

Virginia Tech Transportation Institute

VirginiaTech Transportation Institute

Advancing Transportation through Innovation

Any opinions, findings, and conclusions or recommendations expressed during the workshop are those of the workshop participants and do not necessarily reflect the views of the National Science Foundation or the United States Government

This workshop was supported by the National Science Foundation under grant No. 1744348

VirginiaTech Transportation Institute

Purpose of Human-Technology Frontier

- Looking at the merging/convergence of technology and humans:
  - Understand and advance the human-technology partnership
  - Promote new technologies to augment human performance
  - Understand the risks and benefits of new technologies
  - Foster lifelong and pervasive learning with technology

Questions We Tried to Answer

- Understand, anticipate, and shape the implications of this technology for truck drivers.
  - Augmenting human performance for skill acquisition
  - Improving truck driver quality of life and employer financial metrics
  - Enhancing the economic and social well-being of the country

Other Entities

- The Partnership for Transportation Innovation and Opportunity
  - Truck and Car OEMs
  - Find solutions that ensure Americans benefit from adoption of this technology
    - Developing a well-rounded and data-based understanding of the impact of autonomous vehicles on the future of work
    - Soliciting the expertise, concerns, and aspirations of interested parties
    - Fostering awareness of existing and near-term career opportunities for workers during the transition to an autonomous-vehicle-enabled economy

Overview of Activities

- Brief panel presentations
- Group discussion
- Small group discussion

Advancing Transportation through Innovation

Session 1: Automated Truck Implementation Scenarios and the Role of the Driver

Virginia Tech Transportation Institute

Advancing Transportation through Innovation

Summary

- ❑ Many different scenarios
  - Vision based
  - Sensor based
  - Remote control
- ❑ All scenarios yield reduced operating costs
- ❑ Likely skip to Level 4
- ❑ Tailored to operational domain
- ❑ Possibly "train" system to act more like human drivers in mixed situations
- ❑ Still a need for drivers

7/11/2018

Virginia Tech Transportation Institute

Advancing Transportation through Innovation

Session 2: Regulatory/Insurance Perspectives on Automated Trucks

VirginiaTech
Transportation Institute

Advancing Transportation through Innovation

10

Summary

- ❑FMCSRs to identify unintended or unnecessary barriers to safe deployment
- ❑What is safe enough?
- ❑Do we need to "license" systems?
- ❑Insurance will still be needed given inefficiency in product liability
- ❑Federal, state, and local issues
  - Federal: safety standards for vehicle design, insurance minimums for trucks
  - State: licensing, liability and insurance, law enforcement
  - Local: congestion, shifting freight to off peak hours

7/11/2018

VirginiaTech
Transportation Institute

Advancing Transportation through Innovation

Session 3: Industry Views on use of Automated in the End-to-end Delivery System

Virginia Tech. Transportation Institute

Advancing Transportation through Innovation

12

Summary

- ❑Adoption will depend on operation type
- ❑New jobs created
- ❑Safety and productivity primary reasons
- ❑Larger firms more likely to adopt
- ❑Smaller firms and IOs may lag
- ❑Who will pay for needed infrastructure?

7/11/2018

Virginia Tech. Transportation Institute

Session 4: Trucking Impact on US Economy/Driver Demand

Summary

- Increase in demand for just in time delivery
- Driver shortage
  - Shortage of qualified drivers
  - Do automated trucks reduce need for qualifications?
- Pay issues (based on time)
- S-shaped curve for adoption
- Trucks may replace trains
  - If so, congestion issues

Session 5: Convergence of Automated Trucks and Human Labor – Enhancing the Driver-Truck Interface and Understanding Skill Requirements

VirginiaTech.
Transportation Institute

Summary

- ❑Likely always a need for driver at some level
- ❑Some automation in 3-7 years
  - Private property
  - Highly limited public road operations
- ❑Wide-scale adoption in 10-15 years
  - "exit-to-exit" driverless, transfer hubs at freeway interchanges
- ❑Vision and/or sensor-based systems possibly legal using existing regulations

VirginiaTech.
Transportation Institute

Session 6: Convergence of Automated Trucks and Human Labor – Job Skills Training/Re-training Challenges and Strategies

Summary

- ❑79 percent probability of automation (average is 45 percent)
- ❑Can increase job from "delivery" to "service"
- ❑More local jobs (decreases wages)
- ❑More precarious and gig jobs
- ❑Job quality/pay more of an issue than displacement
- ❑Apprenticeship training model
- ❑Establish a decent minimum wage

AUTOMATED VEHICLES SYMPOSIUM 2018 — USERS. VEHICLES. INFRASTRUCTURE. — Driven by AUVSI and TRB

Breakout Session 16: Safety Assurance of AD

Summary of Key Findings and Lessons Learned

1. **Which level of Safety do Automated Vehicles need to achieve?**
   - **Safety is the main motivation** in the U.S., EU, Jp, G to introduce automated driving
   - **We need to draw comparisons in order to find the right level:**
     => compare to introduction of Airbags, Vaccination, Aviation (much research involved)
   - **Process is important: society needs to be included**
   - **Need for differentiation** between Level 2, testing Level 3+4, deployment of Level 3+4
     => build public confidence + manage safety risk + create transparency
     => simplify the problem first, allow innovation to take place
   - **Diffentiate between time horizons:** in **medium term** expectation to be significantly better than manual driving

AUTOMATED VEHICLES SYMPOSIUM 2018 — USERS. VEHICLES. INFRASTRUCTURE. — Driven by AUVSI and TRB

Breakout Session 16: Safety Assurance of AD

2. **What is the reference for Safety of Automated Vehicles?**
   - **From a societal perspective**
     - Manual driving, as long as ADAS (Level 1 and 2) have a limited market share.
     - From a product liability perspective: significantly better
   - **From a system development perspective**
     - Driver performance models, each may cover various scenarios;
     - Ideally a simuation-ready „driver population model", which may have different parameters for different countries
   - **From an accident analysis perspective**
     - Physics => the automated vehicle should avoid accidents within given physical limits (e.g. friction potential)
3. **Which are the relevant traffic scenarios?**
   - Multiple Data sources are used: Traffic accidents, incidents from FOT and incidents captured by infrastructure sensors
   - How to seperate relevant scenarios from extremely rare corner cases?
     - Include everything and decide by defining the Operational Design Domain
     - Can rare corner cases be designed by simple scenarios and their combination?
   - How do we include cases, where the driver avoids an incident or accident using anticipation and experience (e.g. loose freight)? How do we know, when the system does not know?

Working group
"Safety for the autonomous vehicle"

Autonomous Vehicles
French National Program

Jean-François SENCERIN
Stéphane GERONIMI

PFA | FILIÈRE AUTOMOBILE & MOBILITÉS

AV French national program: a working group dedicated to « safety »

DEVELOPMENT OF
AUTONOMOUS VEHICLES
Strategic Orientations for Public Action
Summary Document

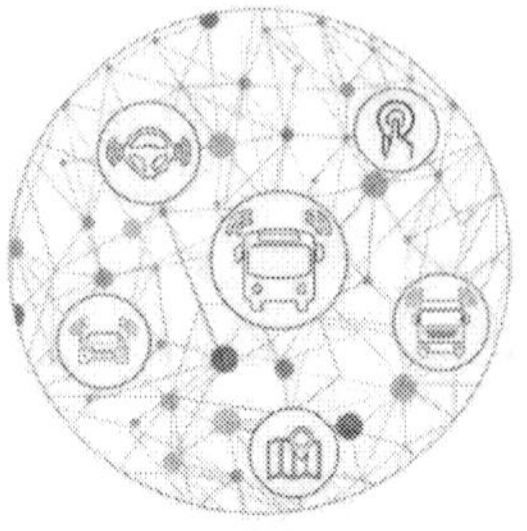

May 2018

PFA | FILIÈRE AUTOMOBILE & MOBILITÉS

\+

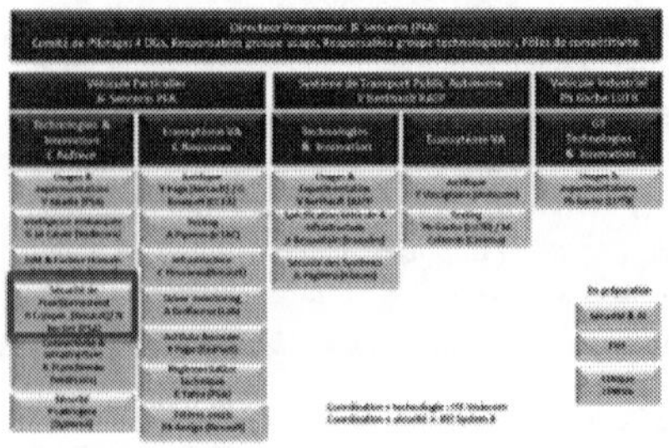

\+

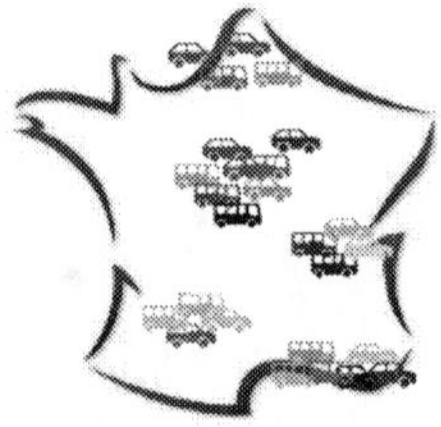

FRENCH SAFETY WORKING GROUP

SAFETY VALIDATION APPROACH FOR AV

3

Working group on « safety for autonomous vehicle »

Scope

- Level 3 and 4
- First step: highways / dual carriageways

Objectives

- Define the general structure of the argumentation for safety for Autonomous vehicle (including methods and tools for validation)
- Bring elements for the revision of the ISO26262 and the new standard PAS 21448

Participants

- PSA Group – Renault - Valeo - Vedecom - SystemX

4

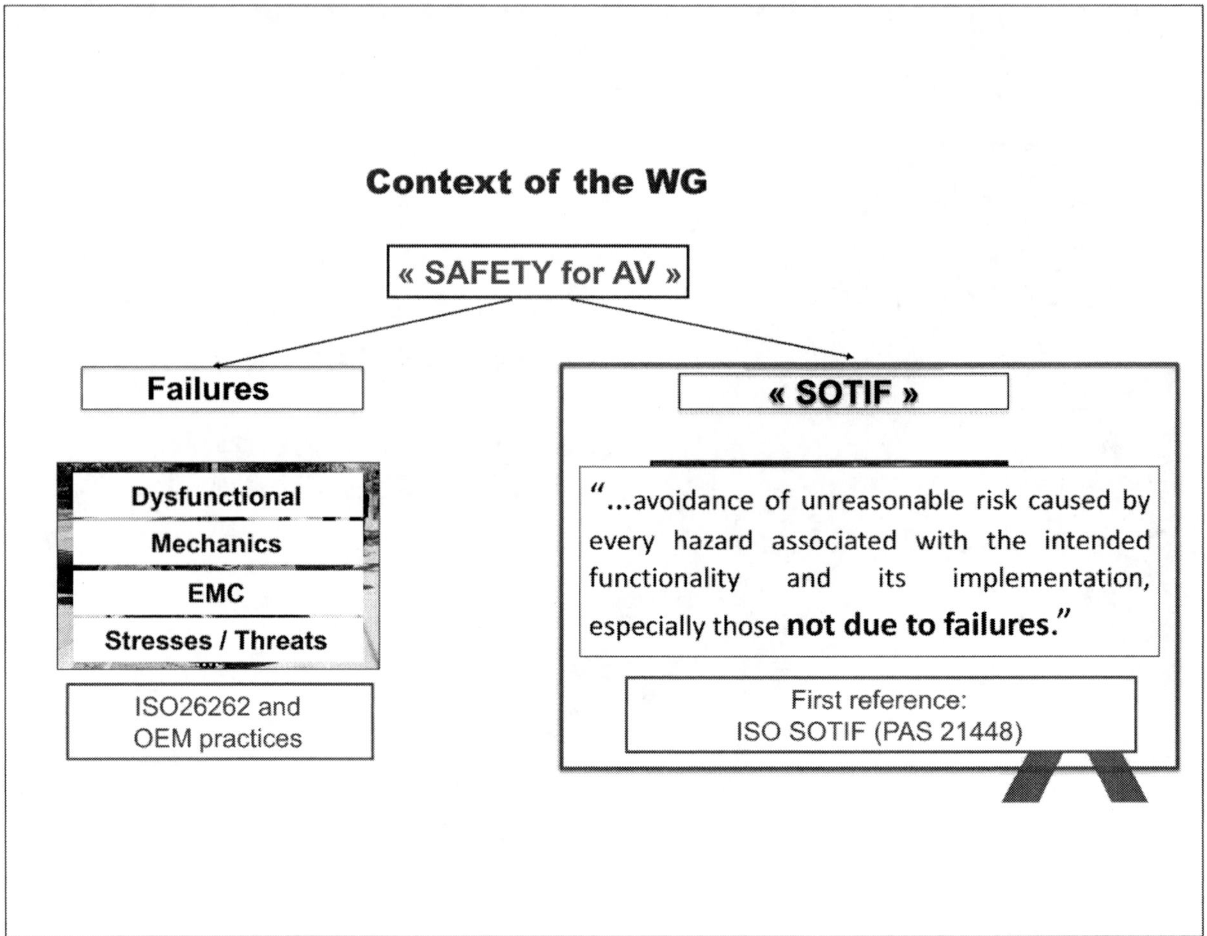
Context of the WG
« SAFETY for AV »
Failures
Dysfunctional
Mechanics
EMC
Stresses / Threats
ISO26262 and OEM practices
« SOTIF »
"...avoidance of unreasonable risk caused by every hazard associated with the intended functionality and its implementation, especially those not due to failures."
First reference: ISO SOTIF (PAS 21448)

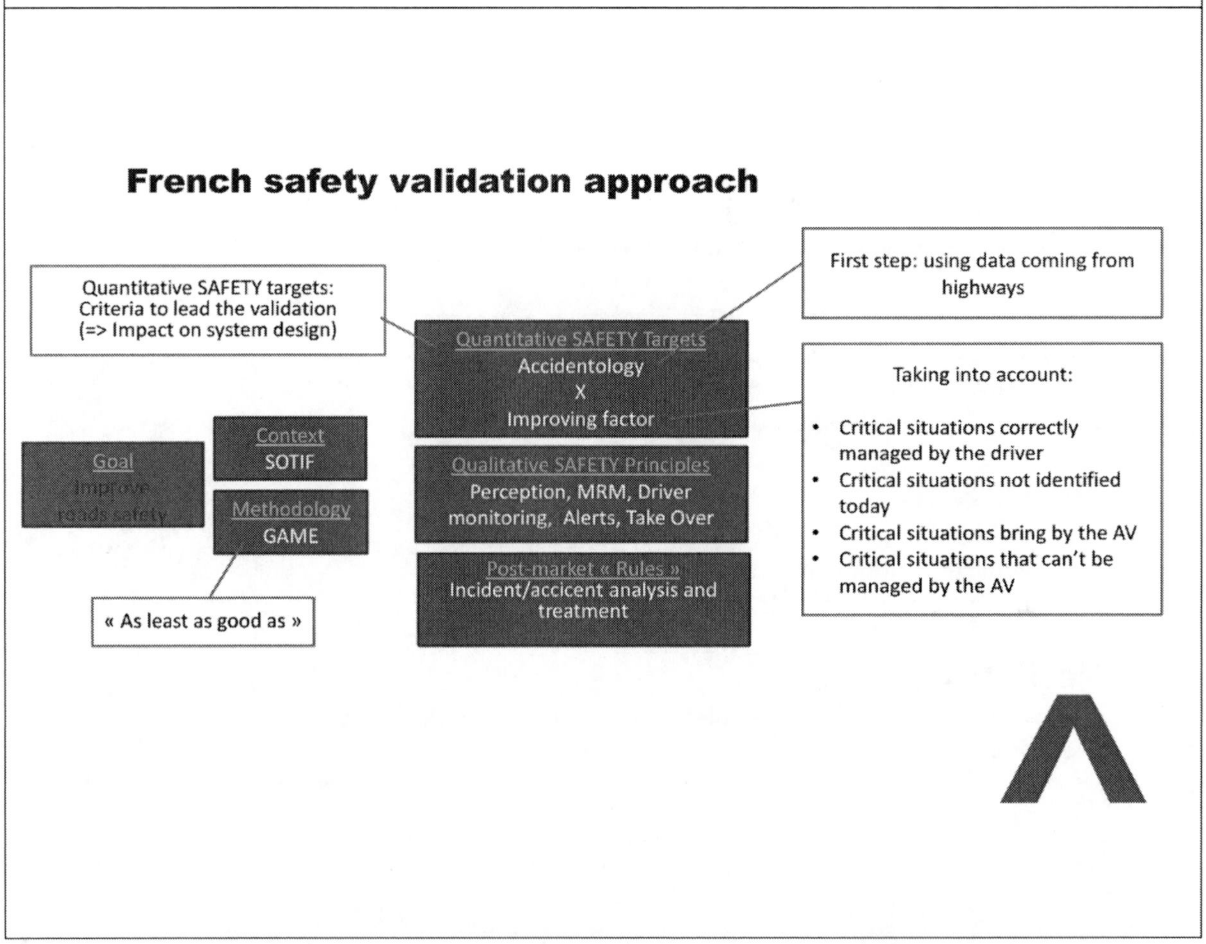
French safety validation approach
Quantitative SAFETY targets: Criteria to lead the validation (=> Impact on system design)
Goal
Improve roads safety
Context
SOTIF
Methodology
GAME
« As least as good as »
Quantitative SAFETY Targets
Accidentology
X
Improving factor
Qualitative SAFETY Principles
Perception, MRM, Driver monitoring, Alerts, Take Over
Post-market « Rules »
Incident/accicent analysis and treatment
First step: using data coming from highways
Taking into account:
• Critical situations correctly managed by the driver
• Critical situations not identified today
• Critical situations bring by the AV
• Critical situations that can't be managed by the AV

French safety validation approach

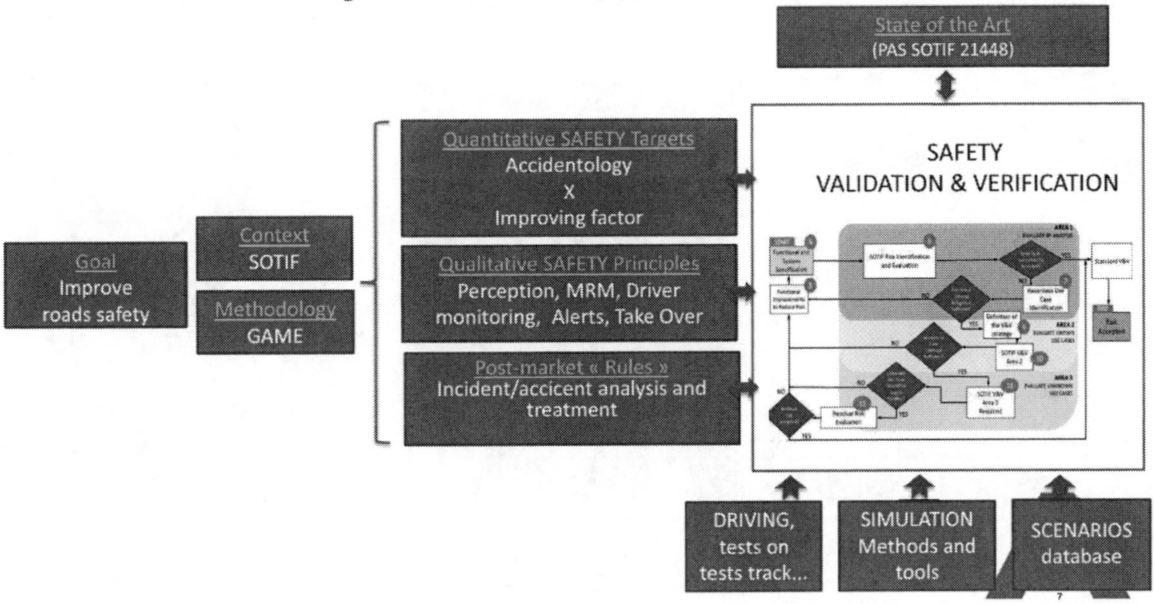

French project « Moove » : Data acquisition / Scenarios identification

Interface example

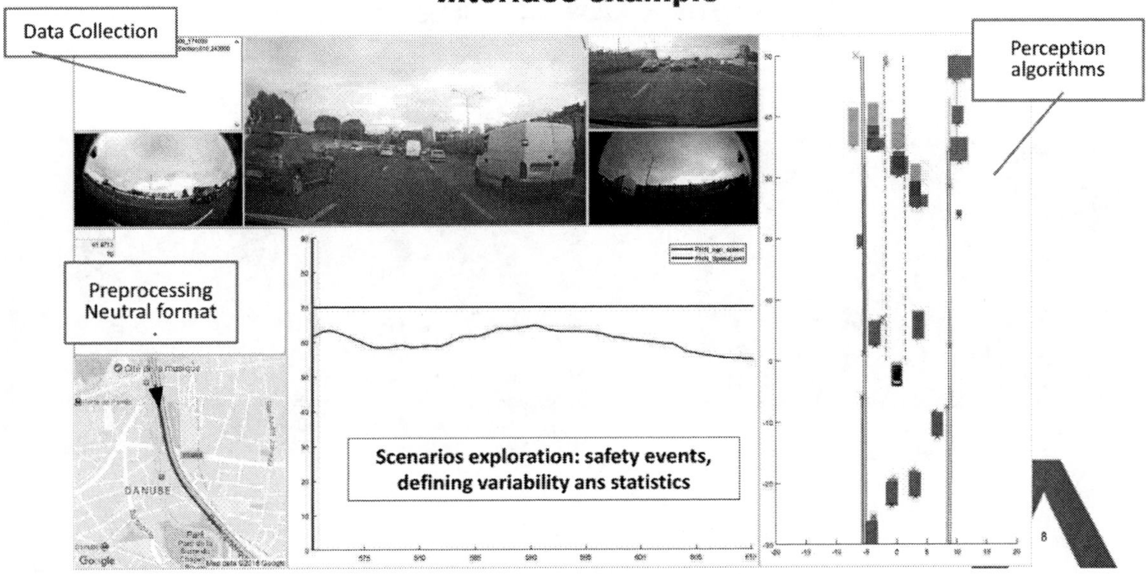

French project « SVA »: Scenarios database / catalog

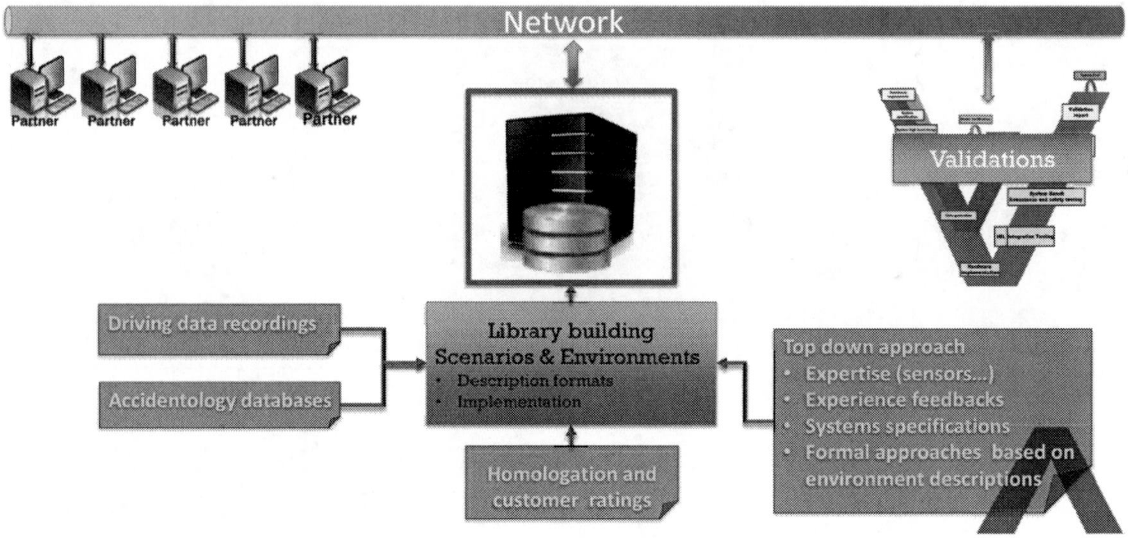

SIMULATION: Methods and tools
Example « Cut-in scenarios »

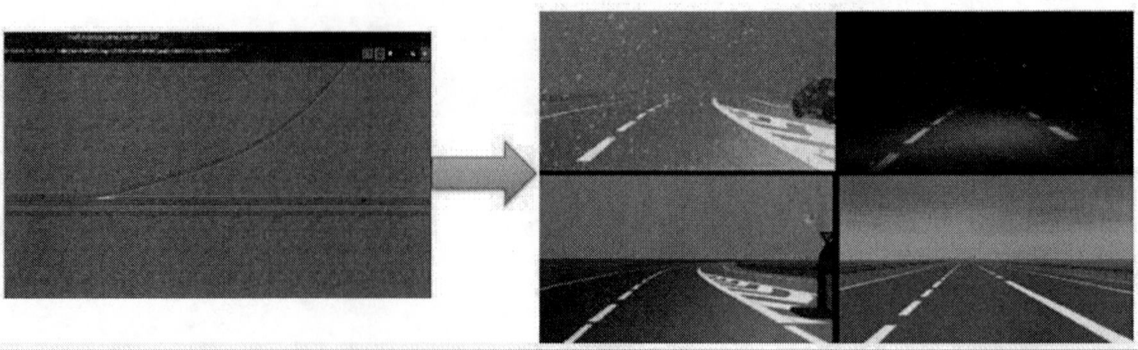

Simulation results

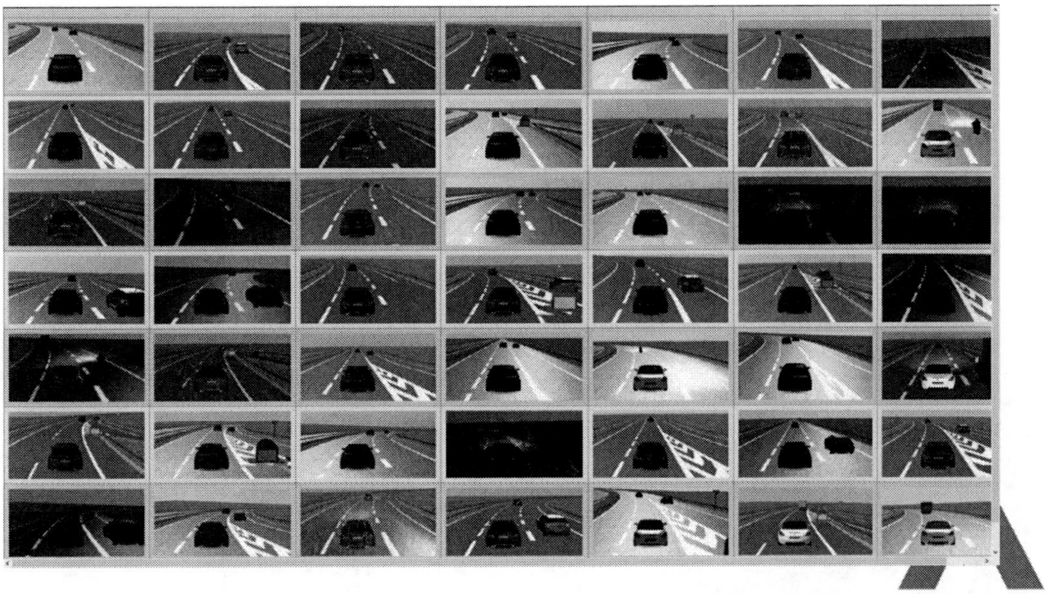

Conclusion: a first step towards « safety » approach

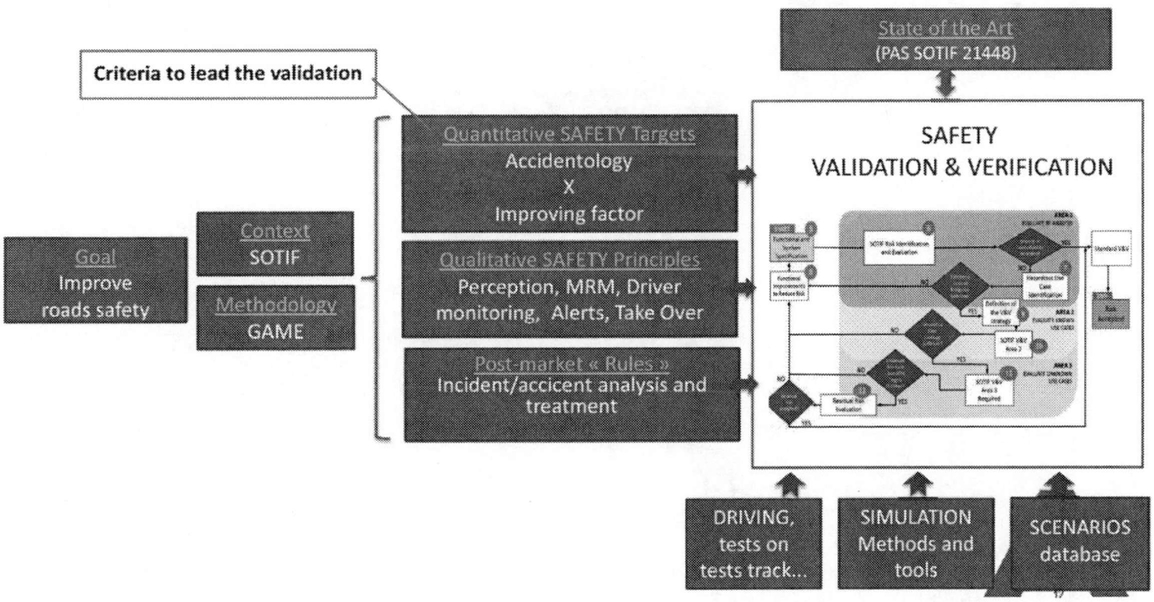

THANK YOU !!

13

VIRES
Simulationstechnologie GmbH
MSC Software Company

Standardization of Data Formats for Increased Efficiency of Simulation Applications

Marius Dupuis
Managing Director, VIRES Simulationstechnologie GmbH

July 11, 2018

PART OF HEXAGON

Introduction

VIRES Simulationstechnologie GmbH

VIRES
Simulationstechnologie GmbH
A Member of the MSC Software Group of Companies

Headquarter:
Grassinger Straße 8
83043 Bad Aibling
Germany

p +49.8061.939093-0
f +49.8061.939093-13
e info@vires.com

Branch Office:
Wiener Straße 12
2700 Wiener Neustadt
Austria

Founded: 1996
Staff: 34+
Management: Marius Dupuis
Wunibald Karl

www.vires.com
www.opencrg.org
www.opendrive.org
www.openscenario.org

Focus on key product: **VIRES Virtual Test Drive®**

VTD

Partner in Standardization Initiatives:

Open*CRG* Open*DRIVE* Open*SCENARIO*

Partner in Research Projects:

PEGASUS ENABLE-S3 + additional ones *(2 starting 2018)*

Partner in: SAE AutoDrive Challenge *(via MSC.Software)*

MSC Software Confidential 3

Motivation

Motivation

Highly Automated Cars Have To Be Safe!

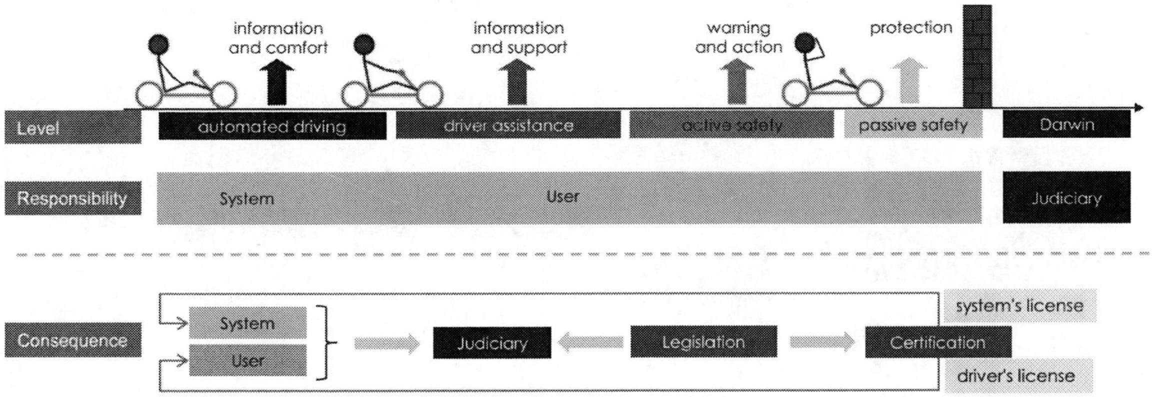

MSC Software Confidential 5

Motivation

Certification Requires Training and Testing

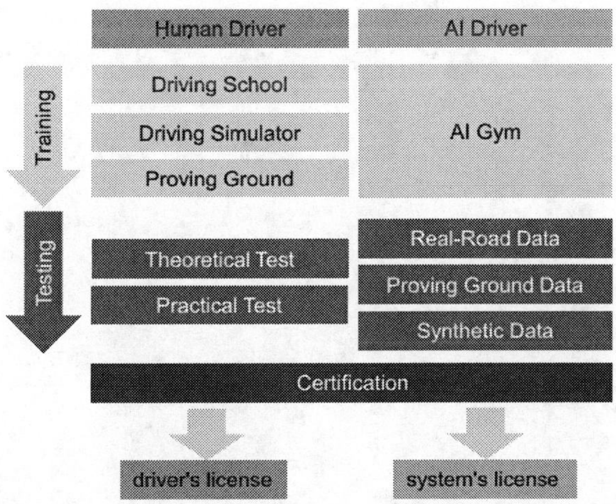

MSC Software Confidential 6

Simulation-Based Certification
Standards and Validated Tools Required!

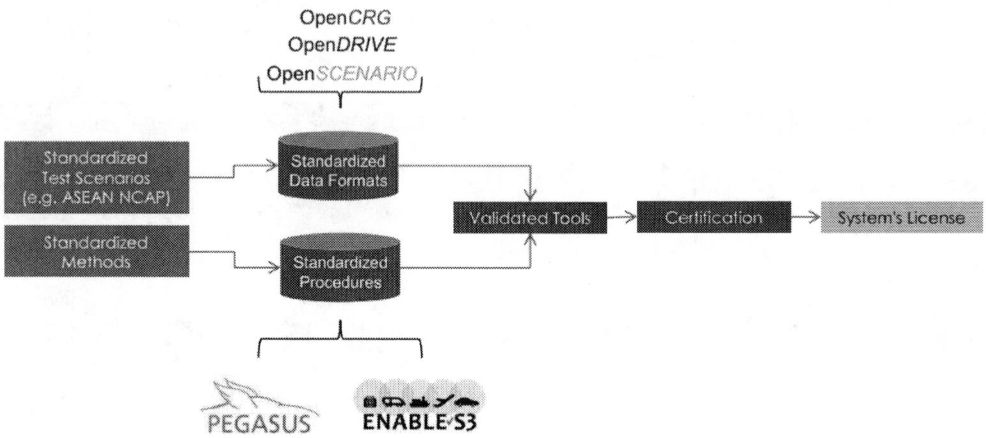

Driving Simulation

It comes in variants...

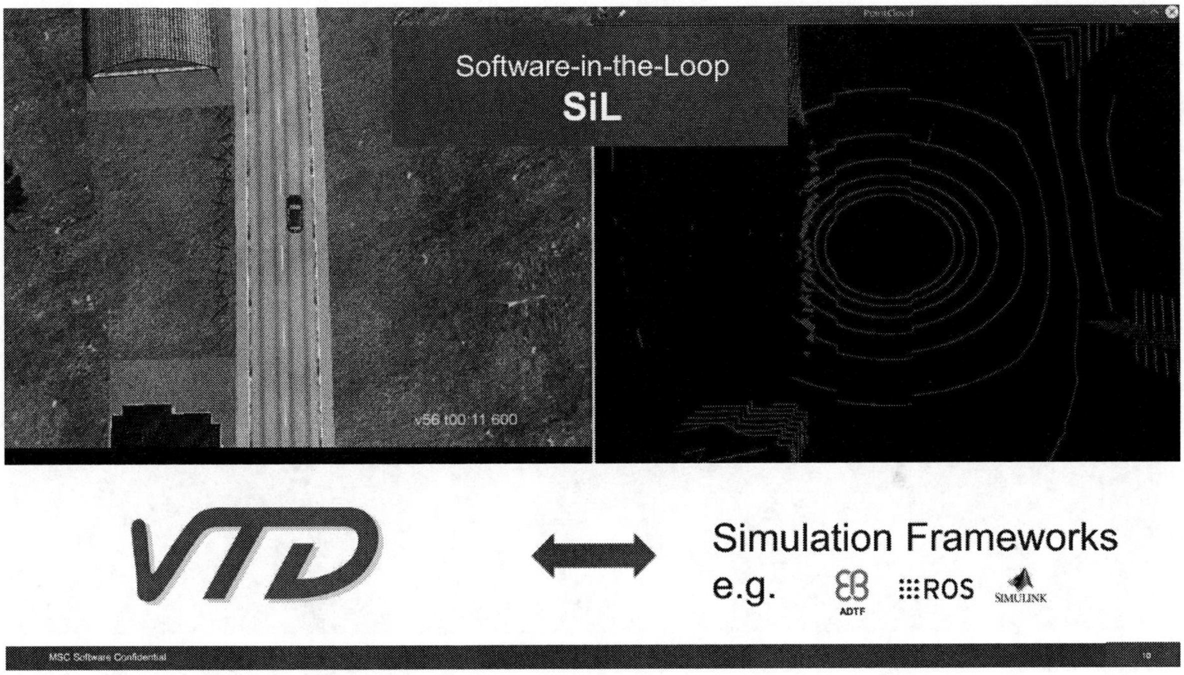
Software-in-the-Loop
SiL
VTD
Simulation Frameworks
e.g.
ADTF
ROS
SIMULINK
MSC Software Confidential
10

ika
Driver-in-the-Loop
DiL
MSC Software Confidential
11

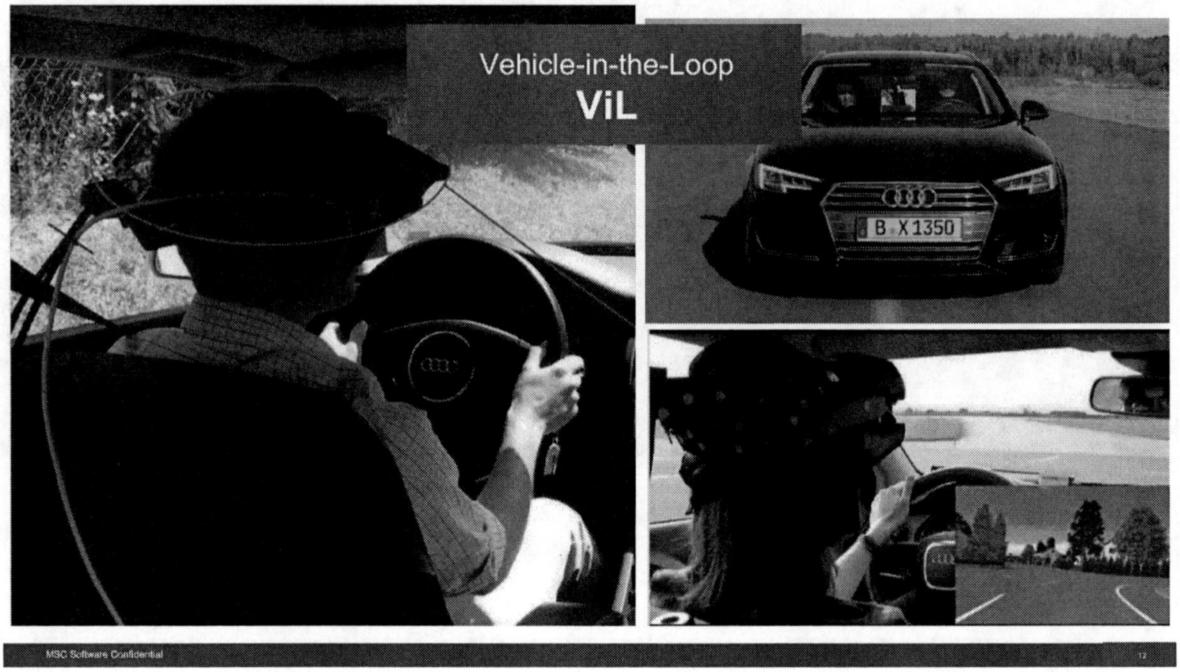
Vehicle-in-the-Loop
ViL
B X 1350
MSC Software Confidential
12

AVL
IPEK FZI
Hardware-in-the-Loop
HiL
Hardware-in-the-Loop-
Test von vorausschauenden
Fahrerassistenzsystemen
Die neue Generation von Fahrerassistenzsystemen verwendet neben physikalischen Umfelddaten auch logische Informationen zur vorausschauenden Berechnung von Fahrsituationen. Diese Informationen stammen aus Navigationsdatenbanken, von Serviceprovidern oder aus der Vernetzung mit anderen Fahrzeugen. Für die Absicherung derartiger Systeme an einem Hardware-in-the-Loop-Prüfstand ist eine Integration dieser Daten in bestehende Testprozesse notwendig. In diesem Artikel wird aufgezeigt, wie die Audi AG gemeinsam mit der Audi Electronics Venture GmbH durch den Einsatz innovativer technischer Lösungen eine Simulation derartiger Szenarien ermöglicht.
MSC Software Confidential
13

Environment Simulation

Standards Make Life a lot Easier...

Standards for Environment Simulation
Application

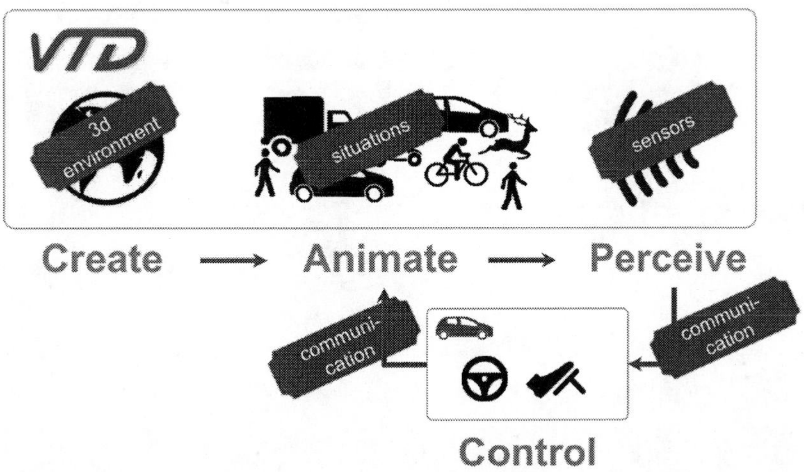

Standards for Environment Simulation
Inspired by VIRES

Open*DRIVE*	Open*CRG*	Open*SCENARIO*

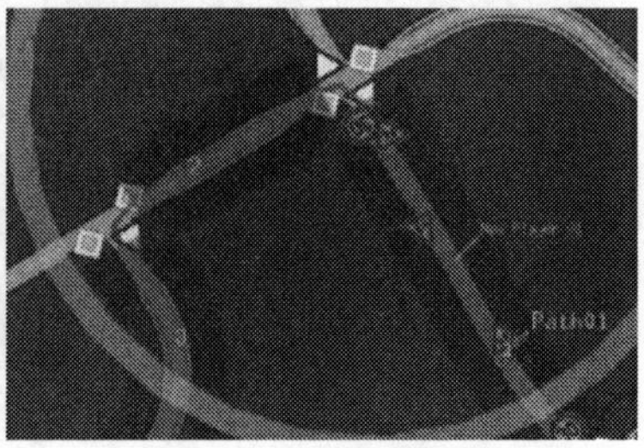

network 2006+	surface 2008+	maneuvers 2016+

Static Content | Dynamic Content

Standards for Environment Simulation

	Open*DRIVE*	Open*CRG*	Open*SCENARIO*
initiators	Daimler VIRES	Audi BMW Daimler Porsche Volkswagen	asc(s Daimler HLRS Opel Porsche VIRES
seed funding	internal	initiators	initiators
type	open format	open source	open format+

Standards for Environment Simulation

	OpenDRIVE	OpenCRG	OpenSCENARIO
			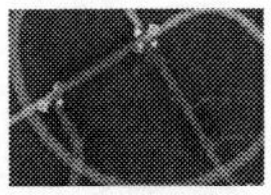
funding	intern	VIRE	intern
management	eam	Daimler	eam
	ASAM	ASAM	ASAM

ASAM's tasks
- collect requirements and write the documents
- run issue tracking and support system
- run website
- run newsletter

Standards for Environment Simulation
12 years in the making...

Our Rulebook

ACT
- **Standards will only gain momentum if someone takes the lead**

CO-OPERATE
- **Invite and include interested parties**

IGNORE
- **Ignore your company's short-term interests**

FOCUS
- **Standardize in batches that can be handled and maintained**

CREATE
- **There's a whole universe left to be standardized!**

Thank You.
VIRES
MSC Software

Virginia Tech Transportation Institute

VTTI Research in Rapid-cycle Mode

AMP

AUTOMATED MOBILITY PARTNERSHIP

Automated Vehicles Symposium 2018

Breakout Session 16: Approaches and Methodologies to Assure Safety of AV

AMP

Virginia Tech Transportation Institute

WHAT CAN YOU LEARN FROM NATURALISTIC DRIVING DATA?

NDD collection techniques were developed by VTTI to capture naturally occurring driver behavior, to the extent practical

- Long data collection periods (up to 3 years) and minimal intervention during data collection
- In situ NDD collection methods allow an unobtrusive, real-world observation of driver behavior and performance
- Naturalistic data collection can be used to estimate prevalence and rates

AMP is leveraging VTTI's access to 70 M miles of continuous NDD, including automated driving systems

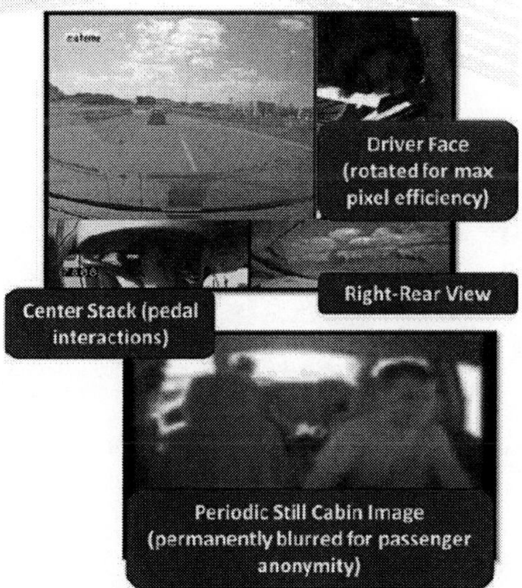

AMP

VirginiaTech. Transportation Institute

CRASHES...

Video Removed

AMP

VirginiaTech. Transportation Institute

CRASHES AVOIDED...

Video Removed

NHTSA
NATIONAL HIGHWAY TRAFFIC SAFETY ADMINISTRATION

Video from NHTSA
Naturalistic Driving Study

AMP

VirginiaTech. Transportation Institute

JUST DRIVING...

Video Removed

AMP

VirginiaTech. Transportation Institute

AMP RESEARCH IN RAPID-CYCLE MODE

AMP
AUTOMATED MOBILITY PARTNERSHIP

An industry partnership promoting the development of tools, techniques, and data resources to support the rapid advancement of automated-vehicle deployment for its members.

AMP

Virginia Tech. Transportation Institute

AMP RESEARCH IN RAPID-CYCLE MODE

AMP Enablers:

- Naturalistic Driving Data (NDD) Scenario Library capturing the complex and chaotic scenarios inherent in real-world driving
- Advanced analytics that measures and characterizes the scenarios
- Tools and techniques that parametrizes the data for reconstruction (virtual and test track)

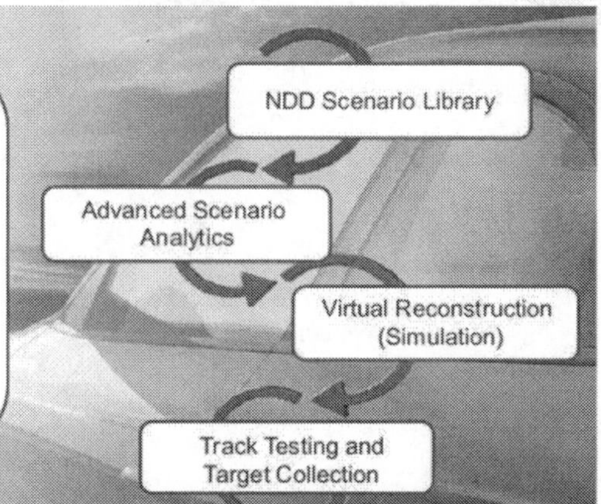

AMP

Virginia Tech. Transportation Institute

ORDERING CASES FOR VIRTUAL RECONSTRUCTION AND ANALYSIS

Epochs and Events

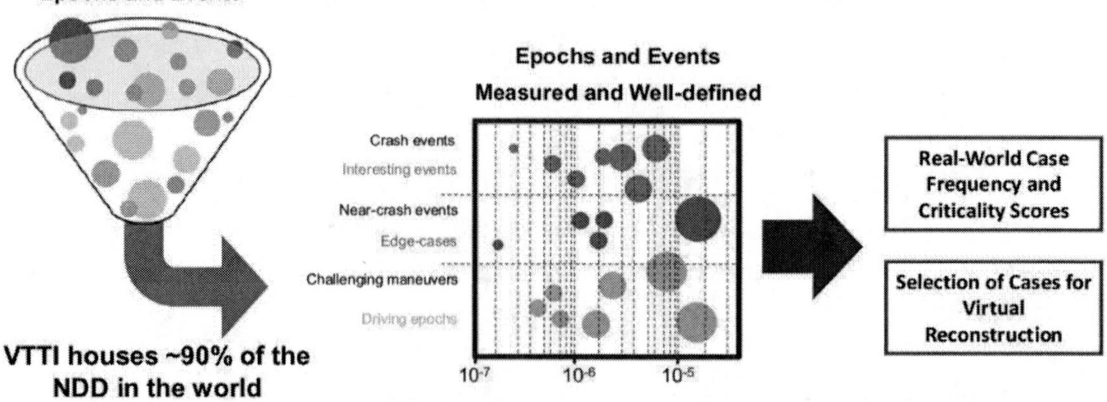

VTTI houses ~90% of the NDD in the world

AMP

Virginia Tech
Transportation Institute

NDD HIGHWAY REVEAL SCENARIO EXAMPLE

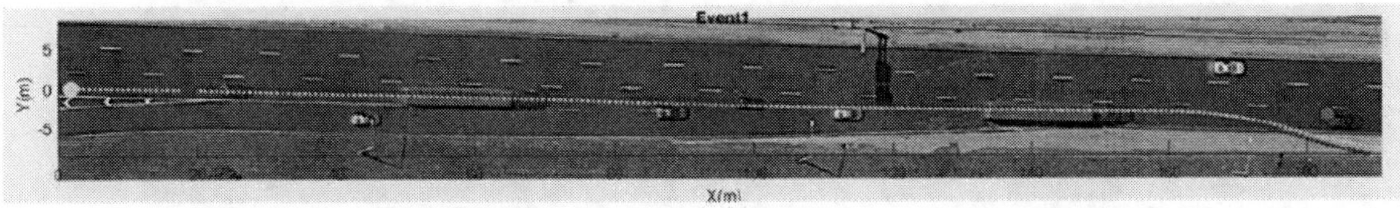

AMP

Virginia Tech
Transportation Institute

NDD HIGHWAY REVEAL SCENARIO EXAMPLE

AMP

VirginiaTech
Transportation Institute

NDD HIGHWAY REVEAL SCENARIO
SIMPLE AUTOBRAKE ALGORITHM EXAMPLE

AMP

VirginiaTech
Transportation Institute

TEST TRACK RECONSTRUCTION EXAMPLE

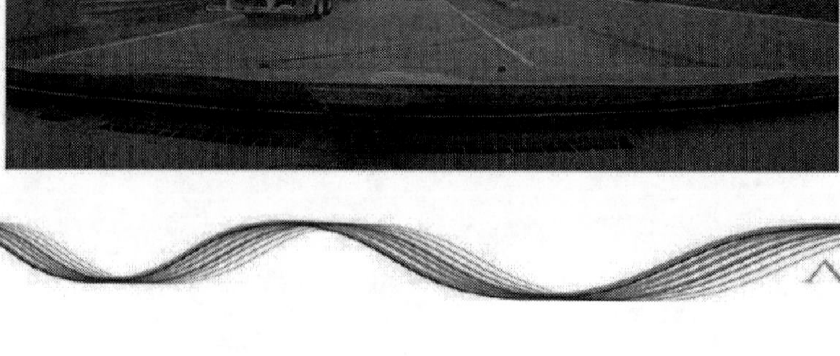

AMP

Virginia Tech.
Transportation Institute

REMOVED STEER / MANUAL BRAKE EXAMPLE

AMP

Virginia Tech.
Transportation Institute

NDD HIGHWAY REVEAL SCENARIO SIMPLE AUTOBRAKE ALGORITHM EXAMPLE

AMP

Virginia Tech. Transportation Institute

IMPORTANCE OF VEHICLE DYNAMICS

AMP

Virginia Tech. Transportation Institute

Questions?

AMP

JAMA
JAPAN AUTOMOBILE MANUFACTURERS ASSOCIATION, INC.

AD Safety Validation Investigation in Japan

AVS Breakout session
11th/July/2018

JAMA AD Safety Validation WG Chair
Satoshi Taniguchi
(satoshi_taniguchi_ad@mail.toyota.co.jp)

page1

➡ 1. Japanese Industry Research Activity Organization
2. AD Safety Validation Strategy
3. Data base
4. Traffic Scenario
5. How safe is safe enough

Japanese Industry Research Activity

JAMA

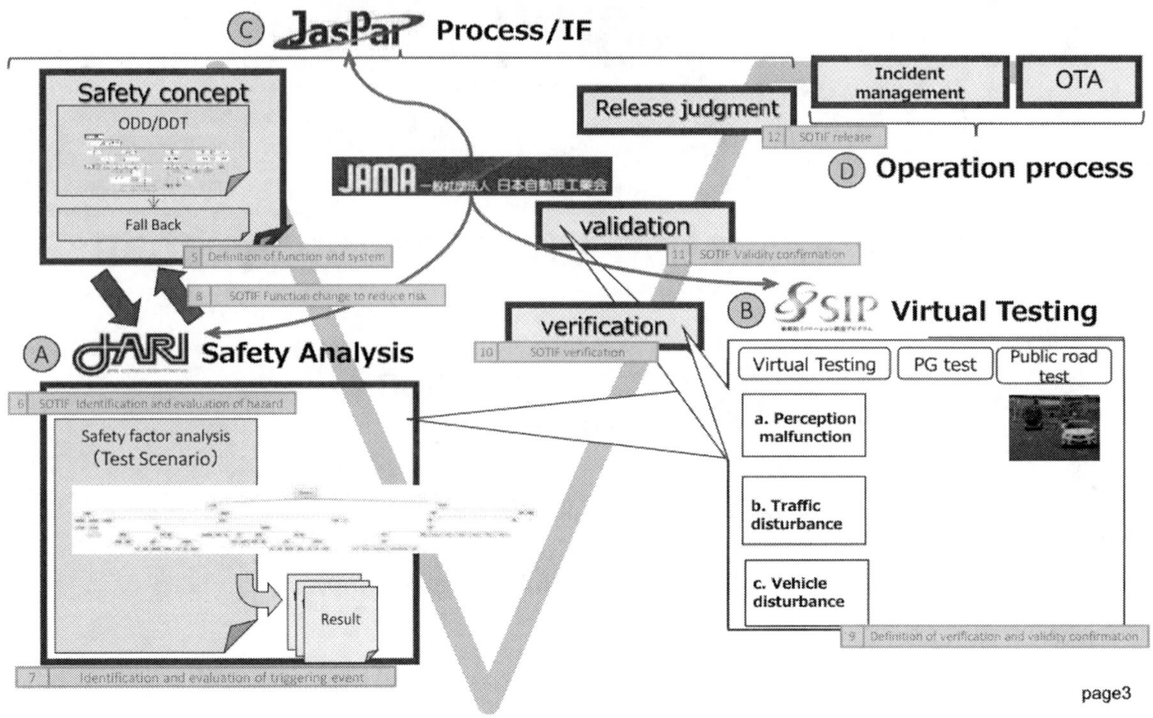

page3

JAMA Structure for AD

JAMA

❶ Regulation : UN-regulation J-industry window

❷ Standard : AD safety evaluation method and criteria investigation

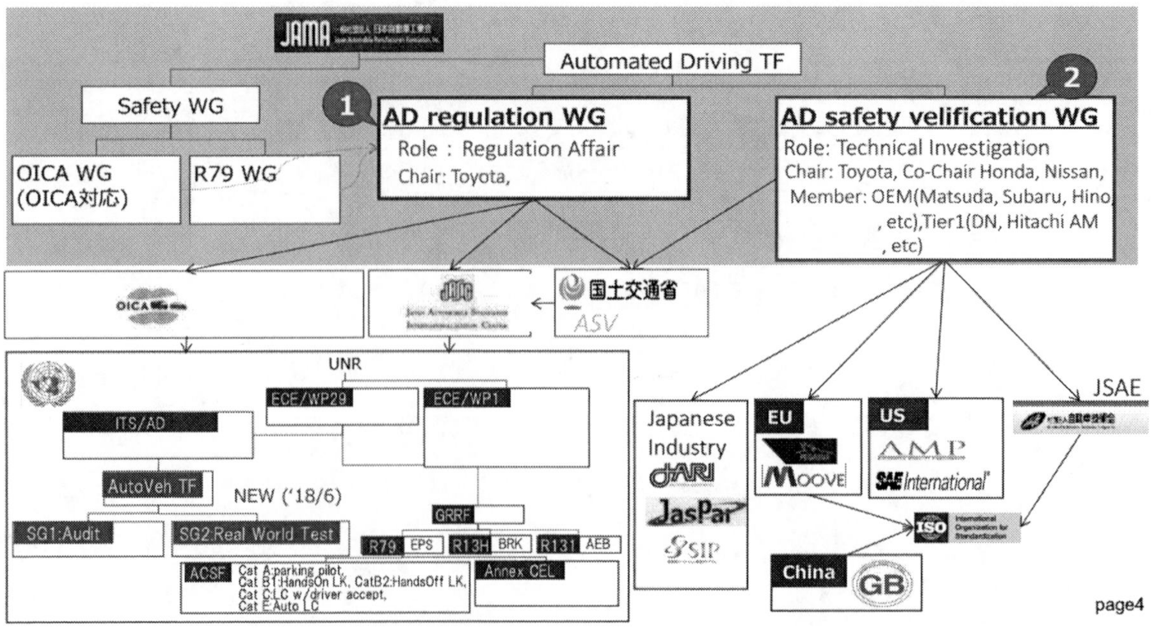

page4

1. Japanese Industry Research Activity Organization
2. AD Safety Validation Strategy
3. Data base
4. Traffic Scenario
5. How safe is safe enough

AD Safety Validation Scenario Catalog

JAMA
JAPAN AUTOMOBILE MANUFACTURERS ASSOCIATION, INC.

AD scenario catalog can be separated from the view point of the physics of AD system, which is perception, judgment, and maneuver in order to ...

- reduce the number and complexity → operational
- cover the all root cause →holistic

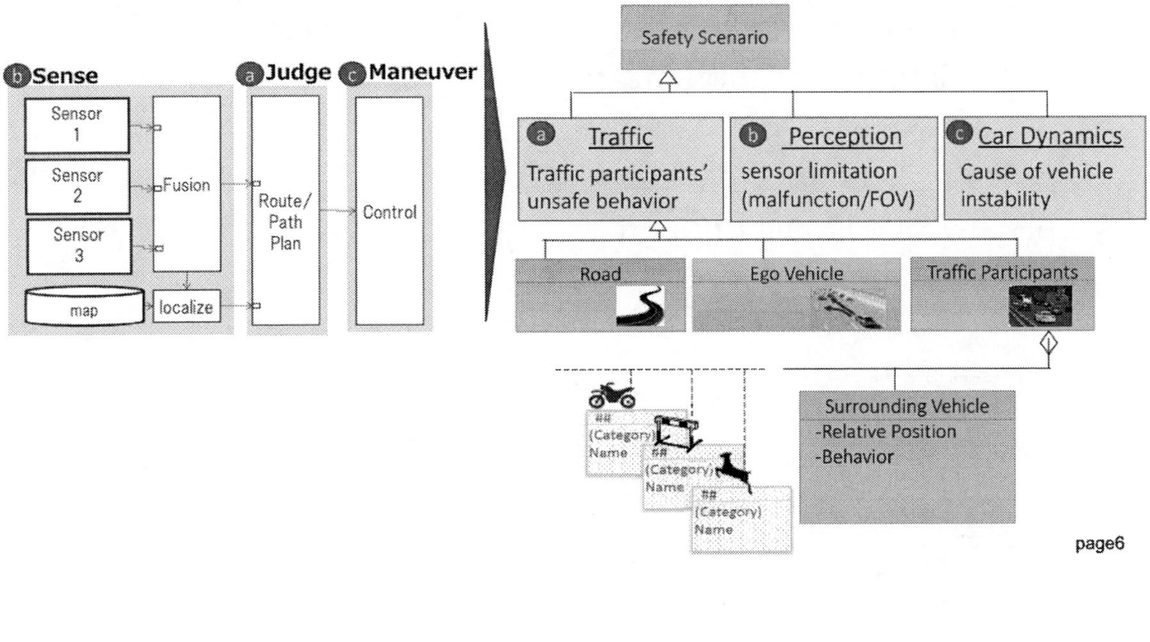

page6

ODD and Testing Scenario

JAMA

- ODD defines the boundary of scenario.
- Each 3 scenario catalog has own ontology based on its physics.
- Testing environment is fitted for each 3 scenario catalog in efficient scale.

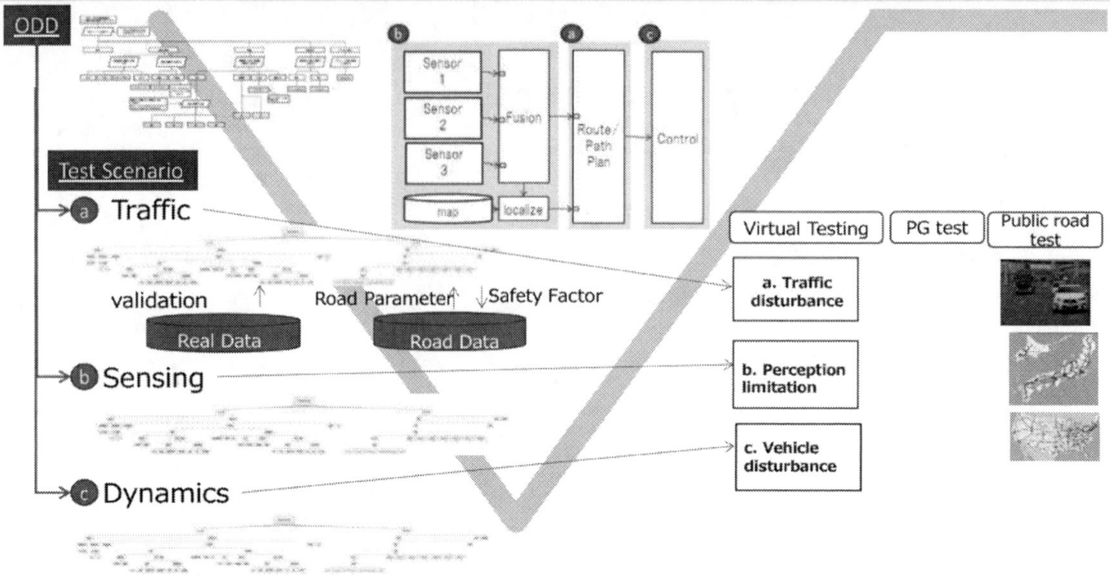

page7

1. Japanese Industry Research Activity Organization
2. AD Safety Validation Strategy
3. Data base
4. Traffic Scenario
5. How safe is safe enough

Data collection scheme

JAMA
JAPAN AUTOMOBILE MANUFACTURERS ASSOCIATION, INC.

① In vehicle collection system
- LIDAR
  (over 100m Around 365 deg.)
- Camera (365 deg.)
- Digital Map matching, GPS

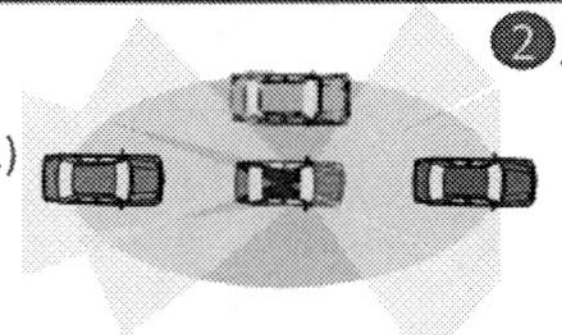

② Fixed location camera
- Camera (4K resolution)
- Lead + 1 can be detected

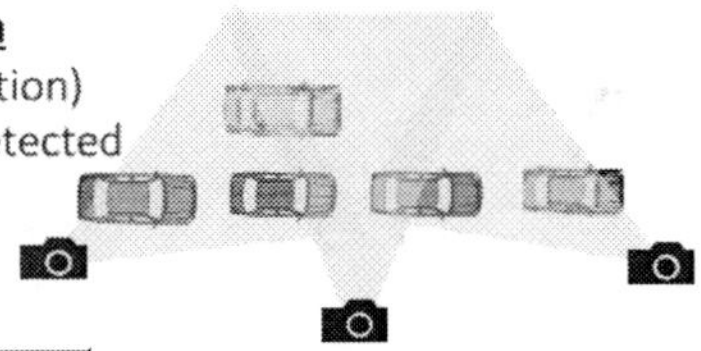

Data Source	TUAT Driving Recorder Data (~2018)	JAMA Driving Recorder data (2008)	Driving Data Base (2017)	On road recognition database (2017)	① Specific Vehicle (2018~)	② Fixed Location Camera (2018~)
◎ ○ △ ×	△ △ ○ △ × Ego × × × ×	△ ○ ○ ○ × Ego × × × ×	△ ○ ◎ ○ ○ Ego ○ △ ○ △	△ ○ ◎ ○ △ Ego △ △ △ △	△ ◎ ◎ ◎ ◎ Ego ◎ ◎ ◎ ◎	○ ○ ○ ○ ○ Ego ○ ○ ○ ○
Express way	Near crash : 255 cases* *At 2018/4/12	Total : (Not counted) Near crash : 23 cases ** **Rear-end crash and Lane change only	TOTAL : 350h Near crash : 15 cases*** *** Estimated Including error	Total : 90h Near crash : Not analyzed	Current planned FY 18 : 1,200 h FY 19 : 4,000 h FY 20 : 0 h Total : 5,200 h	Now under planning
Local way	Near crash : 136,468 cases* *At 2018/4/12	Total : 3,000h Near crash : 506 cases ** ** Rear-end crash and Lane change only	Total : 4,650h Near crash Not analyzed ***Estimated Including error	Total: 1,257h Near crash : Not analyzed	Current planned FY 18 : 0 h FY 19 : 0 h FY 20 : 4,000 h Total : 4,000 h	Now under planning

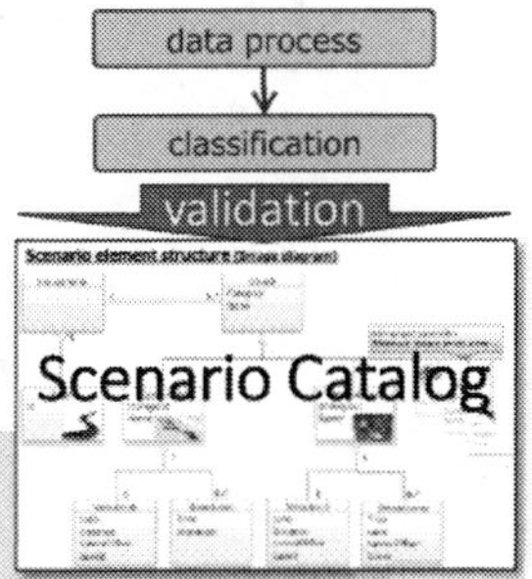

The real world data can be utilized for
- validation of scenario catalog
- definition of scenario parameter boundary

page9

Scenario Data Base

JAMA
JAPAN AUTOMOBILE MANUFACTURERS ASSOCIATION, INC.

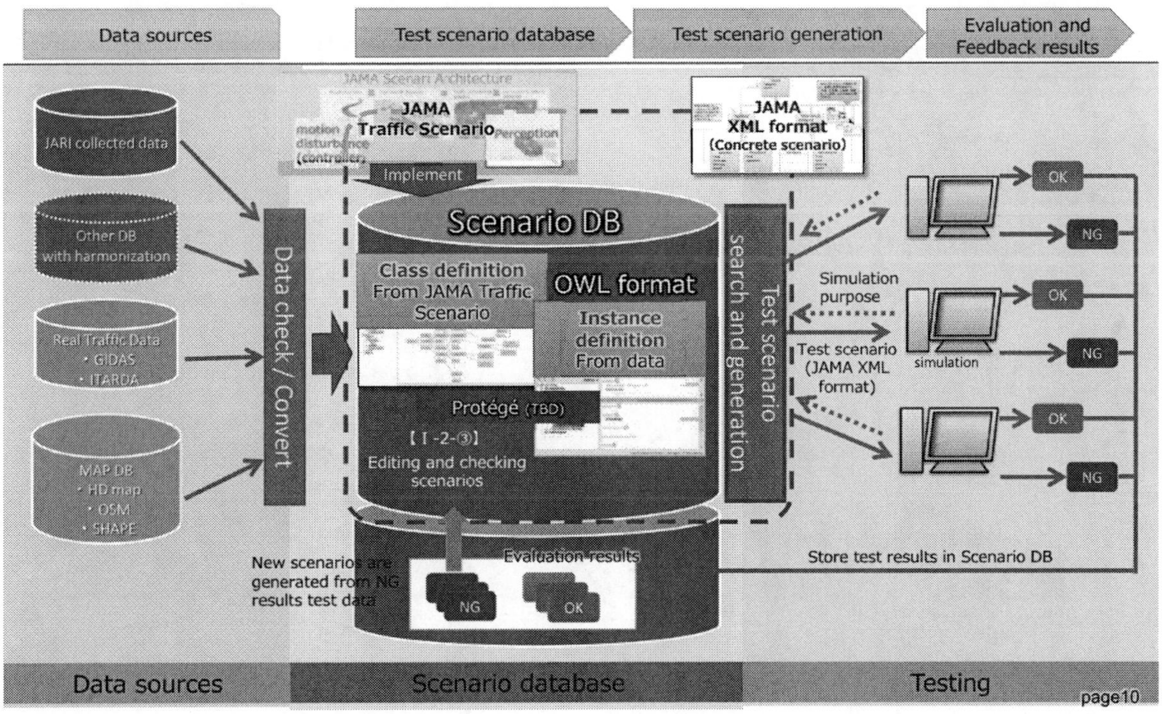

page10

1. Japanese Industry Research Activity Organization
2. AD Safety Validation Strategy
3. Data base
4. Traffic Scenario
5. How safe is safe enough

How to simplify the traffic scenario

JAMA
JAPAN AUTOMOBILE MANUFACTURERS ASSOCIATION, INC.

The scenario ontology which is defined from the view point of safety can make the safety validation holistic and finite.

JAMA's Approach

JAMA separates the traffic scenario to following 4 parts.

Traffic scenario

Road geometry | Ego-vehicle behavior | Traffic participants' position | surrounding vehicle behavior

page12

Road geometry (1/3)

JAMA
JAPAN AUTOMOBILE MANUFACTURERS ASSOCIATION, INC.

Basically, the road is built in accordance with road construction regulation
⇒But in reality severer road shape may be built by many kind of reason
⇒need to define and update the road parameter based on the real world data

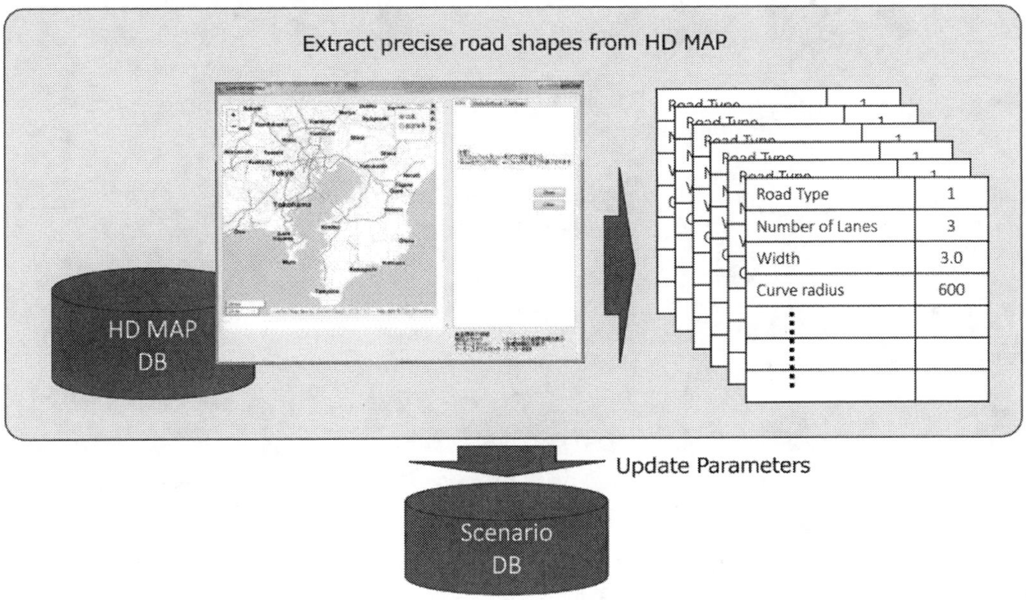

page13

Road geometry (2/3)

JAMA
JAPAN AUTOMOBILE MANUFACTURERS ASSOCIATION, INC.

Total highway length is 9,000km , Tunnel among these is 701km(#743)
⇒In Japan, highway has many difficult situation for AD

page14

Road geometry Critical parameter (3/3)

JAMA
JAPAN AUTOMOBILE MANUFACTURERS ASSOCIATION, INC.

Situation		Perception	Traffic	Vehicle Dynamics
Complicated interchange			Short lane distance of merging and merged	
Blind curve		Curve-R(FOV Limitation)		High Gy
No center zone		Center zone width (non-regulated) Palls		
Narrow tunnel		Reflection noise Shoulder width (non-regulated)		
Bulkhead in marge		Reflection noise Shoulder width (non-regulated)		
Prevention palls for wrong-way		Marge point Palls		

page15

Traffic participants' position and behavior

JAMA
JAPAN AUTOMOBILE MANUFACTURERS ASSOCIATION, INC.

Multiple simple combination of simple sub-scenario can be holistic scenario from safety point of view.
a) Define 8+1 position around ego-vehicle and movement which can invade the ego trajectry.
b) Combine a) into one scenario in order to limit the avoidable trajectory

JAMA's Approach

a

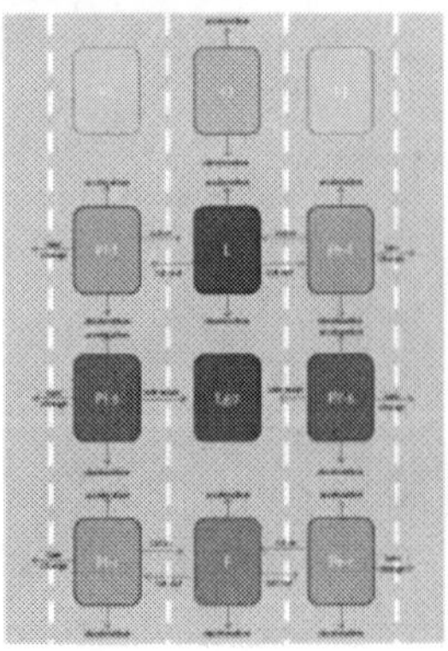

b

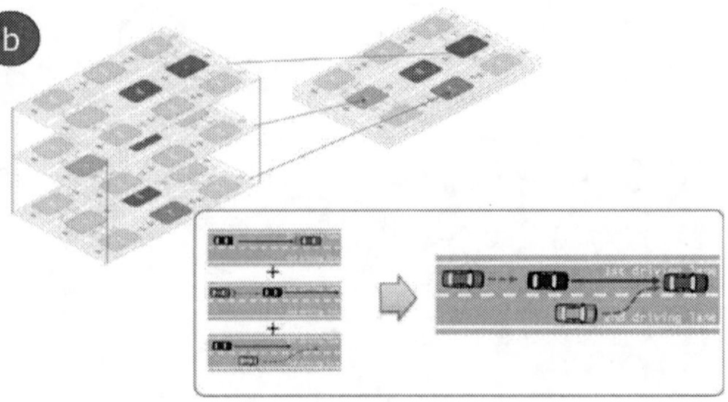

page16

Traffic Scenario Overview

JAMA
Japan Automobile Manufacturers Association, Inc.

Scenario is covered by a combination of ODD definition and surrounding environment definition

Road geometry ✖ Ego-vehicle behavior ✖ Traffic participants' position ✖ surrounding vehicle behavior

Road geometry	Ego-vehicle behavior: Lane keep	Ego-vehicle behavior: Lane change
Main roadway	FreeDriving Following	Lane change Overtaking
Marge	Merging infront	Merge
Branch	-	Branch
Ramp	FreeDriving Following	Lane change Overtaking

Traffic participants' position	surrounding vehicle behavior: Cut in	Cut out	Acceleration	Deceleration (Stop)	Sync
1		○		○	
2			○		
3	○			○	
4					○
5	○		○		
6	○			○	
7					○
8	○		○		

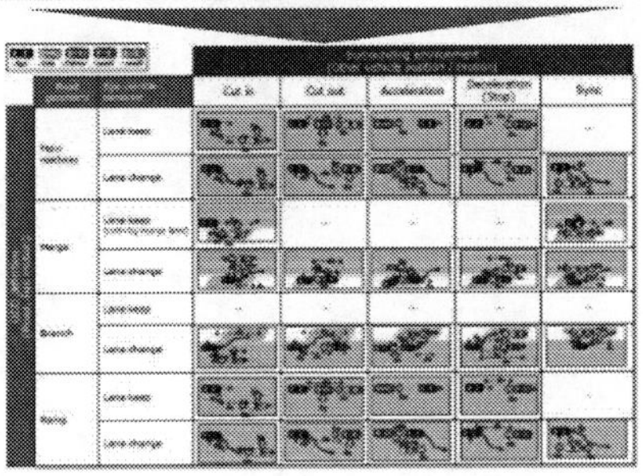

page17

1. Japanese Industry Research Activity Organization
2. AD Safety Validation Strategy
3. Data base
4. Traffic Scenario
5. How safe is safe enough

Two aspects of traffic scenario

JAMA

AD safety performance should be evaluated from the view point of capability of handling 1)Potential risk and 2)unexpected imminent risk.

1) Pre Extreme Condition: piece of mind ⇒『STANDARD』
2) Extreme Condition: potential injury ⇒『REGULATION』

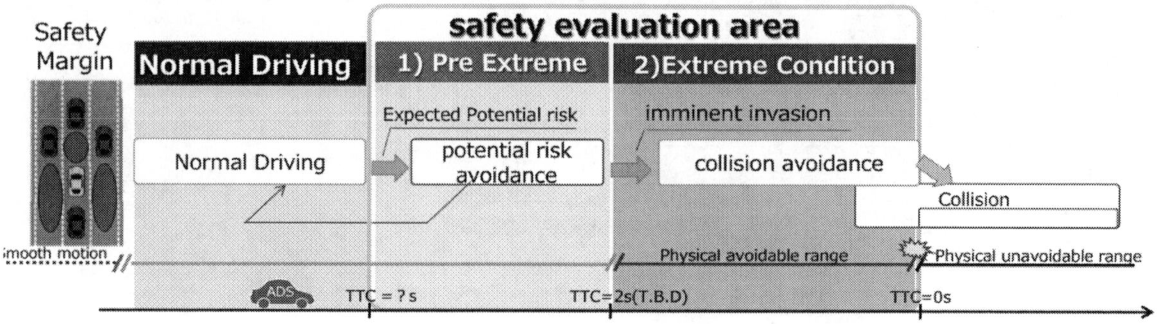

page19

Safety Criteria Candidates

JAMA

Wight each scenario by the exposure rate

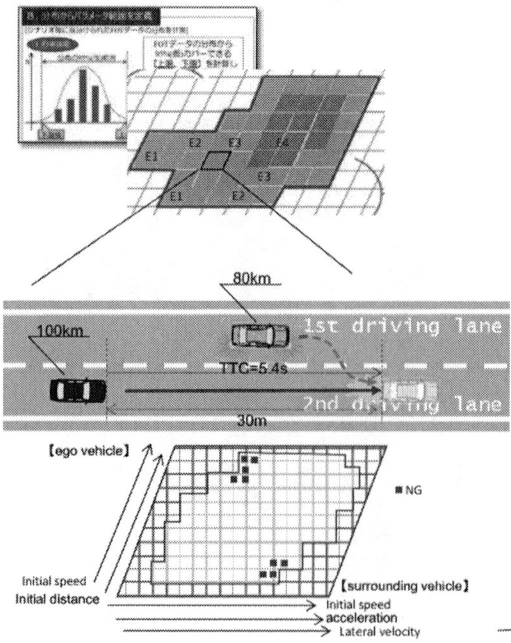

Draft1:better than human driver
comparing the sum of the unavoidable scene exposure rate between ADS model and driver model
・ΣExposureRate_AD < ΣExposureRate_Driver

Draft2: absolute target value of risk
comparing the sum of the unavoidable scene exposure rate of AD and absolute target value, which is used in other industry, like aero space, train, plant.
・ ΣExposureRate_AD $<10^{-9}$

page20

Challenges of Methodology and Criteria

JAMA
Japan Automobile Manufacturers Association, Inc.

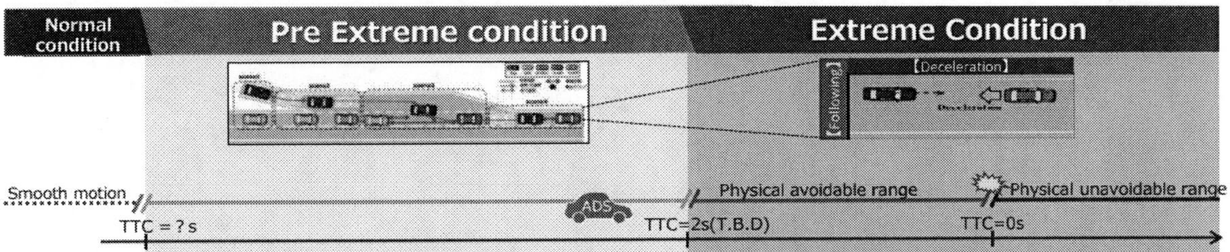

Item	Contents (Pre Extreme)	Possibility	Issues	Contents (Extreme)	Possibility	Issues
Methodology (Other vehicle motion)	Feedback model : Interaction model of ego and other vehicles (VISSIM model, SUMO, etc.) [Other car → Traffic flow; Ego car]	Δ	• How to clarify the range of Pre Extreme scenario and to clarify evaluation range. • What is the range that existing simulation can cover (VISSIM,SUMO, etc.) • What model and method can evaluate the scenarios efficiently.	Feedforward model (Approximation of short time motion) [Other car → Traffic flow; Ego car]	o	• Adequacy of scenario category → validation of traffic flow data • Validation range of initial condition and vehicle motion → limit of driver ~ Physical limit of vehicle?
	Feedforward	Δ				
Metrics (Criteria)	①Comparison with the driver (e.g. : Extreme occurrence rate is smaller than the driver)	Δ	• How to compare the risk of ADS and the driver (How to compare route and ego vehicle motion)	①Comparison with the driver (e.g. : Collision occurrence rate is smaller than the driver)	Δ	How to define global harmonized driver capability (Expert~ Beginner)
	②Absolute value index (e.g. : Extreme occurrence rate is smaller than xx%)	o	(Metrics consistency of Pre-Extreme and Extreme ①or②)	②Absolute value index (e.g. : Collision occurrence rate is smaller than xx%)	×?	Possibility of low social acceptability (How to set the metrics with acceptable reason)

page21

Alternative Traffic Scenario Criteria Draft

JAMA
Japan Automobile Manufacturers Association, Inc.

- There are unavoidable accidents because of the vehicles which don't comply with the traffic rules even if AD drives safely.
- Then the criteria which can judge the responsibility of the AD is needed.

■ Fault Rate Index in Japan

The Japanese lawyers association published "fault rate index for traffic lawsuit" to make judges fair and transparent.
Since first issue (1948), the court and insurance company have been using the criteria.
This implies Japanese society has an common understanding of the criteria.

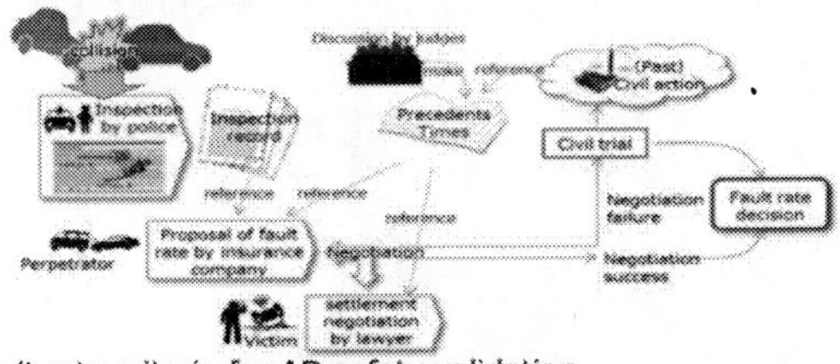

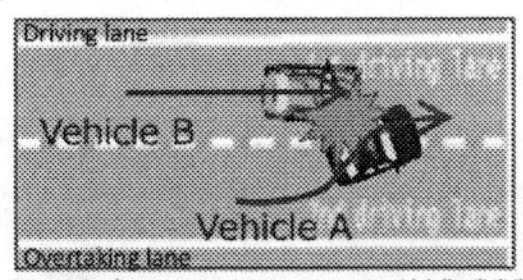

Total damages amount ¥10,000,000
Basic fault rate of lane change
A : B = 70 : 30
Damages burden amount
A : B = ¥7,000,000 : ¥3,000,000

■ Application of fault rate criteria for AD safety validation

【Draft Criteria】
AD is not responsible for primary factor of the accident, which means the fault rate of AD is minimum among accident participants.

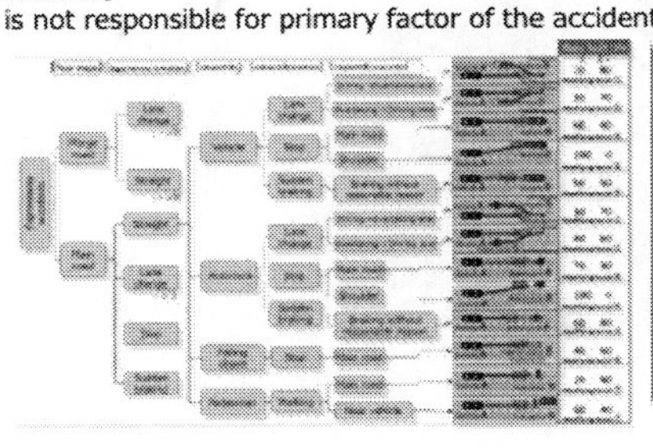

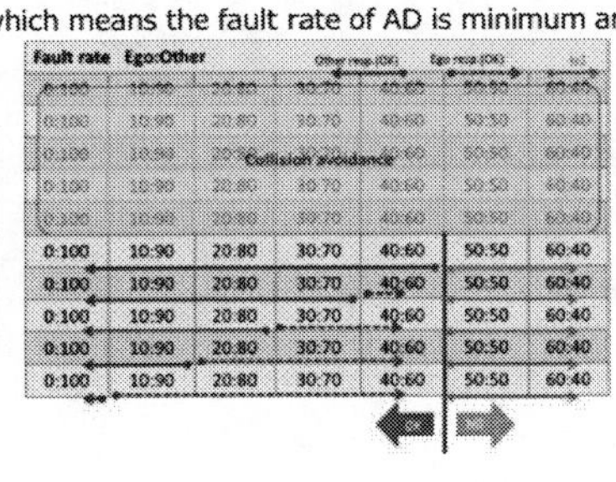

page22

bast

Safety requirements for automated vehicles – a scenario-based approach

Claus Pastor
Department Automotive Engineering
Section Automated Driving

Bundesanstalt für Straßenwesen

History – The Red Flag Act

bast

The Locomotive Act 1865 stipulated that any self-propelled road vehicle had to be preceded by a person walking at least 60 yards ahead, carrying a red flag

Google picture search

Claus Pastor AVS - July, 10th 2018 2

Objectives

- Expectations with regard to Automated and Connected Driving
- Considerations on human driving
- Machines outperforming human drivers ?
- Considerations on scenario-based description
- Performance requirements

Expectations

Ethical Commission – rules for automated & connected driving

Guiding ethical rules for automated and connected driving have been developed by German Ethical Commission (report published in June, 2017).

The primary purpose of [...] automated transport systems is to improve safety for all road users.

The licensing of automated systems is not justifiable unless it promises [...] a positive balance of risks.

[...] technologically unavoidable residual risks do not militate against the introduction of automated driving if the balance of risks is fundamentally positive.

To ensure the safety of all road users, there should be an analysis of hazardous situations that are relevant to erroneous perception and behaviour by the vehicle.

It would appear advisable to hand over relevant scenarios to a central scenario catalogue at a neutral body in order to develop [...]standards, including any acceptance tests.

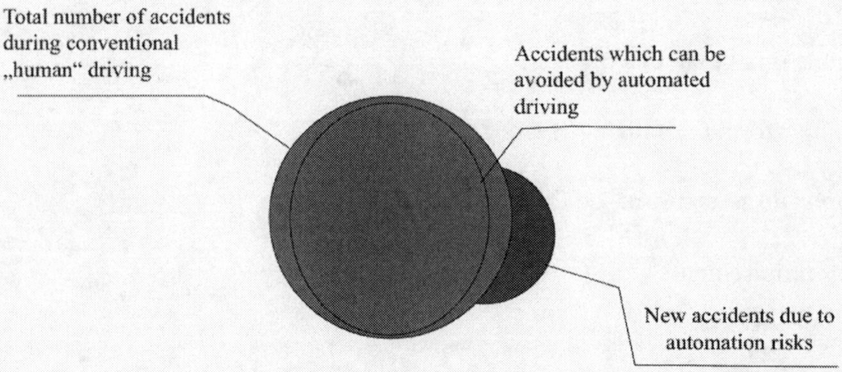

The more information on the traffic situation is available to the driver, the safer the driving status is.

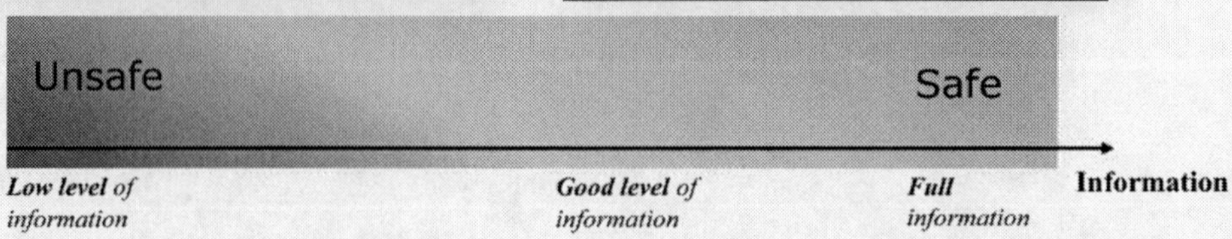

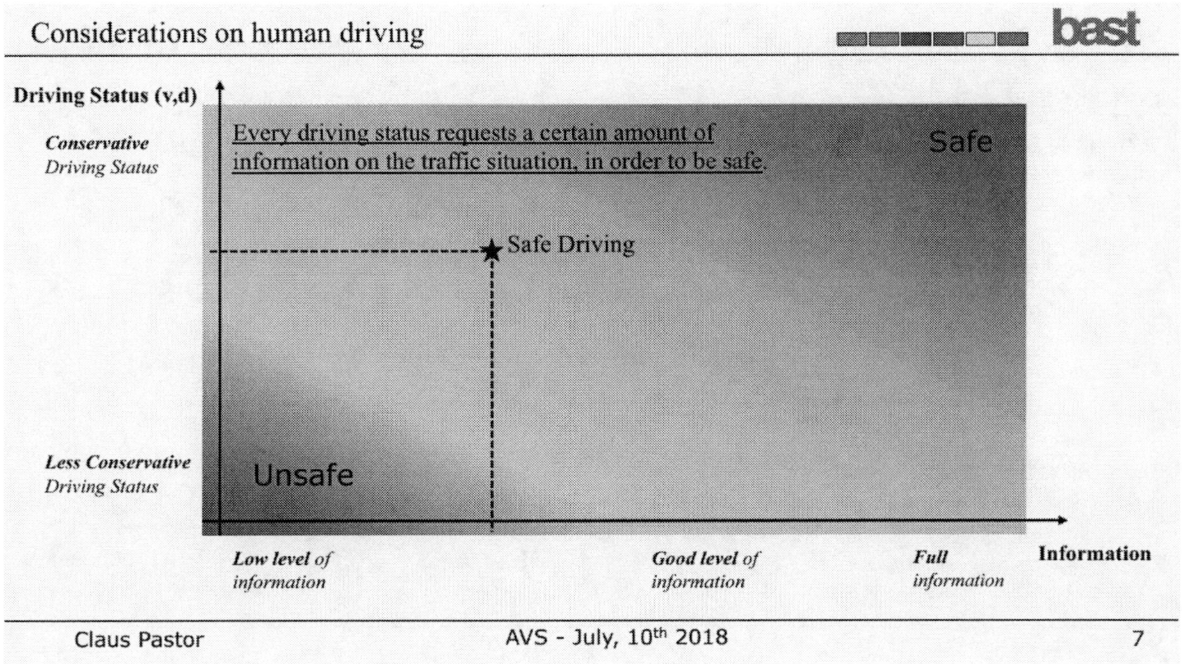
Considerations on human driving
bast
Driving Status (v,d)
Conservative Driving Status
Less Conservative Driving Status
Every driving status requests a certain amount of information on the traffic situation, in order to be safe.
Safe
Safe Driving
Unsafe
Low level of information
Good level of information
Full information
Information
Claus Pastor
AVS - July, 10th 2018
7

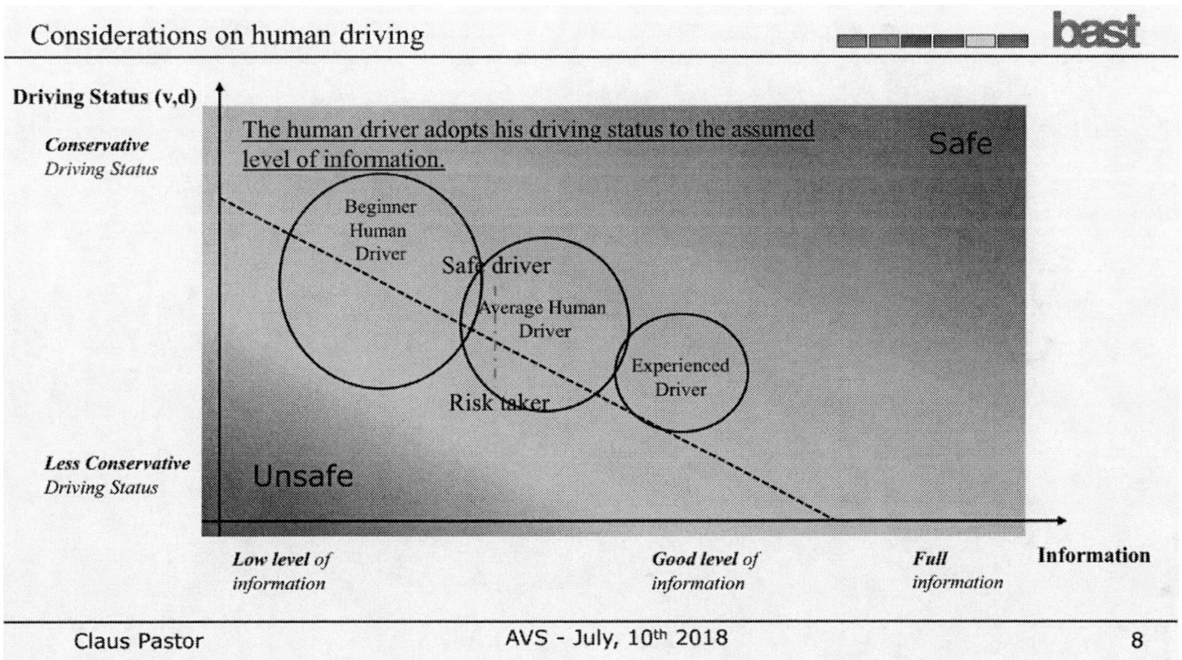
Considerations on human driving
bast
Driving Status (v,d)
Conservative Driving Status
Less Conservative Driving Status
The human driver adopts his driving status to the assumed level of information.
Safe
Beginner Human Driver
Safe driver
Average Human Driver
Experienced Driver
Risk taker
Unsafe
Low level of information
Good level of information
Full information
Information
Claus Pastor
AVS - July, 10th 2018
8

Considerations on human driving

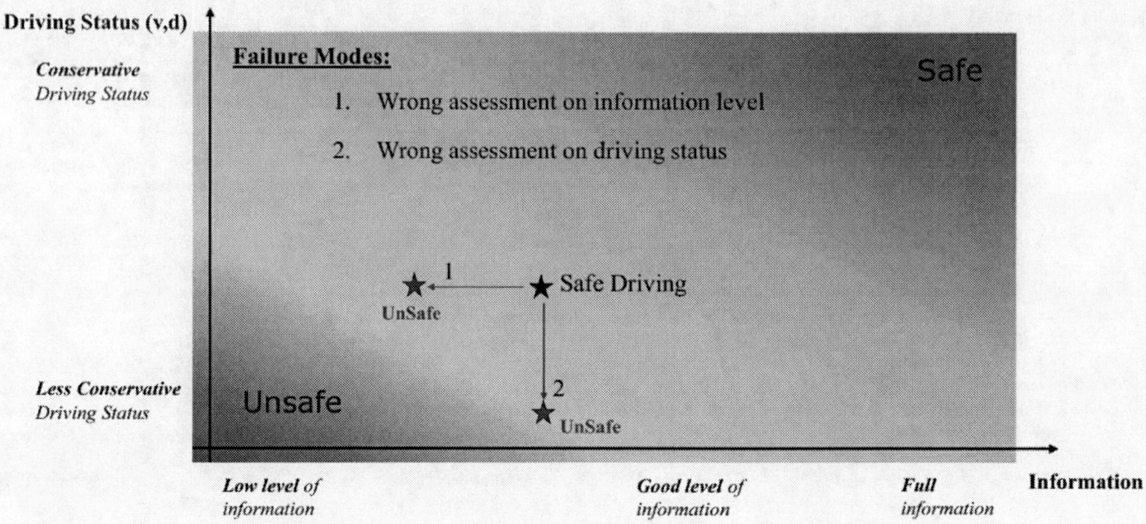

Considerations on human driving

Why do human drivers cause accidents?

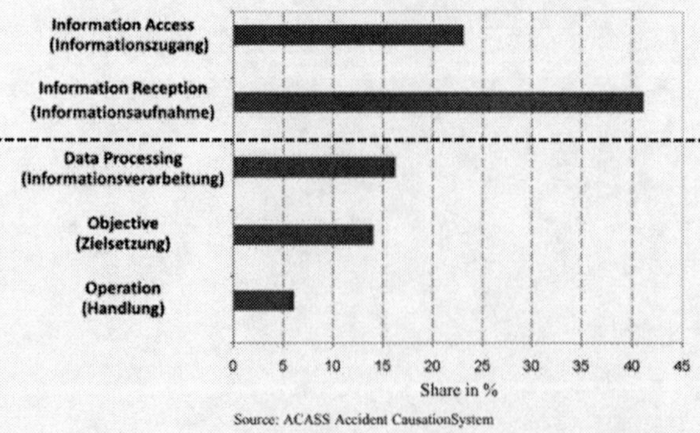

Information related causation
64 %

Driving status related causation
36 %

Machines outperforming human drivers?

Why do machines cause less accidents?

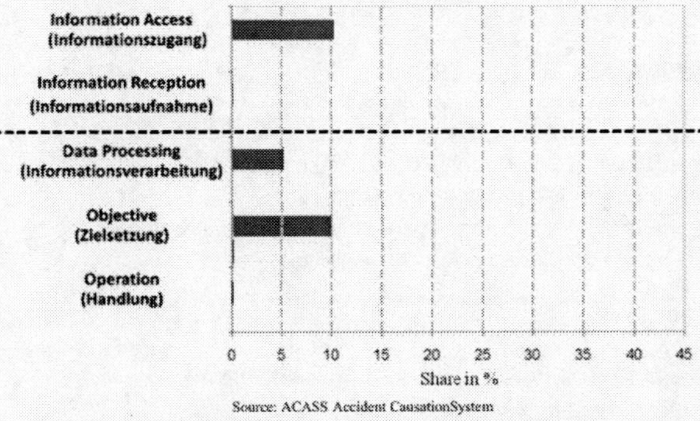

Information related causation

64 % -> 10 %

estimated

Driving status related causation

36 % -> 5 %

estimated

Machines might outperform human drivers by as much as 85%

Claus Pastor AVS - July, 10th 2018 11

Machines outperforming human drivers?

Information related causation (64%)

Issues:

Relevant Information
- unavailable
  (e.g.health issues (e.g.vision); masking by ext. / int. objects; glare)
- available but not captured
  (e.g. distraction; information overload)

Machine perspective
- ~~health issues (e.g.vision, hearing ..)~~
  - but hardware (sensor) failures, misclassification
- masking by external objects (C2C, C2X, improved sensor technology)
- ~~masking by internal objects~~ (sensors outside of vehicle interior)
- weather, ~~low contrast, glare~~ ..(improved sensor technology)
- ~~distraction; information overload~~

Claus Pastor AVS - July, 10th 2018 12

Machines outperforming human drivers?

bast

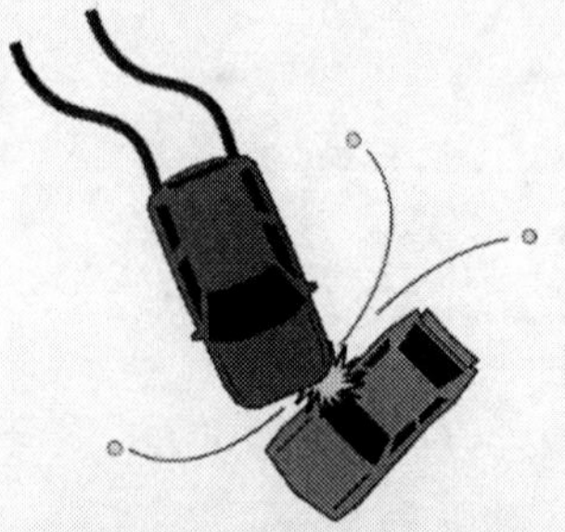

Driving status related causation 36 %

Issues:

Relevant Information available and captured

- but misinterpreted
  (e.g.wrong assessment of speed / distance; traffic rule violation)
- correctly interpreted but wrong decison making
  (e.g. in-appropriate manoeuvre)
- correctly interpreted and right decison making but handling error
  (e.g. confusion of throttle and brake, low braking force)

Machine perspective

- ~~wrong assessment of speed / distance; traffic rule violation~~
- ~~safety-related wrong manoeuvre~~
- ~~handling error (confuse throttle and brake, etc.)~~
  - but transition errors (Level 3)

Machines outperforming human drivers?

bast

Why do human drivers cause accidents?

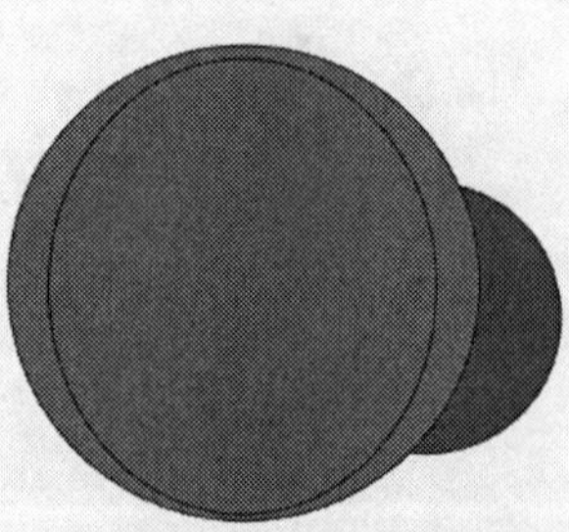

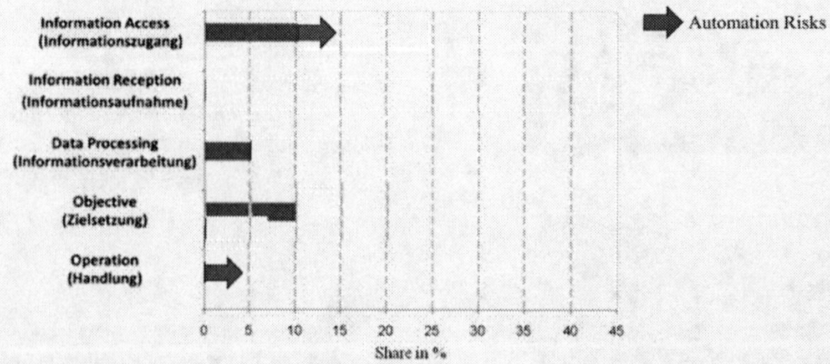

Machines might outperform human drivers by as much as 85%

but automation risks exists – and remain difficult to quantify (by now)

Considerations on scenario-based description

Automation risks -> way out:
Ensuring safe performance by looking into "representative" [critical] traffic scenarios

What is an appropriate „Traffic Scenario Description" for assessing automated driving functions / systems?

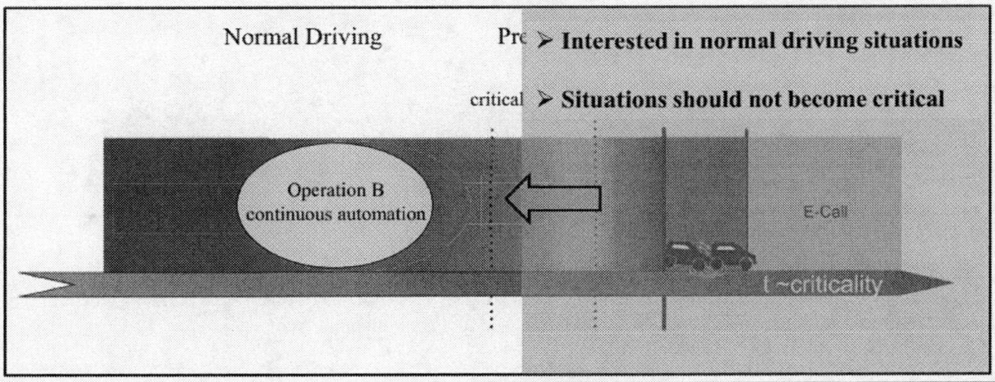

Claus Pastor AVS - July, 10th 2018 15

Example from accident data (GIDAS)

Traffic and Accident Scenario description:
4 passenger cars change consecutively to the deceleration-lane.
A 5th car overtakes all 4 cars and cuts in just in front of the first car in line.

Consequently the 1st, 2nd and 3rd car brake hard. The 4th car does not manage to brake in time and hits the vehicle in front.

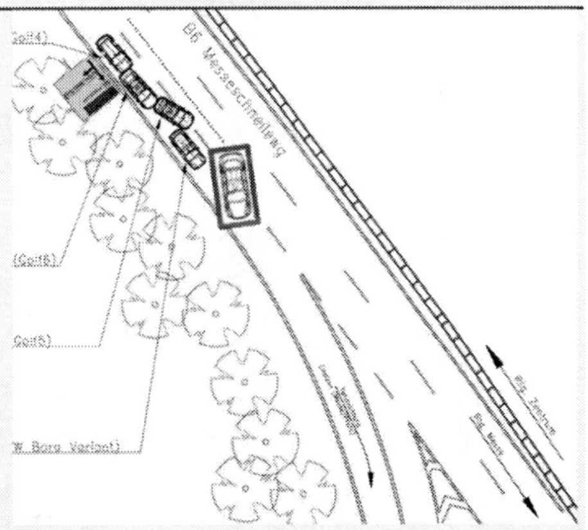

Claus Pastor AVS - July, 10th 2018 16

Considerations on scenario-based description

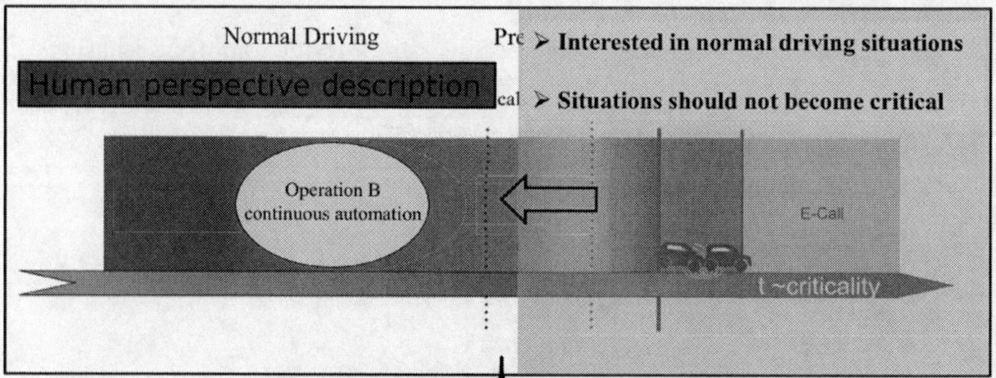

Consequently the 1st, 2nd and 3rd car brake hard.

4 passenger cars change consecutively to the deceleration-lane. A 5th car overtakes all 4 cars and cuts in just in front of the first car in line.	The 4th car does not manage to brake in time and hits the vehicle in front.

Claus Pastor AVS - July, 10th 2018 17

Considerations on scenario-based description

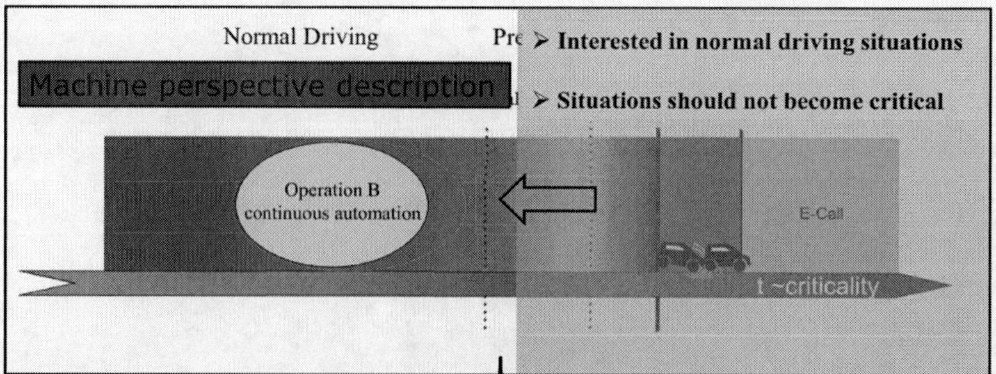

Suddenly the car ahead brakes hard.

The Ego car is following another car into a deceleration lane.	The Ego car does not manage to brake in time and hits the vehicle in front.

Claus Pastor AVS - July, 10th 2018 18

Considerations on scenario-based description

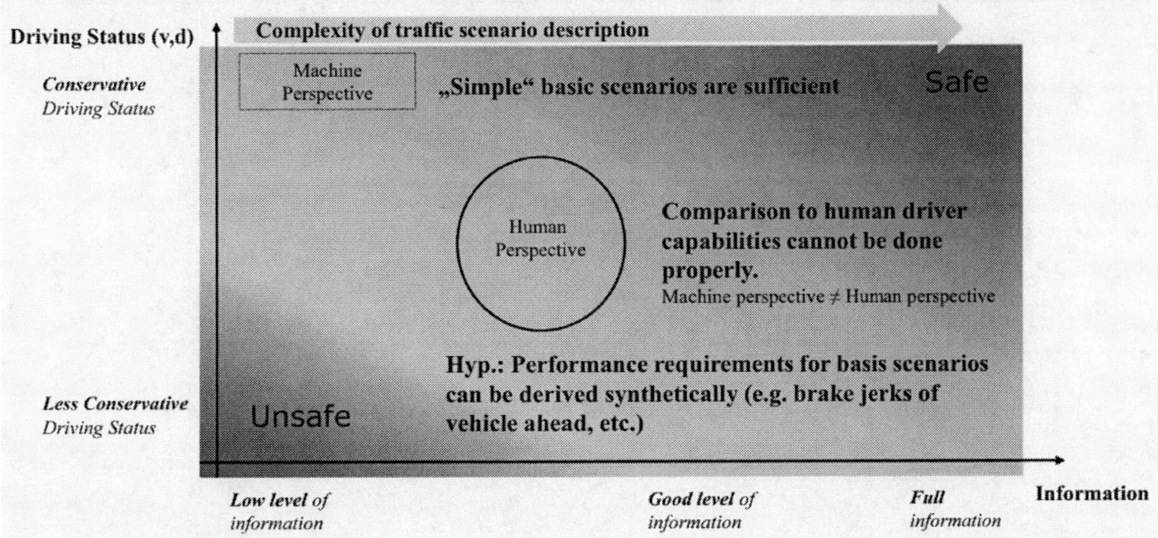

Claus Pastor AVS - July, 10th 2018 19

Considerations on scenario-based description

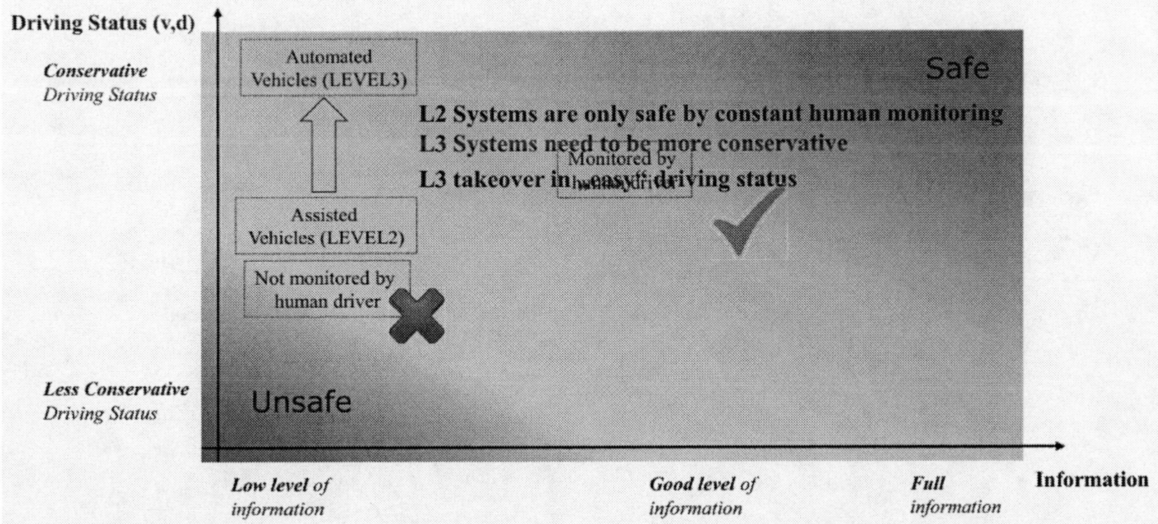

Claus Pastor AVS - July, 10th 2018 20

Considerations on scenario-based description

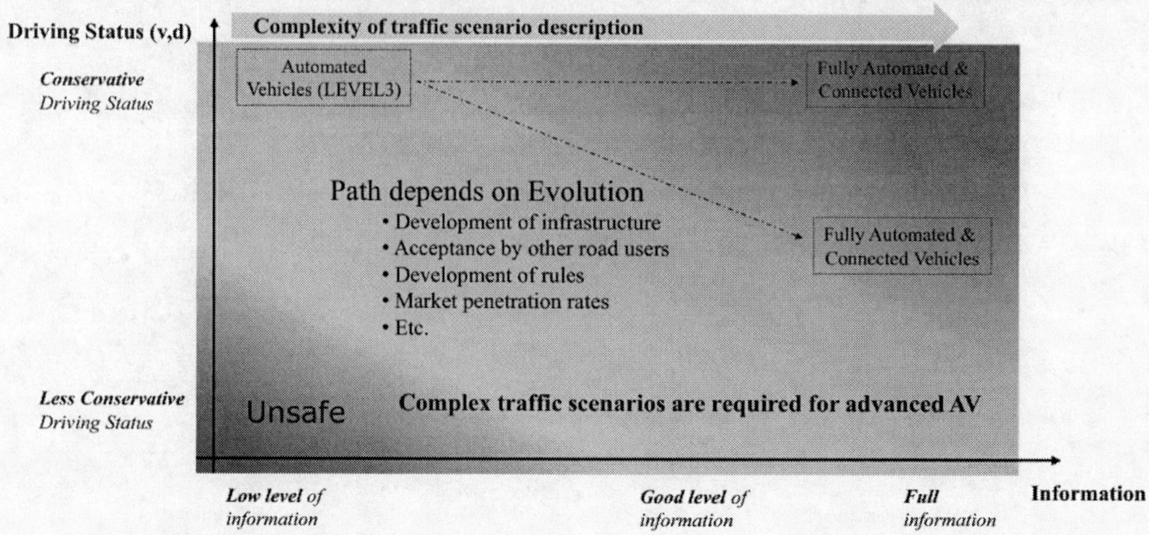

Claus Pastor AVS - July, 10th 2018 21

Summary

- 1st Generation AV can be evaluated by using basic simple scenarios, derived from real world accidents.
- Using simple scenarios, a comparison between human and machine driver performance is invalid.
- Performance requirements for basis scenarios can be derived synthetically.
- Various evolution paths towards next Gen. AV and fully connected driving can be expected.
- To assess smart technologies and connected driving, complex traffic scenarios are required

Claus Pastor AVS - July, 10th 2018 22

Bundesanstalt für Straßenwesen
Federal Highway Research Institute

Claus H. Pastor

Automated Driving

Brüderstraße 53
D-51427 Bergisch Gladbach

Phone + 49 (0)2204 43 5411
Fax + 49 (0)2204 43 5450

pastor@bast.de

INSTITUT FÜR KRAFTFAHRZEUGE
ika | RWTHAACHEN UNIVERSITY

How to Identify the Safety Effect of Automated Driving?

Christian Roesener, Friederike Hennecke, Jan Sauerbier,
Adrian Zlocki, Dirk Kemper, Lutz Eckstein and Markus Oeser

Safety Impact Assessment of Automated Driving
Motivation

INSTITUT FÜR KRAFTFAHRZEUGE
ika | RWTHAACHEN UNIVERSITY

"[..] the licensing of automated systems is not justifiable unless it promises to produce at least a diminution in harm compared with human driving, in other words a positive balance of risks [..]"

„German Ethics Commission on Connected and Automated Driving",
Di Fabio et al.

Safety Impact Assessment of Automated Driving

Framework for Safety Impact Assessment

INSTITUT FÜR KRAFTFAHRZEUGE ika RWTH AACHEN UNIVERSITY

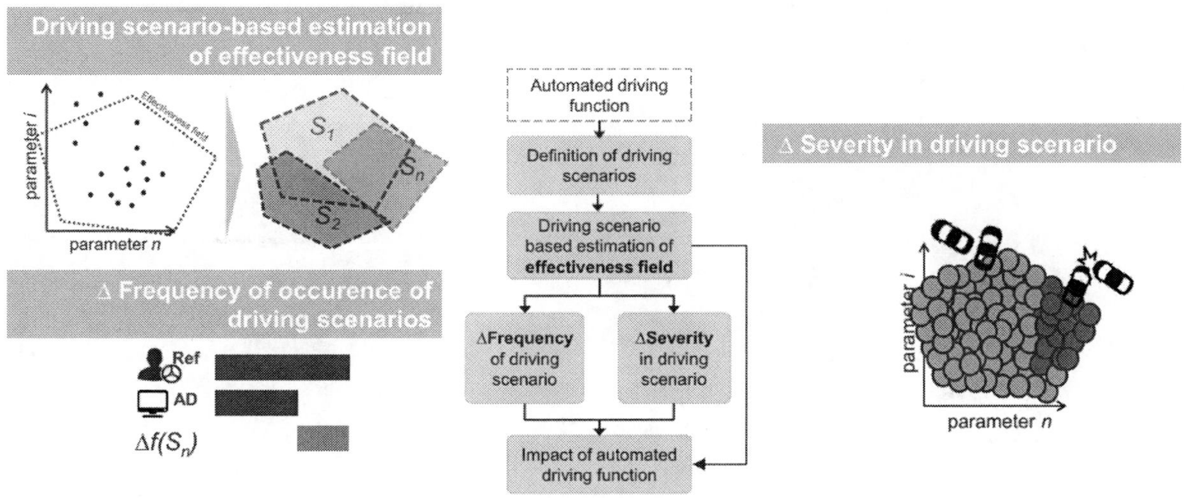

Safety Impact Assessment of Automated Driving

Driving scenario-based estimation of effectiveness field

INSTITUT FÜR KRAFTFAHRZEUGE ika RWTH AACHEN UNIVERSITY

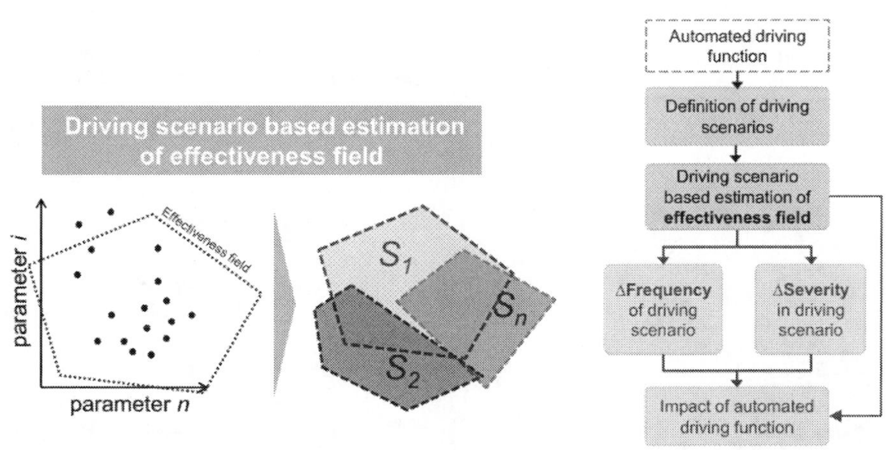

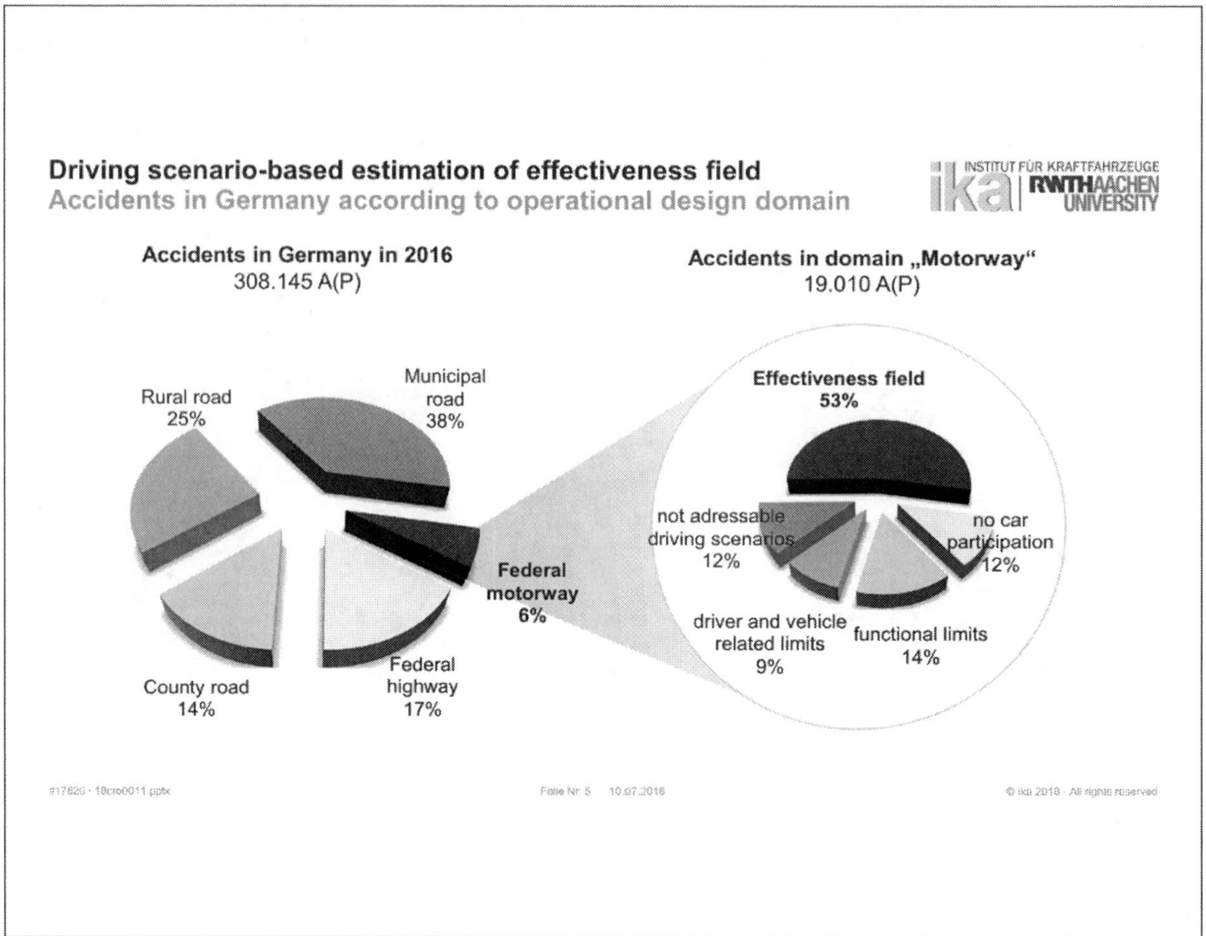
Driving scenario-based estimation of effectiveness field
Accidents in Germany according to operational design domain
INSTITUT FÜR KRAFTFAHRZEUGE
ika
RWTHAACHEN UNIVERSITY
Accidents in Germany in 2016
308.145 A(P)
Accidents in domain „Motorway"
19.010 A(P)
Rural road 25%
Municipal road 38%
Federal motorway 6%
County road 14%
Federal highway 17%
Effectiveness field 53%
not adressable driving scenarios 12%
no car participation 12%
driver and vehicle related limits 9%
functional limits 14%
#17620 · 18cro0011.pptx
Folie Nr. 5 10.07.2018
© ika 2018 · All rights reserved

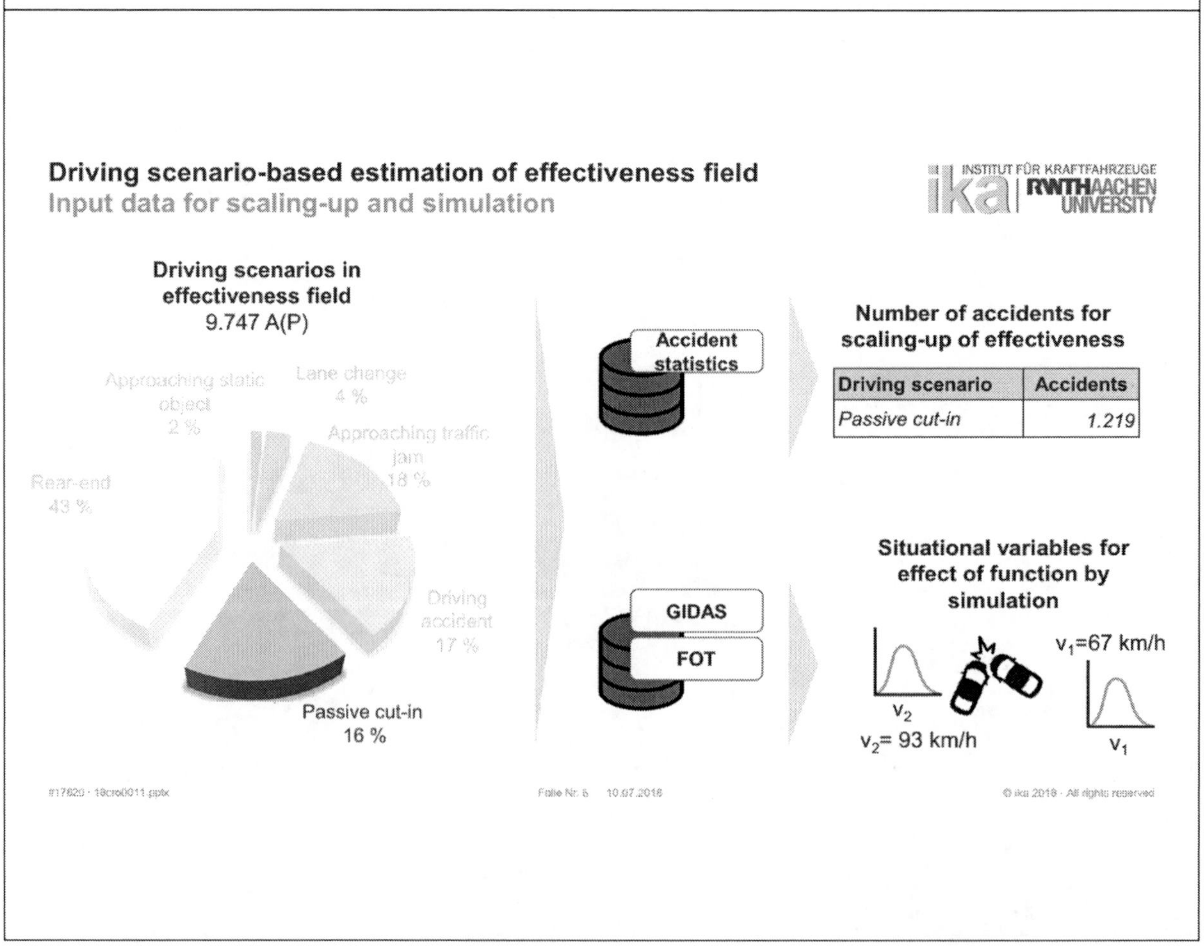
Driving scenario-based estimation of effectiveness field
Input data for scaling-up and simulation
INSTITUT FÜR KRAFTFAHRZEUGE
ika
RWTHAACHEN UNIVERSITY
Driving scenarios in effectiveness field
9.747 A(P)
Approaching static object 2 %
Lane change 4 %
Approaching traffic jam 18 %
Rear-end 43 %
Driving accident 17 %
Passive cut-in 16 %
Accident statistics
GIDAS
FOT
Number of accidents for scaling-up of effectiveness
Driving scenario | Accidents
Passive cut-in | 1.219
Situational variables for effect of function by simulation
v₁=67 km/h
v₂
v₂= 93 km/h
v₁
#17620 · 18cro0011.pptx
Folie Nr. 6 10.07.2018
© ika 2018 · All rights reserved

Safety Impact Assessment of Automated Driving
Identification of Δ frequencies of driving scenarios

INSTITUT FÜR KRAFTFAHRZEUGE ika | RWTH AACHEN UNIVERSITY

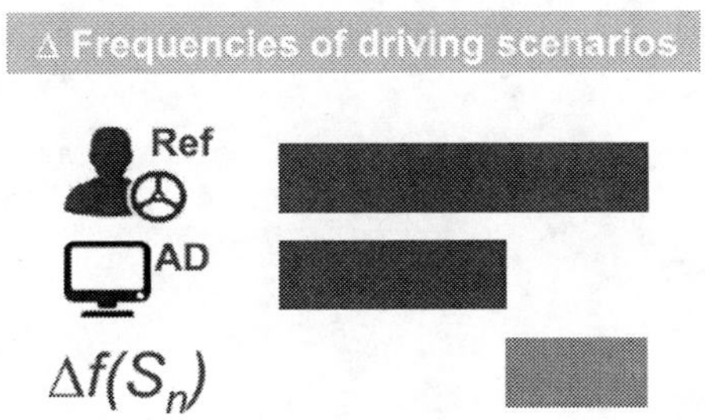

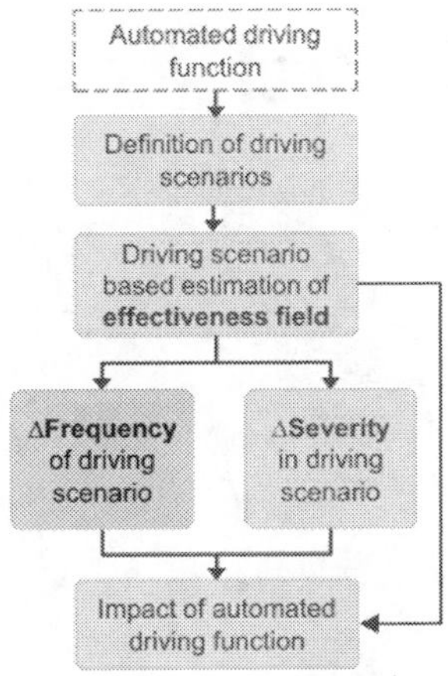

#17620 · 18cro0011.pptx Folie Nr. 7 10.07.2018

Identification of Δ frequencies of driving scenarios
FOT-data

INSTITUT FÜR KRAFTFAHRZEUGE ika | RWTH AACHEN UNIVERSITY

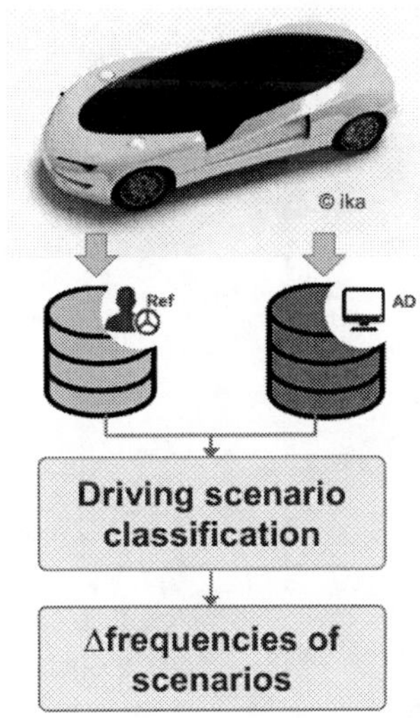

#17620 · 18cro0011.pptx Folie Nr. 8 10.07.2018

Identification of Δ frequencies of driving scenarios
Traffic simulation

INSTITUT FÜR KRAFTFAHRZEUGE ika | RWTH AACHEN UNIVERSITY

#17620 · 18cro0011.pptx Folie Nr. 9 10.07.2018

Safety Impact Assessment of Automated Driving
Identification of Δ severity in driving scenarios

INSTITUT FÜR KRAFTFAHRZEUGE ika | RWTH AACHEN UNIVERSITY

Δ Severity in driving scenarios

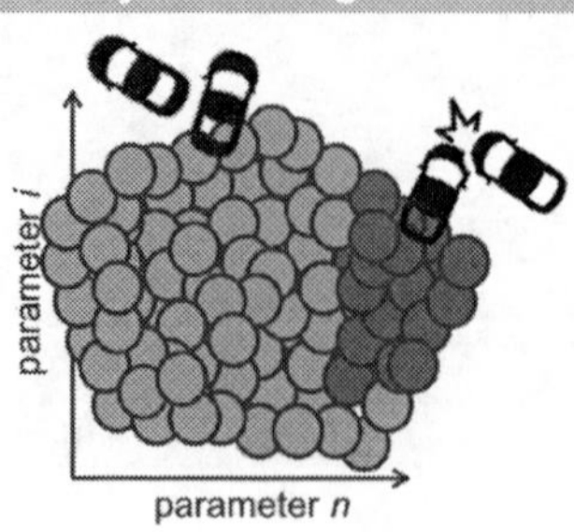

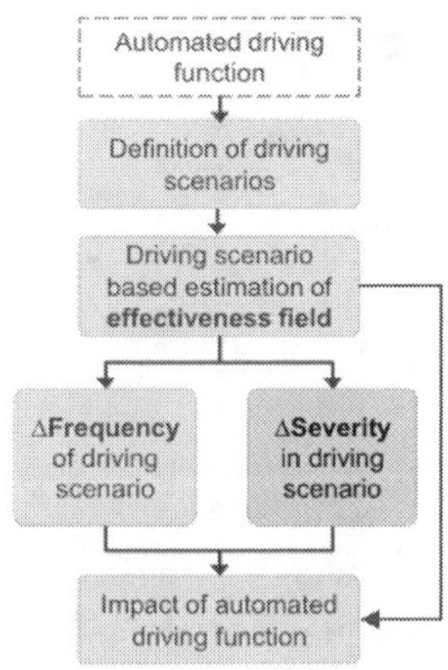

#17620 · 18cro0011.pptx Folie Nr. 10 10.07.2018

Δ Severity in driving scenarios by re-simulation
Simulation framework

INSTITUT FÜR KRAFTFAHRZEUGE
ika | RWTHAACHEN UNIVERSITY

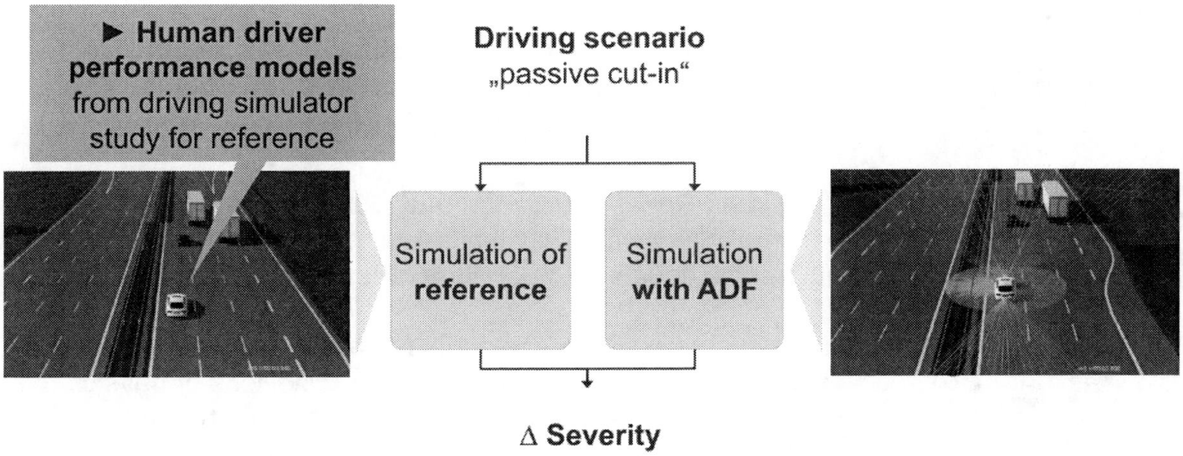

Safety Impact Assessment of Automated Driving
Results

INSTITUT FÜR KRAFTFAHRZEUGE
ika | RWTHAACHEN UNIVERSITY

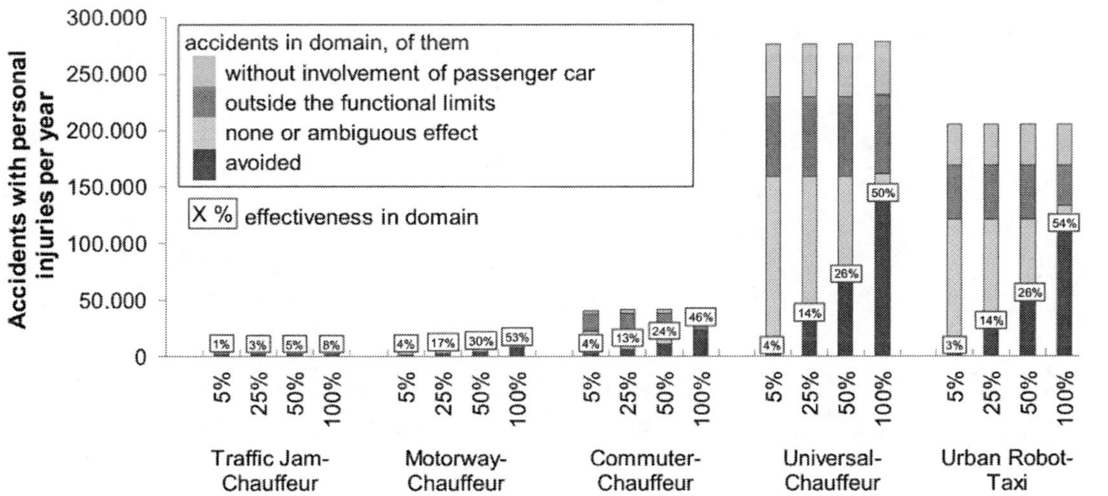

Safety Impact Assessment of Automated Driving
Key results

INSTITUT FÜR KRAFTFAHRZEUGE ika RWTH AACHEN UNIVERSITY

- Motorway-Chauffeur can **reduce 30 % of all accidents** on **German motorways** at a market **penetration of 50 %**. This equals **2 % of all accidents on German roads**.

- The Urban Robot-Taxi **can avoid 26 % of all accidents** with personal injury within city-limits at a market **penetration of 50 %**. This equals **17 % of all accidents on German roads**.

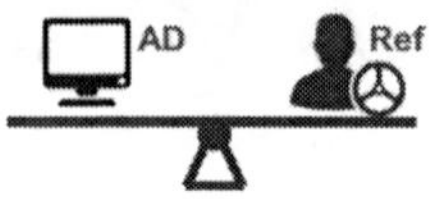

- However, there will be accidents remaining that automated vehicles cannot avoid (due to weather conditions or physics). **But we can show that a human cannot avoid these accidents either.**

Safety Impact Assessment of Automated Driving
Outlook

INSTITUT FÜR KRAFTFAHRZEUGE ika RWTH AACHEN UNIVERSITY

- **Changes in frequencies** of driving/accident scenarios have to be taken into account for safety assessment of AD.

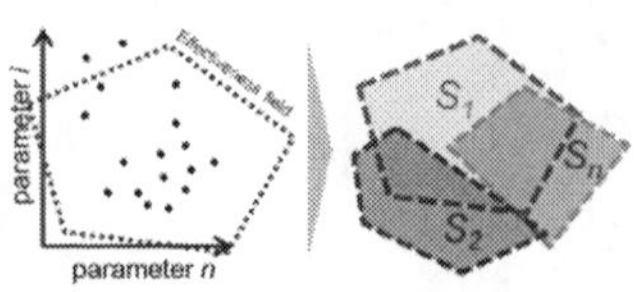

- Methods for **classification of driving scenarios** from FOT-, in-depth accident and other data sources have to be developed.

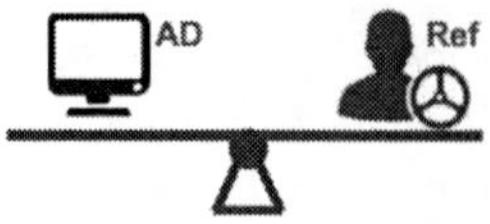

- **Human driver performance** models have to be developed and parametrized from FOT-, in-depth accident and driving simulator studies

Contact

Christian Roesener, Jan Sauerbier,
Adrian Zlocki, Lutz Eckstein

Institute for Automotive Engineering (ika)
RWTH Aachen University
Steinbachstr. 7
52074 Aachen
Germany

Friederike Hennecke, Dirk Kemper,
Markus Oeser

Chair and Institute for Highway Engineering (isac)
RWTH Aachen University
Mies-van-der-Rohe-Straße 1
52074 Aachen
Germany

Phone +49 241 80 23891
Fax +49 241 80 22147

Email christian.roesener@ika.rwth-aachen.de
Internet www.ika.rwth-aachen.de

#17620 · 18cro0011.pptx Slide No. 16 2018/07/10

AUTOMATED VEHICLES SYMPOSIUM 2018
USERS. VEHICLES. INFRASTRUCTURE.
Driven by AUVSI and TRB

Breakout Session 16: Safety Assurance of Automated Driving

- Agenda
- Introduction and report on AVS 2017
- Day 1: State of the Art and Safety Level for Safety Assurance of AV
- **Day 2: Wednesday: Approaches and Methodologies to assure Safety of AV**
- Way Forward

AUTOMATED VEHICLES SYMPOSIUM 2018
USERS. VEHICLES. INFRASTRUCTURE.
Driven by AUVSI and TRB

Breakout Session 16: Agenda

Day 2, Wednesday: Approaches and Methodologies to assure Safety of AV

- ***Introduction*** — ***Lutz Eckstein (ika)***
- *Approaches and Methodologies to assure safety of Automated Vehicles*
  - French approach for safety validation by combination of experiment and simulation — J.-F. Sencerin (Renault)
  - The Automated Mobility Partnership. An approach for utilizing naturalistic driving data to create simulations for evaluating automated systems — T. Dingus (VTTI)
  - AD safety validation investigation in Japan — S. Taniguchi (Toyota, JAMA)
- *Continued after Break*
  - Role of Proving Grounds in Three-Mode AV Validation Methodology — John Maddox (ACM)
  - The Innovations in Simulation that are making AVs safer — C. Partridge (Metamoto)
  - New Ideas on Safety and Certification for full autonomy — P. Perrone (perronerobotics)
  - Standardization of Data Formats for increased efficiency of simulation applications — M. Dupuis (VIRES)
- *Synchronisation of Approaches* — *Lutz Eckstein (ika)*
  - J. Maddox (ACM), S. Taniguchi (Toyota), J.-F. Sencerin (Renault), T. Dingus(VTTI), C. Partridge (Metamoto), P. Perrone (perronerobotics), M. Dupuis (VIRES)
- *Wrap-up & Conclusions* — *Lutz Eckstein (ika)*

AUTOMATED VEHICLES SYMPOSIUM 2018
USERS. VEHICLES. INFRASTRUCTURE.
Driven by AUVSI and TRB

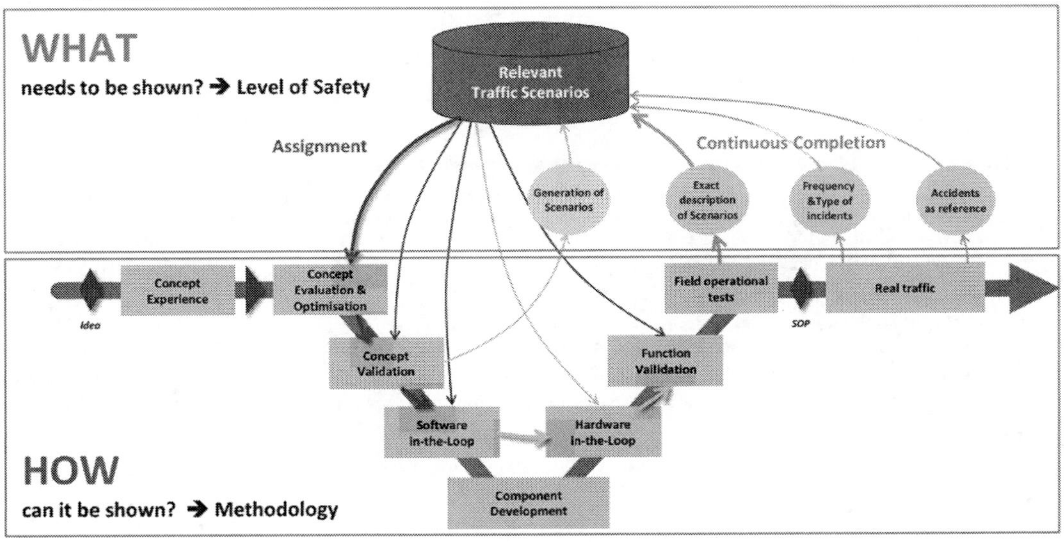

AUTOMATED VEHICLES SYMPOSIUM 2018
USERS. VEHICLES. INFRASTRUCTURE.
Driven by AUVSI and TRB

1. **Which role may simulation play?**
   - With respect to traffic flow, traffic scenarios, system performance, ... sensor performance?
   - For development and optimization by companies vs. certification by authorities?
2. **How can proving grounds contribute?**
   - Which scenarios should be depicted on proving grounds?
   - In which phase of product development? – only for sign-off and certification?
3. **Which role may tests on public roads play?**
   - What are prerequisites in terms of safety?
   - How to deal with updates of the AD functionality?
4. **How can results be combined to find a decision?**
   - Combination of results from different methodologies?
   - Balancing of different scenarios possible?

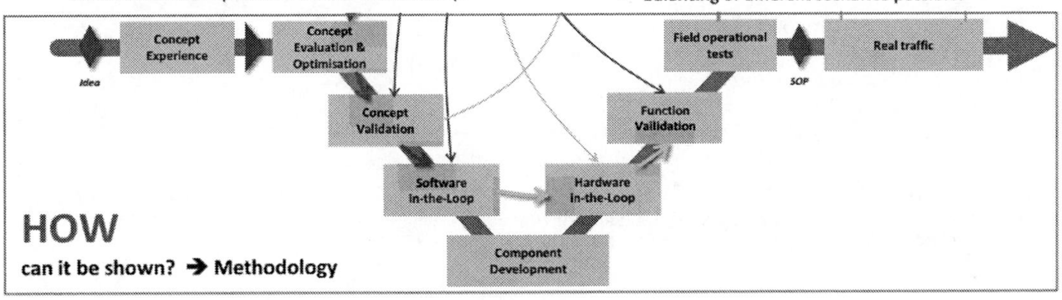

AUTOMATED VEHICLES SYMPOSIUM 2018
USERS. VEHICLES. INFRASTRUCTURE.

Driven by AUVSI and TRB

Breakout Session 16: Agenda

Day 2, Wednesday: Approaches and Methodologies to assure Safety of AV

- *Introduction* — *Lutz Eckstein (ika)*
- ***Approaches and Methodologies to assure safety of Automated Vehicles***
  - French approach for safety validation by combination of experiment and simulation — J.-F. Sencerin (Renault)
  - The Automated Mobility Partnership. An approach for utilizing naturalistic driving data to create simulations for evaluating automated systems — T. Dingus (VTTI)
  - AD safety validation investigation in Japan — S. Taniguchi (Toyota, JAMA)
- *Continued after Break*
  - Role of Proving Grounds in Three-Mode AV Validation Methodology — John Maddox (ACM)
  - The Innovations in Simulation that are making Avs safer — C. Partridge (Metamoto)
  - New Ideas on Safety and Certification for full autonomy — P. Perrone (perronerobotics)
  - Standardization of Data Formats for increased efficiency of simulation applications — M. Dupuis (VIRES)
- *Synchronisation of Approaches* — *Lutz Eckstein (ika)*
  - J. Maddox (ACM), S. Taniguchi (Toyota), J.-F. Sencerin (Renault), T. Dingus(VTTI), C. Partridge (Metamoto), P. Perrone (perronerobotics), M. Dupuis (VIRES)
- *Wrap-up & Conclusions* — *Lutz Eckstein (ika)*

AUTOMATED VEHICLES SYMPOSIUM 2018
USERS. VEHICLES. INFRASTRUCTURE.

Driven by AUVSI and TRB

Breakout Session 16: Agenda

Day 2, Wednesday: Approaches and Methodologies to assure Safety of AV

- *Introduction* — *Lutz Eckstein (ika)*
- *Approaches and Methodologies to assure safety of Automated Vehicles*
  - French approach for safety validation by combination of experiment and simulation — J.-F. Sencerin (Renault)
  - The Automated Mobility Partnership. An approach for utilizing naturalistic driving data to create simulations for evaluating automated systems — T. Dingus (VTTI)
  - AD safety validation investigation in Japan — S. Taniguchi (Toyota, JAMA)
- *Continued after Break*
  - Role of Proving Grounds in Three-Mode AV Validation Methodology — John Maddox (ACM)
  - The Innovations in Simulation that are making Avs safer — C. Partridge (Metamoto)
  - New Ideas on Safety and Certification for full autonomy — P. Perrone (perronerobotics)
  - Standardization of Data Formats for increased efficiency of simulation applications — M. Dupuis (VIRES)
- ***Synchronisation of Approaches*** — ***Lutz Eckstein (ika)***
  - J. Maddox (ACM), S. Taniguchi (Toyota), J.-F. Sencerin (Renault), T. Dingus(VTTI), C. Partridge (Metamoto), P. Perrone (perronerobotics), M. Dupuis (VIRES)
- *Wrap-up & Conclusions* — *Lutz Eckstein (ika)*

AUTOMATED VEHICLES SYMPOSIUM 2018

USERS. VEHICLES. INFRASTRUCTURE.

Driven by AUVSI and TRB

I. Harmonized Definitions and Taxonomy

II. Agreed Format for Scenarios

III. Process for Scenario Sharing

IV. Common Driver Reference Model

V. Agreed Classification of Logical Scenarios

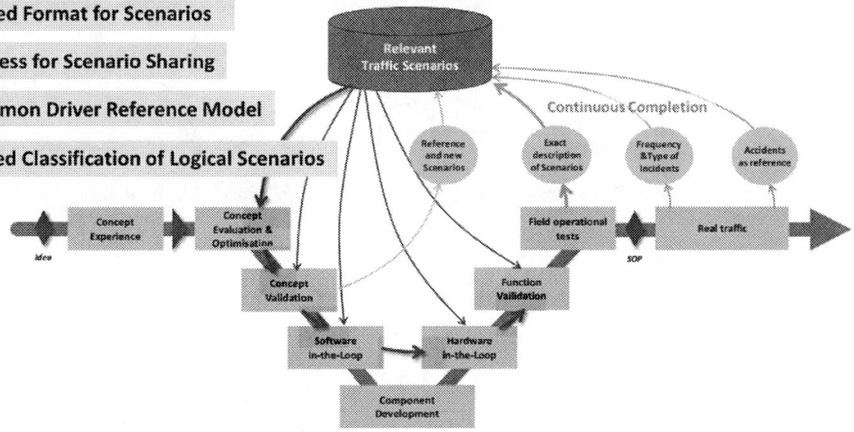

AUTOMATED VEHICLES SYMPOSIUM 2018

USERS. VEHICLES. INFRASTRUCTURE.

Driven by AUVSI and TRB

I. Harmonized Definitions and Taxonomy

II. Agreed Format for Scenarios

III. Process for Scenario Sharing

IV. Common Driver Reference Model

V. Agreed Classification of Logical Scenarios

Status & Process

- SAE already started
- DIN will also start End of August and cooperates with SAE
- ISO Working Group started, implicit harmonization via involved experts.

Definitions

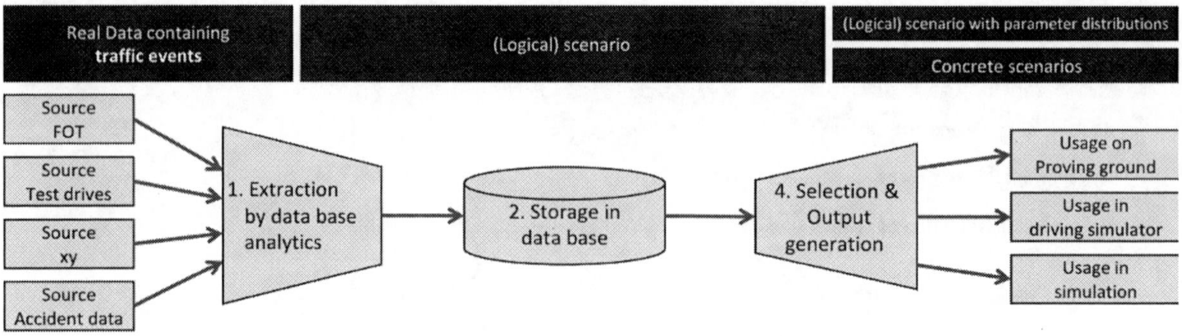

Definitions

- **Definition "Safety relevant traffic event"**

A safety relevant traffic event is a real time sequence in traffic (data) that almost or indeed led to an accident.

- **Definition "Scene"**

A scene is a snapshot that includes the moving and non-moving elements of the traffic environment, the self-representation of all actors and observers and the relations between those elements.

- **Definition "Logical scenario"**

A logical scenario is a model of a conditional sequence of scenes, which starts with a start scene, but does not further specify the behavior of the main actor (subject vehicle) from a defined point in time onwards, so that the result (accident or not) is hypothetical. (class)

- **Definition "Concrete scenario"**

A concrete scenario is a parameterized model of a time sequence of scenes (logical scenario), which begins with a concrete start scene. At a defined point in time, the behavior of the main actor (subject vehicle) is not further specified, so the result is hypothetical, but all other influencing factors are pre-defined by concrete parameters.

AUTOMATED VEHICLES SYMPOSIUM 2018

USERS. VEHICLES. INFRASTRUCTURE.

Driven by AUVSI and TRB

I. Harmonized Definitions and Taxonomy

II. Agreed Format for Scenarios

III. Process for Scenario Sharing

IV. Common Driver Reference Model

V. Agreed Classification of Logical Scenarios

PEGASUS

Scenario Description: Layer model

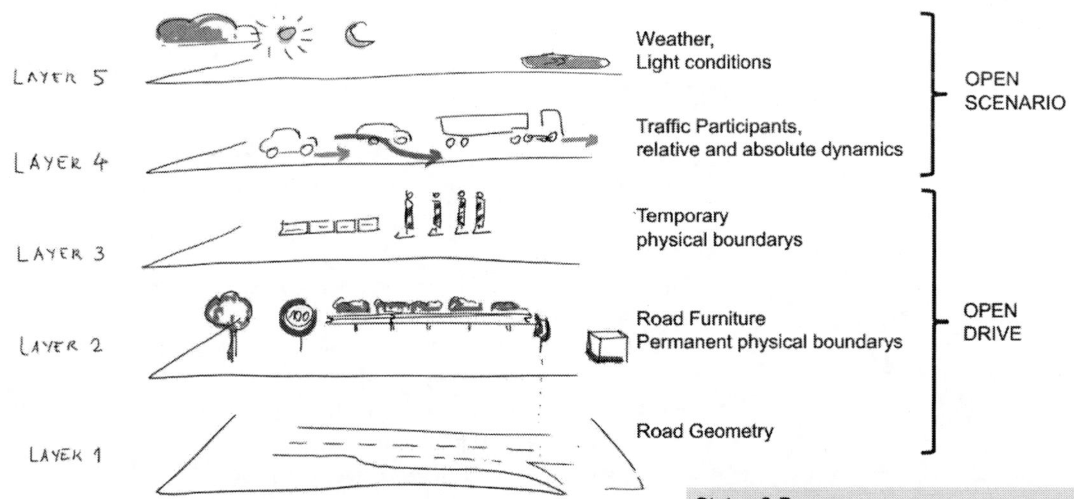

Status & Process
- Both formats have been transferred to ASAM
- First standard version planned for End of 2018.

AUTOMATED VEHICLES SYMPOSIUM 2018

USERS. VEHICLES. INFRASTRUCTURE.

Driven by AUVSI and TRB

I. Harmonized Definitions and Taxonomy

II. Agreed Format for Scenarios

III. Process for Scenario Sharing

IV. Common Driver Reference Model

V. Agreed Classification of Logical Scenarios

AUTOMATED VEHICLES SYMPOSIUM 2018

USERS. VEHICLES. INFRASTRUCTURE.

Driven by AUVSI and TRB

III. Process for Scenario Sharing

MOTIVATION

- Scenarios will differ between countries;
- Societal Consesus, which scenarios an automated vehicle needs to solve, is a national taks, may differ between nations and societies;

PROPOSED APPROACH

- Build and maintain central national databases, which contain an agreed set of scenarios.
- Everybody, who wants to develop an automated vehicle, can access this national database under defined conditions.
- International compatibility on technical level required:
    - Discussion in WP29 startet by BASt
    - Several countries are in direct contact with PEGASUS

I. Harmonized Definitions and Taxonomy

II. Agreed Format for Scenarios

III. Process for Scenario Sharing

IV. Common Driver Reference Model

V. Agreed Classification of Logical Scenarios

AUTOMATED VEHICLES SYMPOSIUM 2018
USERS. VEHICLES. INFRASTRUCTURE.
Driven by AUVSI and TRB

IV. Common Driver Reference Model

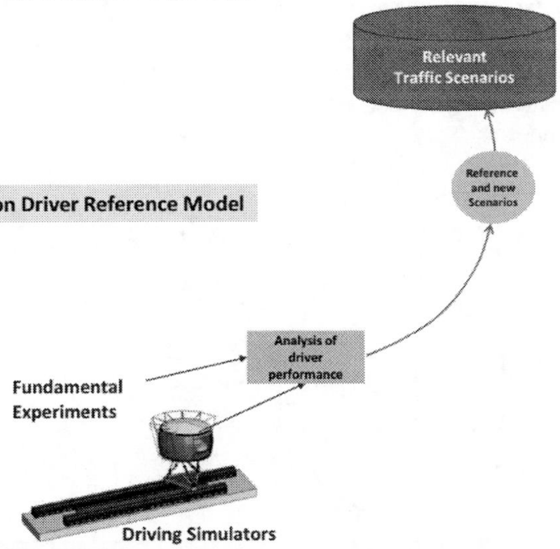

APPROACH

- Define requirements on Driver Reference Model => how to compare AD with human drivers?
- Derive suitable Model and identify Parameters, based on sound scientific experiments.
- Continiously update the parameter distribution of the model based on analysed traffic events.

TO DO

- Agree on requirements – how to compare automated driving with human drivers?
- Agree on Driver Reference Model
- Carry out Experiments to capture Parameters
- Establish a process to share Driver Model and Parameters, which may be market-specific

AUTOMATED VEHICLES SYMPOSIUM 2018

USERS. VEHICLES. INFRASTRUCTURE.

Driven by AUVSI and TRB

I. Harmonized Definitions and Taxonomy

II. Agreed Format for Scenarios

III. Process for Scenario Sharing

IV. Comon Driver Reference Model

V. Agreed Classification of Logical Scenarios

PEGASUS

Classification of logical scenarios (L4)

- Classification of logical scenarios based on accident reconstruction
  → compatible classification between accidents and traffic events (incidents)
  → decades of expertise of accident research is directly taken into account
- Counterfactual approach starting from potential impact (front, side, rear), deriving all possible initial position of the challenger
- Relational Description of layer 4 from the ego vehicle perspective
- Description of relative paths + timing

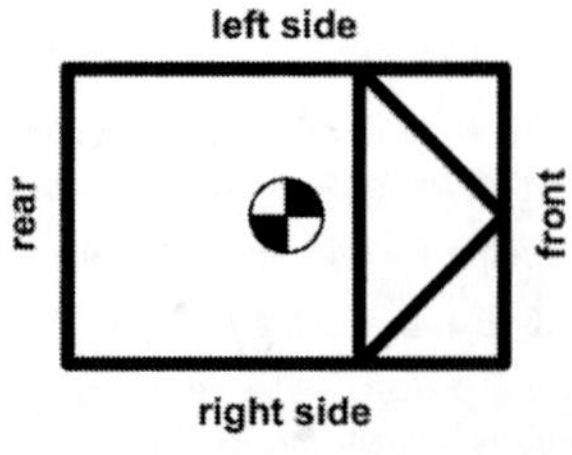

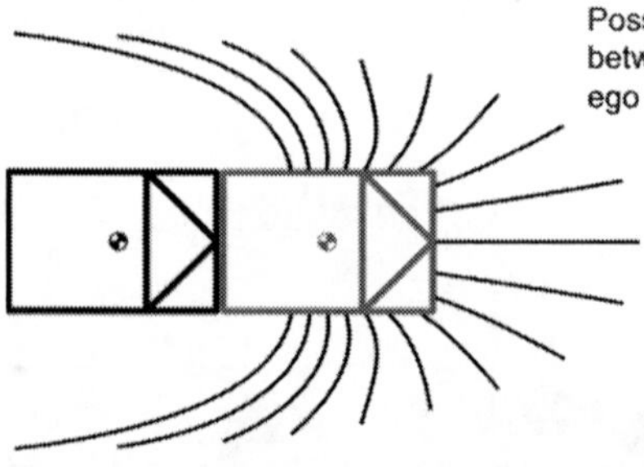

Possible relative paths between challenger and ego vehicle

PEGASUS

Classification of logical scenarios: potential frontal impact

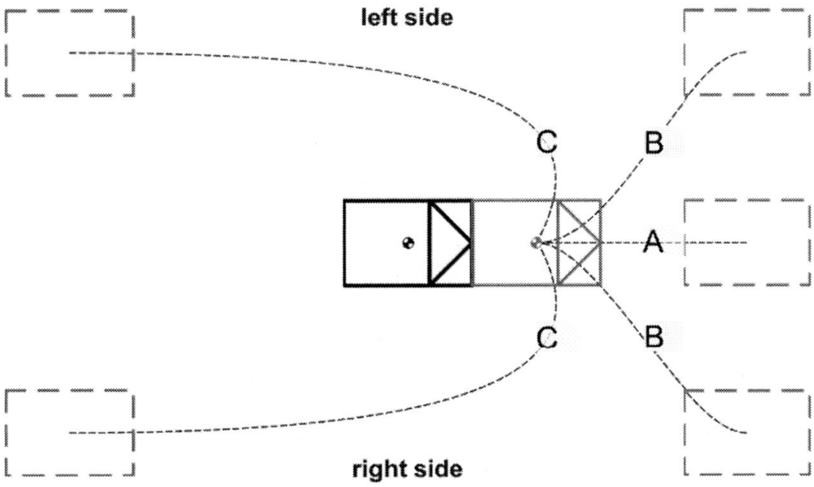

PEGASUS

Classification of logical scenarios: potential rear impact

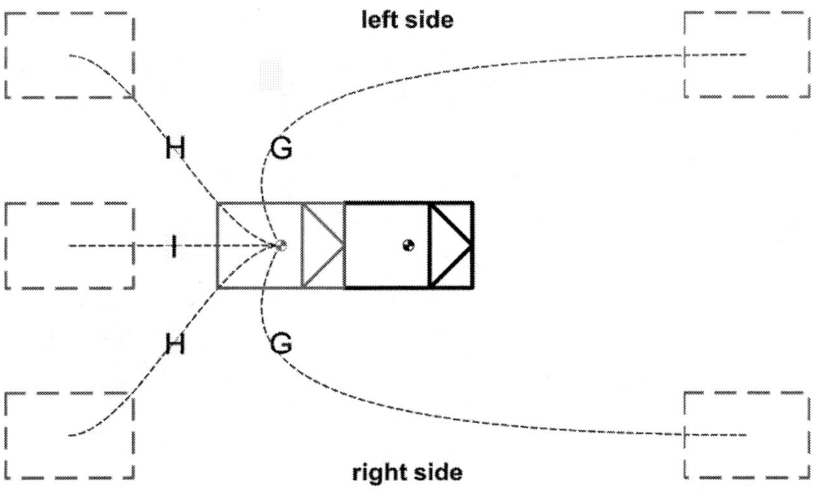

Classification of logical scenarios: potential lateral impact

PEGASUS

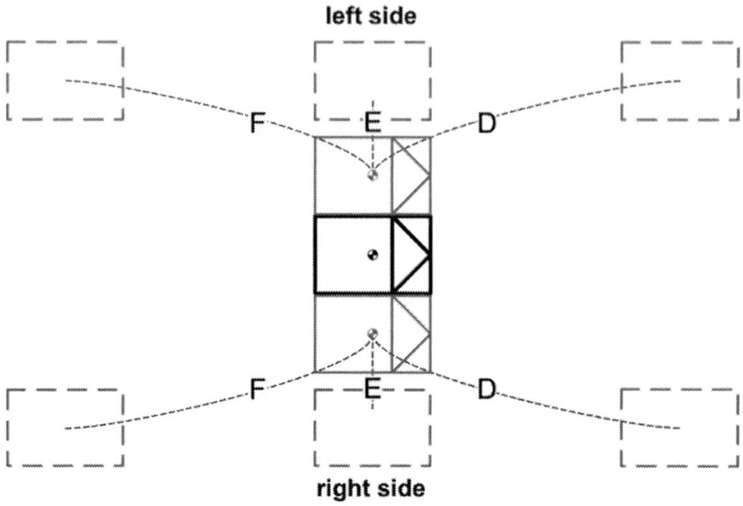

Classification of logical scenarios: Overview

PEGASUS

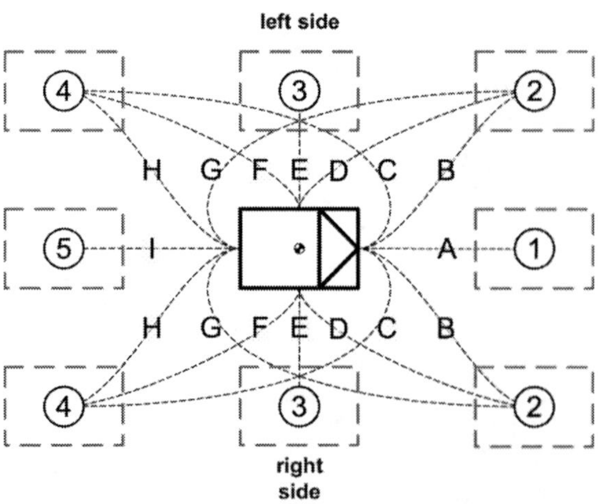

Impact	Initial position	Path	Indication
Front	1	A	Lead vehicle challenger
	2	B	Slower turn into path challenger
	4	C	Overtaking turn into path challenger
Side	2	D	Slower side swipe challenger
	3	E	Side swipe challenger
	4	F	Overtaking side swipe challenger
Rear	2	G	Slower Rear End Challenger
	4	H	Rear end turning into path challenger
	5	I	Rear end challenger

AUTOMATED VEHICLES SYMPOSIUM 2018
USERS. VEHICLES. INFRASTRUCTURE.
Driven by AUVSI and TRB

Breakout Session 16: Agenda

Day 2, Wednesday: Approaches and Methodologies to assure Safety of AV

• *Introduction*	*Lutz Eckstein (ika)*
• ***Approaches and Methodologies to assure safety of Automated Vehicles***	
• French approach for safety validation by combination of experiment and simulation	J.-F. Sencerin (Renault)
• The Automated Mobility Partnership. An approach for utilizing naturalistic driving data to create simulations for evaluating automated systems	T. Dingus (VTTI)
• AD safety validation investigation in Japan	S. Taniguchi (Toyota, JAMA)
• *Continued after Break*	
• Role of Proving Grounds in Three-Mode AV Validation Methodology	John Maddox (ACM)
• The Innovations in Simulation that are making Avs safer	C. Partridge (Metamoto)
• New Ideas on Safety and Certification for full autonomy	P. Perrone (perronerobotics)
• Standardization of Data Formats for increased efficiency of simulation applications	M. Dupuis (VIRES)
• *Synchronisation of Approaches*	*Lutz Eckstein (ika)*
• J. Maddox (ACM), S. Taniguchi (Toyota), J.-F. Sencerin (Renault), T. Dingus(VTTI), C. Partridge (Metamoto), P. Perrone (perronerobotics), M. Dupuis (VIRES)	
• ***Wrap-up & Conclusions***	***Lutz Eckstein (ika)***

AUTOMATED VEHICLES SYMPOSIUM 2018
USERS. VEHICLES. INFRASTRUCTURE.
Driven by AUVSI and TRB

Breakout Session 16: Safety Assurance of AD – Day 2

Summary of Key Findings and Lessons Learned

Speakers & Penalists
J. Maddox (ACM)
S. Taniguchi (Toyota)
J.-F. Sencerin (Renault)
T. Dingus(VTTI),
C. Partridge (Metamoto)
P. Perrone (perronerobotics)
M. Dupuis (VIRES)
Moderator: L. Eckstein (ika)

Approaches and Methodologies to assure Safety of Automated Vehicles

- **Approaches are all based on Scenarios** (U.S., Japan, France and Germany).
- **Scenarios are extracted from different data sources** (FOT, test driving, accident data), complemented by logically derived as well as modified scenarios and stored in a data base.
- **Combination of different Methodologies is necessary** (Simulation, Test tracks, public traffic).
- **Simulation plays a crucial role** in order to cover the vast variety of scenarios and link / combine tests in different instances.
- **Experiments on test tracks are necessary** e.g. for validating simulation, testing improvements, including users ...
- **Functional Safety** needs to cover failure effects as well as Safety of the intended function (Sotif), **Watch-dog concept** is promising to enhance safety as well as security.
- **Urgent need for standardization**, ASAM e.V. addresses OpenDrive, OpenCRG and OpenScenrio Format.

AUTOMATED VEHICLES SYMPOSIUM 2018

USERS. VEHICLES. INFRASTRUCTURE.

Driven by AUVSI and TRB

Breakout Session 16: Safety Assurance of AD – Day 2

Synchronization of Approaches

I. **Harmonized Definitions and Taxonomy**
   - **Draft Definitions** for 'safety relevant traffic event', 'scene', 'logical scenario' and 'concrete scenario'
   - **Proposals from PEGASUS needs to be refined** by SAE and DIN, goal is ISO standard

II. **Common Format for Scenarios**
   - **Proposal of 5-Layer Model** was positively discussed, is depicted by ODR and OSC,
     additional layer to include communication should be added
   - **More detailed discussion needed,** also on different levels of detail (traffic behavior vs. sensor performance)

III. **Process of Scenario Sharing need to be defined**

IV. **Common Driver Reference Model – what are the requirements?**

V. **Agreed Classification / Ontology of Logical Scenarios**
   - **Classification of logical scenarios based on accident reconstruction** was positively discussed, t.b.c.

VI. **Safety Argument**
   - How can results from different methodologies be combined?
   - Balancing of performance in different scenarios?

AUTOMATED VEHICLES SYMPOSIUM 2018
USERS. VEHICLES. INFRASTRUCTURE.
Driven by AUVSI and TRB

Breakout Session 16: Safety Assurance of Automated Driving

Organizers: Prof. Dr. Lutz Eckstein, ika Germany; Dr. Adrian Zlocki, fka Germany

- Agenda
- Introduction and report on AVS 2017
- Day 1, Tuesday: State of the Art and Safety Level for Safety Assurance of AV
- Day 2: Wednesday: Approaches and Methodologies to assure Safety of AV
- Way Forward

AUTOMATED VEHICLES SYMPOSIUM 2018
USERS. VEHICLES. INFRASTRUCTURE.
Driven by AUVSI and TRB

Breakout Session 16: Agenda

Day 1, Tuesday: State of the Art and Safety Level for Safety Assurance of AV

- *Introduction* — *Lutz Eckstein (ika)*
- *State of the Art presentations on ongoing activities and initiatives*
  - Overview on US activities and initiatives — Cem Hatipoglu (NHTSA)
  - Overview on European activities and initiatives — Adrian Zlocki (fka)
  - Overview on Asian activities and initiatives — N. Uchida (JARI)
- *Which level of safety should be aimed at for Safety Assurance of AV*
  - Anticipated Safety Effects of Auotmated Driving — Christian Rösener (ika)
  - U.S. Perspective — Cem Hatipoglu (NHTSA)
  - Safety requirements for automated vehicles – a scenario based Approach — C Pastor (BASt)
- *Discussion on Level of Safety* — *Mike Griffith (FHA)*
  - C. Hatipoglu (NHTSA), N. Uchida (JARI), C. Rösener (ika), C. Pastor (BASt)
- *Wrap-up* — *Lutz Eckstein (ika)*

AUTOMATED VEHICLES SYMPOSIUM 2018
USERS. VEHICLES. INFRASTRUCTURE.
Driven by AUVSI and TRB

Breakout Session 16: Agenda

Day 2, Wednesday: Approaches and Methodologies to assure Safety of AV

- *Introduction* — *Lutz Eckstein (ika)*
- *Approaches and Methodologies to assure safety of Automated Vehicles*
    - French approach for safety validation by combination of experiment and simulation — J.-F. Sencerin (Renault)
    - The Automated Mobility Partnership. An approach for utilizing naturalistic driving data to create simulations for evaluating automated systems — T. Dingus (VTTI)
    - AD safety validation investigation in Japan — S. Taniguchi (Toyota, JAMA)
- *Continued after Break*
    - Role of Proving Grounds in Three-Mode AV Validation Methodology — John Maddox (ACM)
    - The Innovations in Simulation that are making Avs safer — C. Partridge (Metamoto)
    - New Ideas on Safety and Certification for full autonomy — P. Perrone (perronerobotics)
    - Standardization of Data Formats for increased efficiency of simulation applications — M. Dupuis (VIRES)
- *Synchronisation of Approaches* — *Lutz Eckstein (ika)*
    - J. Maddox (ACM), S. Taniguchi (Toyota), J.-F. Sencerin (Renault), T. Dingus(VTTI), C. Partridge (Metamoto), P. Perrone (perronerobotics), M. Dupuis (VIRES)
- *Wrap-up & Conclusions* — *Lutz Eckstein (ika)*

AUTOMATED VEHICLES SYMPOSIUM 2018
USERS. VEHICLES. INFRASTRUCTURE.
Driven by AUVSI and TRB

Breakout Session 16: Safety Assurance of Automated Driving

- Agenda
- **Introduction and report on AVS 2017**
- Day 1, Tuesday: State of the Art and Safety Level for Safety Assurance of AV
- Day 2: Wednesday: Approaches and Methodologies to assure Safety of AV
- Way Forward

AUTOMATED VEHICLES SYMPOSIUM 2018

USERS. VEHICLES. INFRASTRUCTURE.

Driven by AUVSI and TRB

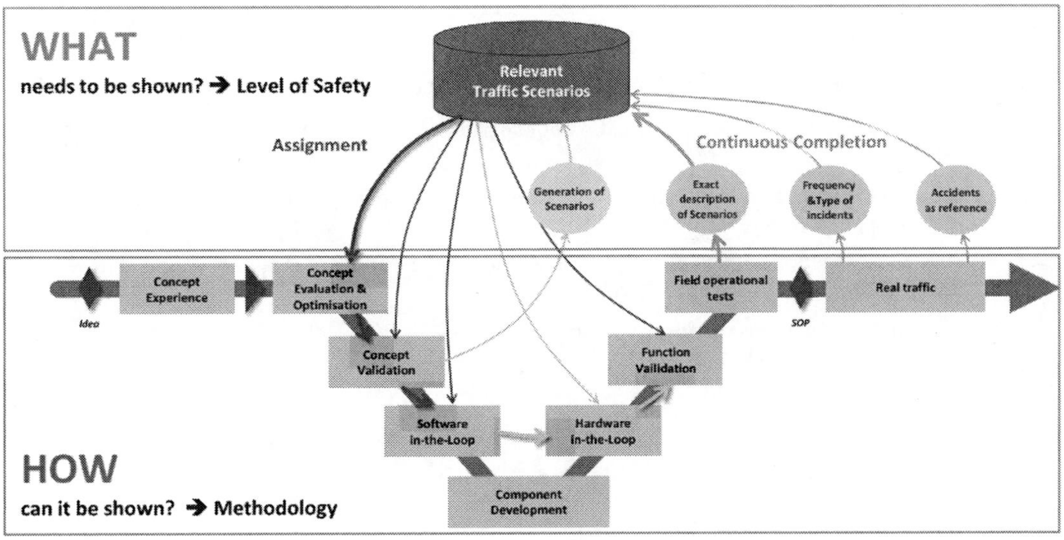

AUTOMATED VEHICLES SYMPOSIUM 2017

USERS. VEHICLES. INFRASTRUCTURE.

Breakout Session 17: Safety Assurance

Summary of Key Findings/Lessons Learned/Issues :

- Database with relevant traffic scenarios is a (the?) key component for the validation & development process.
- It is not possible to evaluate the AV software by physical cases "completely."
- Simulation plays a crucial role on different scales (traffic, vehicles, systems, components)
- Virtual Software-Defined Networking (SDN) for physical resource management is an innovative way of addressing the problem
- Safe validation must include a structured combination of three methodologies (simulation, proving ground testing, on-road testing)
- Cyber security is a concern that needs to be progressively addressed

AUVSI

TRB

AUTOMATED VEHICLES SYMPOSIUM 2017

USERS. VEHICLES. INFRASTRUCTURE.

Breakout Session 17: Safety Assurance

Summary of Key Findings/Lessons Learned/Issues (cont'd):

- Cooperation by all stakeholders will be beneficial, but how to encourage and implement "sharing"
- Customers perception of benefits and consumer understanding of functionalities sometimes are not aligned with reality.
- How to avoid a full verification when a new functional component is added when the lower levels of components are already verified?
- Over-The-Air upgrades may present some challenges, technically as well as with regard to the user, e.g. confuse potentially

AUVSI

TRB

AUTOMATED VEHICLES SYMPOSIUM 2018

USERS. VEHICLES. INFRASTRUCTURE.

Driven by AUVSI and TRB

Breakout Session 16: Safety Assurance of Automated Driving

- Agenda
- Introduction and report on AVS 2017
- **Day 1: State of the Art and Safety Level for Safety Assurance of AV**
- Day 2: Wednesday: Approaches and Methodologies to assure Safety of AV
- Way Forward

WHAT

needs to be shown? → Level of Safety

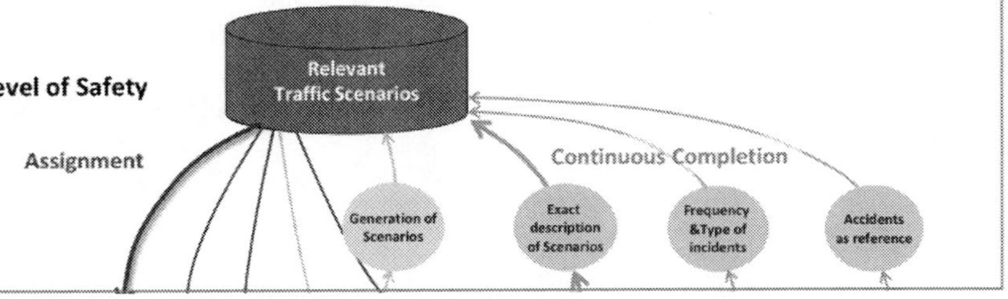

1. **Which level of safety do Automated Vehicles need to achieve?**
   - Initially vs. medium and longterm?
   - as part of public transport vs. private applications?
   - ...
2. **What is the reference?**
   - Manual driving or assisted driving?
   - A „driver population model", which may have different parameters for different countries?
3. **Which are the relevant traffic scenarios?**
   - Data sources and extraction process?
   - Who separates relevant scenarios from extremely rare corner cases? Process?

AUTOMATED VEHICLES SYMPOSIUM 2018
USERS. VEHICLES. INFRASTRUCTURE.
Driven by AUVSI and TRB

Breakout Session 16: Agenda

Day 1, Tuesday: State of the Art and Safety Level for Safety Assurance of AV

- *Introduction* — *Lutz Eckstein (ika)*
- ***State of the Art presentations on ongoing activities and initiatives***
  - Overview on US activities and initiatives — Cem Hatipoglu (NHTSA)
  - Overview on European activities and initiatives — Adrian Zlocki (fka)
  - Overview on Asian activities and initiatives — N. Uchida (JARI)
- *Which level of safety should be aimed at for Safety Assurance of AV*
  - Anticipated Safety Effects of Automated Driving — Christian Rösener (ika)
  - U.S. Perspective — Cem Hatipoglu (NHTSA)
  - Safety requirements for automated vehicles – a scenario based Approach — Claus Pastor (BASt)
- *Discussion on Level of Safety* — *Mike Griffith (FHA)*
  - C. Hatipuglu (NHTSA), N. Uchida (JARI), C. Rösener (ika), C. Pastor (BASt)
- *Wrap-up* — ***Lutz Eckstein (ika)***

AUTOMATED VEHICLES SYMPOSIUM 2018
USERS. VEHICLES. INFRASTRUCTURE.
Driven by AUVSI and TRB

Breakout Session 16: Agenda

Day 1, Tuesday: State of the Art and Safety Level for Safety Assurance of AV

- *Introduction* — *Lutz Eckstein (ika)*
- *State of the Art presentations on ongoing activities and initiatives*
  - *Overview on US activities and initiatives;* — Cem Hatipoglu (NHTSA)
  - Overview on European activities and initiatives — Adrian Zlocki (fka)
  - Overview on Asian activities and initiatives — N. Uchida (JARI)
- ***Which level of safety should be aimed at for Safety Assurance of AV***
  - Anticipated Safety Effects of Automated Driving — Christian Rösener (ika)
  - U.S. Perspective — Cem Hatipoglu (NHTSA)
  - Safety requirements for automated vehicles – a scenario based Approach — Claus Pastor (BASt)
- *Discussion on Level of Safety* — *Mike Griffith (FHA)*
  - C. Hatipuglu (NHTSA), N. Uchida (JARI), C. Rösener (ika), C. Pastor (BASt)
- *Wrap-up* — ***Lutz Eckstein (ika)***

AUTOMATED VEHICLES SYMPOSIUM 2018
USERS. VEHICLES. INFRASTRUCTURE.
Driven by AUVSI and TRB

Breakout Session 16: Agenda

Day 1, Tuesday: State of the Art and Safety Level for Safety Assurance of AV

- *Introduction* — *Lutz Eckstein (ika)*
- *State of the Art presentations on ongoing activities and initiatives*
  - *Overview on US activities and initiatives;* — Cen Hatipuglu (NHTSA)
  - Overview on European activities and initiatives — Adrian Zlocki (fka)
  - Overview on Asian activities and initiatives — N. Uchida (JARI)
- *Which level of safety should be aimed at for Safety Assurance of AV*
  - Anticipated Safety Effects of Auotmated Driving — Christian Rösener (ika)
  - U.S. Perspective — Cen Hatipuglu (NHTSA)
  - Safety requirements for automated vehicles – a scenario based Approach — C Pastor (BASt)
- ***Discussion on Level of Safety*** — ***Mike Griffith (FHA)***
  - C. Hatipoglu (NHTSA), N. Uchida (JARI), C. Rösener (ika), C. Pastor (BASt)
- *Wrap-up* — ***Lutz Eckstein (ika)***

WRAP-UP 1

What needs to be shown?
→ Level of Safety

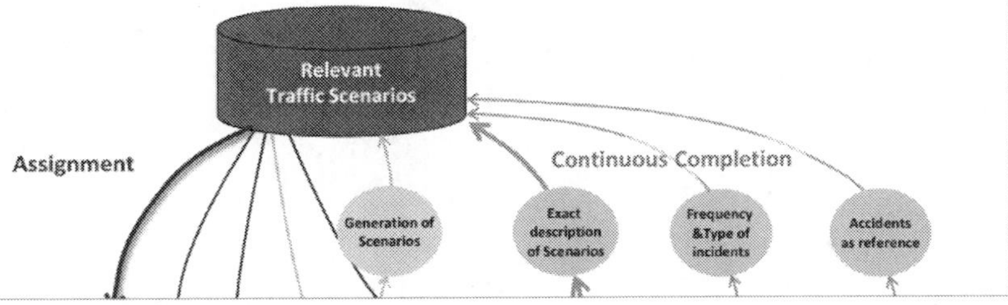

1. **Which level of Safety do Automated Vehicles need to achieve?**
   - **Safety is the main motivation** in the U.S., EU, Jp, G to introduce automated driving
   - **We need to draw comparisons in order to find the right level:**
     => compare to introduction of Airbags, Vaccination, Aviation (much research involved)
   - **Process is important: society needs to be included**
   - **Need for differentiation** between Level 2, testing Level 3+4, deployment of Level 3+4
     => build public confidence + manage safety risk + create transparency
     => simplify the problem first, allow innovation to take place
   - **Diffentiate between time horizons:** in **medium term** significantly better than manual driving

AUTOMATED VEHICLES SYMPOSIUM 2018
USERS. VEHICLES. INFRASTRUCTURE.
Driven by AUVSI and TRB

2. **What is the reference for Safety of Automated Vehicles?**
   - **From a societal perspective**
     - Manual driving, as long as ADAS (Level 1 and 2) have a limited market share.
     - From a product liability perspective: significantly better
   - **From a system development perspective**
     - Driver performance models, each may cover various scenarios;
     - Ideally a simuation-ready „driver population model", which may have different parameters for different countries
   - **From an accident analysis perspective**
     - Physics => the automated vehicle should avoid accidents within given physical limits (e.g. friction potential)
3. **Which are the relevant traffic scenarios?**
   - Multiple Data sources are used: Traffic accidents, incidents from FOT and incidents captured by infrastructure sensors
   - How to seperate relevant scenarios from extremely rare corner cases?
     - Can Rare corner cases be reduced to a simple scenarios or their combination?
     - Include everything and decide by defining the Operational Design Domain
   - How do we include cases, where the driver avoids an incident or accident using anticipation and experience (e.g. loose freight)? How do we know, when the system does not know?

AUTOMATED VEHICLES SYMPOSIUM 2018

USERS. VEHICLES. INFRASTRUCTURE.

Driven by AUVSI and TRB

Breakout Session 16: Safety Assurance of Automated Driving

- Agenda
- Introduction and report on AVS 2017
- Day 1: State of the Art and Safety Level for Safety Assurance of AV
- **Day 2: Wednesday: Approaches and Methodologies to assure Safety of AV**
- Way Forward

AUTOMATED VEHICLES SYMPOSIUM 2018

USERS. VEHICLES. INFRASTRUCTURE.

Driven by AUVSI and TRB

Breakout Session 16: Agenda

Day 2, Wednesday: Approaches and Methodologies to assure Safety of AV

- ***Introduction*** — ***Lutz Eckstein (ika)***
- *Approaches and Methodologies to assure safety of Automated Vehicles*
  - French approach for safety validation by combination of experiment and simulation — J.-F. Sencerin (Renault)
  - The Automated Mobility Partnership. An approach for utilizing naturalistic driving data to create simulations for evaluating automated systems — T. Dingus (VTTI)
  - AD safety validation investigation in Japan — S. Taniguchi (Toyota, JAMA)
- *Continued after Break*
  - Role of Proving Grounds in Three-Mode AV Validation Methodology — John Maddox (ACM)
  - The Innovations in Simulation that are making AVs safer — C. Partridge (Metamoto)
  - New Ideas on Safety and Certification for full autonomy — P. Perrone (perronerobotics)
  - Standardization of Data Formats for increased efficiency of simulation applications — M. Dupuis (VIRES)
- *Synchronisation of Approaches* — *Lutz Eckstein (ika)*
  - J. Maddox (ACM), S. Taniguchi (Toyota), J.-F. Sencerin (Renault), T. Dingus(VTTI), C. Partridge (Metamoto), P. Perrone (perronerobotics), M. Dupuis (VIRES)
- *Wrap-up & Conclusions* — *Lutz Eckstein (ika)*

AUTOMATED VEHICLES SYMPOSIUM 2018
USERS. VEHICLES. INFRASTRUCTURE.
Driven by AUVSI and TRB

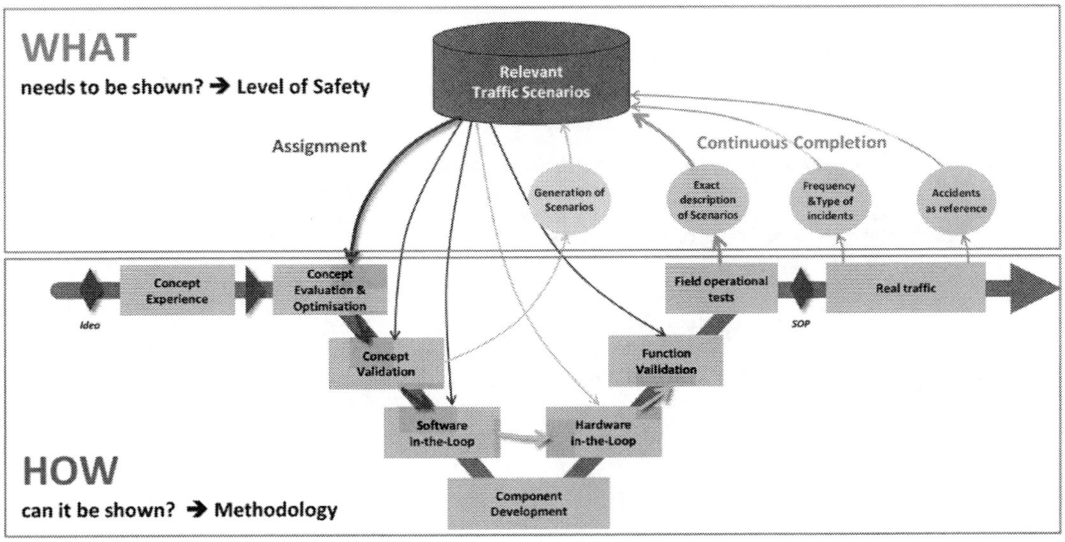

AUTOMATED VEHICLES SYMPOSIUM 2018
USERS. VEHICLES. INFRASTRUCTURE.
Driven by AUVSI and TRB

1. **Which role may simulation play?**
   - With respect to traffic flow, traffic scenarios, system performance, ... sensor performance?
   - For development and optimization by companies vs. certification by authorities?
2. **How can proving grounds contribute?**
   - Which scenarios should be depicted on proving grounds?
   - In which phase of product development? – only for sign-off and certification?
3. **Which role may tests on public roads play?**
   - What are prerequisites in terms of safety?
   - How to deal with updates of the AD functionality?
4. **How can results be combined to find a decision?**
   - Combination of results from different methodologies?
   - Balancing of different scenarios possible?

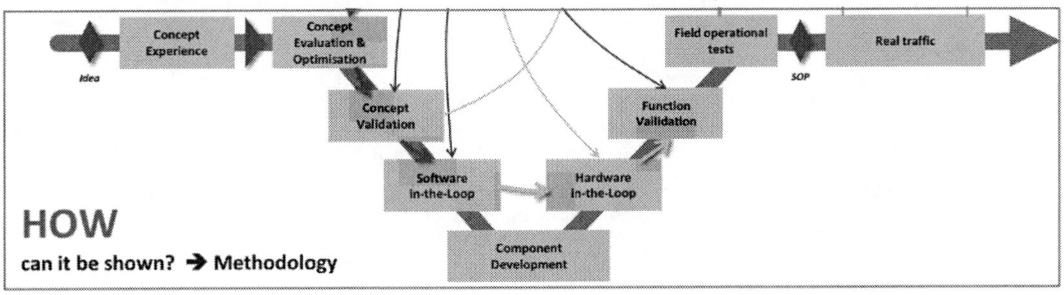

AUTOMATED VEHICLES SYMPOSIUM 2018
USERS. VEHICLES. INFRASTRUCTURE.
Driven by AUVSI and TRB

Breakout Session 16: Agenda

Day 2, Wednesday: Approaches and Methodologies to assure Safety of AV

- *Introduction* — *Lutz Eckstein (ika)*
- ***Approaches and Methodologies to assure safety of Automated Vehicles***
  - French approach for safety validation by combination of experiment and simulation — J.-F. Sencerin (Renault)
  - The Automated Mobility Partnership. An approach for utilizing naturalistic driving data to create simulations for evaluating automated systems — T. Dingus (VTTI)
  - AD safety validation investigation in Japan — S. Taniguchi (Toyota, JAMA)
- *Continued after Break*
  - Role of Proving Grounds in Three-Mode AV Validation Methodology — John Maddox (ACM)
  - The Innovations in Simulation that are making Avs safer — C. Partridge (Metamoto)
  - New Ideas on Safety and Certification for full autonomy — P. Perrone (perronerobotics)
  - Standardization of Data Formats for increased efficiency of simulation applications — M. Dupuis (VIRES)
- *Synchronisation of Approaches* — *Lutz Eckstein (ika)*
  - J. Maddox (ACM), S. Taniguchi (Toyota), J.-F. Sencerin (Renault), T. Dingus(VTTI), C. Partridge (Metamoto), P. Perrone (perronerobotics), M. Dupuis (VIRES)
- *Wrap-up & Conclusions* — *Lutz Eckstein (ika)*

AUTOMATED VEHICLES SYMPOSIUM 2018
USERS. VEHICLES. INFRASTRUCTURE.
Driven by AUVSI and TRB

Breakout Session 16: Agenda

Day 2, Wednesday: Approaches and Methodologies to assure Safety of AV

- *Introduction* — *Lutz Eckstein (ika)*
- *Approaches and Methodologies to assure safety of Automated Vehicles*
  - French approach for safety validation by combination of experiment and simulation — J.-F. Sencerin (Renault)
  - The Automated Mobility Partnership. An approach for utilizing naturalistic driving data to create simulations for evaluating automated systems — T. Dingus (VTTI)
  - AD safety validation investigation in Japan — S. Taniguchi (Toyota, JAMA)
- *Continued after Break*
  - Role of Proving Grounds in Three-Mode AV Validation Methodology — John Maddox (ACM)
  - The Innovations in Simulation that are making Avs safer — C. Partridge (Metamoto)
  - New Ideas on Safety and Certification for full autonomy — P. Perrone (perronerobotics)
  - Standardization of Data Formats for increased efficiency of simulation applications — M. Dupuis (VIRES)
- ***Synchronisation of Approaches*** — ***Lutz Eckstein (ika)***
  - **J. Maddox (ACM), S. Taniguchi (Toyota), J.-F. Sencerin (Renault), T. Dingus(VTTI), C. Partridge (Metamoto), P. Perrone (perronerobotics), M. Dupuis (VIRES)**
- *Wrap-up & Conclusions* — *Lutz Eckstein (ika)*

I. Harmonized Definitions and Taxonomy

II. Agreed Format for Scenarios

III. Process for Scenario Sharing

IV. Common Driver Reference Model

V. Agreed Classification of Logical Scenarios

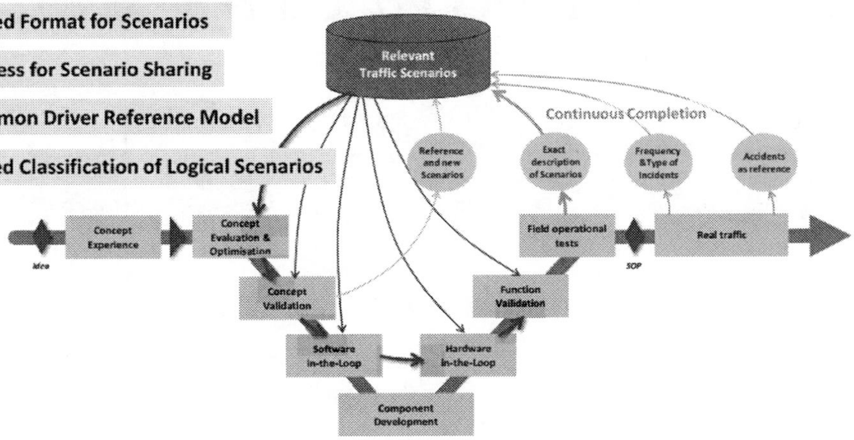

AUTOMATED VEHICLES SYMPOSIUM 2018
USERS. VEHICLES. INFRASTRUCTURE.
Driven by AUVSI and TRB

I. Harmonized Definitions and Taxonomy

II. Agreed Format for Scenarios

III. Process for Scenario Sharing

IV. Common Driver Reference Model

V. Agreed Classification of Logical Scenarios

Status & Process

- SAE already started
- DIN will also start End of August and cooperates with SAE
- ISO Working Group started, implicit harmonization via involved experts.

Definitions

PEGASUS

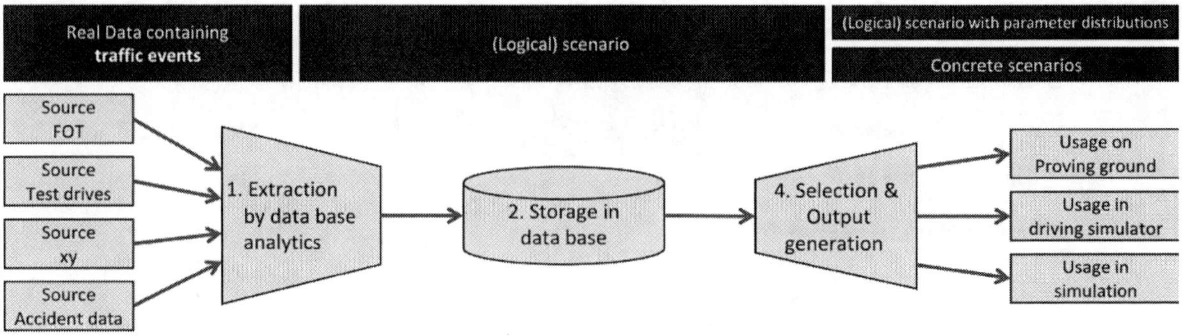

Definitions

PEGASUS

- **Definition "Safety relevant traffic event"**

A safety relevant traffic event is a real time sequence in traffic (data) that almost or indeed led to an accident.

- **Definition "Scene"**

A scene is a snapshot that includes the moving and non-moving elements of the traffic environment, the self-representation of all actors and observers and the relations between those elements.

- **Definition "Logical scenario"**

A logical scenario is a model of a conditional sequence of scenes, which starts with a start scene, but does not further specify the behavior of the main actor (subject vehicle) from a defined point in time onwards, so that the result (accident or not) is hypothetical. (class)

- **Definition "Concrete scenario"**

A concrete scenario is a parameterized model of a time sequence of scenes (logical scenario), which begins with a concrete start scene. At a defined point in time, the behavior of the main actor (subject vehicle) is not further specified, so the result is hypothetical, but all other influencing factors are pre-defined by concrete parameters.

AUTOMATED VEHICLES SYMPOSIUM 2018

USERS. VEHICLES. INFRASTRUCTURE.

Driven by AUVSI and TRB

I. Harmonized Definitions and Taxonomy

II. Agreed Format for Scenarios

III. Process for Scenario Sharing

IV. Common Driver Reference Model

V. Agreed Classification of Logical Scenarios

Scenario Description: Layer model

PEGASUS

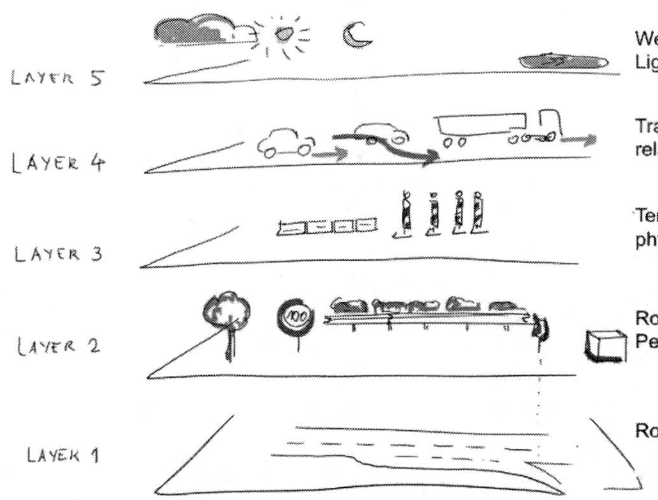

Weather, Light conditions

Traffic Participants, relative and absolute dynamics

Temporary physical boundarys

Road Furniture
Permanent physical boundarys

Road Geometry

OPEN SCENARIO

OPEN DRIVE

Status & Process

- Both formats have been transferred to ASAM
- First standard version planned for End of 2018.

AUTOMATED VEHICLES SYMPOSIUM 2018
USERS. VEHICLES. INFRASTRUCTURE.
Driven by AUVSI and TRB

I. Harmonized Definitions and Taxonomy

II. Agreed Format for Scenarios

III. Process for Scenario Sharing

IV. Common Driver Reference Model

V. Agreed Classification of Logical Scenarios

AUTOMATED VEHICLES SYMPOSIUM 2018
USERS. VEHICLES. INFRASTRUCTURE.
Driven by AUVSI and TRB

III. Process for Scenario Sharing

MOTIVATION

- Scenarios will differ between countries;
- Societal Consesus, which scenarios an automated vehicle needs to solve, is a national taks, may differ between nations and societies;

PROPOSED APPROACH

- Build and maintain central national databases, which contain an agreed set of scenarios.
- Everybody, who wants to develop an automated vehicle, can access this national database under defined conditions.
- International compatibility on technical level required:
  - Discussion in WP29 startet by BASt
  - Several countries are in direct contact with PEGASUS

I. Harmonized Definitions and Taxonomy

II. Agreed Format for Scenarios

III. Process for Scenario Sharing

IV. Common Driver Reference Model

V. Agreed Classification of Logical Scenarios

AUTOMATED VEHICLES SYMPOSIUM 2018
USERS. VEHICLES. INFRASTRUCTURE.
Driven by AUVSI and TRB

IV. Common Driver Reference Model

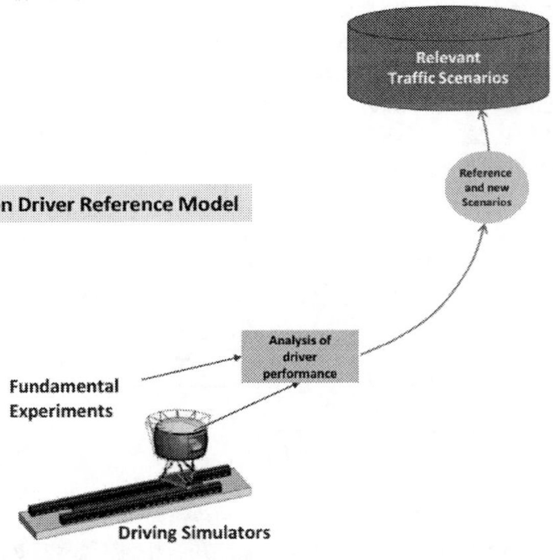

APPROACH

- Define requirements on Driver Reference Model => how to compare AD with human drivers?
- Derive suitable Model and identify Parameters, based on sound scientific experiments.
- Continiously update the parameter distribution of the model based on analysed traffic events.

TO DO

- Agree on requirements – how to compare automated driving with human drivers?
- Agree on Driver Reference Model
- Carry out Experiments to capture Parameters
- Establish a process to share Driver Model and Parameters, which may be market-specific

AUTOMATED VEHICLES SYMPOSIUM 2018

USERS. VEHICLES. INFRASTRUCTURE.

Driven by AUVSI and TRB

I. Harmonized Definitions and Taxonomy

II. Agreed Format for Scenarios

III. Process for Scenario Sharing

IV. Comon Driver Reference Model

V. Agreed Classification of Logical Scenarios

Classification of logical scenarios (L4)

PEGASUS

- Classification of logical scenarios based on accident reconstruction
  - → compatible classification between accidents and traffic events (incidents)
  - → decades of expertise of accident research is directly taken into account
- Counterfactual approach starting from potential impact (front, side, rear), deriving all possible initial position of the challenger
- Relational Description of layer 4 from the ego vehicle perspective
- Description of relative paths + timing

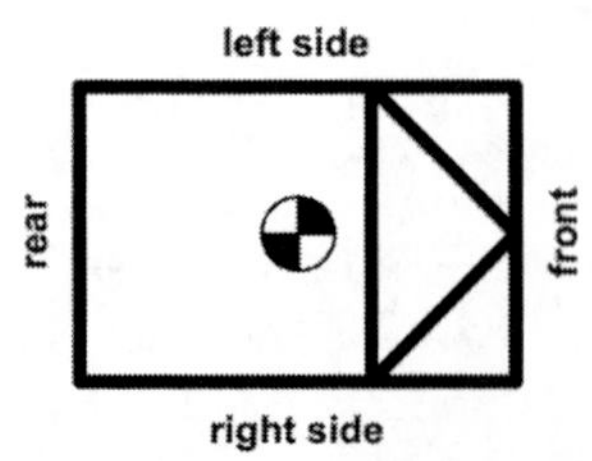

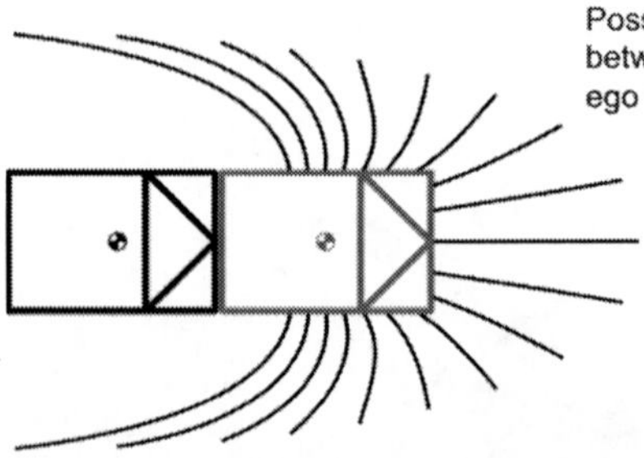

Possible relative paths between challenger and ego vehicle

Classification of logical scenarios: potential frontal impact

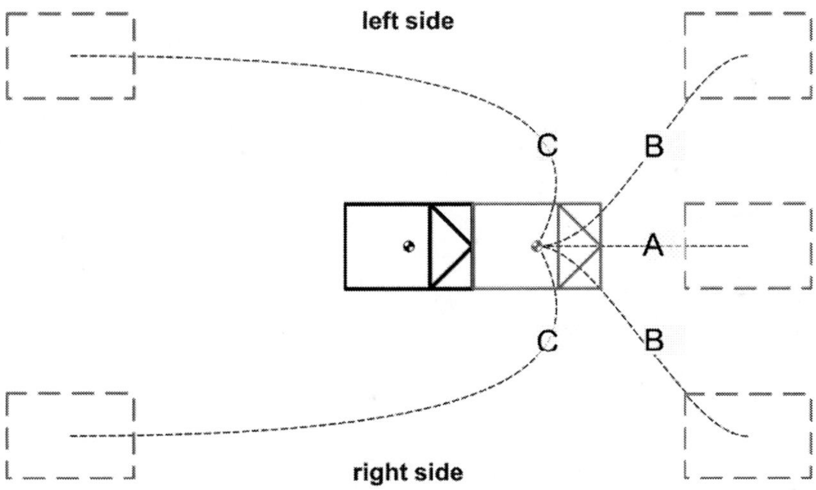

Classification of logical scenarios: potential rear impact

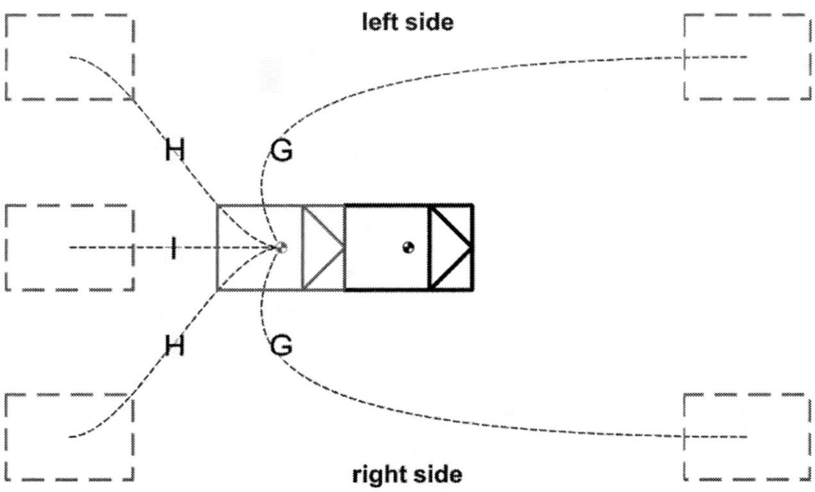

Classification of logical scenarios: potential lateral impact

PEGASUS

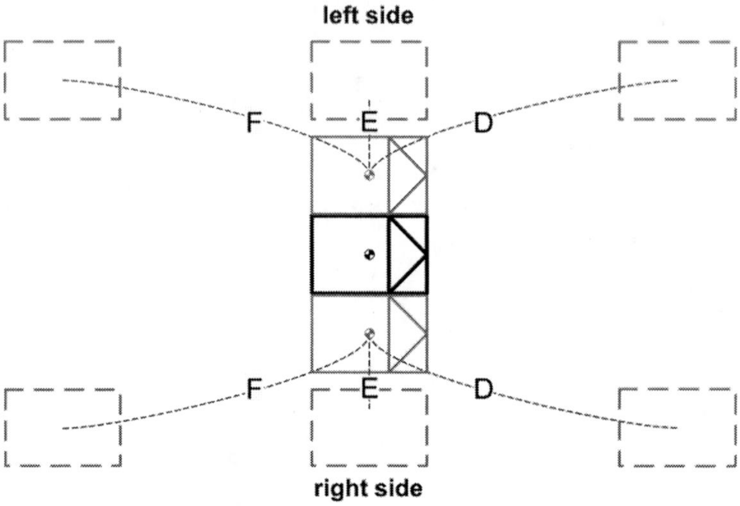

© PEGASUS | Logical scenarios (Layer 4) | July 2018

37

AUTOMATED VEHICLES SYMPOSIUM 2018

USERS. VEHICLES. INFRASTRUCTURE.

Driven by AUVSI and TRB

Breakout Session 16: Agenda

Day 2, Wednesday: Approaches and Methodologies to assure Safety of AV

- *Introduction* — *Lutz Eckstein (ika)*
- ***Approaches and Methodologies to assure safety of Automated Vehicles***
  - French approach for safety validation by combination of experiment and simulation — J.-F. Sencerin (Renault)
  - The Automated Mobility Partnership. An approach for utilizing naturalistic driving data to create simulations for evaluating automated systems — T. Dingus (VTTI)
  - AD safety validation investigation in Japan — S. Taniguchi (Toyota, JAMA)
- *Continued after Break*
  - Role of Proving Grounds in Three-Mode AV Validation Methodology — John Maddox (ACM)
  - The Innovations in Simulation that are making Avs safer — C. Partridge (Metamoto)
  - New Ideas on Safety and Certification for full autonomy — P. Perrone (perronerobotics)
  - Standardization of Data Formats for increased efficiency of simulation applications — M. Dupuis (VIRES)
- *Synchronisation of Approaches* — *Lutz Eckstein (ika)*
  - J. Maddox (ACM), S. Taniguchi (Toyota), J.-F. Sencerin (Renault), T. Dingus(VTTI), C. Partridge (Metamoto), P. Perrone (perronerobotics), M. Dupuis (VIRES)
- ***Wrap-up & Conclusions*** — ***Lutz Eckstein (ika)***

PEGASUS

Classification of logical scenarios: Overview

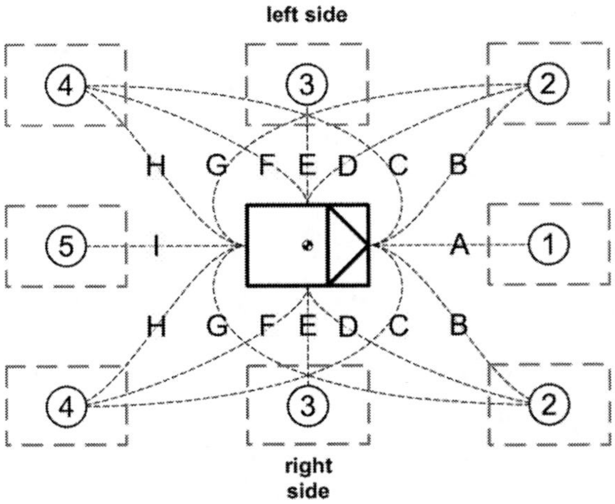

Impact	Initial position	Path	Indication
Front	1	A	Lead vehicle challenger
	2	B	Slower turn into path challenger
	4	C	Overtaking turn into path challenger
Side	2	D	Slower side swipe challenger
	3	E	Side swipe challenger
	4	F	Overtaking side swipe challenger
Rear	2	G	Slower Rear End Challenger
	4	H	Rear end turning into path challenger
	5	I	Rear end challenger

AUTOMATED VEHICLES SYMPOSIUM 2018
USERS. VEHICLES. INFRASTRUCTURE.

Driven by AUVSI and TRB

Breakout Session 16: Safety Assurance of AD – Day 2

Summary of Key Findings and Lessons Learned

Speakers & Penalists
J. Maddox (ACM)
S. Taniguchi (Toyota)
J.-F. Sencerin (Renault)
T. Dingus(VTTI),
C. Partridge (Metamoto)
P. Perrone (perronerobotics)
M. Dupuis (VIRES)
Moderator: L. Eckstein (ika)

Approaches and Methodologies to assure Safety of Automated Vehicles

- **Approaches are all based on Scenarios** (U.S., Japan, France and Germany).
- **Scenarios are extracted from different data sources** (FOT, test driving, accident data), complemented by logically derived as well as modified scenarios and stored in a data base.
- **Combination of different Methodologies is necessary** (Simulation, Test tracks, public traffic).
- **Simulation plays a crucial role** in order to cover the vast variety of scenarios and link / combine tests in different instances.
- **Experiments on test tracks are necessary** e.g. for validating simulation, testing improvements, including users ...
- **Functional Safety** needs to cover failure effects as well as Safety of the intended function (Sotif), **Watch-dog concept** is promising to enhance safety as well as security.
- **Urgent need for standardization**, ASAM e.V. addresses OpenDrive, OpenCRG and OpenScenrio Format.

AUTOMATED VEHICLES SYMPOSIUM 2018

USERS. VEHICLES. INFRASTRUCTURE.

Driven by AUVSI and TRB

Breakout Session 16: Safety Assurance of AD – Day 2

Synchronization of Approaches

I. **Harmonized Definitions and Taxonomy**
   - **Draft Definitions** for 'safety relevant traffic event', 'scene', 'logical scenario' and 'concrete scenario'
   - **Proposals from PEGASUS needs to be refined** by SAE and DIN, goal is ISO standard

II. **Common Format for Scenarios**
   - **Proposal of 5-Layer Model** was positively discussed, is depicted by ODR and OSC, **additional layer to include communication** should be added
   - **More detailed discussion needed,** also on different levels of detail (traffic behavior vs. sensor performance)

III. **Process of Scenario Sharing need to be defined**

IV. **Common Driver Reference Model – what are the requirements?**

V. **Agreed Classification / Ontology of Logical Scenarios**
   - **Classification of logical scenarios based on accident reconstruction** was positively discussed, t.b.c.

VI. **Safety Argument**
   - How can results from different methodologies be combined?
   - Balancing of performance in different scenarios?

NEW IDEAS ON SAFETY
& CERTIFICATION FOR
FULL AUTONOMY
PAUL J. PERRONE
PERRONE
ROBOTICS
JULY 11, 2018

QUICK BACKGROUNDER - PERSPECTIVE
• SAFETY ASSURANCE TECHNIQUES: R&D, SOLUTIONS, & STANDARDS (1992-PRESENT)
• Graduate research – safety assurance methods (1992-1995)
• Railway safety applications - PTS/PTC – FRA standards (1995-1997)
• Automated vehicle testing solutions (2012-2015)
• SAE On Road Automated Driving (ORAD) Committee Chair (2011-2015)
• Started V&V task force within ORAD (2014-2015)
• LARGE SCALE DISTRIBUTED SYSTEMS – SOFTWARE PLATFORMS (1995-2001)
• SOFTWARE PLATFORM FOR ROBOTICS & AVs – GPROS/MAX (2001-PRESENT)
• AV DEVELOPMENT AND APPLICATIONS WITH OEMS & STANDARDS (2004-PRESENT)
• DARPA AV Challenges (2004-2007)
• SAE ORAD Committee Chair (2011-2015)
• Automated vehicle testing solutions (2012-2015)
• Working with automotive and industrial OEMs on AVs (2016-Present)
• APPLIED A.I. TO WORK UNDERTAKEN (1990-PRESENT)
MAX
DARPA
PERRONE
ROBOTICS
2

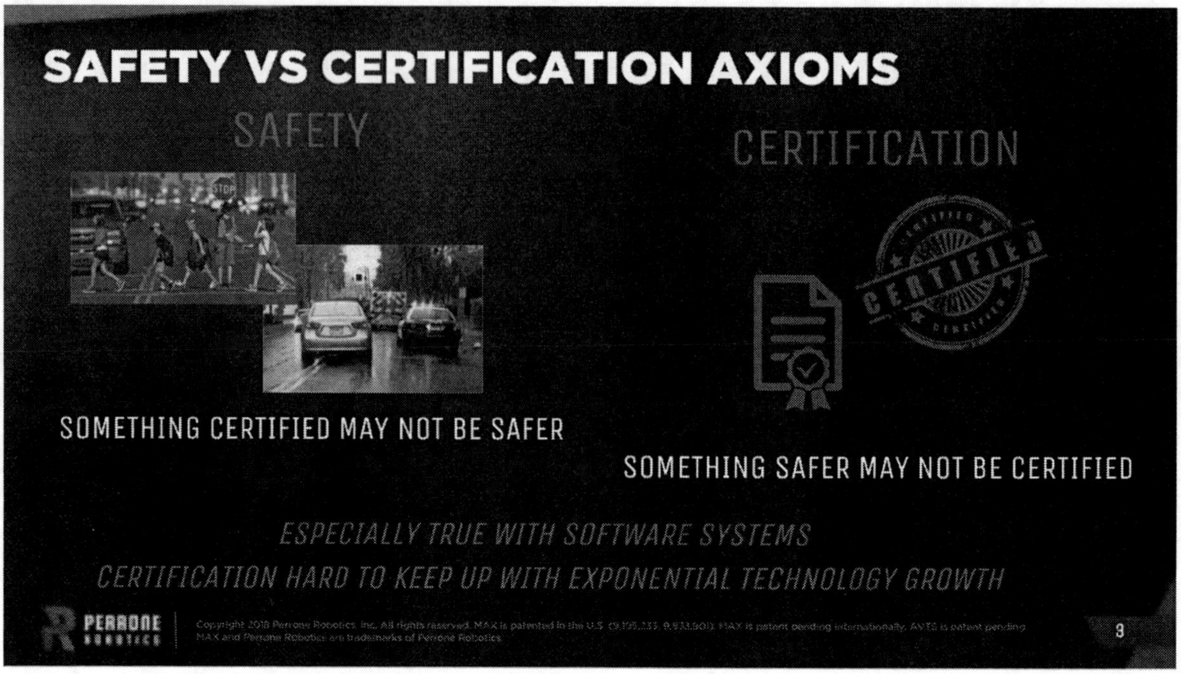
SAFETY VS CERTIFICATION AXIOMS
SAFETY
CERTIFICATION
CERTIFIED
SOMETHING CERTIFIED MAY NOT BE SAFER
SOMETHING SAFER MAY NOT BE CERTIFIED
ESPECIALLY TRUE WITH SOFTWARE SYSTEMS
CERTIFICATION HARD TO KEEP UP WITH EXPONENTIAL TECHNOLOGY GROWTH
PERRONE ROBOTICS
3

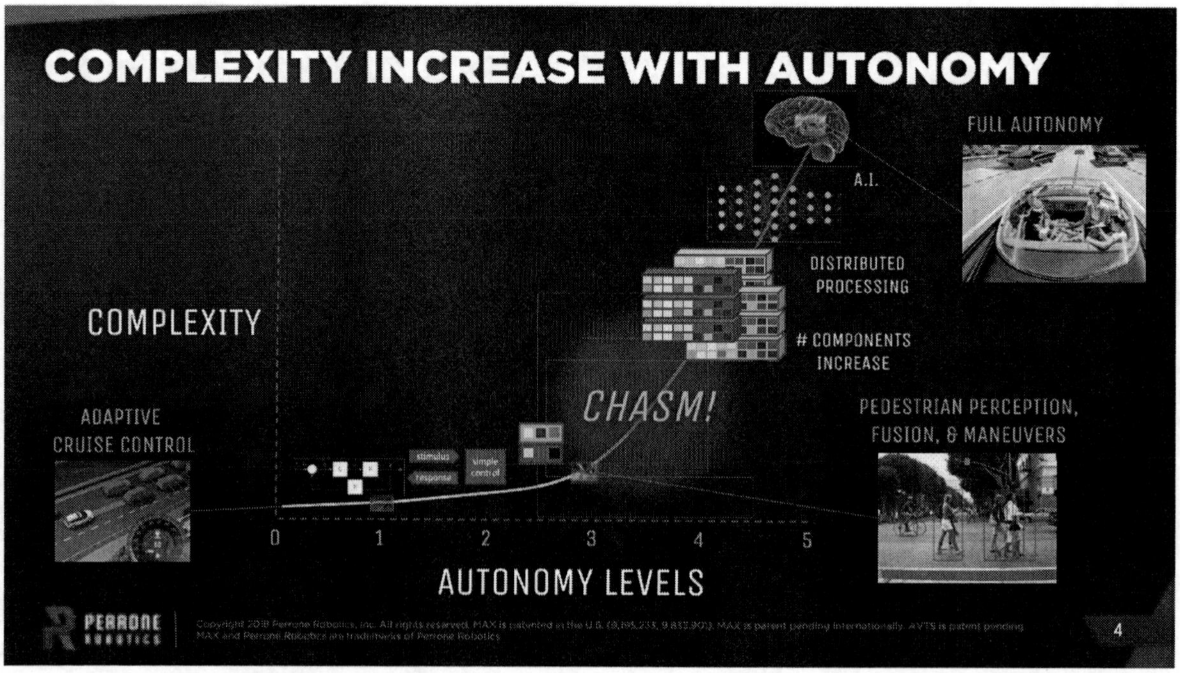
COMPLEXITY INCREASE WITH AUTONOMY
FULL AUTONOMY
A.I.
DISTRIBUTED PROCESSING
COMPLEXITY
COMPONENTS INCREASE
CHASM!
ADAPTIVE CRUISE CONTROL
PEDESTRIAN PERCEPTION, FUSION, & MANEUVERS
stimulus
response
simple control
0
1
2
3
4
5
AUTONOMY LEVELS
PERRONE ROBOTICS
4

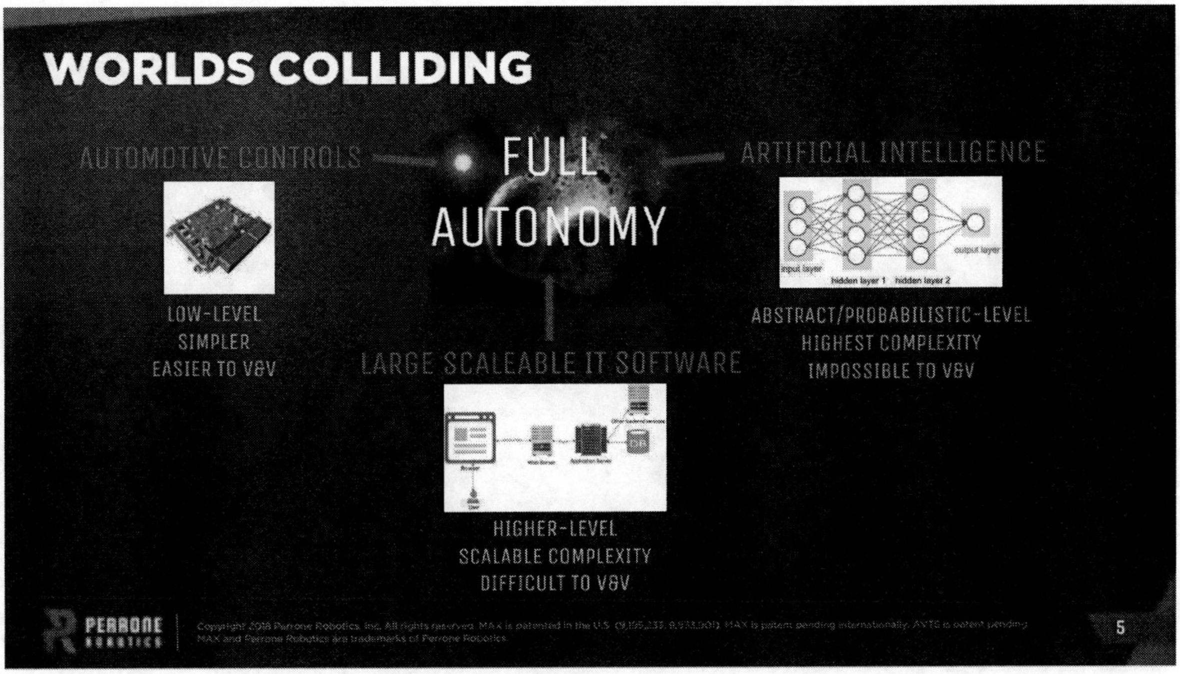
WORLDS COLLIDING
AUTOMOTIVE CONTROLS
FULL AUTONOMY
ARTIFICIAL INTELLIGENCE
LOW-LEVEL
SIMPLER
EASIER TO V&V
ABSTRACT/PROBABILISTIC-LEVEL
HIGHEST COMPLEXITY
IMPOSSIBLE TO V&V
LARGE SCALEABLE IT SOFTWARE
HIGHER-LEVEL
SCALABLE COMPLEXITY
DIFFICULT TO V&V
PERRONE ROBOTICS
MAX and Perrone Robotics are trademarks of Perrone Robotics.
5

LEGACY AUTOMOTIVE STANDARDS
SPECIFIED WITH LOWER LEVEL FUNCTIONALITY IN MIND
ISO 26262
MISRA
THE C PROGRAMMING LANGUAGE
LOWER LEVEL FUNCTIONAL RELEVANCE
LOWER LEVEL PROGRAMMING
ANTIQUATED RELATIVE TO FULL AUTONOMY NEEDS
PERRONE ROBOTICS
6

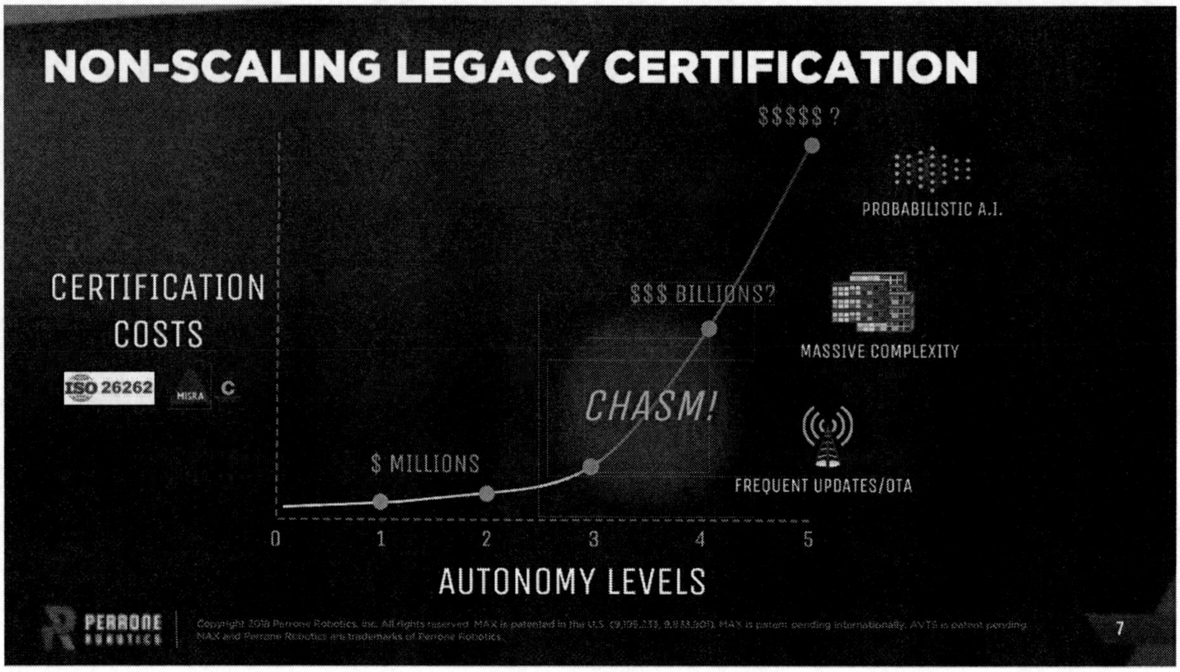
NON-SCALING LEGACY CERTIFICATION
$$$$$?
PROBABILISTIC A.I.
CERTIFICATION COSTS
ISO 26262
MISRA
C
$$$ BILLIONS?
MASSIVE COMPLEXITY
CHASM!
$ MILLIONS
FREQUENT UPDATES/OTA
0 1 2 3 4 5
AUTONOMY LEVELS
PERRONE ROBOTICS
7

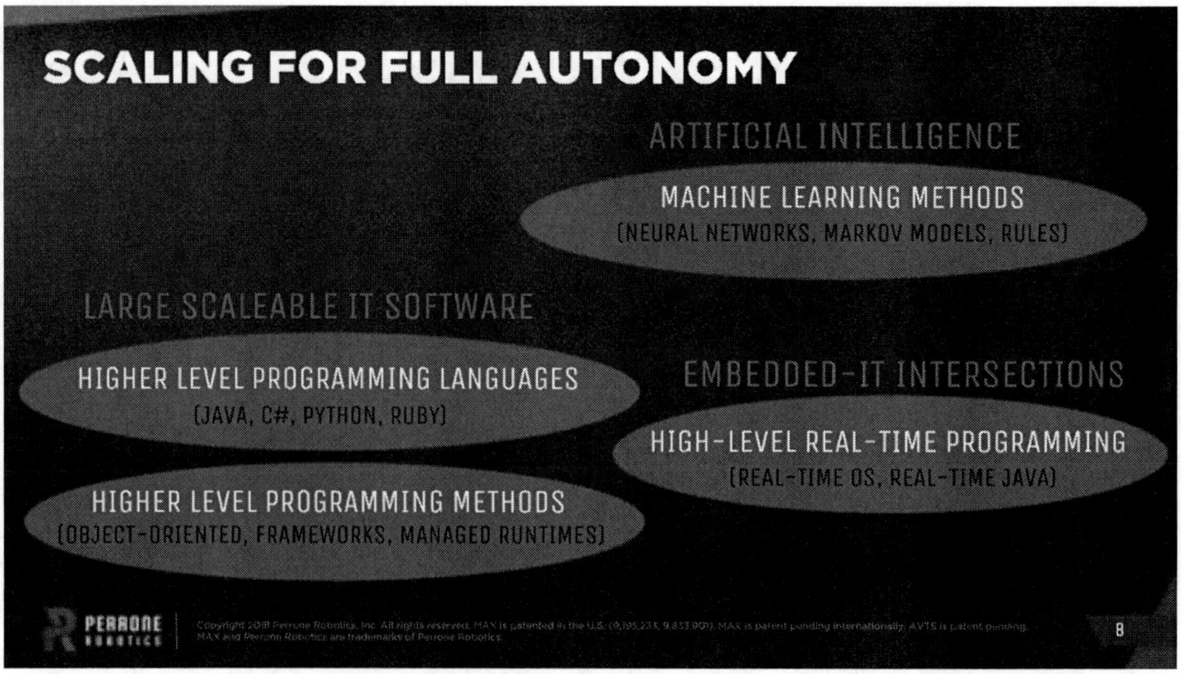
SCALING FOR FULL AUTONOMY
ARTIFICIAL INTELLIGENCE
MACHINE LEARNING METHODS
(NEURAL NETWORKS, MARKOV MODELS, RULES)
LARGE SCALEABLE IT SOFTWARE
HIGHER LEVEL PROGRAMMING LANGUAGES
(JAVA, C#, PYTHON, RUBY)
EMBEDDED-IT INTERSECTIONS
HIGH-LEVEL REAL-TIME PROGRAMMING
(REAL-TIME OS, REAL-TIME JAVA)
HIGHER LEVEL PROGRAMMING METHODS
(OBJECT-ORIENTED, FRAMEWORKS, MANAGED RUNTIMES)
PERRONE ROBOTICS
8

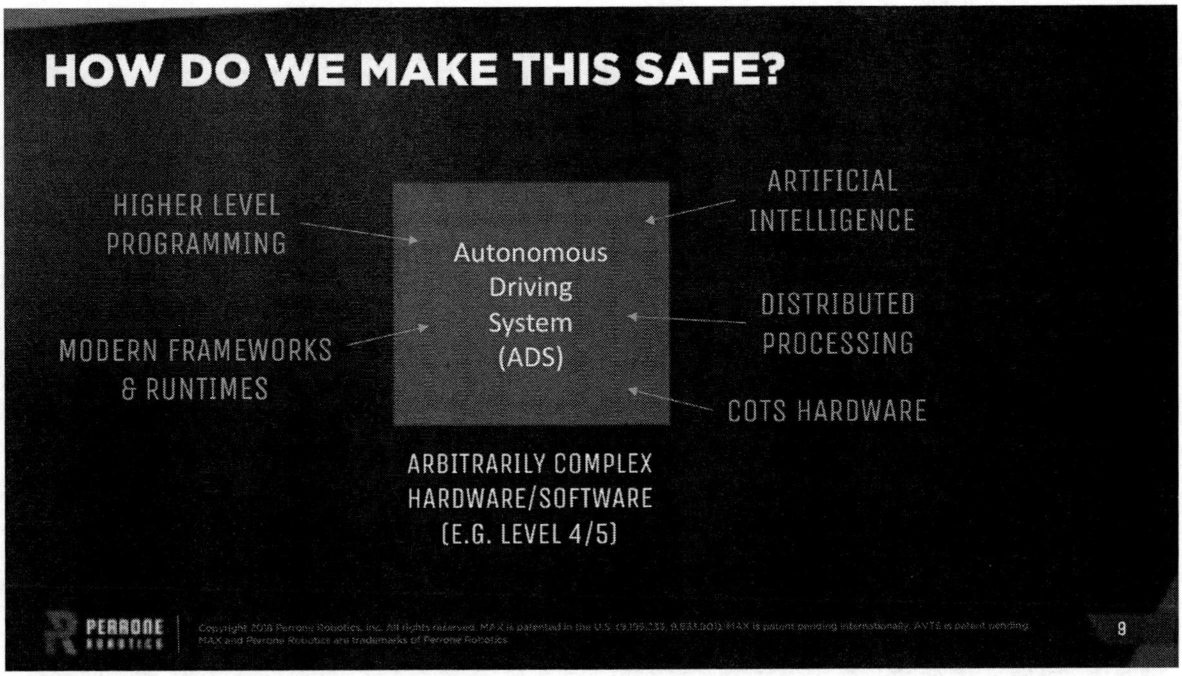
HOW DO WE MAKE THIS SAFE?
HIGHER LEVEL PROGRAMMING
MODERN FRAMEWORKS & RUNTIMES
Autonomous Driving System (ADS)
ARTIFICIAL INTELLIGENCE
DISTRIBUTED PROCESSING
COTS HARDWARE
ARBITRARILY COMPLEX HARDWARE/SOFTWARE (E.G. LEVEL 4/5)
PERRONE ROBOTICS
9

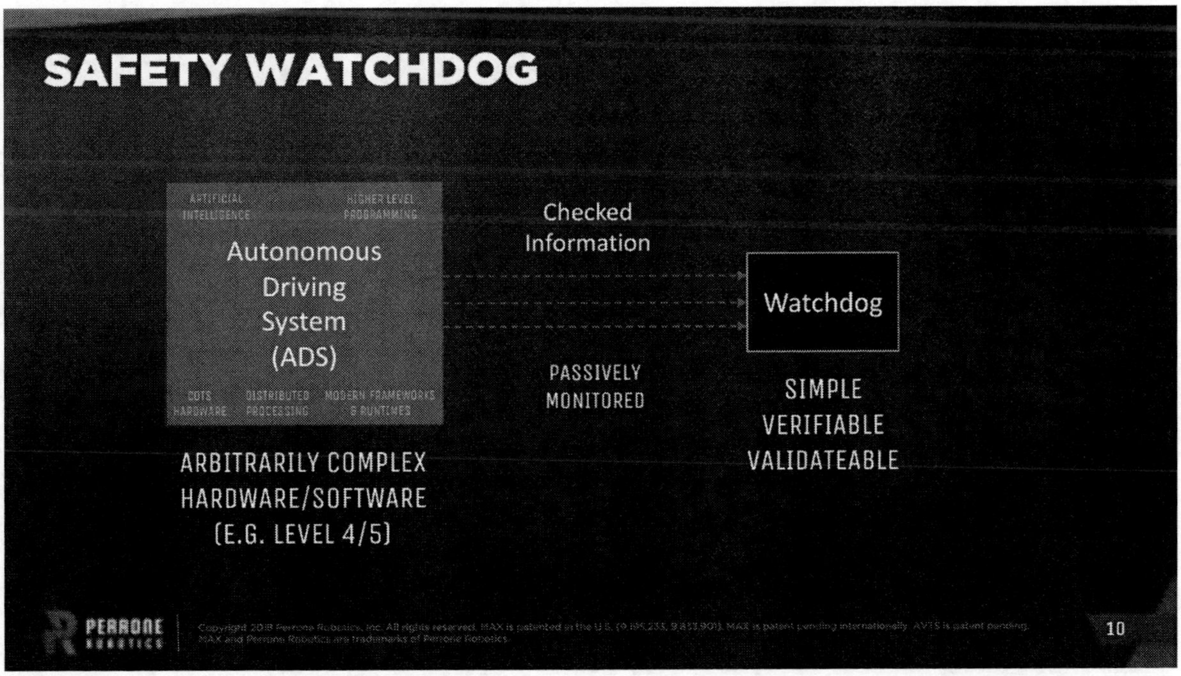
SAFETY WATCHDOG
ARTIFICIAL INTELLIGENCE
HIGHER LEVEL PROGRAMMING
Autonomous Driving System (ADS)
COTS HARDWARE
DISTRIBUTED PROCESSING
MODERN FRAMEWORKS & RUNTIMES
ARBITRARILY COMPLEX HARDWARE/SOFTWARE (E.G. LEVEL 4/5)
Checked Information
PASSIVELY MONITORED
Watchdog
SIMPLE
VERIFIABLE
VALIDATEABLE
PERRONE ROBOTICS
10

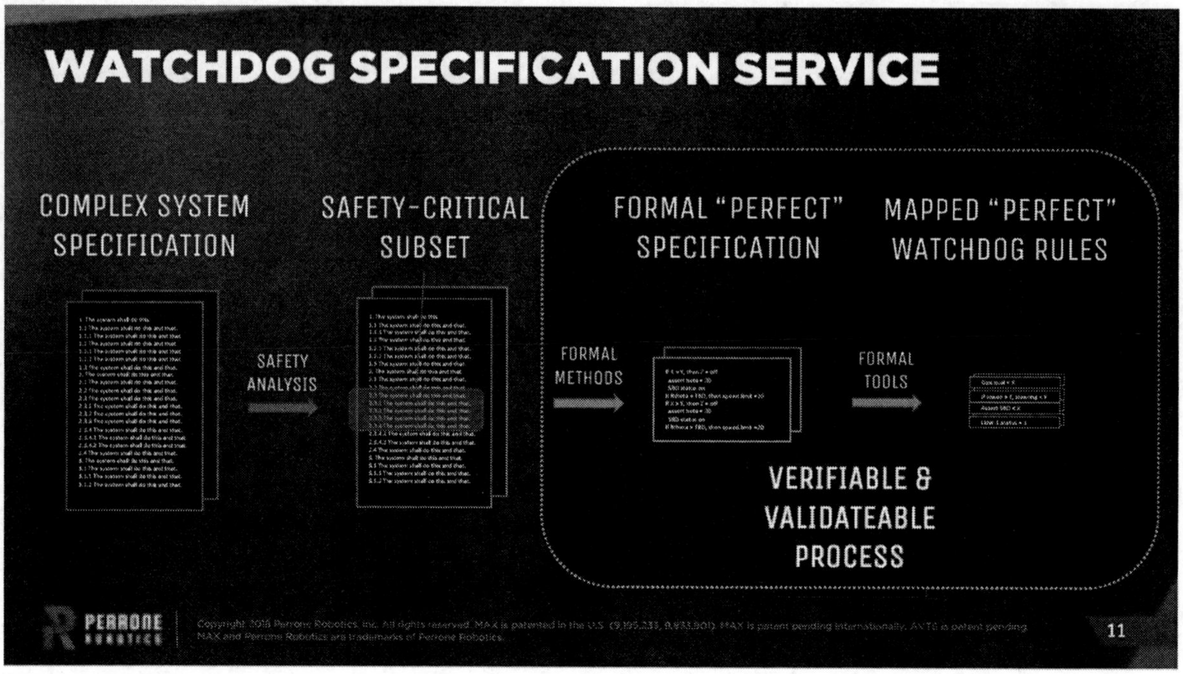
WATCHDOG SPECIFICATION SERVICE
COMPLEX SYSTEM SPECIFICATION
SAFETY ANALYSIS
SAFETY-CRITICAL SUBSET
FORMAL METHODS
FORMAL "PERFECT" SPECIFICATION
FORMAL TOOLS
MAPPED "PERFECT" WATCHDOG RULES
VERIFIABLE & VALIDATEABLE PROCESS
PERRONE ROBOTICS
11

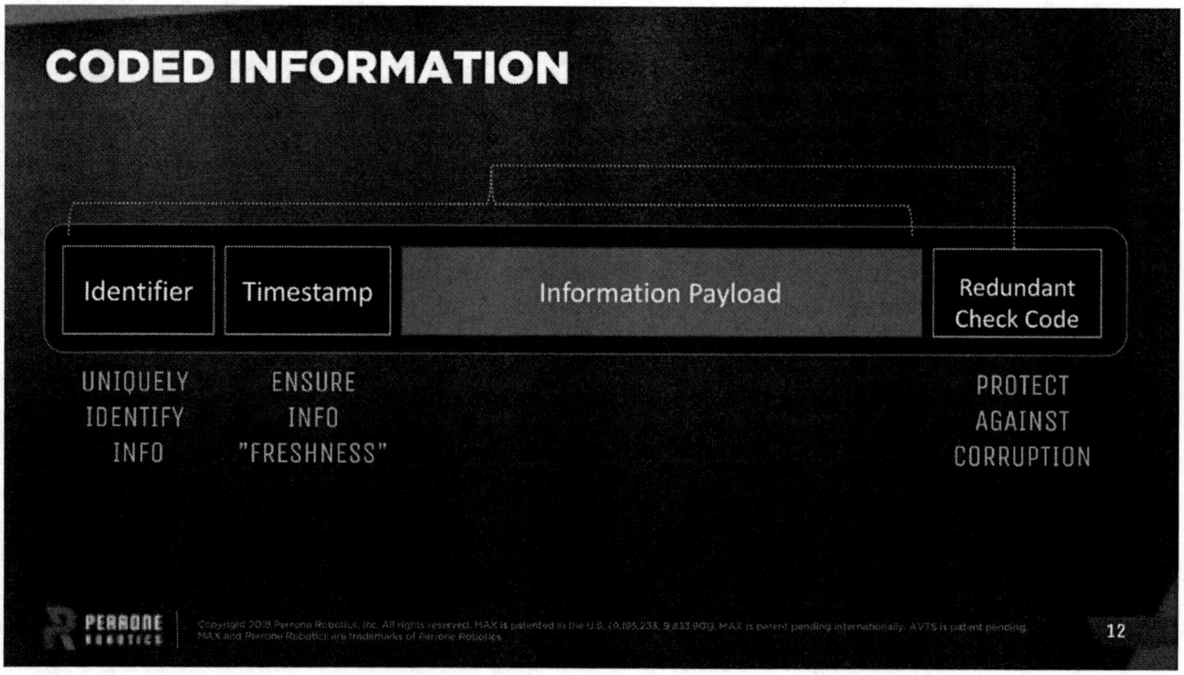
CODED INFORMATION
Identifier
Timestamp
Information Payload
Redundant Check Code
UNIQUELY IDENTIFY INFO
ENSURE INFO "FRESHNESS"
PROTECT AGAINST CORRUPTION
PERRONE ROBOTICS
12

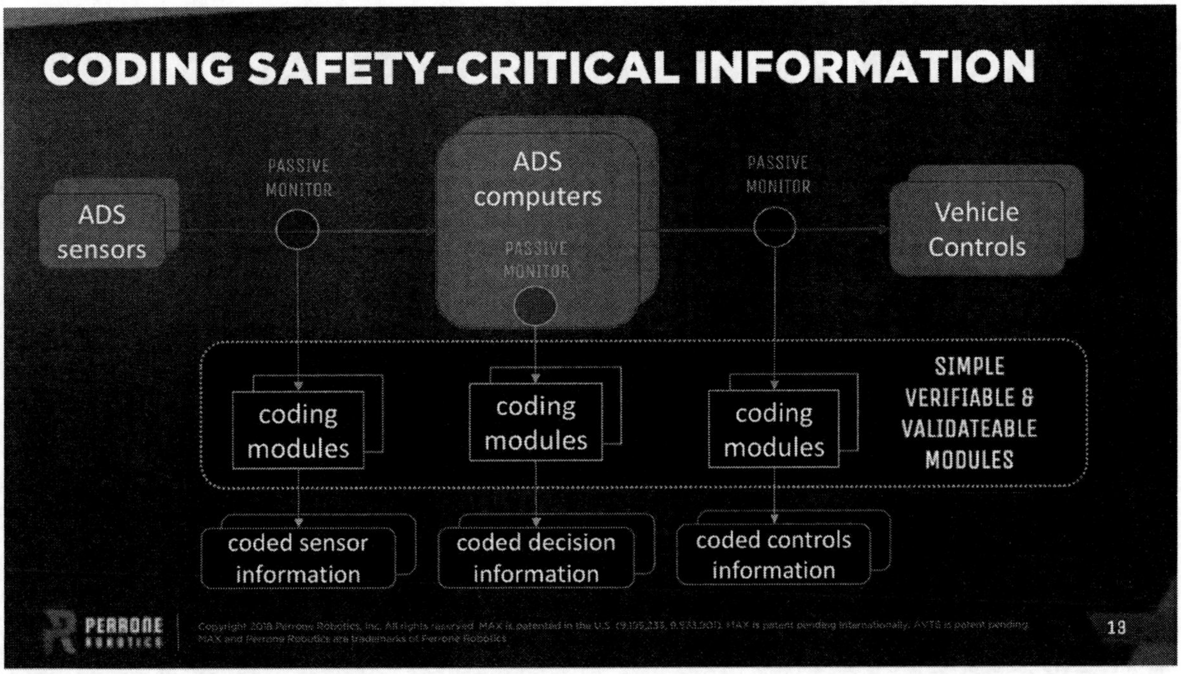
CODING SAFETY-CRITICAL INFORMATION
ADS sensors
PASSIVE MONITOR
ADS computers
PASSIVE MONITOR
PASSIVE MONITOR
Vehicle Controls
coding modules
coding modules
coding modules
SIMPLE VERIFIABLE & VALIDATEABLE MODULES
coded sensor information
coded decision information
coded controls information
PERRONE ROBOTICS
13

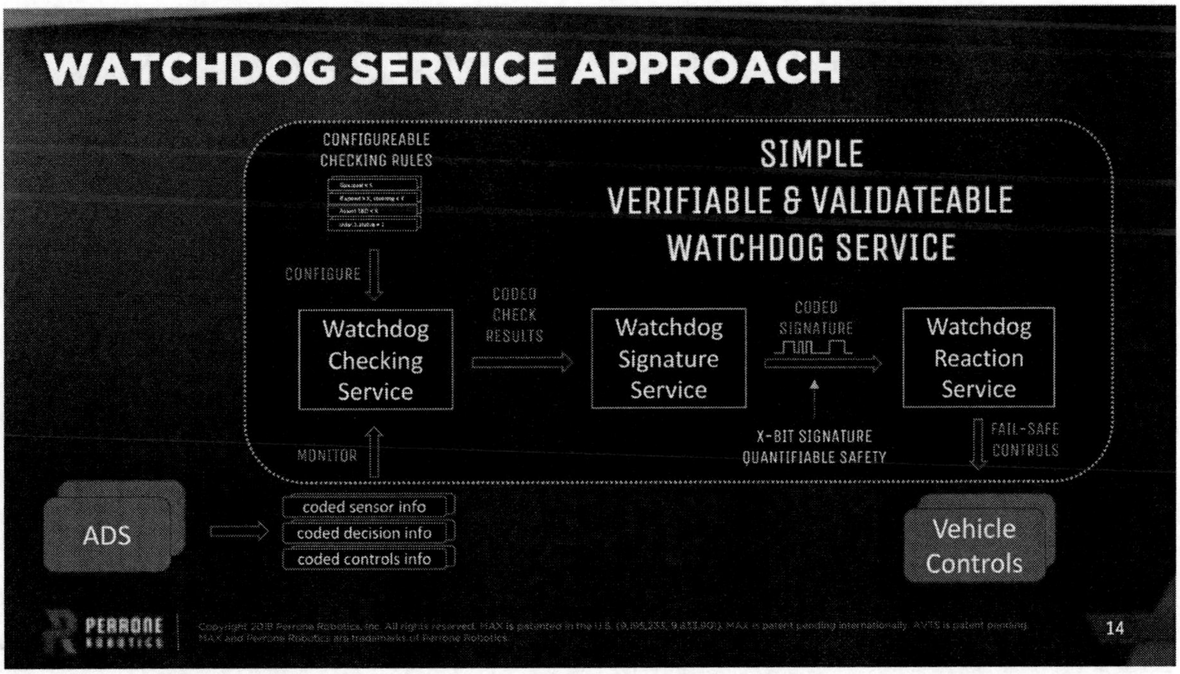
WATCHDOG SERVICE APPROACH
CONFIGUREABLE CHECKING RULES
SIMPLE
VERIFIABLE & VALIDATEABLE
WATCHDOG SERVICE
CONFIGURE
CODED CHECK RESULTS
Watchdog Checking Service
Watchdog Signature Service
CODED SIGNATURE
Watchdog Reaction Service
X-BIT SIGNATURE QUANTIFIABLE SAFETY
FAIL-SAFE CONTROLS
MONITOR
ADS
coded sensor info
coded decision info
coded controls info
Vehicle Controls
PERRONE ROBOTICS
14

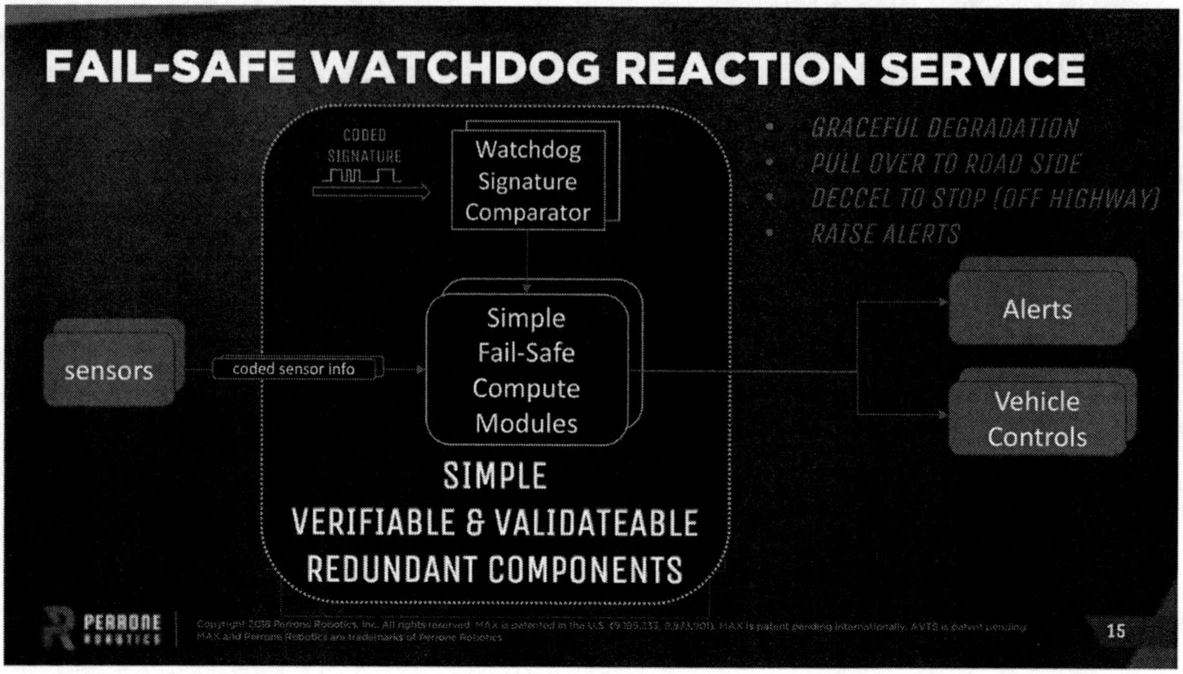

SUMMARY

- CURRENT CERTIFICATION METHODS "DATED" TO DEAL WITH FULL AUTONOMY COMPLEXITY
- HIGHER LEVEL PROGRAMMING LANGUAGES/METHODS (FROM IT) + MODERN FRAMEWORKS/RUNTIMES + AI/ML REQUIRED TO ADDRESS FULL AUTONOMY COMPLEXITY
- NEW APPROACHES NEEDED FOR SAFETY WITH FULL AUTONOMY
- SIMPLE VERIFIABLE/VALIDATEABLE WATCHDOG APPROACH
  - Provable and quantifable safety
  - Bonus: Same approach to safety applies to security

PERRONE ROBOTICS

16

WHITEPAPER FOR THIS TALK
AVAILABLE UPON REQUEST
PAUL@PERRONEROBOTICS.COM
THANK YOU.
QUESTIONS?
WWW.PERRONEROBOTICS.COM
PERRONE ROBOTICS
Copyright 2018 Perrone Robotics, Inc. All rights reserved. MAX is patented in the U.S. (9,195,233, 9,833,901). MAX is patent pending internationally. AVTS is patent pending.
MAX and Perrone Robotics are trademarks of Perrone Robotics.

The Role of Proving Grounds in AV Validation: - A Practical Approach

The American Center for Mobility

WILLOW RUN, YPSILANTI, MICHIGAN

John Maddox
President & CEO
July 2018

Automated Transportation Can Change our Transportation System

Potential for up to 90% reduction in crashes

Reduced congestion, and better use of passengers' time

Reduced emissions & energy consumption

Reduced cost of moving goods

American Center for Mobility
CONNECTED. AUTOMATED. VALIDATED.

Automated Vehicle Validation

- Significant technical and policy challenges remain
  - no established processes yet exist for safe testing, validation, and verification
  - field repairs, updates, and aftermarket solutions must be validated
  - AV workforce issues
- Voluntary standards are needed immediately
  - industry driven, harmonized
  - can inform regulation later
  - vehicle validation, data, simulation, etc...

The American Center for Mobility | Willow Run, Ypsilanti, Michigan

What is the American Center for Mobility?

<u>CAV Proving Ground</u>

- Non-Profit

- Product Development
- Voluntary Standards
- Education

The American Center for Mobility | Willow Run, Ypsilanti, Michigan

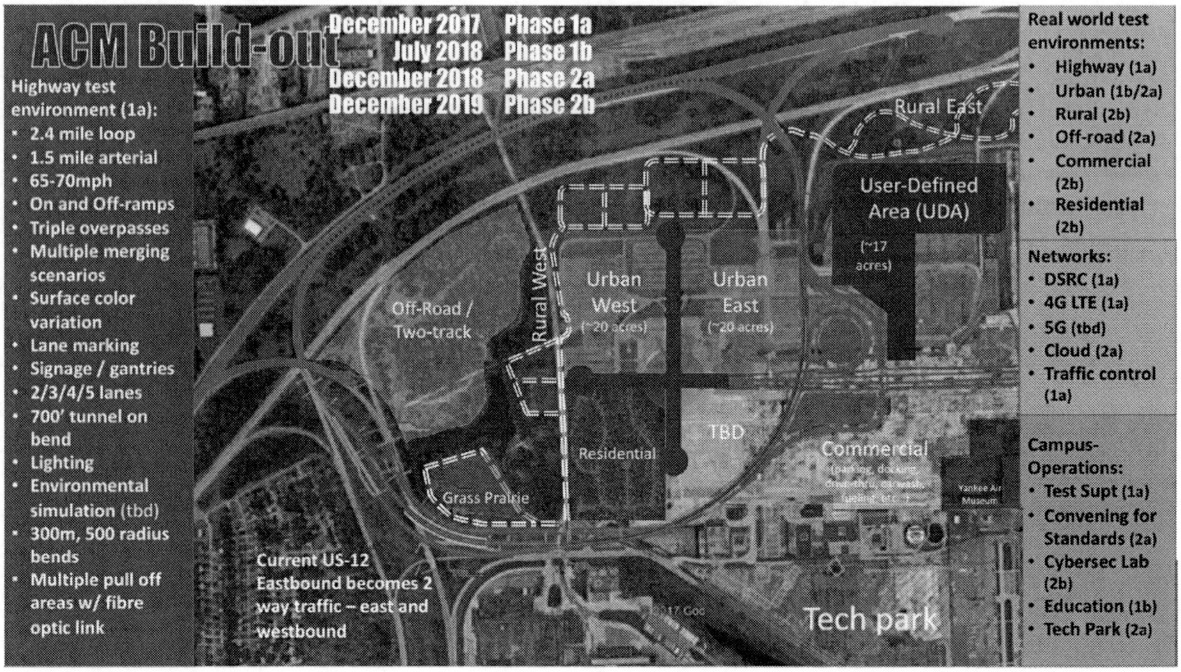
ACM Build-out
December 2017 Phase 1a
July 2018 Phase 1b
December 2018 Phase 2a
December 2019 Phase 2b
Highway test environment (1a):
• 2.4 mile loop
• 1.5 mile arterial
• 65-70mph
• On and Off-ramps
• Triple overpasses
• Multiple merging scenarios
• Surface color variation
• Lane marking
• Signage / gantries
• 2/3/4/5 lanes
• 700' tunnel on bend
• Lighting
• Environmental simulation (tbd)
• 300m, 500 radius bends
• Multiple pull off areas w/ fibre optic link
Rural East
User-Defined Area (UDA)
(~17 acres)
Rural West
Urban West (~20 acres)
Urban East (~20 acres)
Off-Road / Two-track
Residential
TBD
Commercial
Grass Prairie
Yankee Air Museum
Current US-12 Eastbound becomes 2 way traffic – east and westbound
Tech park
Real world test environments:
• Highway (1a)
• Urban (1b/2a)
• Rural (2b)
• Off-road (2a)
• Commercial (2b)
• Residential (2b)
Networks:
• DSRC (1a)
• 4G LTE (1a)
• 5G (tbd)
• Cloud (2a)
• Traffic control (1a)
Campus-Operations:
• Test Supt (1a)
• Convening for Standards (2a)
• Cybersec Lab (2b)
• Education (1b)
• Tech Park (2a)

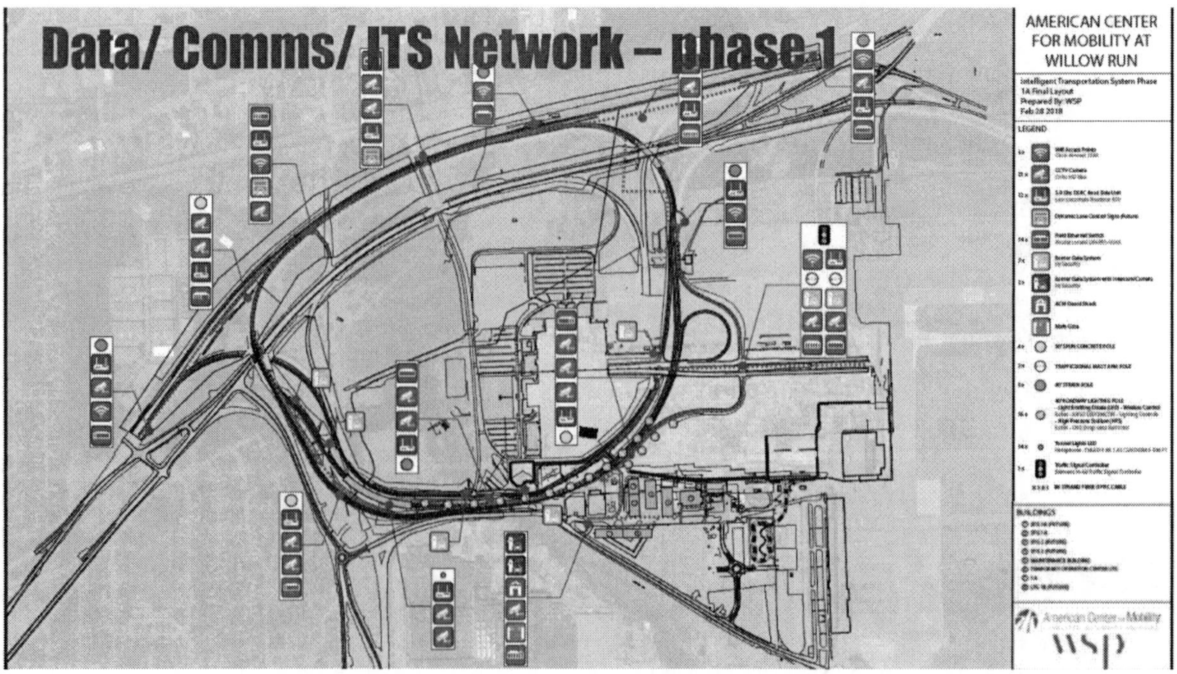
Data/ Comms/ ITS Network – phase 1
AMERICAN CENTER FOR MOBILITY AT WILLOW RUN
Intelligent Transportation System Phase 1A Final Layout
Prepared By: WSP
Feb 28 2018
LEGEND
BUILDINGS
American Center for Mobility
wsp

Data Management & Analytics Platform (DMAP)

- Closed-track data collection
- Public-road data collection
- Simulation tool data collection, and analysis
- Evaluate CAV safety, energy efficiency, advanced control algorithms in controlled and real-world settings
- **ACM will leverage DMAP & Microsoft Azure to seamlessly provide access to data through the cloud**

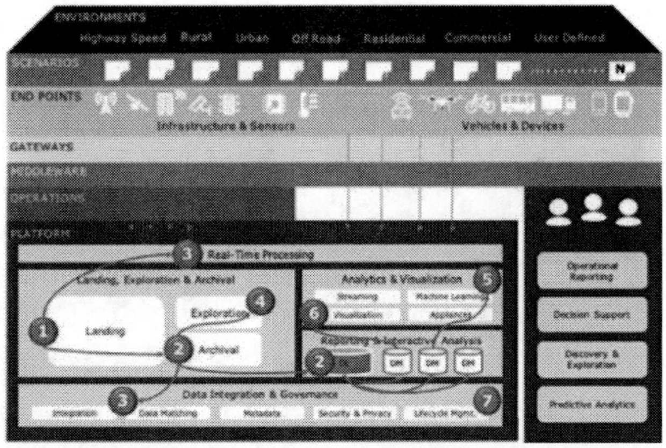

Microsoft

American Center for Mobility
CONNECTED. AUTOMATED. VALIDATED.

Center for Mobility | Willow Run, Ypsilanti, Michigan

5 Environments Open for Testing

- **Highway**
- **Arterial**
- **Boulevard**
- **Tunnel**
- **Intersection**

American Center for Mobility
CONNECTED. AUTOMATED. VALIDATED.

Pillars of Activities

TESTING

STANDARDS

SAE INTERNATIONAL

ite

IEEE

EDUCATION

American Center for Mobility
CONNECTED. AUTOMATED. VALIDATED.

Pillars of Activities

TESTING

STANDARDS

SAE INTERNATIONAL

ite

IEEE

EDUCATION

American Center for Mobility
CONNECTED. AUTOMATED. VALIDATED.

AV Product development: must include a structured combination of five methodologies

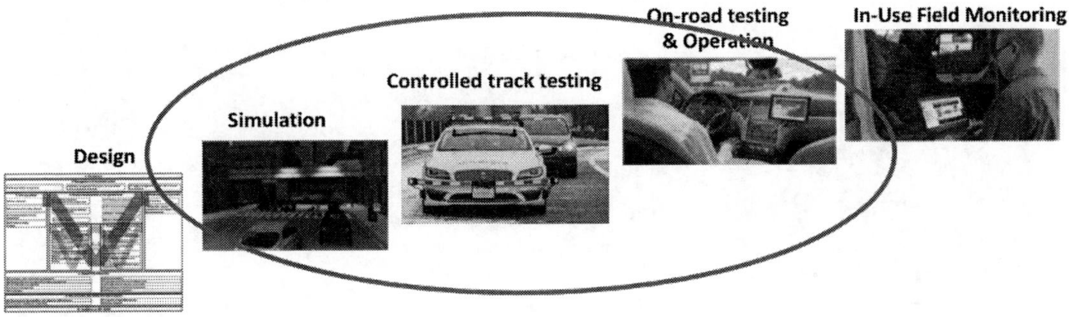

AV Product development & validation: Safe validation must include a structured combination of three methodologies

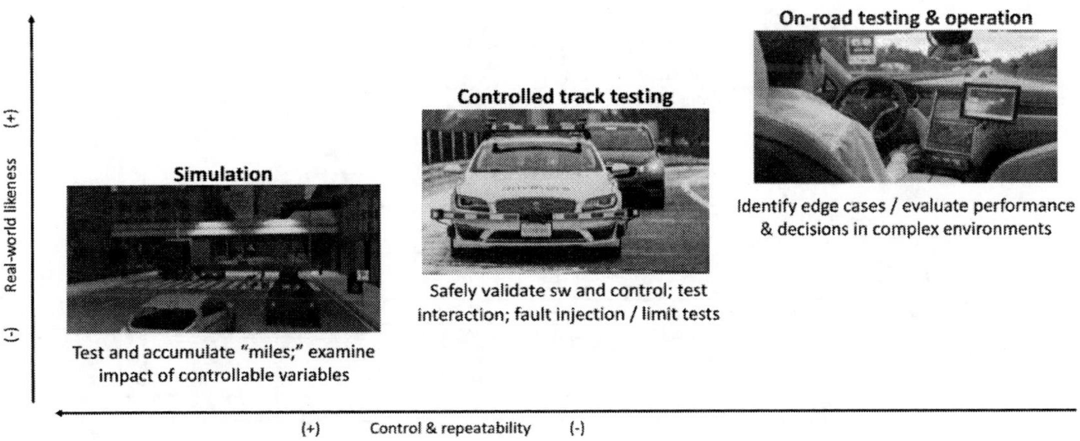

AV Product development & validation: Safe validation must include a structured combination of three methodologies

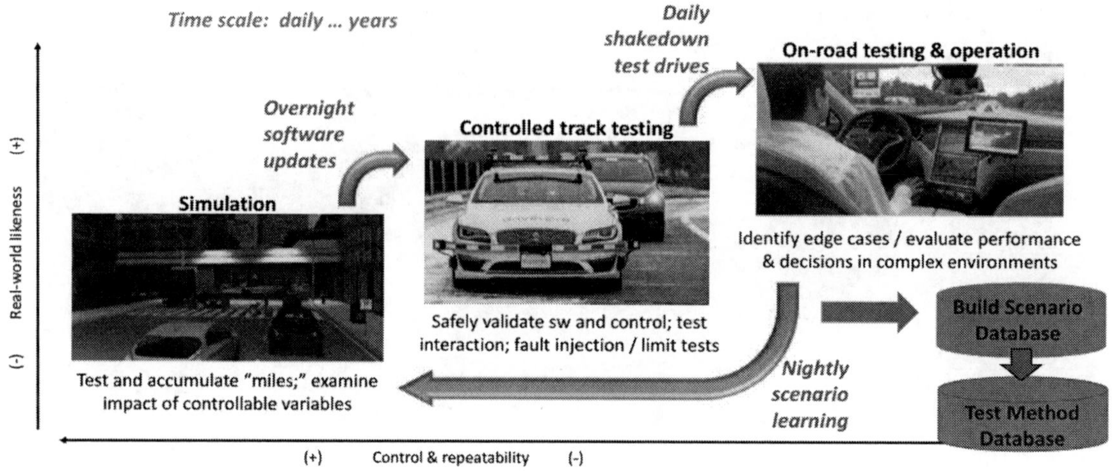

Three methodologies have strengths and weaknesses

		Simulation	Closed Course	On-Road
Normal Driving	Accumulate "miles"	✔	✖	✔✔
	Extrapolation/Fuzzing	✔✔	✔	✔
Edge Case	Identification	✔	✔	✔✔
	Verification	✔	✔✔	✖
Failed State	Verification	✔	✔✔	✖

Outside ODD Verification	✔	✔	✔✔
Software Update Verification	✔	✔✔	✖

Simulation Validation	✖	✔✔	✖
Standard Test Method Development	✔	✔✔	✖
Standard Test Infrastructure/Equipme	✔	✔✔	✖
Customer Evaluation	✖	✔	✔✔
Third Party Evaluation	✖	✔✔	✖

Practical Approach: Develop toolset based on strengths and weaknesses

Pillars of Activities

TESTING | STANDARDS | EDUCATION

SAE INTERNATIONAL

ite

IEEE

American Center for Mobility
CONNECTED. AUTOMATED. VALIDATED.

The challenge is cross-disciplinary

Connected & Automated Vehicles

- Mechanical engineering
- Electrical engineering
- Software engineering
- HMI
- Cybersecurity
- Wireless & Telecoms
- Infrastructure/ civil engineering
- Optics
- Big Data management
- Urban planning
- Behavioral sciences
- Other...

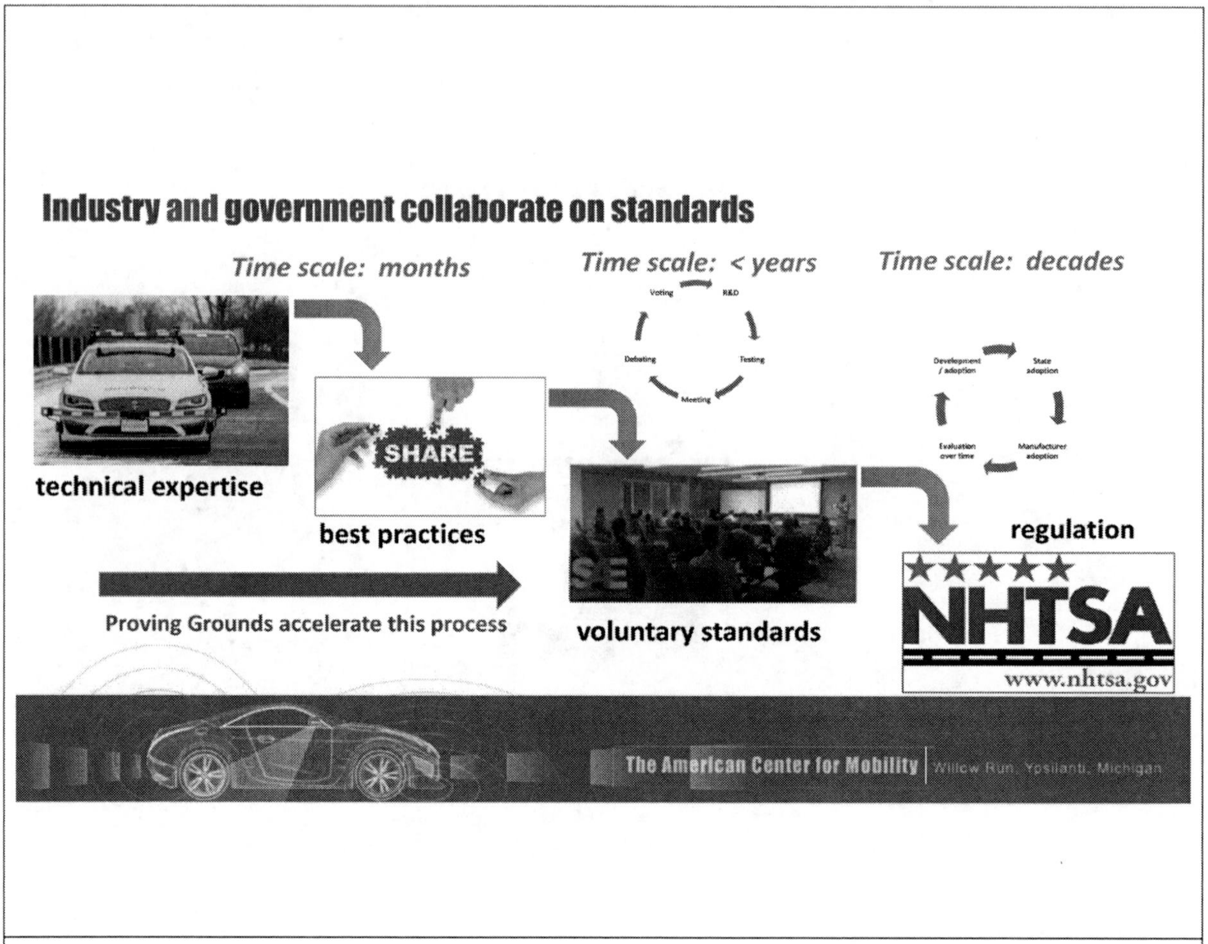
Industry and government collaborate on standards
Time scale: months
Time scale: < years
Time scale: decades
Voting
R&D
Debating
Testing
Meeting
Development / adoption
State adoption
Evaluation over time
Manufacturer adoption
SHARE
technical expertise
best practices
regulation
Proving Grounds accelerate this process
voluntary standards
NHTSA
www.nhtsa.gov
The American Center for Mobility | Willow Run, Ypsilanti, Michigan

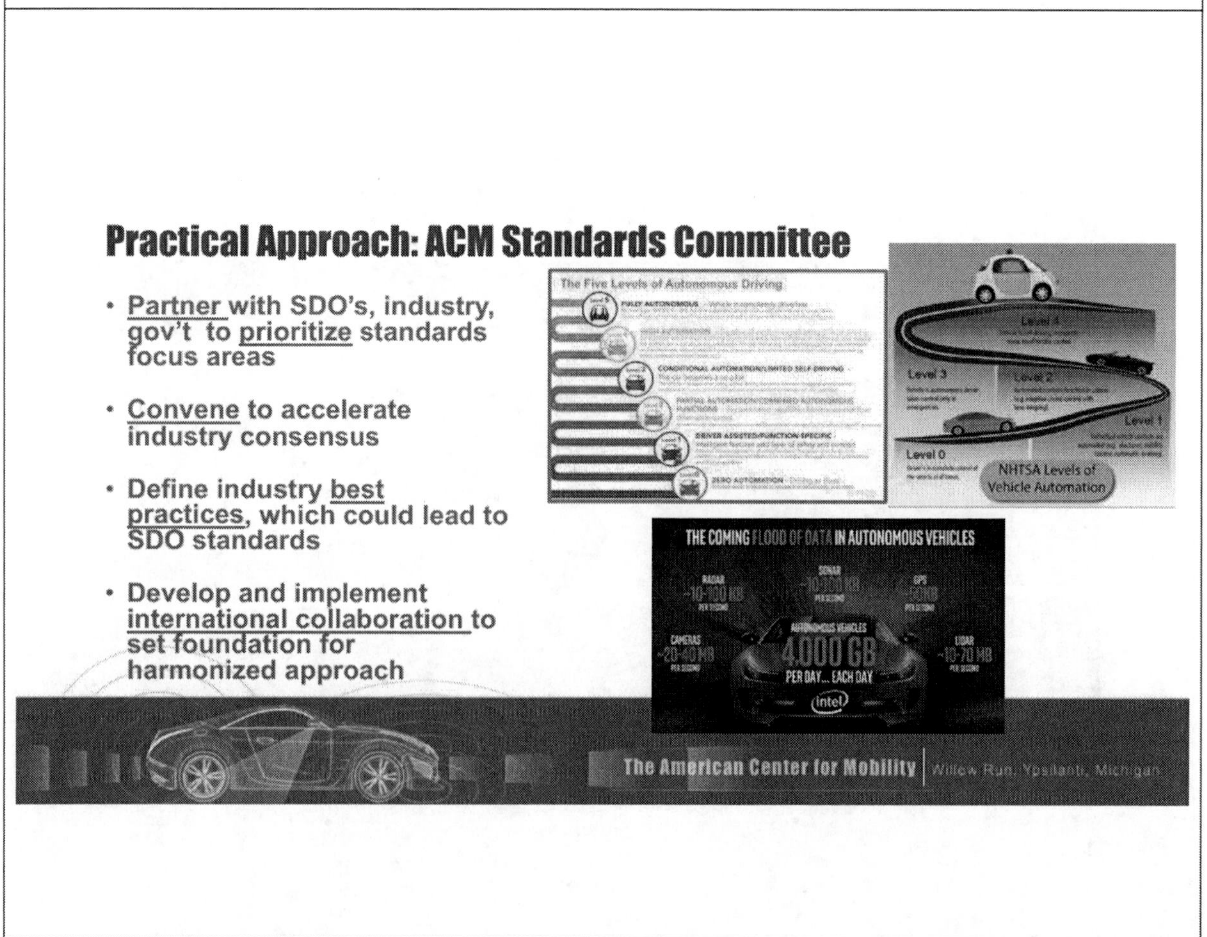
Practical Approach: ACM Standards Committee
• Partner with SDO's, industry, gov't to prioritize standards focus areas
• Convene to accelerate industry consensus
• Define industry best practices, which could lead to SDO standards
• Develop and implement international collaboration to set foundation for harmonized approach
The Five Levels of Autonomous Driving
Level 4
Level 3
Level 2
Level 1
Level 0
NHTSA Levels of Vehicle Automation
THE COMING FLOOD OF DATA IN AUTONOMOUS VEHICLES
4,000 GB
PER DAY... EACH DAY
intel
The American Center for Mobility | Willow Run, Ypsilanti, Michigan

ACM Forms Global Proving Ground Partners, Signs MOU

MOU signing aligns to ACM's strategic goal of ***accelerating standards development focused on CAV testing and data sharing** by*

- Leveraging partnerships w. international AVPGs
    - Formally established International Facilities for Automated and Connect Transportation (IFACT)
- Facilitating safe validation of CAVs

Participating Partners:

- AstaZero AB (Sweden)
- K-City (Republic of Korea)
- Centre of Excellence for Testing and Research for AVs – NTU (CETRAN) (Singapore)

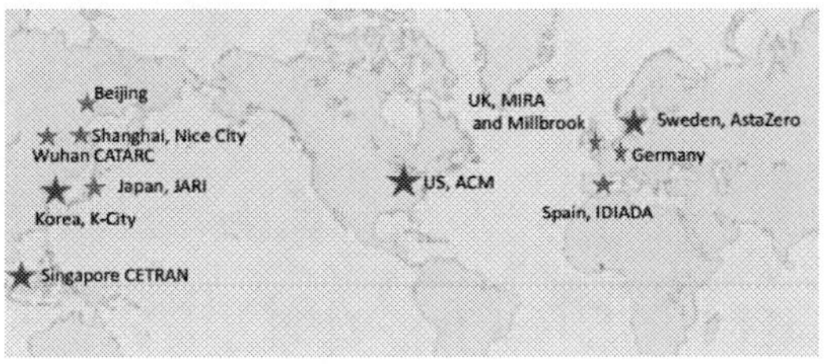

Pillars of Activities

TESTING	STANDARDS	EDUCATION

SAE INTERNATIONAL

ite

IEEE

American Center for Mobility

CONNECTED. AUTOMATED. VALIDATED.

Four Key Education Needs

Workforce

Higher Ed

Public & K-12

R&D

American Center for Mobility
CONNECTED. AUTOMATED. VALIDATED.

Practical Approach: Driver Training & Career Programs at PG's

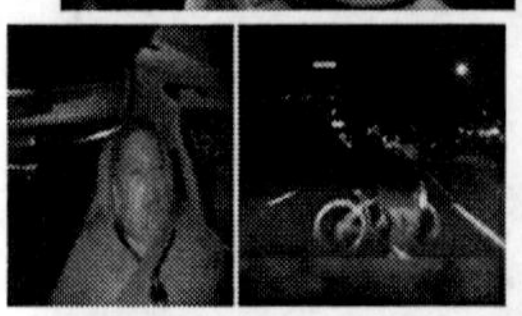

- Rideshare Driver Technology Training
- **AV Safety Driver Training and "Certification"**
- Commercial Driver Technology Demonstrations
- Commercial Driver Technology Adoption Course
- Commercial Driver Technology Certification
- First Responder Driver Technology Training & Certification

American Center for Mobility
CONNECTED. AUTOMATED. VALIDATED.

The Future of our Workforce

- 2016 BLS: over 100,000 jobs in Michigan (2.4% of workforce) include service drivers
- Who will be affected in this transition?: drivers, professionals, manufacturing, others
- Displaced workers will need new career opportunities
- Increasing demand for workers in connected and automated vehicles, fleet management, cyber security, shared vehicles, delivery bot operations etc.
- **HOW CAN WE PREPARE THE WORKFORCE FOR THE NEEDS OF TOMORROW?**

NBC NEWS

Sixth New York City cab driver dies of suicide after struggling financially
by Phil McCausland / Jun.16.2018 / 2:59 PM ET

American Center for Mobility
CONNECTED. AUTOMATED. VALIDATED.

Career Programs Needed – PGs can help

- **CAV Repair and Maintenance Technicians**
- **CAV Operational Monitoring Technicians**
- **Sensor Maintenance Technicians**
- **AV Crash Forensics**
- **"*Ground* Traffic Controller"**
- **Customer Support for OTA Updates**
- First Responder Training
- AV Urban planning
- Paratransit support
- AV Traffic Engineer
- Traffic Signal Repair and Maintenance
- Roadside Unit (RSU) Repair and Maintenance
- ITS Infrastructure Engineering
- Mapping and Geopositioning
- Software & Network Configuration
- Transportation Data Center Analyst
- 3P Application Developer

American Center for Mobility
CONNECTED. AUTOMATED. VALIDATED.

Thank You

intertek
Total Quality. Assured.
Ford
Deloitte.
ADIENT
AT&T
TOYOTA
SIEMENS
HYUNDAI
Visteon®
SUBARU
tass international
Microsoft
American Center for Mobility
CONNECTED. AUTOMATED. VALIDATED.

MEDC
MICHIGAN ECONOMIC DEVELOPMENT CORPORATION
UNIVERSITY OF MICHIGAN
MDOT
Michigan Department of Transportation
planet M®
Michigan. Where big ideas in mobility are born
BUSINESS LEADERS FOR® MICHIGAN
ytown.org
Charter Township of Ypsilanti
SPARK@nn arbor usa

NHTSA

NHTSA's Automated Driving Research Overview

July 2018

What does NHTSA Research do?

2

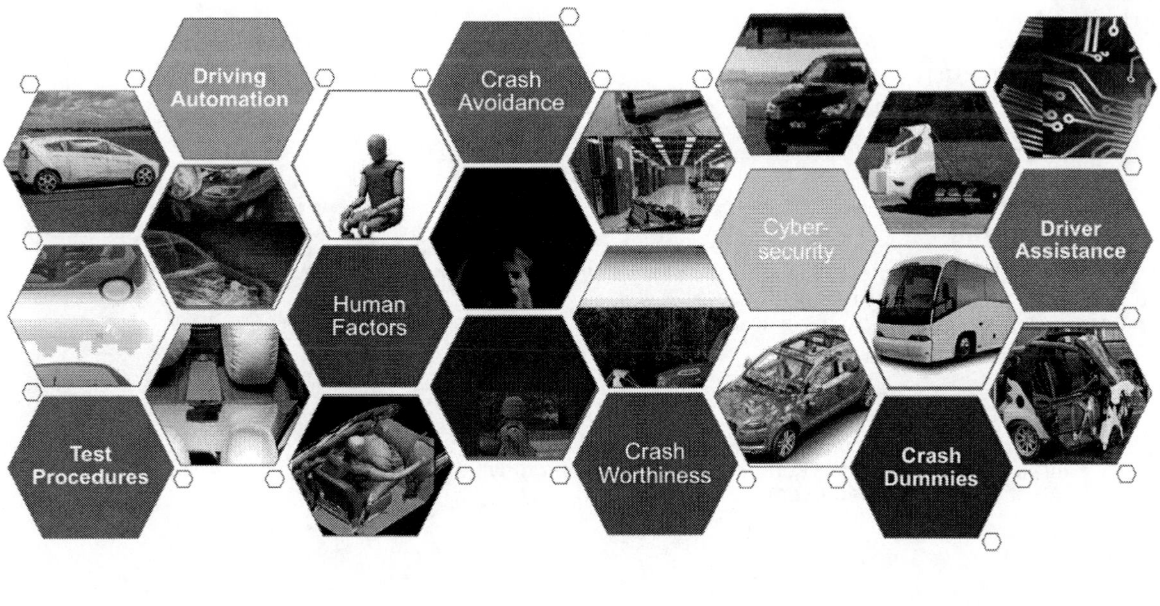

ADS 2.0: A vision for safety

3

Driving System Intent Communication to Shared Road Users

4

- **Background**
  - Drivers communicate with other road users using gestures, eye contact and other complex means. A similar need will likely emerge to communicate Automated Driving Systems' intentions to shared road users
- **Objectives**
  - Describe social interaction strategies used by drivers when communicating to other road users
  - Conduct initial studies to test potential strategies for driving automation systems
- **Deliverables**
  - Report on types of communication needs and potential strategies for effectively communicating automated systems' intention to other road users

SAE Level 2 Naturalistic Driving Study

5

- **Background**
  - Builds on past study on Human Factors Evaluation Of L2/L3 Automated Driving Concepts (published)
  - Numerous SAE Level 2 equipped vehicles deployed
- **Objectives**
  - Develop baseline understanding of naïve drivers':
    - Interaction & use of production driving automation systems
    - Operator responses to different implementations
- **Deliverables**
  - Report on study observations

Driver Expectations for SAE Level 2 Driving Automation Systems

6

- **Background**
  - Vehicles with SAE Level 2 driving automation systems are being deployed
  - However limited market penetration to date
  - Drivers' expectations of SAE Level 2 systems could affect their responses to unexpected situations and safety outcomes
- **Objectives**
  - Evaluate response times for resuming control after surprise events during SAE Level 2 function engagement
  - Evaluate secondary task choices during SAE Level 2 function engagement
- **Deliverables**
  - Report on driver reactions to unexpected errors with SAE Level 2 system performance

On Road Testing of Driving Automation Systems

7

- **Background**
  - Characterize the performance of production Level 2 driving automation systems.
- **Objectives**
  - Real-world data collection mixed roads
  - Observe operational boundaries
  - Measure human / machine interactions (driver monitoring, fallback, misuse/abuse timeout, human-machine interface, etc.)
- **Deliverables**
  - Report documenting test data and system observations

Driving Automation Test Equipment

8

- **Background**
  - Testing aspects of driving automation systems will likely need a highly repeatable test procedure with multiple actors involved that can be programmed to react to the subject vehicle or other stimuli.
- **Objectives**
  - Develop the capabilities to perform objective testing of driving automation systems
    - Procure/develop equipment to test subject vehicles with robotically-controlled principal other vehicles, guided soft targets, pedestrians, and bicyclists
    - Develop scenario/test design software
    - Effort will support scalable scenario
- **Deliverables**
  - Tools to perform objective and repeatable test maneuvers synchronized with multiple vehicles, pedestrians, and bicyclists

Testable Cases and Scenarios Framework for Concept Automated Driving Systems

9

- **Background**
  - Common methods, taxonomy and metrics could benefit the process of validating performance and safety of Automated Driving Systems
- **Objectives**
  - Establish a preliminary framework to describe Automated Driving System test scenarios
- **Deliverables**
  - Identify sample concept driving automation functions
  - Identify the initial set of Operational Design Domain (ODD) scenario attributes
  - Develop a preliminary sample set of objective tests for two concept functions

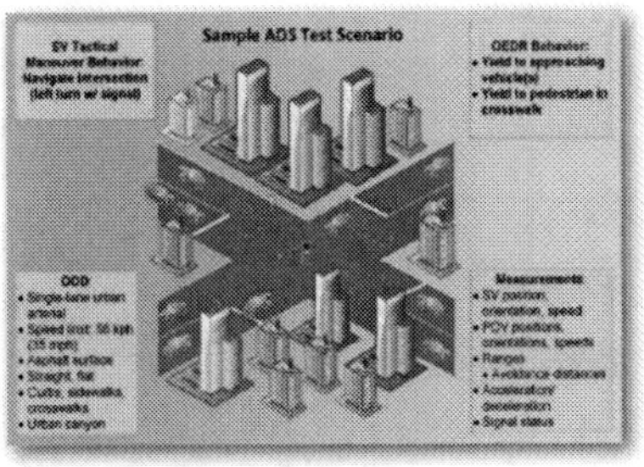

Research on potential FMVSS compliance challenges for vehicles with Automated Driving Systems

10

- **Background**
  - Assessment, evaluation, and approaches to modification of Federal motor vehicle safety standards that may impact compliance of innovative new vehicle designs associated with Automated Driving Systems (SAE Levels 3-5)
- **Objectives**
  - Identify & research standards that are candidates for regulatory translations to facilitate certification of innovative new vehicle designs while still ensuring that standards are objective, practicable, and meet the need for safety
  - Perform test track evaluations needed to support the findings
- **Deliverables**
  - Report on novel concepts
  - Summit Feedback summary

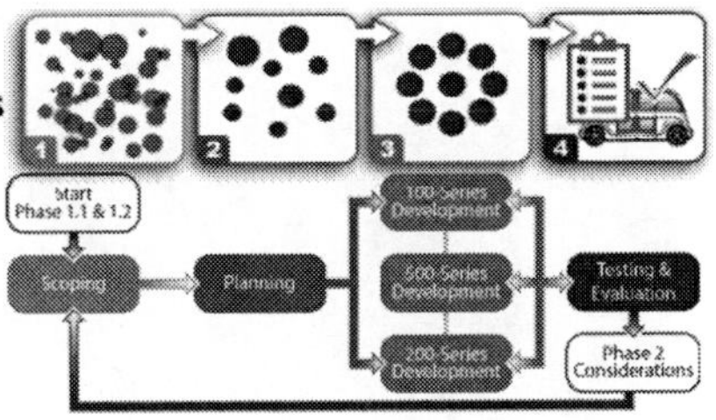

11

General Research Areas under Crash Avoidance

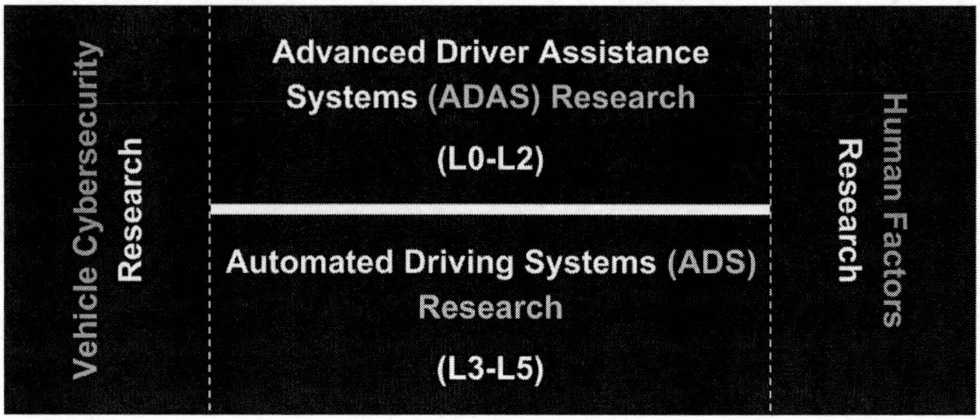

12

Automated Driving Systems (ADS)

Core Research Areas:

1. Support Regulatory Decision on the Removal of Potential Barriers
2. System Safety Performance
3. Functional Safety and ADS Subsystems
4. Crashworthiness

13

Resources

DOT WEBSITES

1. DOT Strategic Plan for FY 2018-2022 - https://www.transportation.gov/administrations/office-policy/dot-strategic-plan-fy2018-2022
2. Research, Development and Technology Strategic Plan - https://ntlsearch.bts.gov/inc/images/kms/StrategicPlan.pdf
3. Federal Register Docket- https://www.federalregister.gov/agencies/national-highway-traffic-safety-administration
4. DOT Research hub- https://researchhub.bts.gov/search

FEDERAL REGULATION DOCKETS

1. Vehicle Automation Topics and Publications: https://www.regulations.gov/docket?D=NHTSA-2014-0070
2. Automotive Cybersecurity Topics and Publications: https://www.regulations.gov/docket?D=NHTSA-2014-0071
3. Automotive Functional Safety and Reliability Topics and Publications: https://www.regulations.gov/docket?D=NHTSA-2014-0092
4. Crash Avoidance Data, located at https://www.regulations.gov/docket?D=NHTSA-2014-0134
5. Docket # NHTSA-2018-0027 - Crash Avoidance and Electronic Systems Safety Research Topics and Publications
6. Heavy Truck Automatic Emergency Braking, link: https://www.regulations.gov/docket?D=NHTSA-2015-0024
7. Crashworthiness and Biomechanics Research Topics and Publications : https://www.regulations.gov/docket?D=NHTSA-2018-0026
8. 2018 NHTSA Research Public Meeting Docket: https://www.regulations.gov/docket?D=NHTSA-2018-0016

AUTOMATED VEHICLES SYMPOSIUM 2018
USERS. VEHICLES. INFRASTRUCTURE.
Driven by AUVSI and TRB

Session # 17B: CAVs' Energy & Emissions Implications

Key Findings & Lessons Learned on Day 2

•*Congrats, Croatia!* (Et bonne chance, France!)

•U.S. **vehicle-miles travelled likely to rise 30%**, so **congestion & emissions are real concerns.**

•Fortunately, **CAV tech improves economics of electrifying** our fleets. We **expect HEVs & PEVs.**

•**Different settings favor different ownership & use:** Privately owned AVs will ultimately cost many households less (at the margin) for longer trips & in low-density landscapes, but shared AVs with dynamic ride-sharing (**SAVs + DRS**) are hard to beat in CBDs with pricey parking.

•***Even with SAEV fleets***, U.S. is **nowhere near meeting GHG targets**. More **electrification** (trucks too!), **right-sizing, credit-based congestion pricing, carbon taxes, & renewable feedstocks are key.** Also, **no empty driving for privately owned AVs** without proper pricing in place.

AUTOMATED VEHICLES SYMPOSIUM 2018
USERS. VEHICLES. INFRASTRUCTURE.

TRB

Session # 17B: Energy & Emissions Implications (2)

Presenters: Corning's Ameya Joshi, SAFE's Amitai Bin-Nun, UC Davis' Austin Brown, UT Austin's Kara Kockelman, UIC's Bo Zou, ANL's Josh Auld, & UMichigan's Morteza Taiebat

Research Needs

•Investigate **trade-offs between public transit vs. MaaS-based SAV fleet use**. And can mass-transit providers eventually compete head-to-head with TNCs?

•Surveys & most simulators are not sufficiently realistic. **More realistic CAV experiences needed,** to better anticipate adoption & use of these new modes.

•**Design energy-minimizing shared fleets** to assign **right-sized & right-ranged vehicles** to **travelers & packages**. How best can **shared fleets** & smart **first-mile, last-mile decisions** protect our planet's climate & our lungs?

•Evaluate **emissions & health implications of CAV electrification** across different power grids. Will future-year coal-plant SO_2 emissions be a notable issue in some locations?

AUTOMATED VEHICLES SYMPOSIUM 2018
USERS. VEHICLES. INFRASTRUCTURE.
Driven by AUVSI and TRB

Breakout Session # 18: Enabling Technologies

Summary of Key Findings and Lessons Learned

- Focus on 5 technologies: **• Communications • Sensors & Sensor Fusion • Positioning • Driver Monitoring • High Performance Computing**
- Format: short expert presentations followed by interactive Q&A with audience
- Outcome:
  - Communications: no decision yet on who is the winner (DSRC or C-V2X/5G) in the "one-takes-it-all" competition. Many open questions around performance, conformance, and interoperability.
  - Sensors & Sensor Fusion: need for more than one sensor (quality, redundancy) but no obvious choice which one(s). Standardization difficult due to market competition and variety of possible approaches.
  - Positioning: high precision localization provided by precision GNSS. real-time kinematic (RTK) systems dramatically improve GNSS localization accuracy. ADS requires GNSS as part of sensor suite
  - Driver Monitoring: not just driver but all occupants, for safer and more personalized ride experience
  - High Performance Computing: (de)centralized, distributed → centralized and hybrid, also depending on software

AUTOMATED VEHICLES SYMPOSIUM 2018
USERS. VEHICLES. INFRASTRUCTURE.
Driven by AUVSI and TRB

Breakout Session # 18: Enabling Technologies

Research Needs

- Communications: (better) comparable test setup/ data analysis ("apple-to-apple")
- Sensors & Sensor Fusion:
  - Object classification in 3D point cloud. Better use of existing data.
  - Development of a common testing protocol (e.g. vision/ visibility spec, measure the recognition of "edge" case scenarios) might be beneficial but will take time to develop. Who takes the lead?
- Driver Monitoring: more quantifiable parameters besides 'nausea'; initial research with physiological responses – next: correlate this with AI-analyzed facial and body tracking
- High Performance Computing: lower power consumption. how to build safety compliant high compute systems

AUTOMATED VEHICLES SYMPOSIUM 2018
USERS. VEHICLES. INFRASTRUCTURE.
Driven by AUVSI and TRB

Breakout Session #19 – What's a Digital Infrastructure Anyway?: Building a Shared Vision for National AV-Readiness

Summary of Key Findings and Lessons Learned

- 2 dynamic panels with a mix of perspectives
  - State - **Scott Marler, Tracy Larkin-Thomason**; Federal - **John Corbin**; European - **Serge van Dam**; Academia - **Nikola Ivanov**; outside perspective - **Jon Neff;** industry perspective - **Valerie Shuman**
- Generated a group definition of digital infrastructure and discussed
  - What are the key components / capability areas of a digital infrastructure?
  - What are the top three national definitions/models that must be in place to enable development of the digital infrastructure?
- Picked apart a preliminary data maturity model for data supporting highly automated vehicles

AUTOMATED VEHICLES SYMPOSIUM 2018
USERS. VEHICLES. INFRASTRUCTURE.
Driven by AUVSI and TRB

Breakout Session #19 – What's a Digital Infrastructure Anyway?: Building a Shared Vision for National AV-Readiness

Next Steps

- White paper
- Findings going into FHWA's National Dialogue on Highway Automation
  - Partnering closely with FHWA on how to feed this work directly into their activities with upcoming workshop in Seattle August 1-2

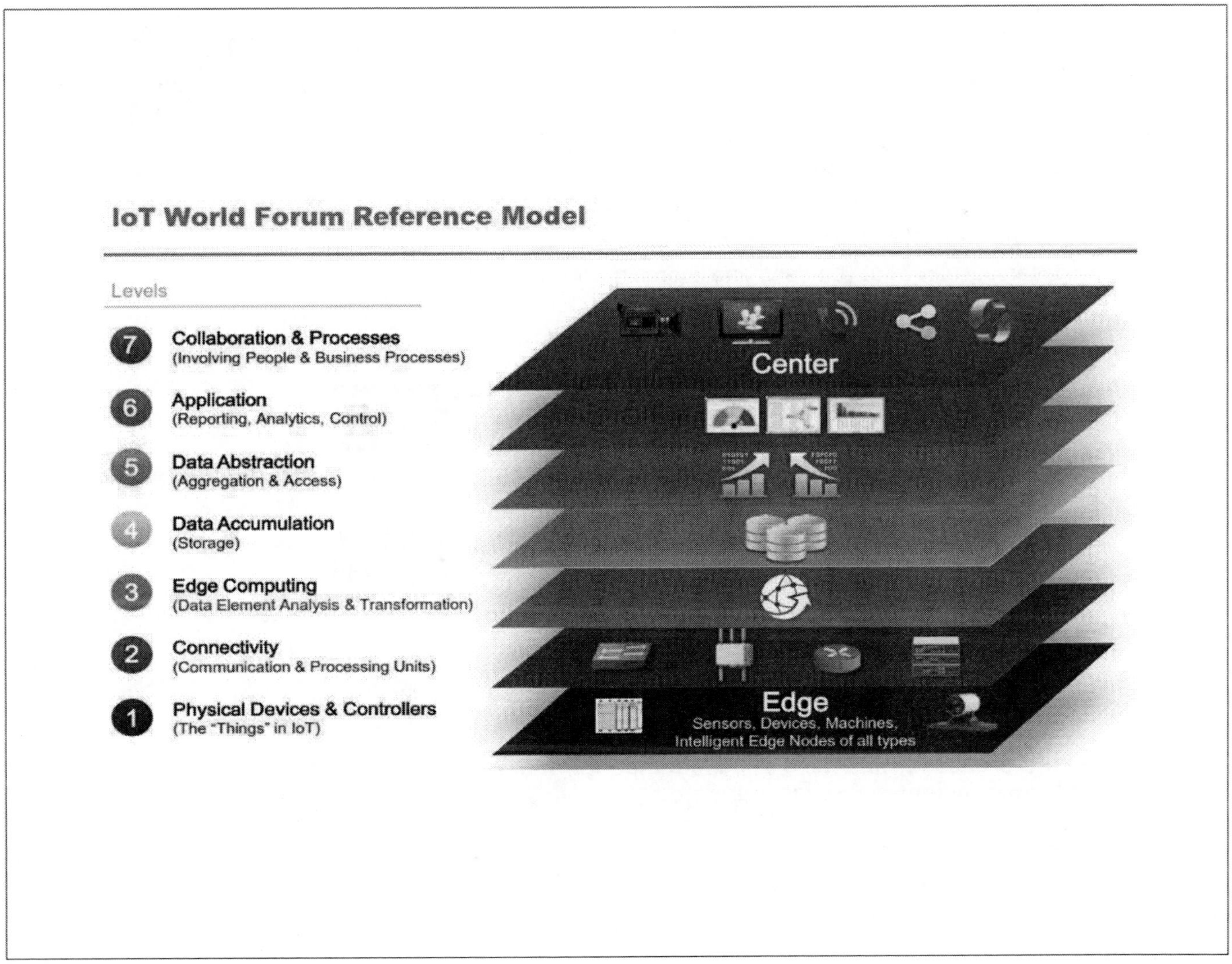
IoT World Forum Reference Model
Levels
7 Collaboration & Processes
(Involving People & Business Processes)
6 Application
(Reporting, Analytics, Control)
5 Data Abstraction
(Aggregation & Access)
4 Data Accumulation
(Storage)
3 Edge Computing
(Data Element Analysis & Transformation)
2 Connectivity
(Communication & Processing Units)
1 Physical Devices & Controllers
(The "Things" in IoT)
Center
Edge
Sensors, Devices, Machines,
Intelligent Edge Nodes of all types

Internet Engineering Task Force

A successful model for developing standards

- IETF has been developing standards for the Internet since 1986.
- The IETF is not a corporation and has no board of directors, no members, and no dues.
- "We believe in rough consensus and running code".
- "Be conservative in what you send and liberal in what you accept".

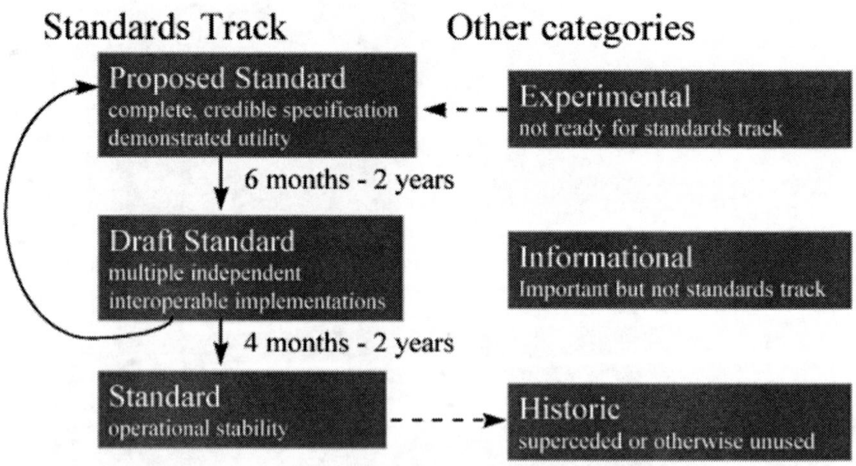

IETF RFC process would need to be modified for automated vehicles to include government-mandated standards, preferably at the beginning and end.

1

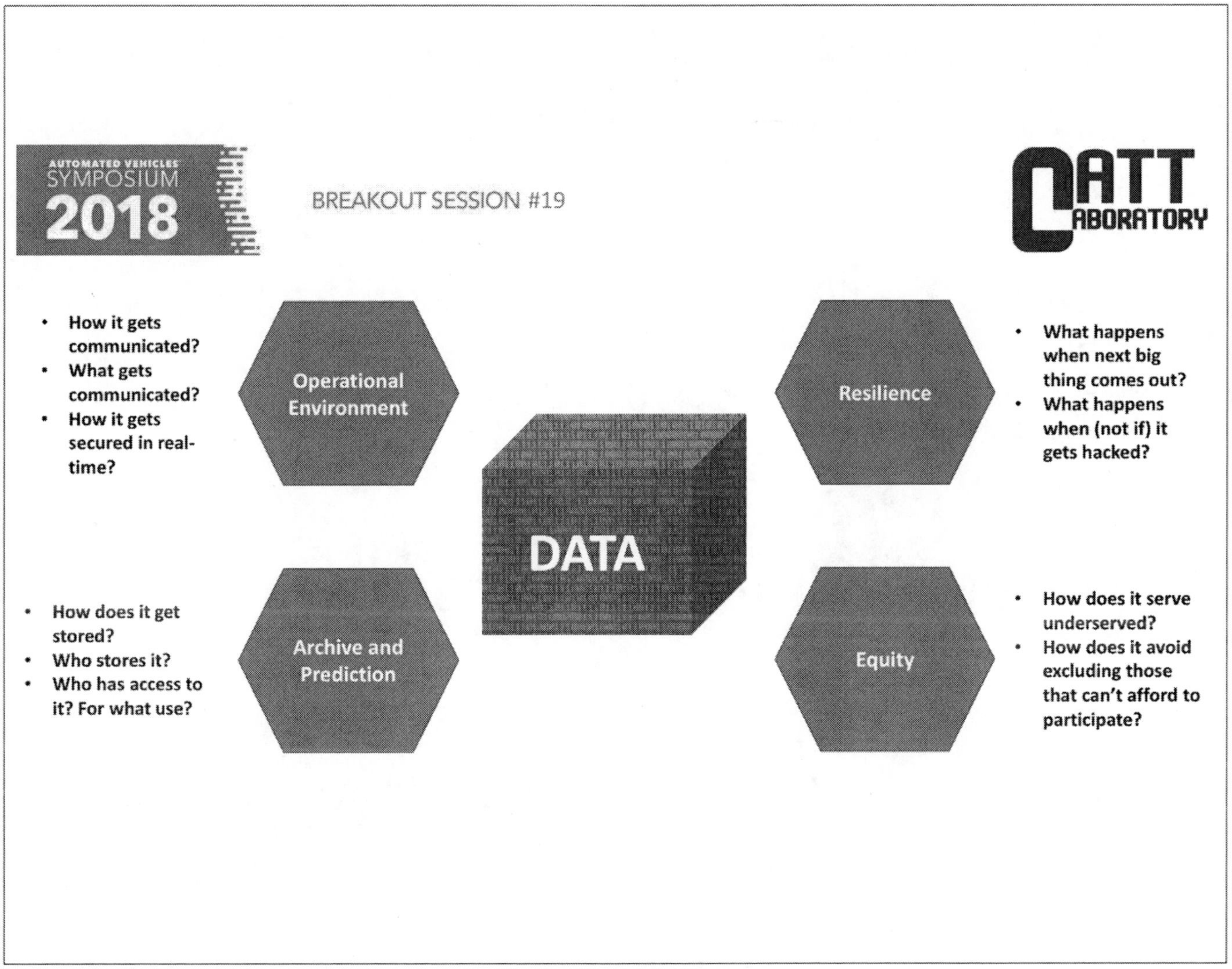
AUTOMATED VEHICLES
SYMPOSIUM
2018
BREAKOUT SESSION #19
CATT LABORATORY
• How it gets communicated?
• What gets communicated?
• How it gets secured in real-time?
Operational Environment
Resilience
• What happens when next big thing comes out?
• What happens when (not if) it gets hacked?
DATA
• How does it get stored?
• Who stores it?
• Who has access to it? For what use?
Archive and Prediction
Equity
• How does it serve underserved?
• How does it avoid excluding those that can't afford to participate?

Frameworks for Digital Infrastructure

1. Bold Vision

2. Data

3. Digital Infrastructure

IOWA DOT

C-ITS developments in Europe

- Big debates on cellular versus short-range: What is hybrid communication?
- Draft legislation (delegated act) on C-ITS
- Progress in testing and harmonization, however:
- OEM ‘soft’ commitment for 2019 deployment…

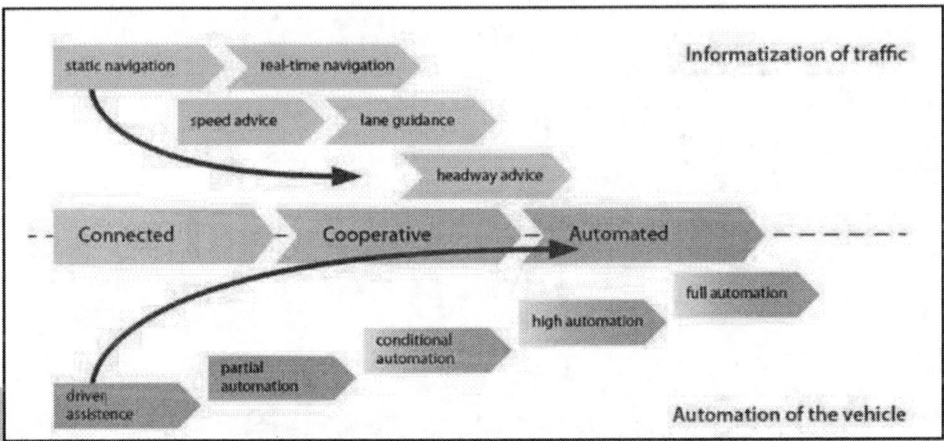

Alignment of Local, State, and National Standards
Protection of the 5.9GHZ Safety Spectrum
Sharing/ Handling of Data Exchange
Digital Infrastructure
Automated Vehicle Symposium
San Francisco
July 10, 2018
Tracy Larkin Thomason, PE
tlarkin@dot.nv.gov
NEVADA DOT

DRAFT

AV Data Maturity Roadmap – Framework Only

	A	B	C	D	E
	Baseline	*Inform*	*Warn*	*Assist*	*Control*
PRESENCE (Availability, Geographic Coverage, Event Coverage)					
TIMELINESS (Start, Update, End, Reporting Period, Latency)					
ACCURACY (Location, Reporting, Error Rate)					
STANDARDS COMPLIANCE					
DATA QUALITY CERTIFICATION					

AUTOMATED VEHICLES SYMPOSIUM 2018

USERS. VEHICLES. INFRASTRUCTURE.

Driven by AUVSI and TRB

Breakout Session # 21
A Call to Action: It's Time for a Transit Renaissance!
Linking Automation, Sharing, and Public Transportation to Advance the Public Good

Summary of Key Findings and Lessons Learned

- Use automation as a new tool to enhance society, our communities, and cities, not just vehicles
- Focus on addressing long-standing transportation problems – it's a chance to rethink land-use, equity, and accessibility.
- Let's move people, not cars!
- Purpose-built vehicles matched with shared mobility could provide new opportunities
- Don't forget about rural applications
- Design vehicles with people not for people

AUTOMATED VEHICLES SYMPOSIUM 2018

USERS. VEHICLES. INFRASTRUCTURE.

Driven by AUVSI and TRB

Breakout Session # 21
A Call to Action: It's Time for a Transit Renaissance!
Linking Automation, Sharing, and Public Transportation to Advance the Public Good

Research Needs

- How will efficiencies in gained in the utilization of automation alleviate challenges and costs associated with paratransit
- How can automation extend the catchment area of and support existing investment in public transportation
- How can existing model of shared mobility pave the way for autonomous vehicles.
- How can we better utilize land use, parking, curb space, accessibility, equity, and mobility through the adoption of automated vehicles.

Robotic Research, LLC

Offering Innovative Engineering Solutions since 2002

Expertise in Autonomous Mobility Across Various Domains, Sizes, and Types

Road Map to Commercial AV Deployments

- Since 2002, RR has been known for autonomous mobility and ancillary technologies for DoD and DHS
- In the last three years, RR has been transitioning these technologies to the commercial market
- Our technology has operated on roads in various locations, such as Fort Hood, Fort Bragg, Greenville, and Afghanistan

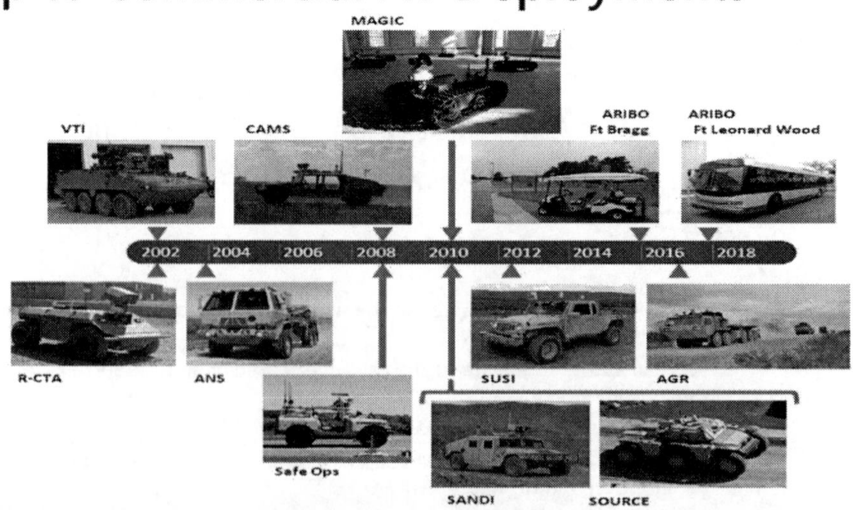

ARIBO – We Have a Problem

ARIBO

- Autonomous electric vehicle navigates roads, intersections, and pedestrian walkways
- On-demand transport system for solders, cadre, and medical staff
- Operational Area: 5 stops 1x1km
- No infrastructure changes required
- 6 intersections, 3 traffic circles
- SAE Level 4 operations

On-demand: Reservation / Scheduling System

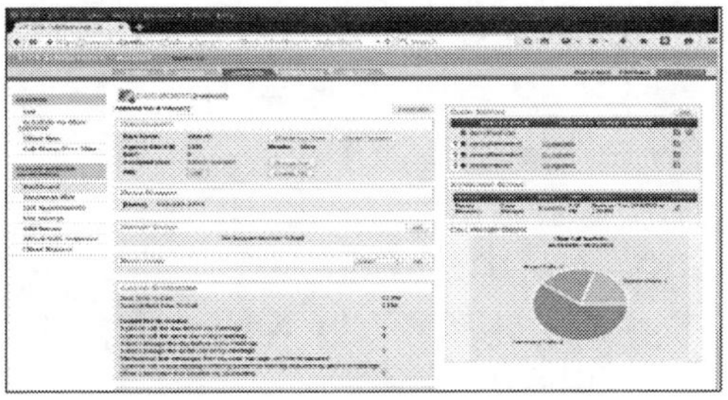

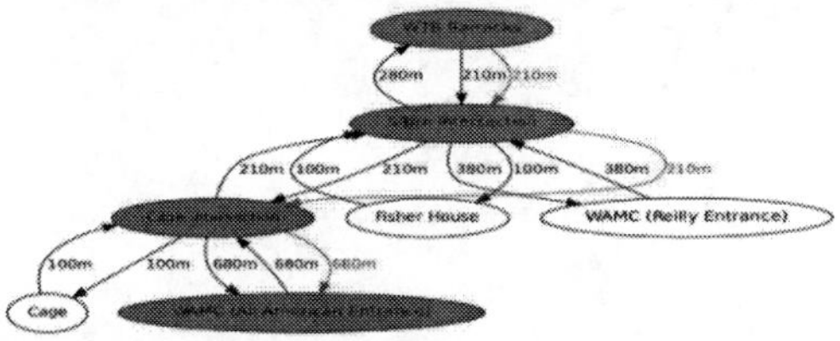

- Smartphone/kiosk reservations
- Dynamic, ride-sharing, on-demand scheduling
- User and admin webpages

Automated Accessibility

- Automated loading and securement system
- Provides passengers with severely reduced mobility the ability to ingress/egress without any human aid
- Rapid loading of wheelchair passengers

Data Collection and Analytics: nSight™

Comprehensive suite of hardware and software that provides the end to end solution from data collection to analysis of manned/unmanned ground vehicles.

Blackbox Recorder
- Vehicle mounted data collection module
- Computational and removable storage capability in one unit
- Expandable storage

Backend Database server and Web server
- Expandable data storage and server backend for multi-year, multi-vehicle data collection with backup
- Suite of automated modules that can evaluate manned or unmanned vehicles
- Extensible API for adding additional analysis modules

Automated Analysis
- Auto-loading of processed data into database
- Event tags and Info Graphics
- Performance Statistics (miles driven, number of event tag occurrences)

After Action Review
- Web UI allows a user to review, compare, and evaluate the vehicles' behavior
- All data is time synchronized and can be viewed simultaneously
- Tags, videos, world model, planned trajectories
- Allows the user to view, add, edit, and delete spatiotemporal tags
- Provides access to raw and processed data

nSight™ After Action Review

Human to AV Comparison

Show the Autonomous shuttle is as safe or safer than the human

- Humans drove the routes while the Autonomy System ran in the background
- Using nSight™, we collected all available data on the shuttle over a 6 Month period
  - Human input, Autonomy data, internal and external video, weather conditions.
  - Roughly 80TB of data
- The result of what the human did and what the robotic system would have done is then post processed and compared.
- Differences between the two then create edge cases

7/26/2018 AVS Breakout #21 - Robotic Research 9

Operational Evaluation: Human vs Robot at Intersections

The Autonomous Shuttle never had lane marking violations or traffic sign violations

Human vs. Robot behavior / % of Observed Situations	Human Driver **Stops** at Intersection (Stopping)	Human Driver **Enters** Intersection (Resuming)
Robot behavior nearly matches human driver behavior	12%	16%
Robot is more aggressive	28%	0%
Robot is more cautious	52%	84%
Robot observes moving object after short delay	8%	0%

7/26/2018 AVS Breakout #21 - Robotic Research 10

nSight™ Enhancements Example

Right-Of-Way Rules

Crosswalks. The system verifies that the autonomous system yields to pedestrians in crosswalks. In particular,

a. it verifies that the crosswalk is detected at sufficient distance to stop,
b. that the pedestrian/s at the crosswalk is detected with sufficient distance to stop,
c. that the pedestrian/s has right of way (if signals are available),
d. that the vehicle stops when the pedestrian has the right of way, and
e. does not stop when the pedestrian does not have the right of way.
f. It also verifies that there is a sufficient safety buffer that is maintained around the pedestrians and
g. the stopping of the vehicle is not abrupt and it follows the demarcation lines if available, and
h. it stops without encroaching into the pedestrian walkway areas even if the lines are not visible.

Measuring safety throughout lifetime: Why?

Human vision and reaction time change with age. The same is true for Ladars, radars, and cameras.

- The light intensity of a laser diode diminishes with time.
- Coatings on the lenses and protective covers get scratched and less transparent over time
- Cameras and laser detectors "burn in" with time.
- The chassis of the vehicle deforms and the internal components become loose and mis-calibrate
- This degradation is unlikely to be do catastrophic
- nSight™ can be used to measure the degradation of the sensors over time

The main point here, is trying to identify when sensors start to degrade. Then by understanding the operational conditions for each AV, what level of sensor degradation is acceptable to continue safe operations.

Olli is Robotic Research Driven

ROBOTIC RESEARCH

AutoDrive™
Driverless Vehicle
Controller

IBM WATSON
Cognitive Rider Experience

7/26/2018 AVS Breakout #21 - Robotic Research 13

IS IT TIME FOR A TRANSIT RENAISSANCE? LINKING SAVS TO ADVANCE PUBLIC GOOD

SUSAN SHAHEEN, PHD
EMAIL: SSHAHEEN@BERKELEY.EDU
TWITTER: SUSANSHAHEEN1
LINKEDIN: SUSAN SHAHEEN

ITSBerkeley 70 YEARS

Berkeley Transportation Sustainability RESEARCH CENTER

SAV Development Updates – Conventional Vehicle SAVs

SAV pilots with conventional vehicles have expanded, with multiple programs planning to deploy commercial SAV services by 2019.

Waymo
Phoenix, AZ

Moved safety engineer to the back seat for its testing with passengers in November 2017.

Cruise/GM
San Francisco, CA

In August 2017, Cruise began offering employees work and non-work trips in their SAVs.

Aptiv/Lyft
Las Vegas, NV

In May 2018, Aptiv deployed 30 SAVs in Las Vegas, NV, available to be summoned by the public through Lyft's ridesourcing app.

SAV Development Updates – Low-Speed SAV Shuttles

In the last year, many new low-speed SAV developments have taken place on private roads, planned communities, university campuses, and local streets.

Optimus Ride
Boston, MA

In January 2018, the company was approved to carry passengers within the Marine Park area of the Boston Waterfront.

May Mobility
Detroit, MI

Launched a fleet of five 6-seater SAVs in June 2018 to replace an existing bus shuttle route in downtown Detroit.

EasyMile/Transdev
Babcock Ranch, FL

Trail rides have taken place since January 2018 and Transdev plans to operate a fixed-route SAV service in the planned community.

SAV Policy Developments

A NUMBER OF POLICY DEVELOPMENTS RELATED TO SAVS HAPPENED OVER THE PAST YEAR, INCLUDING:

- In February 2018, **California approved AV deployment regulations**, which allow companies to apply for permits to test AVs in the state without human backup drivers.
- **Chandler, a suburb of Phoenix, amended its city zoning code** in May 2018. The amendments reduce parking requirements for new developments by up to 40%, if studies can show there will be less demand for parking due to SAVs.
- Only a few states include provisions in their AV legislation that **specifically relate to the regulation of SAV services, at present**
  - Nevada's AB 69 = **3% excise tax** of total SAV fare
  - Tennessee's SB 1561 = **one cent-per-mile tax on AV passenger vehicles**, 2.6 cent-per-mile tax on AV trucks

Stocker and Shaheen, forthcoming

SHARED MOBILITY AND TRANSPORTATION EQUITY

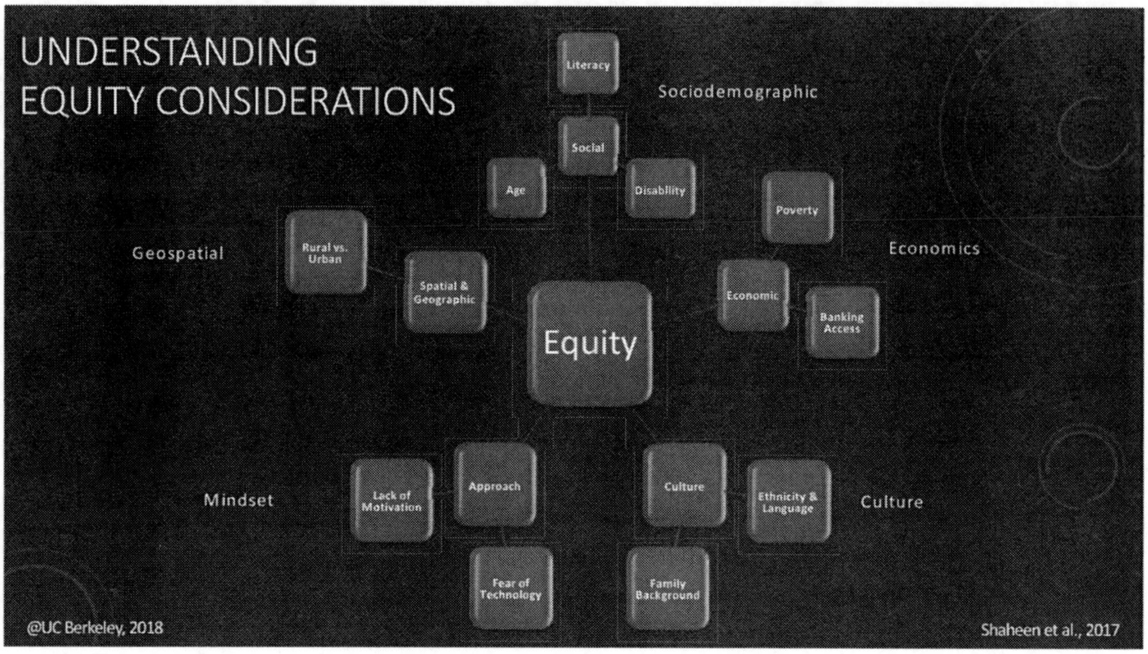

EQUITY CHALLENGES

Some common challenges ...

- **No smartphone**/data access (cannot access)
- Un- and **under- banked** households (cannot pay)
- Low-income users (cannot afford)
- Low-income / minority neighborhood (**lack of service availability**)
- **People with disabilities** (cannot access / lack of service availability)

@UC Berkeley, 2018

Shaheen et al., 2017

SUMMARY OF SHARED MOBILITY USER DEMOGRAPHICS

Mode	Race/Ethnicity	Income	Educational Attainment	Age
Roundtrip Carsharing (N. America)		21% earned >$100K 23% earned <$40K	81% had a 4-year degree or higher	35% ages under 30 31% Ages 30-40
One-Way Carsharing (N. America)	80-87% Caucasian 1-10% Hispanic/Latino 1-5% African American	35-56% earned >$100K 7-17% earned <$35K (US)	72-96% had a 4-year degree or higher (across 5 cities)	48-64% ages under 35 32-41% ages 35-54
P2P Carsharing (N. America)	67% Caucasian 3% Hispanic/Latino 3% African American	30% earned >$100K 21% earned <$35K (US)	86% had a 4-year degree or higher	73% ages under 35 23% ages 35-54
Station-Based (Docked) Bikesharing (N. America Multi-City Studies)	74-92% Caucasian 1-5% Hispanic/Latino 1-2% African American	29-39% earned >$100K 9-26% earned <$35K	55-89% had a 4-year degree or higher (2 studies)	37-54% ages under 35 36-51% ages 35-54
Ridesourcing/TNC (SF Bay Area)		38% earned >$100K 9% earned <$30K	81% had a 4-year degree or higher	73% ages under 35 25% ages 35-54
Microtransit (Kansas City)	89% Caucasian 6% African American 6% Asian American	50% earned >$100K 6% earned <$35K	100% had a 4-year degree or higher	55% ages under 35 39% ages 35-54

@UC Berkeley, 2018

Shaheen et al., 2018

STEPS TO TRANSPORTATION EQUITY FRAMEWORK

- SPATIAL EFFECTS
- TEMPORAL
- ECONOMIC
- PHYSIOLOGICAL
- SOCIAL

@UC Berkeley, 2018

Shaheen et al., 2017

STEPS TO TRANSPORTATION EQUITY FRAMEWORK

SPATIAL EFFECTS

Spatial factors can compromise daily travel needs (e.g., excessively long distances between destinations, lack of public transit within walking distance)

Opportunities:

- Public transit operators and ridesourcing first- and last-mile partnerships
- Microtransit for lower-density areas

@UC Berkeley, 2018

Shaheen et al., 2017

STEPS TO TRANSPORTATION EQUITY FRAMEWORK

TEMPORAL

Travel time barriers can inhibit a user from completing time-sensitive trips, such as arriving at work (e.g. public transit reliability issues, limited operating hours, traffic congestion), school, and medical appointments

Opportunities:

- Dynamic microtransit (on-demand transit)
- Late-night ridesourcing and shuttle services
- Commuter carpooling services

@UC Berkeley, 2018

Shaheen et al., 2017

STEPS TO TRANSPORTATION EQUITY FRAMEWORK

ECONOMIC

Direct costs (e.g., fares, tolls, vehicle ownership costs) and indirect costs (e.g., smartphone, Internet, credit card access) can create economic hardship or preclude users from completing basic travel

Opportunities:

- Shared mobility subsidies for low-income users
- Multiple payment options for shared mobility services
- Multi-modal hubs with Wi-Fi access

@UC Berkeley, 2018 Shaheen et al., 2017

STEPS TO TRANSPORTATION EQUITY FRAMEWORK

PHYSIOLOGICAL

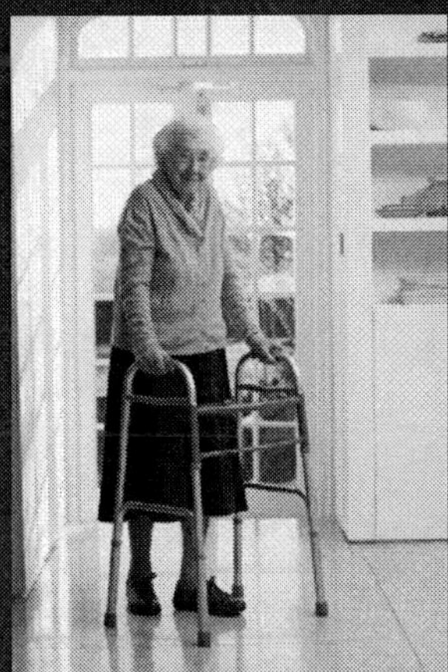

Physical and cognitive limitations can make using standard transportation modes difficult or impossible (e.g., infants, older adults, and disabled)

Opportunities:

- Older adult focused shared mobility services
- Voice activated mobility app features

@UC Berkeley, 2018 Shaheen et al., 2017

STEPS TO TRANSPORTATION EQUITY FRAMEWORK

SOCIAL

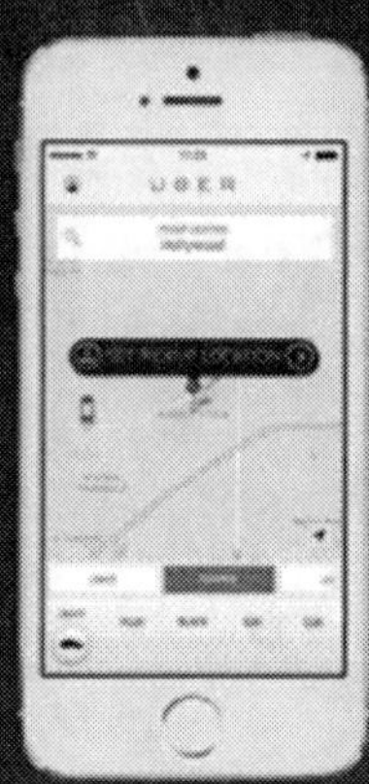

Social, cultural, safety, and language barriers can inhibit a user's comfort in using transportation (e.g. neighborhood crime, poorly targeted marketing, lack of multi-language information) modes and services

Opportunities:

- Ridesourcing app interface that minimizes sociodemographic profiling
- Targeted outreach to low-income and minorities
- App information in user's native language

@UC Berkeley, 2018

Shaheen et al., 2017

Examples:

- Same transportation modes
- Same hours and frequency of service
- Comparable wait times

NEIGHBORHOODS WITH SIMILAR DENSITIES AND BUILT ENVIRONMENTS SHOULD HAVE EQUAL ACCESS & EQUIVALENT LEVEL OF TRANSPORTATION SERVICES

@UC Berkeley, 2018

Shaheen et al., 2017

POLICY QUESTIONS: "THE 3-DS"

- *Divide:* How do we ensure access to all and equivalent level of service among shared mobility providers?
- *Discrimination:* How do we prevent discrimination against protected classes (i.e., minority, people with disabilities, etc.)?
- *Displacement:* How do we make all shared modes affordable and accessible to low-income, digitally impoverished, and un/under-banked users?

@UC Berkeley, 2018

https://www.fhwa.dot.gov/policy/otps/shared_use_mobility_equity_final.pdf

@UC Berkeley, 2018

SHARED MOBILITY AND PUBLIC TRANSPORTATION
DO THE RIGHT MIX

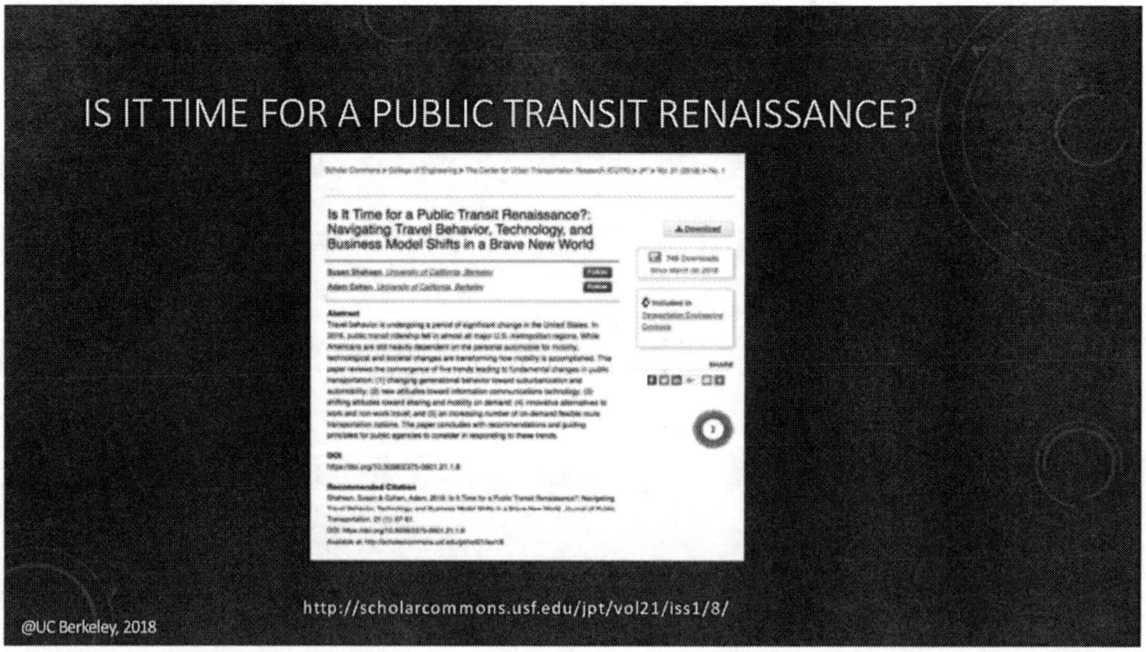
IS IT TIME FOR A PUBLIC TRANSIT RENAISSANCE?
Is It Time for a Public Transit Renaissance?: Navigating Travel Behavior, Technology, and Business Model Shifts in a Brave New World
http://scholarcommons.usf.edu/jpt/vol21/iss1/8/
@UC Berkeley, 2018

RISKS CONFRONTING PUBLIC TRANSPORTATION

- Protecting public transit riders from *crime and terrorism*
- *Demographic shifts* changing modal choice
- *Technological innovation* creating new demand for real-time information services, rider amenities, and on-demand/flexible route services
- *Labor interests & security risks* may slow or limit innovation in public transit
- *Rising transit fares*, as other modes (ridesourcing and driving) are less expensive
- Many public transit systems *in need of maintenance and repair*
- Competition with low fuel prices (auto use)

@UC Berkeley, 2018

Shaheen, 2018

RECENT DECLINES IN PUBLIC TRANSIT RIDERSHIP

UZA Name	Sum of 2015	Sum of 2016	Change
Seattle, WA	178,640,154	185,913,534	4.1%
Houston, TX	83,285,295	85,180,409	2.3%
Milwaukee, WI	40,610,851	41,476,982	2.1%
Detroit, MI	36,734,180	37,079,598	0.9%
New York-Newark, NY-NJ-CT	4,222,700,561	4,241,214,495	0.4%
San Francisco-Oakland, CA	454,952,418	454,906,256	0.0%
Boston, MA-NH-RI	403,464,723	402,554,159	-0.2%
Pittsburgh, PA	63,990,430	63,570,697	-0.7%
Denver-Aurora, CO	101,021,365	99,777,407	-1.2%
Portland, OR-WA	112,440,100	110,985,034	-1.3%
San Antonio, TX	37,983,886	37,290,201	-1.8%
Salt Lake City-West Valley City, UT	44,909,741	43,776,825	-2.5%
Minneapolis-St. Paul, MN-WI	96,636,368	93,716,867	-3.0%
Chicago, IL-IN	623,466,948	603,747,357	-3.2%
Urban Honolulu, HI	68,587,549	66,361,162	-3.2%
Las Vegas-Henderson, NV	72,044,767	69,420,973	-3.6%
Dallas-Fort Worth-Arlington, TX	75,998,371	72,137,725	-5.1%
Baltimore, MD	111,070,976	105,214,371	-5.3%
Atlanta, GA	141,154,134	132,925,293	-5.8%
Philadelphia, PA-NJ-DE-MD	369,644,085	346,276,496	-6.3%
Phoenix-Mesa, AZ	69,525,177	64,898,486	-6.7%
San Diego, CA	94,921,830	88,507,937	-6.8%
St. Louis, MO-IL	47,250,866	44,020,031	-6.8%
Cleveland, OH	46,844,074	43,507,057	-7.1%
Los Angeles-Long Beach-Anaheim, CA	619,459,557	572,589,716	-7.6%
San Jose, CA	44,718,244	40,763,554	-8.8%
Miami, FL	156,449,301	141,556,090	-9.5%
Washington, DC-VA-MD	441,222,366	396,260,838	-10.2%
Austin, TX	32,795,531	28,893,906	-11.9%
San Juan, PR	38,853,326	32,289,221	-16.9%

Increase

No Change

Decrease

@UC Berkeley, 2018

NTD 2017

SUMMARY OF SHARED MOBILITY IMPACTS ON PUBLIC TRANSPORTATION

Mode	Decrease/Increase	Public Transit Impacts
Roundtrip Carsharing (N. America)	Net decrease (-)	For every 5 members that use rail less, 4 ride it more; For every 10 members that use the bus less, 9 ride it more.
One-Way Carsharing (N. America)	Net decrease, although an exception in Seattle (- / +)	In Seattle, where a small percentage of respondents increase their use exceeding the smaller percentage of respondents decreasing their rail use. Across the other four cities, more people report a decrease in their frequency of urban rail and bus use than an increase.
P2P Carsharing (N. America)	Net decrease (-)	Those increasing and decreasing their bus and rail use were closely balanced in number, with 9% increasing bus and 10% decreasing use. Similar effects were found with rail, as 7% reported increasing rail use, while 8% reported decreasing it.
Station-Based (Docked) Bikesharing (N. America Multi-City Studies)	Net increases in bus/rail in small- and medium-sized cities Small net decreases in bus/rail in larger cities (+ / -)	-Small net increases in bus and rail use in small- and medium-size cities (e.g., Minneapolis) -Small net decreases in bus and rail use in larger cities (e.g., Mexico City)
Pooling (Casual Carpooling in Bay Area)	Net decrease (-)	Majority of casual carpoolers were public transit users. In the Bay Area, 75% were casual carpoolers.
Ridesourcing/TNCs (SF Bay Area)	Net decrease (-)	30% competition with public transit, 4% first mile and last mile (destination or origin is public transit stop)

@UC Berkeley, 2018 Shaheen et al., 2018

SHIFTS TO RIDESOURCING PREDOMINANTLY FROM DRIVING, PUBLIC TRANSIT, AND TAXIS

Study Authors Location Survey Year Mode	Rayle et al.* San Francisco, CA 2014	Henao* Denver and Boulder, CO 2016	Gehrke et al.* Boston, MA 2017	Clewlow and Mishra** Seven U.S. Cities***** Two Phases, 2014 – 2016	Feigon and Murphy*** Seven U.S. Cities***** 2016	Hampshire et al.**** Austin, TX 2016
Drive (%)	7	33	18	39	34	45
Public Transit (%)	30	22	42	15	14	3
Taxi (%)	36	10	23	1	8	2
Bike or Walk (%)	9	12	12	23	17	2
Would not have made trip (%)	8	12	5	22	1	-
Carsharing / Car Rental (%)	-	4	-	-	24	4
Other / Other ridesourcing (%)	10	7	-	-	-	42 (another TNC) 2 (other)

Note: Mode replacement findings of these studies employ various methodologies, depending on survey instrument used and analysis methods chosen. Different methodologies can have a notable impact on findings.

@UC Berkeley, 2018 Shaheen et al., 2018

THE RELATIONSHIP BETWEEN SHARED MOBILITY & PUBLIC TRANSIT
First-and-Last Mile Connections
Public Transit Enhancement & Replacement
Late-Night Transportation
Paratransit
Others...
@UC Berkeley, 2018

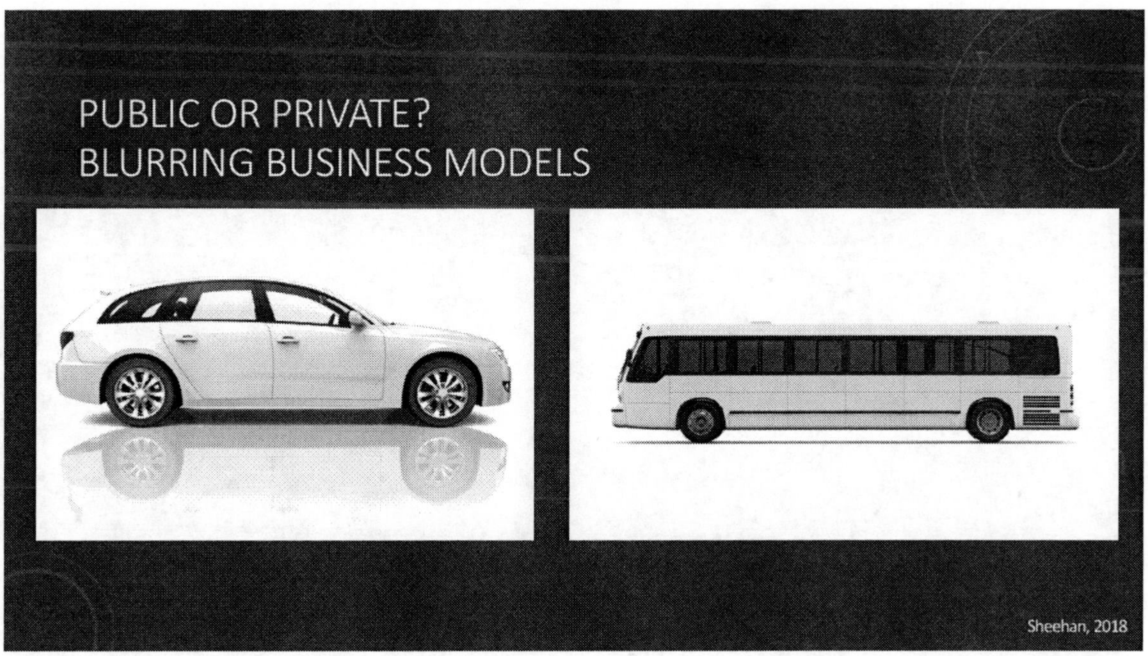
PUBLIC OR PRIVATE?
BLURRING BUSINESS MODELS
Sheehan, 2018

SHARED MOBILITY & PUBLIC TRANSIT

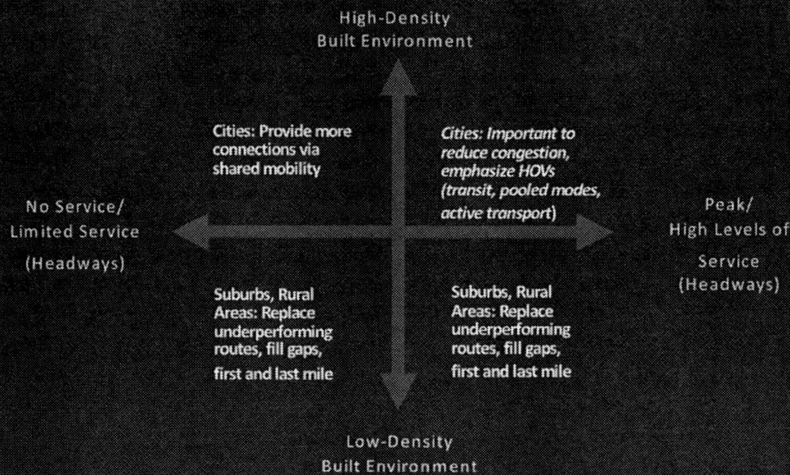

POLICY QUESTIONS

What policies are needed to ensure shared modes complement public transit rather than compete?

- Data sharing
- Integrated routing, booking & payment
- Equitable Access
- Pooled rides (occupancy)
- Peak-hour limitations
- Pricing
- Curb/parking access

INNOVATIVE MOBILITY HIGHLIGHTS, CARSHARING OUTLOOK, AND LATEST RESEARCH
mobility
Home
Research
News
People
About
Contact
Email Address
First Name
Last Name
Subscribe
Last Week In Innovative Mobility
mobility
Subscribe for the latest updates (Innovative Mobility Highlights, Carsharing Outlooks, Policy Briefs, Research Highlights and more) at:
www.innovativemobility.org (bottom of home page)
@UC Berkeley, 2018

RECENT RESOURCES
SHARED MOBILITY
CURRENT PRACTICES AND GUIDING PRINCIPLES
SMARTPHONE APPLICATIONS TO INFLUENCE TRAVEL CHOICES
PRACTICES AND POLICIES
PLANNING FOR SHARED MOBILITY
https://ops.fhwa.dot.gov/publications/fhwahop16022/fhwahop16022.pdf
https://ops.fhwa.dot.gov/publications/fhwahop16023/fhwahop16023.pdf
https://www.planning.org/publications/report/9107556/
@UC Berkeley, 2018

RECENT RESOURCES

https://escholarship.org/uc/item/68g2h1qv

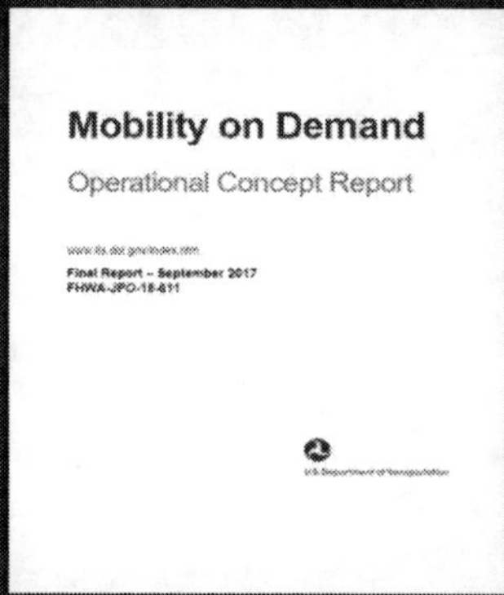

https://rosap.ntl.bts.gov/view/dot/34258

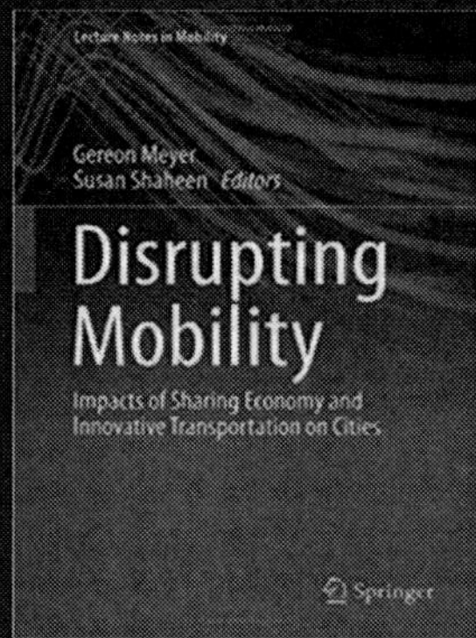

https://www.amazon.com/Disrupting-Mobility-Impacts-Innovative-Transportation/dp/3319516019

ACKNOWLEDGEMENTS

www.its.berkeley.edu/sustainabilitycenter
www.innovativemobility.org

- Shared mobility operators, experts, and industry associations
- Federal Highway Administration
- Mineta Transportation Institute, San Jose State University
- California Department of Transportation
- American Planning Association
- Adam Cohen, Rachel Finson, Elliot Martin, Adam Stocker, Hannah Totte, Mark Jaffee, Marcel Moran, Mikela Hoffman-Stapleton, TSRC, UC Berkeley

tsrc

Susan Shaheen, Ph.D.
Email: sshaheen@berkeley.edu
Twitter: SusanShaheen1
LinkedIn: Susan Shaheen

ITSBerkeley 70 YEARS

AUTOMATED VEHICLES SYMPOSIUM 2018

Driven by: AUVSI & TRB

July 9–12, 2018

Hilton San Francisco Union Square | San Francisco

A Call to Action: It's Time for a Transit Renaissance! Linking Automation, Sharing, and Public Transportation to Advance the Public Good

nadtc
National Aging and Disability Transportation Center

Carol R. Wright Kenderdine
Easterseals, Inc.
Co-Director, NADTC

National Aging & Disability Transportation Center

- Partnership of National Association of Area Agencies on Aging (n4a) and Easterseals, Inc.
- National Technical Assistance Center funded by the U.S. Department of Transportation, Federal Transit Administration
- Promotes the availability & accessibility of transportation options that serve the needs of:
  - Older Adults
  - People with Disabilities
  - Caregivers

www.nadtc.org nadtc 2

What Do We Want & How Do We Get It?

- Improve mobility for people who currently have limited transportation options
- Have autonomous vehicle solutions which are:
  - **Affordable**
  - **Available**
  - **Accessible**
- Shared autonomous vehicles that have diverse payment systems

What Do We Know? Older Adults – The Numbers:

- The number of U.S. residents age 65+ is projected to increase to 72 million in 2030/88 million by 2050 (10,000 people turn 65 every day).
- Nearly 16 million people age 65+ live in communities where public transportation is poor or nonexistent. (That # is expected to grow rapidly as Baby Boomers remain living outside of large cities.)
- 92% of older people intend to "age in place".
- 22% of Baby Boomers are now (or are at risk of becoming) "elder orphans" – individuals who have no family members, friends or neighbors to provide transportation for them.

Rural Mobility – Autonomous Vehicles as a Solution for Targeted Audiences:

- Elderly people who are no longer able to drive. The proportion of people 65+ is higher in rural areas than in urban areas, and their numbers are expected to increase over the next decade.
- Young people who are not yet of driving age living where there is no available public transit.
- People with disabilities who cannot drive or have difficulty accessing public transit.

NOTE: Increased mobility, especially for people in isolated areas or with mobility issued related to age or ability, has a direct impact on personal physical and mental health, as well as quality of life.

Final Thoughts: How Do We Get It?

- Design with those who cannot drive in mind (Include this target audience when designing vehicles)
- Ensure that the Americans with Disabilities Act (ADA) is front and center in the design and rollout of autonomous vehicles
- Remember that transportation is only one part of the complete travel experience. The journey planning must include getting to the vehicle, getting into the vehicle, the trip itself, and getting from the vehicle to the final destination.
- With proper, inclusive design and planning, autonomous vehicles can provide even more people with significantly increased independence and mobility in their lives.

nadtc

NATIONAL AGING AND DISABILITY TRANSPORTATION CENTER

1101 Vermont Avenue NW, Suite 510

Washington, DC 20005

www.nadtc.org

cwright@Easterseals.com

easterseals taking on disability together

n4a advocacy | action | answers on aging

DEPARTMENT OF TRANSPORTATION UNITED STATES OF AMERICA

Questions?

people
not
cars

The New York Times
SHARE
Opinion
Automated Vehicles Can't Save Cities
By Allison Arieff
Contributing opinion writer on architecture and design

This is what streets looked like in the early 1900s.

There are people walking and riding bikes.

There are horses, pushcarts and
even horseless carriages.

It's chaos.
But it works because...

No single mode of
transportation is privileged.

ONLY
Look at that same street today.
It's designed primarily for...
ONLY

cars
and cars
and more cars

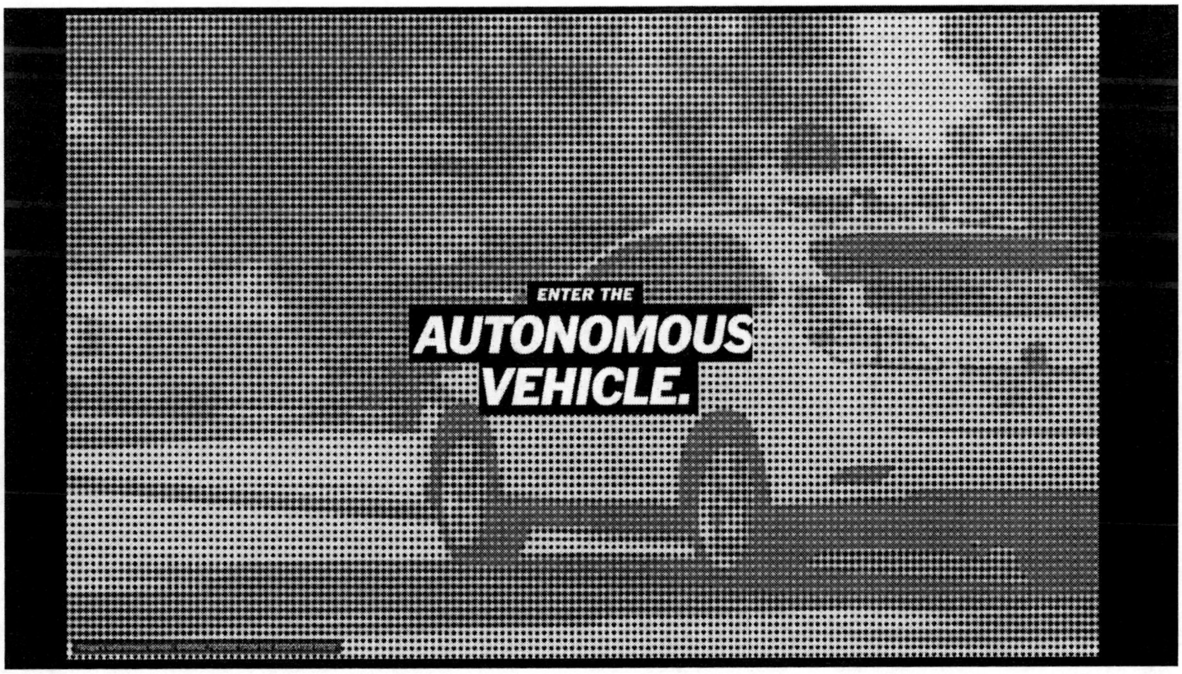
ENTER THE
AUTONOMOUS
VEHICLE.

It's one of the most talked about "solutions" to a host of urban problems. . .

from traffic congestion

to CO_2
emissions and
lack of green
space.

But the reality is that AVs
alone won't save cities
from these problems.
(They could actually make them worse.)

We need to ask ourselves:

Just what is the problem that AVs are meant to solve?

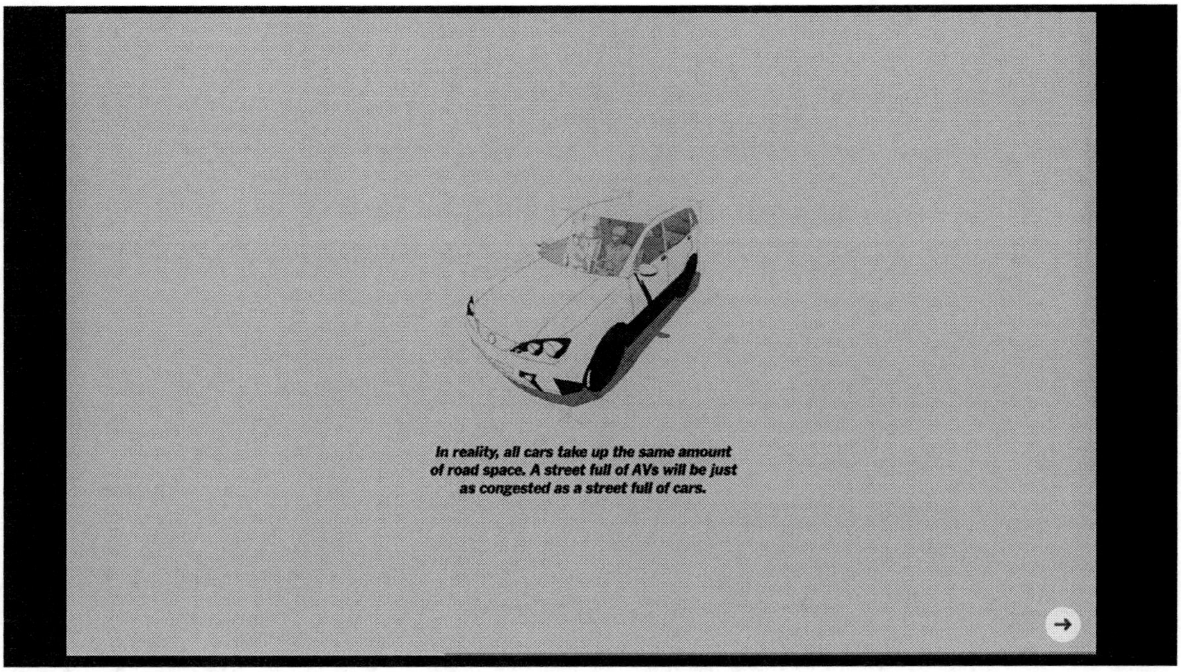
In reality, all cars take up the same amount of road space. A street full of AVs will be just as congested as a street full of cars.

PROMISE #2
AVs could help create more public space because they would eliminate the need for parking structures.

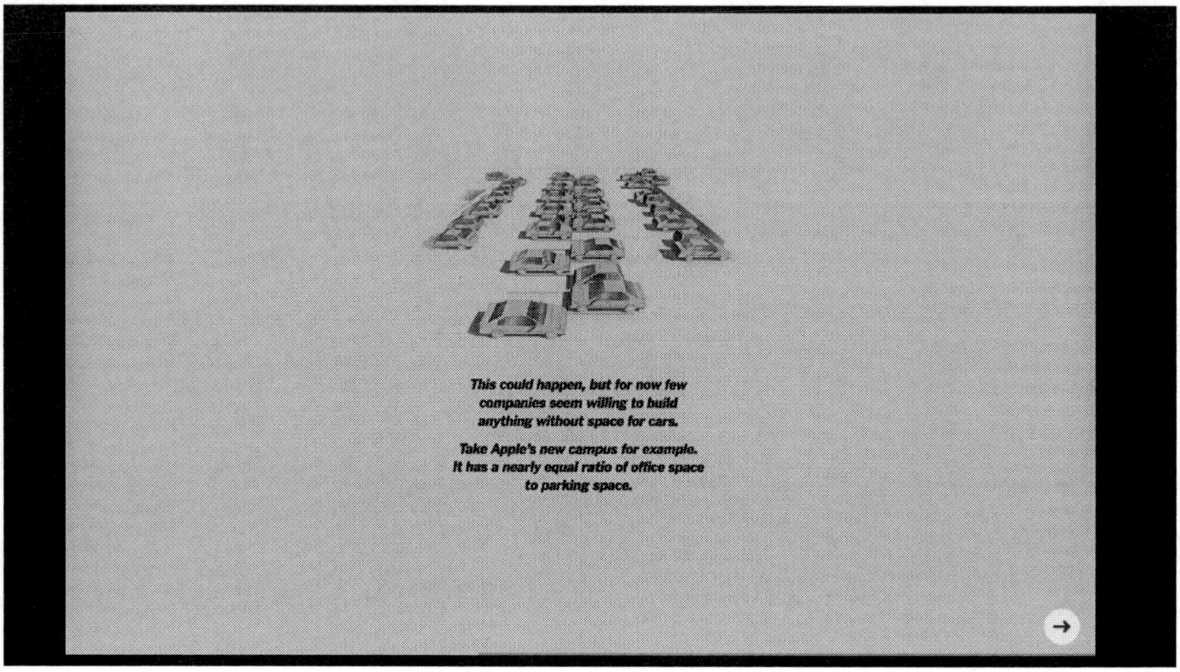
This could happen, but for now few companies seem willing to build anything without space for cars.
Take Apple's new campus for example. It has a nearly equal ratio of office space to parking space.

THAT'S ENOUGH ROOM FOR
11,000 CARS.
Two parking structures on Apple's Cupertino campus. FOOTAGE BY MATTHEW ROBERTS

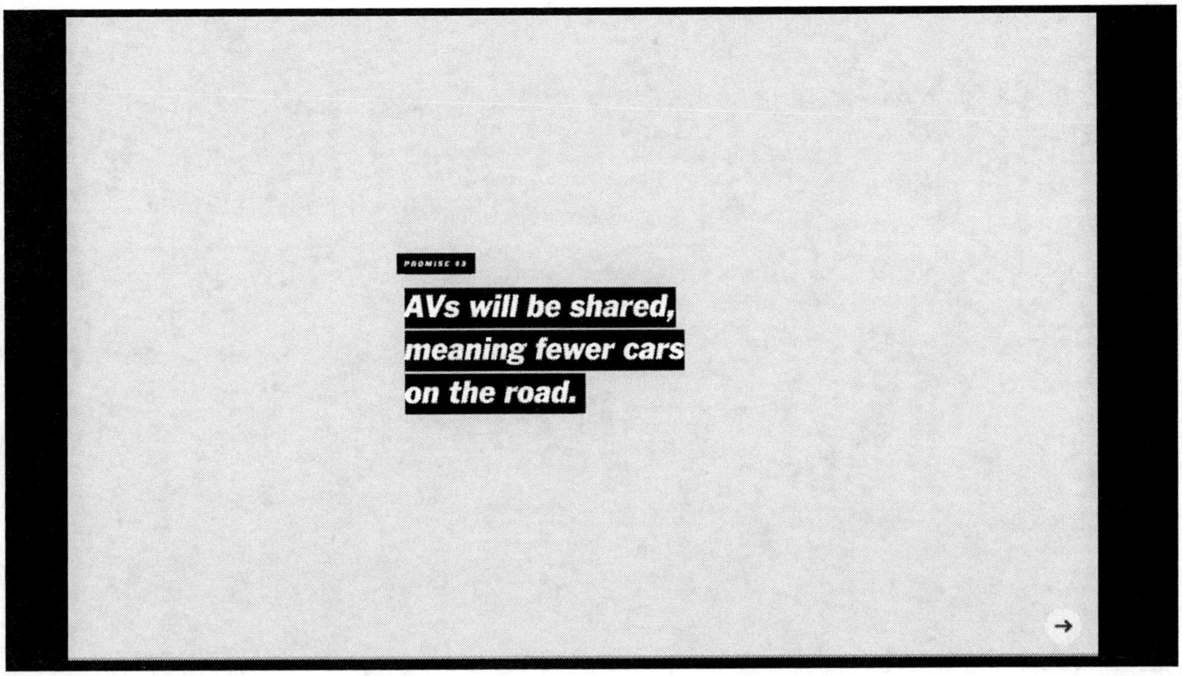
AVs will be shared,
meaning fewer cars
on the road.

AMERICANS LOVE
THEIR CARS.
Getting them to give up private car
ownership to embrace carsharing may be
more complicated than many think.

There are many more promises.
This just scratches the surface.

Our driverless future can be
one where many transportation
problems are solved...

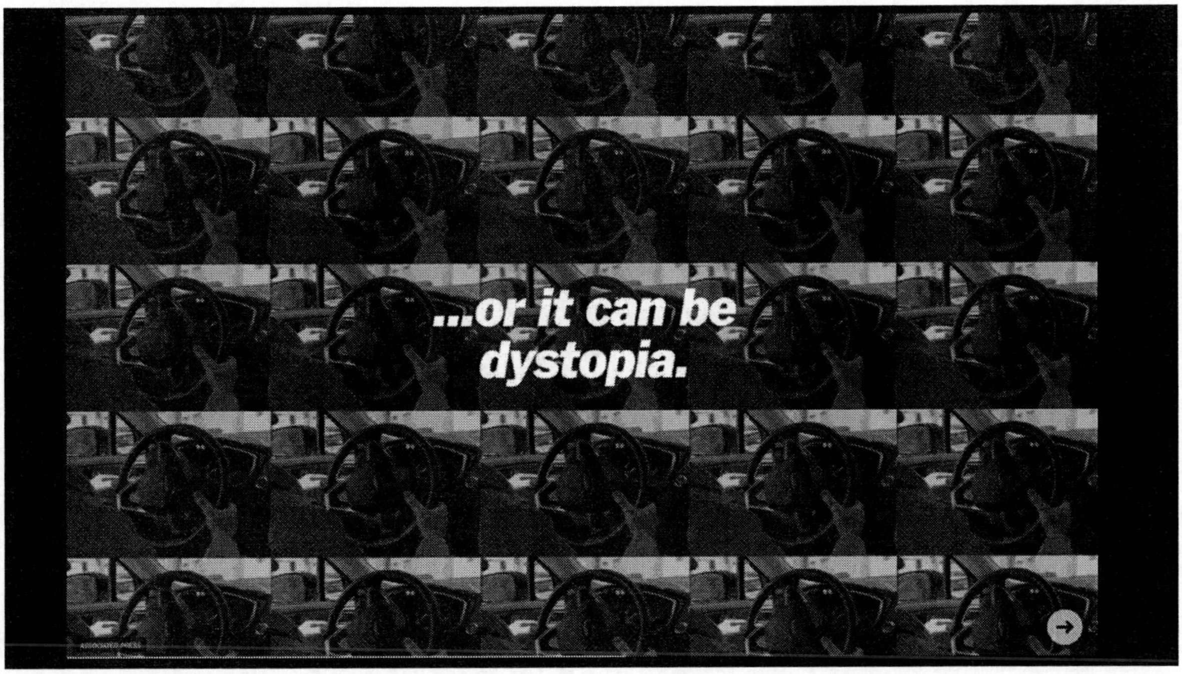
...or it can be
dystopia.

We need to regulate how
AVs will operate in cities
sooner rather than later.
But more importantly, we
need to design streets for...

people
not
cars.

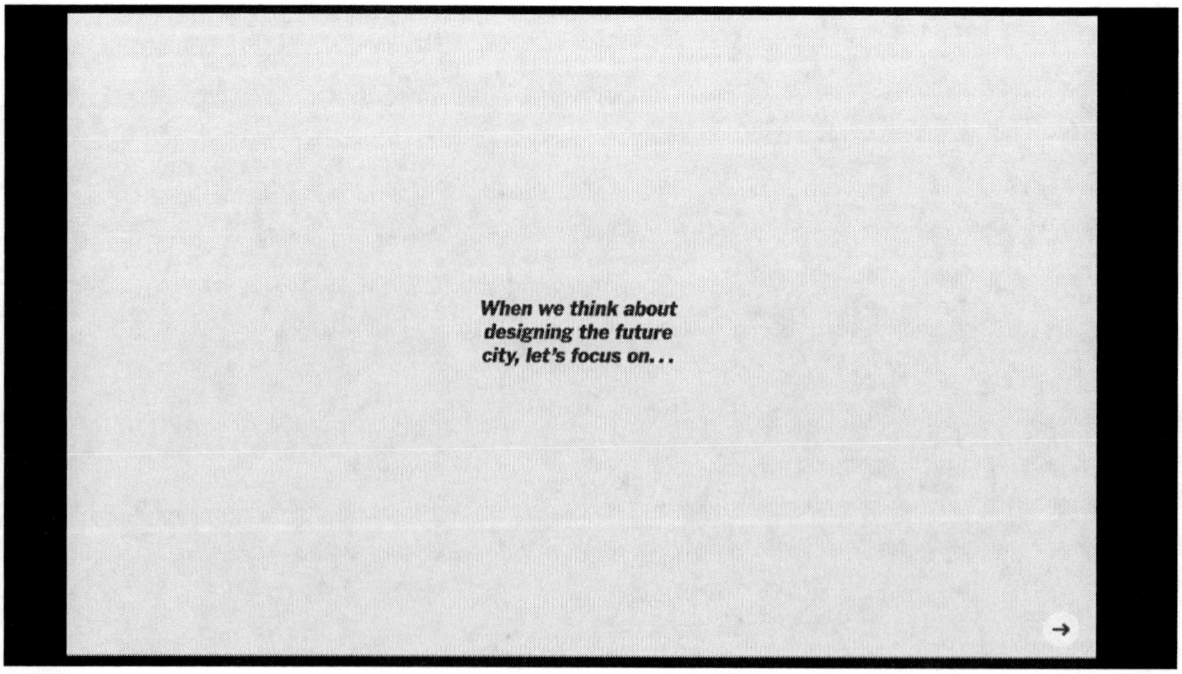
When we think about
designing the future
city, let's focus on...

moving as
many people
as possible.

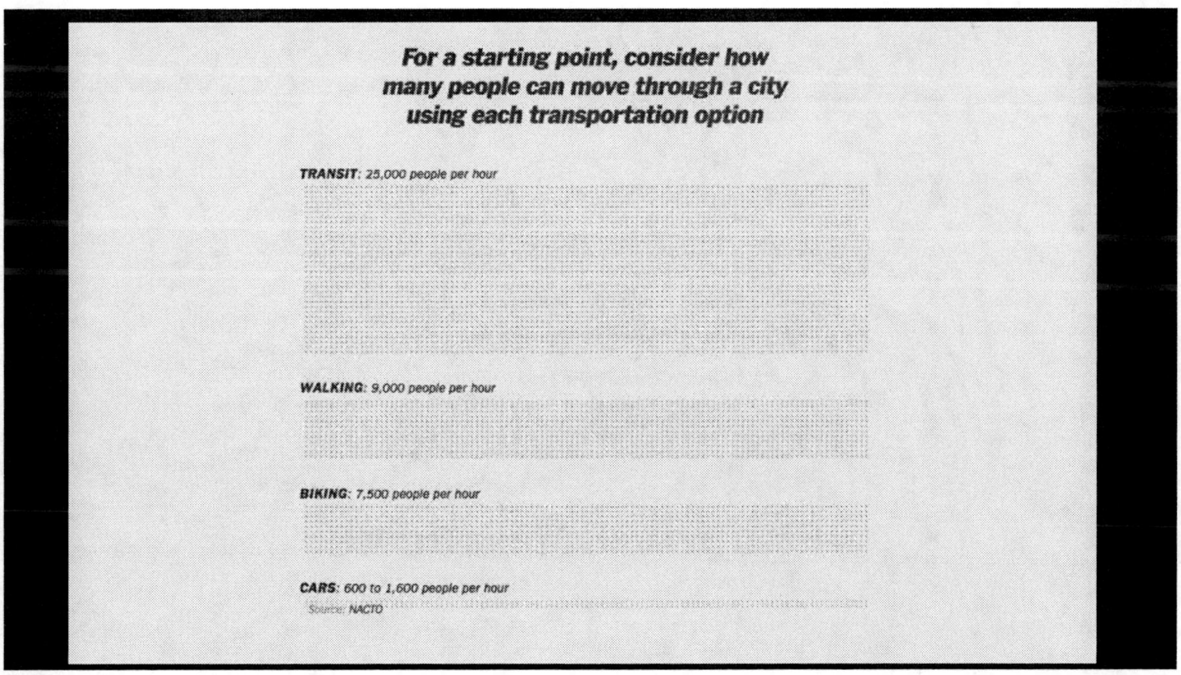
For a starting point, consider how
many people can move through a city
using each transportation option
TRANSIT: 25,000 people per hour
WALKING: 9,000 people per hour
BIKING: 7,500 people per hour
CARS: 600 to 1,600 people per hour
Source: NACTO

Whether they have drivers
or not, cars move the
fewest people per hour –
about 600 to 1,600.

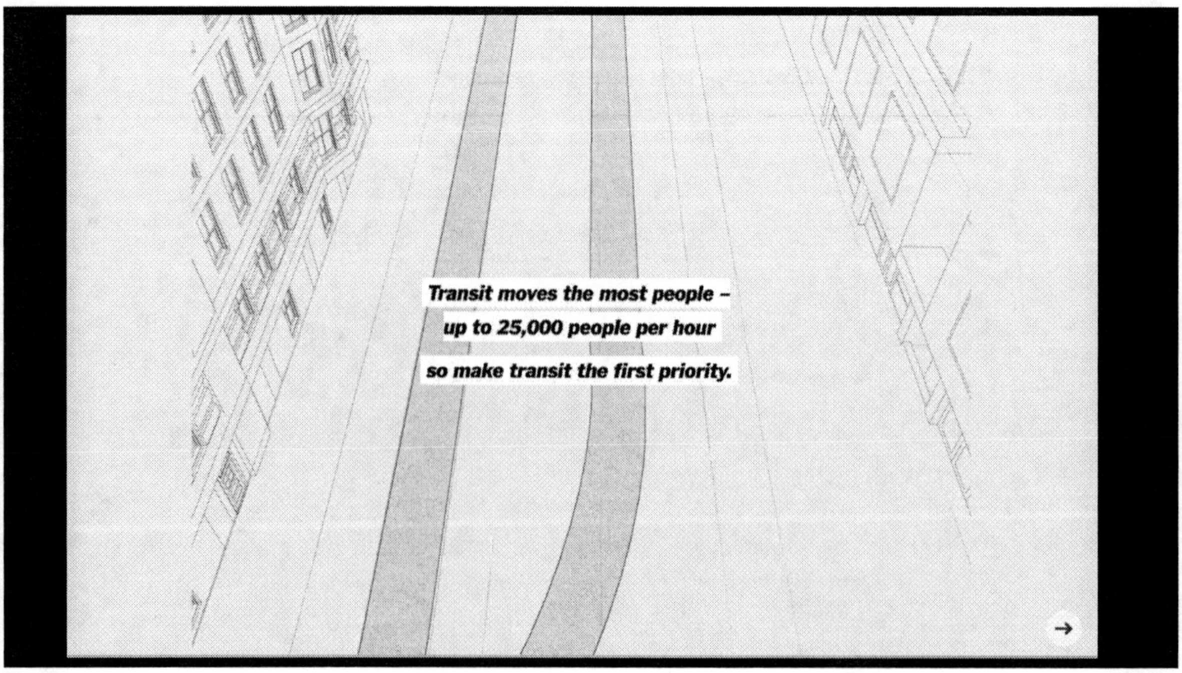
Transit moves the most people –
up to 25,000 people per hour
so make transit the first priority.

Next comes walking, with
9,000 people per hour on
new, larger sidewalks.

Dedicated bus lanes
carry 8,000 passengers
an hour. Two-way bike
lanes carry 7,500 per
hour. Those come next.

When we understand that urban
transportation is about moving...

people
not
cars

our priorities for space and investment become obvious.

SPUR
Ideas + Action for a Better City

Educating the public about the cities they live in with an eye toward creating a community of urban advocates.

10 Strategies for a Better Transportation Future (with or without autonomous vehicles)

1. Don't wait for AVs for safety.
2. Offer mobility to those who need it most.
3. Design streets for fewer cars, not more.
4. Manage curbs and their many uses.
5. Set prices to encourage the conditions we want.

6. Fund infrastructure.
7. Make public transportation better.
8. Deploy flexible transportation.
9. Manage growth, manage sprawl.
10. Try pilots, partnerships and experiments.

By doing these things, we could save many lives for much less money.

Why not do these things first?

Can we create more opportunities for people to move around our cities without cars?

Shouldn't we get the tech safe enough so that you would feel safe having your child as a passenger?

Autonomous or otherwise, a car is still a car.
We need to think about sustainability, land use and policy and people.

Self-driving cars seemed futuristic a century ago; today, I wonder why we are choosing to focus so much attention on cars at all.

Rates of car ownership are decreasing.
Bikeshare, rideshare and carshare programs are gaining in popularity and acceptance, as are transit-oriented communities.

The climate is getting worse by the minute.

So why continue to design and plan for a car-based society?

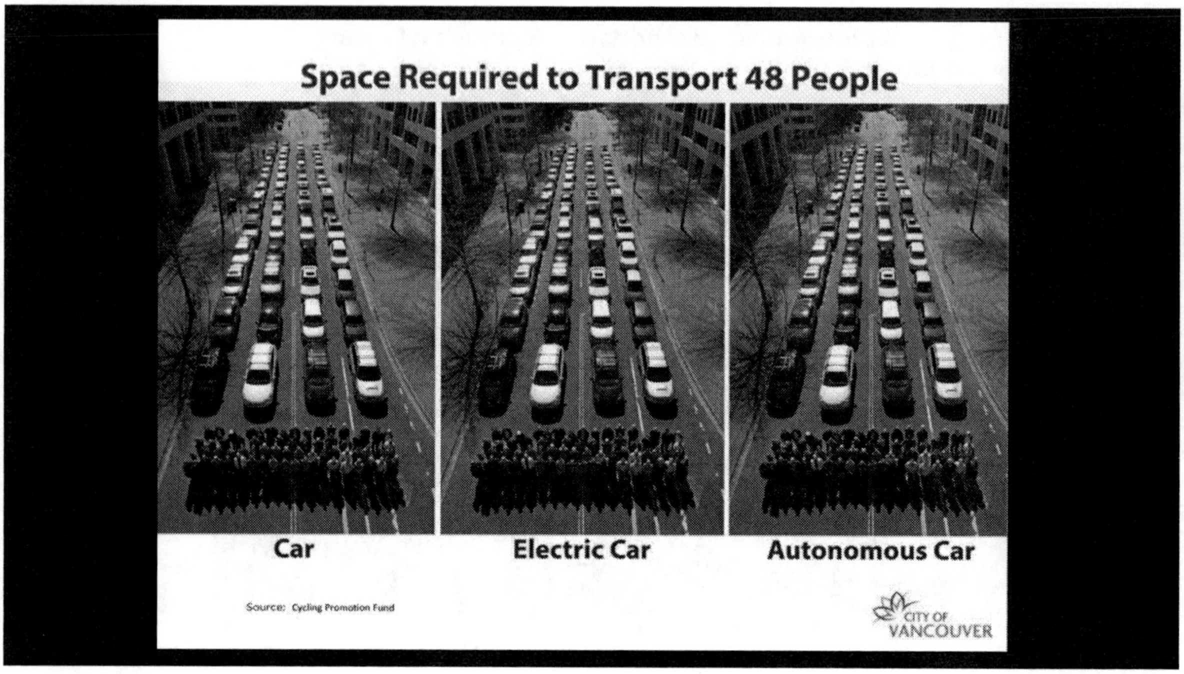

So…in closing:

people
not
cars.

This is what the city of
the future could be

...for
saving
cities.

Thank you.

001
Allison Arieff

002
aarieff@spur.org

003
Twitter: @aarieff

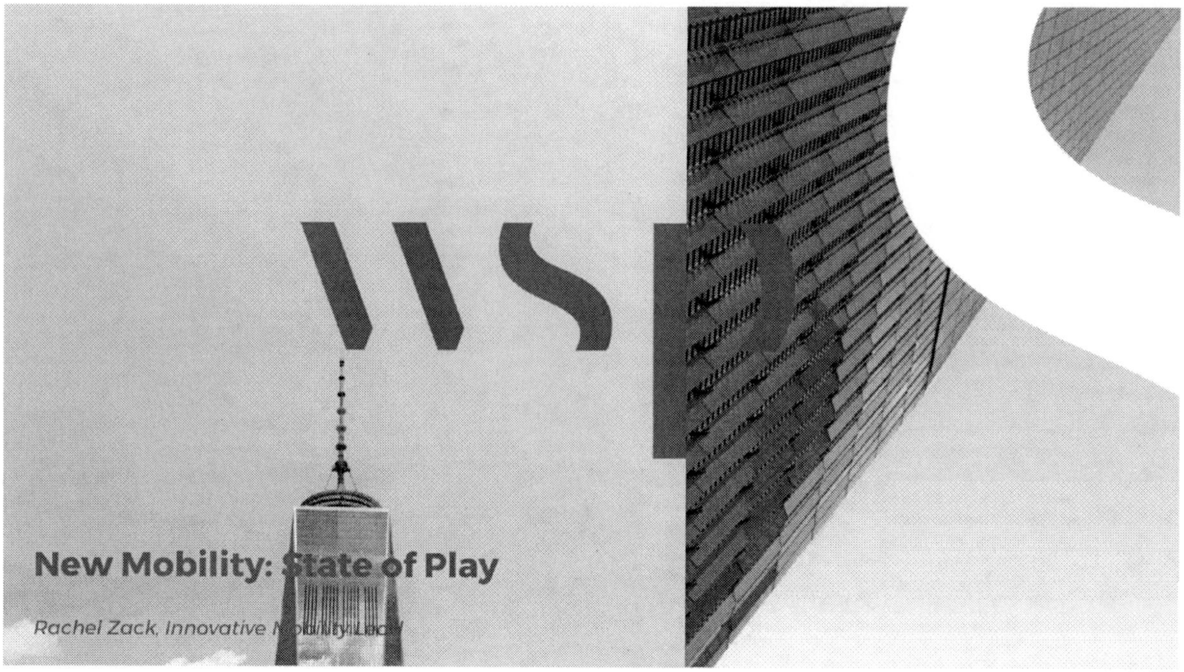

"I think public transport is painful. It sucks. Why do you want to get on something with a lot of other people, that doesn't leave where <u>you</u> want it to leave, doesn't start where <u>you</u> want it to start, doesn't end where <u>you</u> want it to end?"

Emphasis added.

- Elon Musk

wsp

At a glance

State of Play

- Understand – *Model and Regulate*
- Enable – *Action Plans*
- Leverage – *Pilots*

wsp

At a glance

Understand

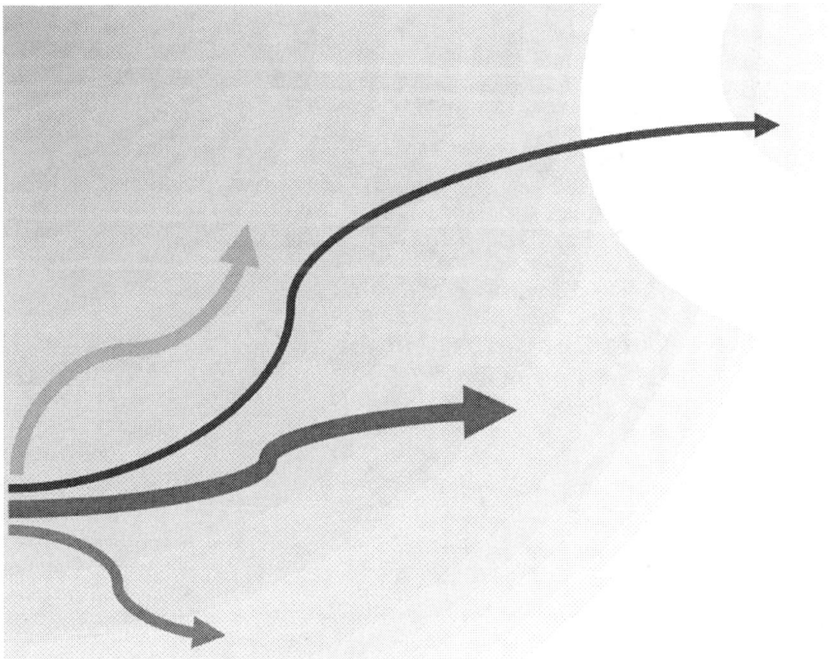

wsp

At a glance

Enable

Welcoming ⟷ Apprehensive

Leverage

Cars Get better Mileage than They Used To

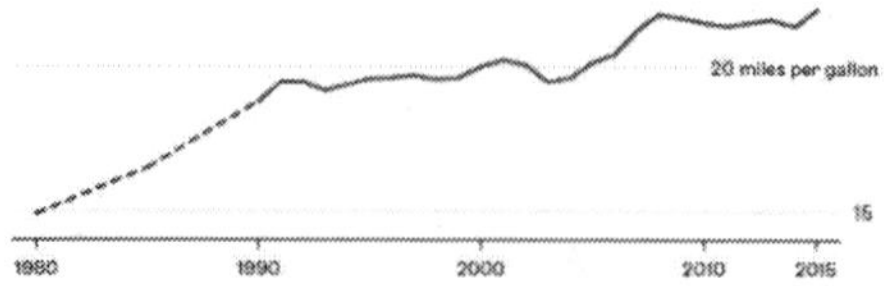

Source: Bureau of Transportation Statistics

Take note: the y-axis does not go to zero. Dotted lines represent years with no data.

Gas Tax Revenue Isn't What it Used to Be

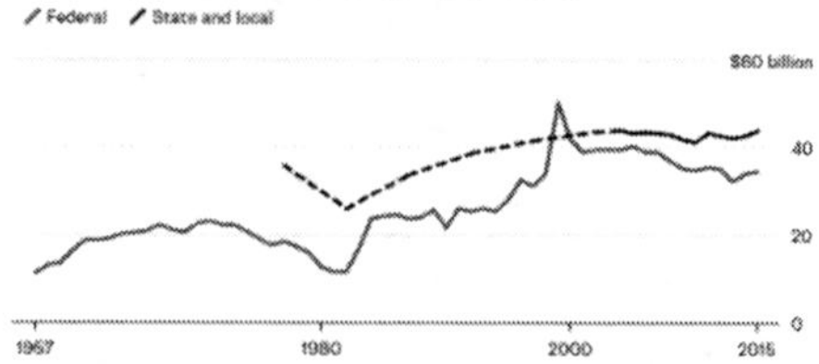

Sources: Federal Highway Administration, Tax Policy Center, Bureau of Labor Statistics

Federal numbers are motor fuel revenues to the Highway Trust Fund; dotted line represents years with no data.

Bloomberg, 2018

Thank you.

Rachel Zack, Innovative Mobility Lead

wsp

Breakout Session #22: Blockchain in the Automated Vehicle Ecosystem

Session Overview

- **Overview:** This session investigated how blockchain has the potential to transform mobility, facilitate transactions, revolutionize cybersecurity, and enable trusted M2M communications without requiring either a trusted third party (e.g., a bank) or a regulatory agency.
- **Expert Panel:** Dale Thompson (FHWA), Rajat Rajbhandari (dexFreight), Chris Ballinger (MOBI), and Karl Wunderlich (Noblis)
- **Session Goals:**
  - Introduce the transportation community to blockchain technology—Blockchain 101
  - Explore potential use cases for blockchain in an automated vehicle ecosystem
  - Take a deep dive into a specific use case for "AV Obstacle Detection and Coordinated Path Planning"

AUTOMATED VEHICLES SYMPOSIUM 2018
USERS. VEHICLES. INFRASTRUCTURE.
Driven by AUVSI and TRB

Breakout Session #22: Blockchain in the Automated Vehicle Ecosystem

Key Observations

lockchain is extremely secure and offers significant opportunities to handle transactions in the mobility space

lliances are forming comprised of many of the world's large automakers, many start-ups, non-profits, governments, transit agencies, and technology companies, all working together to improve mobility services

lockchain in the mobility space is further along than many expected. Several proof of concepts have already been demonstrated; however "Blockchain is a Team Sport" – thus size and scale matter in Blockchain Ecosystems

Examples of Potential Use Cases & Opportunities to Revolutionize the Mobility Space

istributed security credentials management system (SCMS)

hain of custody of vehicle ownership transfers

uto component supply chain tracking

Breakout Session #23

Topic: Optimal integration into the transportation system …

Summary of Key Findings and Lessons Learned

- Connectivity beneficial for both vehicles and infrastructure owner-operator (IOO)
  - E.g., can smooth transition between ODDs, allow additional services
- Automation must be accommodated along with all other current system users
- IOOs and OEMs need to know what to expect from one-another
  - Early agreement on key information flows beneficial
- Broadly accepted stakeholder consensus can obviate need to regulate
  - Still achieve desirable nationwide/cross regional interoperability
- Life-cycle management requires careful planning
  - Vehicle, infrastructure and communication technology lifecycles very different

AUTOMATED VEHICLES SYMPOSIUM 2018
USERS. VEHICLES. INFRASTRUCTURE.
Driven by AUVSI and TRB

Breakout Session #23

How the ITS System Fits Together …

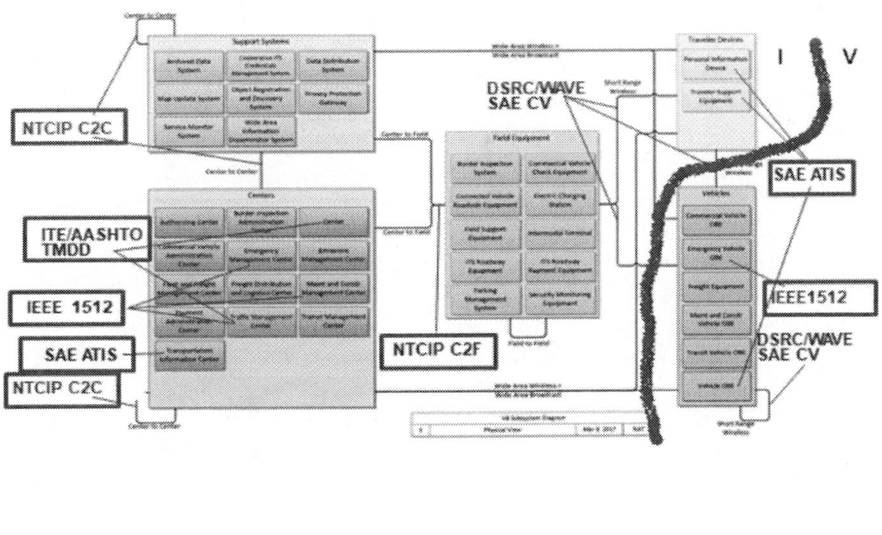

AUTOMATED VEHICLES SYMPOSIUM 2018

USERS. VEHICLES. INFRASTRUCTURE.

Driven by AUVSI and TRB

Breakout Session #23

Research Needs

- Key information flows desired for vehicles
  - Package of regulatory information at jurisdictional boundary
    - Cooperate with EU effort?
  - Real time condition information
    - Roadway striping condition, work zones, snow cover, traffic congestion ...
  - Regional and local differences in expected driver behavior to assist AV integration
- Key information flows desired for infrastructure
  - Vehicle position/speed/wipers/headlights etc.
    - Allows better understanding of network state, queue length, congestion
    - Eventually reduce infrastructure costs (e.g., video, loop detection)?
  - Automation capability of vehicles
    - Support e.g., managed lanes, appropriate warnings, MRC advice
- Stakeholder consensus on an interface architecture, specific standards
  - Many stakeholders beyond obvious groups ... need to engage all
  - Common information standards, but allow for multiple communication media?

Scenario Databases: International Activities

Breakout Session #23

Tuesday, July 10

AVS 2018

Anne Marie Lewis, Ph.D.

Director of Safety & Technology Policy

Alliance of Automobile Manufacturers, Inc.

PEGASUS

North American Viewpoint on Scenario Catalogs

The Role of Scenarios for ADS-Equipped Vehicles

Driving scenarios can play an important role in the development, testing and safety assurance framework for ADS-equipped vehicles.

•Product Development:

- Unusual or "edge case" scenarios
- Ongoing research to support the next iteration of product development

•Product Development, Safety Assurance Conformance and Public Acceptance:

- Crash scenarios
- Naturalistic (crash, near-crash, non-crash "just driving")

Scenario Databases Will Not Be "One size fits all"

Databases must be developed for the desired use case.

•It is not physically possible to test a vehicle in every possible driving scenario using real world, on-road testing.

•Therefore, a combination of simulation, track/lab testing and on-road testing will be necessary to test as many relevant scenarios as possible.

•Additionally, an explanation of the functional safety design process may also be helpful.

Raw Data Should Not Be Shared

There are several reasons why:

•Sharing raw data, such as direct outputs from sensors, is problematic from a legal and business confidentiality perspective

•Sharing raw data is inefficient from the technological perspective.

•Raw data is not useful until it has been distilled down to useful information about the identification, location and trajectory of the specific objects that are in the vicinity of the ADS-equipped vehicle for a given scenario.

Learn from Existing Databases

Where possible, draw upon existing databases of driving scenarios and crash and pre-crash scenarios:

- NHTSA Pre-Crash Scenario Typology for Crash Avoidance Research Categories
- NHTSA Fatality Analysis Reporting System (FARS)
- NHTSA National Automotive Sampling System (NASS)
  - Crash Report Sampling System (CRSS) *(formerly General Estimates System (GES))*
  - Crash Investigation Sampling System (CISS) *(formerly Crashworthiness Data System (CDS))*
- NHTSA National Motor Vehicle Crash Sampling Survey (NMVCCS)
- NHTSA Special Crash Investigations (SCI)
- NHTSA Crash Injury Research (CIREN)
- NHTSA State Data System (SDS)
- Naturalistic (e.g. NHTSA/VTTI 100 car and SHRP2; Berkeley DeepDrive from UC Berkeley PATH)
  - ISO DTR-21974 (Naturalistic Driving Studies – Defining and Annotating – Safety Critical Events)
  - VTTI Data Dictionary

Pre-Crash Scenario Typology for Crash Avoidance Research Categories

Based on the GES database, which includes data elements that could be useful for ADS-equipped vehicle scenarios. For example:

C26 Atmospheric Conditions

Definition: This data element records the prevailing atmospheric conditions that existed at the time of the crash as indicated in the PAR.

Additional Information: This data element identifies up to two values. If more atmospheric conditions were reported, the two conditions that most affect visibili selected. Accident.WEATHER1 and Accident.WEATHER2 are coded data elem Accident.WEATHER is derived from these two.

Prior to 2015, this data element's Locator Code or Data Element Number was C

See *Appendix E: Rules for Derived Data Elements* for an expanded explanation element and how it is derived.

2010-2012	*2013-Later*	
0	0	No Additional Atmospheric Conditions
1	1	Clear
2	2	Rain
3	--	Sleet or Hail *(Freezing Rain or Drizzle)*
--	3	Sleet or Hail
4	4	Snow
5	5	Fog, Smog, Smoke
6	6	Severe Crosswinds
7	7	Blowing Sand, Soil, Dirt
8	8	Other
10	10	Cloudy
11	11	Blowing Snow
--	12	Freezing Rain or Drizzle
98	98	Not Reported
99	99	Unknown

National Automotive Sampling System (NASS) General Estimates System (GES) Analytical User's Manual 1988-2015, US Department of Transportation, National Highway Traffic Safety Administration, August 2016.

Work Towards International Harmonization

In the development of new scenario database(s) for ADS-equipped vehicles, we should work towards international harmonization while recognizing and addressing potential challenges and limitations.

- Challenges/Limitations:
  - Scenarios will not necessarily be comparable internationally - different roadway designs, traffic laws and protocols (e.g. right hand vs. left hand drive), traffic density and patterns, and levels of driver skill.
  - Self-certification vs. type approval motor vehicle safety regime
- Path forward:
  - Develop common methodologies, tools and formats such that relevant information and data can be exchanged and compared. For instance, a similar method(s) for selecting relevant driving scenarios could be used by all.
  - As we have seen with previous motor vehicle safety harmonization efforts, there will likely be some commonalities that can be used in both the self-certification and type approval frameworks.

Anne Marie Lewis, Ph.D.
Director of Safety & Technology Policy
Alliance of Automobile Manufacturers

North America PEGASUS Team Members

- Auto Alliance (Convener)
- ACM
- VTTI
- Booz Allen / SAE
- UMTRI
- TRI
- Metamoto
- PATH / UC Berkeley
- Waymo
- Daimler (US)
- Mazda (US)
- Toyota (US)
- Motus Ventures

Graphic credit: https://xkcd.com/1429/

Access to data is a limiting factor for AV deployment. (It's also a sensitive topic.) One solution: ***data exchanges***.

USDOT is using our *convening power* to understand critical use cases for data exchange and the appropriate federal role to enable them.

See the AV Data Guiding Principles and Framework at https://www.transportation.gov/AV/Data

Outcomes

•Clarity on value of federal government as convener and facilitator

•Priority use cases for data exchange: work zones, scenarios, cybersecurity, others

Next Steps

•Enable voluntary data exchanges as "One DOT" via pilot projects

•Incorporate into AV policies

•Continue conversations – including at AVS (Breakout Session #35) and via AVDX@dot.gov

Photo credit: Dan Morgan

Summary Report available via: https://www.transportation.gov/AV/Data

BRIEFING ON STREETWISE

10 July 2018 | AVS2018 San Francisco | Breakout Session #24

Olaf Op den Camp

TNO innovation for life

AV SAFETY ASSESSMENT FRAMEWORK

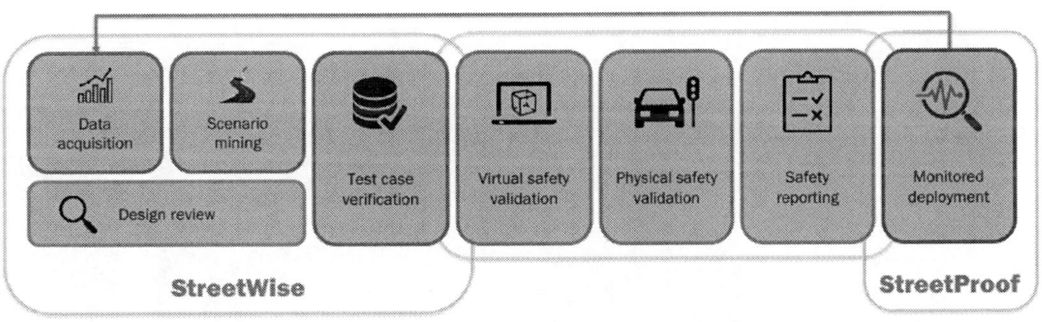

CETRAN ENABLE-S3 PEARS openPASS

TNO innovation for life

STREETWISE PIPELINE

WWW.TNO.NL/STREETWISE

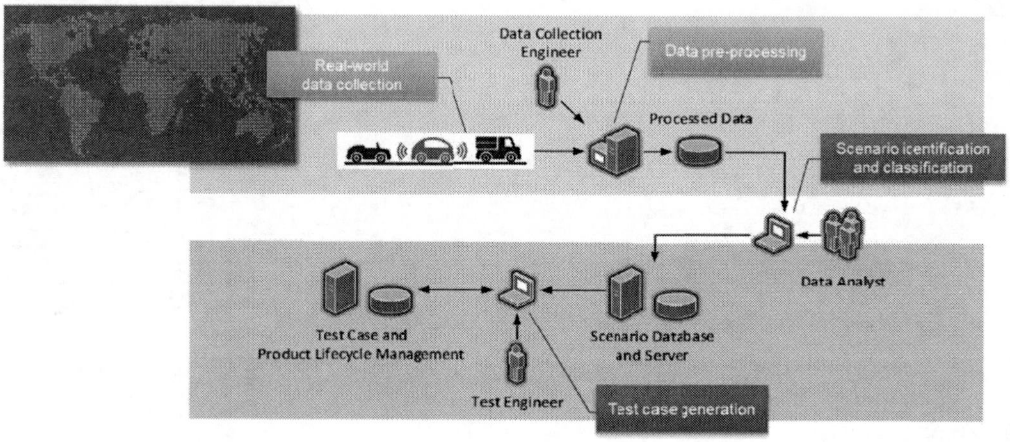

TNO innovation for life

SCENARIO IDENTIFICATION & CLASSICATION

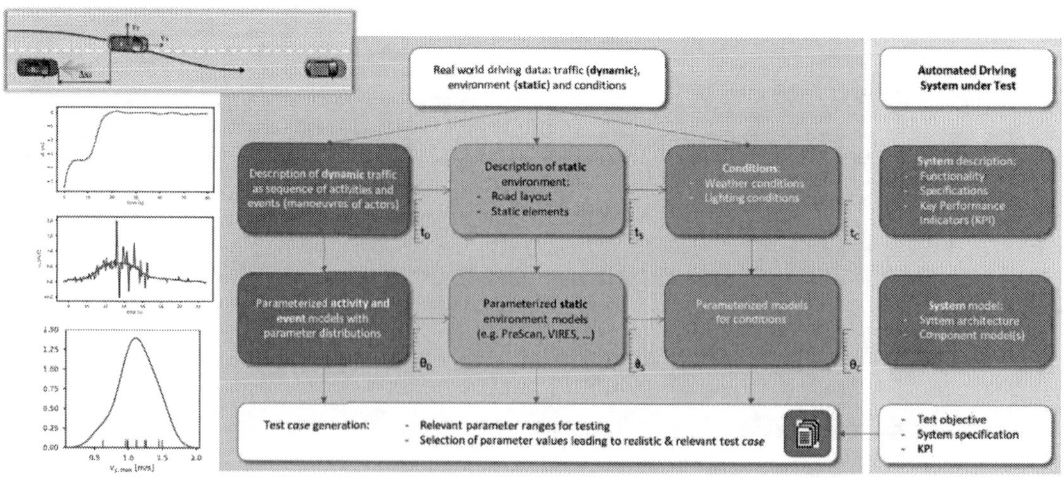

TNO innovation for life

SCENARIO EXAMPLE

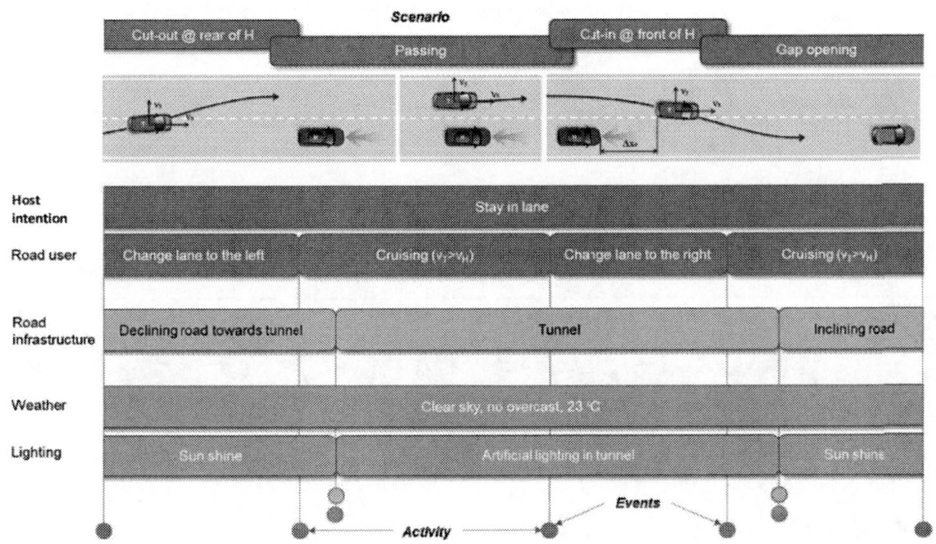

ADS VARIABLE PERFORMANCE TESTING PROJECT

AUVSI AV Symposium 2018

July 10, 2018

SAE INTERNATIONAL

Iterative Activities to Understand the Issue

Date	Action	Purpose
Sep 15, 2017	• NHTSA sends letter to SAE highlighting need for standards for "Variable Performance Testing for Automated Driving Systems	• Highlight NHTSA priorities

Nov 2017- Mar 2018	• Various group meetings with MVC, SAE experts, NHTSA and Industry	• Begin to Plan Approach, beginning with a broad industry workshop
		• Understand two key NHTSA objectives:
		  • Rapidly develop deliverables of impact
		  • Achieve a common performance testing environment with broad industry acceptance

April 9, 2018	• Variable Performance Testing for ADS Workshop	• Engage industry; develop some initial planning
June 19, 2018	• Second Industry Workshop	• Engage a broader industry group • Discuss progress to date and refine

SAE INTERNATIONAL

Program Mission, Goals and Objectives

Mission:

Develop a common approach to variable performance testing for ADS that assures public confidence

Goals and Objectives:

•Performance-based safety assurance testing for ADS

•Rapidly develop initial tasks of a common approach to ADS safety assurance testing; target is calendar 2018

•Use the exemplar Use Case Low-Speed Urban L4 to apply testing framework developed

•Attain broad industry acceptance

Program Initial Tasks and Broad Principles

Initial Tasks

- Terms and definitions, and a taxonomy for common ADS safety assurance testing
- A harmonized CONOPS approach to ADS safety assurance testing

Principles

•Testing approach will include variation to prevent "developing to the test"

•Will use a combination of simulation, track, and road testing

•Safety assurance testing will be representative sample of full development testing it does not re-create development testing

•Testing will be an agnostic common approach, not the approach of any one entity

•Feed appropriate SAE standards

Short-Term Program Actions

1. Small group of technical experts is meeting to scope the program Approach and Concept of Operations
    - Using existing research for baselines, e.g.
        - NHTSA Framework for Automated Driving System Testable Cases and Scenarios
        - CAMP AVR Safety Principles
        - NHTSA "37 Crash Types" research
        - NHTSA Driver Task Analysis
        - SAE J3016 and J3131
2. Begin Initial Tasks (goal: complete initial deliverables in calendar 2018)
3. Outreach and Stakeholder Engagement
    - SAE will use its convening power to ensure broad industry engagement

Approach

SAE ITC
An SAE International Affiliate

SAE INTERNATIONAL

- Develop a broad industry consortium administered by SAE
- Stakeholders will identify overall technical program plan and individual technical tasks

- Small groups of industry stakeholders develop detailed technical scope/schedule/cost
- SAE contracts with technical performers to rapidly develop deliverables guided by small groups

- SAE publishes deliverables
- Deliverables are injected into appropriate SAE Standards Committees
- SAE Committees expedite standards through open, collaborative process

A Framework for Automated Driving System Testable Cases and Scenarios

Eric Thorn, PhD
Southwest Research Institute

7/10/2018

Project Overview

- Objectives:
  - Develop a preliminary objective testing and evaluation approach, which may be used by industry and others to verify and validate the safe operation of ADS features and behaviors
  - Take the first steps of partitioning the ADS performance space as a test framework of independent factors
  - Refine the testing framework through methods of modeling, simulation, track testing, and open road testing
- Outcomes:
  - Identified Tactical Maneuver/Competency behaviors from various sources
  - Identified the ADS Operational Design Domain (ODD)
  - Identified ADS Object and Event Detection and Responses (OEDR)
  - Identified ADS Fail-operational / Fail-safe Mechanisms
  - Developed a Model Framework of Evaluation Factors by cross factoring ODD, OEDR, and Tactical Maneuver/Competency dimensions

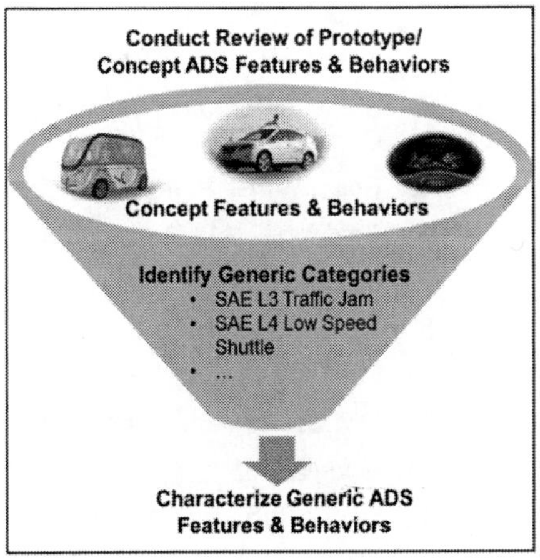

Stakeholder Engagement

- Motivation: Incorporate important perspectives to inform the project framework and validate results.
- Expert Panel: Group of manufacturers, suppliers, and academics that provided detailed input to project deliverables.
- Public Engagement: In-person stakeholder workshop at AVS 2017 to present preliminary findings and gather broad stakeholder perspectives.
- Role in Project: The information gathered during these engagements were aggregated and used to update project deliverables and become part of a publically available final report (pending publication).

NHTSA | VirginiaTech. Transportation Institute | SwRI | Booz | Allen | Hamilton

2

24 Sample Proprietary ADS Features were Identified

1. Audi Traffic Jam Pilot
2. Audi Highway Pilot
3. Auro Self Driving Shuttle
4. Baidu Automated Transportation Network Company (TNC)
5. Bosch Valet Parking
6. CityMobil2 Automated Shuttle
7. Bosch Highway Pilot
8. EZ10 Self Driving Shuttle
9. Ford Automated TNC
10. GM Cruise Automation TNC
11. Google Car
12. Honda Automated Drive
13. Mercedes Highway Pilot Truck
14. Navya Arma Shuttle
15. Nissan Autonomous Drive
16. Olli Local Motors Shuttle
17. Otto Trucking
18. Tesla Self-Drive
19. Toyota Chauffeur
20. Toyota Guardian
21. Uber Automated TNC
22. Varden Labs Self Driving Shuttles
23. Volkswagen I.D. Pilot
24. Volvo IntelliSafe Auto Pilot

NHTSA | VirginiaTech. Transportation Institute | SwRI | Booz | Allen | Hamilton

3

24 Proprietary ADS Features Grouped into 7 Generic ADS Features

Generic ADS Features (Speculated Commercially Available Date) *Selected for focused analysis	Commercially Available? (Y/N)	Speculated Level of Driving Automation (SAE 1-5)	Tactical and Operational Maneuvers (X = demonstrated, ? = speculated)															
			Parking	Maintain Speed	Car Following	Lane Centering	Lane Switching/Overtaking	Enhancing Conspicuity	Merge	Navigate On/Off Ramps	Follow Driving Laws	Navigate Roundabouts	Navigate Intersection	Navigate Crosswalk	Navigate Work Zone	N-Point Turn	U-Turn	Route Planning
L3 Conditional Automated Traffic Jam Drive (2018)	N	3		X	X	X												
L3 Conditional Automated Highway Drive (2020)	N	3		X	X	X	X	X	X		X							X
L4 Highly Automated Low Speed Shuttle (2018)	Y	4	X	X	X	X	X	X	X		X	X	X	X	?	X	X	X
L4 Highly Automated Valet Parking (2020)	N	4	X	X	?	?		X			?		?	?		?	?	X
L4 Highly Automated Highway Drive (2020)	N	4		X	X	X	X	?	X		?							X
L4 Emergency Take-Over (?)	N	4		X	X	X	X	X		X	X							
L4 Highly Automated Vehicle / TNC (2020)	N	4	X	X	X	X	X	X	X	X	X	X	X	X	X	X	X	X

**Note: Information was gathered in 2016

- Tactical/operational maneuvers represent a set of testable capabilities/conditions for ADS

NHTSA | Virginia Tech. Transportation Institute | SwRI | Booz | Allen | Hamilton

Examples of Operational Design Domain (ODD) Elements

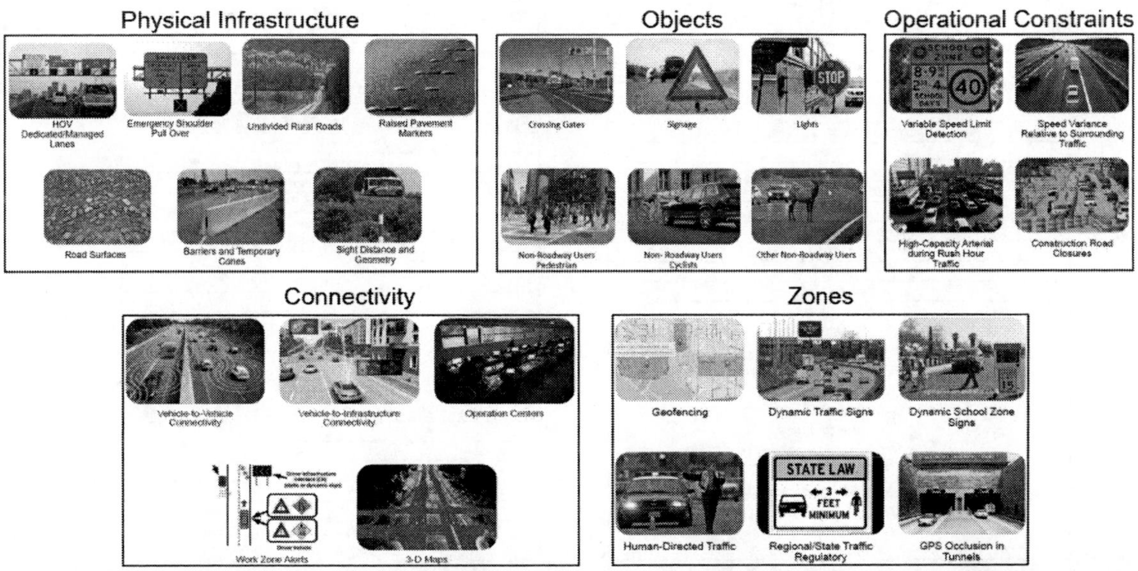

NHTSA | Virginia Tech. Transportation Institute | SwRI | Booz | Allen | Hamilton

Sample ODD Taxonomy

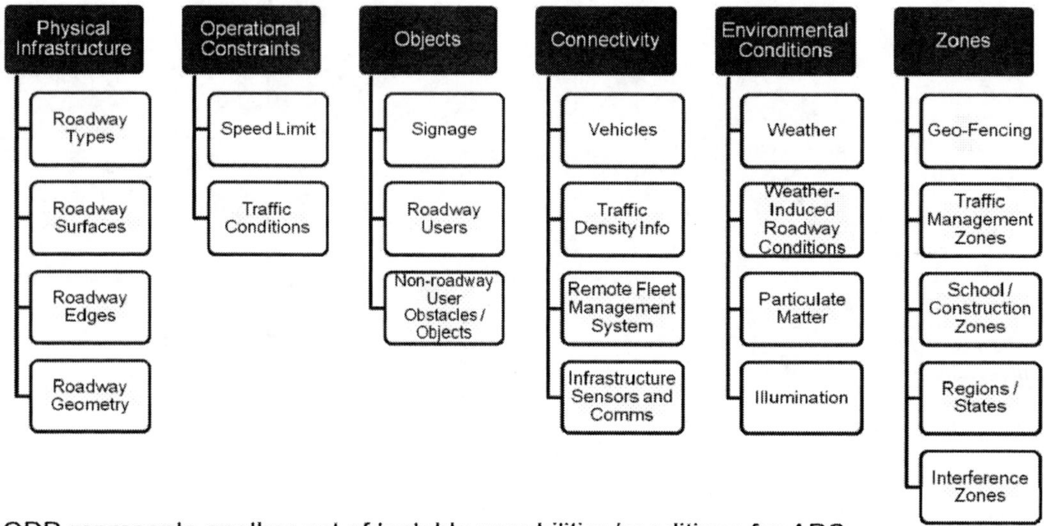

- ODD represents another set of testable capabilities/conditions for ADS

Sample ODD Element Taxonomy Decomposition (Demonstrative, not normative)

PHYSICAL INFRASTRUCTURE
Roadway Types
Divided highway
Undivided highway
Arterial
Urban
Rural
Parking (surface lots, structures, private/public)
Bridges
Multi-lane / single lane
Managed lanes (HOV, HOT, etc.)
On-off ramps
Emergency evacuation routes
One way
Private roads
Reversible lanes
Intersection Types - signaled - U-turns - 4-way vs. 3-way vs. 2-way - stop sign - roundabout - merge lanes - left turn across traffic, one-way to one-way - right turn - multiple turn lane - crosswalk - toll plaza - railroad crossing
Other

ENVIRONMENTAL CONDITIONS
Weather
Wind
Rain
Snow
Sleet
Temperature
Other
Weather-Induced Roadway Conditions
Standing Water
Flooded Roadways
Icy Roads
Snow on Road
Other
Particulate Matter
Fog
Smoke
Smog
Dust/Dirt
Mud
Other
Illumination
Day (Sun: Overhead, Back-lighting and Front-lighting)
Dawn
Dusk
Night
Street lights
Headlights (Regular & High-Beam)
Oncoming vehicle lights (Overhead Lighting, Back-lighting & Front-lighting)
Other

Object and Event Detection and Response (OEDR) Behaviors for ADS Features

- ❑Detect[1] & Respond to Speed Limit Changes
- ❑Detect & Respond to Encroaching, Oncoming Vehicles
- ❑Perform Vehicle Following
- ❑Detect & Respond to Relevant Stopped Vehicles
- ❑Detect & Respond to Relevant Lane Changes / Cut-ins
- ❑Detect & Respond to Relevant Static Obstacles in Lane
- ❑Detect[1] & Navigate Work Zones
- ❑Detect & Respond to Relevant Safety Officials
- ❑Detect[1] & Respond to Relevant Access Restrictions
- ❑Detect[1] & Respond to Relevant Dynamic Traffic Signs
- ❑Detect & Respond to Relevant School Buses
- ❑Detect & Respond to Relevant Emergency Vehicles
- ❑Detect & Respond to Relevant Pedestrians
- ❑Detect & Respond to Relevant Pedalcyclists
- ❑Detect & Respond to Relevant Animals
- ❑Detect & Respond to Relevant Vehicle Cut-out / Reveal
- ❑Detect & Respond to Relevant Vehicle Roadway Entry
- ❑Detect & Respond to ODD Boundary Transition

1. Detection may be implemented in a variety of ways (e.g., using onboard sensing, digital map information, etc.)

- OEDR represents additional testable capabilities of an ADS feature related to perception[1], decision making, and navigation
- Must distinguish between "detect" and "respond" when developing test procedures

Failure Modes and Mitigation Strategies

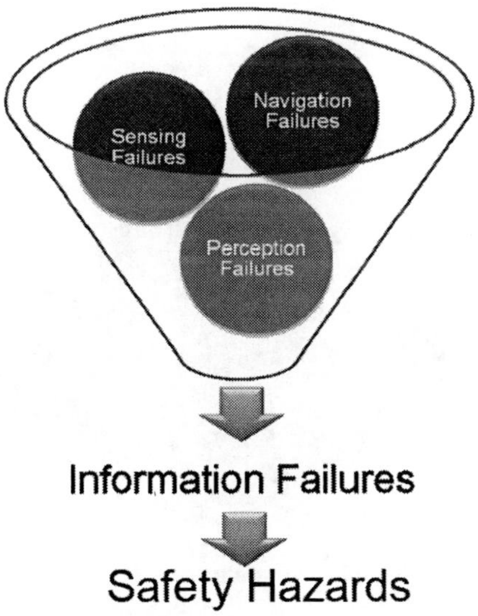

- Fail-safe
  - Request fallback-ready user intervention
  - Stop in lane of travel
  - Move out of travel lane and stop
- Fail-operational
  - Hardware / software redundancies
  - Adaptive compensation
  - Degraded operation(s):
    - Reduced speed
    - Reduced level of automation
    - Reduced ODD
    - Reduced maneuver capabilities
    - Reduced OEDR capabilities

- Failure strategies represent another set of testable capabilities for ADS

Multi-faceted Testing Architecture

- Modeling & Simulation
  - Controlled, predictable, repeatable
  - Opportunity to run large number of simulations very quickly
  - Opportunity to perform sensitivity analysis of ODD and OEDR variability, and identify candidate scenarios for further testing
- Closed-Track Testing
  - Controlled, predictable, repeatable
  - Opportunity to assess full system performance
- Open-Road Testing
  - Uncontrolled, unpredictable
  - Exposure to variety of environmental conditions (e.g., weather) and other ODD elements (e.g., local traffic patterns and infrastructure conditions)
  - Exposure to variety of real-world scenarios that may be difficult to replicate on a test track or simulator

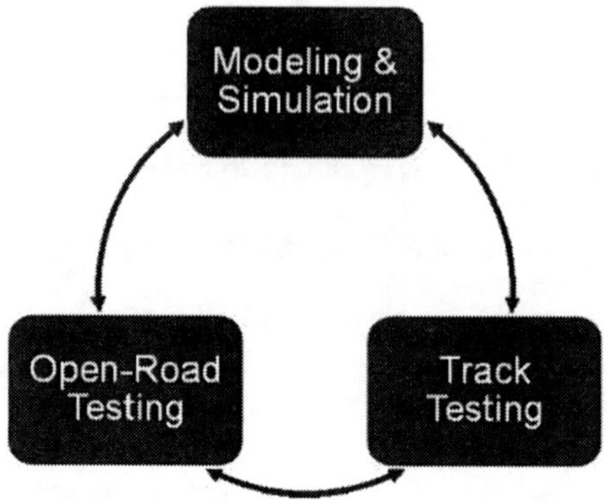

Test Scenario Framework for ADS Features

- Composed of elements from prior tasks (maneuver behaviors, ODD, OEDR, failure modes)
- Develop scenario tests for specific competencies
  - Positive and negative test cases (e.g., school bus with / without traffic control devices activated)
- Full operational scenario can be composed of multiple behavioral competency tests
- Scenarios that can be evaluated in all phases of the testing architecture

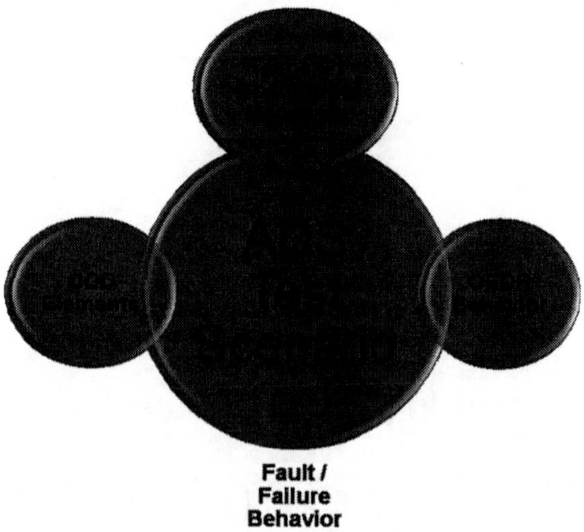

Test Scenario Sample – Detect & Respond to School Bus

- Test scenario description
  - Test subject and purpose
  - Visualization
- Test protocol
  - Parameters, including potential variations in tests
- General procedure
  - Ambient conditions, personnel, facility, equipment
  - Mode: closed course, on-road, simulation
- Test execution and evaluation
  - Performance metrics and evaluation
  - Initial conditions, trial validity

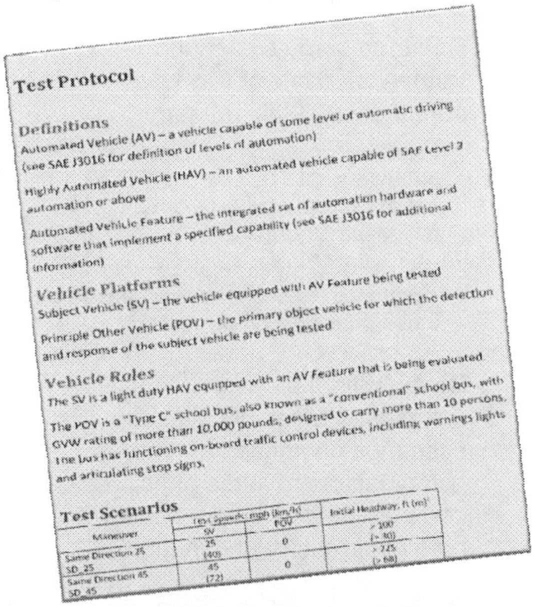

Test Protocol

Definitions

Automated Vehicle (AV) – a vehicle capable of some level of automatic driving (see SAE J3016 for definition of levels of automation)

Highly Automated Vehicle (HAV) – an automated vehicle capable of SAE Level 3 automation or above

Automated Vehicle Feature – the integrated set of automation hardware and software that implement a specified capability (see SAE J3016 for additional information)

Vehicle Platforms

Subject Vehicle (SV) – the vehicle equipped with AV Feature being tested

Principle Other Vehicle (POV) – the primary object vehicle for which the detection and response of the subject vehicle are being tested

Vehicle Roles

The SV is a light duty HAV equipped with an AV Feature that is being evaluated

The POV is a "Type C" school bus, also known as a "conventional" school bus, with GVW rating of more than 10,000 pounds, designed to carry more than 10 persons. The bus has functioning on-board traffic control devices, including warnings lights and articulating stop signs.

Test Scenarios

Maneuver	Test Speeds, mph (km/h) SV	POV	Initial Headway, ft (m)
Same Direction 25 SD_25	25 (40)	0	> 100 (> 30)
Same Direction 45 SD_45	45 (72)	0	> 225 (> 68)

NHTSA | VirginiaTech. Transportation Institute | SwRI | Booz | Allen | Hamilton

12

ADS Testing Challenges (examples)

Related to technology:

- Probabilistic / non-deterministic / statistical algorithms
  - "Best guess"
  - No guarantee of output
  - Output not necessarily repeatable
- Inductive Inference (machine learning)
  - Learning from prior experiences and responses
  - Emergent behavior
- Regression testing
  - Software / hardware / firmware updates (including over-the-air)
  - Learning systems

Related to test execution:

- Complete testing
  - Number of miles and tests to achieve statistical significance
- Scalability
  - Variety and combination of ODD elements
  - Variety and combination of OEDR elements
- Degraded testing
  - Infrastructure (lane markers, signs, etc.)
  - Environmental (rain, snow, lighting, etc.)

NHTSA | VirginiaTech. Transportation Institute | SwRI | Booz | Allen | Hamilton

13

Common Sample ADS Test Scenario Framework

- Common sample scenario framework concept is a key contribution of this research
  - Identified relevant tactical maneuver behaviors, ODD elements, OEDR behaviors, and failure modes / mitigation strategies
  - Specific test procedures can be developed for scenarios composed of these principal components
- Development of a set of tools / techniques to rapidly evaluate scenarios composed of different combinations of these principal components could be valuable to both government and industry

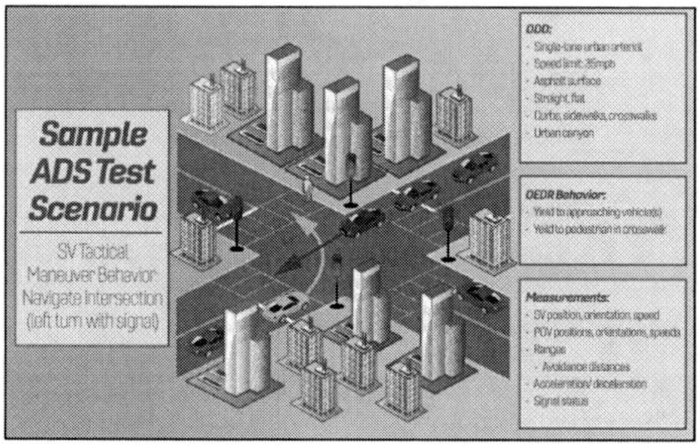

Summary

- Identified the most important independent variables within a four-factor test framework of ODD, OEDR, Tactical Maneuvers/Behavioral Competencies, and Failure Behaviors.
- Next Steps:
  - Identify the most predictive set of dependent variables for parsing L3-L5 scenarios under comparable terms and to build initial models.
  - Utilize simulation, test track, and open road testing to refine the models and to test L3-L5 operational scenarios (e.g., DMV-like driving tests for ADS).
  - Develop an innovative testbed to reliably benchmark scenario performance over the hundreds of complex test cases in the test framework.

Questions?

Eric Thorn, PhD
Southwest Research Institute
ethorn@swri.org

NHTSA VirginiaTech. Transportation Institute SwRI Booz | Allen | Hamilton

16

Breakout Session # 25

Summary of Key Findings and Lessons Learned

- Presentations from Andy Schaudt (VTTI), Jingyi Zhang (Nissan), and Justin Owens (VTTI)
- Research on AV external communications is ongoing on multiple fronts
  - Although most research is not finding changes in behavior in the presence of AV external communication, most people indicate that they would like to see AVs have these types of signals in the future
- ISO technical report to be published later this year on recent work in this space and defining the different types of external communication

AUTOMATED VEHICLES SYMPOSIUM 2018
USERS. VEHICLES. INFRASTRUCTURE.
Driven by AUVSI and TRB

Breakout Session # 25

Research Needs

- New design iterations of signals that attract the attention of pedestrians
  - Auditory may be able to augment (multi-modal)
- Further understand other road user behavior in the presence of AV external communication over a longer period of time
- Effective communication may improve trust and acceptance as part of encouraging safe pedestrian interaction
- Disabled user needs (e.g., last minute adjustment AV parking so user can enter)

Breakout Session 25: The Role of Human Factors in the Design of Automated Vehicle External Communications

Organizers

Andy Schaudt, Virginia Tech Transportation Institute
Justin Owens, Virginia Tech Transportation Institute
Sheldon Russell, Virginia Tech Transportation Institute

Reporters

Sheldon Russell, Virginia Tech Transportation Institute
Andy Schaudt, Virginia Tech Transportation Institute

Session Focus

Research is underway across numerous countries investigating the potential of new external communication methods for highly automated vehicles (HAVs). As drivers are replaced, we lose human-to-human communication between road users such as eye contact and hand gestures. With this loss, and our uncertainty as to whether HAVs will be able to perceive and communicate intent in the same ways that a human can, do we need a new form of HAV external communication? If so, how and what should it communicate to other road users? This session will provide an update on relevant activities in research and standardization.

A total of 60 individuals were in attendance. The format for the session was a series of short presentations followed by a group discussion exercise.

Session Summary

Study of Automated Vehicle External Communication in the Wild

Presenter: Andy Schaudt

The Ford Motor Company in partnership with the Virginia Tech Transportation Institute recently completed a study on the public roadways of Arlington, VA evaluating road user behavior in the presence of candidate automated vehicle external communication signals. The study design and results were presented along with future research questions.

Road User Behavior in the Presence of Automated Vehicle External Communication in Northern California

Presenter Jingyi Zhang, Nissan Research Center

Nissan presented results of a simulator study of communication intent of multiple autonomous vehicles conducted in Northern California. The focus of the study was to understand the effects of external communication on drivers' interaction, understanding, feedback, as well as distraction potential.

Updates on ISO Activity Surrounding External Communication
Presenter: Andy Schaudt

This presentation was an update on the numerous activities currently underway by the Road Vehicle Ergonomics committee within the International Organization for Standardization.

Recent Vulnerable Road User Research and Perspectives
Presenter: Justin Owens

This presentation was focused on human factors considerations for enabling safe interactions between automated vehicles and vulnerable road users. Benefits from AVs for pedestrians (e.g., improved sensors and vehicle response times) were presented, along with challenges for pedestrians. Implications for disabled individuals were also presented.

Exercise:

There have been many ideas of what type of information should be communicated, as well as how it should be, with other road users. This was an open discussion session to pull all those in attendance to discuss what is new that may beneficial, how human factors shape the potential of these novel solutions and unique use cases.

Use Cases:

- Taxi- communicating with riders during pickup
- AV transporting goods, not people
- Delivery AV the deploy droves for targeted package delivery

Types of Communication:

Vehicle State - tones, beeps in reverse
Driving Mode - automation level
Perception - AV tells following vehicles about pedestrians that are detected in the forward path
Recognition & Acknowledgement - Communication to pedestrians that they are recognized
Belief state – Braking distance to pedestrian
Guidance - Tell pedestrian it's safe to cross
Intent - AV intent to yield/stop

Suggested Action Items/Discussion topics

Comment: Bring attention to the system; participant looking up rate was low; so use auditory and/or additional visual signals/placement in order to investigate safety benefits.

Will behavior change over time when in a more AV-immersive environment?

Regardless of behavioral change, does the presence of visual external communication help to improve user trust and acceptance?

Will some of these issues resolve themselves as they become agents? Nonverbal communications are often viewed as a positive as they usually are, but possible they are ambiguous or there are mistakes. Predictability of the two agents will be an issue.

AV External Communication

"I'm with the news dude!"

TRANSPORTATION INSTITUTE
VIRGINIA TECH.
Andy Schaudt
Shane McLaughlin
Vicki Williams
Center for Automated Vehicle Systems

7/11/2018

Ford
John Shutko
Teryn Bray

General Research Question

- Does the introduction of AV external visual communication candidates influence the outcomes of encounters with other road users?
  - Safety concern?
  - Safety benefit?
  - Trust?

2

Ford ONE FORD RANSPORTATION INSTITUTE

Study Design

- Field study evaluating a seemingly driverless vehicle without visual external communication (Baseline encounters) and with visual external communication (Intervention encounters)
- Over 2,350 miles (167 hours) of data collected
  - Baseline: 81.8 hrs
  - Intervention: 85.6 hrs
- Primary dependent variable of interest, "Change in Other Road User Behavior" for those who notice the vehicle
  - Assumption: The majority of events will be road users encountering lighting for the first time (in previous Ford research, takes 2-3 times to understand the signal meaning)
- Assumption going in: Approximately 35% of road users encountered would notice (look at) the vehicle, thus providing approximately 1,000 interactions across different scenarios to sufficiently evaluate system performance
  - Only ~19% of road users encountered noticed the vehicle (looked up); lower sample than anticipated

3

TRANSPORTATION INSTITUTE
VIRGINIA TECH

Yielding Signal

Three External Communication Signals

Driver Concealment

View Inside Hood

AV External Communication Signals

- **Driving (Signal State):** Obeying traffic rules/regulations, including stop/go traffic
  - Jaywalker crosses path of AV, AV will slow and avoid striking pedestrian, but signal still indicating intent to drive.
- **Yielding (Signal State):** Identified something to yield to, intent to yield, actively decelerating, including coming to a stop
  - Only if AV is lead vehicle (e.g. first approaching a red light at intersection, pulling over to side of road to pick up pedestrian)
- **About-to-Drive (Signal Transition):** Only if stopped
  - This signal is only and always activated after "Yielding" signal and AV has come to a complete stop (e.g. signal changes from red to green and AV is about to move forward).

5

TRANSPORTATION INSTITUTE

6

TRANSPORTATION INSTITUTE

Independent Variables

- Signal Condition
  - Baseline
  - Intervention (sequence of Yield, About-to-Drive, Drive)
- Scenario Types
  - Study Vehicle Moving Straight through Intersections
  - Study Vehicle Turning Across Traffic
  - Study Vehicle Turning With Traffic
  - Study Vehicle near High Density of Vulnerable Road Users (Groups)
  - Study Vehicle in Parking Situations (on roadway, parking lots)
  - Study Vehicle Mock Pick-ups and Drop-offs at Airport
- Intersection Type
  - Signalized
  - Signage (signs present only)
- Environment
  - Daytime & Nighttime
  - Clear & Rain

7

TRANSPORTATION INSTITUTE
VIRGINIA TECH

Dependent Variables

- Other Road User
  - Intended path
  - Right-of-way
  - Secondary activities
  - First road user glance t
  - First change in behavio
  - Second change in beha
  - Third change in behavior by other road user
  - Change in behavior by other road user after study vehicle has passed by
  - Emotions observe3d
- Study Vehicle
  - Location
  - Travel direction
  - Speed
- Environment
  - Pedestrian signal
  - Factors affecting path
- Signal Type

– First change in behavior by other road user
– Second change in behavior by other road user
– Third change in behavior by other road user

8

TRANSPORTATION INSTITUTE
VIRGINIA TECH

Collection Meta-data

- Intersections w/ Valid Encounters
  - 67% signal controlled; 33% signage only
- Lighting
  - Predominantly daytime; portion of nighttime but low visibility
- Weather
  - Sunny/partly cloudy = 86.5%; Rain/fog/mist = 9.8%, Overcast = 3.7%
- Lighting Signal Sequence Activations
  - >1,800
- Age
  - Predominantly middle-aged (~94%)
- Gender
  - Male = 57.8%; Female = 42.2%

9

TRANSPORTATION INSTITUTE
VIRGINIA TECH.

Scenarios

Study vehicle moving straight

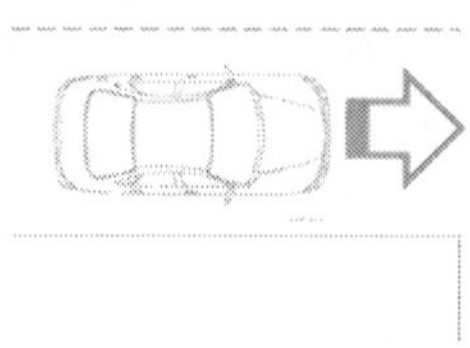

+

Study vehicle turning across

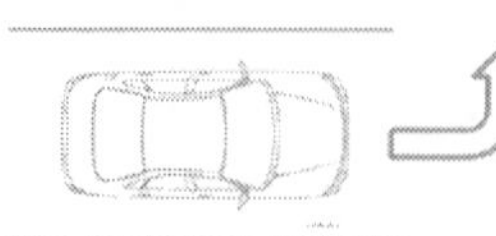

+

Study vehicle Turning with

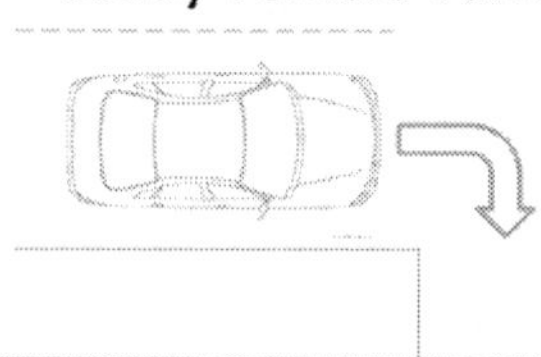

+

Study Vehicle Pickup/Drop-off

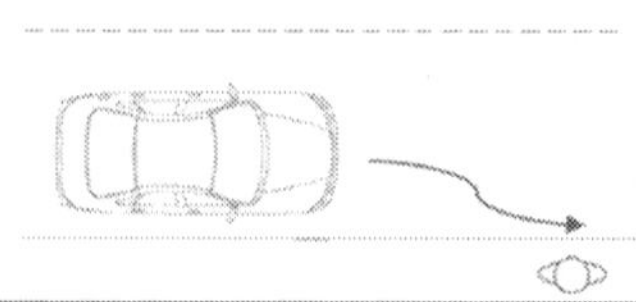

−

Study Vehicle near High Density VRUs

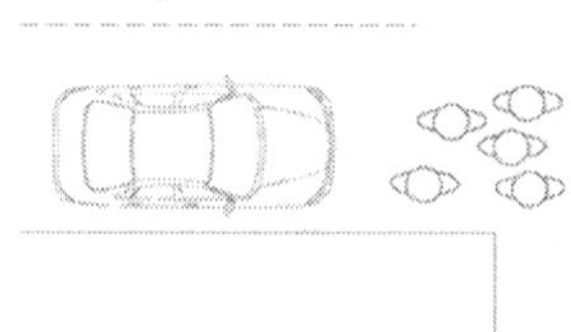

−

Research Questions

A. Do external visual communication signals make the AV more noticeable?
B. Do other road users change their behavior in presence of the signals?
C. What types of other-road-user behaviors are observed in reaction to the signals?
D. What types of emotions are observed by other road users in reaction to the signals?

11

Ford VT TRANSPORTATION INSTITUTE VIRGINIA TECH.

Results Not Available for Distribution at this Time

12

Ford VT TRANSPORTATION INSTITUTE VIRGINIA TECH.

Remaining Questions

- Will behavior change over time when in a more AV-immersive environment?
  - Increase chance of multiple encounters with pedestrians (not just first exposure data)
- Regardless of behavior change, does the presence of visual external communication help to improve road user trust and acceptance of AVs in their environment?
  - If so, there will be tremendous value of such lighting above safety alone

13

Andy Schaudt, aschaudt@vtti.vt.edu
Shane McLaughlin, smclaughlin@vtti.vt.edu
Vicki Williams, vwilliams@vtti.vt.edu

ISO TC22 SC39

Ergonomics

Status Report
AV Exterior Communications
TR 23049
7/11/2018

Andy Schaudt (VTTI, US); John Shutko (Ford, US)

1

Technical Report Currently in Publication Process

Ergonomic aspects of external visual communication from automated vehicles to other road users

- Purpose: guidance for developers of visual external communication systems for automated vehicles
  - Road users need to understand how to safely interact with these vehicles
  - Although there is still some discussion about the need to provide visual external communication, it is important to start working on a common approach.
  - Consistency across the automotive industry is needed to minimize potential road user confusion and establish societal trust with respect to ADS-DVs.
- Objective: propose how ADS-DVs could communicate with other road users via an external communication system.
  - Learnability of these systems is a main focus, limiting the number of signals and ensuring they are distinct and salient yet not distractive, with the aim of their implementation providing a positive impact on societal acceptance and traffic safety.
- The document does not address functionality elements of the ADS-DV external visual communication system itself. Rather, it serves to propose how the system communicates to human users such that it can be learned and understood by society at large.

2

General Outline

3

Section 5 Examples

- **Vehicle State** – e.g. tones and beeps as truck backs up
- **Driving Mode** – e.g. automation level
- **Perception** – e.g. AV tells vehicle following behind about pedestrians ahead
- **Recognition & Acknowledgement** – e.g. communicates that it recognizes a pedestrian
- **Belief State** – e.g. communicates braking distance to encroaching pedestrian
- **Guidance** – e.g. tells pedestrian when safe to cross
- **Intent** – e.g. intends to yield

4

Task Force Meeting in Prague, April 2018

- TF discussed proposals for new work items and agreed to develop and submit 2 NWIPs
  1. AV external communication general design guidance
  2. AV external communication experimental evaluation guidance with coding schemes

5

1 - AV external communication general design guidance

- It was the general consensus of those in attendance in Prague that a NWIP be developed that will produce a "General Design Guidance". Volunteers were collected to work on this NWIP with the target of submitting for ballot and results to be discussed in the fall meeting in Milan.

6

2 - AV external communication experimental evaluation guidance with coding schemes

Draft Scope:

This document proposes and defines terms and variables used for the annotation of video collected from studies evaluating automated vehicle external communication. Regardless of the research approach, the terms and variables defined within will help to provide a common strategy for evaluating other road user behavior in the presence of automated vehicle external communication. Included in this document are the common study types used, recommendations on the video views to be collected, and the important information and variables needed leading up to, during, and after encounters with other road users. The document does not address AV external communication system functionality or design. This document also does not address recommended approaches to analyzing the resulting data.

7

RENAULT NISSAN MITSUBISHI

COMMUNICATING INTENT OF MULTIPLE AUTONOMOUS VEHICLES

A Simulated Study

Jingyi Zhang

July 11th, 2018

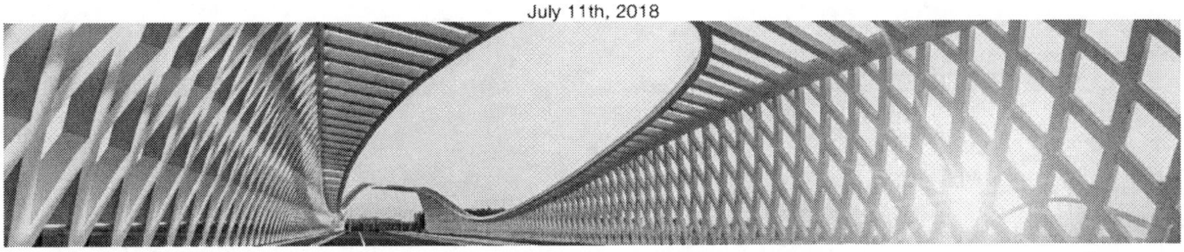

INTENTION INDICATOR

INTENT INDICATION

"I'm going/about to go/waiting."

DETECTION

"I see you."

DRIVER INTERACTING WITH MULTIPLE INTENTION INDICATORS?

Nissan IDS Concept, 2015

2

OBJECTIVE

STUDY DRIVER'S INTERACTION with 1, 3, and 5 autonomous vehicles equipped with 3 different configurations of Intention Indicator respectively in 2 environment settings – 4-way stop sign and midblock uncontrolled crosswalk.

STUDY DRIVER'S UNDERSTANDING of Intention Indicator – Recognition and Interpretation.

STUDY DRIVER'S FEEDBACK on Intention Indicator – Usefulness, Likeability, and Configuration preference.

INVESTIGATE DISTRACTIONS – If Intention Indicator would cause additional distractions in the testing conditions.

3

EXPERIMENTAL DESIGN

33 participants, divided into 4 groups

- 16 female, 17 male, experienced drivers
- Average age = 39.58, SD = 10.01

SECTION I

- 3rd person view video of traffic at T intersection with Intention Indicator animated
- Up to 3 times allowed to watch the video
- Questions being asked after each time: recognition, interpretation
  - *e.g. Able to see the signal? How many states? Can you describe them? What were they doing?*
- Questions being asked at the end of Section I: usefulness, likeability, configuration preference

4

EXPERIMENTAL DESIGN

SECTION II

- Navigate through 9 scenarios as the driver of the car in the video.
- Full instructions of the Intention Indicator configuration at the beginning of the section.
- All AVs were equipped with II.
- Pedal engagement and participants' eye glance behavior were recorded.
- Post-study questionnaire was given at the end of the study.

INDEPENDENT & DEPENDENT VARIABLES

- **Independent Variables** – II Configuration; Scenario Complexity (# of Vehicles, Vehicle Approaching Direction)
- **Dependent Variables** – Questionnaire Questions, Pedal Behavior, Eye-Glance Behavior.

5

EXPERIMENTAL DESIGN

INTENTION INDICATOR CONFIGURATION

	State 1. Going	State 2. About To Go	State 3. Not My Turn
Configuration 1 (Icon – 3-state)	 (solid)	 (blink)	 (solid)
Configuration 2 (Light – 3-state)	Light solid (white)	Light moving backward (white)	Light moving forward (white)

	State 1. Going	State 2. About To Do Something Different
Configuration 3 (Light – 2-state)	Light solid (white)	Light flashing (white)

7

3-STATE LIGHT, 5 AVS, 4-WAY STOP

9

SUMMARY

DISTRACTION

- Intention Indicator would not cause additional distractions.
- When encountering multiple autonomous vehicles simultaneously, tend to improve traffic flow.

ACCEPTANCE

- People like the concept of communicating vehicle intent and/or motion.
- People think that a concept of communicating vehicle intent and/or motion would be useful for autonomous driving in the future (open environment).
- Different types of road users need to be investigated separately

CONFIGURATION

- Preference on color, icon and light vary.
- Light configurations are as easy to be recognized as the icon configuration, less easier to interpret; icon configuration has limited visibility.
- 2-state is a less functional configuration, however simplicity is preferred.

15

Thank you!

Jingyi.Zhang@nissan-usa.com

AUTHOR INDEX

Association for Unmanned Vehicle Systems International
2700 South Quincy Street
Suite 400
Arlington, VA 22206

ISBN 978-1-5108-6840-3

CURRAN ASSOCIATES INC.
proceedings.com

Automated Vehicles Symposium 2018

San Francisco, California, USA
9 - 12 July 2018

Volume 2 of 2

Automated Vehicles Symposium 2018

San Francisco, California, USA
9 - 12 July 2018

Volume 2 of 2

ISBN: 978-1-5108-6840-3

Printed from e-media with permission by:

Curran Associates, Inc.
57 Morehouse Lane
Red Hook, NY 12571

Some format issues inherent in the e-media version may also appear in this print version.

Printed by Curran Associates, Inc. (2018)

For permission requests, please contact Association for Unmanned Vehicle Systems International at the address below.

Association for Unmanned Vehicle Systems International
2700 South Quincy Street
Suite 400
Arlington, VA 22206
USA

Phone: (703) 845-9671
Fax: (703) 845-9679

www.auvsi.org

Additional copies of this publication are available from:

Curran Associates, Inc.
57 Morehouse Lane
Red Hook, NY 12571 USA
Phone: 845-758-0400
Fax: 845-758-2633
Email: curran@proceedings.com
Web: www.proceedings.com

TABLE OF CONTENTS

VOLUME 1

BREAKOUT SESSIONS

VOLUME 2

ENTERPRISE SOLUTION SERIES

PLENARY SESSIONS

AUTOMATED VEHICLES SYMPOSIUM 2018
USERS. VEHICLES. INFRASTRUCTURE.
Driven by AUVSI and TRB

Breakout Session # 26 - Ethical and Social Implications

Summary of Key Findings and Lessons Learned

- Vehicle design can incorporate values, but must explicitly define those values and stakeholders
- Vehicle control should consider the probability that an object is a person, rather than static thresholds
- Consider the special relationship between the passenger and the AV
- Vehicles should have meaningful human control, just as with weapons systems
- Ethics and value design will evolve as technology improves and adoption increases
- Many rights are reliant on access to transportation. AVs should support those rights.
- Machine bias remains an issue, as AVs may classify certain users with higher confidence

AUTOMATED VEHICLES SYMPOSIUM 2018
USERS. VEHICLES. INFRASTRUCTURE.
Driven by AUVSI and TRB

Breakout Session # 26 - Ethical and Social Implications

Research Needs

- Measuring public acceptability of selfish cars
- Methods and metrics to compare various ethical theories
- How to do value sensitive design for policymakers
- What regulatory tools are available to implementing values or ethics
- Strategies for advocating for ethics policies at the state and local level
- What stakeholders to engage with as advocates and at what levels of government

Breakout Session 26: Ethical & Social Implications of Automated Vehicles

Organizers

Noah Goodall, Virginia Transportation Research Council
Patrick Lin, California Polytechnic State University
Filippo Santoni De Sio, TU Delft
Bart van Arem, TU Delft
Jean Pierre Medevielle, TRB International Activities Committee
Katherine Evans, Sorbonne Universités and VeDeCom

Reporters

Noah Goodall, Virginia Transportation Research Council

Session Speakers

Filippo Santoni De Sio, TU Delft
Giulio Mecacci, TU Delft
Sarah Thornton, Stanford University
Bart van Arem, TU Delft
Geoff Keeling, University of Bristol
Katherine Evans, Sorbonne Universités and VeDeCom
Ryan Jenkins, California Polytechnic State University
Nicholas Evans, University of Massachusetts Lowell

Session Focus

This session addressed ethical and social implications of automated vehicles not covered in other breakouts. Panelists discussed the need for human oversight of automated vehicle operation and explored definitions for "meaningful human control" across different operational domains. Panelists then presented approaches to addressed ethical issues from vehicle automation through both engineering and philosophical perspectives.

Session Summary

The session was divided into three panels. Each panelist presented on a topic for 15-20 minutes and took questions from the audience and other panelists. Speakers with philosophy backgrounds submitted position papers prior to the meeting, which were distributed to other panelists but not shared publicly. PowerPoints of most of the presentations are available on the AUVSI website at: http://www.automatedvehiclessymposium.org/

Panel 1 – Meaningful Human Control. The panelists were Filippo Santoni De Sio and Giulio Mecacci, TU Delft.

- Santoni De Sio and Mecacci gave a joint presentation on their recent work in defining meaningful human control for automated vehicle operation and testing. The concept of meaningful human control came out of autonomous weapons systems where there is a strong belief that a human should be somewhere in the decision-process, but how that looks in practice is often undefined. They define two conditions for meaningful human control: tracking and tracing. In tracking, the system should be able to co-vary its behavior with the relevant reasons of the relevant human agents for carrying out or omitting an assignment. Tracing describes a situation where there is at least single human with both the technical knowledge to understand the capabilities of the system as well as some moral responsibility for the system's behavior. Both control and decisions can have short, medium, and long-range components, and the role of control will shift to different agents depending on the time range.

Panel 2 – Engineering Approaches to Automated Vehicle Ethics. The panelists were Sarah Thornton, Stanford University; and Bart van Arem, TU Delft.

- Thornton presented recent work on value-sensitive design (VSD). In one example, a van is stopped on the side of the road in front of a pedestrian crosswalk. An AV wishes to pass the van, but must consider the possibility of a pedestrian stepping out from behind the van just as the AV is passing. The decision of whether to pass and how fast to do involves conflicting values, such as safety, legality, and mobility. VSD has three phases: conceptualization, technical implementation, and empirical analysis. In conceptualization, engineers identify both direct and indirect stakeholders, as well as the relevant values involved in the AV's decision. In the technical implementation phase, an algorithm is developed that aligns with the identified values. For the example scenario, Thornton develops a reward function that measures how well the AV is respecting each value for each stakeholder for a given decision. Finally, engineers use empirical analysis to test the implementation to see if the vehicle behaves in an acceptable way.
- Van Arem discussed a recent undergraduate course component on ethics for intelligent vehicles. Engineering students were asked to identify potential values and stakeholders for a simple truck platooning application. Through this exercise, students quickly understood the ethical components of even fairly mundane driving decisions. Van Arem also discussed recent research on translating these issues algorithmically as risk fields for vehicles on the road, to be used by AVs to avoid potential crashes. The size of risk fields can be adjusted based on vehicle mass (as a surrogate for crash severity) or vulnerability of the user (e.g. giving greater space to pedestrians).

Panel 3 – Philosophical Approaches to Automated Vehicle Ethics. The panelists were Geoff Keeling, University of Bristol; Katherine Evans, Sorbonne Universités and VeDeCom; Ryan Jenkins, California Polytechnic State University; and Nicholas Evans, University of Massachusetts Lowell.

- Keeling argued that autonomous vehicles use classification algorithms that are probabilistic, and that AVs should consider the probability of an objects class in its decisions, rather than simply setting arbitrary thresholds for say a person or other object. He used the example of the recent fatal Uber AV crash in Arizona, where the AV detected the object as a vehicle, then a bicycle, by which time the crash could not be avoided. Given a threshold of, for example, 70% confidence that an object is a person for it to be classified as a person can lead to some illogical results. First, this violates the continuity principle, as a small change in the AV's confidence (increasing from 69% to 70%) results in a disproportionate way that an object is treated. Instead, Keeling advocated a position he calls *uncertain utilitarianism,* whereby an AV should select the crash option that maximizes the utility of the outcome multiplied by the credence or confidence the AV has in the object's class. He considers non-linear relationships between credence and moral weight such as exponential and logarithmic functions, but does not find strong evidence to support them over a simple linear relationship.

- Katherine Evans opened with the quote from a Mercedes representative in 2016 that their cars would generally favor the passenger in a collision, which the company distanced themselves from in a press release five days later. Evans argued that it is morally permissible for an AV to favor its passenger in a collision by a small, non-specified degree. From a technical perspective, the AV is more confident of its own passenger's existence, road user class, and future movements than it is of any object or person outside of the AV. This justifies preferential treatment, as it is unfair to sacrifice a probable person for an improbable person. From an intuitive perspective, there exists a reasonable expectation that an AV, when privately-owned or hired, will not sacrifice its own passengers. This has evidence in the psychological literature, with a concept known as *betrayal aversion*. In surveys, two-thirds of respondents would prefer an airbag with a 2% chance of dying over the life of an airbag, versus one with a 1% chance of dying in a crash plus a 0.01% chance of dying from the result of an accidental airbag deployment. The thought of a product that is designed to protect you instead actively killing you seems to be especially abhorrent. Finally, from a moral perspective, the AV may have a duty to preference the passenger's safety, as the passenger must transfer his decisional autonomy to the AV and therefore has no realistic way to express his intentions in real-time. The AV-passenger relationship is essentially a form of trusteeship, which almost always requires a degree of partiality on the part of the substitute agent. Finally, the wide public acceptance of AVs will likely improve safety for all users, and ensuring that AVs are acceptable to its consumers may be morally preferable.
- Jenkins proposed rule-consequentialism as a decision-making procedure for AVs. Rule-consequentialism is a moral theory that evaluates actions as to how well they adhere to an ideal moral code. The ideal moral code is a set of rules that, were they to be followed by everyone, would maximize good. Codes need not be universal for all vehicles. Just as doctors and police have different moral codes, so should ambulances and police vehicles which would have more leeway to violate traffic laws than, for example, a taxi. Devising specific moral codes for AVs is different that devising ones for humans. First, a computer can handle very complex moral codes, as they are not limited by human memory. On the other hand, AVs will have limited information about many things that people consider morally relevant, for example the age of participants in a trolley-like dilemma. Jenkins proposed the following preliminary ideal code for autonomous vehicles: "Autonomous vehicles out to follow the code of rules such that, if all autonomous vehicles followed that code of rules, the world would be as good as possible." Jenkins favors a machine learning approach to developing specific moral weights for different actions, rather than hard

coding deterministic values. A moral code for AVs should prioritize the following, in order:

- Promote life (prevent deaths)
- Promote well-being (prevent injuries)
- Promote convenience (prevent wasted time and discomfort)

The code could be expanded or altered to include other values such as justice, fairness, and respect for drivers' autonomy.

- Nicholas Evans noted that benefits and harms of AVs are often discussed using collectively term, while the discussion of ethics is dominated by the individual actions of cars. He explored the broader political issues of AVs, through the lens of transportation as a right. While transportation itself is probably not a right, it is derivative of many other rights such as right to freedom of movement and mobility, right to freedom of association, and other civic powers we have as rights. Other rights such as to subsistence and health are also dependent on transportation. In John Rawls' *Theory of Justice,* there are two principles of a scheme of just institutions:
  - Each person is to have an equal right to the most extensive total system of equal basic liberties compatible with a similar system of liberty for all.
  - Social and economic inequalities are to be arranged so they are both to the greatest benefit of the least well-off representative group in society, and attached to offices and positions open to all under conditions of fair equality and opportunity.

  These rules only apply to social institutions and broad rights, not to individual decisions of a single AV. While there are many rights-based transportation services, in practice, these often fail to satisfy the difference principle. For example, road maintenance may be over-funded in wealthy areas and under-funded in others. AVs, of which early versions will likely be privately owned fleets, may worsen this problem through increasing commodification of mobility services.
- The group took questions at the end of the fourth presentation.
  - Audience members pointed out that Keeling's algorithm would formalize any machine bias in a sensing system. For example, machine vision better recognizes lighter skin tones. In a threshold system, this would only benefit lighter-skinned people when the confidence crosses the threshold. In Keeling's system, it would always prioritize lighter-skinned people, although the priority may be rather small.
  - Panelists discussed standards to certify acceptability of an AV. Today's driver's test is far too simple, and instead it should be a battery of tests where an AV must respond correctly in a high percentage of situations, but not 100% of the time. The AV need not agree with itself each time due to differences in machine learning, nor should it need to explain the precise causes for its action. It should, however, be able to justify its actions broadly.
  - Panelists discussed prioritizing passengers vs. pedestrians. They noted bullguards put on the front of trucks are meant to protect bumpers, but are very unsafe for pedestrians. Likewise, vehicle mass has impacts on pedestrian and passenger survivability, yet are ethically permissible.

- The audience asked about ways to create a regulatory framework around ethics. Panelists recommended mobilizing politically, mimicking other groups that have had success in this area, such as the mobility-impaired.

Suggested Action Items

Attendees identified the following recommendations.

- Vehicle design can incorporate values but must explicitly define those values and stakeholders.
- Vehicle control should consider the probability that an object is a person, rather than static thresholds.
- Consider the special relationship between the passenger and the AV.
- Vehicles should have meaningful human control, just as with weapons systems.
- Ethics and value design will evolve as technology improves and adoption increases.
- Many rights are reliant on access to transportation. AVs should support those rights.
- Machine bias remains an issue, as AVs may classify certain users with higher confidence.

Attendees identified the following research needs.

- Measuring public acceptability of selfish cars.
- Methods and metrics to compare various ethical theories.
- How to do value sensitive design for policymakers?
- What regulatory tools are available to implementing values or ethics?
- Strategies for advocating for ethics policies at the state and local level.
- What stakeholders to engage with as advocates and at what levels of government?

Engineering approaches to automated Vehicle ethics

Bart van Arem

TUDelft

Faculty of Civil Engineering and Geosciences
Department of Transport and Planning

Meaningful Human Control over automated driving systems

AV and responsible innovation

AV ethics teaching

AV ethics and safety fields

TUDelft

Meaningful Human Control over automated driving systems

AV ethics and teaching

MSc course Intelligent Vehicles for Safe and Efficient Traffic (40-50 students)

45 minute teaching block

In the US, a Tesla in autopilot mode, crashed into a stationary fire truck. The driver did not take over. The Manual states: "Traffic-Aware Cruise Control cannot detect all objects and may not brake/decelerate for stationary vehicles, especially in situations when you are driving over 50 mph (80 km/h) and a vehicle you are following moves out of your driving path and a stationary vehicle or object is in front of you instead."

Meaningful Human Control over automated driving systems

TU Delft

Interactive classroom question – Feedback Fruits

Which of the following statements do you agree with most?

A	B	C
Manufacturers should not be selling innovative features before they are proven safe.	Not monitoring the autopilot should be fined, just like using hand-held phone.	Public authorities are failing to introduce new type approval and driver license requirements
50%	15%	35%

Meaningful Human Control over automated driving systems

TU Delft

Responsibility of engineers

"Responsibility means being held accountable for your actions and for the effects of your actions. The making of choices, the taking of decisions but also failing to act are all things that we regard as types of actions"

"Active responsibility is responsibility before something has happened. It refers to a duty or task to care for certain state-of-affairs or persons. "

"Passive responsibility is applicable after something (undesirable) has happened."

TU Delft

Meaningful Human Control
over automated driving systems

WePod

- Commissioned by Province of Gelderland "Future of Public Transport in rural areas"
- Political prestige
- Conducted by consortium led by TU Delft
- Low speed automation up to 25 km/h
- Campus Wageningen University and connection to Ede-Wageningen station (11 km)
- 2 vehicles driving in mixed traffic as hospitality service
- Supervised by registered engineer

TU Delft

Meaningful Human Control
over automated driving systems

Active responsibility

Which codes, regulations should the engineer/developer satisfy?

RDW approval, FMEA, ethical approval, research logging (software), road operator approval, ISO standards, ..,

Which professional ideals are involved?

Cool technology, world leadership, sustainability, safety, mobilty4all, …

TU Delft

Meaningful Human Control over automated driving systems

AV and responsible innovation

Human actor responsible and accountable?

Moral philosophy

Responsible AV system design

Cognitive psychology

Traffic engineering

Human actor capable to carry responsibility

TU Delft

Meaningful Human Control over automated driving systems

Truck platooning agreement

Decide who's leading and following.

Discuss responsibilities and agreements for the leader to take control of the following truck.

5 minutes:

- What did you agree upon?
- Which issues are still open?

TU Delft

Meaningful Human Control over automated driving systems

Topics to (dis-)agree upon:

- Activities permitted for follower
- Maximum Take over time
- Manoevre coordination
- Insurance
- Platooning certificate
- Platoon leading license
- Cargo monitoring
- Wireless communication: system integrity, software version, headway, speed, destination?
- Streaming video images leading truck
- …

TU Delft

Meaningful Human Control over automated driving systems

AV ethics and safety fields

1. Traffic scenario data

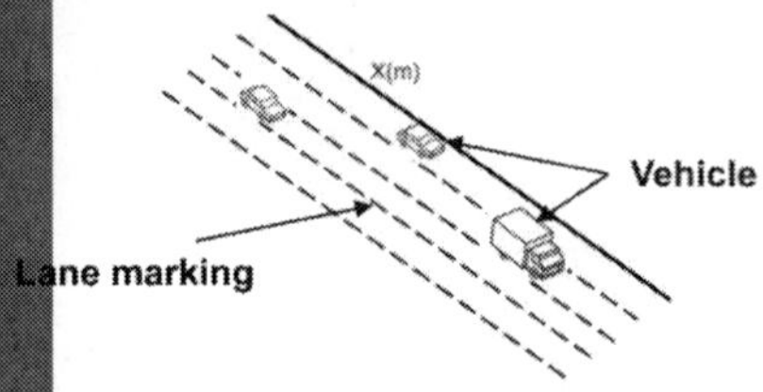

2. Risk field strength estimation

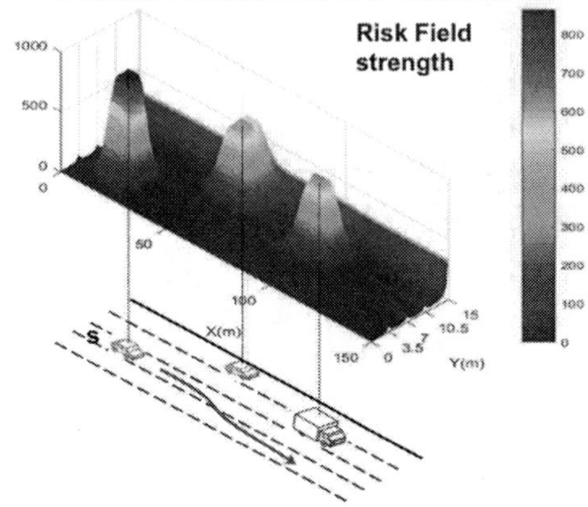

3.Risk estimation

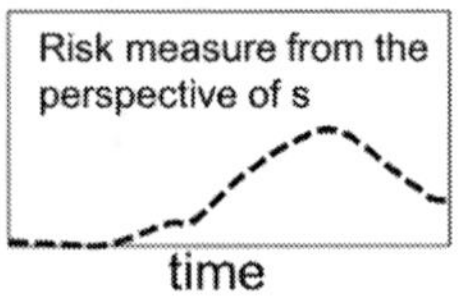

TUDelft

Mullakkal Babu et al (2017) Probabilistic driving risk assessment approach for mixed highway traffic, Poster at proc AVS 2017, DOI: 10.13140/RG.2.2.24191.92323

Meaningful Human Control over automated driving systems

Risk due to the neighbouring vehicle

$$R_n = Collision\_probability(s,n) \times Collision\_severity(s,n)$$

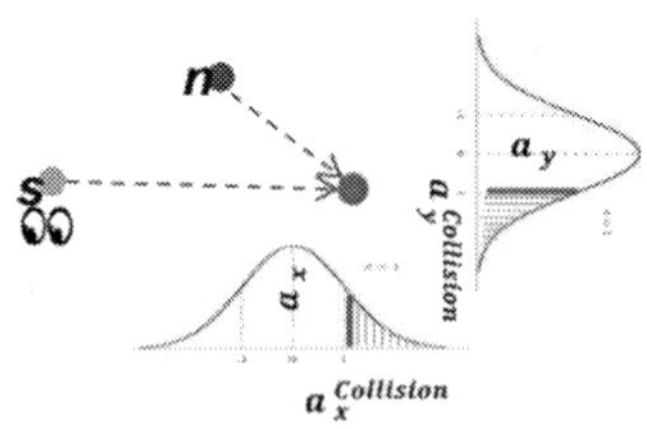

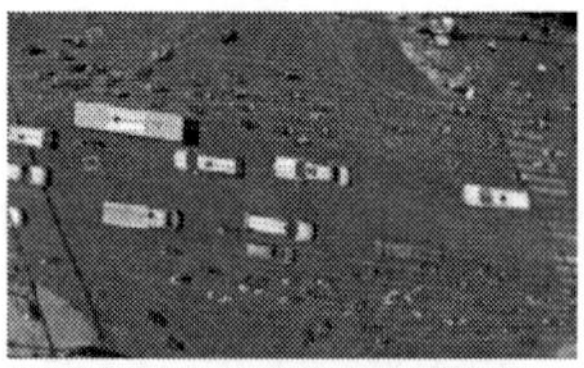

Single-step approach

Collision definition	Angular overlap
Vehicle state prediction	Single time step
Probability estimation	Probability density function

TUDelft

Meaningful Human Control over automated driving systems

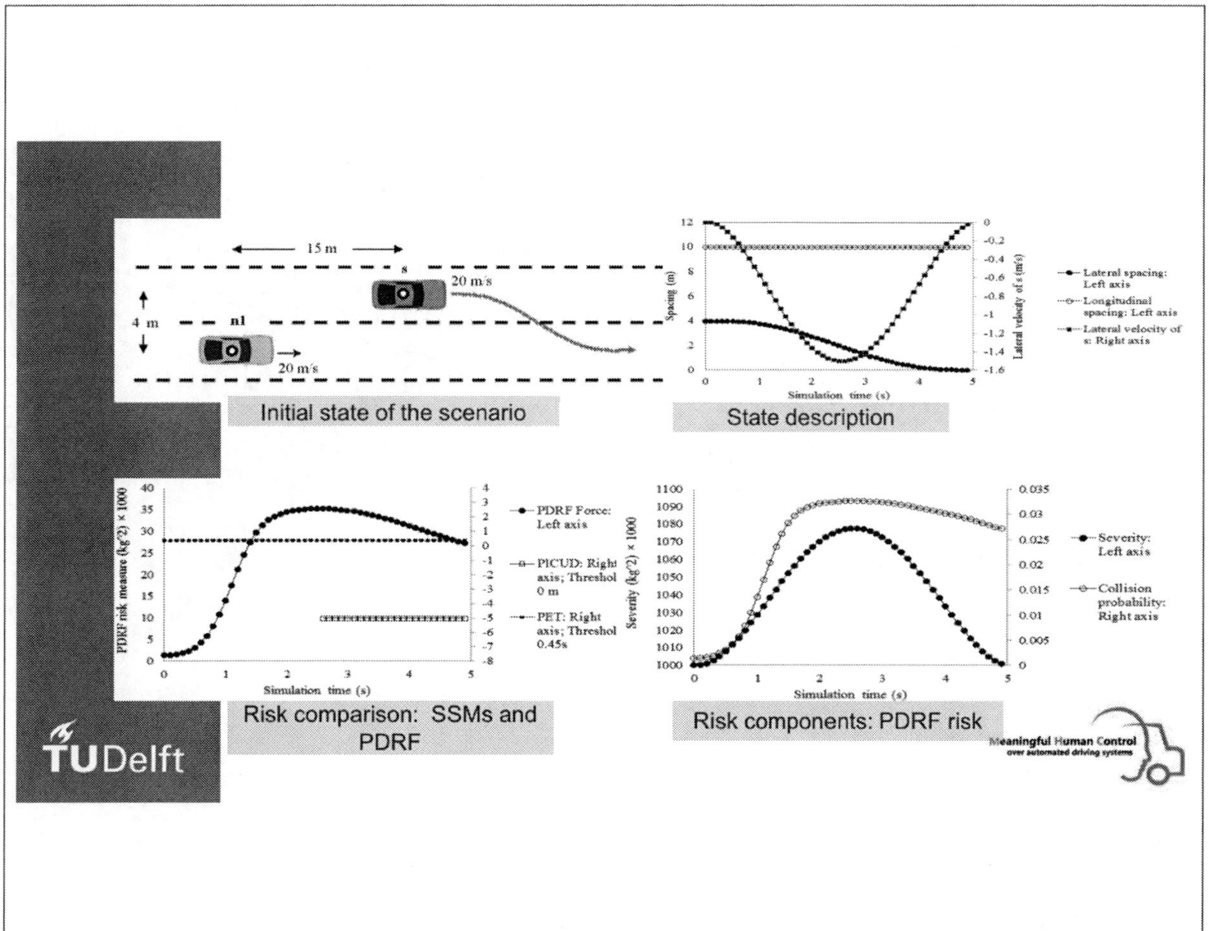
15 m
s
20 m/s
4 m
n1
20 m/s
Initial state of the scenario
Spacing (m)
Lateral velocity of s (m/s)
Simulation time (s)
Lateral spacing: Left axis
Longitudinal spacing: Left axis
Lateral velocity of s: Right axis
State description
PDRF risk measure (kg^2) × 1000
PDRF Force: Left axis
PICUD: Right axis; Threshol 0 m
PET: Right axis; Threshol 0.45s
Simulation time (s)
Risk comparison: SSMs and PDRF
Severity (kg^2) × 1000
Severity: Left axis
Collision probability: Right axis
Simulation time (s)
Risk components: PDRF risk
TUDelft
Meaningful Human Control
over automated driving systems

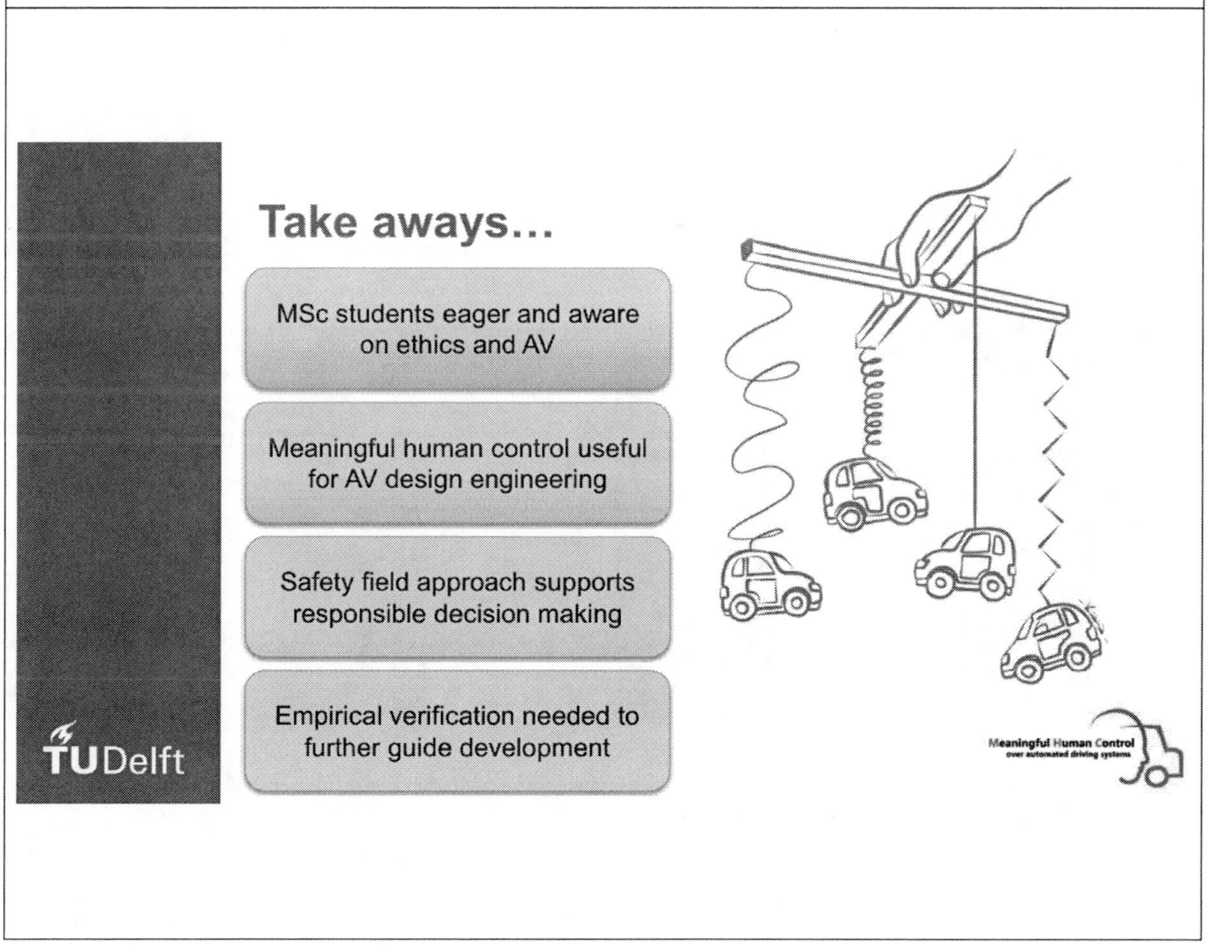
Take aways…
MSc students eager and aware on ethics and AV
Meaningful human control useful for AV design engineering
Safety field approach supports responsible decision making
Empirical verification needed to further guide development
TUDelft
Meaningful Human Control
over automated driving systems

DISTRIBUTIVE JUSTICE AND AUTONOMOUS VEHICLES: FROM INDIVIDUAL TO POLITICAL

NICHOLAS G EVANS
AVS 2018
SAN FRANCISCO

Learning with Purpose

1

UMASS LOWELL

ETHICAL ALGORITHMS IN AUTONOMOUS VEHICLES

NSF AWARD #1734521

Nicholas G Evans
- Philosophy
- University of Massachusetts Lowell

Ryan Jenkins
- Philosophy
- Cal Poly San Luis Obispo

Rocco Casagrande
- Principal
- Gryphon Scientific

Yuanchang Xie
- Civil Engineering
- University of Massachusetts Lowell

Heidi Furey
- Philosophy
- Manhattan College

Ryan Ritterson
- Senior Scientist
- Gryphon Scientific

Learning with Purpose

2

UMASS LOWELL

AUTONOMOUS VEHICLES

FROM INDIVIDUAL TO POLITICAL

- The potential benefits of autonomous vehicles (AVs), particularly at level 5, are often couched in collective terms
  - Safety, efficiency, and environmental benefits
- Likewise, the potential harms of autonomous vehicles are collective in an important sense
  - Inequalities, urban decay, cybersecurity, greater traffic volume, erosion of services
- "Ethics and autonomous vehicles" has typically referred to decisions by individual cars
  - But what about the broader political issues?

Learning with Purpose 3

TRANSPORTATION AS A RIGHT

DO WE HAVE ONE?

- Probably not—at least, not a fundamental right
- But we might have a right to transportation that is *derivative* of other rights
  - Right to freedom of movement and mobility
  - Right to freedom of association
  - Important civic powers we have a matter of rights
- We also have certain welfare rights
  - To subsistence, health, etc.
  - These are highly dependent on transportation
- Transport is an important component in rights, and basic institutions as part of a just society

Learning with Purpose 4

JUST TRANSPORTATION

A RAWLSIAN MODEL

- John Rawls: one of the most prominent 20th century political philosophers
- *Theory of Justice:* Two principles of a scheme of just institutions
  - *1) Each person is to have an equal right to the most extensive total system of equal basic liberties compatible with a similar system of liberty for all.*
  - *2) Social and economic inequalities are to be arranged to they are both*
    - *a) to the greatest benefit of the least well-off representative group in society*
    - *b) attached to offices and positions open to all under conditions of fair equality of opportunity.*
- These do *not* apply to individual decisions, i.e. a "Rawlsian Car"
- Rather, at the level of social institutions, including systems of rights

Learning with Purpose 5

JUST TRANSPORT

HOW DO AUTONOMOUS VEHICLES FEATURE

- There are many services in and around transport that promote rights
  - Transit, assistive and mobility services, road maintenance, highways
- But these do not always (ever?) satisfy the difference principle
  - "Interstate injustice"; diminishing transit funding; selective road maintenance; property rights around transportation hubs
- AVs, improperly implemented, may worsen this problem
  - Uber's foibles, further undermining transit, commodification of mobility services, changes in road infrastructure
- It doesn't have to be this way.

Learning with Purpose 6

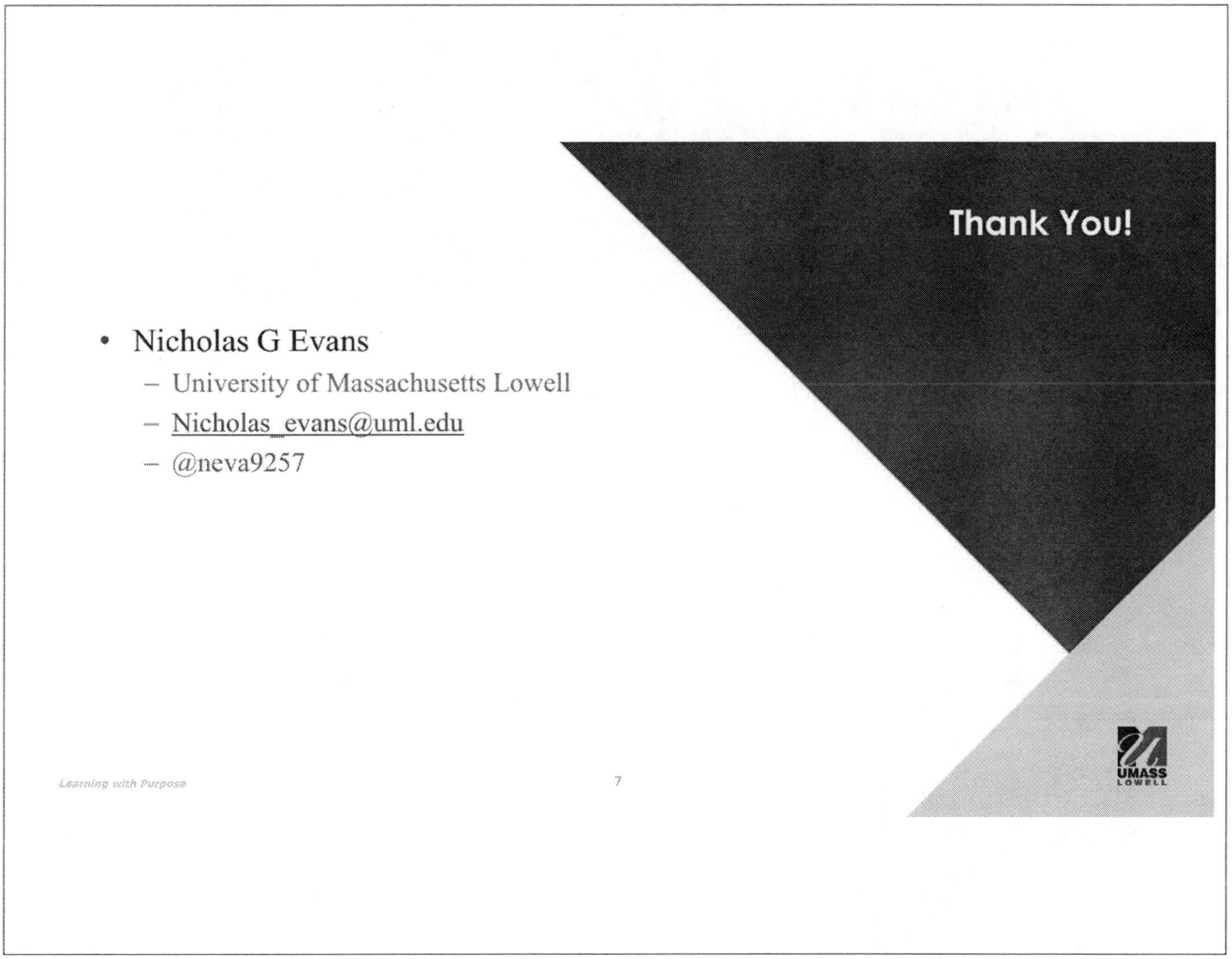

Thank You!
• Nicholas G Evans
– University of Massachusetts Lowell
– Nicholas_evans@uml.edu
– @neva9257
Learning with Purpose
7
UMASS LOWELL

AUTOMATED VEHICLES SYMPOSIUM 2018

University of BRISTOL

Arts & Humanities Research Council

Uncertainty About Persons in Autonomous Vehicle Collisions

Geoff Keeling
University of Bristol

Methodology

- I am concerned with Level 5 vehicles
- I shall examine the kinds of uncertainty which arise in AV collisions
- And then ask what morality requires in dilemmas involving these kinds of uncertainty, drawing on what we already know from other parts of moral philosophy (e.g. the ethics of war)
- I will use trolley cases *carefully*

The Plan

- Uncertainty About Persons
- The Threshold View
- Uncertain Utilitarianism
- Conclusion

Uncertainty About Persons

In *most* moral dilemmas, there is a well-defined set of affected parties

We know who the affected parties are, and what the harms or risks of harm will be to each party conditional on each of the available actions

Is this assumption plausible for AV collisions?

Uncertainty About Persons

- In collisions where (i) harm to at least one person is unavoidable, and (ii) a choice about how to allocate harm between different persons is required, then if there exists a unique Pareto efficient allocation of harm across different persons, then other things being equal, programming a driverless car to bring about the Pareto efficient allocation of harm is justified (Keeling 2018: 422)

Uncertainty About Persons

But let's consider the recent Uber crash in Arizona

The AV detected an unknown object, which was then classified as a vehicle, and then a bicycle, by which time it was too late

Uncertainty About Persons

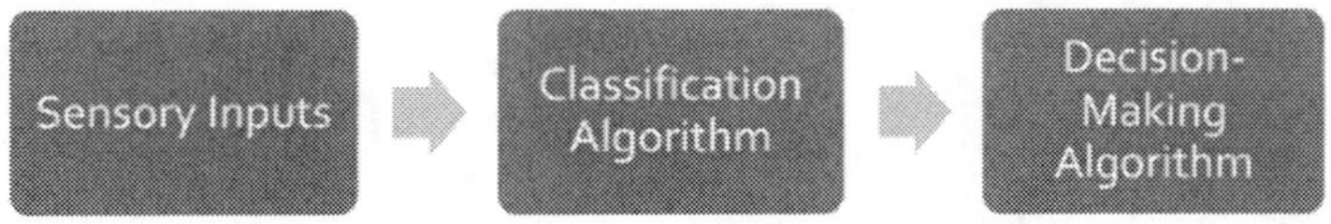

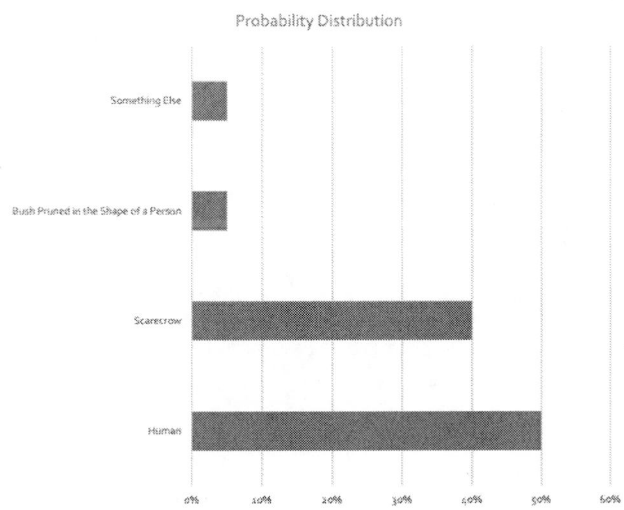

Uncertainty About Persons

The problem is that classification algorithms are **probabilistic**

What does morality require in dilemmas where the AV believes *to some degree* that an object is a person, and *to another degree* that the object is a non-person?

Uncertainty About Persons

- Let $x_1, \ldots, x_n \in E$ be the objects which the AV identifies
- Let $cr: E \to [0,1]$ be a function which maps each $x_i \in E$ to the AV's credence in the proposition that x_i is a person

The Threshold View

- According to the **Threshold View,** the AV should behave as if an object is a person just in case the AV's credence in that object being a person exceeds some threshold
- For any $x_i \in E$, the AV should assign a utility function u_i to x_i which encodes rational and self-interested preferences over collision outcomes which a person in the place of x_i would have if, and only if, $cr(x_i) \geq t$, where t is a real number between 0 and 1

The Threshold View

- The positive case for the Threshold View
- Morally insignificant risks
- Objective and subjective permissibility

The Threshold View

- Consider

  Collision 1: The AV can swerve left or right. To the left is an object, x, and the AV's credence in the proposition that x is a person is equal to the threshold for moral consideration. If the AV swerves left, and x is a person, then x will be severely harmed. To the right are two objects, y and z, and the AV's credence in y and z being persons is slightly below the threshold for moral consideration. If the AV swerves right, and y and z are persons, then y and z will be severely harmed

- Here the Threshold View conflicts with what we can call the

  Continuity principle: a small deviation in the AV's credence in some object being a person does not warrant a large deviation in the degree of moral consideration given to that object.

The Threshold View

- Consider

  Collision 2: The AV can swerve left or right. To the left is a single object, x, and the AV's credence in the proposition that x is a person is very slightly below the threshold for moral consideration. If the AV swerves left, and x is a person, then x will be severely harmed. To the right there are no objects. So, if the AV swerves right, no person will be harmed.

- Here the Threshold View conflicts with what we can call the

  Pareto Principle: if the AV has two options, then if, from the AV's point of view, one option might harm someone and the other is certain to harm nobody, then other things being equal the AV has a decisive moral reason to choose the option which is certain to harm nobody.

Uncertain Utilitarianism

- The AV must choose one act from $a_1, \ldots, a_n \in A$
- Each object $x_i \in E$ is assigned a utility function u_i
- According to **Uncertain Utilitarianism,** the AV should select the action $a \in A$ in a collision which satisfies

$$\underset{a \in A}{\operatorname{argmax}} \sum_{i=1}^{n} cr(x_i) U_{x_i}(a)$$

Uncertain Utilitarianism

- Uncertain Utilitarianism satisfies *Continuity*
- Recall

> *Collision 1:* The AV can swerve left or right. To the left is an object, x, and the AV's credence in the proposition that x is a person is equal to the threshold for moral consideration. If the AV swerves left, and x is a person, then x will be severely harmed. To the right are two objects, y and z, and the AV's credence in y and z being persons is slightly below the threshold for moral consideration. If the AV swerves right, and y and z are persons, then y and z will be severely harmed.

- Uncertain Utilitarianism mandates that the AV swerves left. A small deviation in the AV's credence in some object being a person will not result in a large deviation in the moral consideration given to that object

Uncertain Utilitarianism

- Uncertain Utilitarianism satisfies *Pareto*
- Recall

> *Collision 2:* The AV can swerve left or right. To the left is a single object, x, and the AV's credence in the proposition that x is a person is very slightly below the threshold for moral consideration. If the AV swerves left, and x is a person, then x will be severely harmed. To the right there are no objects. So, if the AV swerves right, no person will be harmed.

- Uncertain Utilitarianism mandates the AV should *swerve right*, because the credence-weighted utility of swerve left is negative, and the credence-weighted utility of swerve right is zero

Uncertain Utilitarianism

- According to uncertain utilitarianism, the AV should select the action $a \in A$ which satisfies

$$\underset{a \in A}{\operatorname{argmax}} \sum_{i=1}^{n} cr(x_i) U_{x_i}(a)$$

- This is one member of a family of views, according to which the AV should select the action $a \in A$ which satisfies

$$\underset{a \in A}{\operatorname{argmax}} \sum_{i=1}^{n} w(x_i) U_{x_i}(a)$$

- Here $w: [0,1] \rightarrow [0,1]$ is a function which takes the AV's credence in each object being a person and outputs a weight for that object between 0 and 1. Uncertain utilitarianism puts $w(x_i) = cr(x_i)$

Uncertain Utilitarianism

Here is $w(x_i) = cr(x_i)^2$

This is a *convex* function

When the AV's credence in some object being a person is low, then the AV should assign that object *even less* weight than uncertain utilitarianism demands

Gains in the AV's credence in some object being a person matter more and more, the higher the AV's credence is

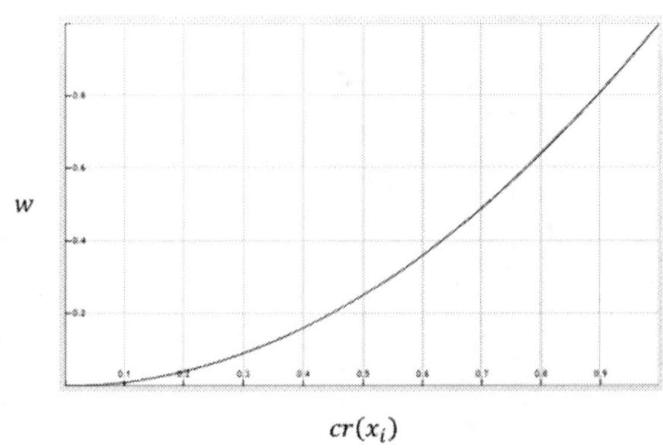

Uncertain Utilitarianism

Here is $w(x_i) = \sqrt{cr(x_i)}$

This is a *concave* function

When the AV's credence in some object being a person is low, then the AV should assign that object *even more* weight than uncertain utilitarianism demands

Gains in the AV's credence in some object being a person matter less and less, the higher the AV's credence is

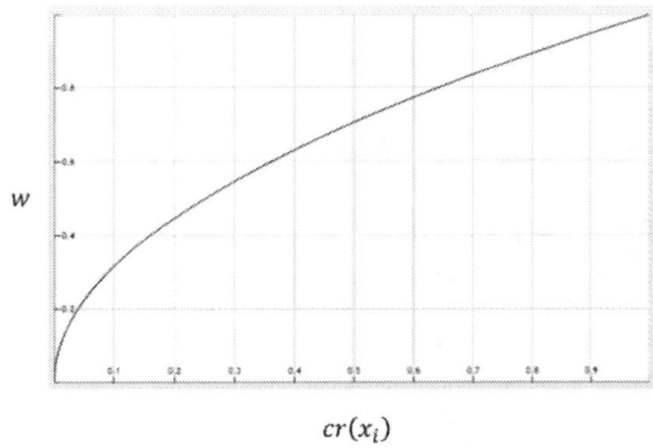

Uncertain Utilitarianism

- These other views face a dilemma
- Uncertain Utilitarianism puts $w(x_i) = cr(x_i)$, because with this weighting function we rationally expect to produce the most utility (minimise harm) in a collision
- These other views must either (i) **reject expected utility theory**; or (ii) appeal to a **moral argument** for why some objects should be given priority over others
- This dilemma is not a knock-down argument, but in my view renders these other weighting functions less attractive

Conclusion

- AV collisions will involve *uncertainty about persons*
- I explored two views about what morality requires in interpersonal moral dilemmas involving uncertainty about persons
- I argued against the Threshold View
- I made a case in favour of Uncertain Utilitarianism

- Thanks for listening!
- Also thanks to Brad Hooker, Richard Pettigrew, Sven Nyholm, Johannes Himmelreich, Farbod Ackhlaghi-Ghaffarokh, Jason Konek, Niall Paterson and Samir Okasha for their comments

AUTOMATED VEHICLES SYMPOSIUM 2018
USERS. VEHICLES. INFRASTRUCTURE.
Driven by AUVSI and TRB

Breakout Session 27

Summary of Key Findings and Lessons Learned

- Cost of urban travel will drop considerably measured on a passenger mile basis
  - This will make it difficult to use congestion pricing tools
  - Varies by cost of travel time, capital costs etc.
- Even a drop below 50 cents a mile will generate major growth in travel demand
- Without respect for AVs, risk of vandalism
- Economic benefits of AV greatly exceed economic cost of job losses
  - Job losses will be delayed
- AV closer to utopia than dystopia (but problems do exist)

AUTOMATED VEHICLES SYMPOSIUM 2018
USERS. VEHICLES. INFRASTRUCTURE.
Driven by AUVSI and TRB

Breakout Session 27

Research Needs

- Explore likely future travel costs, by private car versus shared; urban versus rural
- Analyze impacts of reduced travel cost on PMT and VMT
  - Consider variation by location and nature of impacts
- Explore ways to generate respect for AVs – or ways to monitor vehicle conditions
- Develop business model for options to offset costs of job losses
  - Consider buyouts as well as job training options

AVS18 Proceedings Template for Breakout Sessions [Number and Title of session here]

Organizers

Richard Mudge, Compass Transportation and Technology
Alain Kornhauser, Princeton University
Amitai Bin-Nun, SAFE(Securing America's Future Energy)
Alan Chachich, Volpe National Transportation Systems Center
Jean-Pierre Medevielle
Ellen Partridge, Environmental Law and Policy Center
David Pickeral, Consultant
Jim Scheinman, Maven Ventures
Suna Taymaz, AAA of Northern California

Reporters

Richard Mudge, Compass Transportation and Technoloy
Joshua Cregger, US DOT Volpe Center

Session Speakers

Moderators:
Richard Mudge, Compass Transportation and Technology
Sharon Feigon, Shared Use Mobility Center

Sharks:

- Sven Beiker, Silicon Valley Mobility
- Iain Forbes, UK Ministry of Transportation
- Alain Kornhauser, Princeton University
- Brad Templeton – Singularity University

Speakers:

- Lew Fulton, University of California, Davis
- Alain Kornhauser, Princeton University
- Amitai Bin-Nun, SAFE (Securing America's Future Energy)

Session Focus

The Shark Tank format encourages active debate on topics with important implications for the future of autonomous vehicles. The topics are multi-disciplinary, covering policy, planning, regulations, economics, and of course technology. All modes are considered.

Formal speeches are kept to a minimum. A group of commentators (the sharks) ask provocative questions and may engage in debate among themselves as well as with the presenter. Ample time is allowed for comments and questions from the audience.

This year's topics cover economics (will new jobs stimulated by AV offset job losses— and what new skills will be required); safety (beyond the usual focus on cyber security); planning and regulation (are there practical ways to maximize the benefits of AV and minimize the costs — and is there any agreement about what are costs and benefits); and what are the economic and social implications of dramatic drops in the cost of travel (assuming these forecasts are credible).

PowerPoints of most of the presentations are available on the AUVSI website at: http://www.automatedvehiclessymposium.org/avs2018/program/proceedings.

Session Summary

Part I

Moderator: Richard Mudge, Compass Transportation and Technology

Lew Fulton, University of California, Davis

- **Thesis: In a simple analysis, taking into account all costs, the advantage of a pooled ride goes away due to the hedonic costs of more time spent.**
- Researchers have been looking at fixed and variable costs of different modes. These include "out of pocket" costs (e.g., vehicle purchase, fuel, maintenance, and parking) as well as hedonic costs (e.g., travel time, parking time, and driving stress).
- TNC (MaaS) service (pooled or unpooled) costs more per person-mile traveled than private vehicles, however much of this is driver cost, and automation can make these options competitive with (and slightly cheaper than) private vehicle ownership.
- When hedonic costs are added in, TNC/MaaS options are much more competitive than manned vehicles, though they are more expensive than a private vehicle.
    - There was some discussion as to whether the value of time spent traveling is properly accounted for (it was considered have the price as time spent actively driving for automated and TNC/MaaS options).
- It takes a pretty high price (user fee) to affect user behavior.
    - If the service is highly competitive, prices will be nearly equal to costs.
    - While $3.00 vs $2.00 may not seem like much, for many people, they will care about that difference, especially if it is a daily decision (i.e., a commute).
- Congestion, pollution, greenhouse gas emissions have not been included in the chart, though they could be if a tax/fee was added.
- When pooled trips are operated well and has a high share of all trips, the hedonic costs (time spent waiting) may be lower than assumed.
- What about being in close quarters with strangers? If it is late at night and there is one other user, people may be worried about sharing a ride.
- The big change in urban mobility will be whether people choose to own a car (e.g., cities like Miami and Cleveland may become like Manhattan). TNCs will differentiate and offer different types of service. When the cost of not owning a car is lower than the cost of owning a car, there will be a discrete decision whether or not to own a car (buy or not buy). The significance is much more about whether people will choose to own a car or not.

- Uber/Lyft is not just running an app. It is a city-by-city battle to operate, and they are also spending money on incentives to drivers and users.
    - Part of where these companies are losing money is on advertising.
    - Investors believe that they will make money eventually
    - Uber/Lyft get 25% of the total cost of a ride ($0.50 of $2.00/mile).
    - The price could be much higher than the $2.00, as TNCs are currently losing money (subsidized by investors).
- There are many additional opportunities to subsidize MaaS (e.g., advertising, saving money on parking at residential or commercial buildings).
- Employers pay for transportation (shuttles to work). If companies can get additional benefits to employees at a lower cost, it will have a massive effect on the economics of transportation. The company will pick a specific TNC to provide these rides.
- What about a monthly subscription plan? If it is "all you can ride" for a single subscription cost (no marginal cost to users).
    - Secondary impacts—if it shifts car ownership, it could totally rewrite land use in cities.
    - Mass transportation run by the government may be dead (no one rides buses, trains in favor of shared MaaS).
    - But there may be many people may share the bulk of a commute, and the vehicle will be right-sized (e.g., 30-person bus for the shared portion of 30 commuters' commutes).
- There was some discussion as to whether parking fees would apply to automated MaaS vehicles (of if they would be more diminished).
- There is a difference in the value of time for different people, is that factored in?
    - It is an average for the chart, but there is a paper coming out that varied value of travelers' time.
    - Rural travelers may have a greater portion of mobility disadvantaged travelers, and the only way to afford transportation for those users may be automated MaaS.
    - There is a market for the mobility of those individuals, but it costs $60 a ride currently. In the future with automation, the cost may drop enough to enable a private service rather than an expensive paratransit system.
    - Such a system could also be used for rural deliveries (mobility for goods as well as people).
- There was some discussion on the need for congestion pricing or other methods to mitigate induced demand in dense areas (i.e., cities).
    - Depending on the design of the city, programs such as Waze already distribute traffic in a decentralized way (everyone wants the quickest route based on real-time information).
    - How much more travel can the system handle before congestion becomes an issue again?
    - Ride **sharing** is important to manage mitigation.
    - Big data can aid with strategic pairing up of users.
- Will people be willing to share? Is it like an elevator or sufficiently different?
    - In some places, people only share during rush hour.
    - Elevators do not take detours the way that shared rides do. They just add more stops, so it may not be comparable.

    - Could sharing a ride be positive? For example, will people ride around in vehicles hoping to meet other people?
- How will the rise of telecommuting affect congestion?
    - It will not be 100% of telecommuters, so the charts are still relevant for the travelling population.
- Differentiated vehicles (e.g., Kia vs Mercedes S Class) will cost different amounts.
- If there are already many users who use MaaS options today rather than owning their own vehicle, how will it look once those costs decline by 75% or more?
- Users of services may be more sensitive to costs because they are more obvious than vehicle ownership (where large upfront costs occur and variable costs are relatively low.
- We travel to improve the quality of life.

Alain Kornhauser, Princeton University

- **Thesis: Automated taxis are a great mobility opportunity for moderately dense living environments, but they have no supervision (owner or user) for several hours each day and will only have societal oversight. As a result, they will be the new canvas for artists or vandals.**
- Today public infrastructure, bikeshare bikes, and transit vehicles are already vandalized and covered in graffiti.
- This is one of the most common questions asked by people about small sidewalk delivery robots, but it has not been a problem yet. These systems can catch vandals on camera, sound an alarm, and remote monitors can communicate with them.
- What about people getting sick on the bus?
    - It can be monitored and sent in for cleaning. If you have the rider's credit card, they can be charged appropriately.
    - Other systems have monitoring equipment (e.g., cameras in elevators), but that does not always stop people from misbehaving.
    - It depends on the operator and fleet manager—how do they choose to operate the fleet? If they let it degrade and do not keep it clean, users will drop off.
- How can you ensure that people respect vehicles?
    - They need to feel invested in the system and that they benefit.
    - A challenge of having accountability is that it means that it cannot be anonymous. It may pose a challenge for those without credit cards or smartphones. Wealthy users may have some anonymous options (e.g., company that would bond for the user for a fee).
    - Can the design or decoration of a vehicle help make it locally relevant and help reduce issues?
    - Even shared mobility is an emotional undertaking. Even without ownership, people have an emotional relationship with vehicles.
    - If these vehicles are seen as a symbol of the rich, there will be pushback and vandalism. These vehicles must be for those who are less wealthy and less able to access mobility services.
    - If people are part of a "culture of sharing" then people can feel some sense of ownership and respect. In addition, there can be consequences added in for vandals.
- This technology may not be able to survive a vandalism attack.

- Some transit agencies have taken the "broken windows" strategy of repairing and repainting vehicles as soon as possible after they have been vandalized to demonstrate to would-be vandals that their efforts are futile.
- What about transit drivers, taxi drivers, and delivery drivers who lose jobs? Won't they have an incentive to vandalize vehicles?
    - The number of drivers taken out of a job may be somewhat limited given the need for workers on particularly difficult routes and environments.
- Huge investments to build out a new system can be completely lost (company pulls out) if there is too much vandalism.
    - The public may need to be more involved in "localizing" a solution rather than just replicating the system from other cities.
- How can users be made to leave vehicles if people decide to take up residence in the vehicle?
    - Will we need an autonomous taxi police? Will people be able to write a ticket?
- Today drivers fix many problems on buses—how will those non-driving tasks be handled in automated vehicles?
    - Companies are planning to build service centers for automated vehicles.
    - On-board cameras can monitor to see if a trip to the service center is needed.

Part II

Moderator: Sharon Feigon, Shared Use Mobility Center

Amitai Bin-Nun, SAFE (Securing America's Future Energy)

- **Thesis:** The reason we are having this conversation is because automated vehicles will be impactful (e.g., transforming the way we travel, how we work, what we do) and overall positive.
    - The benefit case is estimated to be on the order ~$800 billion/year by 2050.
    - The job loss case is on the order of $200 billion for the total cost related to job loss as a result of automated vehicles.
- Automated vehicles may lead to prosperity, or may be the flashpoint for culture wars (i.e., benefits may not be distributed evenly).
- Automated vehicles will likely follow the cycle of job creation seen in other technological advance—loses in some areas (drivers), but direct, supplier, and induced jobs in other areas (but these new jobs will not all go to those who lost jobs).
- Automated vehicles could result in a 0.13 percentage point increase in unemployment (peak).
- There are unfilled jobs currently, so automation would help alleviate that need.
    - The shortage is a reflection of low wages—if wages were higher, the shortage would disappear.
    - Everything that needs to get shipped eventually does get shipped, so there must be enough drivers to get items to where they need to go.
- There are three options:
    - Laissez-faire automated vehicle deployment without workforce investment.
    - Throttling automated vehicle deployment to reduce job loss costs.

- Redirect benefits of automation to mitigate the costs to unemployed workers (e.g., retraining programs).
- As an alternative to training programs, there was discussion of taxation or universal basic income (UBI) as a way to redistribute benefits and reduce costs to those who are hurt by automated vehicle deployment.
  - This could take the form of a job buyout.
  - Additional options are available—this conversation should be started early in order to put in pilot programs to have an evidence-based approach to the problem of displaced workers.
  - Maybe there is the need for an annuity from the benefits that can be used to support those who have lost income potential should other policy options fail.
- We have seen technological displacement repeatedly since even before the technological revolution. We have gotten through it so many times before, so why is the situation of automated vehicles different?
  - In previous examples, people who invent/develop technology benefit. Users may save money from lowered costs. If we are investing (policies and transfers) in assisting displaced (or to-be-displaced) workers, who is going to foot the bill for those changes?
  - Previous transitions have not always gone well, but that does not mean that we are locked into the same (lack of) actions that have been used previously.
  - There is a role for companies to anticipate needed skills in the future, communicate with workers those needs, and support them in transitioning to new roles (e.g., through training).
  - No policy is a "silver bullet" and there are roles for multiple stakeholders (i.e., both public and private).
- Automated vehicles are not the only change that will happen over this time period, but other artificial intelligence-driven changes will compound the issue.
- Geographically, the driver jobs are generally spread across the country, though some stated have more drivers than others.
- Truck drivers will be the last ones taken out of vehicles due to the risks of the larger format and the value of the cargo in the vehicle.
  - Railroad consolidation (Conrail in 1970s) bought out three of the five jobs on every train. Those people were then made whole through legislation.
- Assuming that truck automation does happen, the rate of attrition is high, and may be higher than the adoption of automated trucks.
- What about downstream effects (e.g., fewer road construction, as infrastructure can be more efficiently used by automated vehicles).
  - Internet enabled business models could not have been predicted, so there are boundaries on the analysis due to uncertainty in the future.
  - What about congestion disbenefits from automated vehicles? This analysis only considers benefits.
- The key takeaway is not the solution itself, but the tractability of the solution. Next steps are on getting more specific on the solution and these discussions are a good start.
- It is not about what to do specifically, but the point that the benefits are much larger than than the costs.
  - The job of a driver is not high among those kids care about.

- BUT – that might not be the right measure of whether or not a job is desirable.
- Especially given the cultural implications (socioeconomic demographics).

Utopian / Dystopian Discussion

- **Poll: The majority of the audience believes on balance automated vehicles will do more good from harm (utopia over dystopia). Remaining audience members who expressed a view suggested that they may have negligible impacts either way.**
- There was some discussion as to whether there would be energy implications of electrified automated vehicles:
    - Automated vehicles will be able to plug themselves in, so there could be chargers near substations where vehicles could stop, charge, and pull away to allow other vehicles to charge.
    - There may be more demand for energy, but solar panels are the cheapest way to add marginal energy to the grid.
- Automated single-occupancy vehicles may be more efficient than trains and buses.
    - The single occupancy vehicles can be build smaller and more efficient electric vehicles.
    - What about during rush hour when buses and trains are full? Yes, they may be better on that period, but systems are purchased for all hours, so the average occupancy is still a relevant metric.
    - There is no one vehicle type—vehicles can be right-sized for the trip (e.g., 1, 2, 3, 4 person vehicles.
- Road metering and congestion pricing has been successful. It may be possible to convince even wealthy users to pool as they value their time over all else.
- Will it be the free forces of the market that will get us there or will it be regulation?
    - It may be mostly the free market, but there could be a role for regulation.
    - In the United States, congestion charging has been resisted (it gives the impression that the roads are for the rich).
    - The idea of demand-side management could apply to travelers—people could volunteer to not use the roads (or use alternate routes) during high demand periods.
    - In a way, that would be dynamic congestion pricing.
    - In a fleet world, there is an incentive to have shared rides, because it uses the asset better and brings in more income.
    - When we move to very high percentage of TNC service, there will be so much information about people and where they go, competing service will want larger market share, so they will work to reduce costs and offer better (and more distinct/segmented) services. Competitive forces will wind up increasing the efficiency. People will be incentivized to pool rides and the system will have enough information to effectively pool people. The central dispatcher will assign the vehicle along with 100s of other vehicles, so it will be able to manage demand.
    - One option would be to charge per square foot second of road the vehicle takes up (a large bus is charged more, but it divides those costs among many riders, so it will still be cheaper on a per-person basis than a small car).

- The fear of loss of privacy will be a barrier to the deployment of automated vehicles, especially when the government is involved with a user fee, but if TNCs manage this action, riders may not care much.
- For elevators, the charge for the elevator is built into the rent of the building. Would the same model make sense for transportation along roadways? Maybe the property tax structure of properties changes.
- Maybe road pricing is old fashioned – maybe we end up with lifestyle-centric mobility taxes.
- There are no pick-up/drop-off locations on curb. Automated vehicles can "stand" and do not need to park to let people out of their cars.
  - In the south, some cities have taken parking spaces out and replaced them with pick-up/drop-off areas.
  - Geofencing is another option to limit certain areas for pick-up and drop-off activities.
  - The example of the Texas unsignalized intersection with reservations for cars to cross is an example of a high-technology solution that does not take into account practical needs (e.g., pedestrians, non-connected vehicles). Some of these solutions for curb reservation may be similarly impractical.

Suggested Action Items

No action items were explicitly discussed.

Shark Tank II at Automated Vehicles Symposium

Hilton, San Francisco, July 11, 2018

Cost of automated vehicle trips: How low can it go?

Lew Fulton, STEPS Director

ITS UCDAVIS INSTITUTE OF TRANSPORTATION STUDIES

www.steps.ucdavis.edu

Costs of Mobility – Exploration of Hedonic Costs

- Still trying to get a handle on monetary costs of different modes
  - Wide range of fixed and variable costs
  - ICE vs electric and automated vehicles
  - Differences by trip type and region
- But at the same time, we have reason to believe that non-monetary (hedonic/productivity) costs are as important or potentially more important.
  - Even harder to quantify
  - But let's try

UCDAVIS SUSTAINABLE TRANSPORTATION ENERGY PATHWAYS

2

The wide range of costs related to mobility choices

Out-of-pocket Costs

- Vehicle purchase
- Vehicle maintenance
- Fuel
- Insurance
- Cleaning
- Parking
- Driver
- MaaS fees
- Tolls
- Registration-related fees

Hedonic costs

- **Travel time (driving)**
- **Travel time (passenger)**
- **Parking search time**
- **Walking time**
- **Driving stress**
- **Shared trips (e.g. lack of privacy)**
- **EV range, charging anxiety**
- **Car ownership negatives (maintenance, registration, inspections etc.)**
- **Car ownership positives (car pride, guaranteed ride; can leave personal belongings in the car)**
- **Perceived Environmental Cost**

UCDAVIS
SUSTAINABLE TRANSPORTATION ENERGY PATHWAYS

Recap - monetary costs: Comparison of modes

- ***Driven TNC vehicles are premium service, automation makes these competitive***

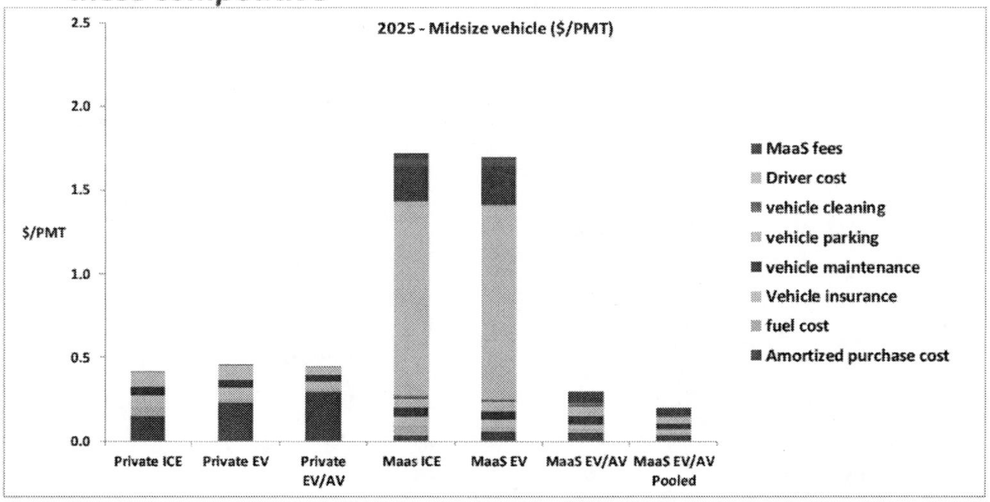

UCDAVIS
SUSTAINABLE TRANSPORTATION ENERGY PATHWAYS

We added a value of time for driving, travelling, parking

- ***Time costs are equal to or in some cases far greater than the out-of-pocket costs***
- ***Pooled mobility advantage disappears***

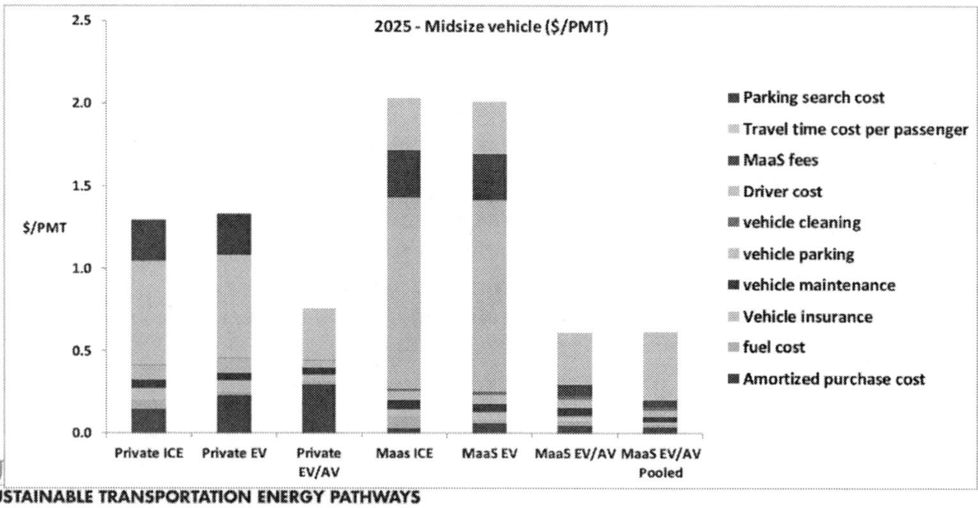

U
SUSTAINABLE TRANSPORTATION ENERGY PATHWAYS

Removed vehicle purchase, insurance costs, kept only variable costs (daily decision)

- Ignore private car purchase, insurance cost
- ***The AV/EV private car becomes cheaper than shared mobility options***

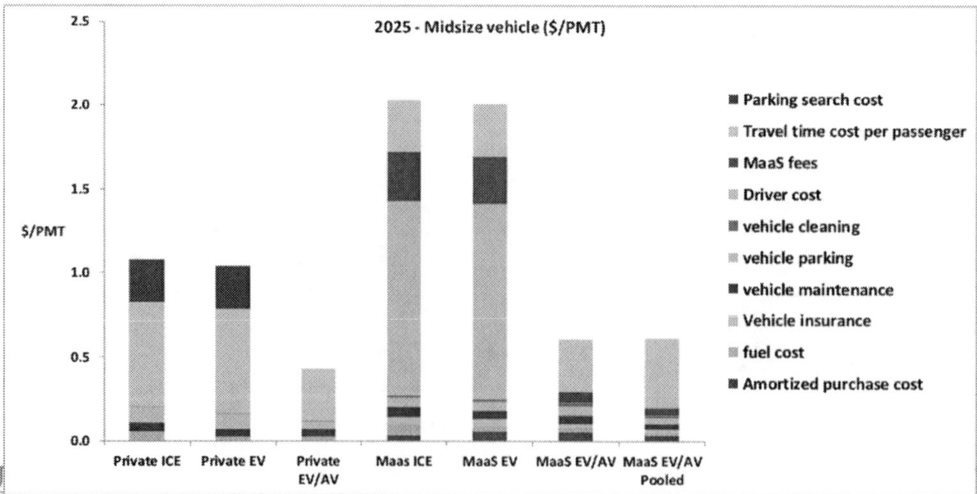

U
SUSTAINABLE TRANSPORTATION ENERGY PATHWAYS

Thank you

Lew Fulton

Director,
Sustainable Transportation Energy Pathways Program
ITS-Davis
lmfulton@ucdavis.edu

UCDAVIS
SUSTAINABLE TRANSPORTATION ENERGY PATHWAYS

7

AUTOMATED VEHICLES SYMPOSIUM 2018

Driven by: AUVSI & TRB

July 9–12, 2018
Hilton San Francisco Union Square | San Francisco

Breakout Session #28

What We Know: Evidence-based Behavioral Studies of Impacts of Automated Vehicle Systems

Session Goals and Objectives

- Forum to discuss issues and challenges in collecting data (evidence) about adoption, factors influencing adoption, and how the adoption might affect mobility and location choices.
  - To reduce uncertainty regarding the potential impacts of automated vehicles on people's activity-travel behavior.
  - To highlight empirical research on behavioral phenomena that could determine the impacts of AVs on activity-travel demand, vehicle ownership, and longer-term location choices in the future.
  - To examine the implications of such studies.

Agenda

- Short presentations of empirical research (Shiftan)
- Interactive discussion (Outwater)
  - What are the key findings from the various behavioral studies?
  - What are the gaps in knowledge to be addressed in future behavioral studies?
  - What are appropriate data and research methodologies to obtain better insight into behavioral processes in the world of AVs?
  - How can the knowledge gained from these behavioral studies be used to inform modeling exercises?
- Key takeaways and next steps (Kolodge and Lenz)

Agenda

- Short presentations of empirical research (Shiftan)
- Interactive discussion (Outwater)
  - What are the key findings from the various behavioral studies?
  - What are the gaps in knowledge to be addressed in future behavioral studies?
  - What are appropriate data and research methodologies to obtain better insight into behavioral processes in the world of AVs?
  - How can the knowledge gained from these behavioral studies be used to inform modeling exercises?
- Key takeaways and next steps (Kolodge and Lenz)

AUTOMATED VEHICLES SYMPOSIUM 2018

BREAKOUT SESSION

BREAKOUT SESSION # 28

What We Know: Evidence-based Behavioral Studies of Impacts of Automated Vehicle Systems

Wednesday, July 11, 1:30 PM – 5:30 PM

Room: Continental 1

Organizers

Johanna Zmud, Texas A&M Transportation Institute
Yoram Shiftan, Technion University
Chandra Bhat, University of Texas at Austin
Ram Pendyala, Arizona State University
Kristin Kolodge, JD Power
Maren Outwater, Resource Systems Group
Barbara Lenz, German Aerospace Center (DLR)
Don Mackenzie, University of Washington
Thomas Stephens, Argonne National Laboratory
Jeremy Raw, FHWA
Scott Smith, Volpe
Josh Peterman, Fehrs and Peers
Zhenhong Lin, Oak Ridge National Laboratory

Session Description

This breakout session aims to serve as a forum that brings together scholars, researchers, and industry professionals who have conducted surveys and collected data and evidence about how people may adopt automated vehicles, factors influencing adoption, and how the adoption of such vehicles might affect various mobility and location choices. The breakout session will feature a number of short presentations of behavioral research studies and key findings from the data gathered. Through this forum, the breakout session organizers envision creating a synthesis or annotated bibliography of behavioral research studies that will serve as a resource to the planning and modeling community.

Goals/Objectives/Outputs

- Begin to reduce uncertainty regarding the potential impacts of automated vehicles and other transformational transportation technologies on people's activity-travel behavior

- Highlight empirical research that sheds light on a number of behavioral phenomena that could determine the potential impacts of automated vehicles on activity-travel demand, vehicle ownership, and longer term location choices in the future
- Examine the implications of such studies

Agenda

1:30 PM – 2:00 PM Rationale and Goals for Session
Introduction to session and attendee self-introductions
Johanna Zmud, Texas A&M Transportation Institute

2:00 PM – 3:15 PM Short presentations of empirical research
5 minute presentations of research – gathered through a call for abstracts.
Yoram Shiftan, Technion

Presenters:

- Farzad Alemi, Georgia Institute of Technology
    - Studying the Adoption of an Autonomous Shuttle in a Pedestrian-Friendly Sustainably-Inspired Community: Before and After Pilot Study in Dubai
- Regina Clewlow, Populus, UC-Davis
    - Survey on Adoption, Utilization and Travel Behavior Impacts of Ride-Hailing Services
- Yoram Shiftan, Technion University
    - Mode Choice in the World of Fully Automated Services
- Stephen Greaves, University of Sydney
    - Driverless Vehicles Down Under: An Empirical Investigation of Consumer Sentiment
- Mustapha Harb, UC-Berkeley
    - Projecting Travelers into a World of Self-Driving Cars: A Naturalistic Experiment for Travel Behavior Implications
- Scott John Hardman, UC-Davis
    - Perceptions and Use Patterns of Autonomous Vehicles amongst Owners of Semi-Autonomous Vehicles
- Viktoriya Kolarova, German Aerospace Center
    - Empirical Evidences on the Impact of Autonomous Driving on the Value of Time Savings and Mode Choice
- Nikhil Menon, University of South Florida
    - Understanding the Influence of Consumers' Perceptions towards Automated Vehicles on their Intended Adoption: A Cluster Analysis
- David R. Keith, MIT Sloan School of Business
    - AVs, Shared Mobility, and the Cost of Driving: Will Lower Costs Reduce the Incentive to Pool?
- Baiba Pudane, Delft University
    - Daily Activity Schedules in the Automated Vehicle-era: A Focus Group Study
- Ipek Sener, Texas A&M Transportation Institute
    - Examining Future Automated Vehicle Usage: A Focus on the Role of Ride-haling
- Zia Wadud, University of Leeds
    - The Potential Use and Usefulness of Travel Time in Fully Automated Vehicles

3:15 PM – 3:45 PM Break

3:45 PM – 5:00 PM Interactive discussion
Following the series of short presentations, participants in the breakout session will engage in an interactive format to document answers to the following questions.

Facilitators: Chandra Bhat, UT-Austin and Maren Outwater, RSG
Note-takers: Jeremy Raw, FHWA and Scott Smith, Volpe

- What are the key findings from the various behavioral studies?
- What are the gaps in knowledge that remain to be addressed in future behavioral studies?
- What are the appropriate data and research methodologies to obtain better insight into behavioral processes in the world of automated vehicles?
- How can the knowledge gained from these behavioral studies be used to inform modeling exercises that aim to simulate the impacts of automated vehicles on future activity-travel demand and location?

5:00 PM – 5:30 PM Conclusion

Key takeaways and next steps
Kristin Kolodge, JD Power and Barbara Lenz, German Aerospace Center (DLR)

Autonomous Vehicles Down Under: An Empirical Investigation of Consumer Sentiment

Presented by Professor Stephen Greaves, University of Sydney

Paper prepared by Brett Smith, Stephen Greaves, Tony Arnold, Doina Olaru, Andrew Collins

THE UNIVERSITY OF SYDNEY

Background/Context

- Australia a relative laggard in the AV space with a generally-held view that consumers are sceptical and not yet ready to embrace this technology (KPMG 2018).
- However, recent evidence suggests this scepticism may not be heterogeneous, with around 2/3rds indicating a 'positive' attitude to AVs and 1/3rd indicating they would consider purchasing an AV if available in 2025 (forthcoming in Smith et al. 2018).
- Raises questions about who is likely to embrace AVs and equally who is not and how personal, situational/contextual characteristics and existing travel patterns and constraints might influence consumers' attitudes towards AVs.

Methodological Approach

- 15 minute online survey of 474 adults from Sydney and Perth drawn from an online panel.
- Attitudes and concerns towards AVs and sharing their own car and AVs.
- Demographics and personality traits (60-item Big Five).
- Cluster analysis based on attitudes and concerns to AVs with a MANOVA to uncover most significant attributes across the groups.
- *Choice task (IATBR paper)*

THE UNIVERSITY OF SYDNEY

THE UNIVERSITY OF WESTERN AUSTRALIA

Consent

Vehicle Sharing Survey

Participation Information Statement

This study is being undertaken by the University of Sydney and University of Western Australia to identify your views on vehicle-sharing schemes. A vehicle-sharing scheme is designed to facilitate the short-term rental of vehicles between you and other drivers for financial reward. Your responses will give us valuable information on views towards sharing vehicles and help inform future transport policy in Australia.

This is an independent research study conducted by The University of Sydney (Professor Stephen Greaves, Dr Andrew Collins) and The University of Western Australia (Dr Brett Smith and A/Professor Doina Olaru). Your responses are voluntarily and anonymous. If you wish to leave the survey you can close your browser at any time.

The survey will take 15 minutes. You will be compensated for your time.

Attitudes Towards Driverless Vehicles

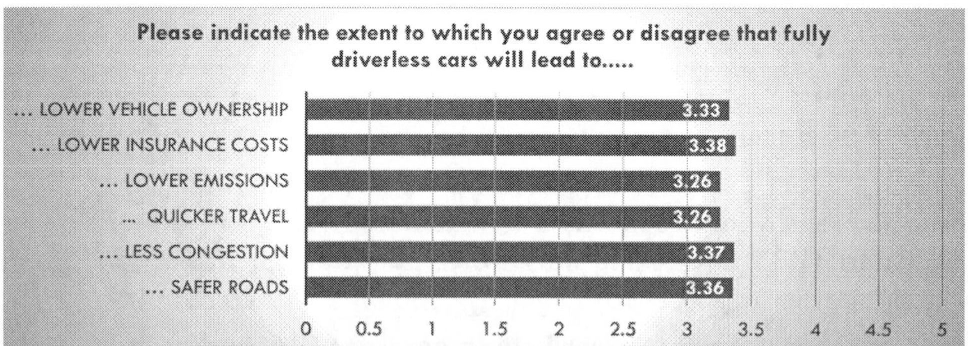

Overall, one-third indicated an interest in purchasing an AV if available in 2025.

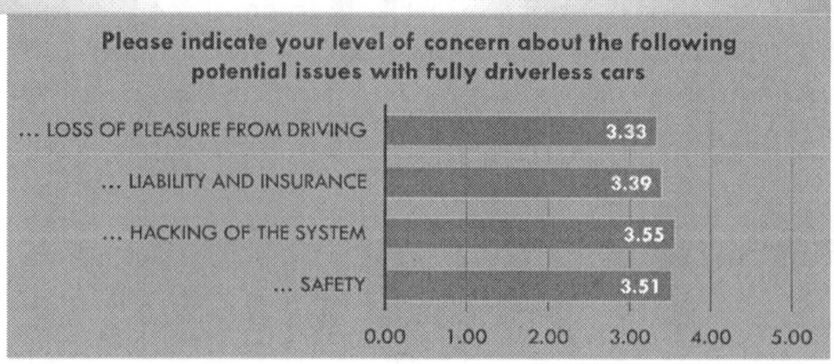

Attitudes Towards Car Sharing

Overall, 25% indicated an interest in sharing their car.

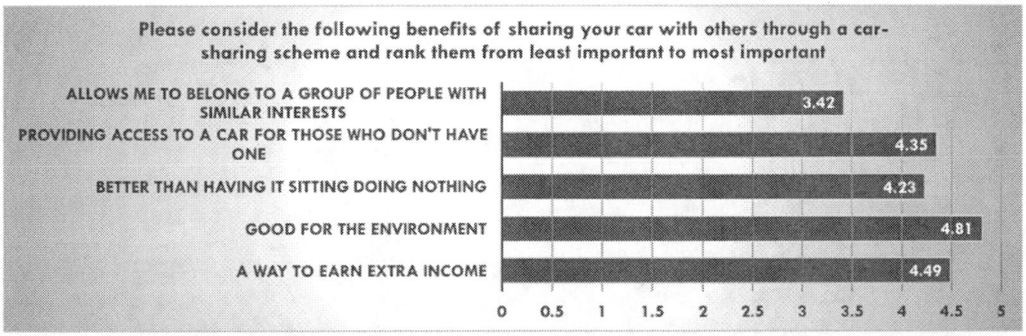

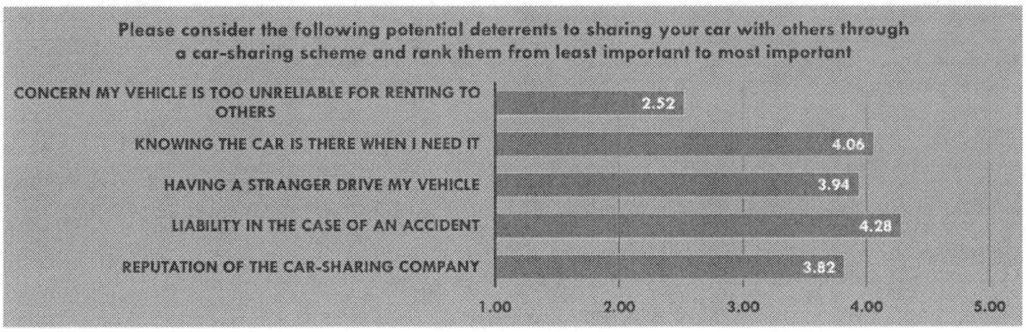

Personality Traits, WTS and Likelihood of AV Purchase

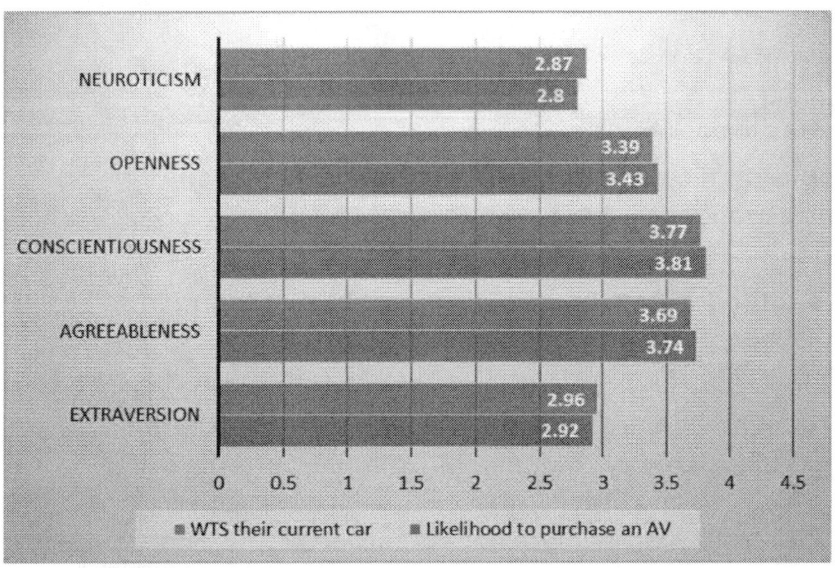

Cluster Results

Cluster Details	AV Optimists	AV Fence-sitters	AV Sceptics	p-value
Cluster Size	148	187	123	
Attitudes towards benefits of AVs	3.85	3.16	2.06	0.000
Concerns about AVs	3.02	3.13	4.36	0.000
Likelihood of Purchasing an AV in 2025	4	3	2	0.000
Profile				
Gender (% male)	58%	35%	42%	0.000
Age	44	42	49	0.000
Car Use				
Allow vehicle to be part of peer-to-peer sharing	2.72	2.36	1.83	0.000
No. of days drive to work	2.7	2.8	3.3	0.025
Human Values Max-Difference Scores				
Stimulation	-0.02	0.07	-0.05	0.028
Hedonism	0.07	0.15	0.06	0.037
Achievement	-0.30	-0.29	-0.43	0.014
Power	-0.47	-0.52	-0.55	0.115
Benevolence	0.39	0.49	0.53	0.001

Discussion & Conclusions

- Australian consumers are divided with roughly equal proportions of 'Optimists' and 'Sceptics' while the majority remain undecided about this technology.
- More favourable attitudes towards AVs are associated with younger, male respondents, those who drive less currently and those more open to enjoying life's pleasures.
- More negative attitudes prevail with older, female respondents, those who drive more, and less open to new experiences.
- The strong relationship between attitudes towards the benefits of AVs and likelihood to purchase (despite higher levels of concern), suggests it is critical to highlight both the benefits of AVs as well as focus on alleviating concerns, which has come under particular scrutiny after the Arizona fatality.

Projecting Travelers into a World of Self-Driving Vehicles: Estimating Travel Behavior Implications via a Naturalistic Experiment

Mustapha Harb – Yu Xiao – Giovanni Circella
Patricia Mokhtarian – Joan Walker

ITSBerkeley Institute of Transportation Studies
ITS UCDAVIS INSTITUTE OF TRANSPORTATION STUDIES
Georgia Tech

1

Motivation:

- A future with fully self-driving vehicles is fast approaching.
- Governments are racing to develop the necessary policies.
- How will fully autonomous vehicles impact travel and activity behavior?

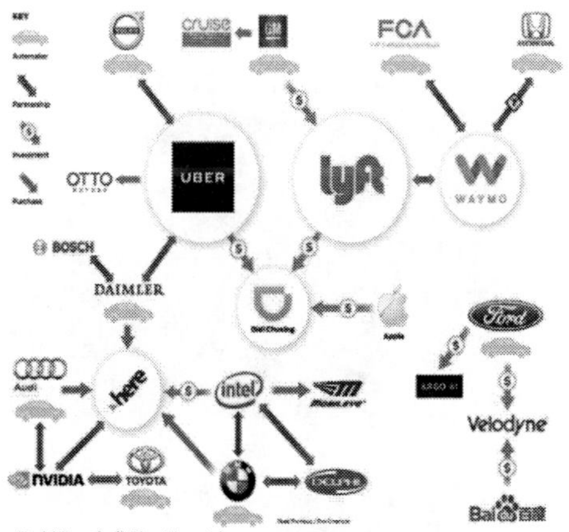

Todd Trumbull, The Chronicle

ITSBerkeley Institute of Transportation Studies
ITS UCDAVIS INSTITUTE OF TRANSPORTATION STUDIES
Georgia Tech

2

Predicting Travel Implications:

- Rich transport behavior literature:
  - Pricing, multitasking, parking, sharing vehicles and rides, mode choice, safety, etc...
- Approaches:

Driving simulators and controlled testbeds

US DOE, Volpe center (August 19, 2014)

Stated preference surveys

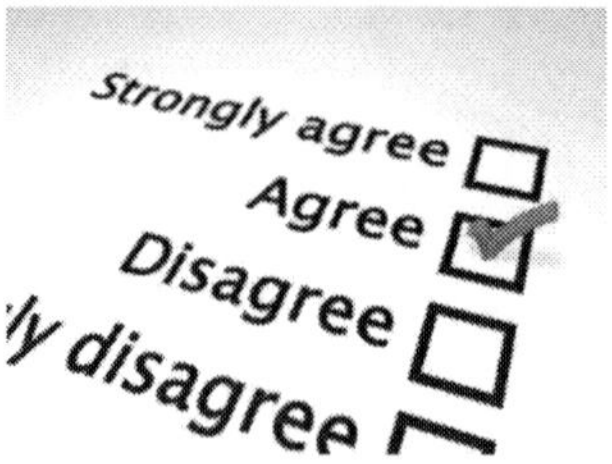

The Apopka voice (July 26, 2016)

Microsimulations and agent based model scenarios

ISI Foundation & ISI Global Science Foundation (April 20, 2016)

3

Naturalistic Experiment:

- FUTURE OF INTEREST: a fully autonomous vehicle

Don't have to drive the car ✓
Full multitasking ✓
No parking worries ✓
Can send on errands ✓

SCENARIO
- *Status quo infrastructure and policies.*
- *Private vehicle ownership.*

- SIMULATION OF FUTURE: a personal driver

- *Subjects use their own household vehicles.*
- *We provide the chauffeur to the household for free.*
- *Chauffeur stays with the car whether in use or not.*

ITSBerkeley Institute of Transportation Studies | ITS UCDAVIS INSTITUTE OF TRANSPORTATION STUDIES | Georgia Tech

4

Naturalistic Experiment:

Recruitment & Onboarding → Travel Week 1: Status quo → Travel Week 2: 60 hours of chauffeur service → Travel Week 3: Status quo → Offboarding

Recruitment & Onboarding	Travel Weeks 1–3	Offboarding
Entrance Online survey	All travel of subjects and vehicles tracked via mobile phone app, vehicle tracking GPS, and subject and chauffeur reports	Exit Online survey

qualtrics

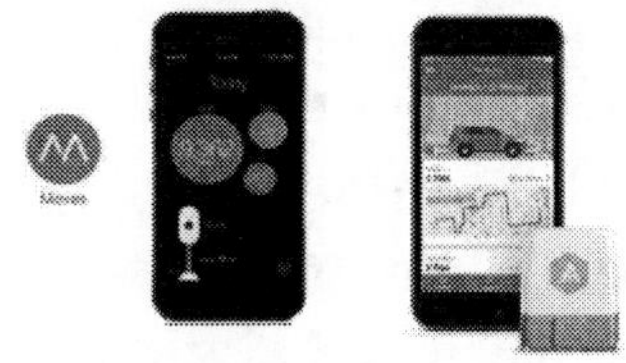

qualtrics

ITSBerkeley Institute of Transportation Studies   ITS UCDAVIS INSTITUTE OF TRANSPORTATION STUDIES   Georgia Tech

5

Summer 2017 Beta Test:

Dense experiment → run a beta test on **13 subjects** before running the real experiment

Five Retirees:

- Free time
- Safety concerns
- Travel restrictions (distance and time of day)

Four Millennials:

- Tech savvy
- Uber/Lyft users
- Active lifestyle

Four Families:

- Parents' schedule is tied to kids' schedule and vice versa

ITSBerkeley Institute of Transportation Studies   ITS UCDAVIS INSTITUTE OF TRANSPORTATION STUDIES   Georgia Tech

6

Beta Test Results – Summary:

(Harb et al., 2017)

- More auto travel
  - 83% increase in VMT
  - 21% of increased VMT were ghost trips
- Change in activity patterns
  - 58% increase in # of vehicle trips
  - 91% increase in # of longer trips (over 20 miles)
  - 88% increase in # of evening trips (after 6 pm)
- Impact on miles walked was not directional
  - 30% decreased (-31% on average), 70% increased (+38% on average)
- Virtually no biking, transit, TNC use in the sample

ITS Berkeley Institute of Transportation Studies | ITS UC DAVIS INSTITUTE OF TRANSPORTATION STUDIES | Georgia Tech

7

Potential Limitations:

Resource limitations ($$$)

- Sample size
- Relatively short treatment/adjustment time (1 week)
- 60 hour chauffeur limit
- Novelty effect

Context limitations

- Private ownership model, not shared
- "Automation" is provided for free

Technology limitations

- Human driver, not robot

8

Quotes From Exit Interviews/Surveys:

- "With all the limitations of the experiment, I definitely felt the benefits of a self-driving car. I noticed that I reach work less tired, I noticed that I can do work on my way back home and not worry much about traffic jams, and I noticed that my commute overall feels more pleasant." - *Millennial*
- "I spend a lot of time in the car driving my kids around to activities. Having a self-driving car would enable me to spend more time on work and would afford my kids more freedom." - *Family*
- "At my age, I am looking forward to the independence a self-driving car will provide as my driving skills decline. I believe self-driving cars will improve safety in driving, a real boon." - *Retiree*

9

Moving Forward:

Fine-tuning of the methodology:

- Issues with chauffeurs
- Tweaking of survey questions
- More diverse sample (e.g. income level, age, neighborhood type, minors in household)

Data collection (Fall 2018): 100 subjects from Sacramento region

10

Thank you for listening! Questions?

11

Detailed VMT Changes

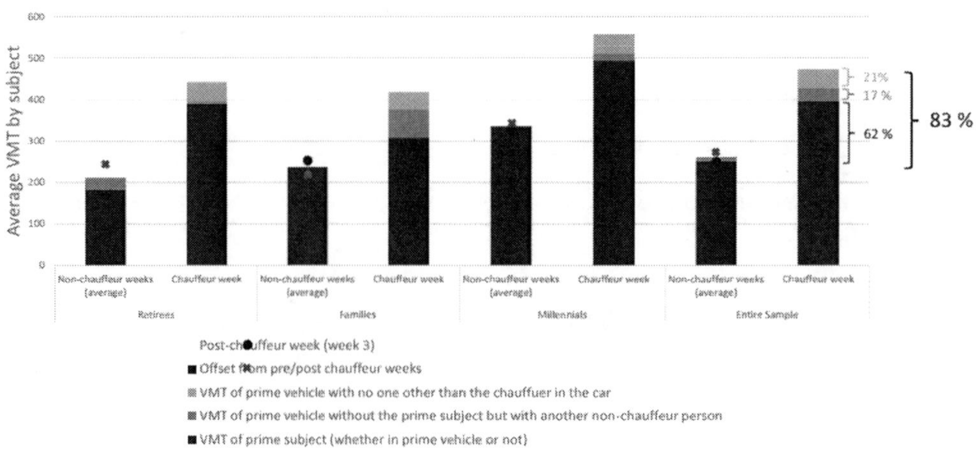

ITSBerkeley Institute of Transportation Studies | ITS UCDAVIS INSTITUTE OF TRANSPORTATION STUDIES | Georgia Tech

12

ITS UCDAVIS INSTITUTE OF TRANSPORTATION STUDIES

3 REVOLUTIONS SHARED·AUTOMATED·ELECTRIC

Testing the Deployment of an Autonomous Shuttle in a Sustainability-Inspired Community in Dubai: The Sustainable City

Farzad Alemi[1]*, Giovanni Circella[1,2], Patricia Mokhtarian[2], Karim El-Jisr[3], Tim Rogmans[3]

[1]University of California, Davis (USA)
[2]Georgia Institute of Technology (USA)
[3]The Sustainable City, Dubai (UAE)

*corresponding author: falemi@ucdavis.edu

AV Symposium
San Francisco, July 2018

The Sustainable City (TSC)

- A 46-hectare zero-net energy development, located in "Dubai Land"
- Mixed-use development:
  - 500 villas
  - 89 apartments
  - Offices
  - Retail space and restaurants
  - Nursery, hotel (coming soon...)
- Home of more than 2,700 residents, comprising high-educated, professional expats
- Built by Diamond Developers (DD) and inspired by UC Davis West Village

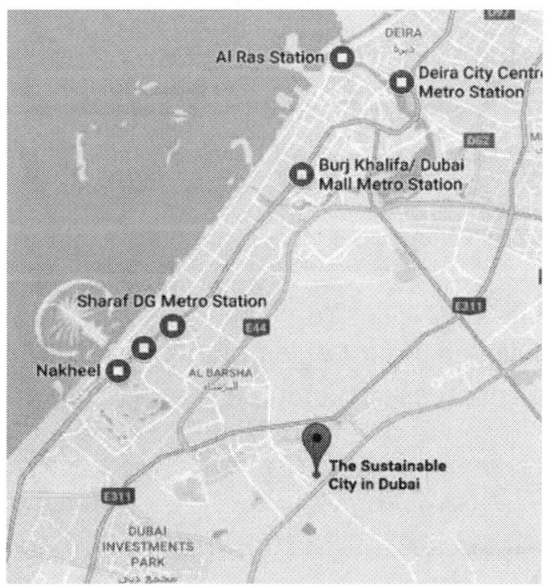

2

Autonomous Shuttle Deployment in TSC

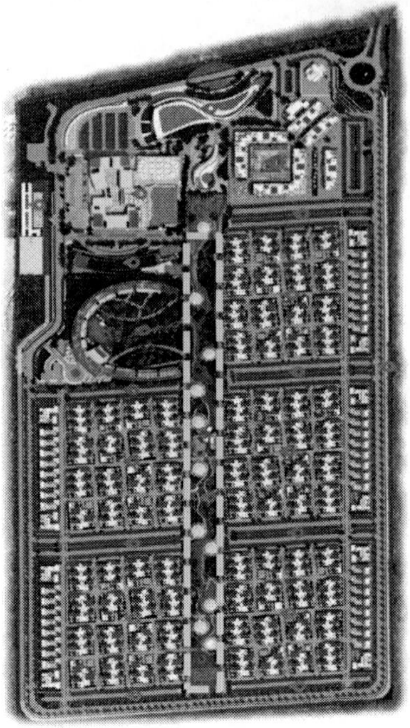

- 25% of the total transportation in Dubai expected to switch to autonomous modes by 2030 (*source*: Dubai RTA).
- In January 2018, the Dubai Road and Transport Authority (Dubai RTA) and TSC signed an MOU to allow the *operation of an autonomous shuttle* inside TSC.

3

Research Design

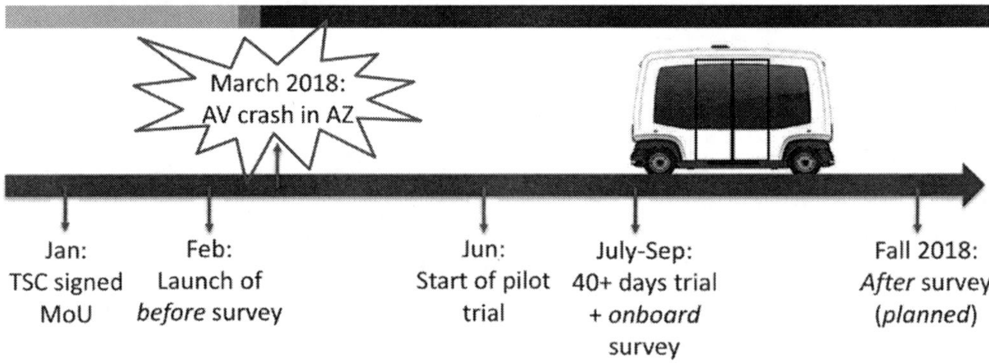

Online *before* survey:

- Designed as part of larger mobility study in TSC
- Distributed via community news and email listserv

Survey content:

1. *Attitudes and Preferences*
2. *Lifestyle*
3. *Current Travel Behavior*
4. *Future Transportation (carsharing, self-driving cars, driverless shuttle)*
5. *Sociodemographic traits*

4

Sociodemographics (*Before* Survey)

Distribution of Key Sociodemographic Traits (N≈108)

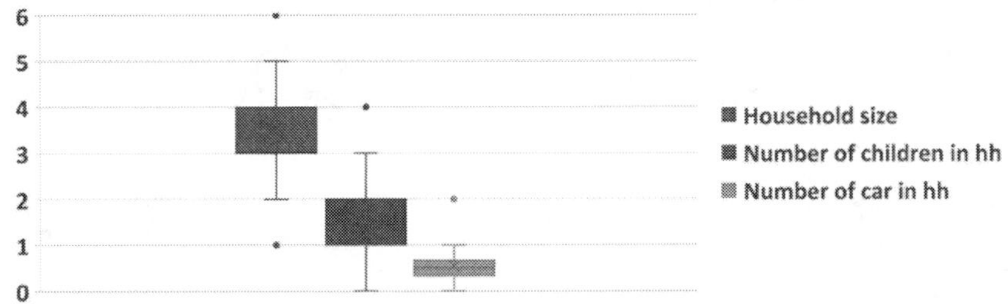

Household Income (Dhs)

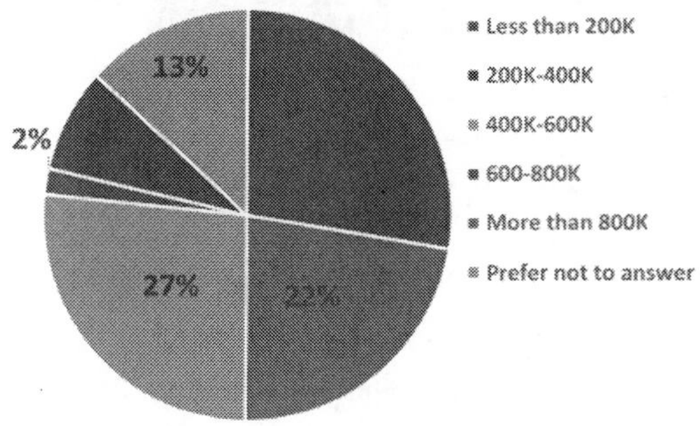

Individual Education

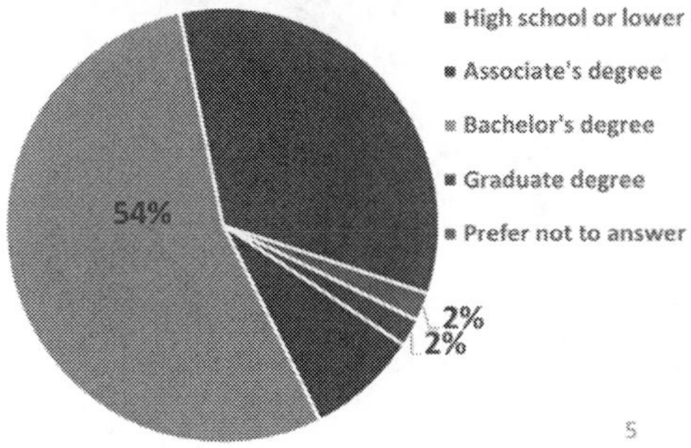

5

Preliminary Results (*Before* Survey)

How Do Residents Travel Inside the Sustainable City?

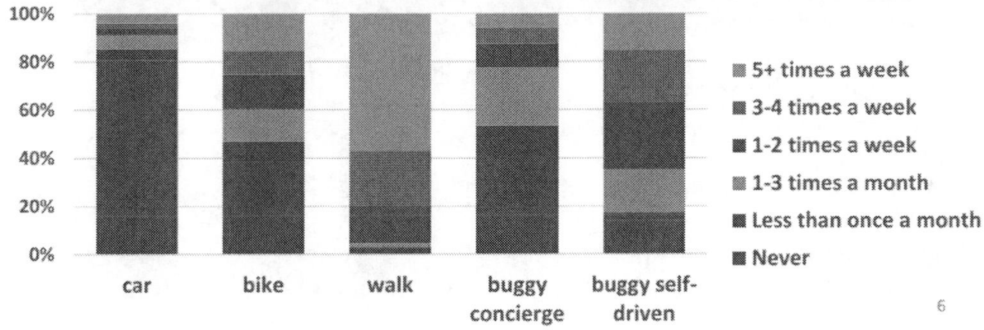

6

Preliminary Results (*Before* Study)-2

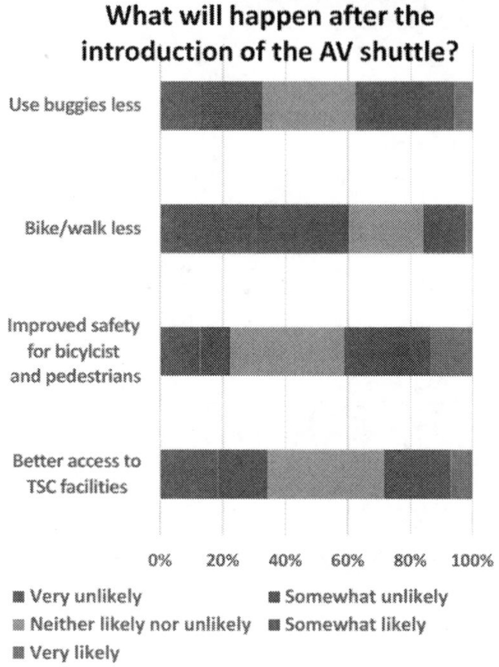

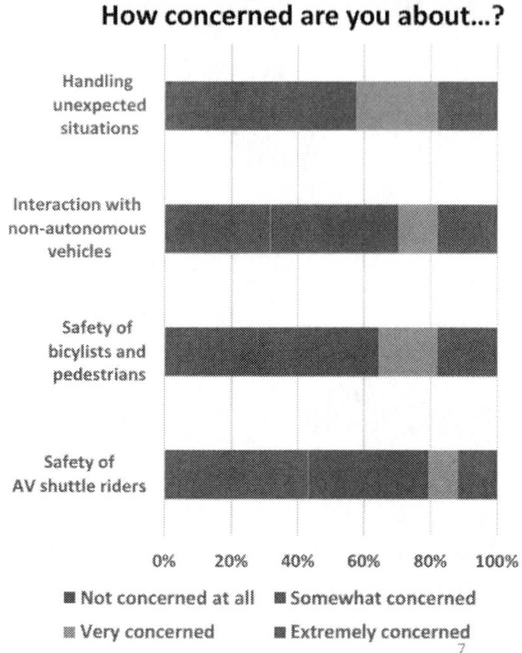

7

Challenges and Lessons Learned

- Modification of Infrastructure
  - New signage
  - Road marking and barriers
  - Change the circulation of traffic
- Backlash from TSC Residents
  - Safety concern
  - Loss of convenience (two-way → one-way paths; right turn only)
  - Reduced speed for EV buggies
  - Attracting visitors from outside
- Operational Challenges
  - Tailing of buggies sharing the road with AV shuttle
  - Weather and battery life
  - Waiting time
  - High sensitivity of sensors (foliage, birds, etc.) and emergency braking

8

Thank you for your attention!

Farzad Alemi, Ph.D.
Institute of Transportation Studies, University of California, Davis
falemi@ucdavis.edu

AUTOMATED VEHICLES SYMPOSIUM 2018
USERS. VEHICLES. INFRASTRUCTURE.

Driven by AUVSI and TRB

Breakout Session 28 :What We Know: Evidence-based Behavioral Studies of Impacts of Automated Vehicle Systems

Organizers :

Johanna Zmud, Texas A&M Transportation Institute
Chandra Bhat, University of Texas at Austin
Kristin Kolodge, JD Power
Barbara Lenz, German Aerospace Center (DLR)
Thomas Stephens, Argonne National Laboratory
Scott Smith, Volpe
Zhenhong Lin, Oak Ridge National Laboratory
Yoram Shiftan, Technion University
Ram Pendyala, Arizona State University
Maren Outwater, Resource Systems Group
Don Mackenzie, University of Washington
Jeremy Raw, FHWA
Josh Peterman, Fehrs and Peers

Presentations

- Farzad Alemi, Georgia Institute of Technology
  - Studying the Adoption of an Autonomous Shuttle in a Pedestrian-Friendly Sustainably-Inspired Community: Before and After Pilot Study in Dubai
- Regina Clewlow, Populus, UC-Davis
  - Survey on Adoption, Utilization and Travel Behavior Impacts of Ride-Hailing Services
- Yoram Shiftan, Technion University
  - The impact of multi-tasking on Value of Time
- Stephen Greaves, University of Sydney
  - Driverless Vehicles Down Under: An Empirical Investigation of Consumer Sentiment
- Mustapha Harb, UC-Berkeley
  - Projecting Travelers into a World of Self-Driving Cars: A Naturalistic Experiment for Travel Behavior Implications
- Scott John Hardman, UC-Davis
  - Perceptions and Use Patterns of Autonomous Vehicles amongst Owners of Semi-Autonomous Vehicles
- Viktoriya Kolarova, German Aerospace Center
  - Empirical Evidences on the Impact of Autonomous Driving on the Value of Time Savings and Mode Choice
- Nikhil Menon, University of South Florida
  - Understanding the Influence of Consumers' Perceptions towards Automated Vehicles on their Intended Adoption: A Cluster Analysis
- David R. Keith, MIT Sloan School of Business
  - AVs, Shared Mobility, and the Cost of Driving: Will Lower Costs Reduce the Incentive to Pool?
- Baiba Pudane, Delft University
  - Daily Activity Schedules in the Automated Vehicle-era: A Focus Group Study
- Ipek Sener, Texas A&M Transportation Institute
  - Examining Future Automated Vehicle Usage: A Focus on the Role of Ride-haling
- Zia Wadud, University of Leeds
  - The Potential Use and Usefulness of Travel Time in Fully Automated Vehicles

Lively Discussion

- What are the key findings from these various behavioral studies?
- What are the gaps in knowledge?
- Research methodologies that are appropriate in this context?
- How can evidence be used to inform models?

- Key takeaways and next steps
  - Kristin Kolodge, JD Power and Barbara Lenz, German Aerospace Center (DLR)

Key Takeaways

- Third year, conversation has progressed
- Presentations demonstrate the amount of work out there

Key findings from all presentations

- Missing consistency among demographic factors
- Caution against future research
- Working towards representative sample size, current research is targeted small sample sizes

Gaps

- Creative new ideas methodologies are needed
  - Chauffeur

Data and research methodologies

- Which data do we use to put into the models
- Which models are still valid – push and pull from that respect
- How does the introduction of new technologies affect use of time and how relate to mobility models

Next Week

- **IATBR – International Association of Travel Behavior Research, Santa Barbara**
- **Workshop on Automation and Self-Driving**

Understanding the Influence of Consumers' Perceptions towards Automated Vehicles on their Intended Adoption: A Cluster Analysis Approach

Nikhil Menon, Yu Zhang, Abdul Pinjari & Fred Mannering

Presentation at the Automated Vehicle Symposium 2018

AUTOMATED VEHICLES SYMPOSIUM 2018

INTRODUCTION

- Web-based stated-preference surveys conducted in 2015
  - University sample (USF Tampa) + non-university sample (members of AAA south) → n=2477
  - Elicited info on consumers' perceptions, their intended adoption, and anticipated impacts of AVs (and SAVs) on travel behavior
- Consumers' perceptions about AVs influence their adoption decisions

Two-step cluster analysis

Autonomous vehicle consumer market segments

2

TWO-STEP CLUSTER ANALYSIS

Description of Autonomous Vehicles Perception Variables	Benefits-Dominated Cluster (N=513)	Uncertain Cluster (N=732)	Well-Informed Cluster (N=811)	Concerns-Dominated Cluster (N=602)
Fewer traffic crashes and increased roadway safety	**4.65**	3.08	**4.14**	2.47
Less stressful driving experience	**4.62**	2.89	**4.21**	2.27
Less traffic congestion	**4.18**	2.46	3.35	1.89
More productive (than driving) use of travel time	**4.57**	2.97	**4.24**	2.57
Increased fuel efficiency	**4.21**	3.07	3.85	2.69
Safety of the vehicle occupants and other road users such as pedestrians, bicyclists	2.35	3.43	**4.26**	**4.43**
System/equipment failure or AV system hacking	2.77	3.48	**4.4**	**4.73**
Performance in (or response to) unexpected traffic situations, poor weather conditions	2.82	3.49	**4.44**	**4.64**
Difficulty in determining who is liable in the event of a crash	2.46	3.15	3.63	**4.52**
Privacy risks from data tracking on my travel locations and speed	2.68	3.07	3.67	**4.59**
Loss in human driving skill over time	2.46	3.34	3.45	**4.49**
Likelihood of adopting autonomous vehicles when they become available in the market	4.24	2.64	3.30	1.74

Two-step cluster analysis

Autonomous vehicle consumer market segments

- Benefits-dominated
- Concerns-dominated
- Uncertain
- Well-Informed

Multinomial logit model

Probability of respondents belonging to a particular market segment

MARKET SEGMENT PROBABILITIES

- Multinomial Logit Model
  - determines characteristics that make respondents more/less likely {one of the market segments}

$$MS_{in} = \beta_n X_{in} + \varepsilon_{in},$$

  - error term → generalized extreme value distributed
  - extreme value Type 1 – most common EV distribution – chosen for computational convenience – similar to normal dist. (Washington et al., 2011)

$$P_n(i) = \frac{EXP[\boldsymbol{\beta}_i \boldsymbol{X}_{in}]}{\sum_{\forall I} EXP(\boldsymbol{\beta}_I \boldsymbol{X}_{In})}$$

  - estimation of parameters by maximum-likelihood
- Likelihood ratio tests
  - male vs female → $X^2 = -2[LL(\boldsymbol{\beta}_{total}) - LL(\boldsymbol{\beta}_{male}) - LL(\boldsymbol{\beta}_{female})]$ ×
  - university vs non-university → $X^2 = -2[LL(\boldsymbol{\beta}_{total}) - LL(\boldsymbol{\beta}_{uni}) - LL(\boldsymbol{\beta}_{non\text{-}uni})]$ ×

GENERAL MAKE UP OF AV CLUSTERS

BENEFITS-DOMINATED SEGMENT

Parameter	Effect
Male	↑
University Respondents	↑
Commute Distance (20+ mi one-way)	↑
Daily Travel Time (45 mins or less)	↑
Parking Time (10+ mins)	↑

WELL-INFORMED SEGMENT

Parameter	Effect
Millennial GG	↑
Commute time (60+ mins one-way)	↑

LEGEND

Parameter	Effect
Positive Fixed Parameter	↑
Negative Fixed Parameter	↓

UNCERTAIN SEGMENT

Parameter	Effect
Generation-X Respondents GG	↓
Licensed driver (3+ drivers)	↓
Non-Commuter	↑

CONCERNS-DOMINATED SEGMENT

Parameter	Effect
Baby Boomer GG	↑
Household Income ($150,000+)	↓
Graduate	↓
Drive Alone	↓
Major Injury Severity	↓
Vehicle Ownership (3+ vehicles)	↑
New Vehicle Purchase (Own/Lease)	↓

Two-step cluster analysis

Autonomous vehicle consumer market segments

- Benefits-dominated
- Concerns-dominated
- Uncertain
- Well-Informed

Multinomial logit model

Probability of respondents belonging to a particular market segment → make up

Ordered probit (with random parameter) model

Factors influencing adoption of autonomous vehicles

CUTR USF Automated Vehicle Symposium 2018 7

AV ADOPTION LIKELIHOODS

- Ordered probability modeling approach (from *extremely unlikely* to *extremely likely*)
  - +**β** →increase in **X**, increases the probability of *extremely likely*, decreases the probability of *extremely unlikely;* interior categories – average marginal effects

- Unobserved heterogeneity → explanatory variables varying across observations → normally distributed random parameters – simulated MLE – 1000 Halton draws

- Likelihood ratio tests
  - **cluster 1 vs cluster 2 vs cluster 3 vs cluster 4 →** $-2[LL(\boldsymbol{\beta}_{total}) - LL(\boldsymbol{\beta}_{cluster1}) - LL(\boldsymbol{\beta}_{cluster2}) - LL(\boldsymbol{\beta}_{cluster3}) - LL(\boldsymbol{\beta}_{cluster4})]$ ✓
  - **university vs non-university models →** $X^2 = -2[LL(\boldsymbol{\beta}_{total}) - LL(\boldsymbol{\beta}_{uni}) - LL(\boldsymbol{\beta}_{non-uni})]$ ✗

CUTR USF Automated Vehicle Symposium 2018 8

DIFFERENCES IN ADOPTION BEHAVIOR

BENEFITS-DOMINATED SEGMENT

Parameter	Effect
Millennials	⇧
Baby Boomers	⇧
Great Generation	⬇
Hispanic/Black	⇧
Household Income ($100,000+)	⬆
Non-Commuter	⇧
Commute Distance (< 5 miles)	⇩
Commute time (45+ minutes)	⇧
Parking Time (< 5 minutes)	⇩
Vehicle Ownership (> 1 vehicle)	⬇
New Vehicle Purchase (Own/Lease)	⬆
Crash Involvement	⬆

Gender and belonging to the university population had no influence in the adoption decisions for this market segment

LEGEND

Parameter	Effect
Positive Random Parameter	⇧
Positive Fixed Parameter	⬆
Negative Random Parameter	⇩
Negative Fixed Parameter	⬇

DIFFERENCES IN ADOPTION BEHAVIOR

Current vehicle ownership has no influence in the adoption decisions for this market segment

CONCERNS-DOMINATED SEGMENT

Parameter	Effect
Male	⬇
University Respondent	⬆
Great Generation	⬆
White	⬇
Household Income (< $50,000)	⬇
Daily Travel Time (90+ minutes)	⬇
Parking Time (< 5 minutes)	⬇
Used Vehicle Purchase (Own/Lease)	⬇

LEGEND

Parameter	Effect
Positive Random Parameter	⇧
Positive Fixed Parameter	⬆
Negative Random Parameter	⇩
Negative Fixed Parameter	⬇

DIFFERENCES IN ADOPTION BEHAVIOR

All respondent demographics except household income insignificant in the adoption decisions for this market segment

UNCERTAIN SEGMENT

Parameter	Effect
Household Income (< $50,000)	⬇
Drive Alone	⬆
Commute Distance (< 5 miles)	⬇
Daily Travel Time (< 30 minutes)	⬆
Vehicle Ownership (3+ vehicles)	⬇

LEGEND

Parameter	Effect
Positive Random Parameter	⇧
Positive Fixed Parameter	⬆
Negative Random Parameter	⇩
Negative Fixed Parameter	⬇

DIFFERENCES IN ADOPTION BEHAVIOR

WELL-INFORMED SEGMENT

Parameter	Effect
Male	⇧
University Respondent	⬇
Baby Boomers	⬇
Great Generation	⬇
White	⇩
Household Income ($100,000+)	⇧
Two-person Household	⇩
Commute time (45+ minutes)	⇧
Vehicle Ownership (Zero Vehicles)	⇧
New Vehicle Purchase (Own/Lease)	⬆

Most current travel characteristics except longer commute times and parking insignificant in the adoption decisions for this market segment

LEGEND

Parameter	Effect
Positive Random Parameter	⇧
Positive Fixed Parameter	⬆
Negative Random Parameter	⇩
Negative Fixed Parameter	⬇

THANK YOU

Speaker

Nikhil Menon

Postdoctoral Research Scholar
Center for Urban Transportation Research (CUTR)
University of South Florida
4202 E. Fowler Ave., CUT 100, Tampa, FL, 33620-5375
Tel: (813) 974-1832, E-mail: nikhilmenon@cutr.usf.edu
Web: http://nikhilmenon.in

CUTR USF

13

Empirical evidenced on the impact of autonomous driving on value of travel time savings and mode choices

Viktoriya Kolarova
German Aerospace Center

DLR

Estimation of VTTS for autonomous driving

Viktoriya Kolarova, Felix Steck, Francisco Bahamonde-Birke

Focus of the first study:

- How might autonomous driving affect VTTS?

Methodology:

- Online survey; N ~ 500
- Commuting, shopping, leisure trips; long-distance trips
- 2 Stated Choice experiments (8 decision situations each)
- Mixed Logit Model has been estimated (to explore heterogeneity)

Survey design:

1. Reference trip: ____ km
2. SC Exp. 1: Currenltly available modes of transport
3. Introduction of AD (privat vs. shared AV) in short videos

4. SC Exp. 2: Future available modes of transport

Please mark which of the following transportation modes would you choose.

	Option 1	Option 2	Option 3	Option 4	Option 5
Mode of transport	Walk	Bicycle	Public transport	Private AV	Driverless taxi (SAV)
Trip duration	1 h 09 min	22 min	15 min	18 min	16 min
Access / egress time		2 min	5 min		
Waiting time			10 min	2 min	5 min
Ridesharing					no
Costs			2,25 €	1,05 €	2,15 €
	Availible time: no Total trip time: 1 h 09 min	Availible time: no Total trip time: 24 min	Availible time: 10 min Total trip time: 30 min	Availible time: up to 11 min Total trip time: 18 min	Availible time: 11 min Total trip time: 21 min
	○	○	○	○	○

DLR

Estimation of VTTS for autonomous driving

Viktoriya Kolarova, Felix Steck, Francisco Bahamonde-Birke

Main findings:

- **VTTS reduction of 34%** when driving autonomously vs. driving manually, however, only for commuting trips
- In-vehicle time in an **AV is perceived similar as using public transport** (similar results for long-distance trips and train)

- Riding **a privately owned AV** in automated mode is **more attractive than using a driverless taxi**
- **No differences** were found **for using the vehicle alone vs. sharing the ride**

DLR

Outlook: Explaining VTTS for autonomous driving

Viktoriya Kolarova, Advisors: Prof. B. Lenz, Prof. E. Cherchi

Focus of the second study:

- How can the changes in VTTS be explained? (focus on psychological factors)

Methodology:

- A conceptual model (basis: behavior theories, theoretical considerations on positive utility of travel, empirical results from studies on acceptance of AD, and results from FGD
- SC experiment *(build upon the first study)* + measurement of psychological constructs
- Hybrid choice model

Status:

- Pre-test: July/August 2018
- Online survey: Autumn 2018

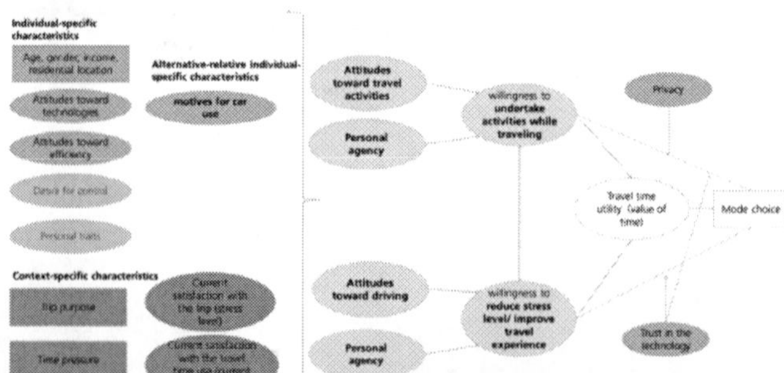

DLR

Related works

- Trommer, S., Kolarova, V., Fraedrich, E., Kröger, L., Kickhöfer, B., Kuhnihof, T., Lenz, B., Phleps, P. (2016). Autonomous driving – The impact of Vehicle Automation on Mobility Behavior. Final Report. Ifmo – institute of mobility research.

- Steck, F., Kolarova, V., Bahamonde-Birke, F., Trommer, S., and Lenz, B. (2018). How Autonomous Driving May Affect the Value of Travel Time Savings for Commuting. In: Transportation Research Record: Journal of the Transport Research Board.

- Cyganski, R., Heinrichs, M., von Schmidt, A., and Krajzewicz, D. (2018). Simulation of automated transport offers for the city of Brunswick. In: Procedia Computer Science 130: 872-879.

- Kolarova, V., Steck, F., Bahamonde-Birke, F. Assessing the effect of autonomous driving on value of travel time savings: a comparison between current and future preferences. *(forthcomming)*

DLR

UNIVERSITY OF LEEDS

The Use and Usefulness of Travel Time in Fully Automated Vehicles

Zia Wadud

Institute for Transport Studies

With Fuad Y Huda

AVS 2018, San Francisco

Any *revealed evidence* on activities in cars *now*?

UNIVERSITY OF LEEDS

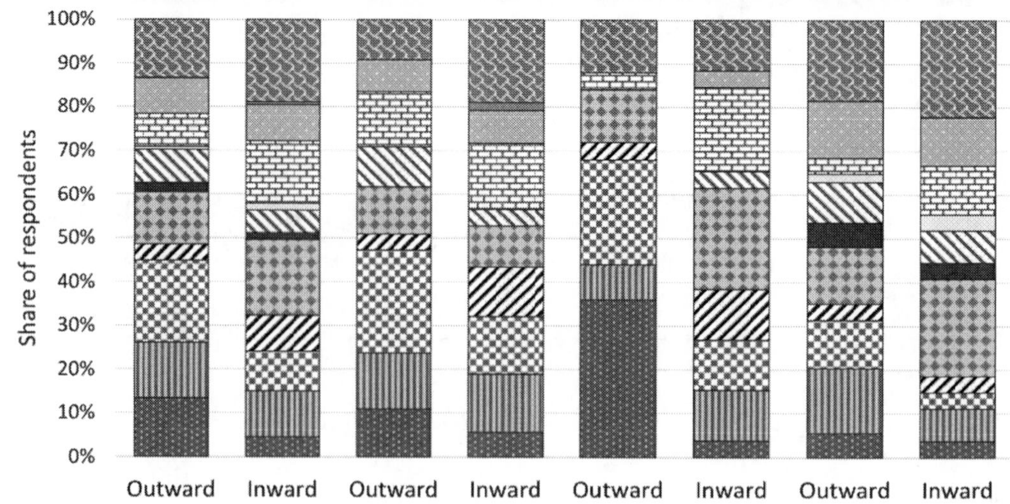

Respondents use chauffeur driven cars in Bangladesh

Type of activities depend on trip purpose & direction

What are the *Intended* activities in *FAVs*?

UNIVERSITY OF LEEDS

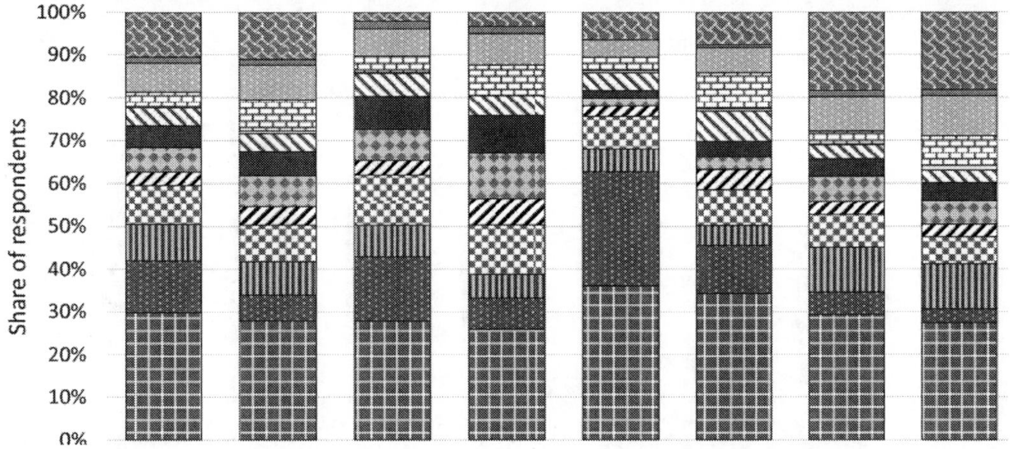

Type of activities depend on trip purpose & direction

Good correlation between revealed activities in chauffeured cars and intended activities in future FAVs

Similarity with literature on travel time use in trains

Will time be *more useful* in FAVs?

UNIVERSITY OF LEEDS

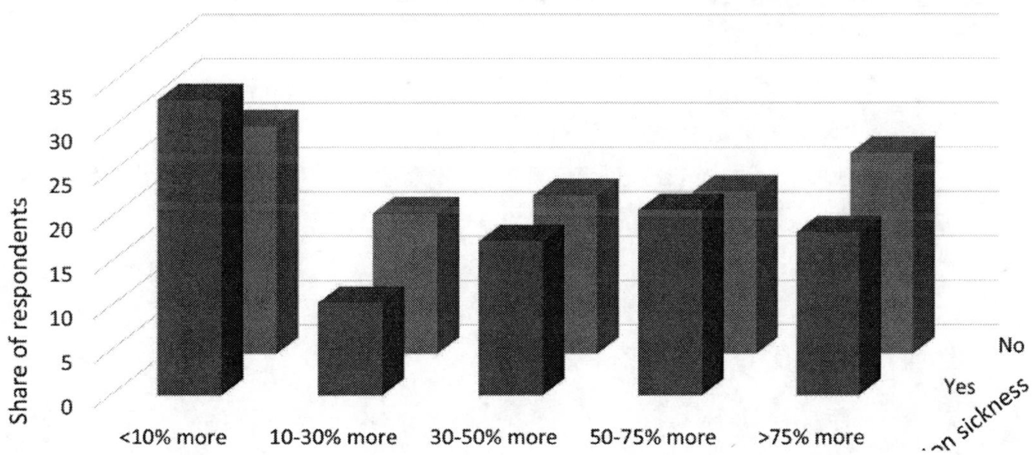

Respondents believe that their time to be more useful in FAVs

Motion sickness does not massively change the usefulness of time in FAVs

-- but activities change

Does *interest to use* FAVs depend on usefulness of time?

UNIVERSITY OF LEEDS

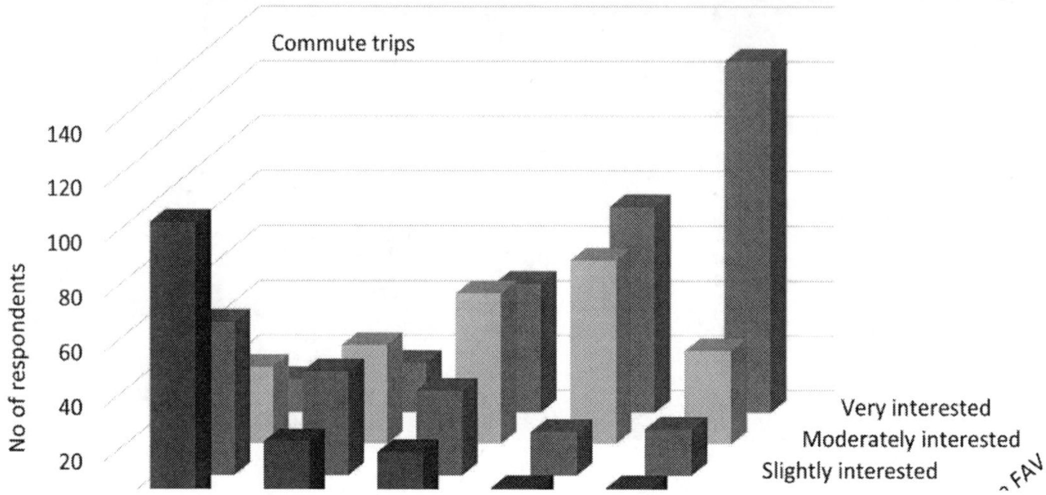

Strong correlation between usefulness of time and interest to use FAV

-- except for women and parents

UNIVERSITY OF LEEDS

Thank you

Z.Wadud@leeds.ac.uk

The use and usefulness of travel time in fully automated vehicles, 2018 (review)

Fully automated vehicles: A cost of ownership analysis to inform early adoption, 2017

Help or Hindrance? The travel, energy and carbon impacts of highly automated vehicles, 2016

Daily Activity Schedules in the Automated Vehicle-era:
A Focus Group Study

For Automated Vehicle Symposium, July 2018, San Francisco

Baiba Pudāne
Michał Rataj
Eric Molin
Niek Mouter
Sander van Cranenburgh
Hadi Asghari
Caspar Chorus

Spatial and Transport Impacts of Automated Driving

TU Delft

1

Intuition:
AVs may cause re-arrangements in activity schedules

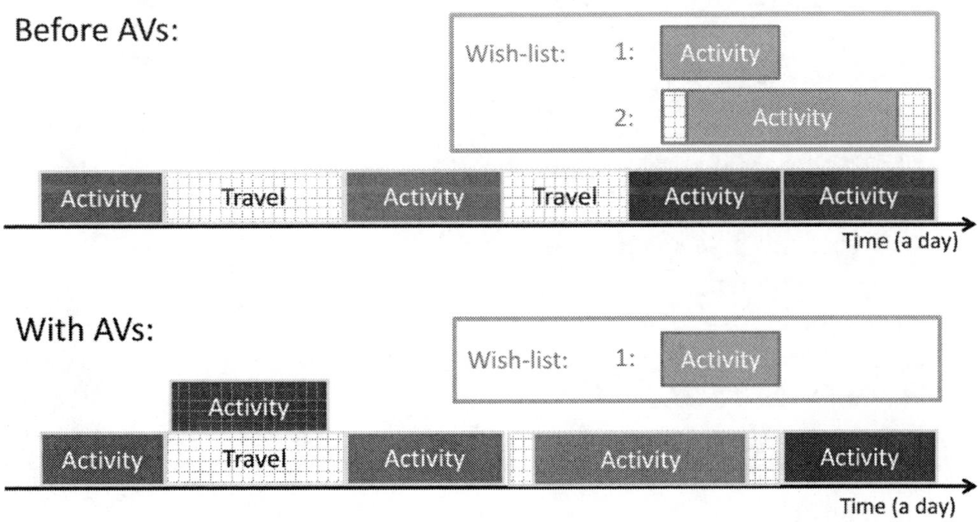

TU Delft

2

Why is that important?

State of practice for travel behaviour modelling with AVs:

Automated vehicles

↓ More pleasant travel | ↓ New feasible on-board activities

↓

Lower VoT

↓

More travel

Many activity re-arrangements do not fit in this picture:

- Activity 'transfers'
- Saved time
- Potential of less travel

↓

Solution:
Activity based modelling / Time-use models

Current work:
Discuss in focus groups activities during travel and their influence on daily activity schedules

TUDelft

3

Finding 1/3: reflecting on travel in AV, pleasure from travelling ≠ feasibility of activities

Characteristics of travel in AV	Influence of the characteristics on ...	
	pleasure from travelling in AV	feasibility of on-board activities
a) Fully automated driving	+/–	+
b) Availability, little planning needed	+	0
c) Travel continuity	+	+
d) Comfort	+	+
e) Equipment, storage possibilities	0	+
f) Privacy, isolation	+/–	+
g) Predictability, reliability of travel time	+/–	+
h) Longitudinal and lateral movement, position of the traveller	–	–

TUDelft

4

Finding 2/3: activities during travel: feasible ≠ desired and performed

Time-inflexible activities *Time pressure*

Saved time available for re-arrangements; most re-arrangements could be expected

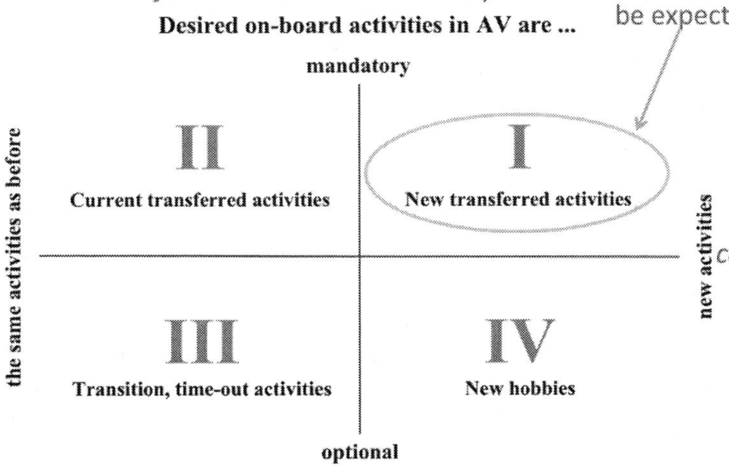

Short trips

Useful anti-activity

Activities ideally compatible with AVs

TUDelft

5

Finding 3/3: impacts of feasible activities on time pressure

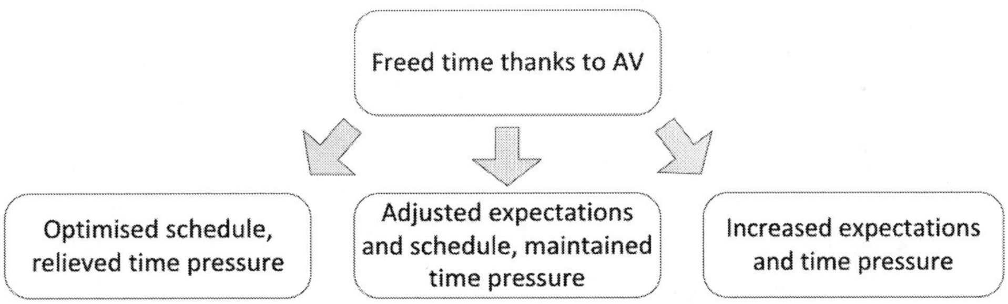

Saved time for more important things, more efficient and relaxed

Disappearing time savings

Expected to work during travel

TUDelft

Thank you!

6

Additional information

TU Delft

7

Focus group study:
investigate potential activity re-arrangements

When: Sep-Nov 2017

How many: 5 groups, 27 participants in total

Who: TU Delft students and employees (1 group),
Dutch working population who commute with car or PT (4 groups)

Scope: fully automated vehicles, mostly privately used. 100% safe, secure, available, customisable

Core questions:

1. Imagine travelling in an AV. What are the **pros and cons** compared to your current travel?
2. What would you like to **do while travelling** and why?
3. Would you **transfer to AV activities** that you currently perform at home / work / ...?
4. Would you change anything in your **daily habits**?
5. Would you travel **further / more often** to perform activities? Would you like to **move**?

TU Delft

8

MIT MANAGEMENT SLOAN SCHOOL

AVs, Shared Mobility, and the Cost of Driving: Will Lower Costs Reduce the Incentive to Pool?

David Keith and Sergey Naumov
System Dynamics Group
MIT Sloan School of Management

Automated Vehicles Symposium, San Francisco, July 2018

Research Questions

- Many believe that the attractiveness of AVs will induce additional demand for driving, worsening congestion and environmental impacts *all else being equal*.
- A countermeasure gaining increasing attention is pooling (e.g. Sperling 2017), serving the same number of passenger miles with fewer vehicle miles.
- However, consumer adoption of pooling today is modest.

Question 1: What are consumer preferences for the attributes of pooled mobility services?

- A key attraction of pooling for consumers today is lower trip prices (relative to private trips).
- However, it is anticipated that the cost of driving in ridesharing fleets may fall from about $2/mile today to $0.25/mile or less with AVs (Fulton, 2018).

Question 2: Will lower driving costs reduce consumer use of pooling?

Experimental Design

- Sample: 300 respondents from Amazon MTurk
- Random assignment into 2 experimental conditions:
  - Using a ride-hailing service to travel to the airport
  - Using a ride-hailing service to go shopping
- Respondents were asked to choose between:
  - A trip in a private vehicle, a service we called MyRider (similar to UberX or Lyft); or
  - A trip in a pooled vehicle, a service we called RiderPool (similar to UberPOOL or Lyft Line).
- Attributes/Levels:
  - Pick-up Time: 2, 5, and 8 min
  - Travel Time: MyRider: 10, 15, and 20 min, RiderPool: 8-12,12-18, and 16-24 min
  - Trip Price: $3, $5.25, $7.50, $10, $15, $20
- Post-Experiment Survey: Demographics, Vehicle Ownership, Ride-hailing Usage.

Results

	Utility coefficients			
	(1)	(2)	(3)	(4)
RiderPool:(intercept)	−0.796*** (0.063)	−0.838*** (0.263)	−0.788*** (0.137)	−0.692*** (0.165)
PickUp Time	−0.050*** (0.014)	−0.050*** (0.014)	−0.051*** (0.014)	−0.051*** (0.014)
Ride Price	−0.240*** (0.011)	−0.240*** (0.011)	−0.244*** (0.011)	−0.244*** (0.011)
Expected Travel Time	−0.071*** (0.008)	−0.073*** (0.012)	−0.072*** (0.009)	−0.072*** (0.009)
RiderPool: Maximum Travel Time		0.002 (0.014)		
RiderPool: Age > 35 years			0.206** (0.105)	0.212** (0.106)
RiderPool: Female			0.111 (0.103)	0.105 (0.104)
RiderPool: Income ('000s)			−0.007*** (0.002)	−0.007*** (0.002)
RiderPool: Bachelor's or Graduate Degree			0.320*** (0.106)	0.319*** (0.106)
RiderPool: Party Democrat			0.207* (0.113)	0.194* (0.114)
RiderPool: Party Republican			−0.126 (0.146)	−0.119 (0.147)
RiderPool: Party Other			0.135 (0.342)	0.123 (0.342)
RiderPool: 1+ Children in Household				0.024 (0.106)
RiderPool: # of Vehicles per Adult in Household				−0.165 (0.111)
RiderPool: Rushing to Destination				0.048 (0.098)
Observations	2,480	2,480	2,480	2,480
R^2	0.240	0.240	0.249	0.250
Log Likelihood	−1,292.352	−1,292.338	−1,275.676	−1,274.484
LR Test	814.455*** (df = 4)	814.482*** (df = 5)	847.806*** (df = 11)	850.190*** (df = 14)

Note: *p<0.1; **p<0.05; ***p<0.01

Simulation of the Falling Cost of Driving

Utility Attribute	MyRider	RiderPool
Initial Trip Price	$11.25	$6.75
Pick Up Time	3 min	7 min
Travel Time	16 min	16-24 min

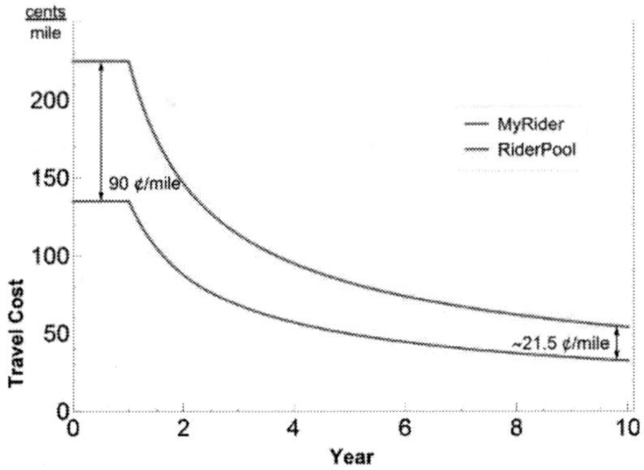

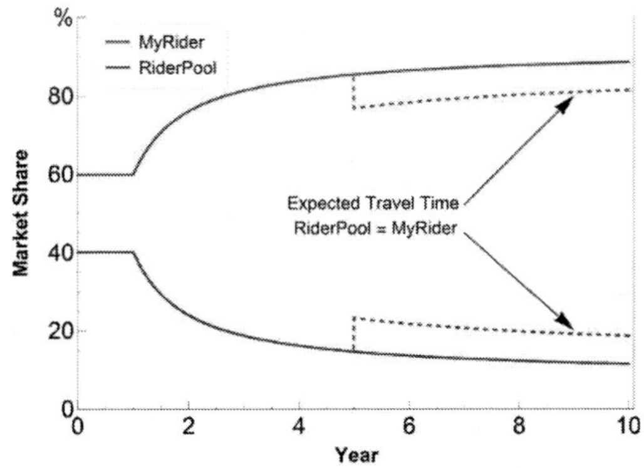

Discussion

- As expected, we find that:
  - Consumers prefer private rides, controlling for ride attributes & demographics
  - Consumers prefer cheaper rides
- Therefore, if the cost of driving falls with AVs, people are likely to pool less
- Widespread adoption of pooling may require policies such as HOV lanes or a VMT tax to overcome the falling unattractiveness of pooling.

- Improvements in our next experiment include:
  - Increase in sample size
  - Representative sample of US population
  - Strengthen nudge invoking time pressure
  - Introduce further variation in travel time uncertainty for pooled rides

Thank You

MIT MANAGEMENT SLOAN SCHOOL

- Questions?
- Contact us:
  - David Keith dkeith@mit.edu
  - Sergey Naumov snaumov@mit.edu

Examining Future Automated Vehicle Usage: A Focus on the Role of Ride-Hailing

Presented by

Ipek Nese Sener, Ph.D.

AUTOMATED VEHICLES SYMPOSIUM 2018

Breakout Session 28:
What We Know: Evidence-based Behavioral Studies of Impacts of Automated Vehicle Systems

July 11, 2018
San Francisco

Texas A&M Transportation Institute

Measuring Intent to Use

- Future acceptance and use are highly uncertain.
  - New technology, no experience, preferences change.

Track & Monitor

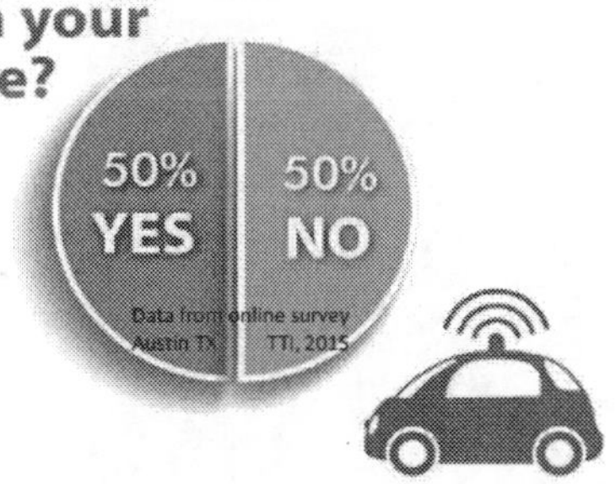

TTI's prior research:

- Austin – 2015 survey (Sponsor: TTI PRC)
- Houston, Dallas, Waco – 2016 survey (Sponsor: TxDOT)

Texas A&M Transportation Institute

Intent to Use & Technology Adoption

- Technology adoption was a very significant factor on intent to use self-driving vehicles.
  (Zmud et al. 2016, Sener et al. 2017)

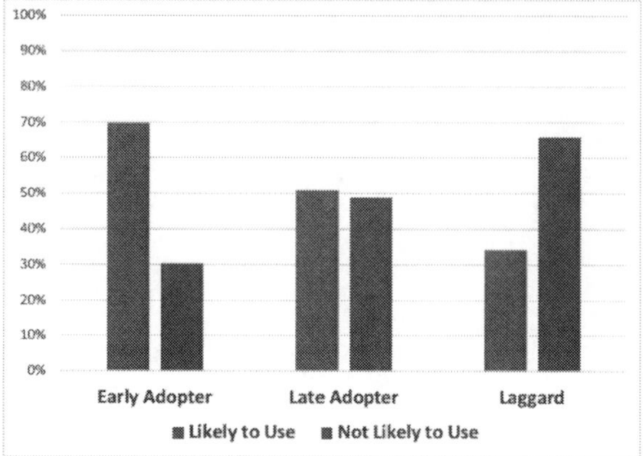

- Would persons using innovative mobility solutions be early adopters of automated vehicles?

Texas A&M Transportation Institute

The Role of Ride-Hailing
"Intent to Use: 2017 survey"

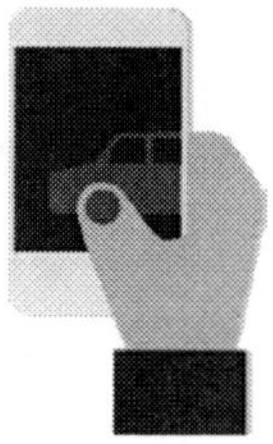

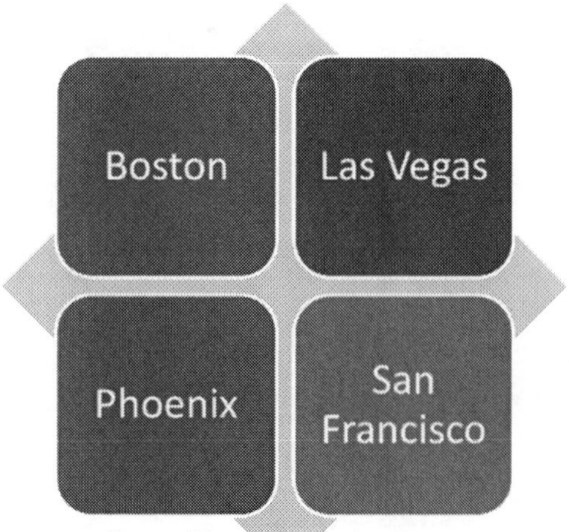

Four cities already testing.

- Target sample:
  - Non-ride-hailing user
  - New ride-hailing user
  - Long-term ride-hailing user

- Focus on three platforms:
  - Privately owned vehicle
  - Car-sharing service
  - Ride-hailing service

Texas A&M Transportation Institute

Intent to Use Self-Driving Vehicles

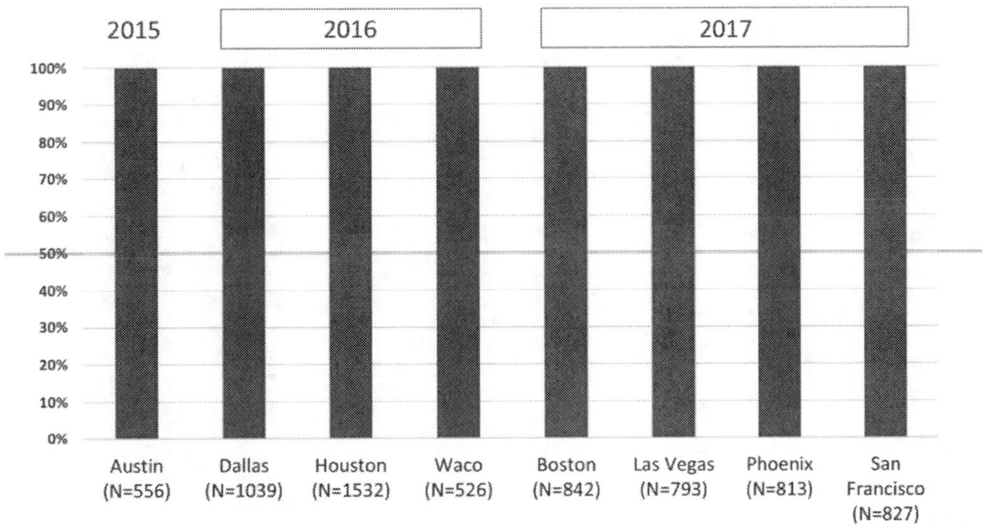

Texas A&M Transportation Institute

Intent to Use by Ride-Hailing Use

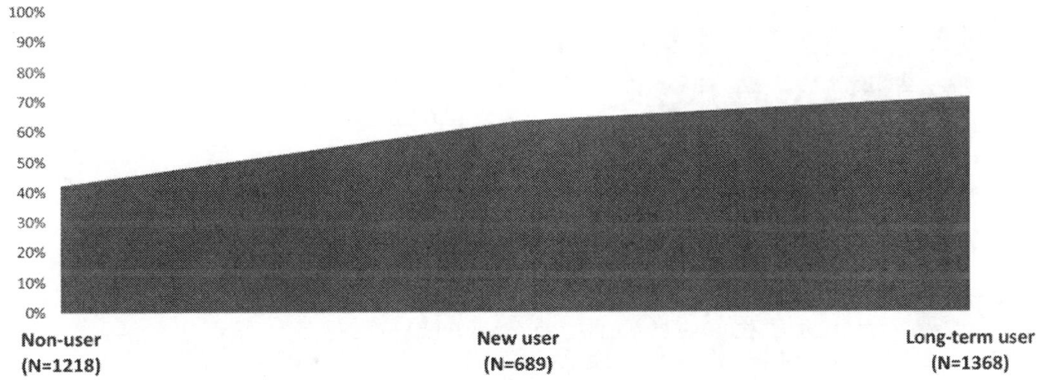

Ride-hailing users were more likely to use self-driving vehicles than non-users.

Acceptance/likely usage of self-driving vehicles increased with ride-hailing experience.

Texas A&M Transportation Institute

Models of Intent to Use

- Three distinct models of likely use across different types.
- Several variables were influential.
  - Individual and household demographics
  - Residential characteristics
  - Technology and shared mobility use factors
  - Attitudes and perceptions

Variables \ Self-driving vehicle type		POV	CSS	RHS
Age				
Female				
Unable to drive (medically)				
Household income >= $100,000				
No motorized vehicle in household				
Neighborhood density				
Base: Boston	Phoenix			
	Las Vegas			
	San Francisco			
Base: Non-ride-hailing user	New user			
	Long-term user			
Car-share membership				
Frequent Internet shopping				
Not concerned with data privacy				
Very concerned with data privacy				
Late/last to adopt technology				
AVs- Fun & interesting				
AVs- Not safe				

POV=Privately owned vehicle
CSS= Car-sharing service
RHS=Ride-hailing service

Positive impact | Negative impact | No impact

Texas A&M Transportation Institute

EXAMINING FUTURE AUTOMATED VEHICLE USAGE
A FOCUS ON THE ROLE OF RIDE HAILING

- Sponsor:
  - (2017) Lyft
- Full Report
  - http://tti.tamu.edu/documents/TTI-2018-2.pdf
- Study Brief
  - http://tti.tamu.edu/documents/TTI-2018-2-brief.pdf

- TTI's (other) relevant research:
  - https://trid.trb.org/view/1496491
  - https://www.sciencedirect.com/science/article/pii/S2352146517305884
  - https://trrjournalonline.trb.org/doi/abs/10.3141/2565-07
- Sponsors:
  - (2015) TTI Policy Research Center
  - (2016) Texas Department of Transportation
  - (2018) AAA Foundation for Traffic Safety --- ongoing
- TTI Contact:
  - Ipek N. Sener, i-sener@tti.tamu.edu
  - Johanna Zmud, j-zmud@tti.tamu.edu

Texas A&M Transportation Institute

REGINA CLEWLOW, PH.D.
CEO & CO-FOUNDER

POPULUS

www.populus.ai

MOBILITY SERVICES HAVE RAPIDLY EVOLVED IN CITIES

POPULUS
www.populus.ai

LEGACY TRANSPORTATION PLAYERS ARE IN THE DARK

COMPREHENSIVE, REPRESENTATIVE DATA FROM MAJOR U.S. METROPOLITAN AREAS

ATTITUDES | DEMOGRAPHICS | VEHICLE OWNERSHIP | MOBILITY SERVICES | BEHAVIOR CHANGE

POPULUS
www.populus.ai

MODE SUBSTITUTION FOR UBER AND LYFT

What people would do if Uber or Lyft didn't exist? (Based on representative data from 7 U.S. cities from 2016)

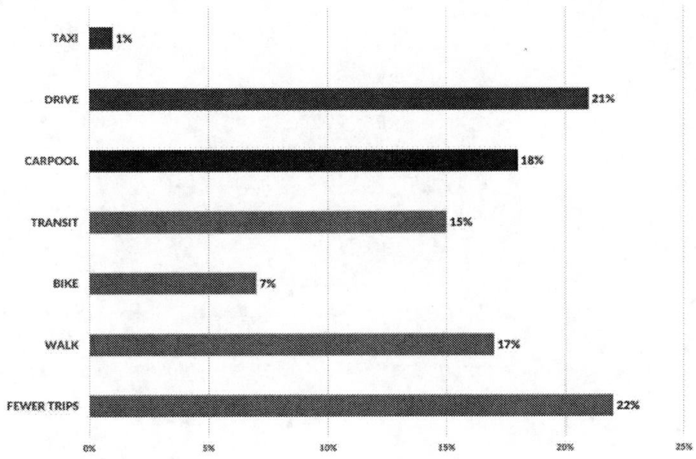

(Source: Populus, adapted from Clewlow & Mishra, 2017)

POPULUS
www.populus.ai

NEARLY 50% OF PEOPLE IN THE BAY AREA NOW USE UBER OR LYFT

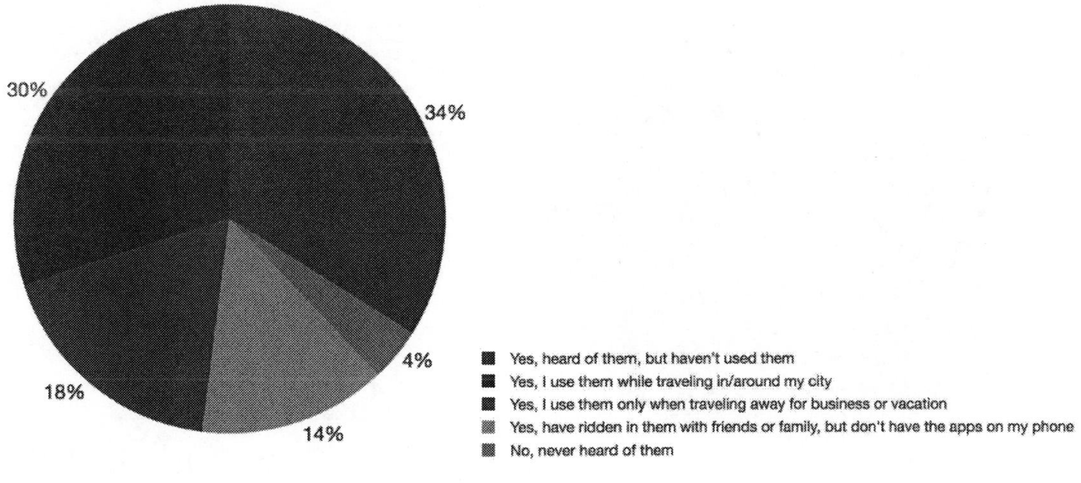

AMONG FREQUENT UBER/ LYFT USERS, 36% OF THEIR LAST TRIPS WOULD HAVE BEEN MADE BY TRANSIT

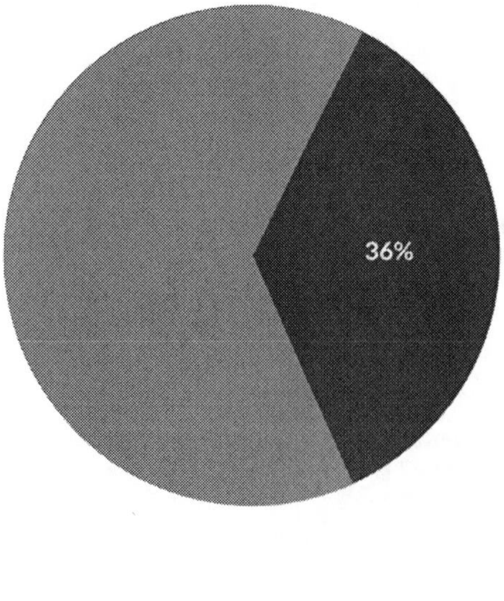

POPULUS
www.populus.ai

47% OF RECENT UBER OR LYFT TRIPS IN THE BAY AREA WOULDN'T HAVE BEEN MADE AT ALL OR BY TRANSIT

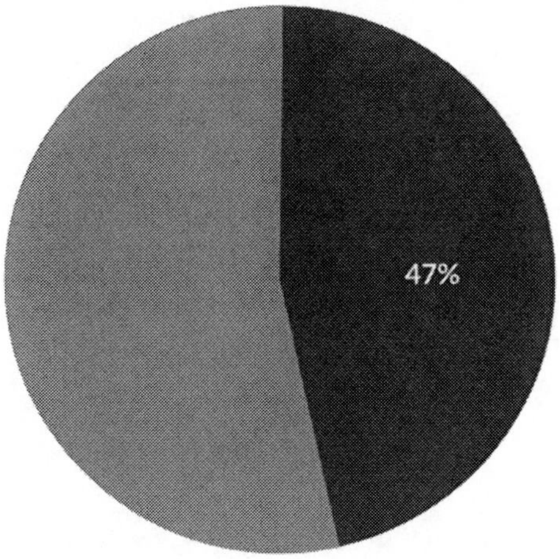

Another 11% wouldn't have made their most recent trip at all if Uber/Lyft wasn't available.

POPULUS
www.populus.ai

THANK YOU

REGINA CLEWLOW, PH.D.
CEO & CO-FOUNDER
REGINA@POPULUS.AI

POPULUS

AS FEATURED IN

CNN tech
FORTUNE
The Atlantic
Vox
The New York Times
WIRED
THE VERGE
San Francisco Chronicle SFGate
Chicago Tribune
The Economist
The Washington Post

HOWEVER, WE ALSO FIND THAT RIDE-HAILING USERS ARE GETTING RID OF VEHICLES: 11% IN THE BAY AREA

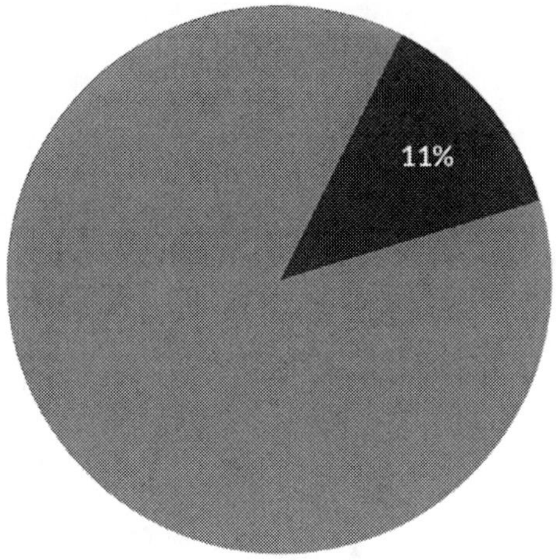

POPULUS
www.populus.ai

A FURTHER 10% OF PEOPLE WHO USE UBER/LYFT NEVER HAD A CAR TO BEGIN WITH

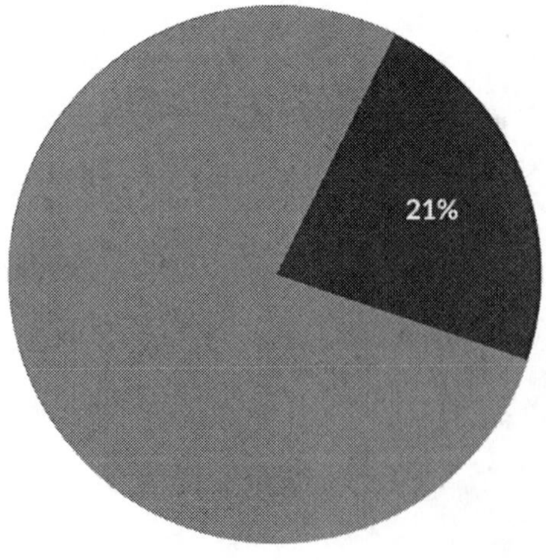

POPULUS
www.populus.ai

UCDAVIS
PLUG-IN HYBRID & ELECTRIC VEHICLE RESEARCH CENTER
of the Institute of Transportation Studies

Use Patterns of Tesla BEVs with Autopilot

26 July 2018

Scott Hardman
Rosaria Berliner
Gil Tal

UCDAVIS
PLUG-IN HYBRID & ELECTRIC VEHICLE RESEARCH CENTER
of the Institute of Transportation Studies

Introduction

- Questionnaire survey of 2715 BEV owners in 36 states
- Topics include:
  - Purchase motivations and impact of incentives for PEVs
  - Knowledge, Perceptions, and Purchase Intentions of AVs
  - Perceptions and use of TNCs (uber/lyft)
- This presentation:
  - Use patterns of Tesla BEVs with Autopilot enabled

UC DAVIS
PLUG-IN HYBRID & ELECTRIC VEHICLE RESEARCH CENTER
of the Institute of Transportation Studies

Introduction to Autopilot

- Partial Automation (L2) nearing Conditional Automation (L3)
- Tesla vehicles sold after October 2016 have hardware
- Not all vehicles have software enabled
  - $5,000 option

26 July 2018 3

UC DAVIS
PLUG-IN HYBRID & ELECTRIC VEHICLE RESEARCH CENTER
of the Institute of Transportation Studies

Use patterns of semi-automated vehicles

- 96% of Tesla with autopilot owners use it. 63% use it on their commute.

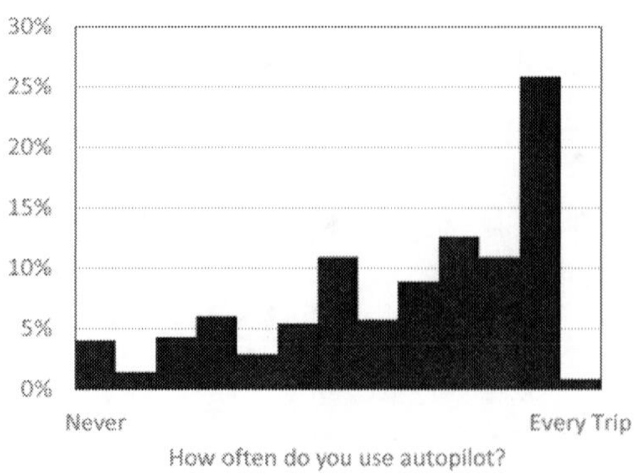

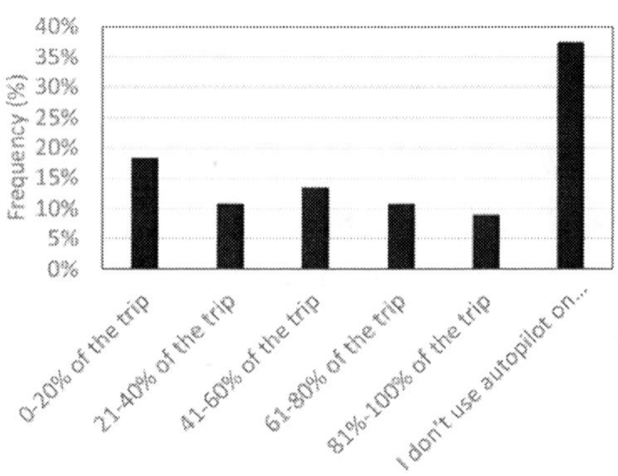

26 July 2018 4

UCDAVIS
PLUG-IN HYBRID & ELECTRIC VEHICLE RESEARCH CENTER
of the Institute of Transportation Studies

Mean Annual VMT of PEVs

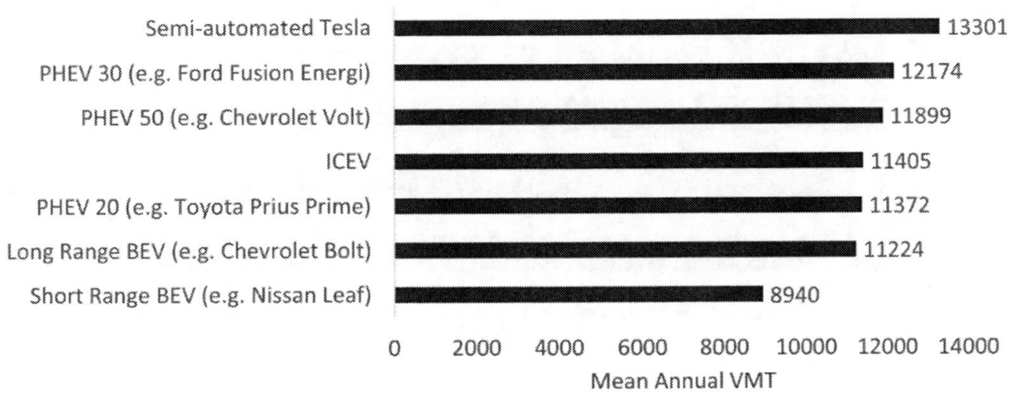

26 July 2018 5

UCDAVIS
PLUG-IN HYBRID & ELECTRIC VEHICLE RESEARCH CENTER
of the Institute of Transportation Studies

Mean Tesla Autopilot and US Annual VMT by Age

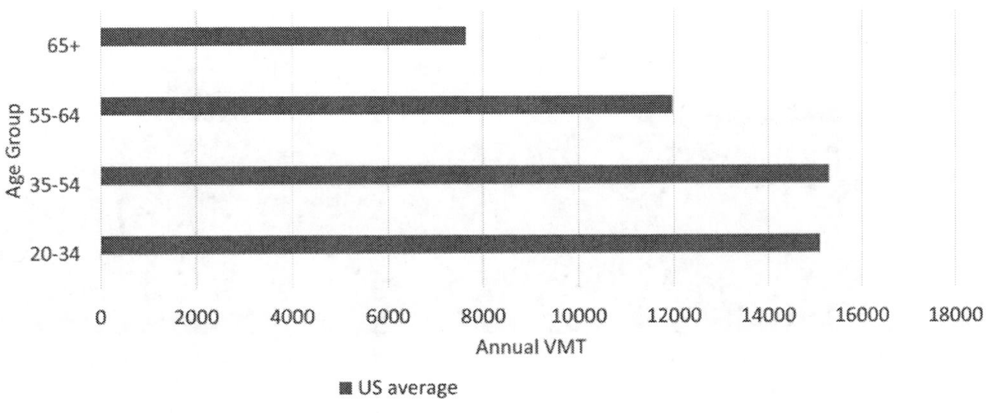

26 July 2018 6

UCDAVIS
PLUG-IN HYBRID & ELECTRIC VEHICLE RESEARCH CENTER
of the Institute of Transportation Studies

Mean Tesla Autopilot and US Annual VMT by Age

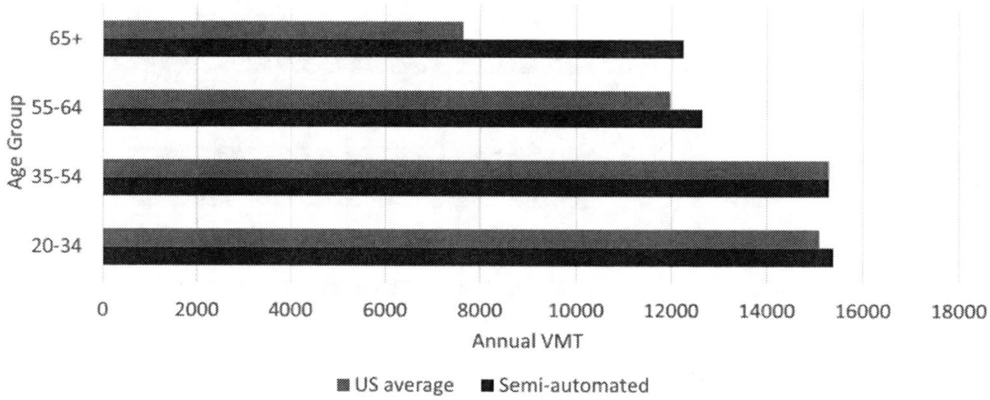

26 July 2018

7

UCDAVIS
PLUG-IN HYBRID & ELECTRIC VEHICLE RESEARCH CENTER
of the Institute of Transportation Studies

Tesla Autopilot VMT and Household Income

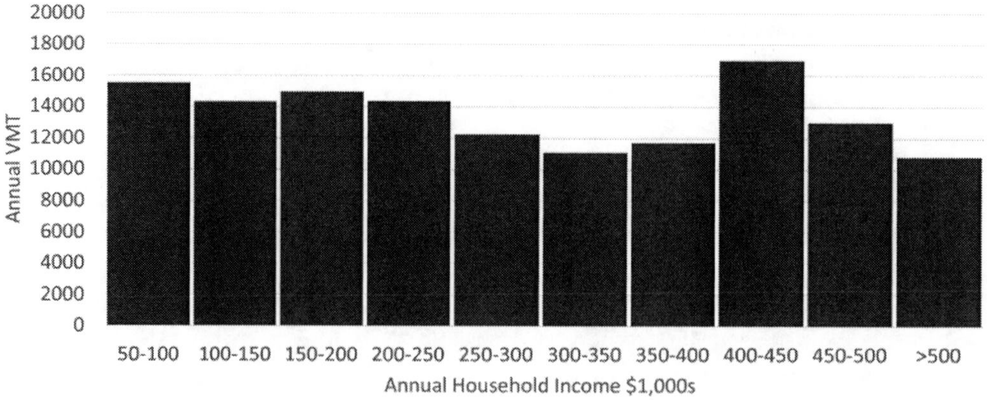

26 July 2018

8

UCDAVIS
PLUG-IN HYBRID & ELECTRIC VEHICLE RESEARCH CENTER
of the Institute of Transportation Studies

Summary and Future Work

Summary

- Tesla BEVs with autopilot appear to have highest VMT of all PEVs
- Hard to say how VMT will change with semi-autonomous vehicles, more work needed

Future Work

- More analysis including modelling to understand trends
- Qualitative interview with Tesla Autopilot users

26 July 2018 9

UCDAVIS
PLUG-IN HYBRID & ELECTRIC VEHICLE RESEARCH CENTER
of the Institute of Transportation Studies

Thanks

Scott Hardman
shardman@ucdavis.edu
Plug-in Hybrid & Electric Vehicle Research Center

UCDAVIS
PLUG-IN HYBRID & ELECTRIC VEHICLE RESEARCH CENTER
of the Institute of Transportation Studies

Autopilot use by road type

- Mostly on freeways and rural roads

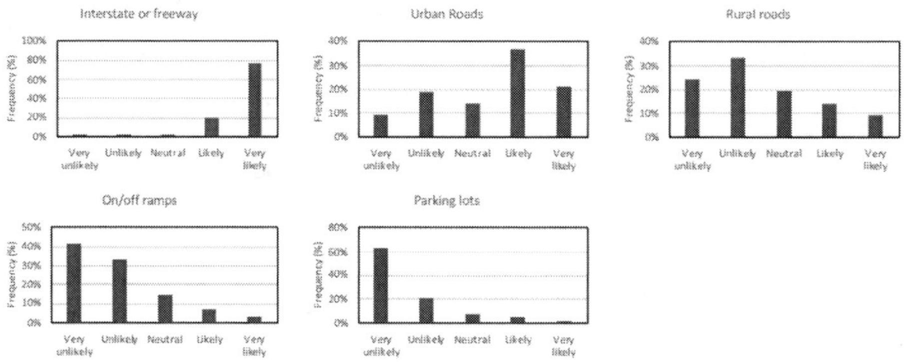

26 July 2018 11

UCDAVIS
PLUG-IN HYBRID & ELECTRIC VEHICLE RESEARCH CENTER
of the Institute of Transportation Studies

Autopilot use by weather conditions

- Mostly in clear weather conditions

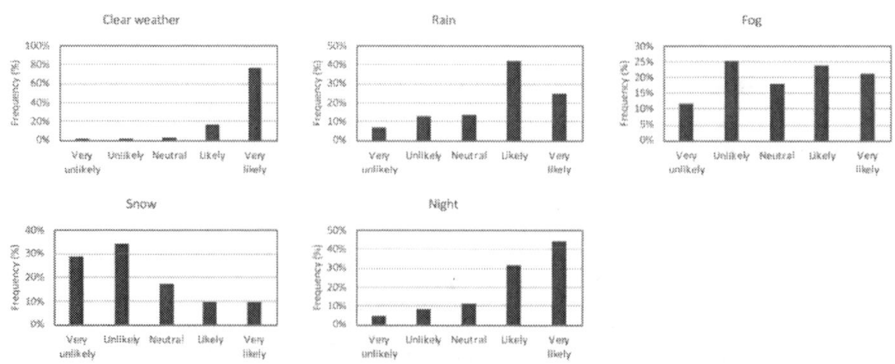

26 July 2018 12

Autopilot use by traffic conditions

- Not largely impacted by traffic conditions

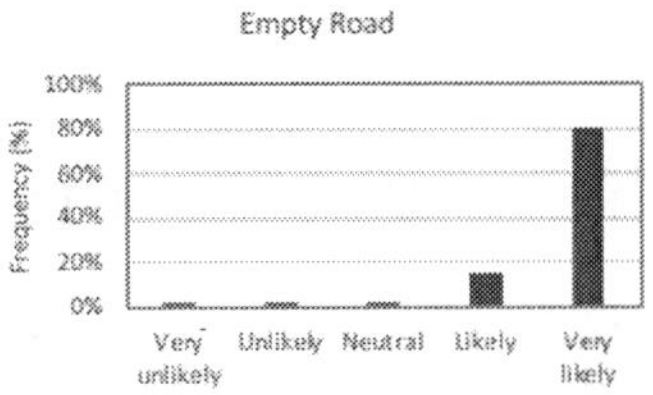

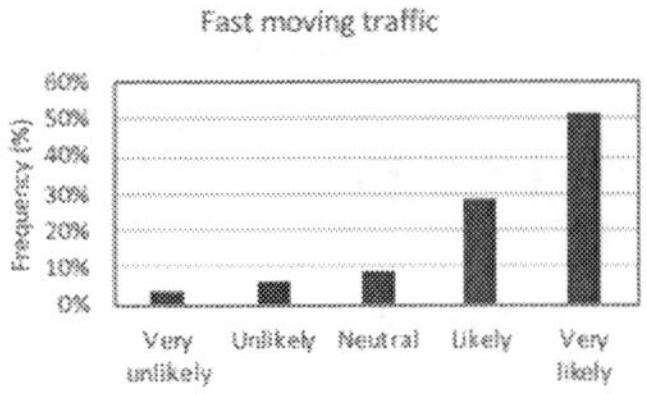

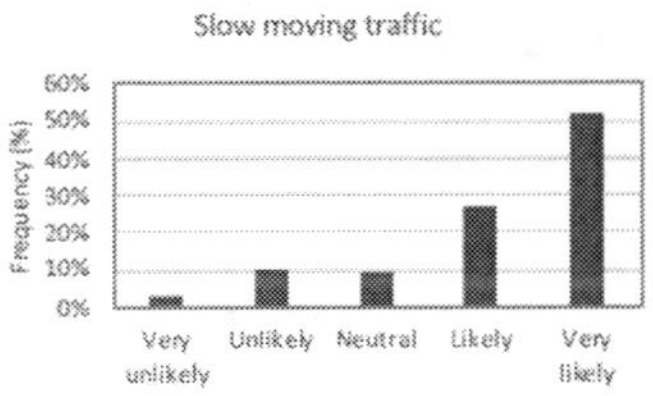

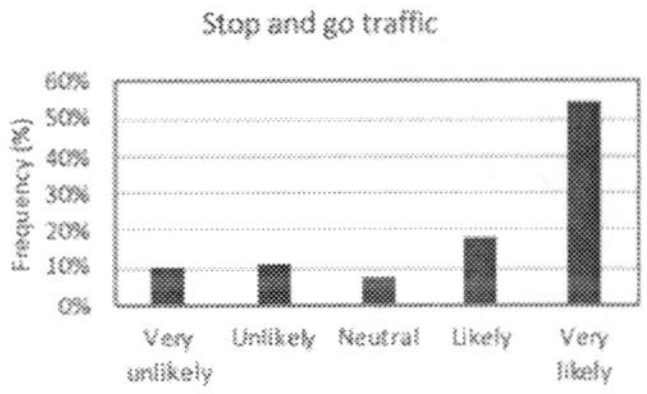

What are the key findings from these various behavioral studies?

- No consistencies in terms of demographics – impact on adoption or acceptance or not?
    - Gender?
    - Difficult to collect the representative sample of drivers
    - Looking at attitudes and concerns
    - Elderly more concerned and less likely
    - Male, young and educated more willing to share
    - Need overrides trust
    - Attitudes are not stable – so careful when using as predictive variables
- Heterogeneity in Populus data
- Missing is a broader framework for lifecycle changes
- Method of using a driver as an analog for self-driving vehicle?
    - Driver important for pooling or not?
- Consistent value of time reduced
- Zero occupant trips? Add to VMT.
    - Chauffeur driven cars do that all the time.
    - Impacts of vehicles per HH
    -
- Public education – increasing AV adoption with familiarity
- Bias towards urban populations – how to integrate other areas
    - Rural areas? Where are the future problems? And funding goes to address urban problems?

- Don't have observed behavior so finding ways to try to capture data on experience.
- Vastly underestimate the uses for cars beyond travel from here to there (lockers, style, performance)
- Will the behaviors change or the attitudes?
- Driver assist technologies slip into privately owned vehicles – acceptance is increasing
- Anecdotes into data?
- Bias towards commute travels – induced travel for long-distance holiday, weekend travels
- Funding issue – Self-driving cars not a priority no need to understand now
    - Understand impacts today to develop policies, decisions
- More VMT, increased congestion, less transit
- Pooling is viable because it is cheaper, as cost of autonomous driving decreases incentive to pool goes away
- People are concerned about safety and liability
- Lack of knowledge by public
    - Methodological challenge – seem to have conventional or self-driving nothing in-between – may lead to inconsistency of result
    - Naming conventions across OEMS is not consistent
- With methods we are using we can only present so much in a clearly consistent way, respondents also bring into the study their personal experiences, understanding, etc.
- Complex to model – we don't know what "multitasking" will look like in future

What are the gaps in knowledge?

- Longer term choices – location decision, land development
- Individuals less than 18 – teens or children – who may be important future consumers—asking parents about children
- We don't know much about adoption likelihood of vulnerable populations, disabled, elderly
- Impact on labor market – and concentrated wealth into smaller group of people –
- How much different models of adoption will co-exist
    - ADAS and higher levels of automation
    - Automated fleets
    - What are the regions, people, etc., that will use one versus other model
- Small samples that are not necessarily representative – transportation agencies need larger samples, robust data for decisions
- How to ask about future services in ways that will be accurate and make sense to respondents?
- Now AV ride-sharing or private—what about first/last mile solutions? More multimodal trips?
- Uncertainty related to commercial actors in the market – what about impact of subsidies on mobility choices.
- Decision of whether people give us car ownership in lieu of ride-hail or ride share vehicle?
- Right representative sample – representative for what? Need many small sample? More careful in comparing results based on samples we're talking about
- Data repository project – with meta data on samples, sample sizes

- Influence of all these new activities performed during travel – on activity-travel behavior
    - Impact on traffic flows – changes in peak hours
    - What is the fate of airlines?
- Linkage between automation and its own value – sustainable environment
    - How will streets look? What happens to demand?
- Acceptance to riding in automated car
- Autopilot on tesla – you can turn it on and off – autonomous features people turn them on and off – make its way into the system and still allow some control
- Cost and time not being everything in choice – may be exacerbated with automated vehicles –
- What is the ideal experience from the consumer point of view – what might the ideal experience be? Mental model
- Solving a problem for the consumer – safety is not a problem for individuals. Model resistance of people, why they resist adoption?
- People owning vehicles because of convenience – how the symbolic relationship people have with their vehicle may change?
- What technology – already handed over automation to that affects our lives? Giving up locus of control
    - Public transport in the long distance train still the symbolic attribution is missing
    - Telecommuting – didn't have to go to specific place – what people did with extra time?
    - AV vs public transport – privacy, storage, personalization possibilities; comparison not appropriate

Research methodologies that are appropriate in this context?

- Stated choice – we don't have those experiences; the more that we can define the experience through videos or simulations (chauffeur) it is better
    - How much bias the respondent brings into that survey – given the level of detail we provide – impact of "instrumentation"
    - Respondents may be unfamiliar choice models may not work that well – simulation or agent based models better? Diffusion of innovations (Everett Rogers) – adoption of autonomous vehicles
    - Address attitudes and perceptions in the models
    - Insights from focus groups and qualitative interviews - but they are limited in sample sizes
    - How does stated choice work in light of social agreement?
- Menu of options in terms of research methods
    - Ensemble modeling – two or three different methodologies applied to same questions
    - How to use cell phones or other technologies to get larger samples
    - Data collection part requires more of our attention
    - Problems with imputation when using passive data
    - Be careful with too much modeling – planning community and decision makers don't understand.
    - Public sector results expected to published – private sector expectation that results will be proprietary. Hiring their own modelers and survey firms.
    - OEMs are collecting real world data – right now. Could we get that data?

How can evidence be used to inform models?

- Distinguishing mandatory activities from other activities during travel; new activities
- Benefit from new VoTs rather than making assumptions
- Zero occupant trips added to modeling framework
- Activity based models can handle the future?
- MPOs or other state agencies – we have to find a way to help them with what they have.
- Exciting time to do research – challenging time to do models.
- Classic paradigm – how does that play in this world. Not easy
- The marginal cost to the user of privately owned vehicle as opposed to all of these new modes; better ways to transfer that knowledge to the public
- More research on social trends and generations with all of the innovation and technology – at what point in the lifecycle is that generation.

Key Takeaways

- Third year, conversation has progressed
- Presentations demonstrate the amount of work out there

Key findings from all presentations

- Missing consistency among demographic factors
- Caution against future research
- Working towards representative sample size, current research is targeted small sample sizes

Gaps

- Creative new ideas methodologies are needed
  - Chauffeur

Data and research methodologies

- Which data do we use to put into the models
- Which models are still valid – push and pull from that respect
- How does the introduction of new technologies affect use of time and how relate to mobility models

Knowledge gained to stimulate impacts

- Linking to actual world

What do we do next?

- Come a long way still much for us to do
- Collaboration – does it help with respect to sample size and funding
- Becoming more systematic in research – recognizing the gaps that have been identified –
- It doesn't help if we are all doing research on the same topic
- Zero occupancy trips – theory for a couple of years – coming to fruition what can we do to continue to formulate and quantify that problem
- Link between time use community and mobility research community. Bring into the conversation.

1

Technion
Israel Institute of Technology

The Impact of Multitasking On Travel Behavior

Bar Bergman, Yoram Shiftan
Transportation Research Institute

AV 2018, San Francisco

2

Methodology

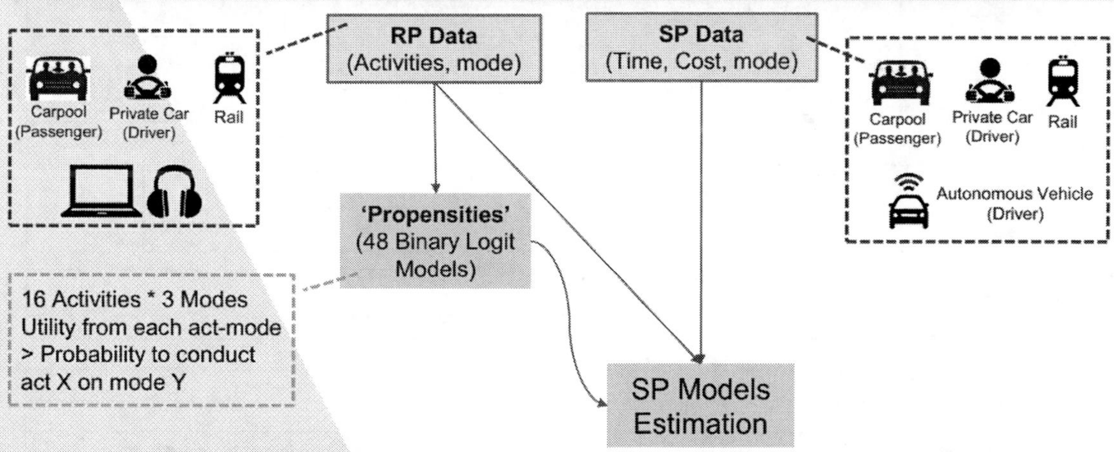

4

Data

- **16 Activities**

MNL	3A-M4			4A-M4			
	PC	CP	RAIL	PC	CP	RAIL	AV
Time	-0.085*** (-7.64)	-0.106*** (-13.00)	-0.09*** (-11.18)	-0.086*** (-6.65)	-0.0817*** (-9.15)	-0.078*** (-10.47)	-0.0825*** (-7.86)
Cost	-0.033*** (-4.54)	-0.052*** (-9.25)	-0.097*** (-6.63)	-0.016*** (-2.19)	-0.026*** (-3.72)	-0.032** (-1.98)	-0.026*** (-5.17)
Prop_Eat	-	**2.47*** (7.37)**	**0.68*** (2.42)**	-	**1.51*** (5.88)**	**0.88*** (3.26)**	-
Prop_Laptop	-	-	**1.16*** (2.96)**	-	-	**1.24*** (3.60)**	-
Prop_Read	-	-	-	-	**-1.07*** (-3.02)**	-	-
Prop_Rest	-	**0.86*** (3.21)**	-	-	**1.36*** (6.33)**	-	-
Prop_Call	**1.09*** (3.26)**	-	-	**1.20*** (3.54)**	-	-	-
Constant	**Base**	**6.13*** (3.39)**	**3.33* (1.91)**	**Base**	**4.20** (2.12)**	**3.83* (1.81)**	**4.71*** (2.59)**
VOT (NIS/hr)	~ 154	~ 123	~ 55	~317	~188	~145	~189
No. of Parameters	29			31			
Null log-likelihood	-1421.604			-1795.251			
Cte log-likelihood	-1225.77			-1526.926			
Final log-likelihood	-814.417			-1053.546			
Rho-square	0.427			0.413			
Adjusted rho-square	0.407			0.396			

6

Value of Time (VOT)

MNL	3A-M4			4A-M4			
	PC	CP	RAIL	PC	CP	RAIL	AV
VOT (NIS/hr)	~ 154	~ 123	~ 55	~317	~188	~145	~189

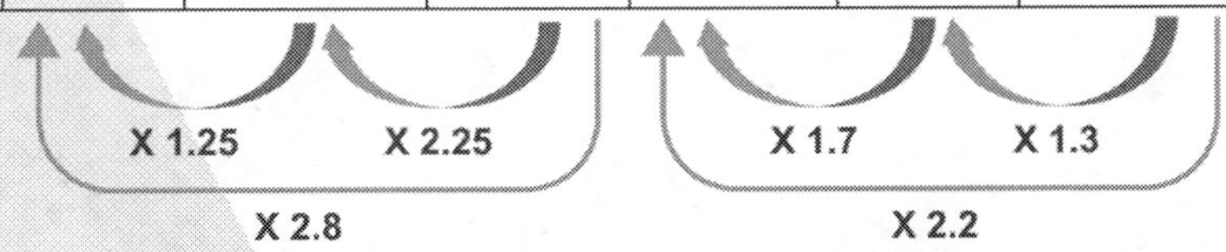

7

Value of Time (VOT)

	PC	CP	Rail
VOT [Nis/hr] (High propensity)	42	37	34
VOT [Nis/hr] (Low propensity)	117	150	70

THANKS!

AUTOMATED VEHICLES SYMPOSIUM 2018 — USERS. VEHICLES. INFRASTRUCTURE. — Driven by AUVSI and TRB

Breakout Session #29: Integrating Automated Vehicles on City Streets: An Interactive Planning Workshop

Summary of Key Findings and Lessons Learned

- Our expert panel: Ellie Casson (**Waymo**), Danie Beukman (**Cruise Automation**), Alisyn Malek (**May Mobility**), Stephen Zoepf (**Stanford**), Andrea Broaddus (**San Jose State University**), Kris Hansen and Justin Johnson (**3M**), and Billy Riggs (**University of San Francisco**)
- Cities need to evolve **from auto-centric to people-centric**. In cities, AVs will always be secondary to people.
- Transportation and the electric grid will be intertwined. We need to plan for **renewable energy** systems for mobility.
- Today, the curb is critical; tomorrow, the curb may be irrelevant with new **flexible designs** that are not yet possible today.
- Accessible, **universal design** is a challenge for both vehicles and streets. This challenge needs to be tackled and resolved.

AUTOMATED VEHICLES SYMPOSIUM 2018 — USERS. VEHICLES. INFRASTRUCTURE. — Driven by AUVSI and TRB

Breakout Session #29 Integrating Automated Vehicles on City Streets: An Interactive Planning Workshop

Top 5 Policy Recommendations

1. Design for clarity, **design** for people
2. Open doors to **pilot** new ideas
3. Evolve **policy** faster, more collaboratively
4. Advance mapping efforts with **digital** infrastructure inventories
5. Rethink infrastructure **finance** to consider dynamic pricing

AUTOMATED VEHICLES SYMPOSIUM 2018

USERS. VEHICLES. INFRASTRUCTURE.

Driven by AUVSI and TRB

Breakout Session # 30
Accelerating Market Penetration of CAVs by Leveraging Infrastructure

Summary of Key Findings and Lessons Learned

- Quantifying the effect of smart infrastructure on the penetration of CAVs and costs
- Determination of the liability between OEMs, data carriers, and AV infrastructure service providers, even multiple service providers - who is going to be the main responsible for decision making?
- Addition of infrastructure developers into the law suits
- The need for retrofitting of infrastructure and more often maintenance of infrastructure
- Public willingness to pay for infrastructure and liability costs

Presenters: Swaminathan Gopalswamy, Scott Shogan, Aybike Ongel, Greg Larson, James Sayer, Peter Polit, Shawn Brovold, Joe Gustavus, Tina Georgieva , Roger Berg

AUTOMATED VEHICLES SYMPOSIUM 2018

USERS. VEHICLES. INFRASTRUCTURE.

Driven by AUVSI and TRB

Breakout Session # 30
Accelerating Market Penetration of CAVs by Leveraging Infrastructure

Research Needs

- Cost/benefit assessment for smart infrastructure vs. sensors on the vehicle, including initial costs, maintenance costs, and operational feasibility
- The vulnerability of smart infrastructure to cyber attacks, compared to that of the sensors on the vehicle
- The effects of reliability and performance of communication technology on the use of infrastructure, e.g. DSRC, 4G-LTE, 5G, etc.
- Determination of optimum share of situational awareness between sensors on the road and sensors on the vehicle
- The effects of smart infrastructure on equity

AUTOMATED VEHICLES SYMPOSIUM 2018
USERS. VEHICLES. INFRASTRUCTURE.
Driven by AUVSI and TRB

Breakout Session #30
Accelerating CAV Market Penetration by Leveraging Infrastructure

Discussion Topics:

- Low-Level Automation on Managed Lanes
- Ramp Metering for CAV's
- CAV Operations on Signalized Corridors

AUTOMATED VEHICLES SYMPOSIUM 2018
USERS. VEHICLES. INFRASTRUCTURE.
Driven by AUVSI and TRB

Breakout Session #30
Accelerating CAV Market Penetration by Leveraging Infrastructure

Low-Level Automation in Managed Lanes:

- Level 2, or even just Level 1 automation on separated facilities
    - Allow access to Managed Lanes as a enticement to equip vehicles
- CACC, dependent on V2V communications (DSRC?) for safe close-following
    - Could also be "platooning", which is fixed-distance following
- Speed controls sent by operators to vehicles using I2V for speed harmonization?
- Peak Demand Hours only! Normal operation during off-peak or low demand

AUTOMATED VEHICLES SYMPOSIUM 2018 — USERS. VEHICLES. INFRASTRUCTURE. — Driven by AUVSI and TRB

Breakout Session #30
Accelerating CAV Market Penetration by Leveraging Infrastructure

Ramp Metering for CAV's:

- Vehicles on the freeway broadcast their position (BSM) using V2I (DSRC?)
- Algorithms in the Ramp Metering System identify gaps in traffic and release vehicles accordingly
- Waiting vehicle types are known (car, bus, truck) to account for proper acceleration time
- In the longer term, gaps in freeway traffic could be opened automatically to accommodate more entering vehicles
- "Bus-on-Shoulder" operations?

AUTOMATED VEHICLES SYMPOSIUM 2018 — USERS. VEHICLES. INFRASTRUCTURE. — Driven by AUVSI and TRB

Breakout Session #30
Accelerating CAV Market Penetration by Leveraging Infrastructure

CAV Operations on Signalized Corridors:

- "Traffic Optimization on Signalized Corridors" (TOSCo)
- "Group" vehicles together, and schedule green time for the entire group to move through the intersection
- Cross-traffic is also scheduled to use the space between groups to cross the corridor
- Also moves groups of stopped vehicles through intersections more smoothly than with manual driving
- Works best on one-way corridors (fewer turning movements)

Organization Committee and Panelist Profiles

Swami Gopalswamy, Organization Committee Chair, and Panel Speaker:

Swami Gopalswamy, Ph.D. is the Director of the Connected Autonomous Safe Transportation (CAST) Program and the co-director of the Center for Autonomous Vehicles and Sensor Systems (CANVASS) at Texas A&M University, College Station, with joint appointment at Texas A&M Transportation Institute (TTI). Swami Gopalswamy has over two decades of experience in Model Based Systems Engineering and Control System development working in the automotive industry, with a special interest in safety and systems reliability for mechatronic systems. His current focus is to look at new techno-business paradigms for autonomous transportation that can improve safety and reliability, increase trustworthiness, and reduce liability, all towards accelerating the adoption of autonomous vehicles

Scott Shogan, Organization Committee, and Panel Moderator:

Scott Shogan is the Connected/Automated Vehicle Market Leader for WSP USA, a practice which is broadly defined as the intersection of CAV technologies, vehicle electrification, and shared mobility. Scott has over two decades of experience in traffic operations, transportation planning, and emerging transportation technologies. He has served in leadership roles in transformative projects such as the Safety Pilot Model Deployment, the American Center for Mobility, and CAV strategic planning for a number of state DOTs, including Michigan, Minnesota and Utah.

Aybike Ongel, Organization Committee, and Panel Speaker:

Aybike Ongel is the principal investigator at TUMCREATE in Singapore. She is heading the research group Individual Mobility Vehicles and Services. Her research at TUMCREATE focuses on developing autonomous electric vehicle concepts for public transportation

Glenn Havinoviski, Organization Committee, and Break-out Session Reporter:

Glenn N. Havinoviski, PE serves as Associate VP, Transportation Systems for Iteris, Inc. and over the last 34 years has emerged as one of the world's foremost experts in ITS development and transportation systems management. He has served both as a practice builder and a practice leader, providing project management and technical leadership for ITS and traffic management projects both in the US and abroad. For Iteris, he has been involved in connected and automated vehicle program planning activities, and has led firmwide efforts in the area of managed lanes, active traffic management systems, and integrated corridor management, along with real-time traveler information, road pricing and transit signal priority. He has led or served in a key role for projects in 18 U.S. states, Mexico, Eastern Europe, the Middle East and Asia. For three years Glenn served a core instructor in training workshops for the International Road Federation (IRF), providing background to engineers in the Middle East on ITS and integrated mobility strategies. Mr. Havinoviski is a member of the TRB standing committee on Congestion Pricing, is a former President of ITS Virginia and served as US Expert to ISO Technical Committee 204 Working Group 1 (ITS Architecture and Data Dictionaries). He is currently a registered Professional Engineer in Virginia, Maryland, Delaware, Florida, North Carolina, and the District of

Columbia. Glenn graduated from the University of Wisconsin-Milwaukee in 1984 with BS in Civil Engineering.

Joseph Gustavus, Panel Speaker:

Joseph D. Gustavus is a Miller Canfield attorney representing multinational clients and clients primarily in the automotive, defense, aerospace, software and information technology sectors. As a member of the Miller Canfield Automotive Group, Joe represents automotive suppliers in negotiation and drafting of software license, IT service, component sourcing, and technology development agreements with their automotive OEM and Tier 1 customers. Among others, he successfully negotiated transactions for foreign and domestic automotive clients to supply automotive embedded software, Electronic Control Units (ECUs), automotive telematics systems, and development and modeling software for embedded systems.

Tina Georgieva, Panel Speaker:

Tina Georgieva's current practice focuses on providing legal counsel to clients in support of autonomous vehicle technology development and deployment. She also counsels clients on product safety regulatory compliance matters, including those involving the National Highway Traffic Safety Administration. In that capacity, she collaborates closely with clients to implement internal programs to ensure compliance with the National Traffic and Motor Vehicle Safety Act, Federal Motor Vehicle Safety Standards and TREAD Act and coordinate recalls and voluntary product replacement programs. During her career, she has defended the interests of automotive clients in complex product liability litigation across Canada and the United States.

Greg Larson, Panel Speaker:

Greg Larson, P.E., Chief, Office of Traffic Operations Research, Division of Research, Innovation and System Information, California Department of Transportation (Caltrans)

Greg Larson has worked for Caltrans for more than 27 years, starting as a Research Engineer at the Transportation Laboratory, then as a Senior Electronics Engineer, and currently as an Engineering Manager. He is assigned as the Chief of the Office of Traffic Operations Research in the Division of Research, Innovation and System Information. He is responsible for managing and overseeing the work of a professional technical staff performing research in the area of Intelligent Transportation Systems, with the HQ Division of Traffic Operations and various districts as their primary customers. He also serves as Chair of the Strategic Initiatives Technical Working Group of the Cooperative Automated Transportation (CAT) Coalition, and Vice Chair of the Connected and Automated Vehicle Working Group of the AASHTO Committee on Transportation System Operations.

Greg earned Bachelor and Master of Science degrees in Electrical Engineering from California State University, Sacramento, and is a licensed Professional Electrical Engineer in the State of California.

Roger Berg, Panel Speaker:

Roger Berg is Vice President of DENSO's North American Research and Development group. He is responsible for DENSO's regional Research and Development of connected and automated vehicles, and cyber security at their technology center in southern California.

Berg has experience in military, consumer and automotive electronics from previous engineering and executive positions at Northrop Grumman, Motorola and Sony combined with his 17 years at DENSO.

Berg earned a BSEE from the University of Illinois in Urbana, Ill., and a MSEE from Illinois Institute of Technology in Chicago. He is the inventor or co-inventor on eight U.S. and international patents. He is a long standing member of SAE and IEEE, and in December of 2016 he was voted one of Automotive News' "60 people driving the self-driving movement".

Shawn Brovold, Panel Speaker:

Shawn brings a unique perspective having worked as an engineer in both the civil transportation and automotive industry. While at Ford Motor Company, he worked in active safety research, and as part of the Crash Avoidance Metrics Partnership (CAMP). With CAMP, he worked alongside multiple automotive OEMs on the development of safety applications based on V2V and V2I as part the USDOT connected vehicle efforts. Previously, he was a partner in Testquip LLC, and worked with Universities and State DOTs on the development of novel pavement testing equipment. In his current role, he is researching ways the infrastructure will support safe travel in the advent of connected and automated vehicles, through a combination of advanced materials and digital solutions.

Jim Sayer, Panel Speaker:

Jim Sayer is the Director of, and a research Scientist at, the University of Michigan Transportation Research Institute. He served as the PI of the Connected Vehicle Safety Pilot Model Deployment, and currently directs the Ann Arbor Connected Vehicle Test Environment. He also conceived of, designed and oversaw the construction of Mcity, a test facility at the University of Michigan dedicated to the development and evaluation of infrastructure and vehicle-based connected and automated technology.

James R. Sayer, PhD is Director of the University of Michigan Transportation Research Institute (UMTRI); Research Scientist in UMTRI's Human Factors Group; and Adjunct Professor of Civil and Environmental Engineering. He has conducted both basic and translational research in the areas of driver assistance and advanced safety systems development, naturalistic driving behavior, driver distraction, driver vision, and pedestrian conspicuity since 1993.

Jim currently serves as the project manager of the Connected Vehicle Safety Pilot Model Deployment, a U.S. Department of Transportation-sponsored program to demonstrate connected-vehicle technologies in a real-world, multimodal environment. The results of the Safety Pilot Model Deployment will be used by the U.S. DOT to determine driver acceptance for, and evaluate the feasibility, scalability, security and device interoperability of connected-vehicle technologies. In the area of driver assistance systems, Jim has contributed to the development, evaluation, and deployment of adaptive cruise control, collision warning, and collision avoidance systems in both passenger cars and commercial trucks (including simulator, test-track, on-road, and field-operational testing). He has overseen the conduct of four field operational tests involving several hundred drivers, accumulating over 1.2 million miles of naturalistic driving data. In the area of driver distraction. He has conducted research on the effects of cell phones and other secondary behaviors on driving performance, the frequency of secondary tasks, and driver self-regulation. His research interests related to driver vision include the effects of hydrophobic and hydrophilic glass coatings, window tinting, and defrosters/defoggers on visual performance and driving behavior. On the topic of pedestrian conspicuity. Jim has performed multiple research studies concerning the effects of retroreflective markings and safety garment design on the detection of pedestrians, emergency responders and road construction workers.

Peter Polit, Panel Speaker:

Peter Polit is the Vice President and General Manager of Connected Vehicle Strategy and Partnerships at SiriusXM. With over 25 years of industry experience managing both startups and large mature organizations, Peter has been at SiriusXM for five years managing functions ranging from Infotainment to Connected Vehicle Product Management to current focus on Strategy and Partnerships. Peter's background is in business management and development of digital technologies including video and image compression and processing, statistical multiplex, and IoT including connected home and connected vehicle.

Local Roots 》 **Global** Reach

2018 Automated Vehicle Symposium

July 11, 2018, San Francisco

Breakout Session:
Accelerating CAV Market Penetration by Leveraging Infrastructure

A Legal Perspective

Miller Canfield, Joseph Gustavus

MILLER CANFIELD
millercanfield.com

I. Nomenclature

Terms for easy reference. . .

- **AV**-Autonomous (or Automated) Vehicles.
- **SIETC**-Special Infrastructure Enabled Traffic Corridors – AV traffic corridors using MSSPs.
- **MSSPs** -Multi-Sensor Smart Packs –devices monitoring SIETCs for AV purposes.
- **SA Data**-Situational Awareness Data-the AV environmental factors (outside AVs) monitored and collected by MSSPs along SIETCs.
- **SCs**-Smart Connects-devices embedded in or alongside MSSPs that transmit SA Data to AVs in that SIETC via a SA Data Carrier.
- **SA Data Carrier**-the private data carrier transmitting DA Data to AVs using the 5G Network or other sanctioned wireless SA Data transmission system.

MILLER CANFIELD
millercanfield.com 2

II. Government Immunity

- **Governmental (Sovereign) Immunity Doctrine** holds that government and its agencies will be immune from liability, except for very limited cases.
  - Doctrine likely applied in the AV context to federal, state and local governments.

III. Recent State AV-Related Laws Passed

- To date, 29 states have enacted legislation related to AVCs.
- The Michigan Example:
  - **MI PA 332 of 2016**
    - Permits manufacturer R&D use of AVs on MI roads using employees.
    - Provides immunity to AV manufacturers for TP modifications without consent.
  - **MI PA 332 of 2016**
    - Allows only OEMs to participate in SAVE projects.
    - Permits deployment of on-demand self-driving taxis.
    - Hold OEMs completely responsible therefore.

IV. Utilities and Toll Booth Operators

- An SIETC Operator may be analogous to a public utility (PU), such as an electric company.
- SIETC Operator furnishing SIETC Service to a municipality or AV drivers is manifestly a public service or public utility corporation, akin to a PU
- Conceivable that SIETC Operators would be subject, in part, to PU-type law
- PU liability not governed by contracts principles
- In most cases, strict product liability doctrine not applicable to PUs
- However, PU duty owed is determined by negligence test of foreseeability of harm i.e., no liability for unforeseeable failures to anticipate and guard against harm
- In the case of electric PUs, National Electric Safety Code
  - given force of law in some states
  - in other states, merely a floor below which PU safeguards cannot go

MILLER CANFIELD 5
millercanfield.com

V. Current Contracting Structure with Embedded Software and Telematicsclature

- For software embedded within **Electronic Control Units (ECUs)**
  - Brunt of liability is borne by OEMs
  - Tier 1 Supplier ECU developers secondarily so
  - EUS software and components indemnify customers and their customers
    - Warranty claims
    - Voluntary and government-mandated recalls
    - Product liability
    - Class action suits
- For **Telematics Systems** used in fleet tracking and vehicle diagnosis
  - Fleet tracking system providers contract with fleet owners directly
  - in most cases, leaving fleet owners to contract separately for wireless data carrier service

MILLER CANFIELD 6
millercanfield.com

VI. Potential Legal Structure for AV Market

- Hybrid between PU law and contracting law.

MILLER CANFIELD 7
millercanfield.com

Our Local and International Presence Means People on the Ground Assisting with Your Complex Transactions and Litigation

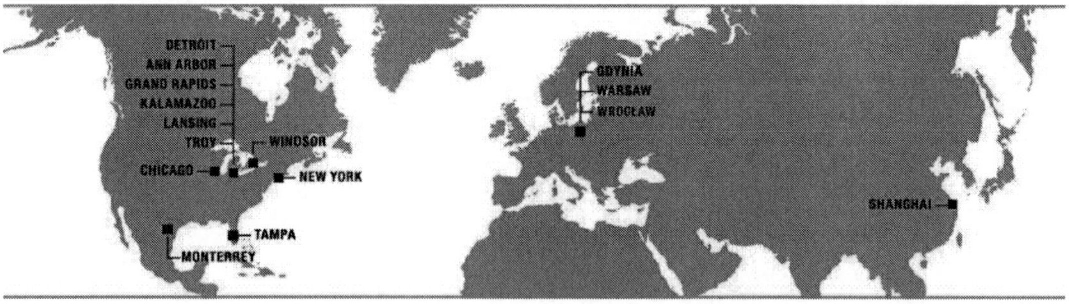

160+ Year reputation and legal leadership Established 1852 in Detroit, Michigan

300+ Lawyers and paralegals

165th Largest U.S. law firm according to the *National Law Journal*

13 "Lawyers of the Year" by *Best Lawyers*

135+ Lawyers in leading referral guides, *Chambers USA*, *Best Lawyers*, *Super Lawyers* and *European Legal Experts*

20+ Lawyers are Fellows of prestigious American Colleges, including *Bond Counsel, Employee Benefits, Mortgage Attorneys, Real Estate Lawyers, Trial Lawyers, Trust and Estate Counsel, College of Labor and Employment Lawyers,* and the *Litigation Counsel of America*

■ The designated firm in Michigan and Poland in the Employment Law Alliance, a network of top labor and employment lawyers in the world – a powerful resource for clients facing local or global issues

MILLER CANFIELD 8
millercanfield.com

Joseph D. Gustavus
Senior Principal
Miller Canfield

+1 (248) 267-3317
Gustavus@millercanfield.com

UNITED STATES ■ CANADA ■ MEXICO ■ POLAND ■ CHINA

MILLER CANFIELD
millercanfield.com

Solutions that will serve a spectrum of vehicles

Human and CAVs will coexist on roadways for many years

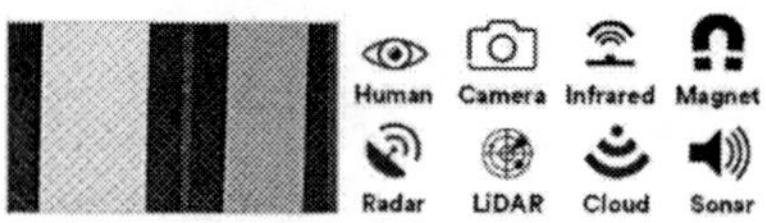

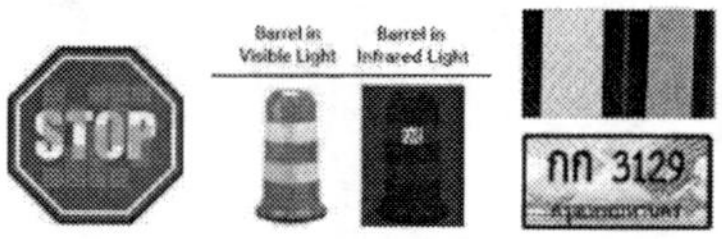

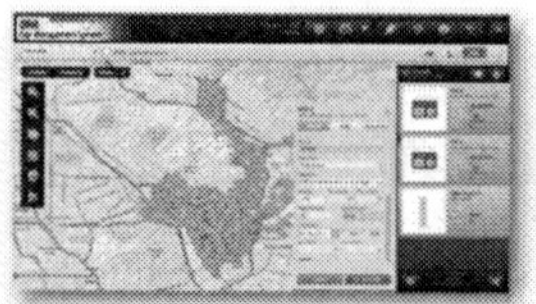

Material Solutions	Material-Data Solutions	Information System Solutions

• Robust portfolio of solutions
• Goal: Read by all sensor modalities in poor weather
• Enabling:
  ◦ safe human & vehicle navigation
  ◦ cyber-physical security
	• Clear, standardized navigational markers
	• Facilitate I2V info exchange
	• Providing:
	  ◦ Safety system with multiple layers of info
	  ◦ cyber-physical security
		• Aid in roadway and work zone safety and in assessing infrastructure* quality
		• Suite of information solutions
		• Enabling safe and secure navigation by humans and CAVs

Engineered context for simplified decision making

Localization *Boundaries* *Uniqueness*

Classification *Guidance*

Data exchange to improve safety

* Infrastructure means anything that DOTs spend on and have responsibility to ensure promotes safety

Confidential.

3M

1

Local Roots » Global Reach

2018 Automated Vehicle Symposium

July 11, 2018, San Francisco

Breakout Session:
Accelerating CAV Market Penetration by Leveraging Infrastructure

A Legal Perspective

Miller Canfield, Tina Georgieva

MILLER CANFIELD
millercanfield.com

"Design Defect" and Autonomous Vehicles

Question to be answered:

Will the introduction of AVs result in a departure from the traditional presentation of a "design defect" product liability case?

MILLER CANFIELD 2
millercanfield.com

Survey of Top Legal Practitioners

Both sides agree: costs will increase exponentially

"Attempting to exempt technologies from tort liability will not occur." – Plaintiffs' Bar

---

Beginning to appreciate the complexities and the associated costs of a traditional "design defect" case in the AV context

---

Attorneys from both sides referenced the *Toyota UA* litigation as a representation of what AV litigation will entail

Traditional method of litigating "design defect" cases is not well-suited to resolving claims directed at AV

NHTSA's Role?

NHTSA'S MISSION

"Save lives, prevent injuries and reduce economic costs due to road traffic crashes, through education, research, safety standards and enforcement activity."

Tina Georgieva
Senior Attorney
Miller Canfield

+1 (248) 267-3251
Georgieva@millercanfield.com

Thank you!

UNITED STATES ■ CANADA ■ MEXICO ■ POLAND ■ CHINA

MILLER CANFIELD
millercanfield.com

A New Paradigm for Autonomous Vehicles: Accelerating deployment through a new separation of responsibilities between automotive OEMs, infrastructure operators, and end-users

Accelerating Market Penetration of CAVs by Leveraging Infrastructure

Breakout Session#30, Automated Vehicle Symposium 2018, San Francisco, CA

Swami Gopalswamy

Director, Connected Autonomous Safe Transportation (CAST) Program,
Department of Mechanical Engineering, Texas A&M University, College Station

Joint appointment with Texas A&M Transportation Institute (TTI)

Concept introduced at the Intelligent Vehicle Symposium 2018

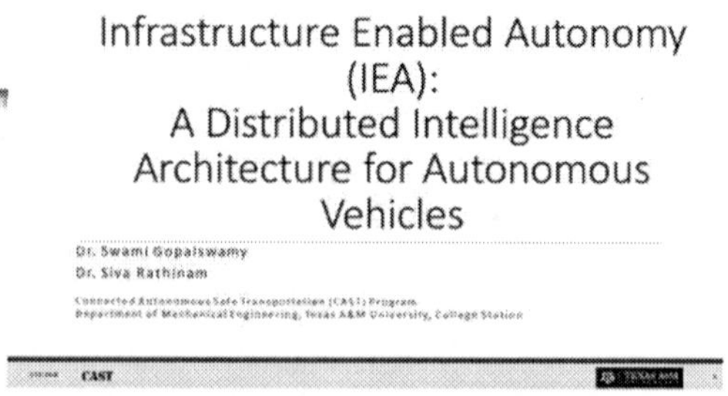

Current Responsibility Distribution

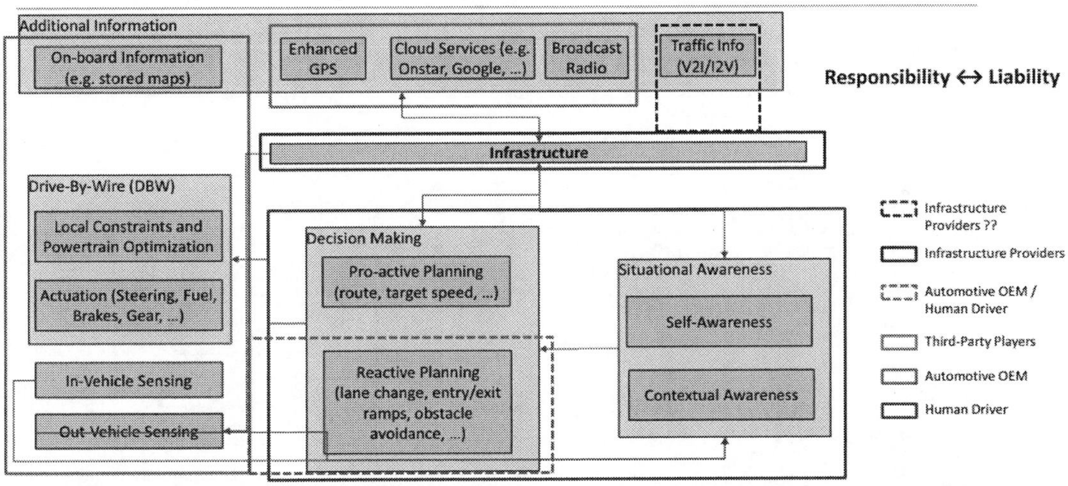

Needed Responsibility Distribution per Current AV Paradigm

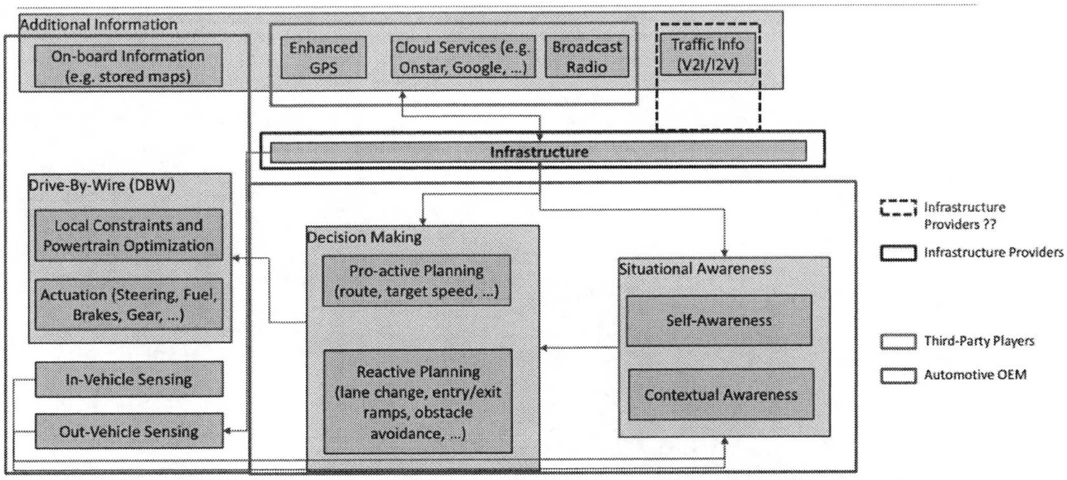

Proposed Responsibility (Re)Distribution per **New** AV Paradigm

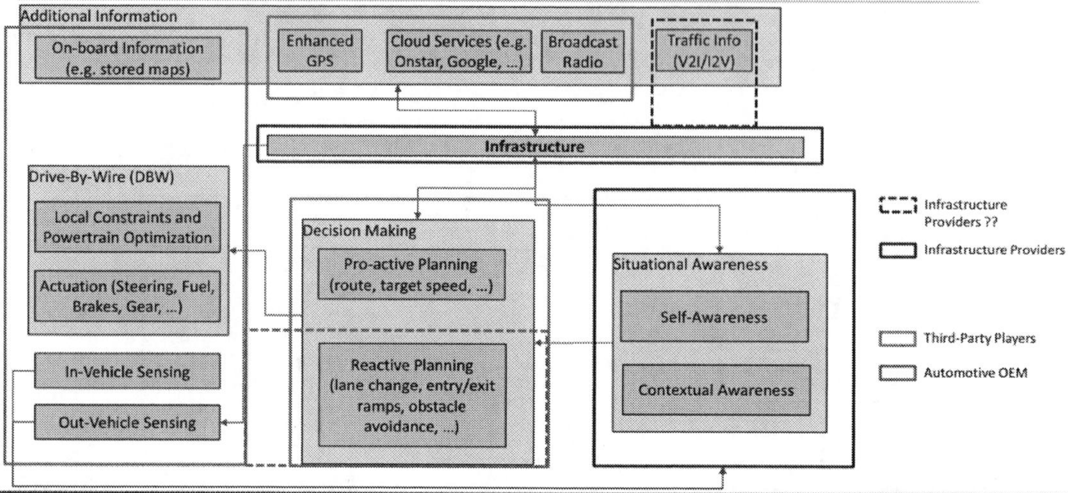

Concept Overview Schematic

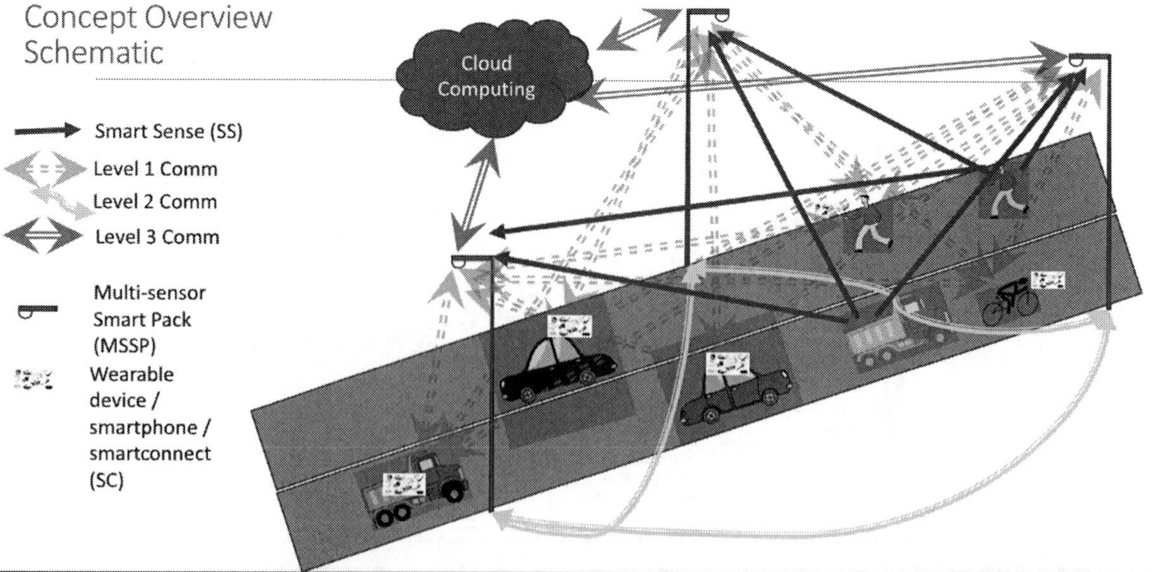

Core Concept

Separation of issues (expertise):

- Automakers shall focus on the drive-by-wire capabilities and make it even more perfect
- The Infrastructure shall become smart and enable awareness
  - "On-Demand Automation" enabled through mutual consent of driver and infrastructure

Smart-Infrastructure enabled Awareness

- Higher reliability and precision because:
  - Precise base-geolocation using "fixed" MSSP – independent of weather, cloud cover, (GPS Denied environment) etc
  - Occlusions overcome!
  - Sensor fusion across multiple neighbors
  - increased time and processing power (leveraging the cloud) for detection, classification, and probabilistic motion estimation of objects
  - Long Term memory of environment (e.g. pot holes, rocks, trees)

Deployment Model – Autonomy as a Service

Example Deployment Scenario

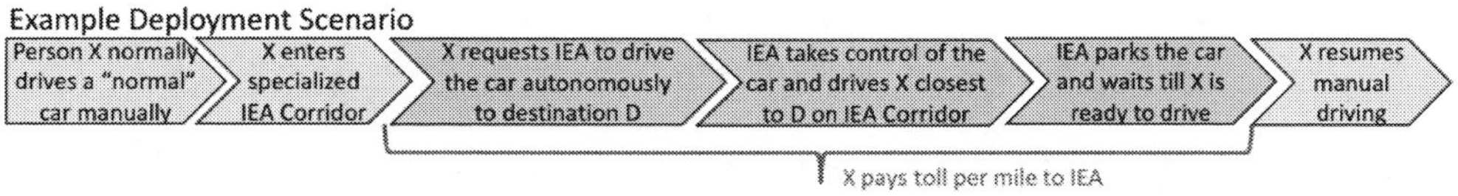

User pays only for usage of Autonomy (Autonomy as a Service)

Opportunity for monetization of Autonomy

- Private Infrastructure Players – avoid the traditional barriers with public sector infrastructure investments

Value Proposition for a Large Ecosystem

Automotive OEMs
- Focus effort and energy on core competency, incur no additional liabilities, still grow revenues!

Infrastructure Operators (e.g. toll-booth operators, cell phone operators, etc.)
- New market and opportunities for monetization of their services, breathing new life into a business that is getting more and more commoditized.

Device Makers
- Ripe business opportunities for entrepreneurs / businesses to manufacture and install MSSPs and SCs on the infrastructure.

SC Application Makers
- SC not just for driving the automobile. It can be used to communicate warnings and diagnostics to non-automated entities such as pedestrians and bicyclists and manually driven cars.
- Many opportunities for entrepreneurs / businesses to develop new "apps"– either working with the smartphones or the native SCs.

Law enforcement, Infrastructure Maintenance, and Traffic Management:
- The presence of the infrastructure and sensing capabilities provides opportunities for newer applications leveraging the infrastructure to support traffic management, infrastructure maintenance, and law enforcement.

Society at Large
- The new paradigm will accelerate the penetration of automated driving overall, and correspondingly accelerate the reaping of the societal, environmental and economic benefits expected from autonomous vehicles in general

BACKUP

Why Autonomous Vehicles

Table 2: Estimates of Annual Economic Benefits from AVs in the United States

	10%	50%	90%
Crash Cost Savings from AVs			
Lives Saved (per year)	1,100	9,600	21,700
Fewer Crashes	211,000	1,880,000	4,220,000
Economic Cost Savings	$5.5 B	$48.8 B	$109.7 B
Comprehensive Cost Savings	$17.7 B	$158.1 B	$355.4 B
Economic Cost Savings per AV	$430	$770	$960
Comprehensive Cost Savings per AV	$1,390	$2,480	$3,100
Congestion Benefits			
Travel Time Savings (M Hours)	756	1680	2772
Fuel Savings (M Gallons)	102	224	724
Total Savings	$16.8 B	$37.4 B	$63.0 B
Savings per AV	$1,320	$590	$550

Great Benefits

Source: Eno Report, October 2013

How can we accelerate achieving High Penetrations of Autonomous Vehicle Usage ?

Automotive Embedded Software Landscape

Safety a Growing Concern for many Automotive OEMs

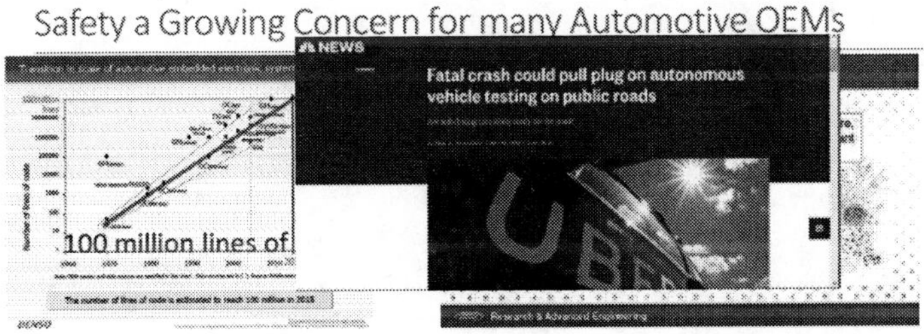

Dramatically amplified impact of failures of mechatronic systems – when machines, electronics and the society (humans) intersect

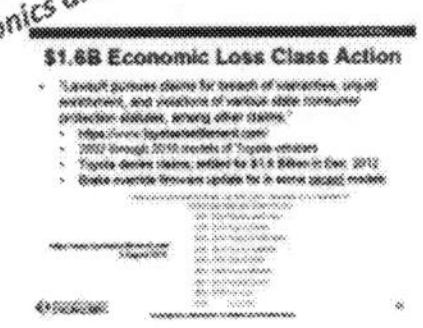

Dramatic Failures Led by Automotive, But Across Industries

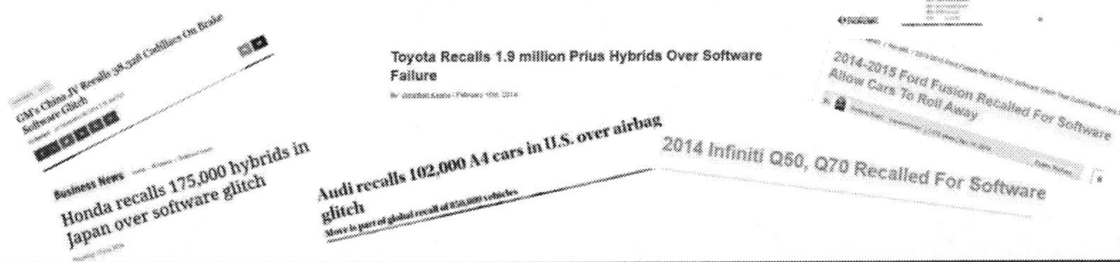

Infrastructure Enabled Autonomy

Autonomy-as-a-Service – A New Techno-Business Paradigm for CAVs

TTI/ CAST MEEN: Dr. Swami Gopalswamy

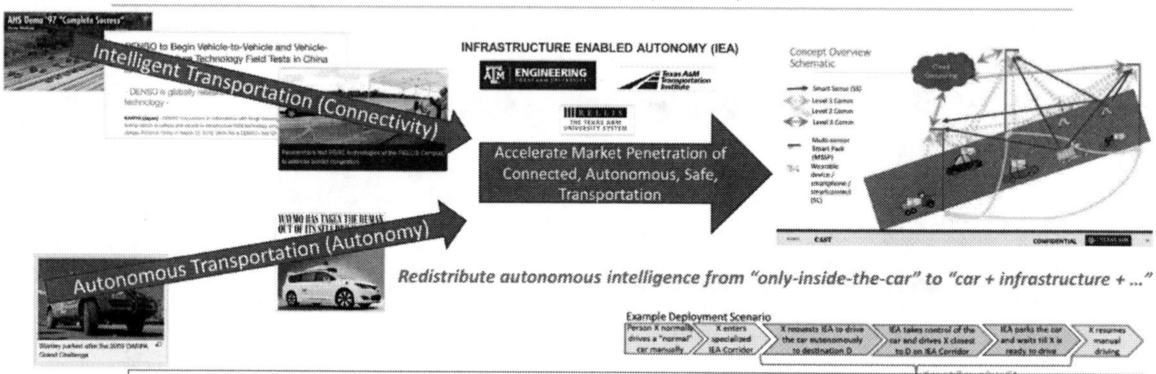

- "Autonomy as a Service" – new value proposition for the end customer
- Reduced Liability for Automotive OEM
- New Business Opportunities for Private Infrastructure Operators, Third Party Players, and automotive OEMs
- Integration with First Responders and Law Enforcement

Functional Schematic of IEA

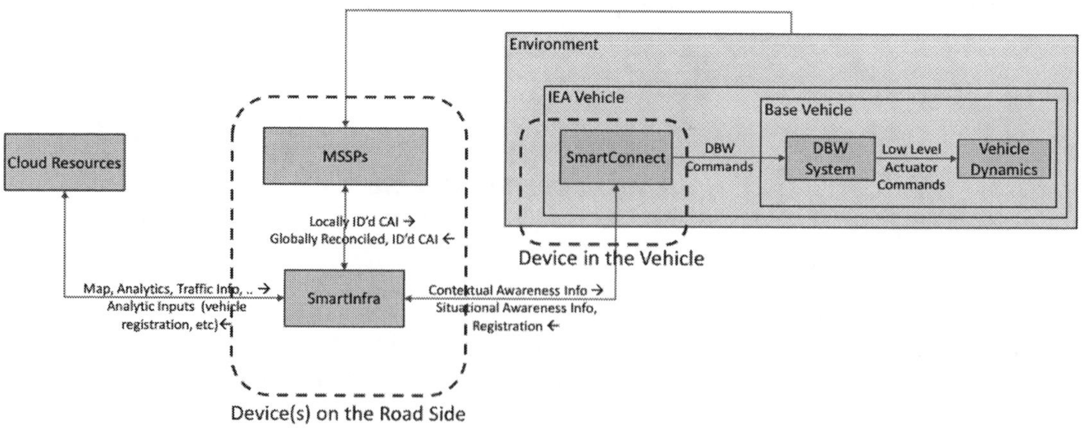

Local Roots » Global Reach

2018 Automated Vehicle Symposium

July 11, 2018, San Francisco

Breakout Session:
Accelerating CAV Market Penetration by Leveraging Infrastructure

A Legal Perspective

Miller Canfield, Joseph Gustavus

MILLER CANFIELD
millercanfield.com

I. Nomenclature

Terms for easy reference. . .

- **AV**-Autonomous (or Automated) Vehicles.
- **SIETC**-Special Infrastructure Enabled Traffic Corridors – AV traffic corridors using MSSPs and operated by AV Infrastructure Operators
- **MSSPs** -Multi-Sensor Smart Packs –devices monitoring SIETCs for AV purposes.
- **SA Data**-Situational Awareness Data-the AV environmental factors (outside AVs) monitored and collected by MSSPs along SIETCs.
- **SCs**-Smart Connects-devices embedded in or alongside MSSPs that transmit SA Data to AVs in that SIETC via a SA Data Carrier.
- **SA Data Carrier**-the private data carrier transmitting DA Data to AVs using the 5G Network or other sanctioned wireless SA Data transmission system.

MILLER CANFIELD 2
millercanfield.com

II. Government Immunity

- **Governmental (Sovereign) Immunity Doctrine** holds that government and its agencies will be immune from liability, except for very limited cases.
  - Doctrine likely applied in the AV context to federal, state and local governments.

MILLER CANFIELD
millercanfield.com
3

III. Recent State AV-Related Laws Passed

- To date, 29 states have enacted legislation related to AVCs.
- The Michigan Example:
  - **MI PA 332 of 2016**
    - Permits manufacturer R&D use of AVs on MI roads using employees.
    - Provides immunity to AV manufacturers for TP modifications without consent.
  - **MI PA 332 of 2016**
    - Allows only OEMs to participate in SAVE projects.
    - Permits deployment of on-demand self-driving taxis.
    - Hold OEMs completely responsible therefore.

MILLER CANFIELD
millercanfield.com
4

IV. Utilities and Toll Booth Operators

- An SIETC Operator may be analogous to a public utility (PU), such as an electric company.
- SIETC Operator furnishing SIETC Service to a municipality or AV drivers is manifestly a public service or public utility corporation, akin to a PU
- Conceivable that SIETC Operators would be subject, in part, to PU-type law
- PU liability not governed by contracts principles
- In most cases, strict product liability doctrine not applicable to PUs
- However, PU duty owed is determined by negligence test of foreseeability of harm i.e., no liability for unforeseeable failures to anticipate and guard against harm
- In the case of electric PUs, National Electric Safety Code
  - given force of law in some states
  - in other states, merely a floor below which PU safeguards cannot go

MILLER CANFIELD 5
millercanfield.com

V. Current Contracting Structure with Embedded Software and Telematicsclature

- For software embedded within **Electronic Control Units (ECUs)**
  - Brunt of liability is borne by OEMs
  - Tier 1 Supplier ECU developers secondarily so
  - EUS software and components indemnify customers and their customers
    - Warranty claims
    - Voluntary and government-mandated recalls
    - Product liability
    - Class action suits
- For **Telematics Systems** used in fleet tracking and vehicle diagnosis
  - Fleet tracking system providers contract with fleet owners directly
  - in most cases, leaving fleet owners to contract separately for wireless data carrier service
  - Wireless data carriers take on no liability for this arrangement
  - What result for SA Data Carriers?

MILLER CANFIELD 6
millercanfield.com

VI. Potential Legal Structure for AV Market

- Hybrid between PU law, Federal Regulation and Contract Law
- Federal Regulation applied to AV industry players beyond just OEMs
- Contract Law
  - Licensed Application Scope
  - Warranty Scope
  - Customer Satisfaction Campaigns and Voluntary Recalls
  - Government Mandated Recalls
  - Indemnification
  - Limitation of Liability
- Key AV Industry Players and Their Potential Liability Exposure
  - AV OEMs and AV Outfitters
  - AV Infrastructure Operators
  - AV Data Carriers for SA Data and AV Data
- Decision Making Algorithm for AV Navigation
  - Who owns responsibility for it?

MILLER CANFIELD 7
millercanfield.com

Our Local and International Presence Means People on the Ground Assisting with Your Complex Transactions and Litigation

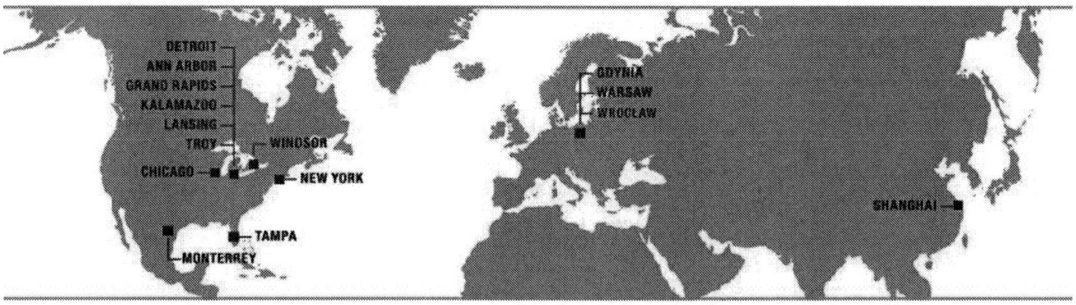

160+ Year reputation and legal leadership Established 1852 in Detroit, Michigan

300+ Lawyers and paralegals

165th Largest U.S. law firm according to the *National Law Journal*

13 "Lawyers of the Year" by *Best Lawyers*

135+ Lawyers in leading referral guides, *Chambers USA*, *Best Lawyers*, *Super Lawyers* and *European Legal Experts*

20+ Lawyers are Fellows of prestigious American Colleges, including *Bond Counsel, Employee Benefits, Mortgage Attorneys, Real Estate Lawyers, Trial Lawyers, Trust and Estate Counsel, College of Labor and Employment Lawyers,* and the *Litigation Counsel of America*

■ The designated firm in Michigan and Poland in the Employment Law Alliance, a network of top labor and employment lawyers in the world – a powerful resource for clients facing local or global issues

MILLER CANFIELD 8
millercanfield.com

Joseph D. Gustavus
Senior Principal
Miller Canfield

+1 (248) 267-3317
Gustavus@millercanfield.com

UNITED STATES ■ CANADA ■ MEXICO ■ POLAND ■ CHINA

MILLER CANFIELD
millercanfield.com

AUTOMATED VEHICLES SYMPOSIUM 2018
USERS. VEHICLES. INFRASTRUCTURE.
Driven by AUVSI and TRB

Breakout Session #30
Accelerating CAV Market Penetration by Leveraging Infrastructure

Discussion Topics:

- Low-Level Automation on Managed Lanes
- Ramp Metering for CAV's
- CAV Operations on Signalized Corridors

AUTOMATED VEHICLES SYMPOSIUM 2018
USERS. VEHICLES. INFRASTRUCTURE.
Driven by AUVSI and TRB

Breakout Session #30
Accelerating CAV Market Penetration by Leveraging Infrastructure

Low-Level Automation in Managed Lanes:

- Level 2, or even just Level 1 automation on separated facilities
  - Allow access to Managed Lanes as a enticement to equip vehicles
- CACC, dependent on V2V communications (DSRC?) for safe close-following
  - Could also be "platooning", which is fixed-distance following
- Speed controls sent by operators to vehicles using I2V for speed harmonization?
- Peak Demand Hours only! Normal operation during off-peak or low demand

AUTOMATED VEHICLES SYMPOSIUM 2018
USERS. VEHICLES. INFRASTRUCTURE.
Driven by AUVSI and TRB

Breakout Session #30
Accelerating CAV Market Penetration by Leveraging Infrastructure

Ramp Metering for CAV's:

- Vehicles on the freeway broadcast their position (BSM) using V2I (DSRC?)
- Algorithms in the Ramp Metering System identify gaps in traffic and release vehicles accordingly
- Waiting vehicle types are known (car, bus, truck) to account for proper acceleration time
- In the longer term, gaps in freeway traffic could be opened automatically to accommodate more entering vehicles
- "Bus-on-Shoulder" operations?

AUTOMATED VEHICLES SYMPOSIUM 2018
USERS. VEHICLES. INFRASTRUCTURE.
Driven by AUVSI and TRB

Breakout Session #30
Accelerating CAV Market Penetration by Leveraging Infrastructure

CAV Operations on Signalized Corridors:

- "Traffic Optimization on Signalized Corridors" (TOSCo)
- "Group" vehicles together, and schedule green time for the entire group to move through the intersection
- Cross-traffic is also scheduled to use the space between groups to cross the corridor
- Also moves groups of stopped vehicles through intersections more smoothly than with manual driving
- Works best on one-way corridors (fewer turning movements)

TUMCREATE

Impacts of Autonomous and Connected Vehicles on the Physical Infrastructure

Aybike Ongel, TUMCREATE

John Harvey, University of California Davis

AVS 2018 San Francisco

07.11.2018

TUMCREATE

Roadmap for Deployment of AVs in Singapore

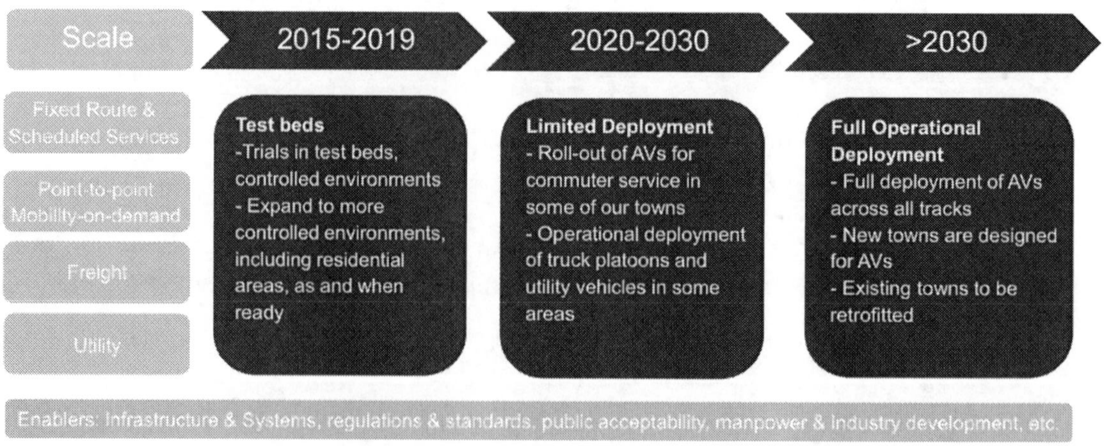

2

TUMCREATE

Singapore Smart Nation Sensor Platform

- Public agencies to optimise their sensor deployment needs and facilitate the sharing of resources, sensor data and data insights amongst agencies
- Agencies to share common sensor communications backhaul as well as the data collected
- So far… CCTVs in public areas, environmental sensors and connectivity infrastructure
- Future… collaborate with the Land Transport Authority (LTA) to leverage its lamppost infrastructure to test the feasibility of deploying a shared network for wired and wireless sensors

3

TUMCREATE

TUMCREATE's mission: Improve the travel experience for people of Singapore

The Mission of TUMCREATE is to seek the ultimate public transport system for the people of Singapore. Our innovative road transport solutions will **provide high comfort and a positive travel experience**, best protection of the environment and maximum benefit to the society and the economy.

www.tum-create.edu.sg

4

TUMCREATE

The DART (Dynamic Autonomous Road Transit)

A Semi-Rapid Transit System for Singapore

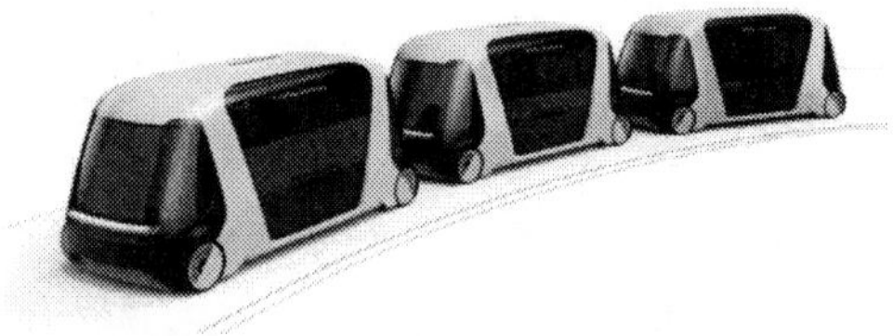

System components:
DART – MODULE: Autonomous modular electric vehicles that can dynamically platoon
DART – INFRASTRUCTURE: Facilities for charging, road monitoring, stops boarding and alighting
DART – POWER: Intelligent power supply and power management for the electric vehicles
DART – CONNECT: Inter-vehicle and infrastructure communication mechanisms
DART – CENTRAL: Optimal System Control through intelligent information exchange

5

TUMCREATE

Features of the DART System

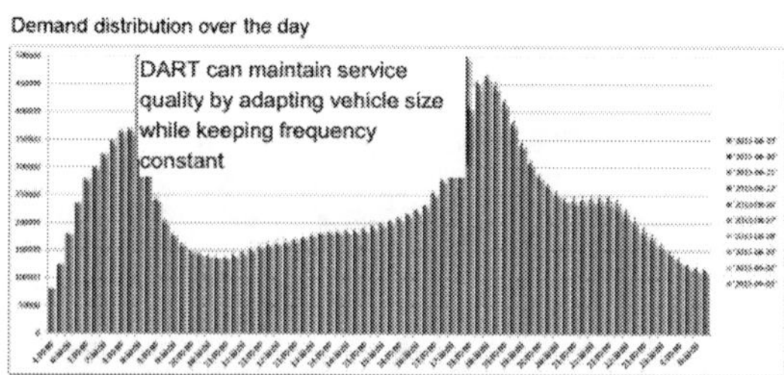

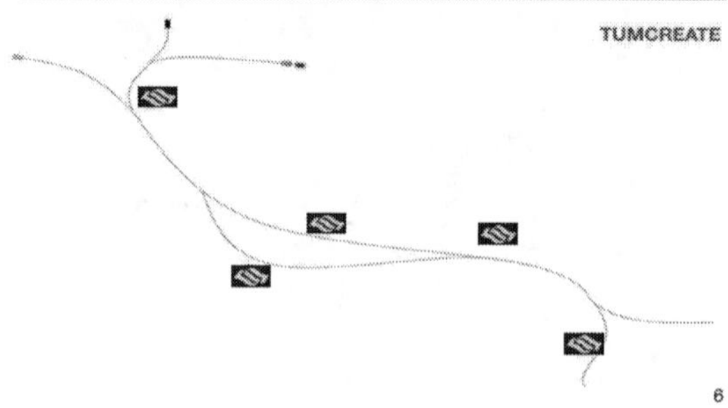

1. **Versatile operation** using modular autonomous vehicles -platooning
2. **Flexible capacity** during peak hours using dynamic platoons of up to 10 x 30-passenger modules
3. **Virtual Right of Way** between stops for faster travel
4. **Cleaner urban environment** by using electric vehicles with zero local emissions

6

TUMCREATE

Modular Autonomous Vehicles

Max axle load:
Front: 7 t
Drive: 12 t

Max axle load:
Front: 7 t
Center: 9
Drive: 12 t

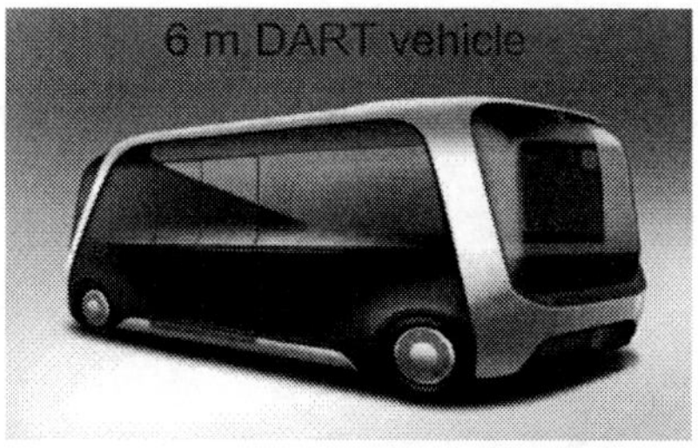

Axle load:
Front: 3.5 t
Drive: 3.5 t

ESALs for 12 m 2 axle bus = 5.6
ESALs for 12 m 3 axle bus = 7.2
ESALs for 6 m DART= 0.07

7

TUMCREATE

Independent Steer-by-wire Steering

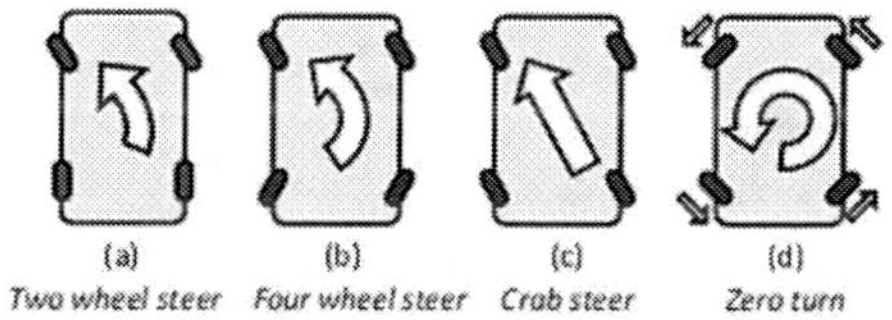

https://contest.techbriefs.com/2015/entries/automotive-transportation/5236

Many modes can be achieved:
- Front wheel steering
- All wheel steering
- Lateral motion
- Zero point turning

8

TUMCREATE

Independent Steer-by-wire Steering – cont'd

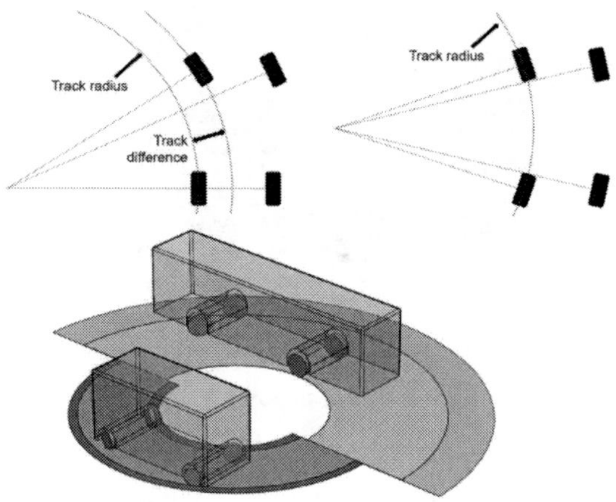

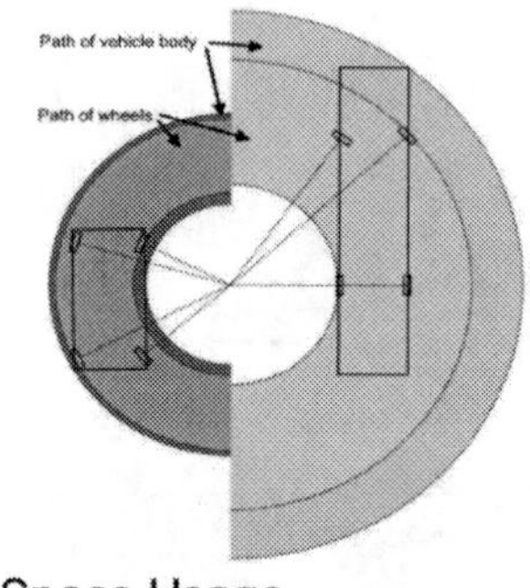

- Reduce Road Space Usage
- Smaller Turning Radius (11 m → 6.5 m)

9

TUMCREATE

Lane Keeping

- Smaller lane widths
- Less wander
- Regular maintenance of road surfaces
- Regular maintenance of road markings
- Signage quality
- Improved drainage

10

TUMCREATE

Conclusions

- Platoon → reduced headway → increased capacity → increased traffic volume → reduced pavement life
  → higher speeds → faster cracking damage for asphalt pavements from shorter rest periods
- Drive-by-wire (4WIS) → increased steering angle → reduced turning circle → increased shear stresses at speed → reduced pavement life from rutting
- Modular systems → reduced weight → reduced stresses → increased pavement life
- Sensing technologies → less wander → reduced lane width → increased rutting → decreased pavement life
  → more frequent surface maintenance/road markings
  → better road surface drainage
  → signage quality

11

Why Autonomous Vehicles

Table 2: Estimates of Annual Economic Benefits from AVs in the United States

	10%	50%	90%
Crash Cost Savings from AVs			
Lives Saved (per year)	1,100	9,600	21,700
Fewer Crashes	211,000	1,880,000	4,220,000
Economic Cost Savings	\$5.5 B	\$48.8 B	\$109.7 B
Comprehensive Cost Savings	\$17.7 B	\$158.1 B	\$355.4 B
Economic Cost Savings per AV	\$430	\$770	\$960
Comprehensive Cost Savings per AV	\$1,390	\$2,480	\$3,100
Congestion Benefits			
Travel Time Savings (M Hours)	756	1680	2772
Fuel Savings (M Gallons)	102	224	724
Total Savings	\$16.8 B	\$37.4 B	\$63.0 B
Savings per AV	\$1,320	\$590	\$550

Great Benefits

Source: Eno Report, October 2013

How can we accelerate achieving High Penetrations of Autonomous Vehicle Usage ?

TUMCREATE

Impacts of Autonomous and Connected Vehicles on the Physical Infrastructure

Aybike Ongel, TUMCREATE

John Harvey, University of California Davis

AVS 2018 San Francisco

07.11.2018

TUMCREATE

Roadmap for Deployment of AVs in Singapore

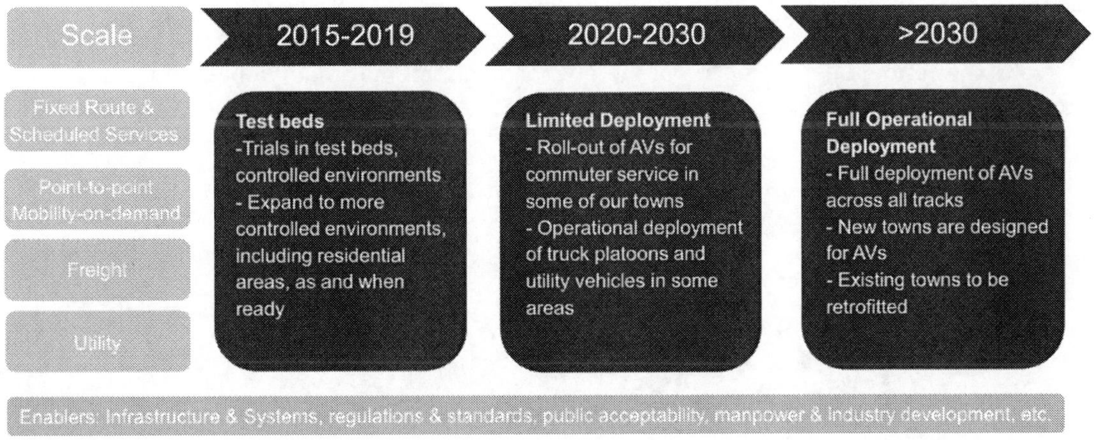

2

TUMCREATE

Singapore Smart Nation Sensor Platform

- Public agencies to optimise their sensor deployment needs and facilitate the sharing of resources, sensor data and data insights amongst agencies
- Agencies to share common sensor communications backhaul as well as the data collected
- So far… CCTVs in public areas, environmental sensors and connectivity infrastructure
- Future… collaborate with the Land Transport Authority (LTA) to leverage its lamppost infrastructure to test the feasibility of deploying a shared network for wired and wireless sensors

3

TUMCREATE

TUMCREATE's mission: Improve the travel experience for people of Singapore

The Mission of TUMCREATE is to seek the ultimate public transport system for the people of Singapore. Our innovative road transport solutions will **provide high comfort and a positive travel experience**, best protection of the environment and maximum benefit to the society and the economy.

4

TUMCREATE

The DART (Dynamic Autonomous Road Transit)

A Semi-Rapid Transit System for Singapore

System components:
DART – MODULE: Autonomous modular electric vehicles that can dynamically platoon
DART – INFRASTRUCTURE: Facilities for charging, road monitoring, stops boarding and alighting
DART – POWER: Intelligent power supply and power management for the electric vehicles
DART – CONNECT: Inter-vehicle and infrastructure communication mechanisms
DART – CENTRAL: Optimal System Control through intelligent information exchange

5

TUMCREATE

Features of the DART System

1. **Versatile operation** using modular autonomous vehicles -platooning
2. **Flexible capacity** during peak hours using dynamic platoons of up to 10 x 30-passenger modules
3. **Virtual Right of Way** between stops for faster travel
4. **Cleaner urban environment** by using electric vehicles with zero local emissions

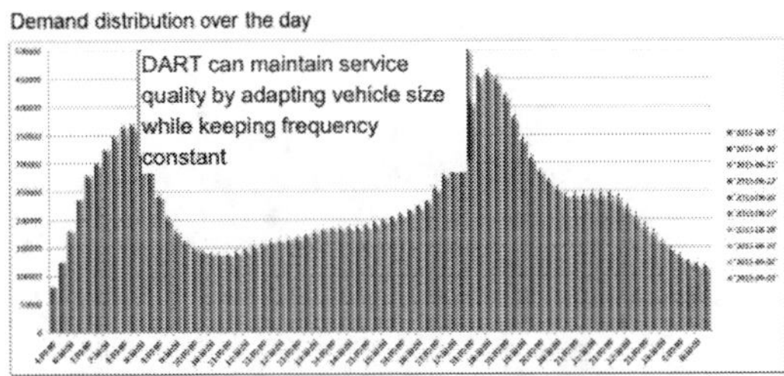

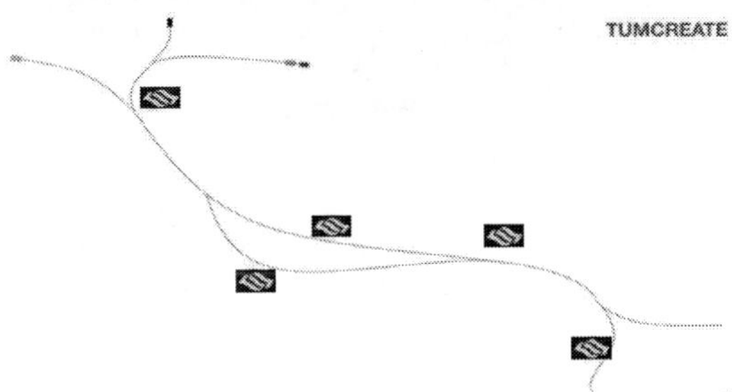

6

TUMCREATE

Modular Autonomous Vehicles

Max axle load:
Front: 7 t
Drive: 12 t

Max axle load:
Front: 7 t
Center: 9
Drive: 12 t

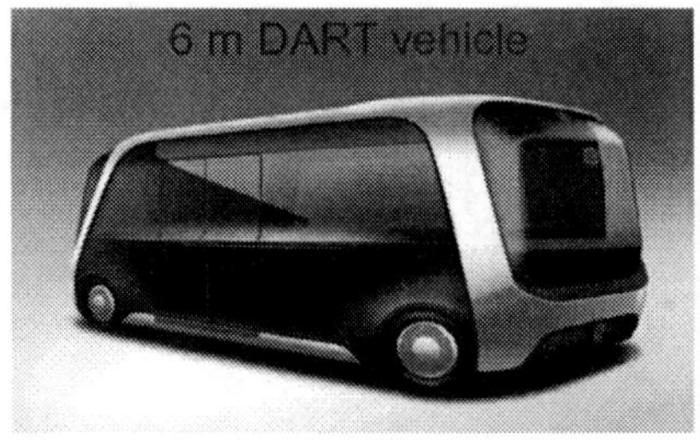

Axle load:
Front: 3.5 t
Drive: 3.5 t

ESALs for 12 m 2 axle bus = 5.6
ESALs for 12 m 3 axle bus = 7.2
ESALs for 6 m DART= 0.07

7

TUMCREATE

Independent Steer-by-wire Steering

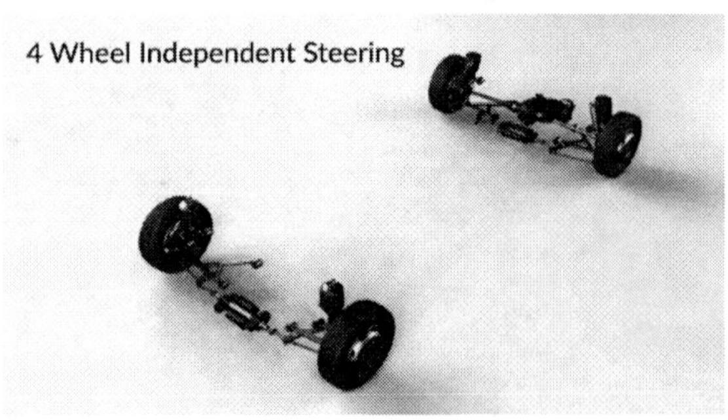

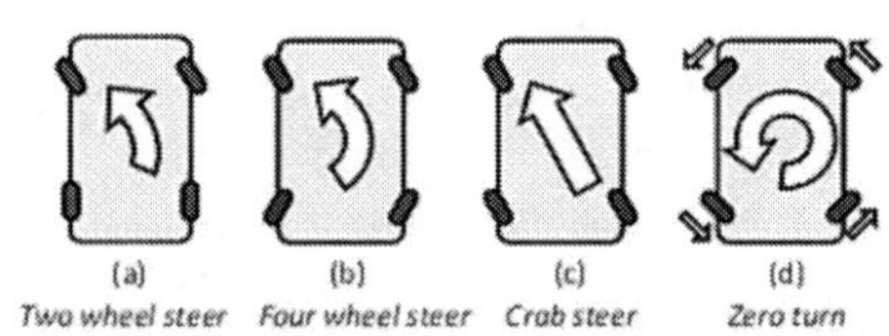

https://contest.techbriefs.com/2015/entries/automotive-transportation/5236

Many modes can be achieved:
- Front wheel steering
- All wheel steering
- Lateral motion
- Zero point turning

8

TUMCREATE

Independent Steer-by-wire Steering – cont'd

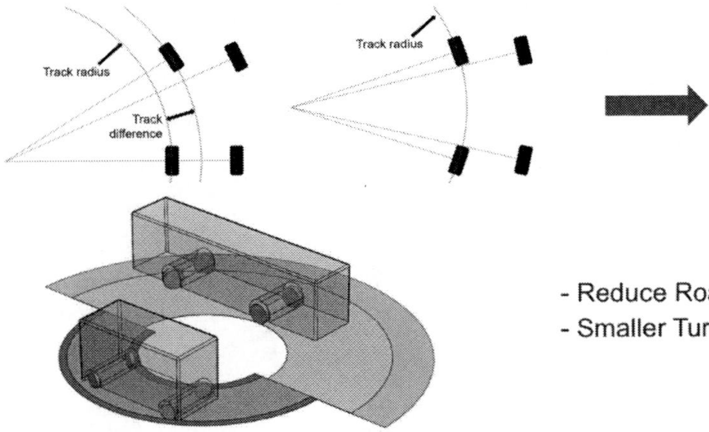

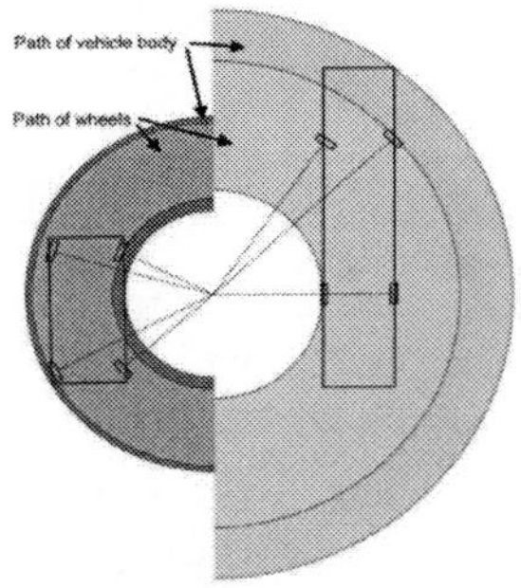

- Reduce Road Space Usage
- Smaller Turning Radius (11 m → 6.5 m)

9

TUMCREATE

Obstacle Detection

A. Bódis-Szomorú, T. Dabóczi, and Z. Fazekas, "A Lane Detection Algorithm Based on Wide-Baseline Stereo Vision for Advanced Driver Assistance," IEEE Instrum. Meas. Technol. Conf., 2009

- Smaller lane widths
- Less wander
- Regular maintenance of road surfaces
- Regular maintenance of road markings
- Signage quality
- Improved drainage

10

TUMCREATE

Conclusions

- Platoon → reduced headway → increased capacity → increased traffic volume → reduced pavement life

  → higher speeds → faster cracking damage for asphalt pavements from shorter rest periods
- Drive-by-wire (4WIS) → increased steering angle → reduced turning circle → increased shear stresses at speed → reduced pavement life from rutting
- Modular systems → reduced weight → reduced stresses → increased pavement life
- Sensing technologies → less wander → reduced lane width → increased rutting → decreased pavement life

  → more frequent surface maintenance/road markings

  → better road surface drainage

  → signage quality

11

Solutions that will serve a spectrum of vehicles

Human and CAVs will coexist on roadways for many years

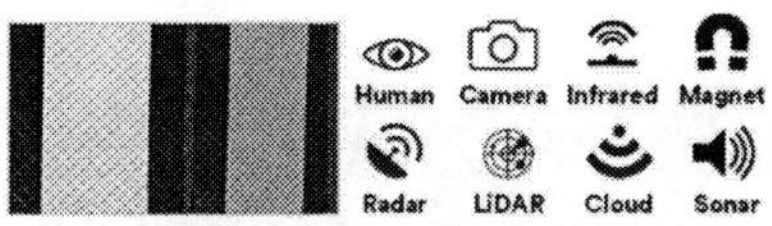

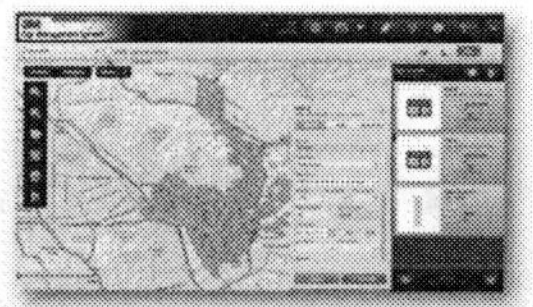

Material Solutions

- Robust portfolio of solutions
- Goal: Read by all sensor modalities in poor weather
- Enabling:
  - safe human & vehicle navigation
  - cyber-physical security

Material-Data Solutions

- Clear, standardized navigational markers
- Facilitate I2V info exchange
- Providing:
  - Safety system with multiple layers of info
  - cyber-physical security

Information System Solutions

- Aid in roadway and work zone safety and in assessing infrastructure* quality
- Suite of information solutions
- Enabling safe and secure navigation by humans and CAVs

Engineered context for simplified decision making

Localization *Boundaries* *Uniqueness*
Classification *Guidance*

Data exchange to improve safety

* Infrastructure means anything that DOTs spend on and have responsibility to ensure promotes safety

3M

UNIVERSITY of SOUTH FLORIDA

Autonomous Vehicle Longitudinal Trajectory Control in Mixed Traffic with Machine Learning

Yu (Crystal) Wang
Dr. Xiaopeng Li

CATS (Connected & Autonomous Transportation Systems) Lab
Department of Civil and Environmental Engineering
University of South Florida

7/11/2018

2018 AVS Breakout Session:
Artificial Intelligence and Deep Machine Learning Tools and Algorithms for Automated Vehicles: The State of the Art and Practice

1

UNIVERSITY of SOUTH FLORIDA

Introduction

- Connected Automated Vehicles (CAV) – A cue for traffic oscillation

Sources of photo: online

2

Introduction

- Stop-and-go traffic brings lots of adverse impacts

Video Presentation in Original e-media

Sources of video: online

Introduction

- Human-driven Vehicles V.S. Connected Automated Vehicles (CAV)
  - Connected → Information sharing
  - Automated → Precise control

Sources of photo: online

Introduction

- Existing control methods
  - Car-following model: ACC, CACC (Milanés, Shladover, 2014; Xiao, Wang, Arem, 2017)
  - Only rely on individual vehicle sensors, not utilize traffic information (Ahn, Rakha, Park, 2013; Yang, Jin, 2014)
  - Separate prediction from control (Wang, Zhang,Yi, 2017)
- Proposed control method
  - Adaptive controller
  - Data-driven control
  - Reinforcement learning

UNIVERSITY of SOUTH FLORIDA

Controller design

- Problem statement

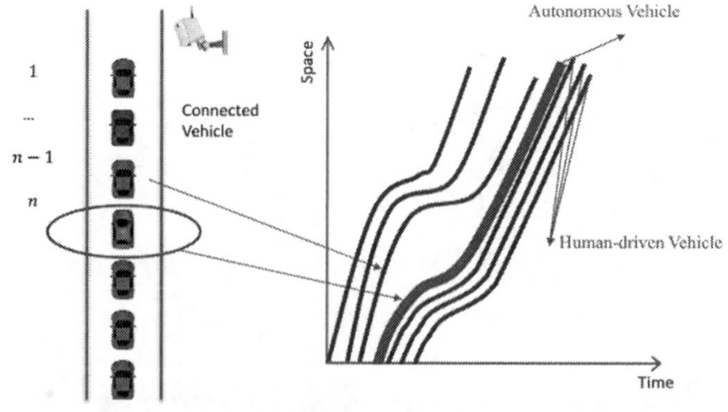

Data

- Field tests setup
  - Objective: Collect complete traffic oscillation data on highway with high resolution GPS devices
  - Location: 5 km highway section of China National Highway G202 (Lanxi direction) in Harbin, Heilongjiang, China
  - By Professor An's team and Dr. Li's team

UNIVERSITY of SOUTH FLORIDA

Data

- Field tests setup
  - 12 vehicles with the identical size and style (i.e., Kia K5)
  - High resolution GPS device, called GPS-RTK based STAR-RTK-M9
    - The data measurement errors are within ± 1 m for location and ± 1 km/h for speed.
    - The sampling frequency is 20 Hz (i.e., collecting 20 data per second).

Sources of photo: online

UNIVERSITY of SOUTH FLORIDA

Data

- Field tests setup
  - Bi-direction and four lanes highway
  - Black line: the field tests segment in google map
  - Blue line: the moving direction
  - No traffic lights
  - No overtaking/ lane changing

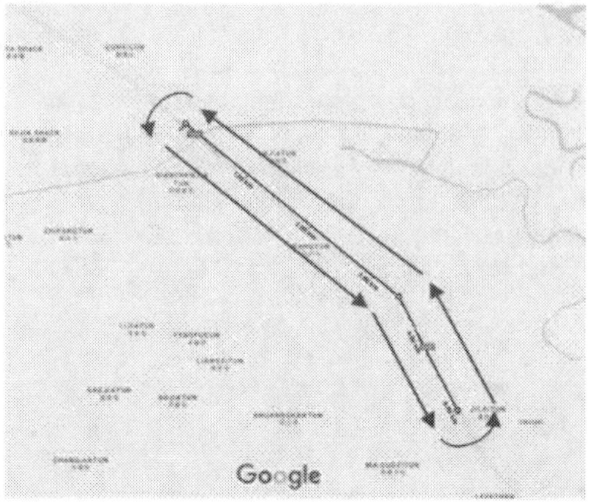

Sources of photo: online

9

UNIVERSITY of SOUTH FLORIDA

Data

- Field tests setup
  - The first vehicle is required to drive with pre-designed speed and accelerate/decelerate

	Time period (s)	Designed acceleration (km/h/s)	Speed range (km/h)	Average speed (km/h)
Case 1	30	1	10	40
Case 2	30	2	10	40
Case 3	30	1	20	40
Case 4	30	2	20	40
Case 5	120	1	10	40
Case 6	120	2	10	40
Case 7	120	1	20	40
Case 8	120	2	20	40
Case 9	30	1	10	70
Case 10	30	2	10	70
Case 11	30	1	20	70
Case 12	30	2	20	70
Case 13	120	1	10	70
Case 14	120	2	10	70
Case 15	120	1	20	70
Case 16	120	2	20	70

10

UNIVERSITY of SOUTH FLORIDA

Data

- Field tests results
  - Speed of the leading vehicle

Time period 120
Average speed 40
Speed range 10

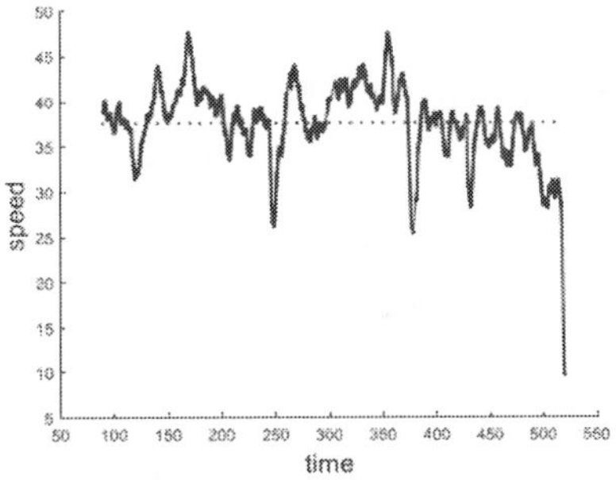

Case 5

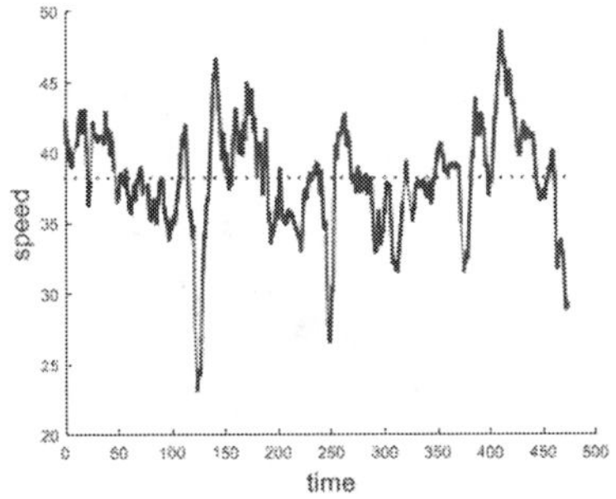

Case 6

Time period 120
Average speed 40
Speed range 10

11

UNIVERSITY of SOUTH FLORIDA

Data

- Field tests results
  - Lead vehicle speed trajectory

Time period 30
Average speed 70
Speed range 20

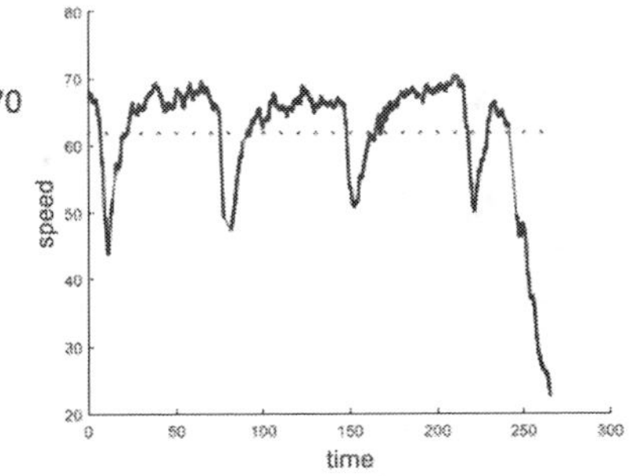

Case 11

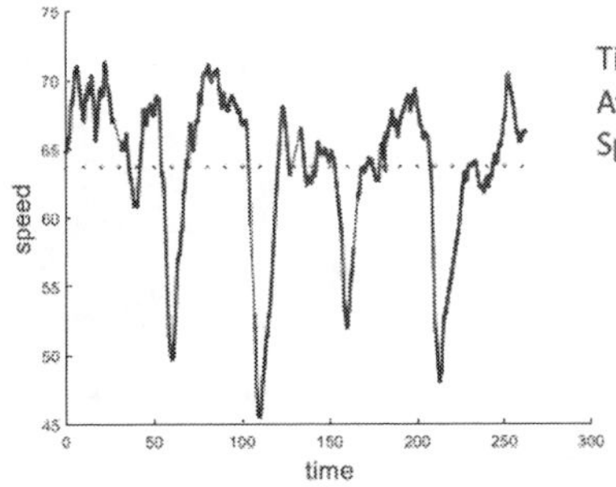

Case 12

Time period 30
Average speed 70
Speed range 20

12

UNIVERSITY of SOUTH FLORIDA

Data

- Field tests results

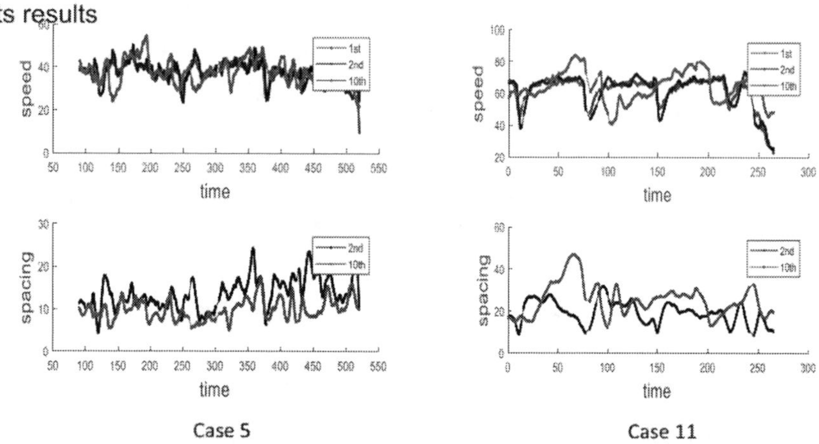

UNIVERSITY of SOUTH FLORIDA

Controllers

- Car-following model based controller
- Neural network based controller (ongoing)

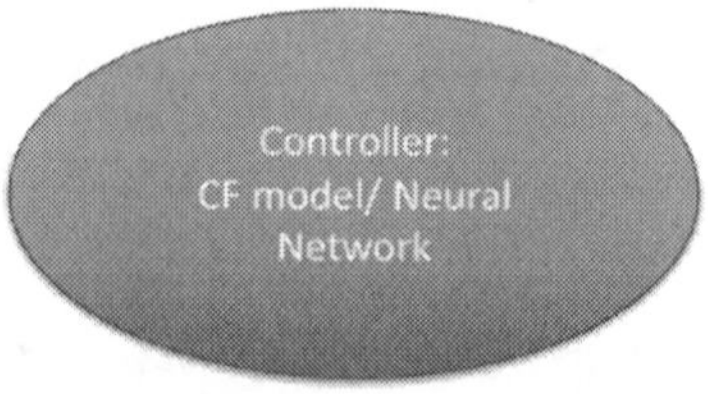

UNIVERSITY of SOUTH FLORIDA

Car-following model based controller

- Controller:

$$a[t,n] = f(\,w[0] * s[t - tau, n] + w[1] * v[t - tau, n] + w[2] * vd[t - tau, n] + w[3] * a[t - tau, n - 1] + b)$$

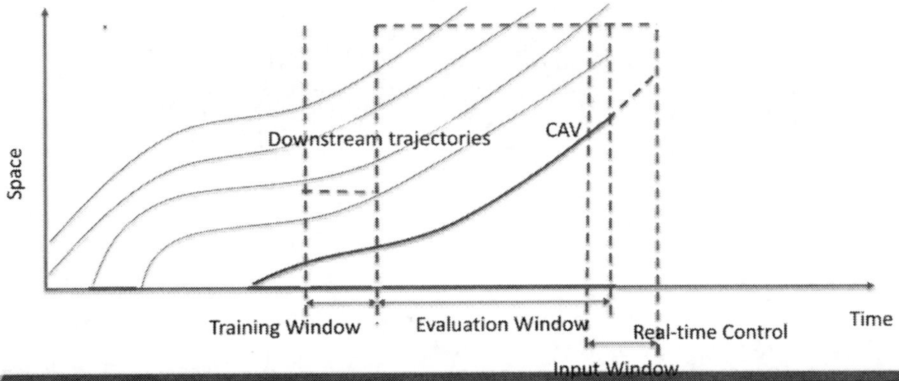

UNIVERSITY of SOUTH FLORIDA

Car-following model based controller

- Controller:

$$a[t,n] = f(\,w[0] * s[t - tau, n] + w[1] * v[t - tau, n] + w[2] * vd[t - tau, n] + w[3] * a[t - tau, n - 1] + b)$$

- Input data structure:
  - (depth, width, height) = (4, 2, 1)
  - 4 variables: spacing, speed, speed difference, acceleration
  - 2 vehicles: preceding vehicle, controlled vehicle
- Perceptron
- Rectify: safety constraint
- Optimize: Gradient decent
- Objective function:

$$min \sum 0.001 * gap - 10 * fuel\ consumption - 1 * reverse\_time\_to\_collision$$

UNIVERSITY of SOUTH FLORIDA

Car-following model based controller

- Results

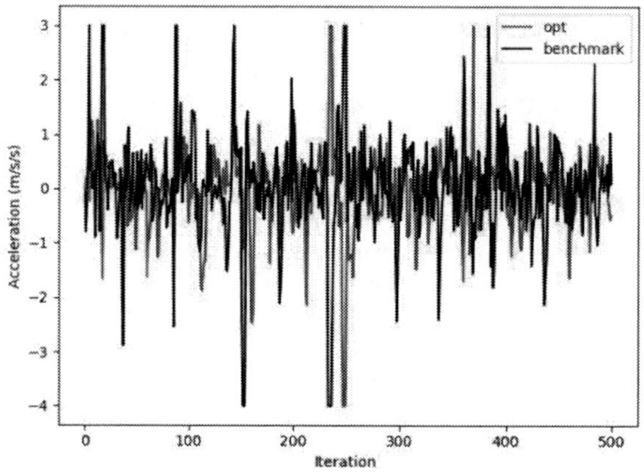

17

UNIVERSITY of SOUTH FLORIDA

Car-following model based controller

- Results

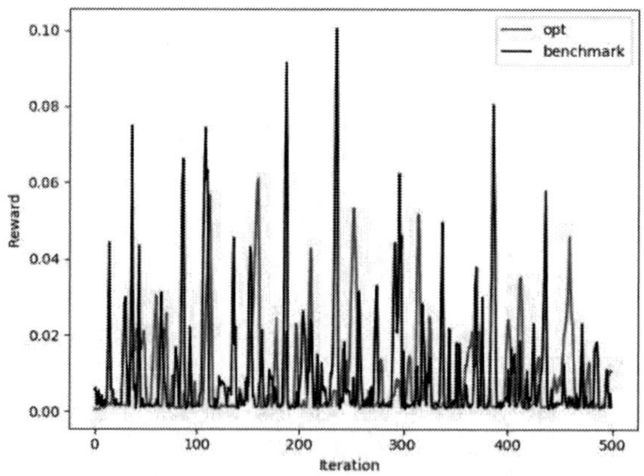

Performance:
improved by 18.3%
Computation time:
0.4s on CPU

18

UNIVERSITY of SOUTH FLORIDA

Car-following model based controller

- Controller:

$$a[t,n] = f(\sum_{0}^{t}\sum_{0}^{n}\sum_{0}^{p} w[t,n,p] * inputdata[t,n,p] + b)$$

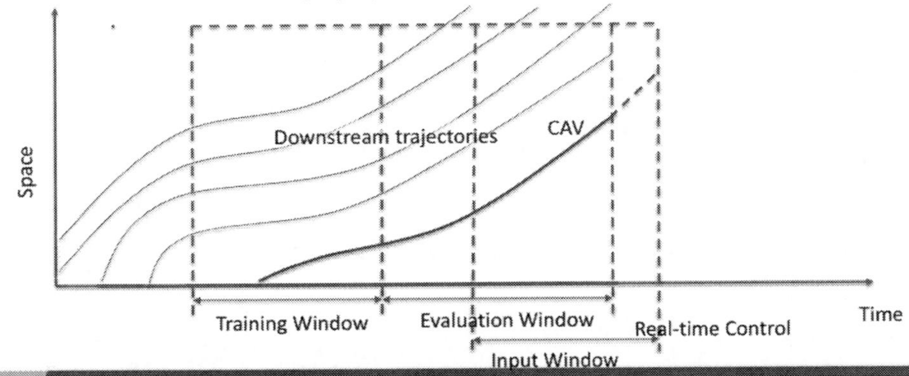

19

UNIVERSITY of SOUTH FLORIDA

Car-following model based controller

- Data-driven controller

$$a[t,n] = f(\sum_{0}^{t}\sum_{0}^{n}\sum_{0}^{p} w[t,n,p] * inputdata[t,n,p] + b)$$

- Input data structure:
  - (depth, width, height) = (p, n, t)
  - P= 4 variables: spacing, speed, speed difference, acceleration
  - N=12 vehicles: 11 preceding vehicles, controlled vehicle
  - T= 35 seconds: enough to capture the one oscillation period

20

UNIVERSITY of SOUTH FLORIDA

Neural network based controller

- 2 convolution layers

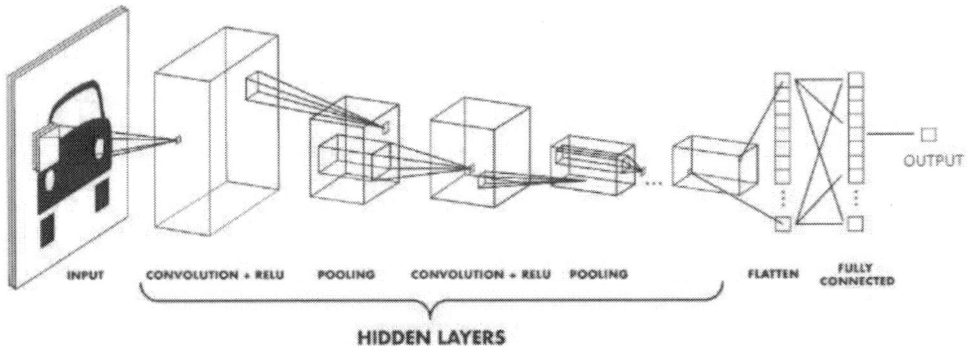

21

UNIVERSITY of SOUTH FLORIDA

Neural network based controller

- Parameter setting:
  - Filter for the first CNN layer: 8 3*3 filters at stride 1
  - Filter for the first pooling layer: 2*2 filters at stride 2
  - Filter for the second CNN layer: 16 3*3 filters at stride 1
  - Filter for the second pooling layer: 2*2 filters at stride 2
  - Fully connected: 8 neurons

22

UNIVERSITY of SOUTH FLORIDA

Future research plans

- Mutli-CAV cooperative control
- Include information of neighborhood vehicles
- Pure CAV data

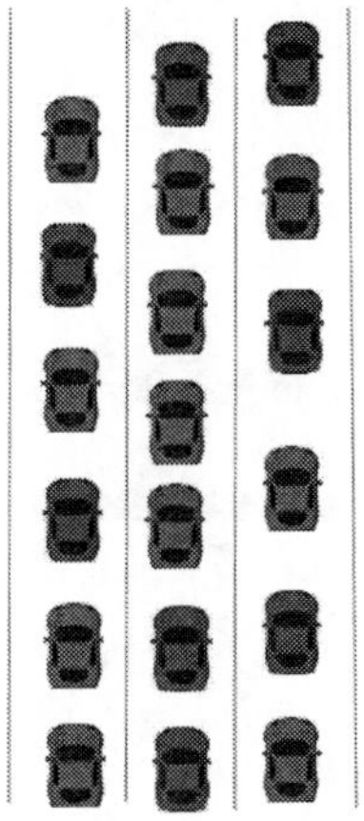

23

UNIVERSITY of SOUTH FLORIDA

Thank You!

Crystal Wang

24

The Potential Role of AI on the Impacts of Partial Penetrations of Connected and Automated Vehicles

Jackeline Rios Torres[1], Andreas Malikopoulos[2]

1. Oak Ridge National Laboratory- riostorresj@ornl.gov
2. University of Delaware

Automated Vehicles Symposium 2018
July 11, 2018

Acknowledgements

- DOE SMART Mobility Initiative
- Laboratory Directed Research and Development Program of the Oak Ridge National Laboratory
  - Eugene P. Wigner Fellowship
- ARPA-E Nextcar program

2

Automobiles have revolutionized the way we move…

Late 1800's

Early 1900's

Now

And they are key in our society's development….

OAK RIDGE National Laboratory | NATIONAL TRANSPORTATION RESEARCH CENTER

But unintended consequences came along…

> 2.2 million non fatal injuries

~ 37,500 deaths/year →

X 100

1.7 billion metric tons of CO^2

~70% of total petroleum consumption

A driver spends ~40 hours/year stuck in traffic
→ ~ US$160 billons/year

Sources:
National Highway Traffic Safety Administration (NHTSA), Fatal Traffic Crash Data," 2016
Texas A&M Transportation Institute, 2015 Urban Mobility Report
U.S. Department of Transportation. National Transportation Statistics, 2012.

OAK RIDGE National Laboratory | NATIONAL TRANSPORTATION RESEARCH CENTER

Current trends are not helping!

- Population is growing
- Life expectancy is higher
- Will millennials be less interested in driving?
- New transportation business models
- New mobility concepts
- …

UBER lyft

OAK RIDGE National Laboratory | NATIONAL TRANSPORTATION RESEARCH CENTER

5

Potential changes in energy consumption due to vehicle automation

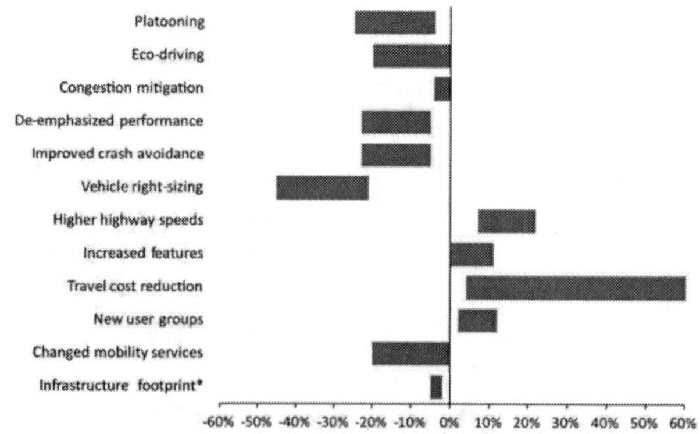

Understanding the implications of the technology is critical!

Zia Wadud, Don MacKenzie, Paul Leiby, "Help or hindrance? The travel, energy and carbon impacts of highly automated vehicles", Transportation Research Part A, Volume 86, April 2016, Pages 1-18

OAK RIDGE National Laboratory | NATIONAL TRANSPORTATION RESEARCH CENTER

Outline

- Exploring potential benefits and implications of Connected and Automated Vehicles (CAVs) coordination
  - Optimal coordination
  - Full CAVs penetration
  - Partial CAVs penetration
- Challenges and Opportunities in the area of Artificial Intelligence

7

Merging roads are common sources of traffic incidents and bottlenecks

8

Can we optimally coordinate the merging process?

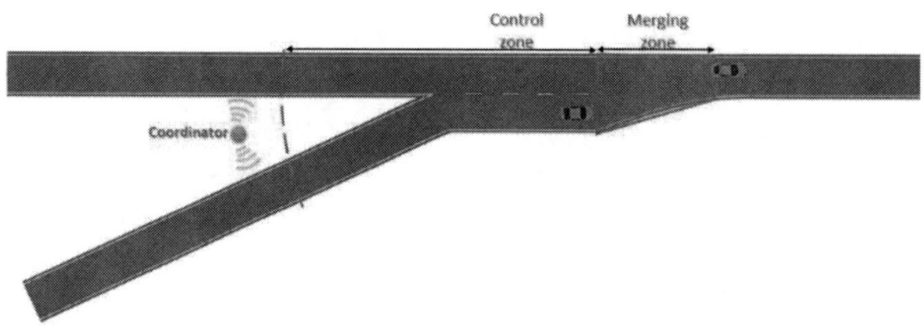

We aim at coordinating the vehicles on the merging scenario while reducing the energy consumption

OAK RIDGE National Laboratory | NATIONAL TRANSPORTATION RESEARCH CENTER

9

Optimal coordination framework

- Based on optimal control
- Energy efficient
- Closed-form
- Implementable in real time

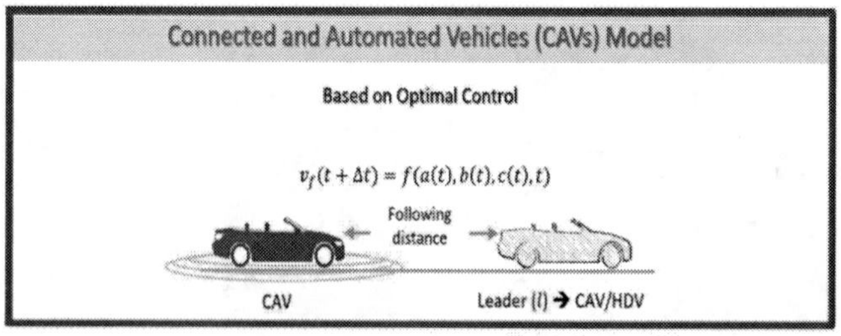

Assumption:

- Vehicles share their state with each other

We aim to compute the optimal acceleration/deceleration of each CAV

J. Rios-Torres and A. A. Malikopoulos, "A Survey on the Coordination of Connected and Automated Vehicles at Intersections and Merging at Highway On-Ramps," IEEE Trans. Intell. Transp. Syst., pp. 1–12, 2016.

J. Rios-Torres and A. A. Malikopoulos, "Automated and Cooperative Vehicle Merging at Highway On-Ramps," in IEEE Transactions on Intelligent Transportation Systems, vol. 18, no. 4, pp. 780-789, April 2017. doi: 10.1109/TITS.2016.2587582

OAK RIDGE National Laboratory | NATIONAL TRANSPORTATION RESEARCH CENTER

Problem formulation

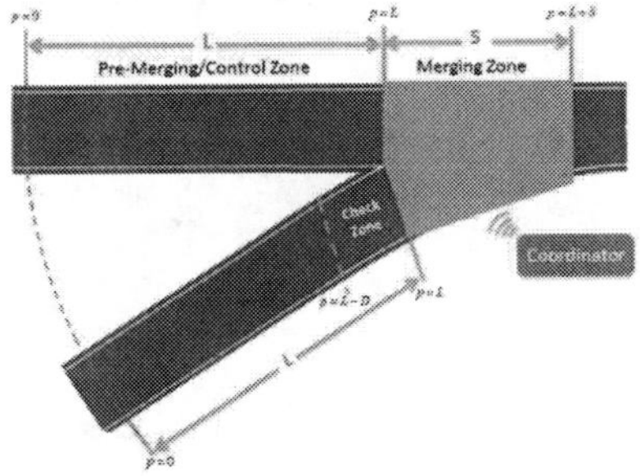

$$\min_{u_i} J = \min_{u_i} \frac{1}{2} \sum_{i=1}^{n} \int_{t_i^0}^{t_i^f} u_i^2 dt$$

Subject to:

Vehicle dynamics

$$\dot{p}_i = v_i$$
$$\dot{v}_i = u_i,$$

Safety Constraints

$$u_i \in R_i$$
$$R_i \triangleq \{u_i(t) \in [u_{\min}, u_{\max}] \mid p_i(t) \le p_k(t) - \delta,$$
$$v_i(t) \in [v_{\min}, v_{\max}], \forall i \in \mathcal{N}(t), |\mathcal{N}(t)| > 1, \forall t \in [t_i^0, t_i^f]\},$$

*Where R_i is the control interval, δ a safe headway distance and **k** the leader of vehicle **i**.*

Hamiltonian analysis allows to find the analytical solution

* *J. Rios-Torres* and A. A. Malikopoulos, "Automated and Cooperative Vehicle Merging at Highway On-Ramps," in IEEE Transactions on Intelligent Transportation Systems, vol. 18, no. 4, pp. 780-789, April 2017. doi: 10.1109/TITS.2016.2587582

OAK RIDGE National Laboratory | NATIONAL TRANSPORTATION RESEARCH CENTER

Analytical closed-form solution

We obtain the optimal control input and states as a function of time and some vehicle specific constants a, b, c, d:

$$u_i^*(t) = a_i t + b_i \longrightarrow \text{Optimal Acceleration}$$
$$v_i^*(t) = \frac{1}{2} a_i t^2 + b_i t + c_i \longrightarrow \text{Optimal Speed}$$
$$p_i^*(t) = \frac{1}{6} a_i t^3 + \frac{1}{2} b_i t^2 + c_i t + d_i \longrightarrow \text{Optimal Position}$$

* *J. Rios-Torres* and A. A. Malikopoulos, "Automated and Cooperative Vehicle Merging at Highway On-Ramps," in IEEE Transactions on Intelligent Transportation Systems, vol. 18, no. 4, pp. 780-789, April 2017. doi: 10.1109/TITS.2016.2587582

OAK RIDGE National Laboratory | NATIONAL TRANSPORTATION RESEARCH CENTER

Optimal control problem solution

The system of equations allows the implementation of a real-time optimal control

$$\begin{bmatrix} \frac{1}{6}(t)^3 & \frac{1}{2}(t)^2 & t & 1 \\ \frac{1}{2}(t)^2 & t & 1 & 0 \\ \frac{1}{6}(t_i^f)^3 & \frac{1}{2}(t_i^f)^2 & t_i^f & 1 \\ \frac{1}{2}(t_i^f)^2 & t_i^f & 1 & 0 \end{bmatrix} \begin{bmatrix} a_i \\ b_i \\ c_i \\ d_i \end{bmatrix} = \begin{bmatrix} p_i(t) \\ v_i(t) \\ p_i(t_i^f) \\ v_i(t_i^f) \end{bmatrix}$$

$$t_i^f = t_i^{out} \begin{cases} \leq t_{i+1}^{in} & \text{→Different road} \\ \leq t_{i+1}^{\delta} & \text{→Same road} \end{cases}$$

The system of equations can be solved at each instant of time

CLOSED LOOP SOLUTION!!!

* *J. Rios-Torres* and A. A. Malikopoulos, "Automated and Cooperative Vehicle Merging at Highway On-Ramps," in IEEE Transactions on Intelligent Transportation Systems, vol. 18, no. 4, pp. 780-789, April 2017. doi: 10.1109/TITS.2016.2587582

OAK RIDGE National Laboratory | NATIONAL TRANSPORTATION RESEARCH CENTER

Optimal coordination increases efficiency in diverse traffic conditions (full market penetration)

- Significant fuel consumption savings in all traffic conditions
- Travel time savings are significant in moderate/high traffic conditions
- Time value > Fuel value

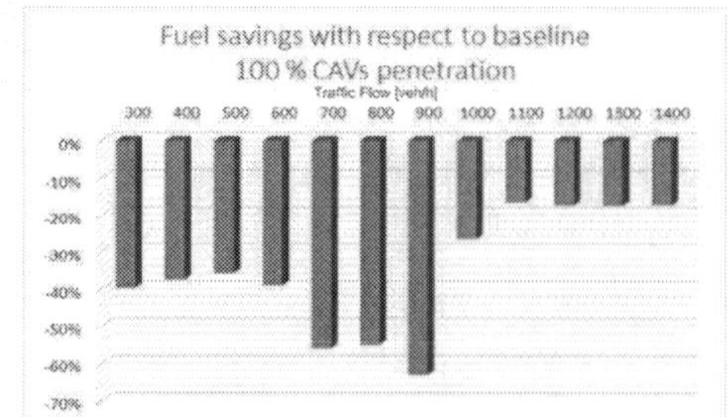

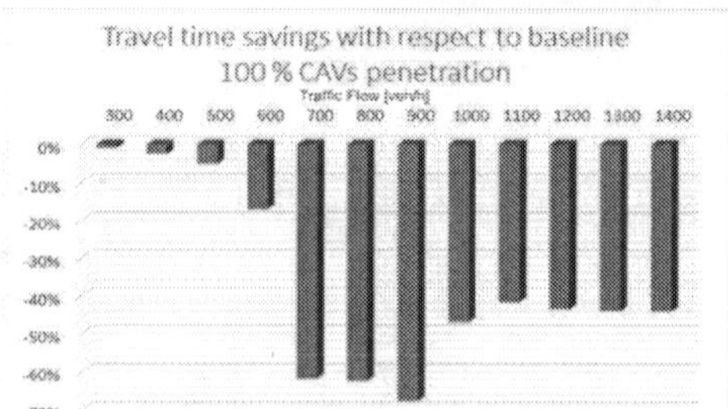

Rios-Torres, J; A. A. Malikopoulos, "Automated and Cooperative Vehicle Merging at Highway On-Ramps," in IEEE Transactions on Intelligent Transportation Systems , no.99, pp.1-10, doi: 10.1109/TITS.2016.2587582
Rios-Torres, J., and Malikopoulos, A.A., "A Survey on the Coordination of Connected and Automated Vehicles at Intersections and Merging at Highway On-Ramps," IEEE Trans. Intel. Trans. Syst., Vol. 18, 5, pp. 1066-1077, 2017

14

OAK RIDGE National Laboratory | NATIONAL TRANSPORTATION RESEARCH CENTER

What about partial market penetration?

Main Assumptions

- Human driver model: Gipps car following model
- Homogeneous traffic
  - Conventional engine-powered vehicles
- CAVs optimal control is disengaged if a desired safety gap is violated (When having a human-driven vehicle as leader)

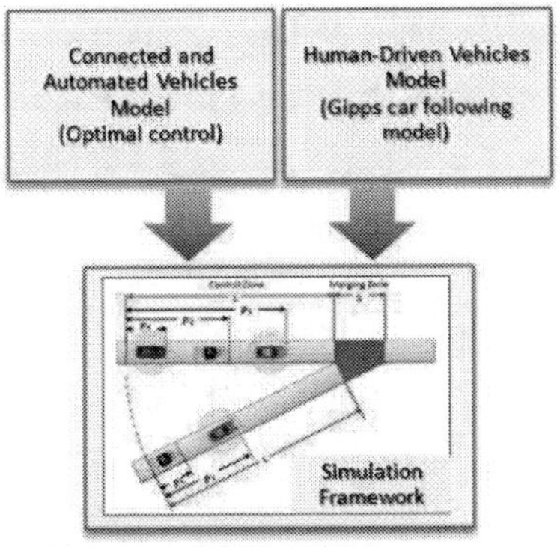

Different trends with partial market penetration

- Fuel savings for low traffic demand
- Fuel increases for medium and high traffic demand

Low traffic

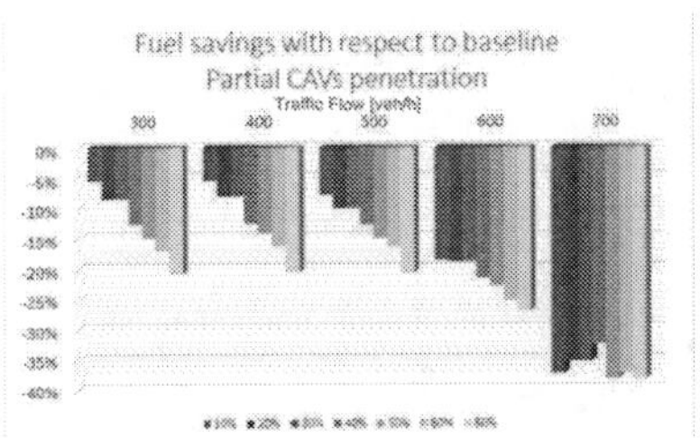

Medium traffic

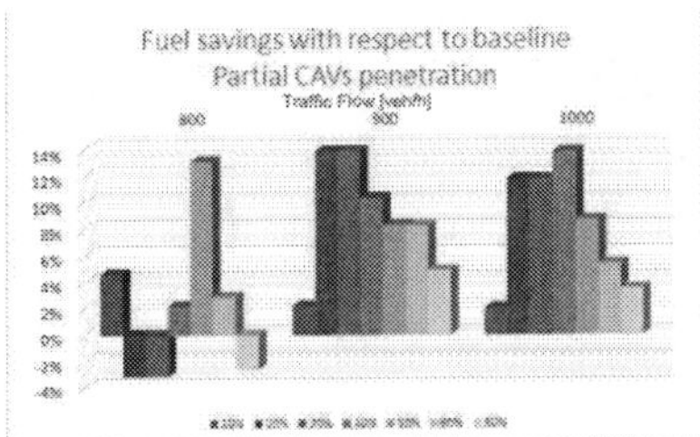

High traffic

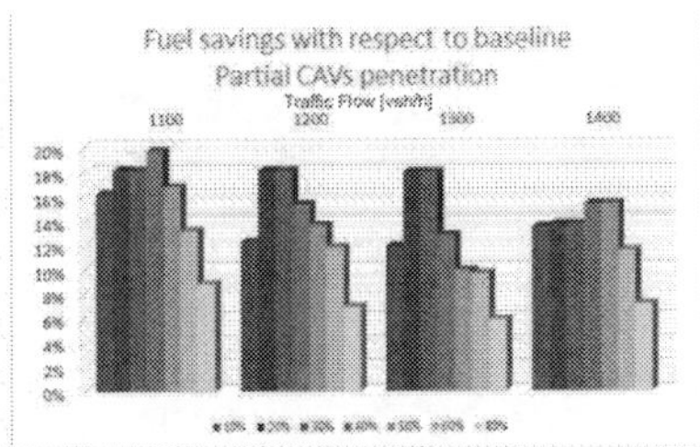

Different trends with partial market penetration (Cont.)

- Remains almost constant for very low traffic demand
- Travel time savings observed for most scenarios in moderate traffic
- Travel time savings at high traffic demand when there is more than 40% CAVs

Low traffic

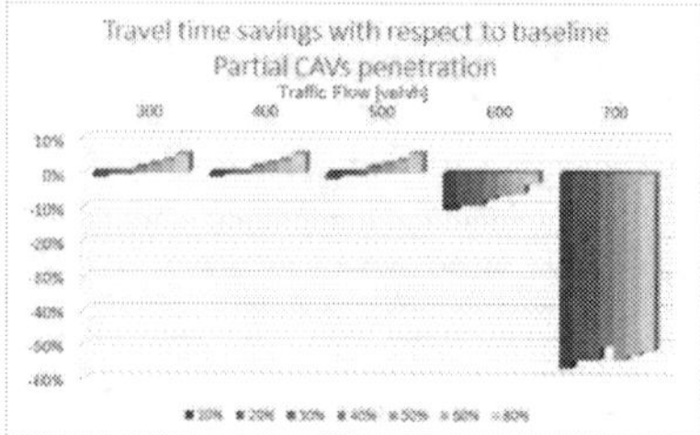

Medium traffic

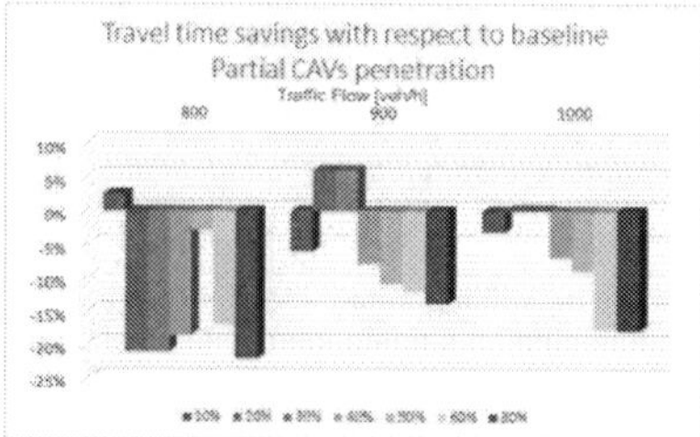

High traffic

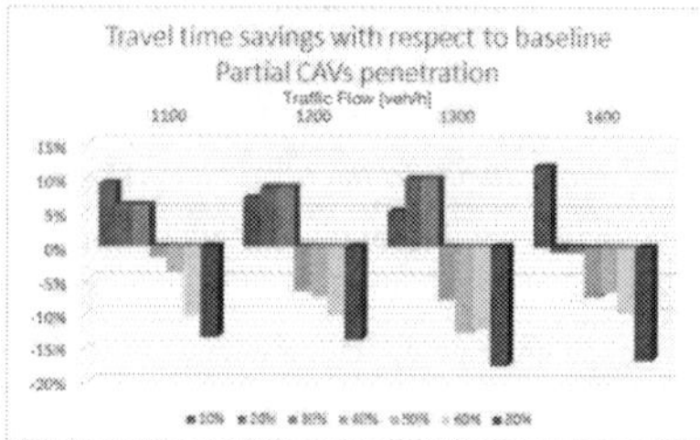

17

OAK RIDGE National Laboratory | NATIONAL TRANSPORTATION RESEARCH CENTER

Outline

- Exploring potential benefits and implications of Connected and Automated Vehicles (CAVs) coordination
  - Optimal coordination
  - Full CAVs penetration
  - Partial CAVs penetration
- Challenges and Opportunities in the area of Artificial Intelligence

18

OAK RIDGE National Laboratory | NATIONAL TRANSPORTATION RESEARCH CENTER

It seems we are facing a new transition period

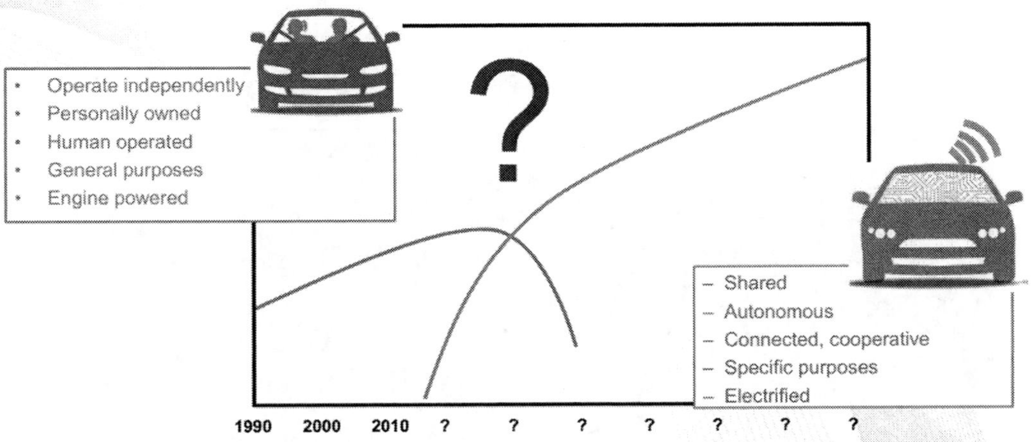

19

OAK RIDGE National Laboratory | NATIONAL TRANSPORTATION RESEARCH CENTER

The main focus has been on safety… but efficiency has become a concern too

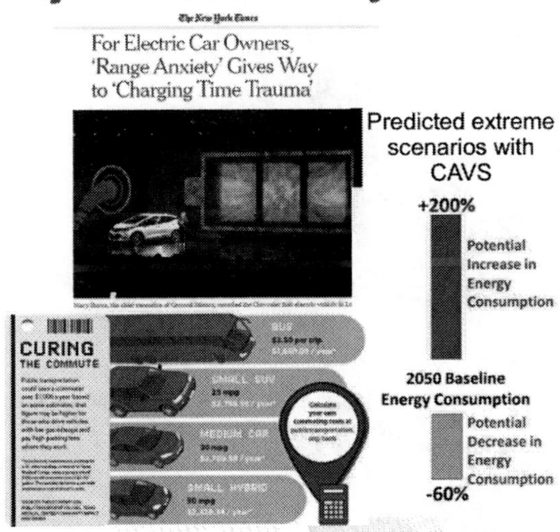

Sources:
Estimated Bounds and Important Factors for Fuel Use and Consumer Costs of Connected and Automated Vehicles (http://www.nrel.gov/docs/fy17osti/67216.pdf)
Transportation Energy Databook, ed. 36

20

OAK RIDGE National Laboratory | NATIONAL TRANSPORTATION RESEARCH CENTER

AI has been key to improve vehicle perception

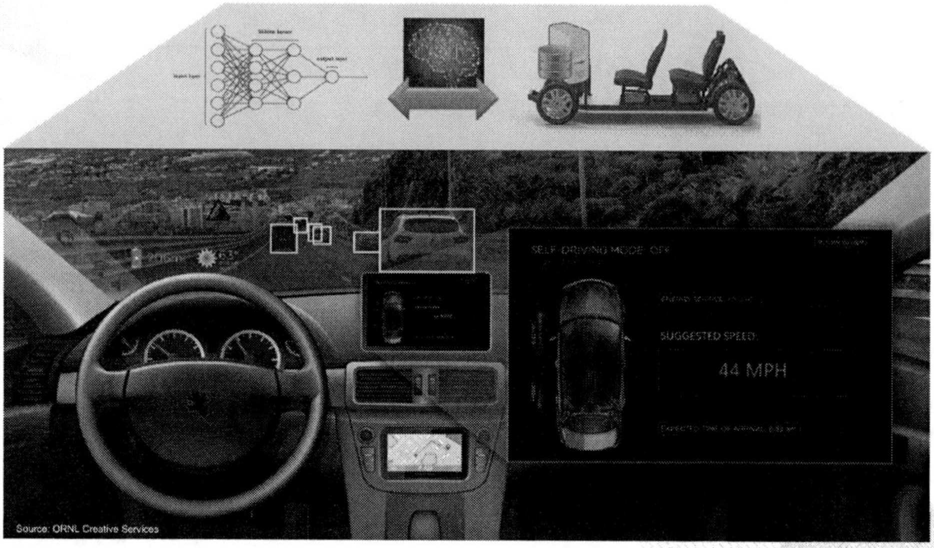

21 NATIONAL TRANSPORTATION RESEARCH CENTER

Can AI help improving mobility energy productivity at different levels?

Source:
https://www.electronicsweekly.com

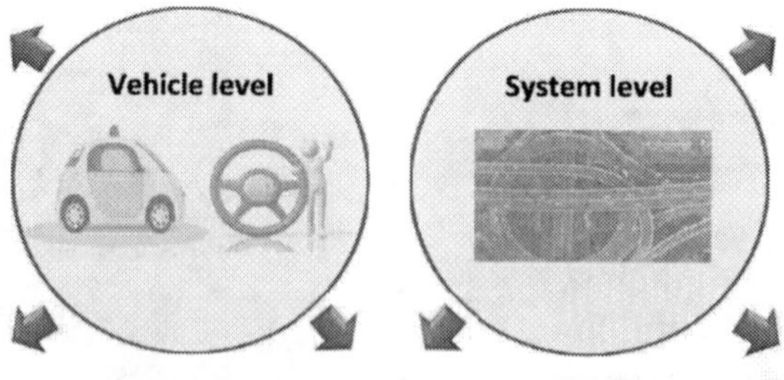

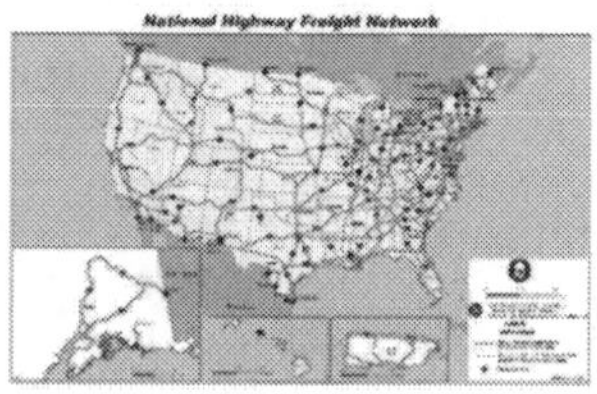

Source:
https://faf.ornl.gov/faf4/

Source:
https://www.sciencedaily.com

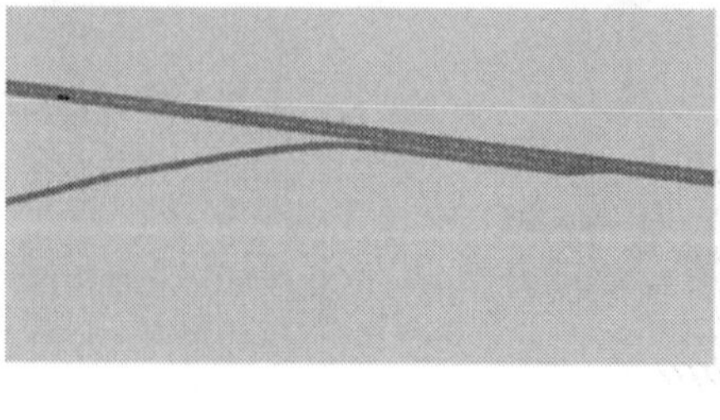

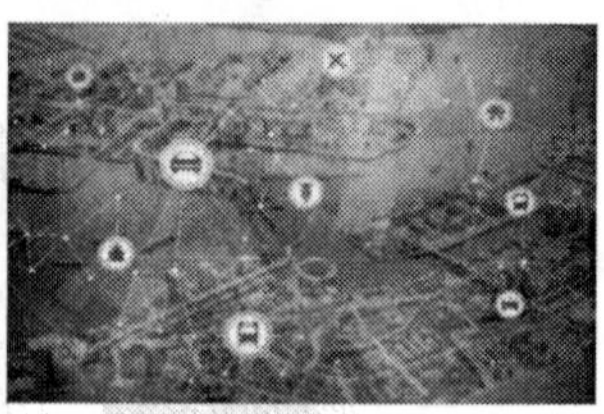

22 OAK RIDGE National Laboratory | NATIONAL TRANSPORTATION RESEARCH CENTER

How smart can our vehicles become?

23

OAK RIDGE National Laboratory | NATIONAL TRANSPORTATION RESEARCH CENTER

Thanks!
Gracias!

Jackeline Rios-Torres
riostorresj@ornl.gov

Application of Deep Learning for Characterization of Driver Engagement in Secondary Tasks

Osama A. Osman, Ph.D.
Research Associate, Center for Sustainable Mobility

Hesham A. Rakha, Ph.D., P.Eng.
Director, Center for Sustainable Mobility

Samuel Reynolds Pritchard Professor of Engineering, Charles E. Via, Jr. Dept. of Civil & Environmental Engineering

Courtesy Professor, Bradley Dept. of Electrical and Computer Engineering

VIRGINIA TECH TRANSPORTATION INSTITUTE

2

Center for Sustainable Mobility CAV Research

- Algorithmic development of CAV systems
  - Freeway traffic control and Speed Harmonization
  - CACC algorithmic development, implementation and field testing
  - Eco-CACC on freeways and arterials
  - CAV control at intersections
  - CAV-enabled network control and management
  - Mobility as a service, Smart Cities, IoT
  - **Driving Safety**

7/26/2018 *Advancing Transportation Through Innovation* VIRGINIA TECH TRANSPORTATION INSTITUTE

3

Secondary Task Engagement

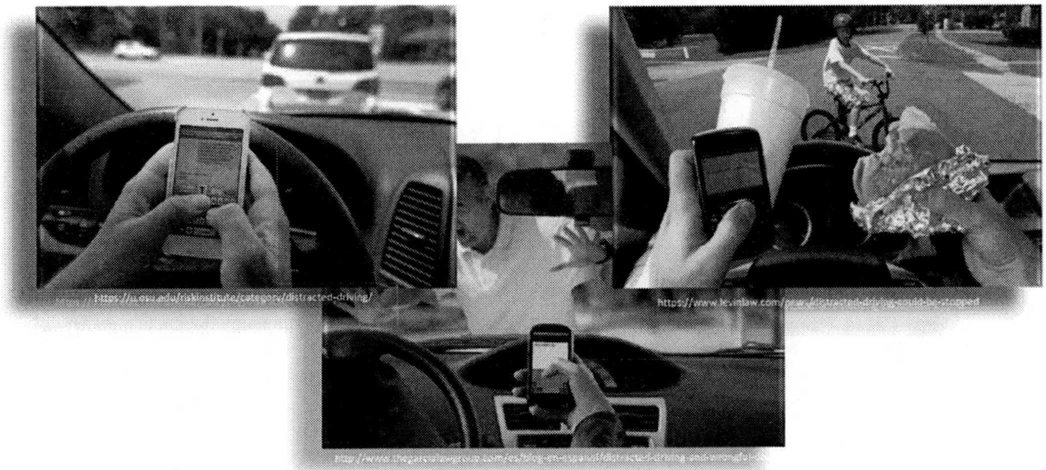

7/26/2018 Advancing Transportation Through Innovation VIRGINIA TECH TRANSPORTATION INSTITUTE

4

Distracted Driving and Driving Safety

- Distracted Driving:
  - Any activity that diverts the driver's attention away from the primary task of driving
    - *Manual* (eating, adjusting the entertainment system, grooming)
    - *Auditory/Visual* (crying baby, passenger conversation, rubber-necking)
    - *Cognitive* (using cell phone, navigation system, reading)
- Secondary tasks increase driver reaction times
  - Lack of attention or perceptual blindness
- Distracted driving is a problem that demands innovative solutions
  - e.g. sensing technologies to detect such instances – CAV technology
- Choose the computational tool to identify secondary task involvement
  - Machine Learning vs. Deep Learning

7/26/2018 Advancing Transportation Through Innovation VIRGINIA TECH TRANSPORTATION INSTITUTE

5

Impact of Distracted Driving

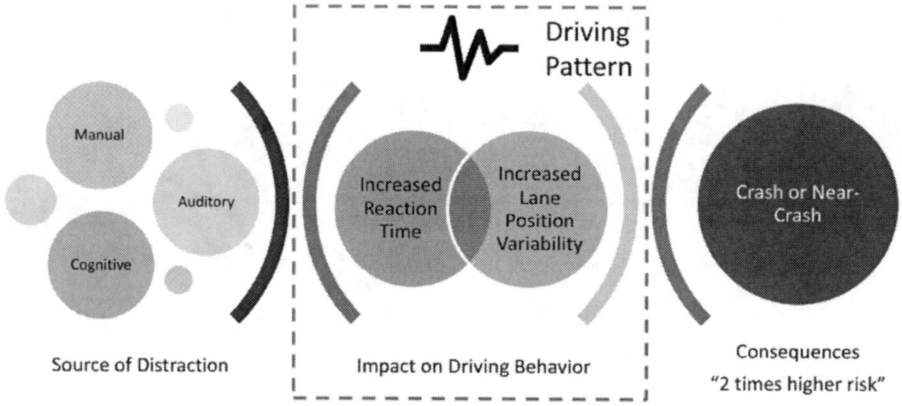

7/26/2018 Advancing Transportation Through Innovation VIRGINIA TECH TRANSPORTATION INSTITUTE

6

Methods

- SHRP2 NDS Vehicle Kinematics Data
  - Longitudinal speed, lateral acc., longitudinal acc., pedal position, yaw rate, and headway
  - Calling, texting, conversation and no secondary task engagement
- Deep Learning (Feed Forward Deep Neural Networks)
- Filters:
  - No crash or near crashes
  - Unimpaired
  - Only one secondary task at a time
- 110 60-second events per task:
  - Up to 60 seconds long for each secondary task within the event
  - ~30,000 data points (≈ 50 minutes) for each of the tasks

7/26/2018 Advancing Transportation Through Innovation VIRGINIA TECH TRANSPORTATION INSTITUTE

7

DNN Model

- Input layer consists of 12 neurons
  - Mean and variance of the 6 kinematics data
- Three hidden layers
  - Layer 1 consists of 15 neurons
  - Layer 2 consists of 20 neurons
  - Layer 3 consists of 10 neurons
- Output layer consists of 4 neurons
  - Three secondary tasks and base "no secondary task"

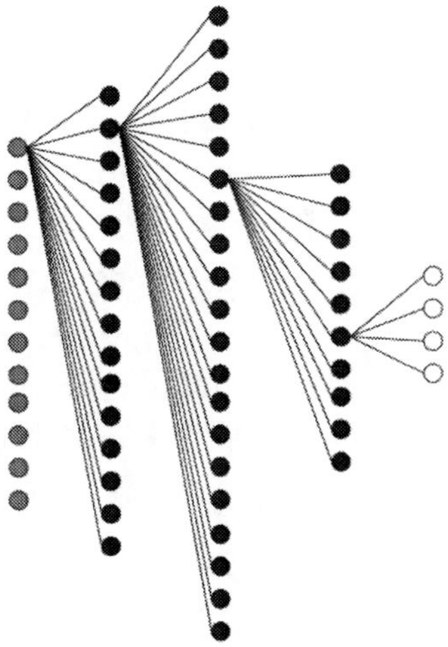

7/26/2018 Advancing Transportation Through Innovation VIRGINIA TECH TRANSPORTATION INSTITUTE

8

DNN Model

- Transfer function
  - ReLU (Rectified Linear Unit) function was used for all layers (except output layer)
    - ReLU is more efficient for DNNs compared to other transfer functions
  - Softmax function was used for the output layer
    - Softmax function performs better when doing multiclass classifications
- Layer batch normalization
  - Normalize the output of every layer before feeding it to the next layer
- Learning rate adjustment
  - Reduce by 10% when validation loss did not decrease by 0.05 in 50+ epochs

7/26/2018 Advancing Transportation Through Innovation VIRGINIA TECH TRANSPORTATION INSTITUTE

9

DNN Model

- Layer dropout
  - Drop 20% of output features in a layer before feeding into the next layer
    - Avoid fitting patterns to insignificant noise in data
- Weight regularization
  - Multiply weights of insignificant features by 0.01, and
  - Allow features with consistently significant impact in the classification process to have larger weights
- Stop training
  - Stop training when the validation loss does not decrease by 0.05 for 500 epochs

7/26/2018 *Advancing Transportation Through Innovation* VIRGINIA TECH TRANSPORTATION INSTITUTE

10

DNN Model Development

- Divided data into three datasets:
  - Training – 60%
  - Validation – 20%
    - Used to ensure no overfitting in the model while training the model
  - Testing – 20%
    - Ensures model is not over-fitted and tested on independent data
      - Could be different drivers

7/26/2018 *Advancing Transportation Through Innovation* VIRGINIA TECH TRANSPORTATION INSTITUTE

11

DNN Model Results

- Overall classification accuracy = 97.3%
- Confusion matrix

		Target			
		Calling	Texting	Interacting with Others	No Secondary Task Engagement
Predictions	Calling	**100.0**	0.1	0.0	0.0
	Texting	0.0	**95.8**	1.7	2.7
	Interacting with Others	0.0	0.5	**95.8**	0.4
	No Secondary Task Engagement	0.0	2.5	2.5	**96.9**

12

DNN (Headway vs. No Headway)

With Headway

- Overall classification accuracy = 97.3%
- Confusion matrix

		Target			
		Calling	Texting	Interacting	Base Case
Predictions	Calling	**100.0**	0.1	0.0	0.0
	Texting	0.0	**95.8**	1.7	2.7
	Interacting	0.0	0.5	**95.8**	0.4
	Base Case	0.0	2.5	2.5	**96.9**

Without Headway

- Overall classification accuracy = 96.9%
- Confusion matrix

		Target			
		Calling	Texting	Interacting	Base Case
Predictions	Calling	**100.0**	0.1	0.0	0.0
	Texting	0.0	**96.1**	2.6	3.0
	Interacting	0.0	0.5	**93.9**	0.4
	Base Case	0.0	2.2	3.5	**96.6**

13

DNN (Headway vs. No Headway)

With Headway

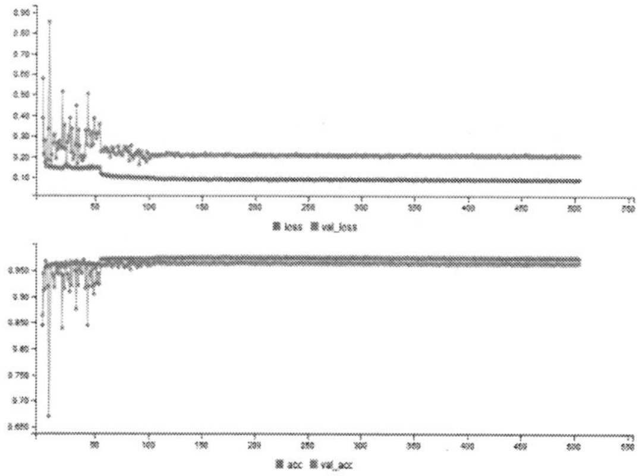

Without Headway

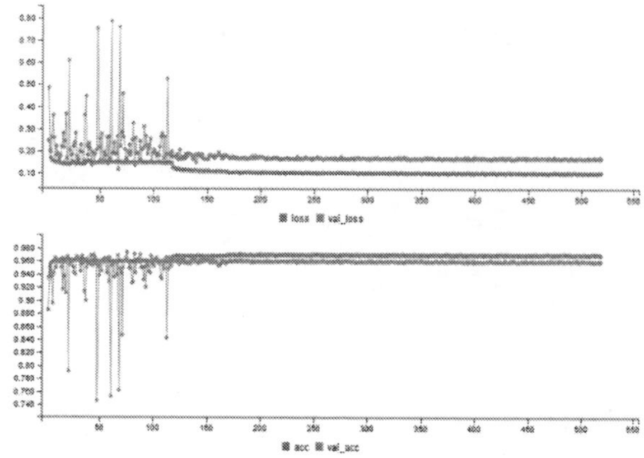

7/26/2018 *Advancing Transportation Through Innovation* VIRGINIA TECH TRANSPORTATION INSTITUTE

14

DNN vs. Ensemble ML

Without Headway

- Overall classification accuracy = 96.9%
- Confusion matrix

		Target			
		Calling	Texting	Interacting	Base Case
Predictions	Calling	**100.0**	0.1	0.0	0.0
	Texting	0.0	**96.1**	2.6	3.0
	Interacting	0.0	0.5	**93.9**	0.4
	Base Case	0.0	2.2	3.5	**96.6**

Osman, O. A., M. Hajij, S. Karbalaieali, and S. Ishak. *A Bi-Level Methodology for Identification of Types of Secondary Tasks from Observed Driving Behavior Data: Application of Ensemble Tree Machine Learning Algorithms on SHRP2 NDS Data.* Presented at the 97th Annual Meeting of the Transportation Research Board, Washington, D.C., USA, January 2018.

- Single Step:

Method	KNN	SVM Nonlinear	Decision Tree (DT)	Random Forest (RF)	AdaBoost	Gaussian NB
Accuracy (%)	64.2	64.2	75.2	60.4	62.1	56.0

- Bi-level (hierarchical)
  - Secondary Task Detection

Method	KNN	SVM Nonlinear	DT	RF	AdaBoost	Gaussian NB
Accuracy (%)	86.6	87.8	99.8	96.0	91.3	66.3
Secondary Task Accuracy (%)	88.0	88.3	99.7	95.1	95.1	93.1
Normal Driving Accuracy (%)	86.2	87.7	99.5	95.2	90.0	53.2

7/26/2018 *Advancing Transportation Through Innovation* VIRGINIA TECH TRANSPORTATION INSTITUTE

15

DNN vs. Ensemble ML

Without Headway

- Overall classification accuracy = 96.9%
- Confusion matrix

		Target			
		Calling	Texting	Interacting	Base Case
Predictions	Calling	**100.0**	0.1	0.0	0.0
	Texting	0.0	**96.1**	2.6	3.0
	Interacting	0.0	0.5	**93.9**	0.4
	Base Case	0.0	2.2	3.5	**96.6**

- Bi-level (hierarchical)
  - Overall Model Performance

Method	KNN	SVM Nonlinear	DT	RF	AdaBoost	Gaussian NB
Overall Accuracy (%)	66.8	64.4	79.7	82.2	60.0	55.5
Texting Accuracy (%)	37.2	20.1	73.0	77.2	3.1	2.1
Calling Accuracy (%)	33.2	16.3	76.8	82.2	3.0	6.2
Conversation Accuracy (%)	91.5	94.9	85.1	93.1	99.5	94.2

- Overall model accuracy = 82.03%

16

Conclusions

- DNNs proved to be a good tool for classifying driver secondary tasks:
  - Single model
  - Higher accuracy (97%) compared to other ML techniques (82%) for RF
- DNN training is an art in addition to a science

17

Future Work

- Investigate the inclusion of video data to enhance the DNN performance
- Include other secondary tasks in the mix
- Consider and identify the interaction of secondary tasks
- Study the impact of secondary tasks on driver longitudinal and lateral movement behavior
- Implement modified behavior models in traffic simulation tools

7/26/2018 *Advancing Transportation Through Innovation* VIRGINIA TECH TRANSPORTATION INSTITUTE

Questions?

Hesham A. Rakha
hrakha@vtti.vt.edu 540-231-1505

VIRGINIA TECH TRANSPORTATION INSTITUTE

DNN vs. Ensemble ML

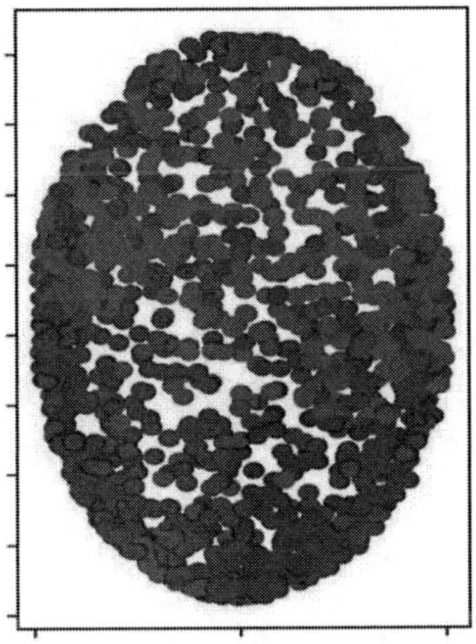

Deep Learning-based Human-Driven Vehicle Trajectory Prediction and its Application for Platoon Control of Connected and Autonomous Vehicles

Lei Lin[1] Siyuan Gong[1] Tao Li[2] Srinivas Peeta[3]

[1]NEXTRANS
Purdue University

[2]Department of Computer Science
Purdue University

[3]Lyles School of Civil Engineering
Purdue University

Automated Vehicles Symposium, 2018

1/30

Outline

1. Problem Description
2. Motivation
   - Mixed Traffic Flow
   - Data-driven Model
3. Data Set and Preprocessing
4. Model Building
5. Results
   - LSTM Prediction Performances
   - CAV Platoon Control Performances

2/30

1. Problem Description
2. Motivation
3. Data Set and Preprocessing
4. Model Building
5. Results

3/30

Problem Description

- Predict the leading Human-driven Vehicle's (HDV) trajectory for platoon control of CAVs.

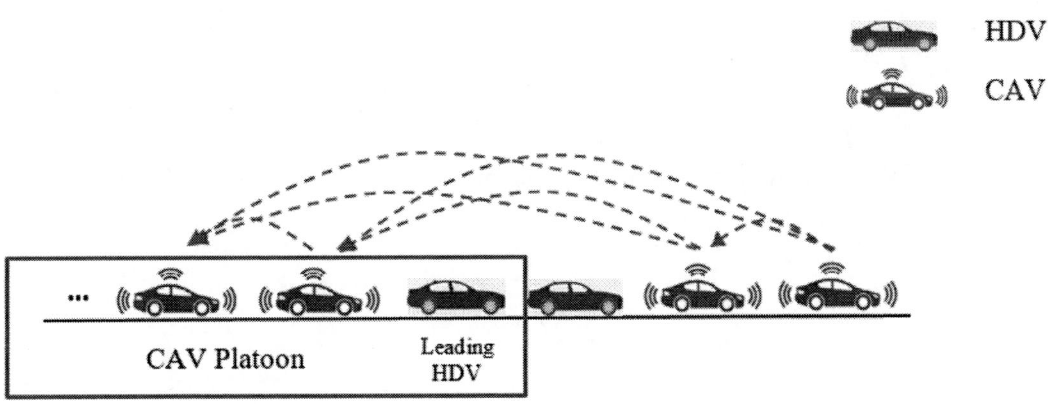

4/30

1. Problem Description
2. Motivation
   - Mixed Traffic Flow
   - Data-driven Model
3. Data Set and Preprocessing
4. Model Building
5. Results

5/30

Motivation-Mixed Traffic Flow

- Long transition period to a fully Connected and Autonomous Vehicle (CAV) environment.
  - All new vehicles will have the connectivity function until 2025 (GSMA, 2013).
  - 75% of vehicles will be autonomous until 2040 (IEEE, 2012).

6/30

Motivation-Data-driven Model

- Most vehicle trajectory prediction models are parsimonious.
  - Provide analytical conclusions and fast simulation speed.
  - Limited model flexibility and accuracy.
  - For example, car following models in traffic simulation studies rely on artificial parameters in empirical equations.
- Trajectory data from CAVs in the communication range can help to predict the HDV's trajectory.

Data Set and and Preprocessing

- Next Generation Simulation (NGSIM) Data Set:
  - 500 meters long, 4:00 PM-4:15 PM, 04/13/2005, I-80.
  - Lateral and longitudinal locations, VehID, Time Step, Velocity, Acceleration, Preceding VehID, LaneID by 10 Hz.
  - Reconstructed data were downloaded (Punzo et al., 2011).

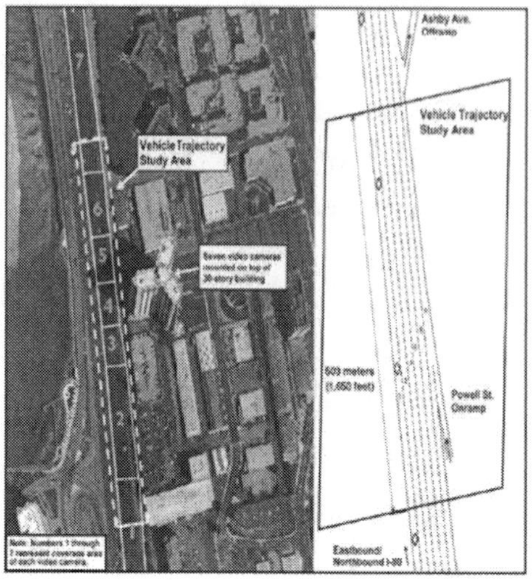

1

[1] https://www.fhwa.dot.gov/publications/research/operations/06137/index.cfm

Data Set and and Preprocessing

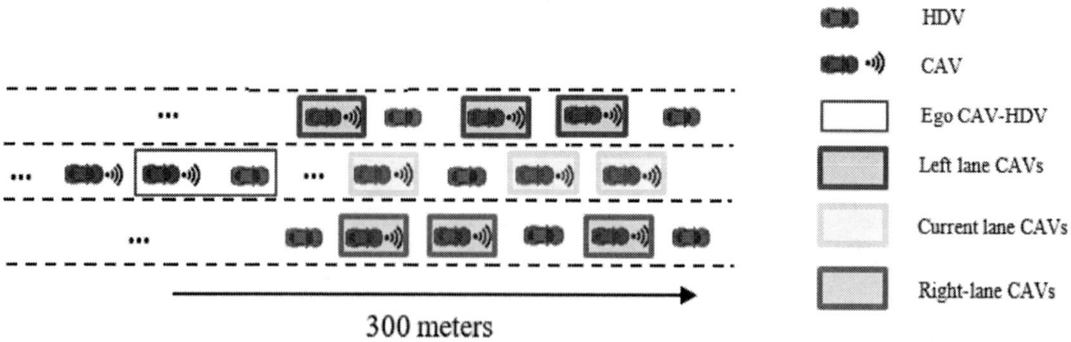

- Data Preprocessing Steps:
  - CAV Market Penetration Rate: 50%.
  - For lane 2, 3, and 4, find the Ego CAV and HDV pairs, and for each pair, find 3 CAVs in the front 300 meters from left, current, and right lanes separately.
  - Record the time step, lane location, speed, acceleration rate, vehicle length for the Ego CAV, the HDV and 9 CAVs.

1 Problem Description

2 Motivation

3 Data Set and Preprocessing

4 Model Building

5 Results

Model Building

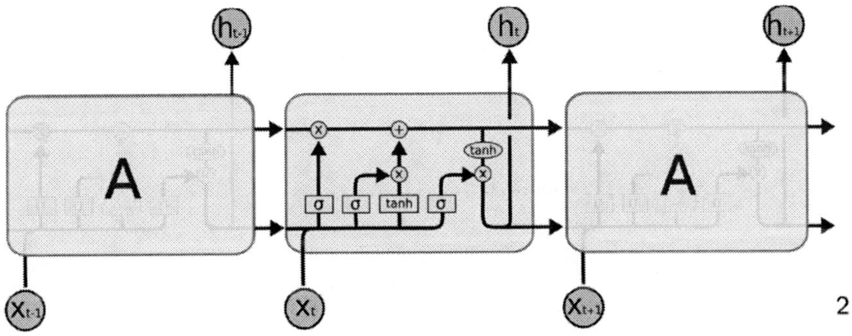

2

- The Long Short-term Memory Model (LSTM).
  - Designed for Sequencial Data.
  - Consists of LSTM Cells, each cell has an input gate, forget gate and output gate.
  - Automatically memorizing relevant spatial and temporal features extracted from previous vehicle states.

[2] http://colah.github.io/posts/2015-08-Understanding-LSTMs/

Model Building

- Experimental Environment:
  - GTX 1080, Tensorflow.
- LSTM Architecture:
  - Learning rate: 0.001.
  - LSTM cell number:100.
  - Mini-batch size: 100.
  - Hidden units in each cell: 400.
  - Early stopping threshold: 100.
- Data Set:
  - Training dataset size: 29376 X 7750.
  - Validation dataset size: 9792 X 7750.
  - Testing dataset size: 9792 X 7750.

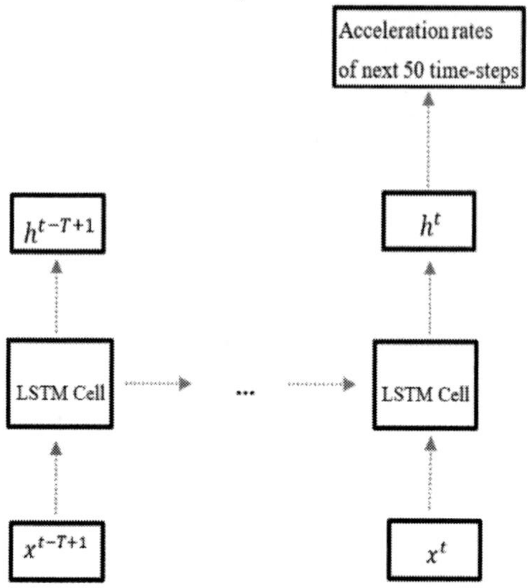

$T = 100$,
$x^{t-i+1} \in \mathbb{R}^{77}$, $i = 1, \ldots, 100$,
7 features × (9 CAVs + 1 Ego CAV + 1 HDV).

13/30

1. Problem Description
2. Motivation
3. Data Set and Preprocessing
4. Model Building
5. **Results**
   - LSTM Prediction Performances
   - CAV Platoon Control Performances

14/30

Results-LSTM Prediction Performances

Table 1: RMSE of Testing Data Sest.

Number of CAVs	Prediction Steps 10	50
3 CAVs	0.67	0.72
9 CAVs	0.71	0.72

Table 2: Normalized RMSE of Testing Data Sest. (1-RMSE/MEAN)

Number of CAVs	Prediction Steps 10	50
3 CAVs	95.40%	95.11%
9 CAVs	95.19%	95.01%

Results-LSTM Prediction Performances

RMSE by HDV ID (268 HDVs)

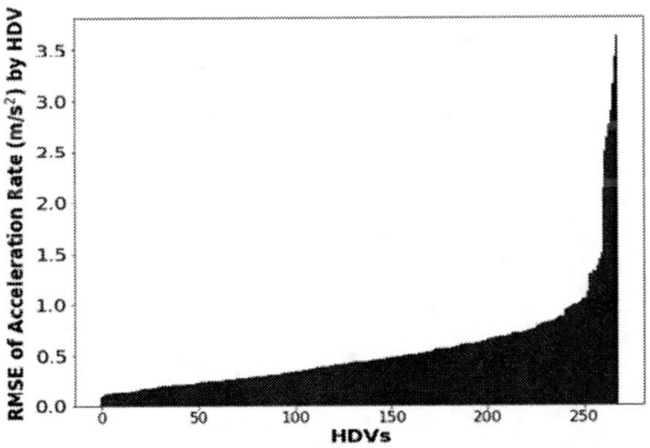

10 Prediction Steps with 3 CAVs

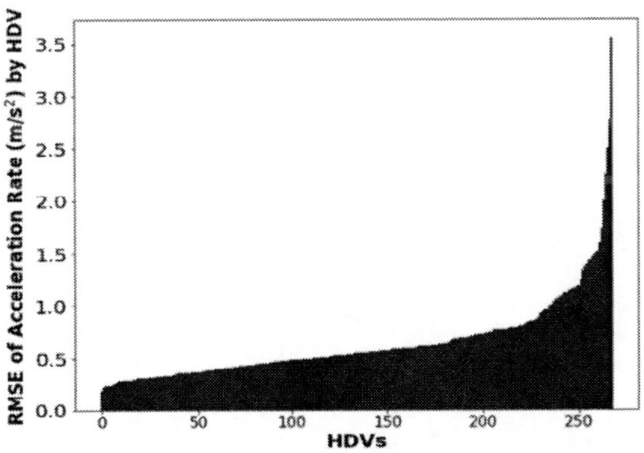

50 Prediction Steps with 3 CAVs

Results-LSTM Prediction Performances

The Minimum and Maximum RMSEs for one Round of Prediction (10 Prediction Steps with 3 CAVs)

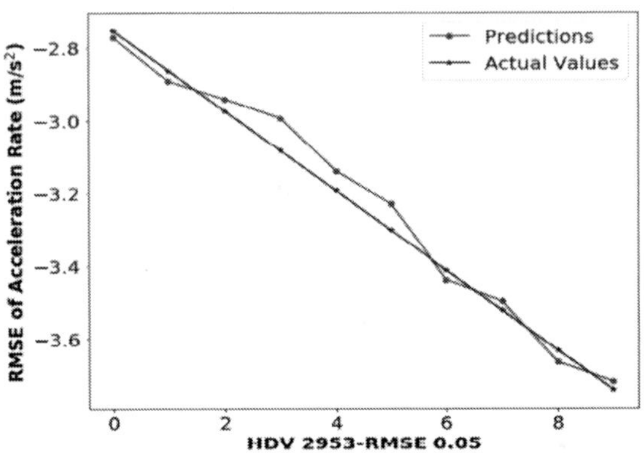

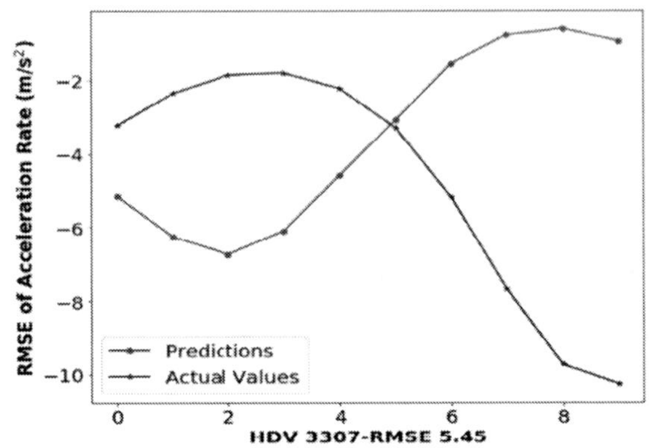

Results-LSTM Prediction Performances

The Minimum and Maximum RMSEs for one Round of Prediction (50 Prediction Steps with 3 CAVs)

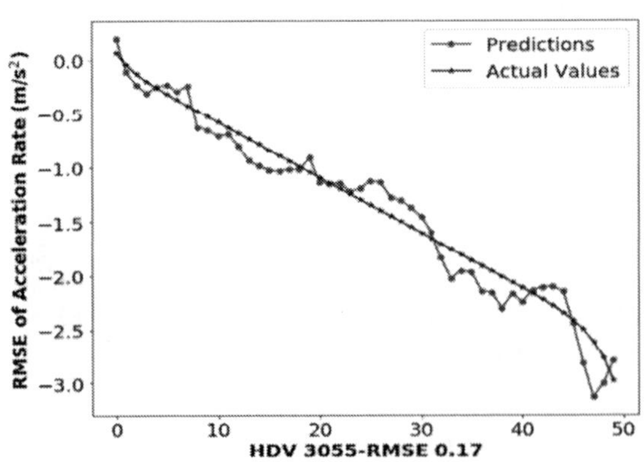

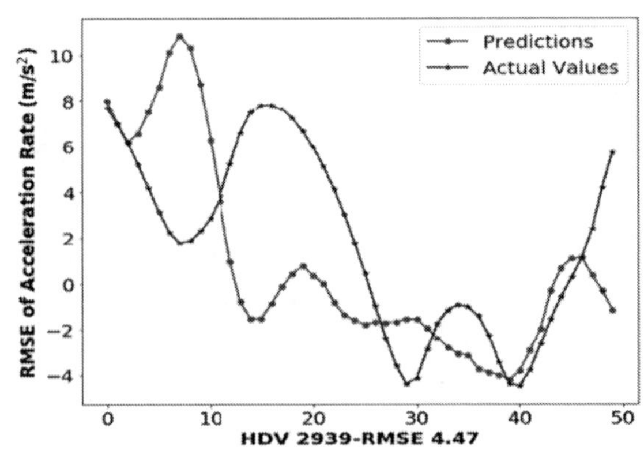

Results-CAV Platoon Control Performances

Select two representative HDVs, assume each is followed by a platoon of 9 CAVs, apply the P-step Model Predictive Control (MPC) (Gong and Du, 2018):

Table 3: RMSEs of Two HDVs.

HDV	Prediction Steps	
	10	50
2973	0.56	0.64
3307	0.94	0.94

Results-CAV Platoon Control Performances

Impact of MPC on the speed of the CAV platoon following HDV 2973:

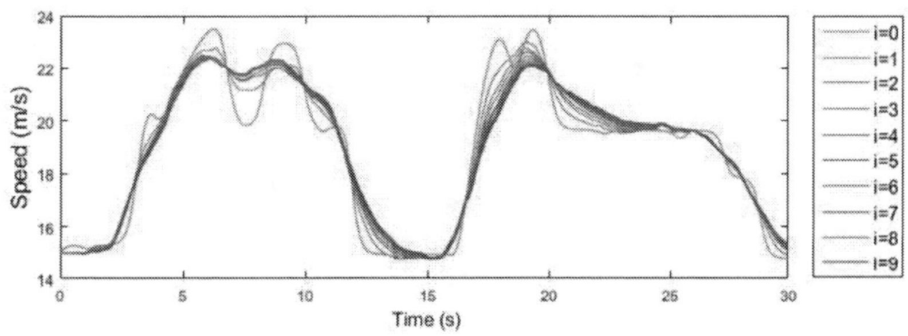

10-step predictions

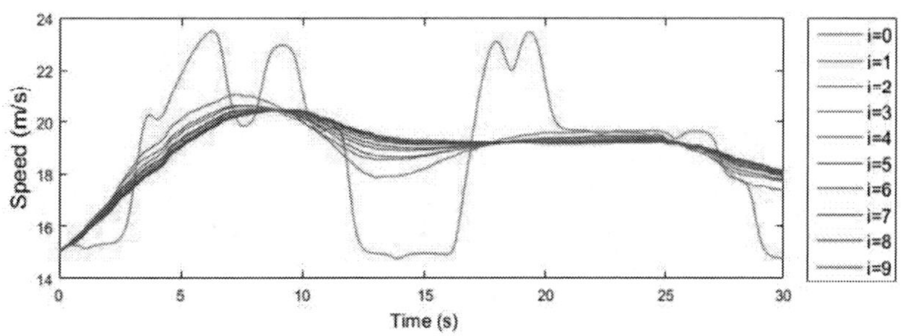

50-step predictions

Results-CAV Platoon Control Performances

Impact on the acceleration rate of the CAV platoon following HDV 2973:

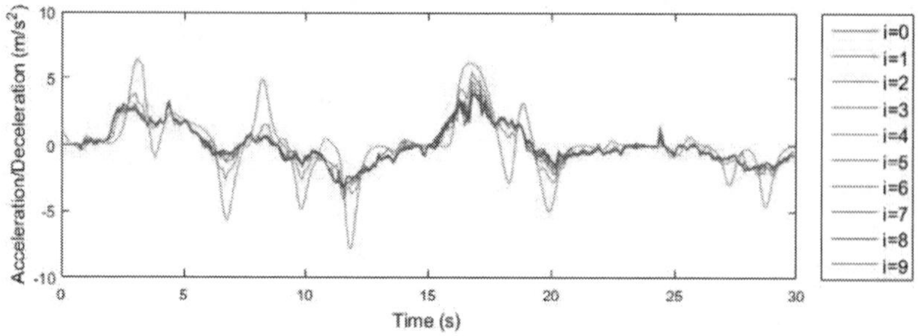

10-step predictions

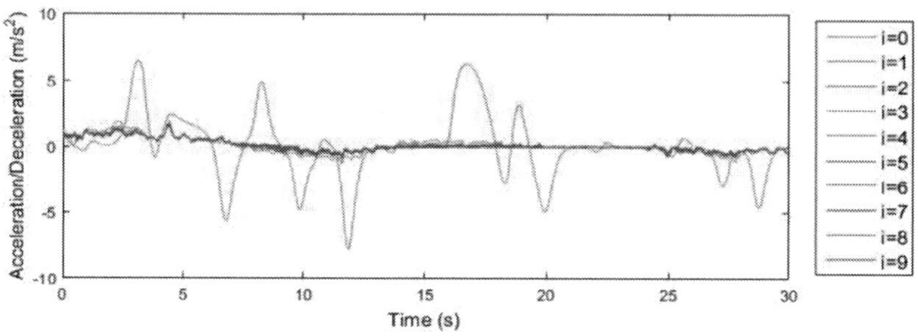

50-step predictions

21/30

Results-CAV Platoon Control Performances

Impact on the spacing of the CAV platoon following HDV 2973:

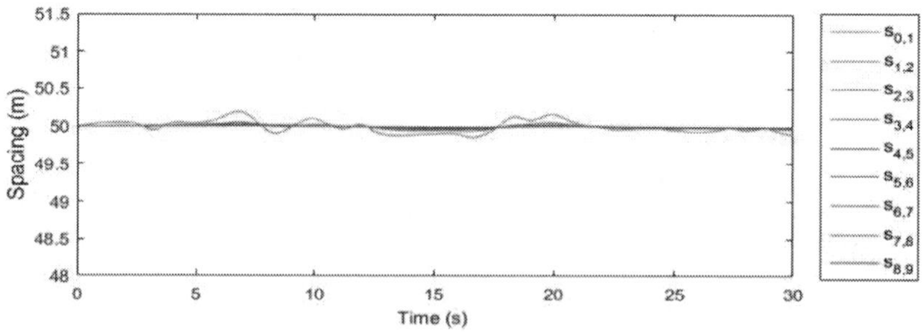

10-step predictions

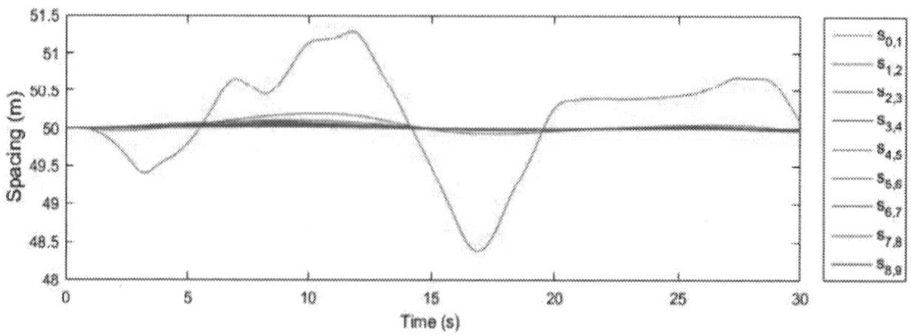

50-step predictions

22/30

Results-CAV Platoon Control Performances

Select two representative HDVs, assume each is followed by a platoon of 9 CAVs, apply the P-step Model Predictive Control (MPC) (Gong and Du, 2018):

Table 4: RMSEs of Two HDVs.

HDV	Prediction Steps	
	10	50
2973	0.56	0.64
3307	0.94	0.94

23/30

Results-CAV Platoon Control Performances

Impact of MPC on the speed of the CAV platoon following HDV 3307:

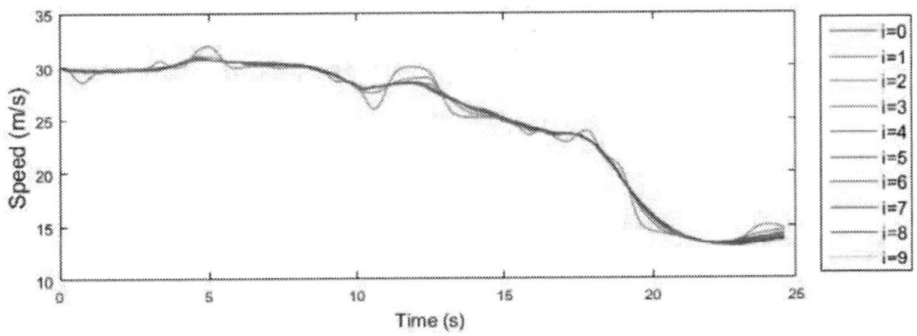

10-step predictions

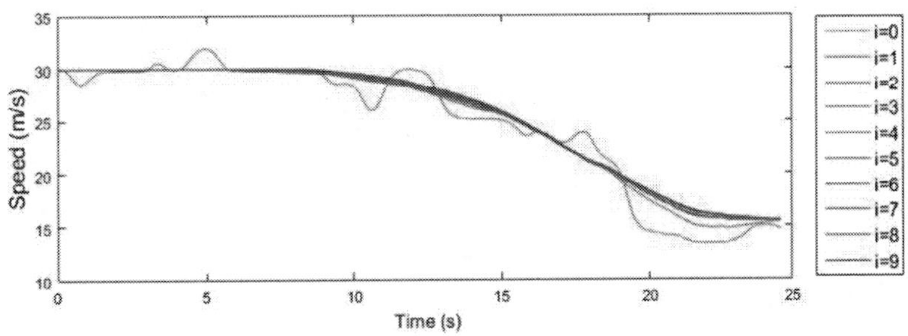

50-step predictions

24/30

Results-CAV Platoon Control Performances

Impact on the acceleration rate of the CAV platoon following HDV 3307:

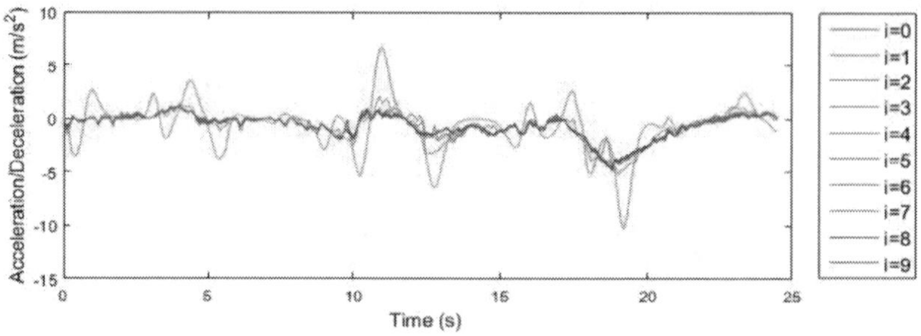

10-step predictions

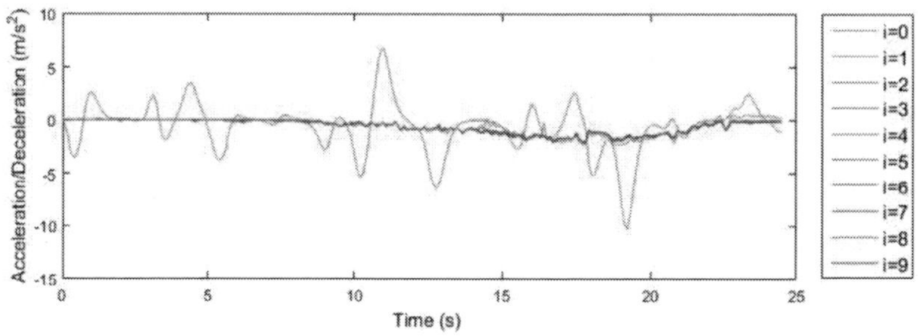

50-step predictions

25/30

Results-CAV Platoon Control Performances

Impact on the spacing of the CAV platoon following HDV 3307:

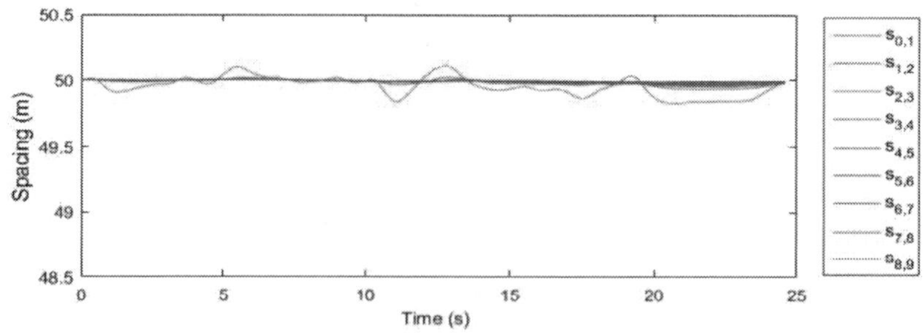

10-step predictions

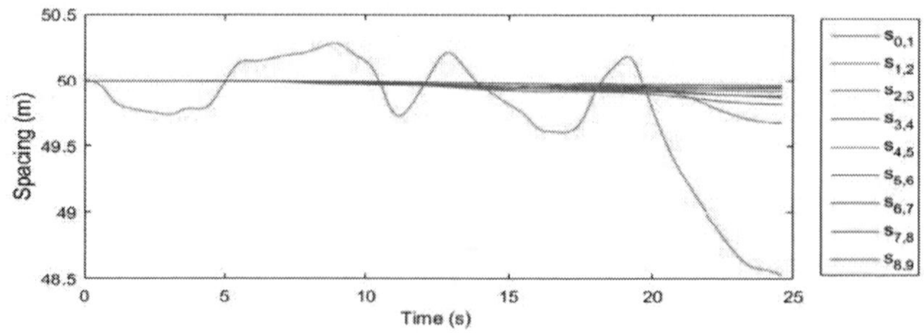

50-step predictions

26/30

Summary

- Contributions
  - Utilized information from multiple vehicles to predict the HDV's acceleration rates.
  - Applied the state-of-the-art LSTM deep learning model for multiple-step-ahead predictions.
  - Based on our predictions, MPC Control can ensure both transient traffic smoothness and asymptotic stability of the CAV platoon.
- Limitations
  - Assumption of "connected".
  - Assumption of "automated".

27/30

Next Steps

- Further clean the dataset.
- Build the CNN-LSTM model.
- Test other platoon control algorithms.
- Evaluate the performances under different CAV MPRs, CV communication ranges and delay and so on.
- Predict a distribution of the acceleration rate and capture the uncertainty.

28/30

References

- GSMA, 2013. Eevery New Car Connected by 2025 as Embedded Mobile Mobile Technology Drives Growth of Connected Car Market.
- IEEE, 2012. https://www.ieee.org/about/news/2012/5september-2-2012.html (accessed 6.1.18).
- Punzo, V., Borzacchiello, M.T. and Ciuffo, B., 2011. On the assessment of vehicle trajectory data accuracy and application to the Next Generation SIMulation (NGSIM) program data. Transportation Research Part C: Emerging Technologies, 19(6), pp.1243-1262.
- Siyuan Gong and Lili Du. 2018. Cooperative Platoon Control for a Mixed Traffic Flow Including Human Drive Vehicles and Connected and Autonomous Vehicles. Transportation Research Part B: Methodological. Accepted.

Thanks for your attentions! Any questions?

UNC CHARLOTTE

Optimal Variable Speed Limit Control in an Autonomous Vehicle Environment for Relieving Freeway Congestion

Wei (David) Fan, Ph.D., P.E.
Director, USDOT CAMMSE University Transportation Center
Professor, Department of Civil and Environmental Engineering
Miao Yu, Ph.D. Candidate
The University of North Carolina at Charlotte
Charlotte, NC 28223

CAMMSE

Automated Vehicles Symposium 2018
Golden Gate 2, Hilton San Francisco Union Square
Wednesday, July 11, 2018

Center for Advanced Multimodal Mobility Solutions and Education (CAMMSE)

- CAMMSE, a five year $ 7.7 million multi-campus Tier I University Transportation Center funded by USDOT as authorized by the FAST Act
- A consortium of five universities
  - University of North Carolina at Charlotte (Lead)
  - University of Texas at Austin
  - University of Connecticut
  - Washington State University
  - Texas Southern University

CAMMSE

UNC CHARLOTTE

Presentation Outline

- Introduction
- Design of Control Model
- Solution Algorithm
- Case Study
- Numerical Results
- Conclusion

Introduction

- Traffic congestion has become a major problem around the world
- Active traffic management (ATM) is a scheme which can be used to relieve congestion and improve traffic flows on freeways
- Variable speed limit (VSL) belongs to the ATM strategy, which enables one to change the posted speed limits dynamically on the basis of the real-time traffic and/or weather conditions
- VSL has been widely implemented around the world
  - Germany, England, Sweden, and the United States

Introduction – Cont.

- Emerging technologies have been developing during recent years
  - e.g., autonomous vehicles (AVs)
- Enhanced outcomes can be achieved through integrating VSL control with AVs
  - e.g., reduced total travel time and fuel consumption
- In this study, an integrated VSL control strategy with CACC on the basis of cell transmission model (CTM) that explicitly considers mixed traffic flows including both trucks and cars

Design of Control Model

- The fundamental diagram (FD) is simplified as having a triangular relationship between flow and density
- When modeling mixed traffic flows, other classes of vehicles are converted to the passenger car equivalents (pce)
- A dynamic pce value that involves physical characteristics of vehicles and prevailing speeds on freeways is used (van Lint et al. 2008)

$$\eta_{i,j}(k) = \eta\left(sd_j, HW_j, v_{i,j}(k)\right) = \frac{sd_j + HW_j v_{i,j}(k)}{sd_{car} + HW_{car} v_{i,car}(k)}$$

Design of Control Model – Cont.

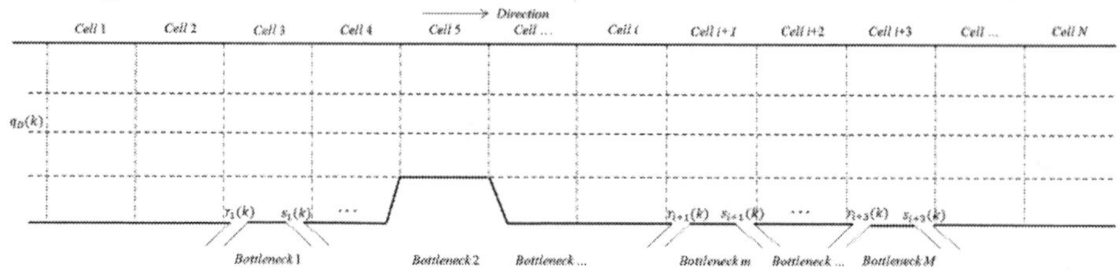

An Illustration of a Freeway Stretch with Multiple Bottlenecks

- More than one bottleneck might be activated
- For simplicity purpose, the following assumptions are made:
  - Under free flow traffic conditions, average speeds of trucks are less than those of cars; while in congested traffic conditions, trucks' speeds equal to cars' speeds;
  - Traffic flow parameters, such as the jam density, might be different at different bottlenecks. However, it is assumed that the values of such parameters on the cells upstream of the nearest bottleneck are the same.

Design of Control Model – Cont.

- To model the capacity drop phenomenon at bottlenecks, a discontinuous FD is used

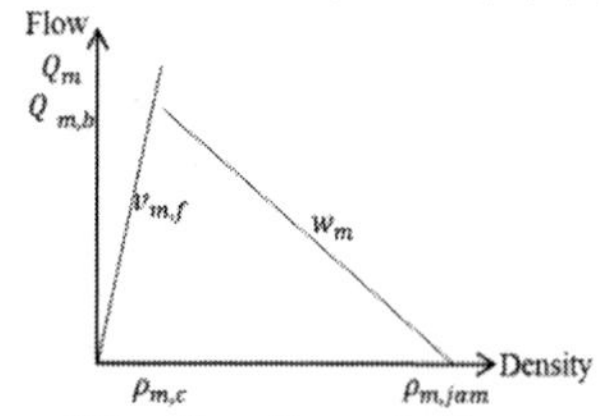

FD with a Capacity Drop

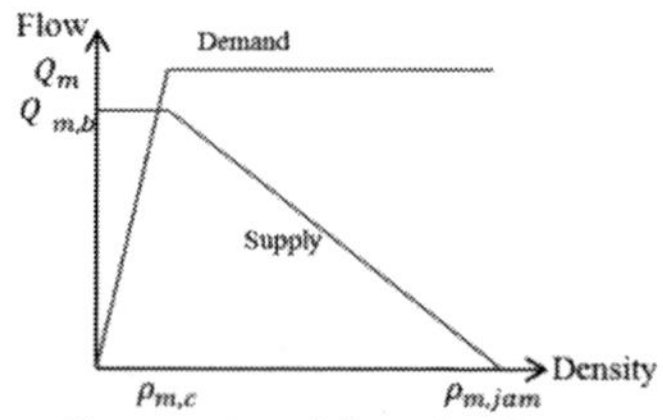

Demand and Supply Lines

- To model mixed traffic flows, a combined FD is used

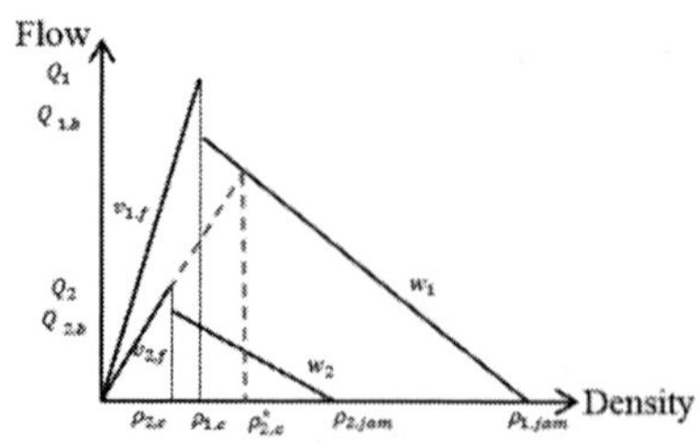

$$\rho_{2,c}^{*} = \frac{w_1 \rho_{1,jam}}{w_1 - v_{2,f}}$$

FD with Capacity Drops for Two Vehicle Classes

Design of Control Model – Cont.

- The average space mean speed of vehicle type *j* on cell *i* during time interval *k* is determined according to the following traffic conditions.

(1). If $\sum_{j=1}^{J} \frac{\rho_{i,j}(k)}{\rho_{j,m,c}} \leq 1$, all the vehicle types are in free flow conditions. The average speed of vehicle type *j* on cell *i* during time interval *k* is $v_{i,j}(k) = min(v_{j,m,f}, u_i(k))$.

(2). If $\sum_{j=1}^{J} \frac{\rho_{i,j}(k)}{\rho_{j,m,c}} > 1$ and $\sum_{j=1}^{j'} \frac{\rho_{i,j}(k)}{\rho^{*}_{j'+1,m,c}} + \sum_{j=j'+1}^{J} \frac{\rho_{i,j}(k)}{\rho_{j,m,c}} \leq 1$, vehicle type 1… j' are in congested traffic conditions, and vehicle type $j'+1$… *J* are in free flow traffic conditions. The average speeds of vehicle class 1 … j' on cell *i* during time interval *k* are $v_{i,j}(k) = \frac{w_m(\rho_{m,jam} - E\rho_i(k))}{E\rho_i(k)}$. The average speeds of vehicle type $j'+1$… *J* on cell *i* during time interval *k* are estimated by $v_{i,j}(k) = min(v_{j,m,f}, u_i(k))$.

(3). If $\sum_{j=1}^{J-1} \frac{\rho_{i,j}(k)}{\rho^{*}_{J,m,c}} + \frac{\rho_{i,J}(k)}{\rho_{J,m,c}} > 1$ and $\sum_{j=1}^{J} \frac{\rho_{i,j}(k)}{\rho_{j,m,jam}} \leq 1$, all the vehicle types are in congested traffic conditions, where $\rho_{j,m,jam} = \frac{\rho_{m,jam}}{\eta_{j,jam}}$, $\eta_{j,jam} = \frac{sd_j}{sd_{car}}$. The average speed of vehicle type *j* on cell *i* during time interval *k* is estimated by $v_{i,j}(k) = \frac{w_m(\rho_{m,jam} - E\rho_i(k))}{E\rho_i(k)}$.

Design of Control Model – Cont.

- The intelligent driver model (IDM) developed by Treiber et al. (2000) is adopted to model the car-following characteristics
- In the IDM, the acceleration $a(k)$ during time interval *k* can be computed

$$a(k) = a\left[1 - \left(\frac{v(k)}{v_0}\right)^4 - \left(\frac{s^*(k)}{s(k)}\right)^2\right] \qquad s^*(k) = max\left(0, s_0 + v * HW + \frac{v(k)\Delta v(k)}{2\sqrt{ab}}\right)$$

- An AV is formulated by adopting the IDM with its headway being smaller than the human-driven vehicle's
- If an AV is following another AV, a smaller headway will be used (0.6s)
- If an AV is following a human-driven vehicle, the vehicle will be acting as a regular AV (1.1s)

Design of Control Model – Cont.

- Objective function
  - Minimizing the sum of TTT spent by all types of vehicles, total speed variation (TSV), and the penalty function
- ***min***

$$J = w_1 T \sum_{k=1}^{T_p} \sum_{i=1}^{N} \sum_{j=1}^{J} \rho_{i,j}(k) l_i + w_2 \sum_{j=1}^{J} \sum_{k=1}^{T_p} \sum_{s=1}^{S} \sum_{i=vsl_{s-1}+1}^{vsl_s} \left[u_i(k) - \left(\frac{vsl_s - i}{N_s} v_{vsl_{s-1},j}(k) + \frac{i - vsl_{s-1}}{N_s} v_{vsl_s,j}(k) \right) \right]^2 + \sum_{s=1}^{S-1} \left(u_{vsl_s}(k) - u_{vsl_s+1}(k) \right)$$

  - The first term of the objective function is the TTT spent by all types of vehicles on the studied freeway corridor
  - The second term is the TSV between speed limits on cell *i* and the traveling speeds of vehicle type *j* on the most upstream and most downstream cells
  - The third term is a penalty function used to represent the speed difference between two consecutive VSL control subsystems
- ***s.t.*** $v_{min} \le u_i(k) \le v_{max}$

  $u_i(k) \in \boldsymbol{V}$ V= {15, 20, 25, 30, 35, 40, 45, 50, 55, 60, 65, 70 mph}

  $|u_i(k+1) - u_i(k)| \le 10$ $|u_i(k) - u_{i-1}(k)| \le 10$

Solution Algorithm

- Genetic algorithm (GA), is selected to optimize the variable speed limits
- Two modules are included
  - GA and VISSIM simulation
- The modified CTM is used to predict the traffic states

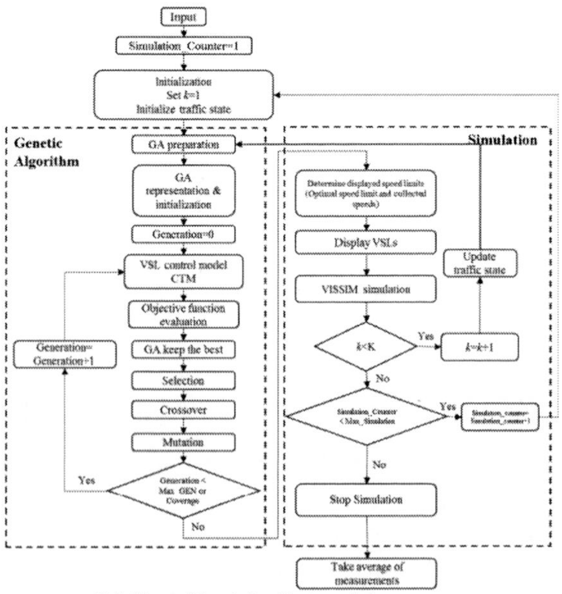

GA Flow Chart for Determining Optimal Speed Limit Set

Case Study

- A real-world freeway corridor is selected
- The studying period is from 5:30 am to 9:00 am on weekdays
- The field data is aggregated into 5-min counts
- The length of the selected freeway corridor is about 5 miles

Map of the Case Study

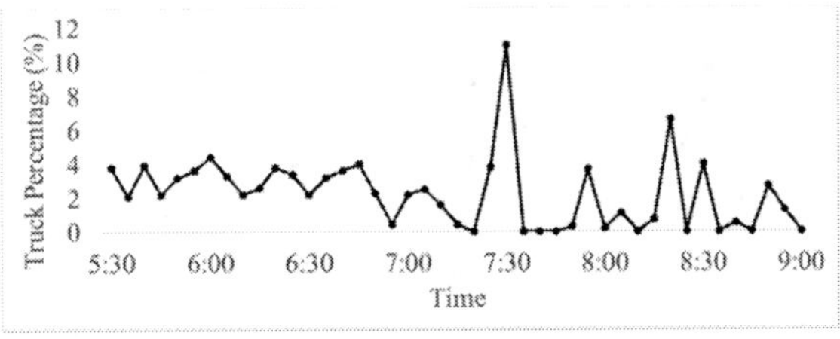

Truck Percentage vs. Study Period

Case Study – Cont.

- A preliminary analysis is performed to identify the positions of bottlenecks (Fan and Gong 2017; Gong and Fan 2017; Gong and Fan 2018)
- Five bottlenecks
  - Detectors 1, 5, 7, 9, and 14
- Three VSL control subsystems are deployed in this study

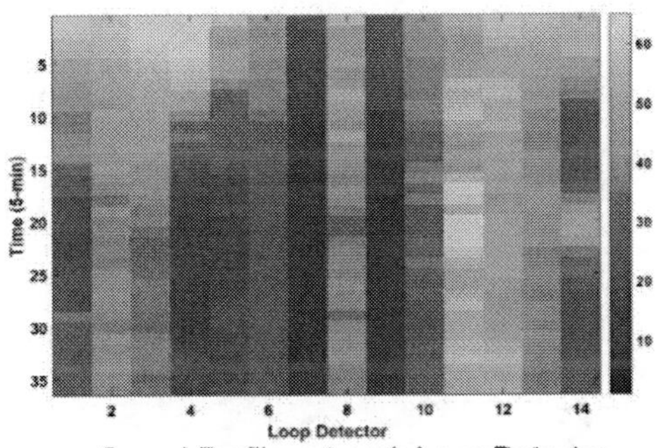

Speed Profiles at each Loop Detector

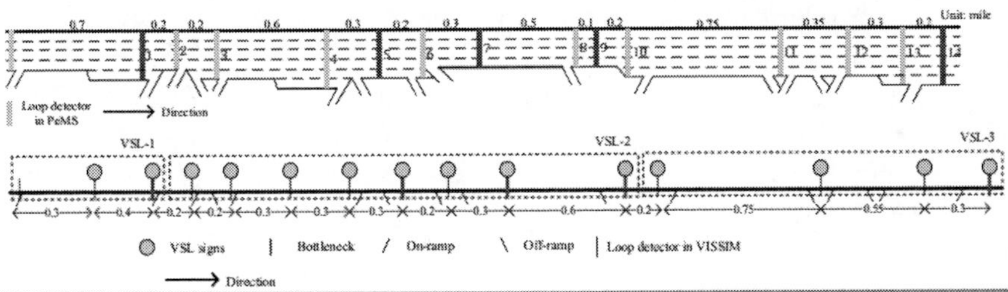

Numerical Results

- Three types of vehicles (i.e., human-driven cars, trucks, and autonomous cars) are included
- Traffic parameters need to be computed (Dervisoglu et al. 2008)
  - E.g., capacity, jam density, and shock wave speed

Computation Results of the CTM at each Bottleneck

Parameters		Bottleneck 1	Bottleneck 2	Bottleneck 3	Bottleneck 4	Bottleneck 5
Capacity (pce/h/lane)		2232	1749	1797	1733	1702
Drop Capacity (pce/h/lane)		2023	1517	1669	1528	1630
Magnitude of Capacity Drop (%)		10.35	15.28	7.66	13.45	4.41
Shock Wave Speed (mph)		10.99	7.26	8.67	8.62	9.24
Critical Density (pce/mile/lane)		34.34	26.33	27.33	26.6	25.66
Jam density (pce/mile/lane)		214.39	235.3	219.86	203.81	202.08
Car (human-driven and AVs)	Free Flow Speed (mph)	64.99	66.42	65.74	65.16	66.32
	Critical Density (veh/mile/lane)	34.34	26.33	27.33	26.6	25.66
Truck	Free Flow Speed (mph)	59.99	61.42	60.74	60.16	61.32
	Critical Density (veh/mile/lane)	20.52	15.12	15.45	15.20	14.45

Numerical Results – Cont.

- To obtain a close match between the collected and simulated traffic data, driver behavior parameters of VISSM are calibrated (Yu and Fan 2017)
  - e.g., standstill distance (CC0) and headway time (CC1)
- Parameters that are used to model the car-following characteristics of the AVs are selected on the basis of existing studies

The IDM's Parameter Value

Vehicle Types	HW_j	a	b	s_0
Human-driven vehicle	1.6 s	3.28 ft/s^2	-6.56 ft/s^2	4.13 ft
AV follows a human-driven vehicle	1.1 s	3.28 ft/s^2	-6.56 ft/s^2	0
AV follows an AV	0.6 s	3.28 ft/s^2	-6.56 ft/s^2	0
References	Treiber et al. 2000; Shladover et al. 2012; Milanés and Shladover 2014; Khondaker and Kattan 2015; Grumert et al. 2015; Li et al. 2017			

Numerical Results – Cont.

- A 3.5-hour simulation with a 30-minute (from 5:30 am – 6:00 am) warm up period is conducted
- The speed limit set that minimizes the objective function over a given prediction horizon (i.e., T_p=5 min)
- The speed limit changes every minute (i.e., T_c=1min)
- The discrete time step used in the control model is T=10s
- w_1=0.9 and w_2=0.1 are selected for the simulation
- Various scenarios are designed in this study

Simulation Scenarios and Descriptions

Scenarios	Description
Scenario 1	With 100% human-driven vehicles and without VSL control
Scenario 2	With 10% AVs and without VSL control
Scenario 3	With 100% human-driven vehicles, VSL control, and the CTM without considering mixed traffic flows
Scenario 4	With 100% human-driven vehicles, VSL control, and the extended CTM
Scenario 5	With 10% AVs and VSL control, and the extended CTM

Numerical Results – Cont.

Performance Comparison under Different Scenarios

Scenario	TTT (veh-h)	Average delays (s)	Average number of stops	Emission (g)			Improvement (%)					
				CO_2	NOx	Particulate	TTT	Delays	Number of stops	CO_2	NOx	Particulate
Scenario 1	8140.51	400.76	67.58	650.33	1734.2	1951	-	-	-	-	-	-
Scenario 2	7988.12	385.75	61.77	641.59	1730.57	1950.45	1.87	3.75	8.59	1.34	0.21	0.03
Scenario 3	5469.65	170.59	26.99	608.91	1585.7	1851.3	32.81	57.43	60.06	6.37	8.56	5.11
Scenario 4	5337.68	158.71	25.74	605	1583.5	1846.12	34.43	60.4	61.91	6.97	8.69	5.38
Scenario 5	5328.65	139.81	25.33	600.32	1578.54	1838.54	34.54	65.11	62.52	7.69	8.98	5.76

- Simulation results under the five designed scenarios
  - TTT, average delays, average number of stops, and emission
- Scenario 1: without control
- Scenarios 3 and 4
  - Examining whether the extended CTM outperforms the CTM without considering mixed traffic flows
- Scenario 4 – VSL control
- Scenario 5 – VSL control in an AV environment

Numerical Results – Cont.

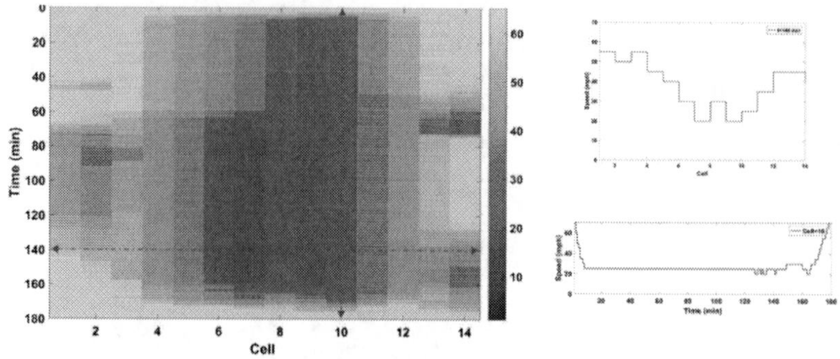

Contour of Speed Limit under Scenario 4

- Speed harmonization impact of VSL control
  - Speed differences among the adjacent cells are noticeably reduced
  - The vehicle speeds at the most congested bottlenecks begin to recover at the end of the simulation
  - The gradual change of color indicates that a smoother transition of speeds among cells has been achieved

Numerical Results – Cont.

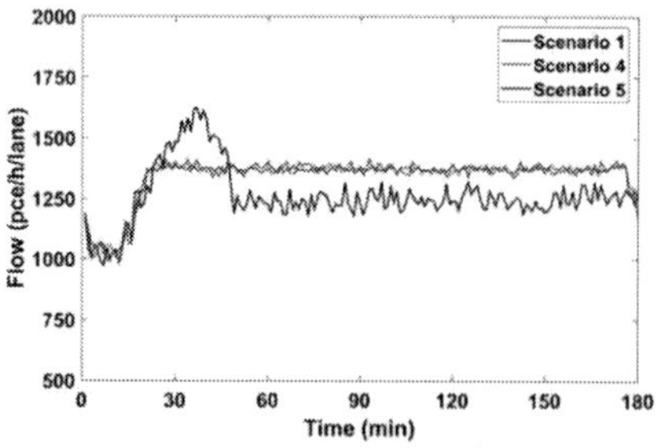

Flow Profiles at Bottleneck 2

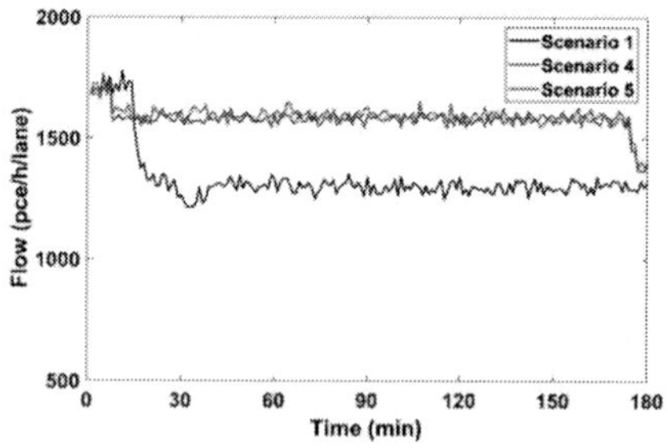

Flow Profiles at Bottleneck 3

- Equilibrium flow (pce/h/lane) profiles during the entire simulation period at bottlenecks 2 and 3 under scenario 1, scenario 4, and scenario 5
  - When traffic demands are greater than the bottleneck capacity, under scenario 1, a drop in flow at the bottleneck can be observed
  - Under scenarios 4 and 5, the equilibrium flow with VSL control remains steady and a higher discharge value is achieved compared to that without VSL control

Conclusion

- A proof-of-concept study on developing a VSL control strategy with CACC in an AV environment for a freeway corridor is performed
- The VSL control is developed on the basis of the extended CTM which considers the capacity drop phenomenon at the bottleneck
- The proposed VSL control model takes the mixed traffic flow (including human-driven cars, trucks and AVs) into consideration
- A real-world freeway corridor is selected to examine the developed control strategy
- The simulation results demonstrate that the developed VSL control can be used to greatly enhance the operational efficiency, improve safety, and reduce the emissions of greenhouse gases
- The VSL control in an AV environment outperforms the VSL control without CACC

QUESTIONS AND COMMENTS?

THANK YOU!

UF | Transportation Institute
UNIVERSITY *of* FLORIDA

Distributed Computation based Constrained Model Predictive Control (MPC) for a Mixed Flow Platoon

Lili Du, Associated Professor,
University of Florida

lilidu@ufl.edu

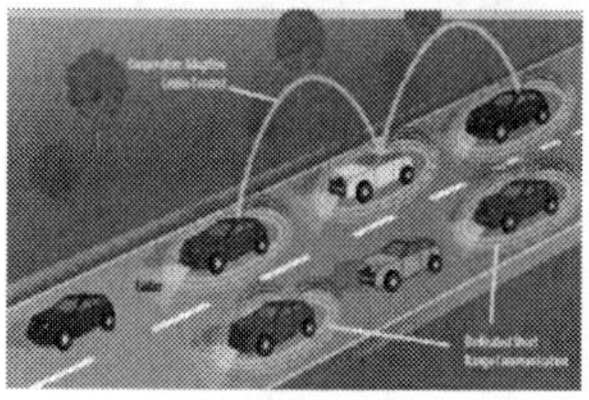

AVS 2018 - Automated Vehicles Symposium, San Francisco, 7/11, 2018

1

UF | Transportation Institute
UNIVERSITY *of* FLORIDA

Research Background

Cooperative adaptive cruise control (CACC)

- Automatically and cooperatively adjusts vehicle speed
- To maintain a stable and safe distance/time headway

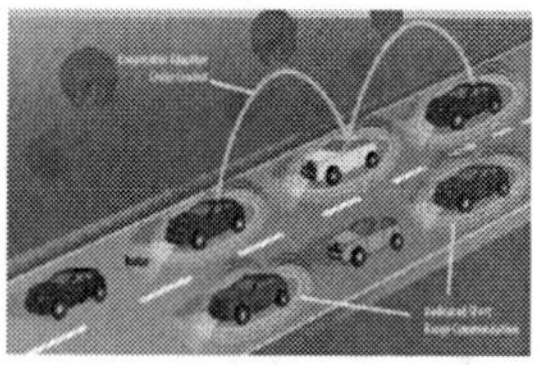

Research gaps and motivation

- CACC mainly focuses on a pure connected vehicle platoon
  - Human-drive vehicles (HDVs) will impair the advantages
- CACC conducts localized control
  - Mainly focuses on neighborhood safety and efficiency
  - Cannot guarantee the performance of the entire platoon
- CACC mainly uses local communication capability
  - Do not fully take advantage of communication capability

Cooperative System Optimal Platoon Control For Mixed Flow

2

Cooperative System Optimal Platoon Control For Mixed Flow

UF | Transportation Institute UNIVERSITY of FLORIDA

Research Objective

- Develop a cooperative platoon control for a mixed flow (CAV and HDV)
- Ensure entire platoon traffic smoothness, efficiency, and stability
- Use model predictive control (MPC) and distributed computation

Research Assumptions

- A sample platoon: a HDV platoon (H) sandwiched by two well-connected CAV platoons (A_1 and A_2).
- The movements of HDVs adjacent to CAVs can be detected.
- Global information exchange is granted in the CAV platoon

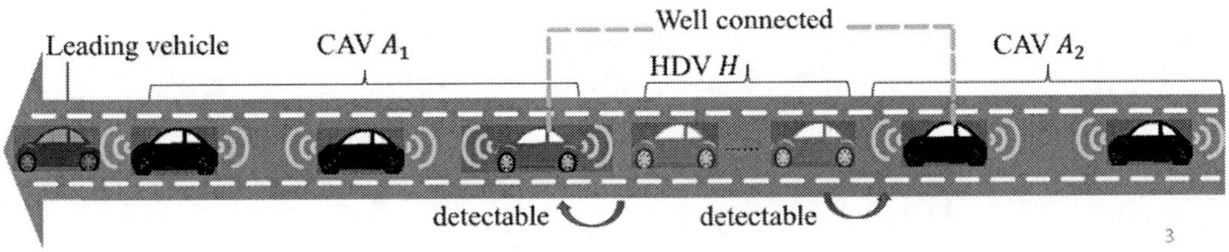

3

Main Idea: Model Predictive Control

UF | Transportation Institute UNIVERSITY of FLORIDA

Control: acceleration/deceleration of the following vehicles at each time step

Procedure: repeat very Δt=1s

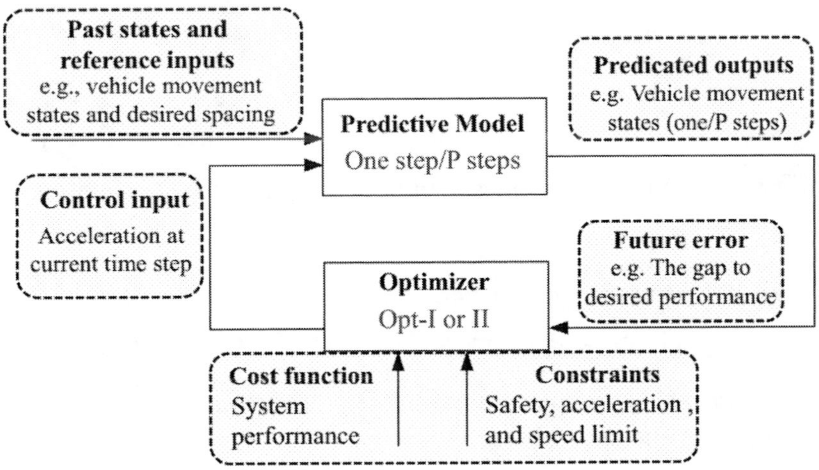

5

Main Idea: Cooperative Platoon Control

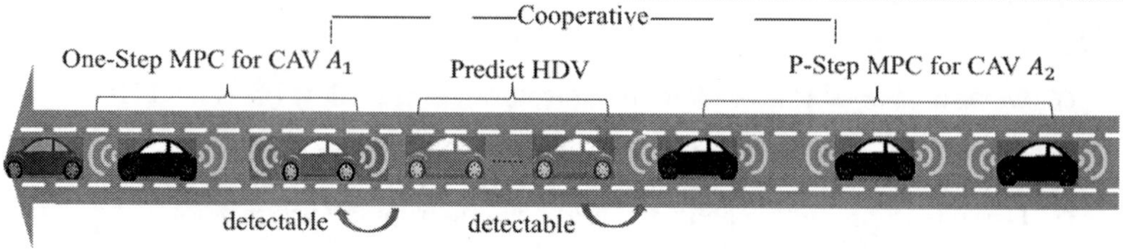

- ☐ One-step MPC (Gong, et al., 2016) for CAV platoon A_1 -- Optimizer-I
  - The movement of its leading vehicle is unpredictable
- ☐ P-step MPC for CAV platoon A_2 -- Optimizer-II
  - The movement of its immediate leading HDV h_n is predictable in P-step
- ☐ The One-Step and P-Step MPCs for platoons A_1 and A_2 are cooperative
  - The movement of the leading HDV for A_2 responds to platoon A_1
  - A_2 cooperatively responds to the predicted movement of the HDV h_n by knowing the movement of A_1

5

Research Challenges

- ☐ Develop and solve 1-step MPC for A_1: addressed by (Gong et al., 2016)

  Siyuan Gong, Jinglai Shen, Lili Du (2016). Distributed computation based car-following control integrating optimal system performance for a platoon of autonomous vehicles. Transportation Research Part B: Methodological, 94, 314-334

- ☐ Develop P-step MPC for A_2 responding to the prediction of HDV movement
  - Keep the traffic smoothness, riding comfort, and string stability.
  - How to determine P?
  - Need to learn the movement dynamics of the adjacent HDVs (h_n and h_1)
- ☐ Solve the optimizers in MPCs : distributed vs. centralized algorithms?
  - Centralized algorithms are not suitable due to platoon topology changes over time and lack of roadside computation infrastructure
  - Distributed algorithms face convergence and computation challenges

6

Vehicle Dynamics

Transportation Institute
UNIVERSITY *of* FLORIDA

CAV dyanmics: Kinematics model

$$x_{i,j}(p+1) = x_{i,j}(p) + \Delta t v_i(p) + \frac{\Delta t^2}{2} u_{ij}(p), \qquad v_{i,j}(p+1) = v_{i,j}(p) + \Delta t u_{i,j}(p)$$

HDV dynamics: follows Newell's linear car-following model (Newell 2002)

$$x_1(j\Delta t + \tau_1) = x_m(j) - d_1 \qquad x_n\left(j\Delta t + \sum_{h=2}^{n} \tau_h\right) = x_1(j) - \sum_{h=2}^{n} d_h$$

- The HDV n (*the leading vehicle of A_2*) responds to CAVs in A_1 with a spatial-temporal delay
- $P = 1 + \left\lfloor \frac{\sum_{h=1}^{n} \tau_h}{\Delta t} \right\rfloor$ ∘ ○ ○ Estimate P?
- A_2 is able to predict the movement of HDV n in the prediction horizon with P control steps, by knowing A_1's movement
- The trajectories of the preceding and following HDVs follow the same shape with a spatial-temporal delay.

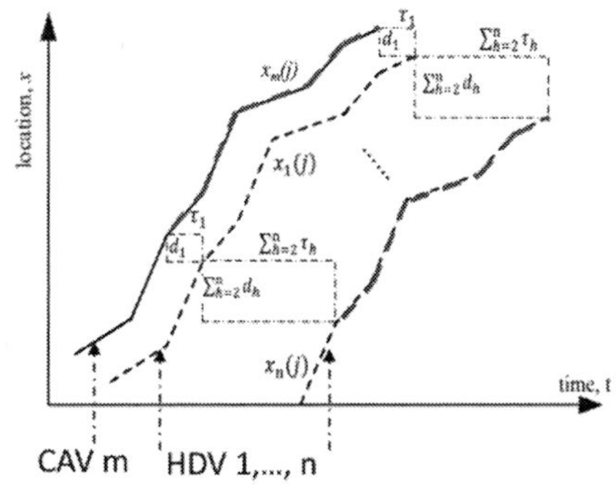

Learn HDV Movement: Curve Matching

Transportation Institute
UNIVERSITY *of* FLORIDA

- Main Ideas (Zhang, 1994):
  - (i) Present the trajectories of the preceding and following vehicles by point sets.
  - (ii) Iteratively move a point on one trajectory to match the closest point on the other by a spatial-temporal motion vector.
  - (iii) The best motion vector is selected to minimize the average distance between all paired points.
  - (iv) Move the trajectory by the best motion vector
  - (v) Repeat the Step (ii) to (iv) until the motion vector becomes small enough
- Find the delay by summing the motion vectors.

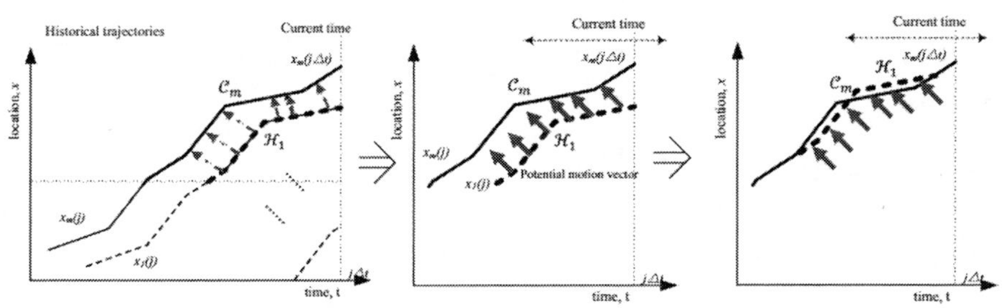

Optimizer II in P-Step MPC for CAV-A_2

UF | Transportation Institute UNIVERSITY of FLORIDA

Min	$\Gamma_2 = \frac{1}{2}\sum_{p=0}^{P}\left\{\left[z_{2,j}^T(p)(Q_\varphi Q_z)z_{2,j}(p) + \left(z_{2,j}'(p)\right)^T (Q_\varphi Q_{z'})z_{2,j}'(p)\right]\right\} + \frac{\Delta t^2}{2}\lVert u_j \rVert_2^2$ Spacing error / Speed fluctuations / Acceleration input	
subject to:	$a_{min,i} \le u_{i,j}(p-1) \le a_{max,i},\ i \in A_2,\ p \in P$	Acceleration limit
	$v_{min} \le v_{i,j}(p) \le v_{max},\ i \in A_2, p \in P$	Speed limit
	$x_{i-1,j}(p) - x_{i,j}(p) \ge L_i + v_{i,j}(p)\,\Delta t - \frac{\left[v_{i,j}(p) - v_{min}\right]^2}{2a_{min,i}}, i = m+2, \dots, \lvert A \rvert, p \in P$ $x_{n,j}(p) - x_{m+1,j}(p) \ge L_{m+1} + v_{m+1,j}(p)\,\Delta t - \frac{\left[v_{m+1,j}(p) - v_{min}\right]^2}{2a_{min,i}},\quad p \in P$	Safety

- Objective: minimizes traffic oscillation using mild control
- Improve Traffic smoothness, Riding comfort, Energy saving, Emission reduction
- Three constraints ensure safety, speed, and acceleration limit.
- The weights (Q_φ, Q_z and $Q_{z'}$) are carefully choosen to keep string stability.

9

Properties of the Optimizer II and MPC

UF | Transportation Institute UNIVERSITY of FLORIDA

- **Lemma 1** (Unique optimal solution, Bazaraa et al., 2006))
  - The objective function is strictly convex ⟺Hessian matrix is positive definite.
  - The constraint set is convex, compact, and nonempty.
- **Lemma 2** (Slater's CQ, Bazaraa et al., 2006))
  - If the leading vehicle $v_{n,j}(p) > v_{min}$, the constraint set always has an interior point solution.
  - It ensures the strong duality for a primal-dual optimization problem.
- **Lemma 3** (Sequential feasibility)
  - If the initial states $\left(x_{i,j}(0), v_{i,j}(0)\right)_{i=m+1}^{\lvert A \rvert}$ and $v_{n,j}(p)$ are feasible (satisfy all constraints), the constraint set is always nonempty.
- **Lemma 4** (Asymptotic stability)
  - The unconstraint P-step MPC is asymptotic stable according to the control theorem developed by *(Löfberg, 2001).*

10

Dual Based Distributed Algorithm

UF | Transportation Institute UNIVERSITY of FLORIDA

- The compact format for Optimizer-II

$$\textbf{Min} \quad \Gamma_2(u) := \frac{1}{2}u^T Hu + c^T u + \gamma$$

$$\textbf{s.t.} \quad \begin{cases} u_{i,p} \in \chi_{i,p}, & i \in A_2;\ p \in P \\ \bar{\varrho}_{i,p}(u) \le 0, & i \in A_2;\ p \in P \\ \underline{\varrho}_{i,p}(u) \le 0, & i \in A_2;\ p \in P \\ g_{i,p}(u) \le 0, & i \in A_2;\ p \in P \end{cases}$$

- $\chi_{i,p}$: acceleration limits- box
- $\underline{\varrho}_{i,p}(u)/\bar{\varrho}_{i,p}(u)$: speed limit- coupled
- $g_{i,p}(u)$: safety constraint- coupled

- Cannot decompose directly to solve it in distributed manner.
- Developing Dual based Distributed Algorithm (Koshal et al., 2011)

- The Lagrangian function is $\mathcal{L}(u,\lambda) = \Gamma_2(u) + \lambda(\bar{\varrho}(u), \underline{\varrho}(u), g(u))^T$
- The primal and associated dual problems:

  Primal $\quad inf_{u\in\chi} sup_{\lambda\in\mathbb{R}_+^{3P|A_2|}} L(u,\lambda)$ $\qquad$ **Dual** $\quad sup_{\lambda\in\mathbb{R}_+^{3P|A_2|}} inf_{u\in\chi} L(u,\lambda)$

- The strong duality holds since Slater's constraint qualification holds.
- Transfer the constraint of $\lambda \ge 0$ to a box constraint set ϕ.

11

Dual based Distributed Algorithm (DBDA-I) for Optimizer-I

UF | Transportation Institute UNIVERSITY of FLORIDA

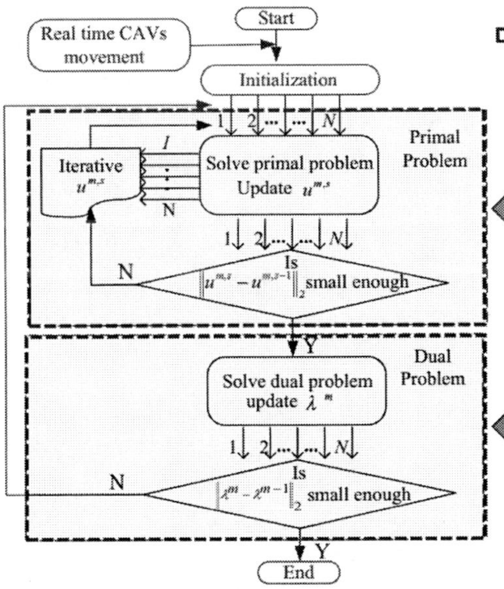

- Each CAV repeat steps 1-4 until converge
  1. Gradient project algorithms solves Primal
     - Given λ^m and u^m, update $u_i^{m,s+1}$

     $$u_i^{m,s+1} = \prod\nolimits_{\chi_i} \left(u_i^{m,s} - \xi \nabla_{u_i}\mathcal{L}(u^{m,s}, \lambda^m)\right)$$
  2. Synchronize to obtain u^{m+1}
  3. Gradient project algorithms solves Dual
     - Given u^{m+1} and λ^m, update λ^{m+1}

     $$\lambda_i^{m+1} = \prod\nolimits_{\phi_i} (\lambda_i^m + \theta[\rho_i(u^{m+1}) - \varepsilon\lambda_i^m])$$
  4. Synchronize to obtain λ^{m+1}
- $\xi \& \theta$ are the step lengths; they affect the convergence.
- Not applicable for the p-step MPC

- *ε is the regularization term to remove degenerate cases*

12

Modified Dual Based Distributed Algorithm (DBDA-II)

UF | Transportation Institute UNIVERSITY of FLORIDA

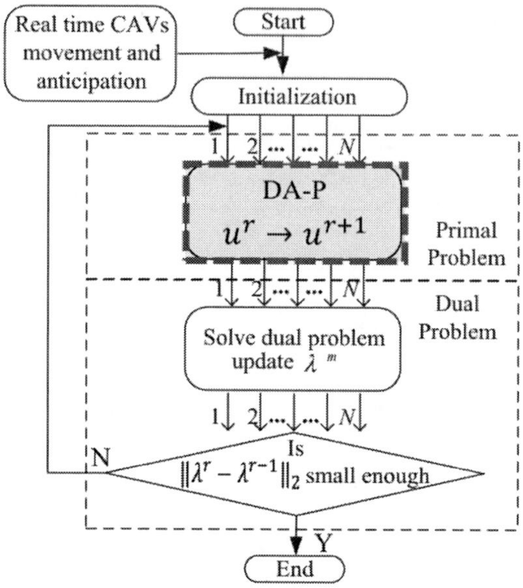

Alternatively solve P and D problems

- Primal Problem has unique optimal solution
- Primal Problem is solved by DA-P algorithm rather than the gradient projection algorithm
- Both DA-P algorithm and the gradient projection algorithm can be implemented in a distributed computation manner with necessary synchronization

13

Numerical Experiment

UF | Transportation Institute UNIVERSITY of FLORIDA

Test Platoon

- Platoon (15): 1 free leading vehicles, A_1 (4 CAVs) , H_1 (5 HDVs), A_2 (5 CAVs).
- The movement of the leading vehicle is obtained from the NGSIM field data.
- Reaction delays and safe spacing of HDVs obtained from existing research.

Experiment Scenarios

- Cooperative platoon control: A_1--one-step MPC and A_2--P-step MPC.
- Non-cooperative platoon control: A_1 and A_2--One-step MPC independently.
- CACC in literature (Schakel et al., 2010).
- No control

Test objectives

- Test the performance of the algorithms: DBDA-II and curve matching.
- Test the performance of the cooperative platoon control.

14

UF | Transportation Institute
UNIVERSITY of FLORIDA

Numerical Experiments

Computation performance DBDA-I vs DBDA-II

Platoon	Algorithm	Computation time	
		Mean (sec)	Variance
A_1 (1-step MPC)	DBDA-II	0.0024	0.000041
	DBDA-I	0.0037	0.000047
A_2 (P-step MPC)	DBDA-II	0.0128	0.000179
	DBDA-I	0.3598	0.027836

- For Optimizer-I (A_1), both DBDA-I and –II converge in very short time.
- For Optimizer-II (A_2), both DBDA-I and –II need more computation time.
- DBDA-II is applicable but DBDA-I not. DBDA-II outperforms DBDA-I.

15

UF | Transportation Institute
UNIVERSITY of FLORIDA

Numerical Experiment

Performance of the Curve Matching Algorithm

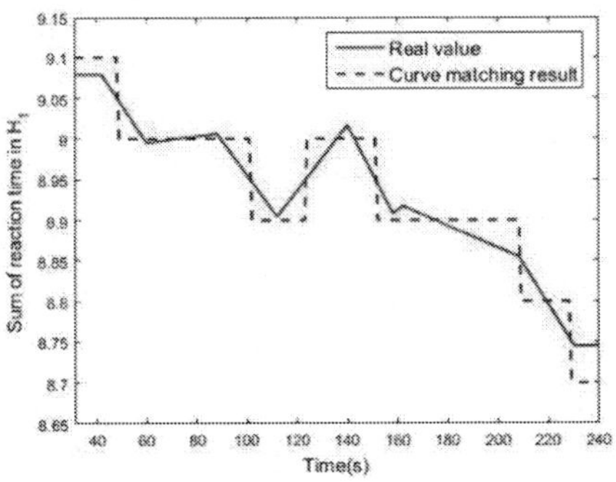

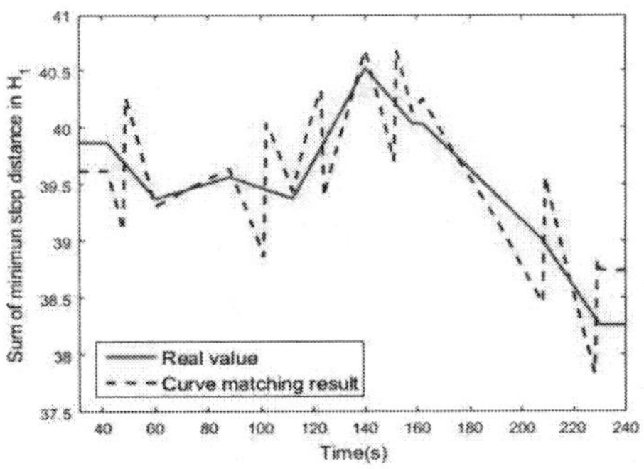

- The aggregated error of the reaction time and the minimum stop distance are less than 0.1s and 1m, respectively.
- The results show that computation time is 0.0021-0.0784s.
- The curve matching algorithm satisfies the online applications.

16

Numerical Experiments

UF | Transportation Institute UNIVERSITY of FLORIDA

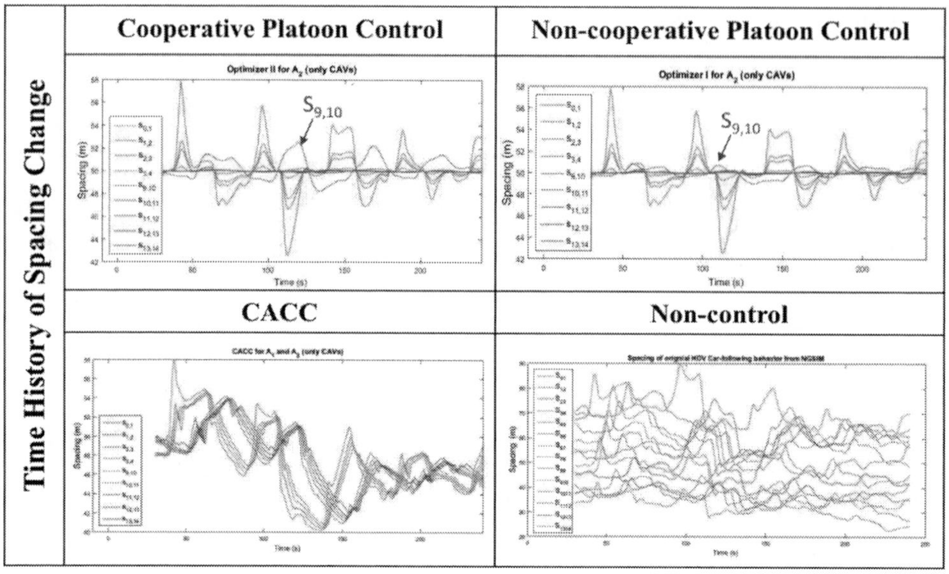

Vehicle 9: the last HDV in H; vehicle 10: the first CAV in A2

17

Numerical Experiment

UF | Transportation Institute UNIVERSITY of FLORIDA

- The cooperative platoon control mitigates speed fluctuation more efficiently.

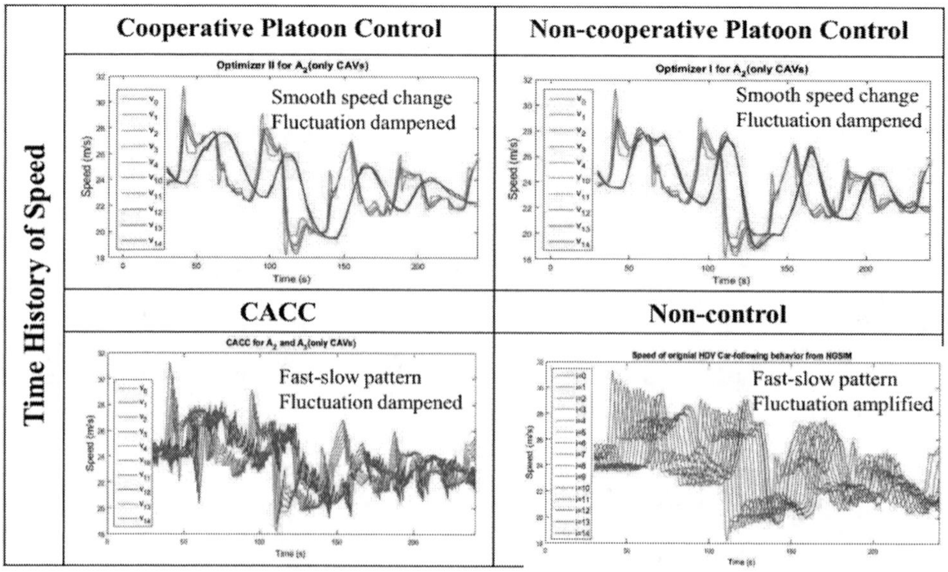

18

Numerical Experiment

UF | Transportation Institute
UNIVERSITY *of* FLORIDA

- The cooperative platoon control generates the mildest control input
- It improves riding comfort and reduce emission and the fuel consumption

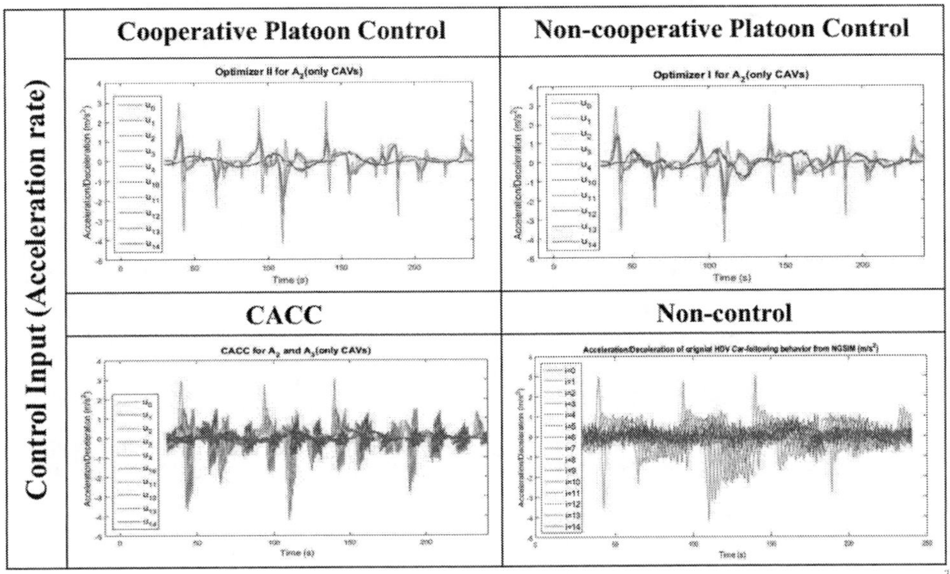

19

Summary

UF | Transportation Institute
UNIVERSITY *of* FLORIDA

- This research develops a cooperative platoon control for a mixed flow platoon.
  - The curve matching learning algorithm combined with Newell's car-following model is used to predict HDV movement.
  - The one-step and P-step constrained MPC are developed to cooperatively control the mixed flow platoon.
  - A modified dual-based distributed algorithm is developed to solve the MPC according to the special properties and structure of the optimizers.
  - The stability of the control scheme is studied for the unconstrained linear closed-loop system.
- Numerical experiments based on field data demonstrate
  - The cooperative platoon control effectively dampens the propagation of traffic fluctuation/oscillation along a platoon
  - It outperforms the scheme independently conduct system optimal control by one-step MPC for each CAV platoon.
  - Both cooperative or non-cooperative system optimal platoon control performs better than CACC scheme (Schakel et al., 2010).

20

UF | Transportation Institute
UNIVERSITY of FLORIDA

Thank you and Questions

lilidu@ulf.edu

University of Florida

Acknowledgments

- The presented research is partially supported by the National Science Foundation awards CMMI-1436786 and CMMI-1554559.
- The co-author
  - Siyuan Gong: my previous Ph.D. student.
  - Jinglai Shen, Associate Professor, UMBC

21

Connected Signals®
SEE THE LIGHT

Data Sharing Panel
Automated Vehicles Symposium

Matt Ginsberg

1

Liability

3

Privacy
FT
4

Looking for the BAD sets...
?
iteris
1

AUTOMATED VEHICLES® SYMPOSIUM 2018

OEM/DOT Dialogue on Dedicated Lanes, Work Zones and Shared Data

(Breakout #32)

AUTOMATED VEHICLES® SYMPOSIUM 2018

OEM/DOT Dialogue on Work Zones

(Breakout #32)

Moderator: Dr. Christopher Poe, Texas A&M Transportation Institute

Agenda

- Moderator / Challenges for Auto Industry 
- Overview of USDOT Connected Work Zone Grant 
- Perspective from IOO 
- Perspective from Smart Work Zone Vendor 

3

Challenges for Automotive Industry

Dr. Christopher Poe, Texas A&M Transportation Institute

4

Work Zone Industry Issues

- Owner
  - Roadway Owner vs. Contractor
- User
  - Traffic Engineer vs. OEM/Supplier vs. Public
- Terminology
  - Work Zone vs. Lane Closure
  - Lane Closure vs. Lane Shift
- Planned vs. Actual Real-time Conditions
- Differences in Use of the Manual on Uniform Traffic Control Devices (MUTCD)

AUTOMATED VEHICLES SYMPOSIUM 2018

5

Automotive Industry Issues

- Auto Industry AV Solution
- Sensing
  - Machine Vision
  - Radar
  - Lidar
  - High Definition Maps

AUTOMATED VEHICLES SYMPOSIUM 2018

6

Needs for National Deployment of AVs

- Super Cruise^{TM} Map

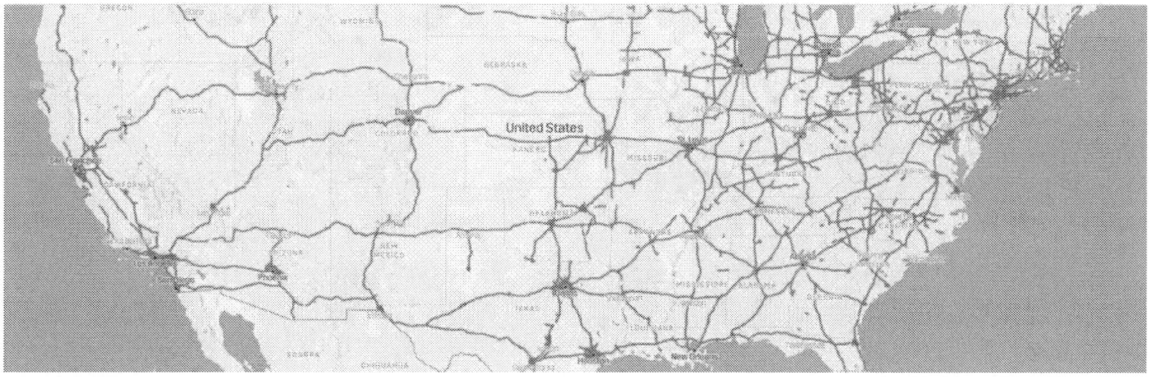

AUTOMATED VEHICLES SYMPOSIUM 2018

7

Automotive Industry Issues (cont.)

- Auto / Infrastructure Industry Shared Solution
- Low-level Automation (Levels 1-2)
  - Turn back to driver
- High-level Automation (Levels 3-5)
  - Navigate work zone

AUTOMATED VEHICLES SYMPOSIUM 2018

8

USDOT Connected Work Zone Grant

Mr. Paul Pisano, Federal Highway Administration

AUTOMATED VEHICLES SYMPOSIUM 2018

9

Connected Work Zone – Project Overview

Goal:

- Test the Reduced Speed Zone Warning / Lane Closure (RSZW/LC) application

Scope:

- Evaluate work zone map data developed by the local agency against the in-vehicle application performance
- Refine/update/adapt the application to variations in map generation capabilities and work zone configurations

Deliverables:

- Guidance document for deployment of the RSZW/LC application
- Refined RSZW/LC application

Application Evaluation Site:

- TxDOT I-35 corridor work zone in Temple, TX

Task Duration:

- May 2017 to March 2019

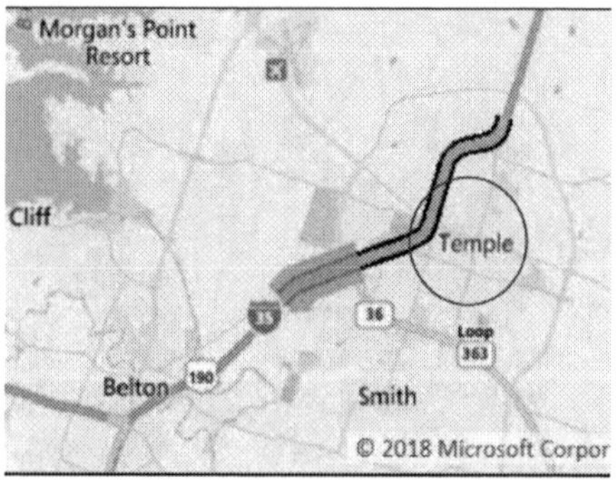

AUTOMATED VEHICLES SYMPOSIUM 2018

10

Connected Work Zone - Project Status

- **Completed:**
  - Established multi-agency working relationships
    - Includes on-going technical information exchange process
    - Includes initial collaboration with TxDOT districts for ultimate work zone placement
  - Developed two-tiered approach (high and low fidelity scenario)
  - Multi-agency site visit (March 2018)
- **Ongoing:**
  - Integrating mapping information with closure information
  - Refining deployment plans
- **Next steps:**
  - Baseline information needs between high and lower fidelity scenarios
  - Establish procedure for locating mapping data
  - District and contractor look-ahead/coordination
  - Equipment deployment and testing

AUTOMATED VEHICLES SYMPOSIUM 2018

11

Perspective from IOO

Mr. Robert Brydia, PMP, Texas A&M Transportation Institute

AUTOMATED VEHICLES SYMPOSIUM 2018

12

How the Agencies Fit Together

ITS Lonestar Intelligent Transportation System

SwRI SOUTHWEST RESEARCH INSTITUTE

CAMP LLC — Vehicle to Infrastructure (V2I) Consortium

HONDA NISSAN KIA Ford SUBARU GM VOLKSWAGEN VOLVO

Texas Department of Transportation

Texas A&M Transportation Institute

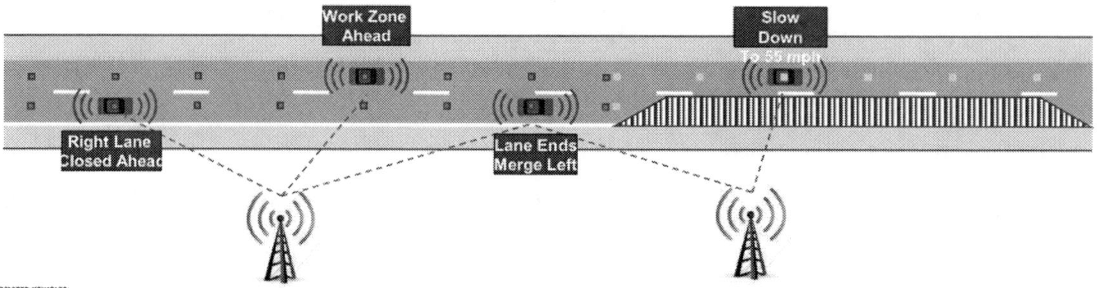

AUTOMATED VEHICLES SYMPOSIUM 2018

13

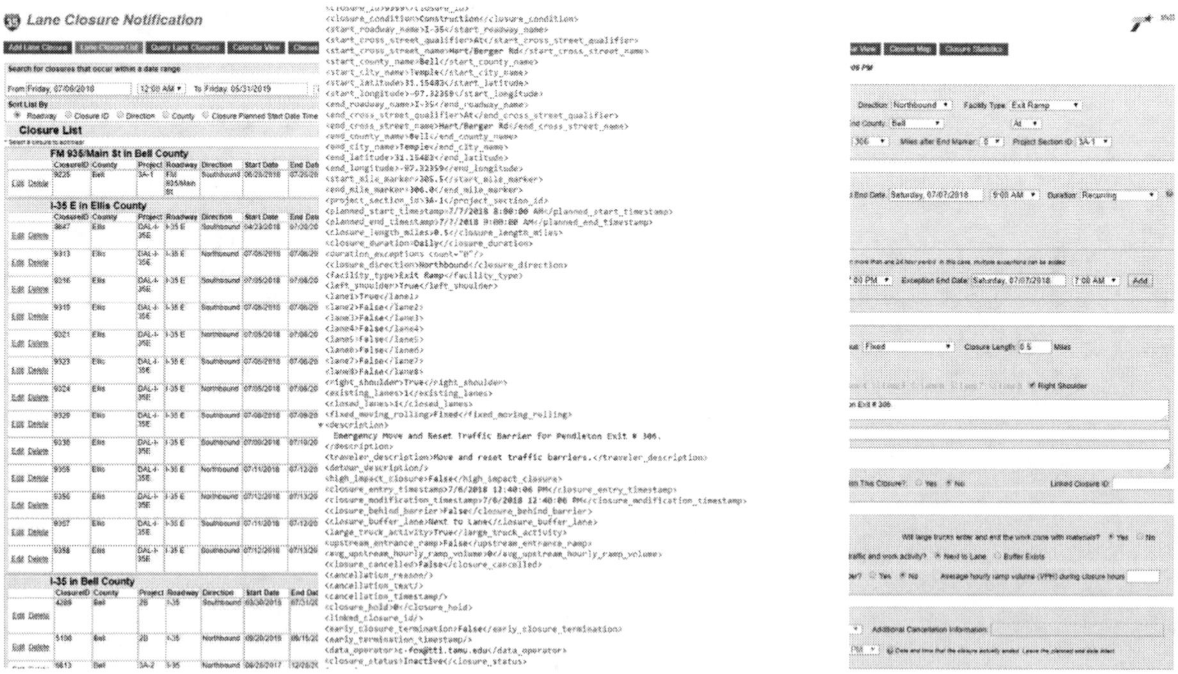

The Need for Linkages

Contractor coordination for lane closure data

- Appropriate detail
- In advance
- Procedures for emergency situations

Multiple systems / endpoints within the IOO

- Traveler information
- ITS
- TMC operations

Work zone mapping to lane closure

- Work zone mapping not yet commonplace
- Correlations of data are not automatic

AUTOMATED VEHICLES SYMPOSIUM 2018

15

Mapping: Two-Tiered Solution

High fidelity scenario

- Detailed lane-level mapping of the roadway and work zone is possible
- Reference point (beginning of lane closure taper) can be accurately defined
- Full information load for RSZW/LC application is supported

Lower fidelity scenario

- Less detailed mapping of the roadway and work zone
- Reference point is estimated
- Reduced information load for RSZW/LC application is supported

Additionally...

- CAMP has developed multiple mapping procedures which can be used to support the above fidelity points

AUTOMATED VEHICLES SYMPOSIUM 2018

16

Perspective from Smart Work Zone Vendor

Mr. Ross Sheckler, iCone Products

AUTOMATED VEHICLES SYMPOSIUM 2018

17

Principle #1 Drivers will do the unexpected

Courtesy of Greg Driskell, President of Professional Pavement Products

AUTOMATED VEHICLES SYMPOSIUM 2018

18

Each Day...

Construction and Maintenance forces 10,000+ Map Changes

- Flaggers
- Lane closures
- Road closures
- Cross-overs
- Moving operations
- Emergency response

AUTOMATED VEHICLES SYMPOSIUM 2018

10,000+ Worksites

Construction/Maintenance Traffic Control is Implemented by Fallible Human Beings

AUTOMATED VEHICLES SYMPOSIUM 2018

New tools you should know about...

- Connected Arrow Boards ➡ Lane closure
- Connected Work Trucks ➡ Hazard
- Connected Delineators ➡ Work zone boundary
- Connected Flaggers ➡ Alternating Single Lane
- Worker Beacon ➡ Active work zone regs.

AUTOMATED VEHICLES SYMPOSIUM 2018

21

Breakout Session # 33
Policy Implications of AVs: Pricing

Summary of Key Findings and Lessons Learned

- 4 scenarios: Technology Wins, Muddling Through, Environment First, Global Chaos
- Pricing of AVs may be necessary to produce/replace infrastructure funding and address externalities of driving (e.g., crashes, congestion, pollution)
- We need to imagine a different future, rather than using today as a reference point
- Political challenges > research challenges
- Travel is often viewed as a right that should be free, but that's not the case in reality
    - Opportunity to make costs transparent and to introduce pricing in new regulations for AVs
- Different levels of detail and inclusion of different tools, in terms of congestion pricing, can be used to target specific externalities that result from congestion related to AVs
- The challenge is to match the tool to the externality
- Distance tax, vehicle weight tax, both interesting options that could be considered if VMT increases to the point where congestion is untenable

AUTOMATED VEHICLES SYMPOSIUM 2017
USERS. VEHICLES. INFRASTRUCTURE.

Breakout Session # 33
Policy Implications of AVs

Research Needs

- Occupancy pricing: vehicle occupancy verification → currently expensive to implement. Can smart phone or in-vehicle systems help verify?
- MaaS systems or built-in vehicle "wallet", where payment for express lanes is included in the overall fee for the service
    - Would this kind of business model make travel costs more transparent and consumers more likely to accept the costs?
- Value of Time → how does it change in an AV, what are the implications for VMT?
- If data collection is brokered by a third party, how would that affect consumer's willingness to release location and other data? Can the blockchain be applied to this to facilitate user trust?
- Could major disruptions like global geopolitical or environmental crises, leaps in technology push pricing or other regs?

AUVSI
TRB

AUTOMATED VEHICLES SYMPOSIUM 2018

Driven by: AUVSI & TRB

#34 A License to Drive: How to Allow Automated Vehicles on Public Roads

1:30 PM – 1:45 PM Setting the Scene 1
Dutch Connected & Automated Driving policy – Edwin Nas (Ministry of Infrastructure & Water Management)
1:45 PM – 2:05 PM Setting the Scene 2
Operational Design Domain Framework in Relation to Deployment Challenges – Tom Alkim (RWS)
2:10 PM – 2:25 PM First Thoughts on a Driver's License for Automated Vehicles
Initial Ideas Regarding an Assessment Framework for AV-Capabilities – Gerben Feddes (RDW)
2:25 PM – 2:45 PM First Thoughts on a Driver's License for Automated Vehicles
Scenario-based safety validation of automated driving: Streetwise – Maurice Kwakkernaat (TNO)
2:45 PM – 3:15 PM Safety Assurance of Automated Road Vehicles
Australian Approach for Safety Assurance of Automated Road Vehicles - Marcus Burke (NTC)
3:15 PM – 3:45 PM Break
3:45 PM – 4:30 PM Interactive Session
Moderator: Serge van Dam (RWS)
4:30 PM – 5:15 PM Panel: A License to Drive, Are We on the Right Track to Allow Automated Vehicles on the Road?
Panel: Marcus Burke, Edwin Nas, Gerben Feddes, Maurice Kwakkernaat, Steve Shladover (UC Berkeley, PATH program)
Moderator: Serge van Dam (RWS)
5:15 PM – 5:30 PM wrap-up and conclusion

AUTOMATED VEHICLES SYMPOSIUM 2018

Driven by: AUVSI & TRB

#34 A License to Drive: How to Allow Automated Vehicles on Public Roads

"To be accepted on the road an AV developer should be forced to be driven by it's creation first. In mixed traffic, on the backseat..."

#34 A License to Drive:
How to Allow Automated Vehicles on Public Roads

Statements - 1

- Do you think we can assess an AV's functionality in advance? **37,5%**
- Can we trust AV developers to self-certify? **17,5%**
- Do you think a "driver's license" for an AV is feasible? **65%**
- Are you in favor of a scenario-based approach (**10%**) or a skill-based (**20%**) approach? Combination **40%**
- Are you in favor of mandatory sharing of edge case data? **75%**

#34 A License to Drive:
How to Allow Automated Vehicles on Public Roads

Statements - 2

AUTOMATED VEHICLES SYMPOSIUM 2018
Driven by: AUVSI & TRB

- Do you think we can quantify something as subjective as acceptable driving behaviour? **25%**
- Connectivity is necessary to make AV work **50%**
- Do you think every country (or state) needs a specific part of the procedure to account for cultural/infrastructural differences? **55%**
- Where do you think the first ODD will be for which an AV can acquire a driver's license? Highways **50%**, Rural **12,5%**, City **10%**

#34 A License to Drive:
How to Allow Automated Vehicles on Public Roads

Statement 1

AUTOMATED VEHICLES SYMPOSIUM 2018
Driven by: AUVSI & TRB

Do you think we can assess an AV's functionality in advance?

37,5% (15/40)

#34 A License to Drive:
How to Allow Automated Vehicles on Public Roads

Statement 2

AUTOMATED VEHICLES SYMPOSIUM 2018
Driven by: AUVSI & TRB

Can we trust AV developers to self-certify?

17,5% (7/40)

#34 A License to Drive:
How to Allow Automated Vehicles on Public Roads

Statement 3

AUTOMATED VEHICLES SYMPOSIUM 2018

Driven by: AUVSI & TRB

Do you think a "driver's license" for an AV is feasible?

65 % (26/40)

#34 A License to Drive:
How to Allow Automated Vehicles on Public Roads

Statement 4

AUTOMATED VEHICLES SYMPOSIUM 2018

Driven by: AUVSI & TRB

Are you in favor of a scenario-based approach or a skill-based approach?

Scenario-based	10% (4/40)
Skill-based	20% (8/40)
Combination	40% (16/40)

#34 A License to Drive:
How to Allow Automated Vehicles on Public Roads

Statement 5

AUTOMATED VEHICLES SYMPOSIUM 2018
Driven by: AUVSI & TRB

Are you in favor of mandatory sharing of edge case data?

75% (30/40)

#34 A License to Drive:
How to Allow Automated Vehicles on Public Roads

Statement 6

AUTOMATED VEHICLES SYMPOSIUM 2018
Driven by: AUVSI & TRB

Do you think we can quantify something as subjective as acceptable driving behaviour?

25% (10/40)

#34 A License to Drive:
How to Allow Automated Vehicles on Public Roads

Statement 6A

AUTOMATED VEHICLES SYMPOSIUM 2018
Driven by: AUVSI & TRB

Connectivity is necessary to make AV work

50% (20/40)

#34 A License to Drive:
How to Allow Automated Vehicles on Public Roads

Statement 7

AUTOMATED VEHICLES SYMPOSIUM 2018
Driven by: AUVSI & TRB

Do you think every country (or state) needs a specific part of the procedure to account for cultural / infrastructural differences?

55% (22/40)

AUTOMATED VEHICLES SYMPOSIUM 2018
Driven by: AUVSI & TRB

#34 A License to Drive:
How to Allow Automated Vehicles on Public Roads

Statement 8

Where do you think the first ODD will be for which an AV can acquire a driver's license?

Highways	50% (20/40)
Rural	12,5% (5/40)
City	10% (4/40)

Ministry of Infrastructure
and Water Management

The Netherlands

Driving the future: The vehicle gets a Driving License?

Edwin Nas
Dep. Teamleader Smart Mobility/
Connected & Automated Driving

Ministry of Infrastructure and Water Management
The Netherlands

24 July 2018

2 Ministry of Infrastructure and Water Management 24 July 2018

Dutch comprehensive approach - ACCESS

Future mobility will be:

- Automated
- Connected
- Cooperative
- Electric and
- Shared

For Society

National approach

- Open road testing with driver possible nationwide (2015)
  - One stop shop for exemption at rdw.nl/its
  - Driver in or outside the vehicle is within the law
  - No legal reference to holding the steering wheel
  - Distance keeping is functional: "safe distance"

- New experimental law nationwide (per 2018)
  - Driving on open road without a driver inside the vehicle
  - New approach for the term 'driver'
  - New testing possibilities for highly and fully autonomous

National approach

- Open road testing with driver possible nationwide (2015)
  - One stop shop for exemption at rdw.nl/its
  - Driver in or outside the vehicle is within the law
  - No legal reference to holding the steering wheel
  - Distance keeping is functional: "safe distance"

- New experimental law nationwide (per 2018)
  - Driving on open road without a driver inside the vehicle
  - New approach for the term 'driver'
  - New testing possibilities for highly and fully autonomous

5 Ministry of Infrastructure and the Environment 24 July 2018

LIVING LAB NL

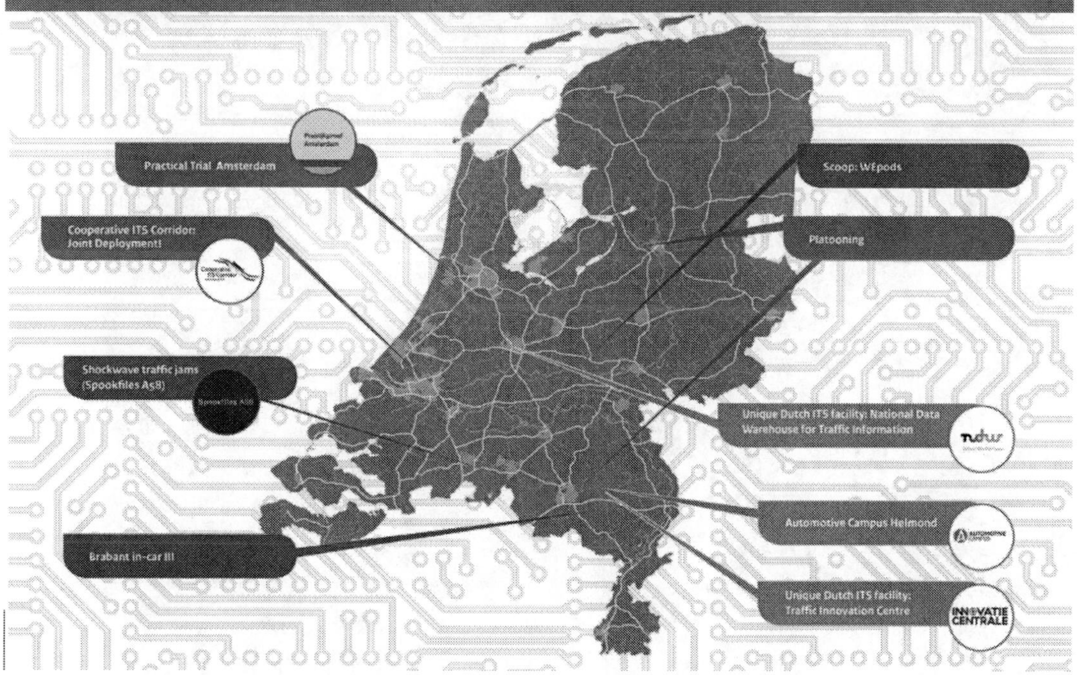

7 Ministry of Infrastructure and the Environment 24 July 2018

8 Ministry of Infrastructure and Water Management 24 July 2018

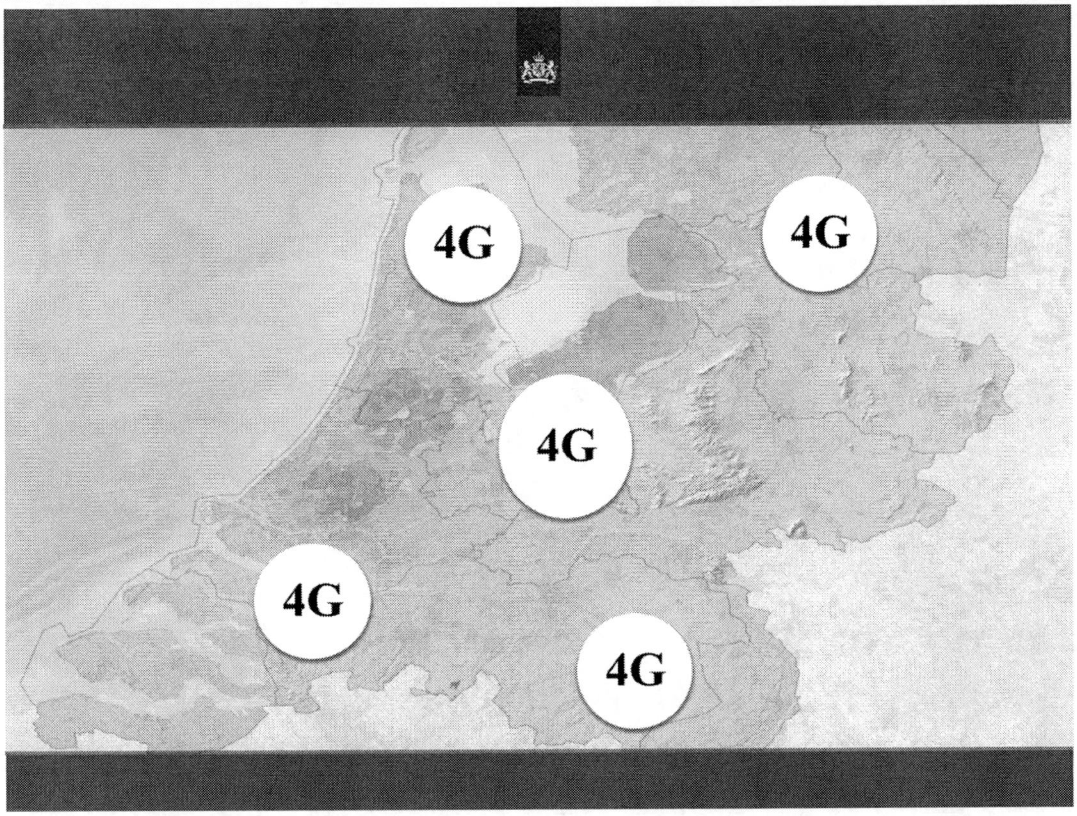
4G
4G
4G
4G
4G

48
50
10
Ministry of Infrastructure and Water Management
24 July 2018

+

=

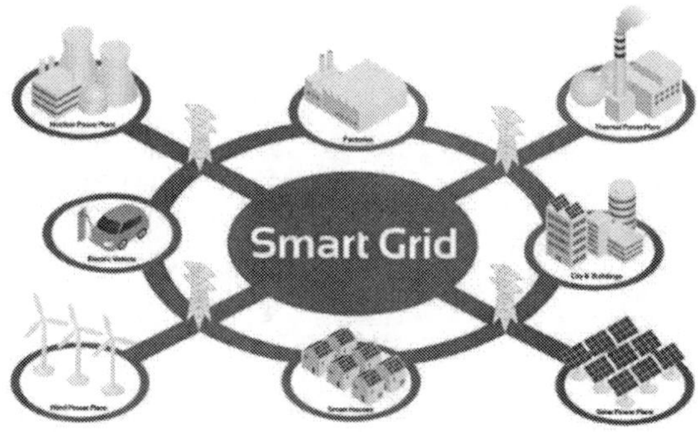

11 Ministry of Infrastructure and Water Management 24 July 2018

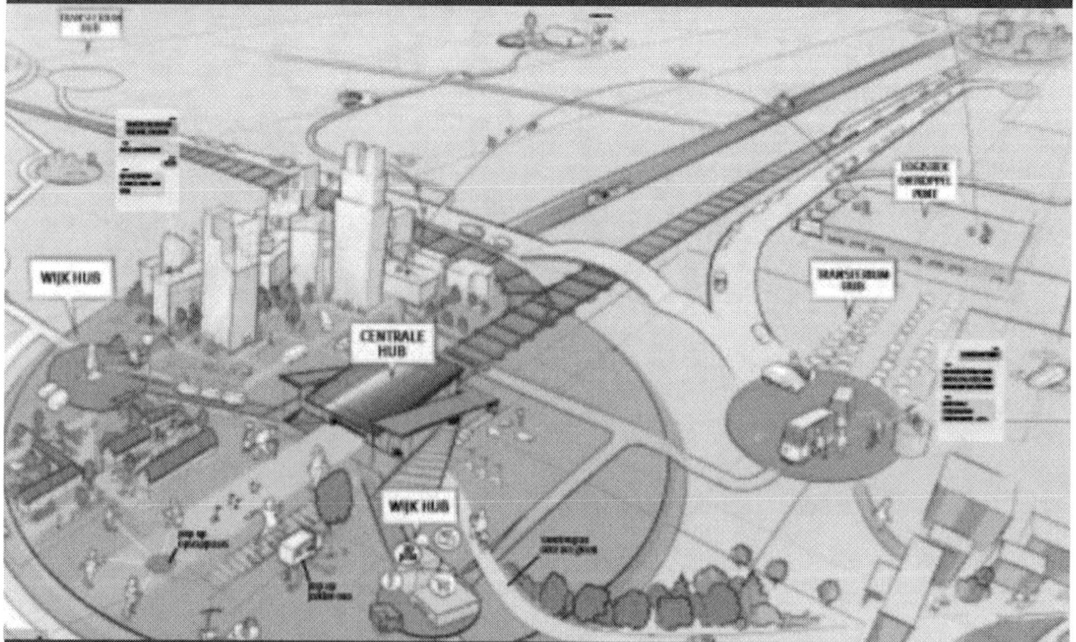

12 Ministry of Infrastructure and Water Management 24 July 2018

Make the (EU) type approval system future proof

And get this show on the road

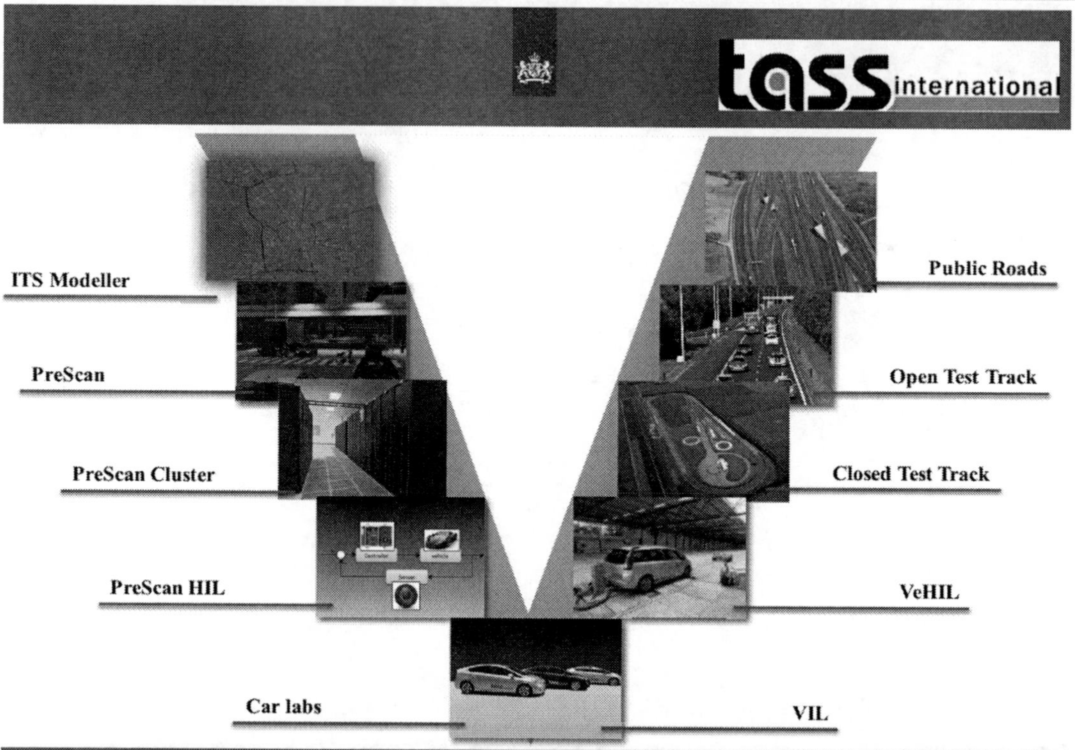

License to drive

- National cooperation with Ministry (DoT), RDW (vehicle authority), RWS (road authority) and CBR (driving license authority)

- Functional assessment of AV driving functions
- Functional safety confrontation based on human experience
- And human driving behaviour

- Creating trust and additional safety

- Process to raise the bar for open road testing...
- ... And national and EU type approval

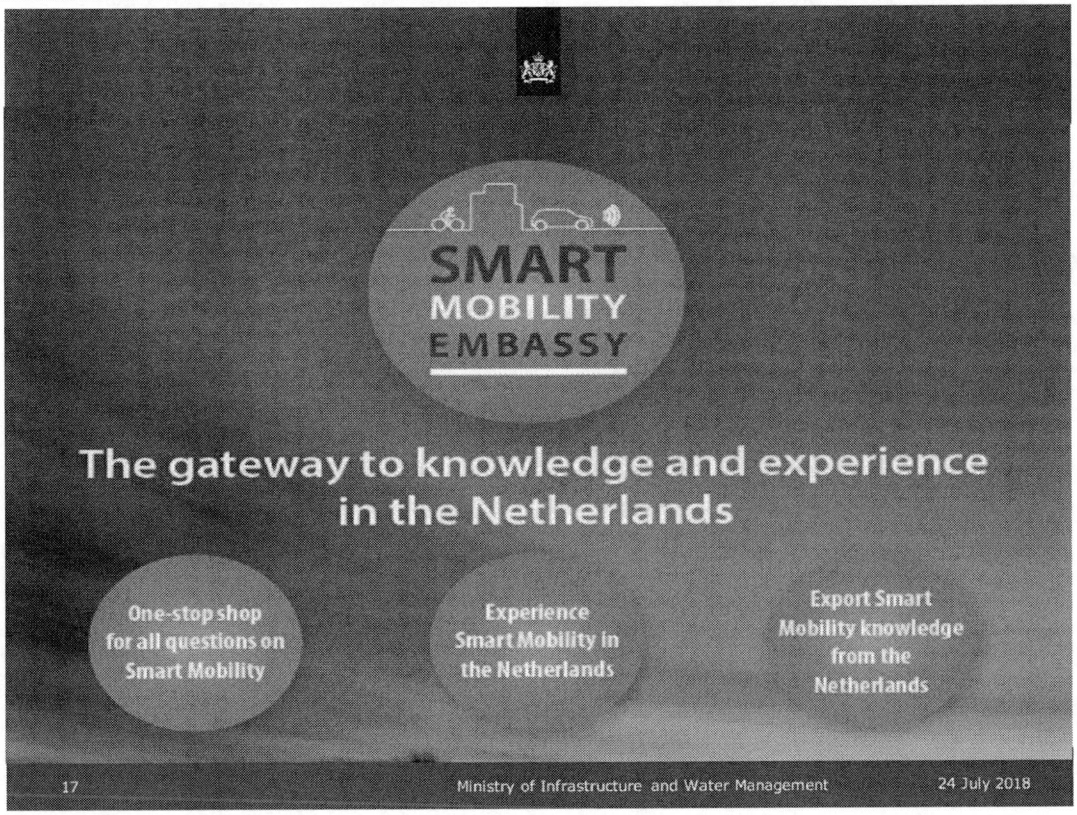
SMART
MOBILITY
EMBASSY
The gateway to knowledge and experience in the Netherlands
One-stop shop for all questions on Smart Mobility
Experience Smart Mobility in the Netherlands
Export Smart Mobility knowledge from the Netherlands
17
Ministry of Infrastructure and Water Management
24 July 2018

More info:
Edwin Nas – edwin.nas@minienm.nl
+31 6 15 35 94 11
18
Ministry of Infrastructure and Water Management
24 July 2018

AUTOMATED VEHICLES SYMPOSIUM 2018

Driven by: AUVSI & TRB

Rijkswaterstaat
Ministerie van Infrastructuur en Waterstaat

The ODD framework

AVS2018 Breakout session #34

"A license to drive, how to allow AVs on public roads"

11 July 2018, San Francisco

Tom Alkim
Senior Advisor
Connected & Automated Driving

Intertraffic 2018

1. "I want to ensure our **infrastructure is ready for connected and automated driving**. I want to sit down with other road operators and the automotive and telecom sector. To discuss what infrastructure we'll need in order for the new generation of vehicles to deliver the greatest gains to society."
2. "I want to **enable the next generation of vehicles to actually take the road**. So I'm creating a legal framework for automated driving. Laying down requirements for reliability and safety that cars must meet before they can hit the road. **A driving licence for self-driving cars**, if you like. Not for the driver – but for the car itself!"
3. "I want to **create room for new transport concepts and make sure people can use them easily**. For instance, 'mobility as a service' – transport that's available when you need it, to wherever you want to go, using your choice of transport modes"

https://www.government.nl/documents/speeches/2018/03/26/speech-by-cora-van-nieuwenhuizen-minister-of-infrastructure-and-water-management-at-the-opening-of-intertraffic-amsterdam-20-march-2018

Past the hype?

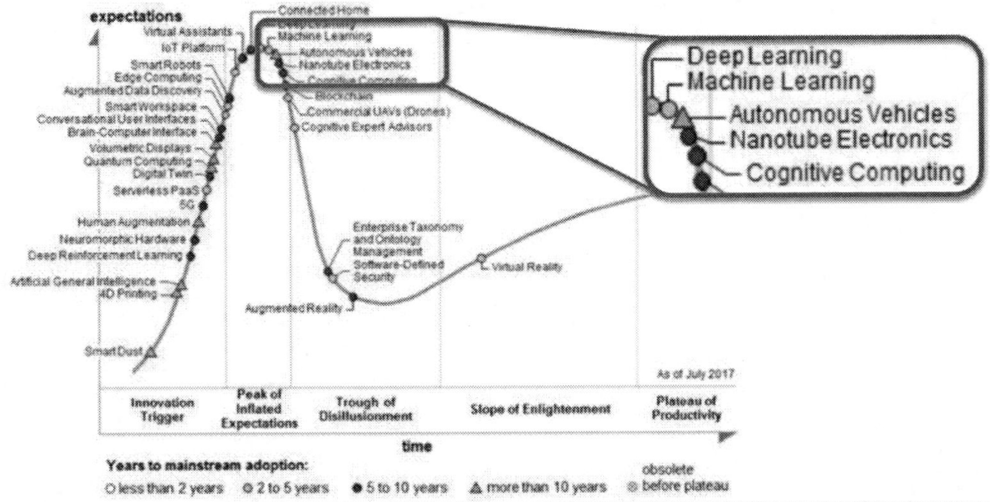

ODD framework

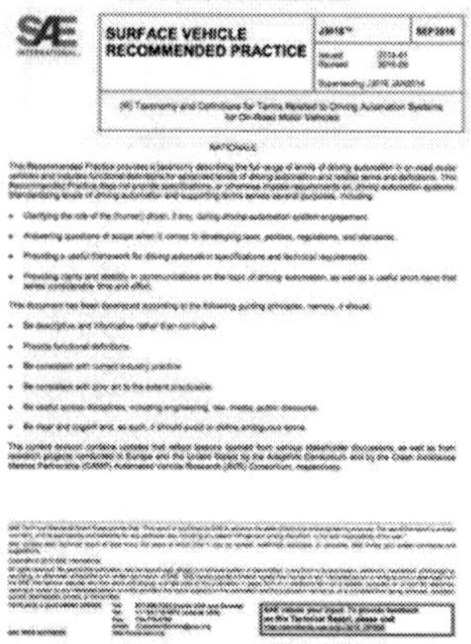

Level	Name	Narrative definition	DDT: Sustained lateral and longitudinal vehicle motion control	DDT: OEDR	DDT fallback	ODD
Driver performs part or all of the DDT						
0	No Driving Automation	The performance by the *driver* of the entire *DDT*, even when enhanced by *active safety systems*.	*Driver*	*Driver*	*Driver*	n/a
1	Driver Assistance	The *sustained* and ODD-specific execution by a *driving automation system* of either the *lateral* or the *longitudinal vehicle motion control* subtask of the DDT (but not both simultaneously) with the expectation that the *driver* performs the remainder of the *DDT*.	*Driver and System*	*Driver*	*Driver*	Limited
2	Partial Driving Automation	The *sustained* and ODD-specific execution by a *driving automation system* of both the *lateral* and *longitudinal vehicle motion control* subtasks of the *DDT* with the expectation that the *driver* completes the *OEDR* subtask and supervises the *driving automation system*.	**System**	*Driver*	*Driver*	Limited
ADS ("System") performs the entire DDT (while engaged)						
3	Conditional Driving Automation	The *sustained* and ODD-specific performance by an *ADS* of the entire DDT with the expectation that the *DDT fallback-ready user* is receptive to *ADS*-issued *requests to intervene*, as well as to *DDT performance-relevant system failures* in other vehicle systems, and will respond appropriately.	*System*	***System***	*Fallback-ready user (becomes the driver during fallback)*	Limited
4	High Driving Automation	The *sustained* and ODD-specific performance by an *ADS* of the entire *DDT* and *DDT fallback* without any expectation that a user will respond to a *request to intervene*.	*System*	*System*	***System***	Limited
5	Full Driving Automation	The *sustained* and unconditional (i.e., not ODD-specific) performance by an *ADS* of the entire *DDT* and *DDT fallback* without any expectation that a user will respond to a *request to intervene*.	*System*	*System*	*System*	**Unlimited**

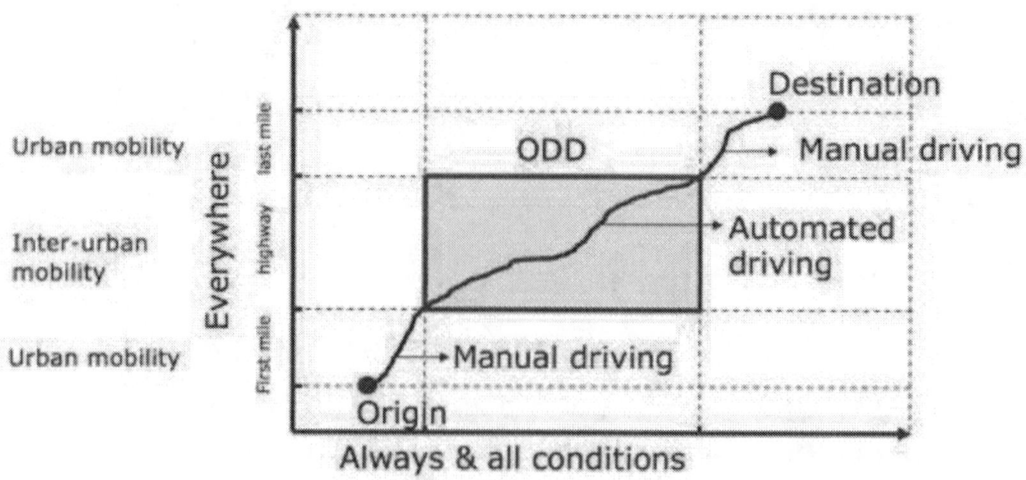

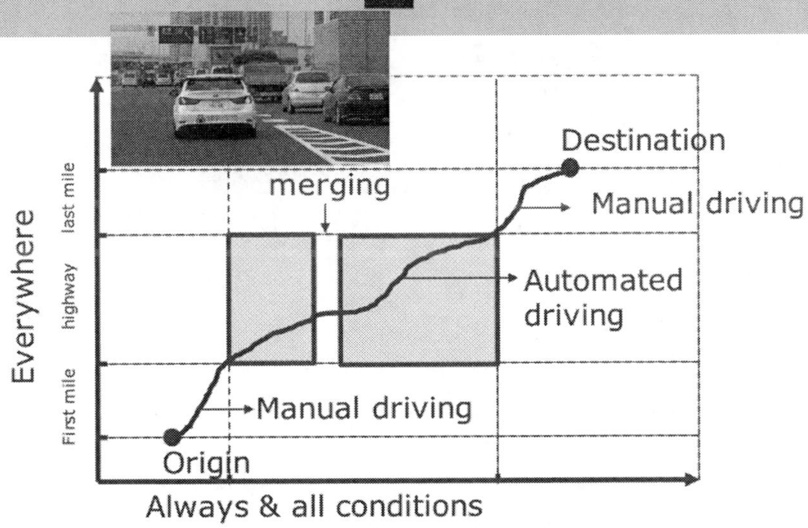
Destination
merging
Manual driving
Automated driving
Manual driving
Origin
Everywhere
First mile
highway
last mile
Always & all conditions

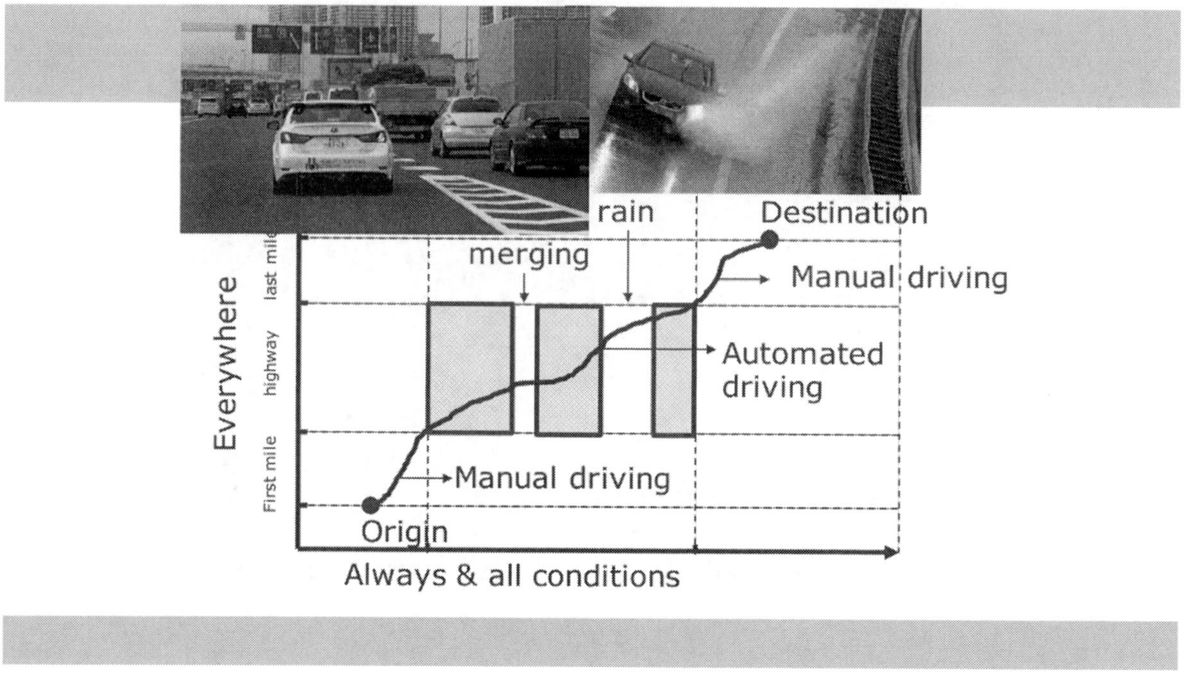
rain
Destination
merging
Manual driving
Automated driving
Manual driving
Origin
Everywhere
First mile
highway
last mile
Always & all conditions

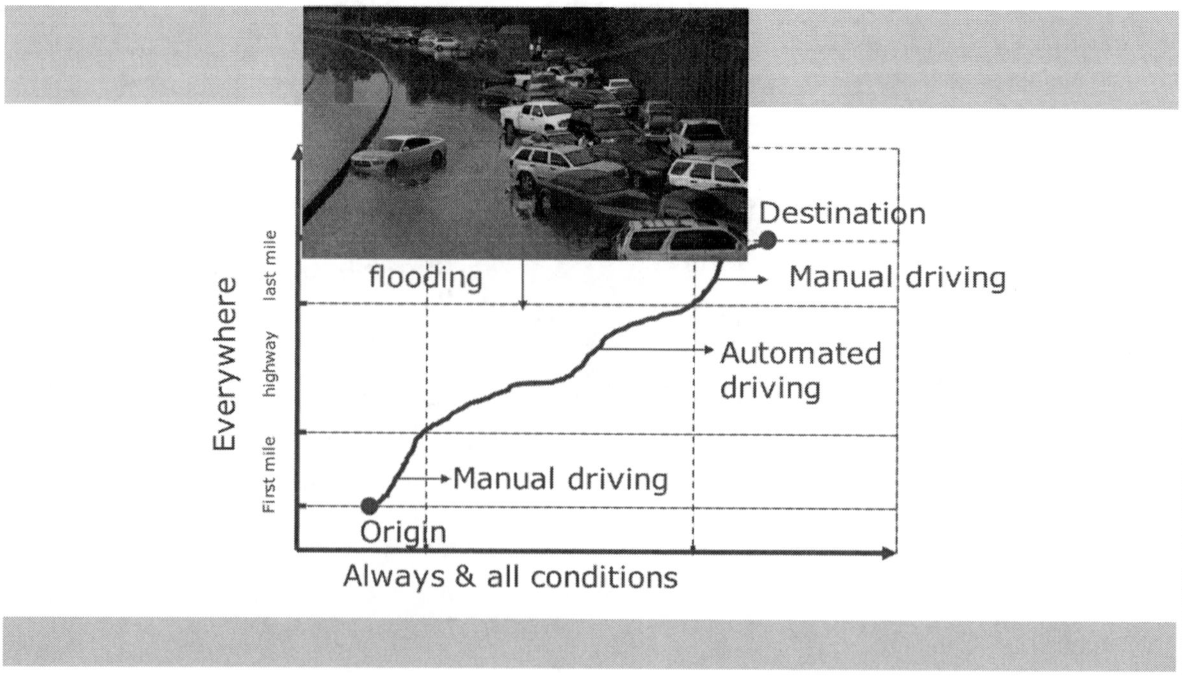
Destination
flooding
Manual driving
Automated driving
Manual driving
Origin
Everywhere
First mile
highway
last mile
Always & all conditions

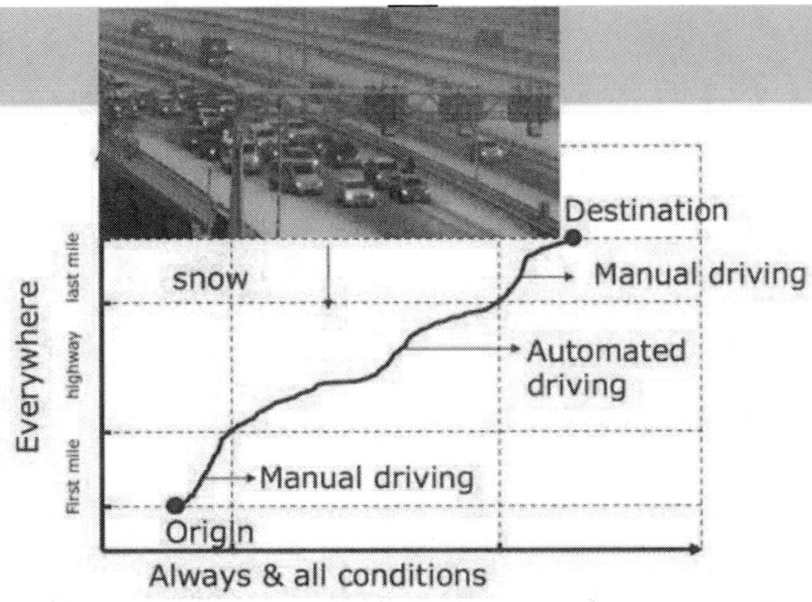
Destination
snow
Manual driving
Automated driving
Manual driving
Origin
Everywhere
last mile
highway
First mile
Always & all conditions

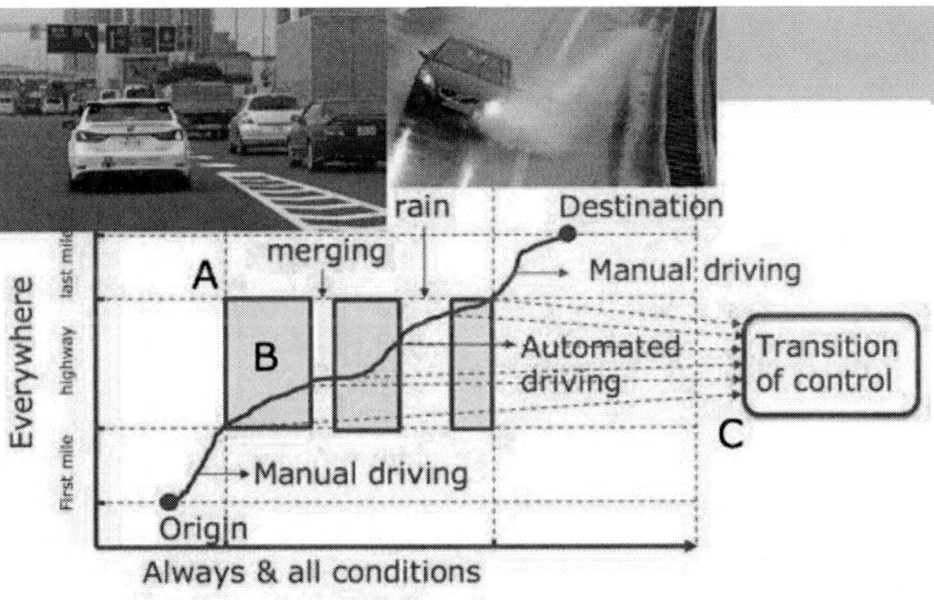
rain
Destination
merging
A
Manual driving
B
Automated driving
Transition of control
C
Manual driving
Origin
Everywhere
last mile
highway
First mile
Always & all conditions

Important research areas

A - How to optimize the ODD
- Various stakeholders involved, what roles and responsibilities?
- How to organise a constructive dialogue?

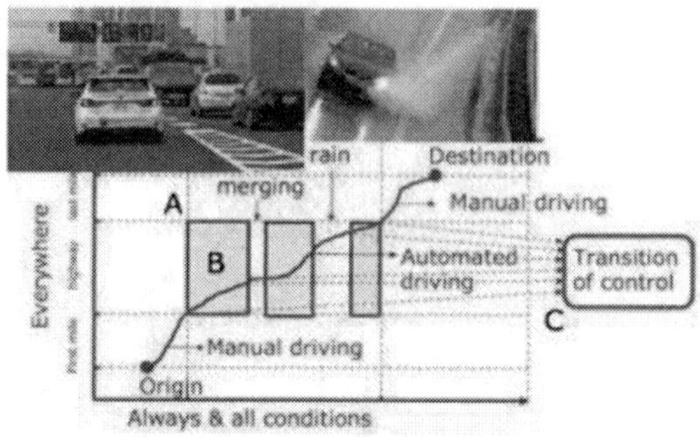

"I want to ensure our **infrastructure is ready for connected and automated driving.** I want to sit down with other road operators and the automotive and telecom sector. To discuss what infrastructure we'll need in order for the new generation of vehicles to deliver the greatest gains to society."

Important research areas

C - Transition of Control (ToC)
- Will happen a lot
- Needs to be safe and swift
- Difference between expected and un-expected ToCs

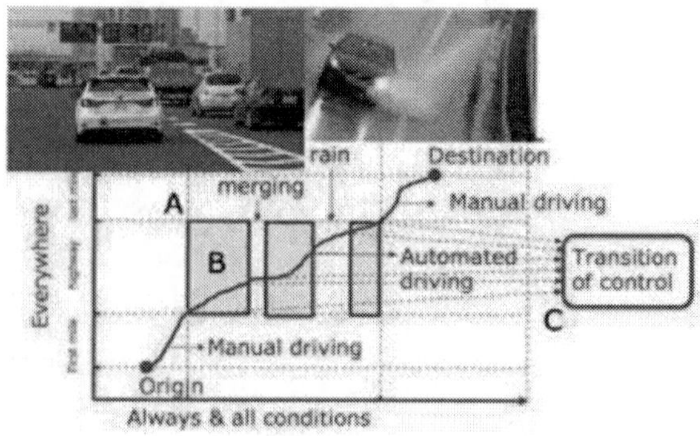

Important research areas

B – Acceptable behaviour within ODD

- detection – context – acceptable behaviour
- Who determines? Multiple perspective (OEM – road operator – driver/passenger)
- How to assess vehicle behaviour? ("Driver's licence"?)
- Mixed traffic challenge

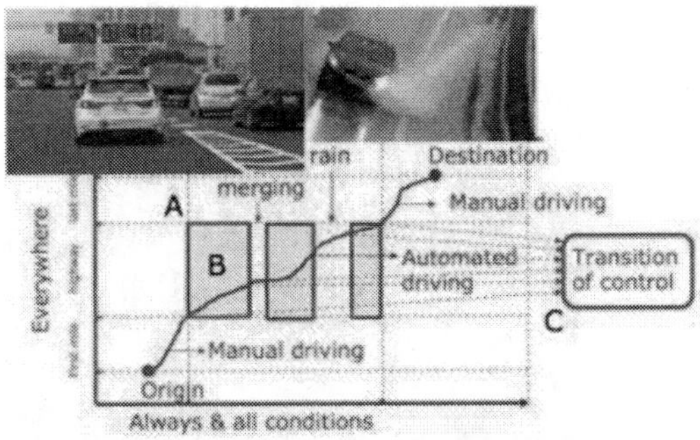

"I want to **enable the next generation of vehicles to actually take the road**. So I'm creating a legal framework for automated driving. Laying down requirements for reliability and safety that cars must meet before they can hit the road. **A driving licence for self-driving cars**, if you like. Not for the driver – but for the car itself!"

Context

Context

Context

Appropriate manoeuvre

Appropriate manoeuvre

Appropriate manoeuvre

Concrete examples as input for constructive dialogue

- Infrastructure design / layout

- Detection errors / mistakes

- Unexpected situations / edge cases

- Adverse weather conditions

Joint stakeholder workshop

discuss in a multi-stakeholder setting:

- Optimization of the ODD
- "Appropriate" behavior of the automated vehicle within the ODD, specifically for mixed traffic conditions.
- Explore how to set up a constructive dialogue with the developers of automated vehicles and other stakeholders to see what each can contribute from it's own role and responsibility.

EU EIP
European ITS Platform

L3Pilot

SAVE THE DATE!

Joint (EU EIP 4.2 / L3Pilot) stakeholder workshop on

The impacts of automated driving, how to maximize the benefits

President Hotel, Athens
25 October 2018 14.00 – 18.00

Objectives:
Automated vehicles are coming to you. But what does that actually mean, when will they come, in what form, and how can we maximize the benefits to the society? Automated vehicles have a vast potential to impact the traffic flow and improve safety, throughput and reduce emissions. Whatever the impact, it depends a lot on factors related to the actual capabilities of the vehicle and the environment in which it is driving. In general one would expect that the more a vehicle is able to drive in automated mode, the more benefits it will bring. However, there is a limited Operational Design Domain (ODD) in which the automated vehicle is capable of functioning as it is intended by the manufacturer.

In this workshop we will discuss in a multi-stakeholder setting how the ODD can be optimized (make it as large and/or as continuous as possible) and what the "appropriate" behaviour of the automated vehicle is within the ODD, specifically for mixed traffic conditions. We would also like to explore how to set up a constructive dialogue with the developers of automated vehicles and other stakeholders to see what each can contribute from its own perspective in the definition of the respective roles and responsibilities.

Who should attend?
All relevant stakeholders are welcome (e.g. automotive OEMs, equipment suppliers, telecom industry, road operators, local and regional authorities, governments, research institutes). Workshop attendance is free of charge.

Why a Joint Workshop?
Coordination and cooperation between the public and private sectors, in particular road operators and the automotive OEMs is needed so that a common approach and a road map can be developed for the introduction of automated vehicles in Europe and to maximise benefits thereof. This workshop, which follows directly the L3Pilot project meetings in the same venue, will offer an important first step on this road!

For more information on the Workshop, please contact: Tom Alkim, tom.alkim@rws.nl or Elias Demirtzis, elias.demirtzis@eict.de

See you in Athens!

About EU EIP
The EU ITS Platform (EU EIP) is the place where National Ministries, Road Authorities, Road Operators and partners from the private and public sectors of almost all EU Member States and neighbouring countries, cooperate in order to foster, accelerate and optimize current and future ITS deployments in Europe in a harmonized way.

About L3Pilot
Automated driving technology has matured to a level motivating a final phase of road tests which can answer key questions before market introduction of the systems. The European research project L3Pilot www.l3pilot.eu tests the viability of automated driving as a safe and efficient means of transportation on public roads. It will focus on large-scale piloting of SAE Level 3 functions, with additional assessment of some Level 4 functions.

Co-financed by the European Union
Trans-European Transport Network (TEN-T)

www.its-platform.eu

RDW

AUTOMATED VEHICLES SYMPOSIUM 2018

USERS. VEHICLES. INFRASTRUCTURE.

What if technology takes over all driving tasks?

Gerben Feddes

11-07-2018

Agenda

1. De RDW at a glance
2. Vehicle Driving License: the software as legal driver
3. Skills versus Scenarios

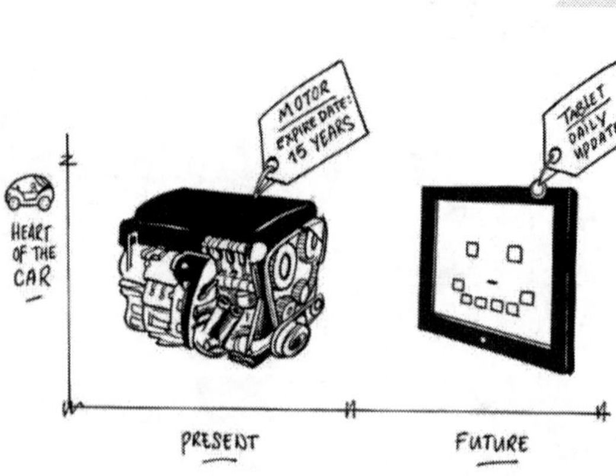

RDW

1. The RDW at a glance

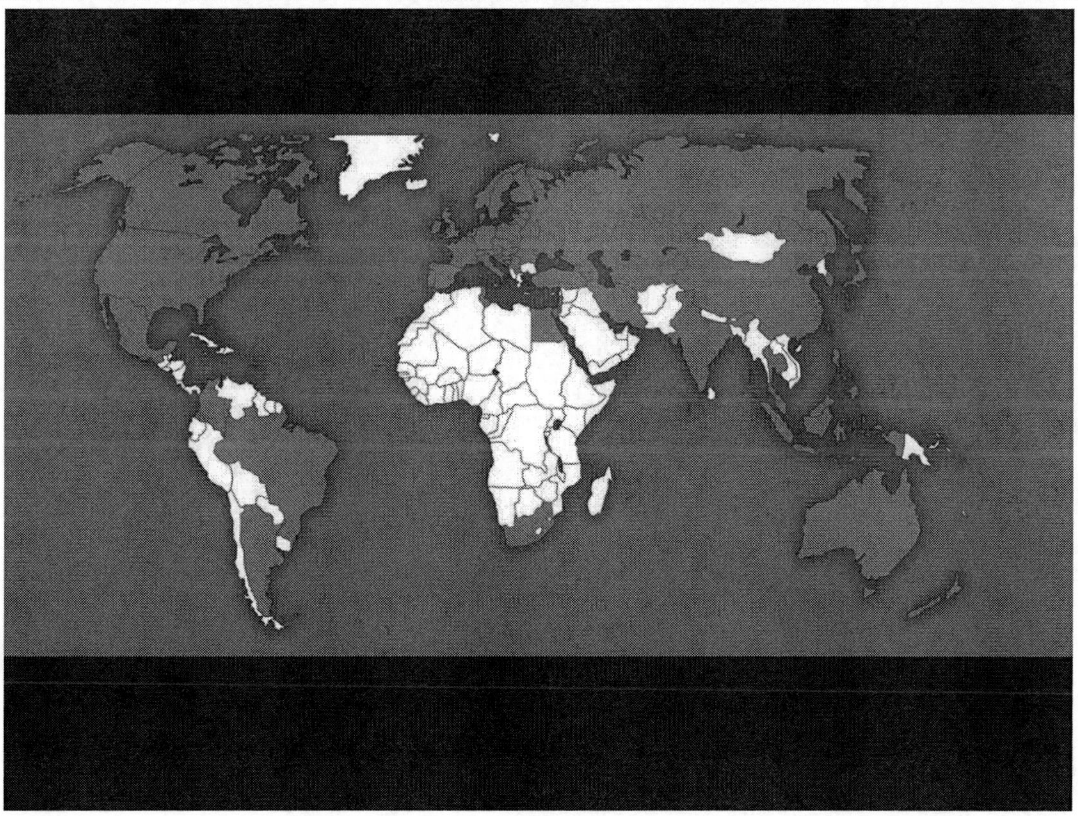

KPMG
Autonomous readiness index 2018

Overall rank	Country	Total score	Policy and legislation		Technology & innovation		Infrastructure		Consumer acceptance	
			Rank	Score	Rank	Score	Rank	Score	Rank	Score
1	The Netherlands	27.73	3	7.89	4	5.46	1	7.89	2	6.49
2	Singapore	26.08	1	8.49	8	4.26	2	6.72	1	6.63
3	United States	24.75	10	6.38	1	6.97	7	5.84	4	5.56
4	Sweden	24.73	8	6.83	2	6.44	6	6.04	6	5.41
5	United Kingdom	23.99	4	7.56	5	5.28	10	5.31	3	5.84
6	Germany	22.74	5	7.33	3	6.15	12	5.17	12	4.09
7	Canada	22.81	7	7.12	6	4.97	11	5.22	7	5.30
8	United Arab Emirates	20.89	6	7.26	14	2.71	5	6.12	8	4.79
9	New Zealand	20.75	2	7.92	12	3.26	16	4.14	5	5.43
10	South Korea	20.71	14	5.78	9	4.24	4	6.32	11	4.38
11	Japan	20.28	12	5.93	7	4.79	3	6.55	16	3.01
12	Austria	20.00	9	6.73	11	3.69	8	5.66	13	3.91
13	France	19.44	13	5.92	10	4.03	13	4.94	10	4.55
14	Australia	19.40	11	6.01	13	3.18	9	5.43	9	4.78
15	Spain	14.58	15	4.95	16	2.21	14	4.69	17	2.72
16	China	13.94	16	4.38	15	2.25	15	4.18	15	3.13
17	Brazil	7.17	20	0.93	18	0.86	19	1.89	14	3.49
18	Russia	7.09	17	2.58	20	0.52	20	1.64	18	2.35
19	Mexico	6.51	19	1.16	17	1.01	17	2.34	19	2.00
20	India	6.14	18	1.41	19	0.54	18	2.28	20	1.91

RDW

3. Vehicle Driving License

Gerben Feddes, senior advisor for intelligent mobility at the Netherlands' vehicle authority, RDW, wants to take testing even further. "In the future, a piece of software might legally be the driver of a production car, making the human driver a passenger in their own car. But where would the knowledge come from to drive that car safely? We believe it should come from the same people who devise and administer driving tests for humans. We're suggesting that maybe there should be a driver's license for cars."

[Above] Ford is testing its third generation of autonomous technologies at sites including the Mcity 32-acre proving ground in Michigan, USA

[Left] Ford has also started testing its technology in snowy conditions

30% wanting the latest technology

84% of those who reject semi-autonomous features think the technology won't live up to their driving skills; **60%** think the technology is too new; **57%** don't want to pay for it; **50%** know too little about it; and **45%** find it annoying

23% of female drivers and **12%** of male drivers rejected the technology, at least partly for being too complicated to use

*Data from an American Automobile Association survey of 1,800 US drivers, published in March 2016

Constant change

Feddes also says the evolving nature of software will necessitate radical reform of homologation procedures. "With software being constantly updated, a car is an ever-changing vehicle and there is no point in one-off admittance – you need to apply performance-based requirements," he says. "A car has to perform in a certain way and it's up to the manufacturer to produce acceptable means of compliance. We've learned from aviation and drone legislation – we're moving away from the how and beginning to focus on the what."

Another issue to settle is who is liable if an accident does happen. Where decisions leading to a crash are made by a machine, can some of the blame be laid at the manufacturer's door? With regard to its current, partially automated systems, Mercedes-Benz says "no". The company issued a statement in April 2016, as part of the Daimler Sustainability Report, which puts the onus on the consumer: "The legal situation in Germany and many other countries is clear: with regard to current, partially automated systems, the driver remains responsible. Although systems such as Lane Keeping Assist in the new E-Class provide support, the driver must still control the vehicle." However, the report did concede that "manufacturers are responsible for damages from product defects".

"WITH SOFTWARE BEING CONSTANTLY UPDATED, A CAR IS AN EVER-CHANGING VEHICLE AND THERE IS NO POINT IN ONE-OFF ADMITTANCE"

Gerben Feddes, *senior advisor for intelligent mobility at the Netherlands' vehicle authority, RDW*

Vehicle driving license

Cora van Nieuwenhuizen, minister of Infrastructure and Water management, during the **Intertraffic 2018 Amsterdam**:

"I'm going to create legal framework for automated driving. Laying down requirements for reliability and safety that cars must meet before they can hit the road. A driving licence for self-driving cars, if you like. Not for the driver – but for the car itself!"

RDW

Vehicle driving license

Assumptions (1/2):

- For SAE level 4 and 5 systems: the system is the legal driver
- Human drivers will be on the road for the coming years, so the automated vehicle has to act like a human
- It's about showing safe and predictable driving behavior related to human performance
- It's a traffic system approach (vehicle-infrastructure-behavior)
- Automated systems will have a stepped admission to public roads (complexity)
- From Compliance Based Standards – To: Performance Based Standards (ensure room for innovation)

RDW

Vehicle driving license

Assumptions (2/2):

- Driving simulators for humans can speed up the assessment process
- From: risk mitigation in the vehicle – To: risk mitigation on a system level

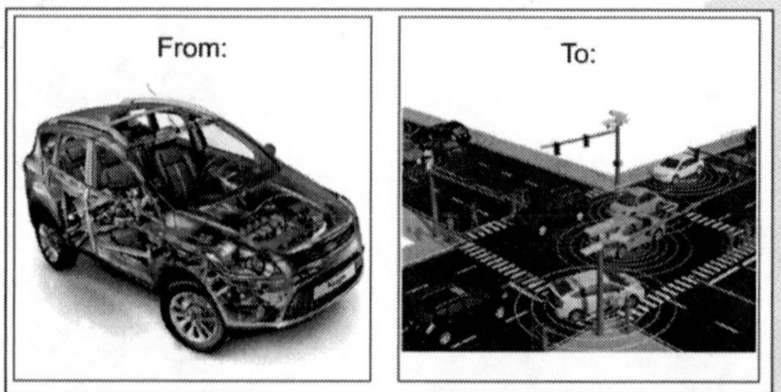

RDW

A driving license for an automated vehicle

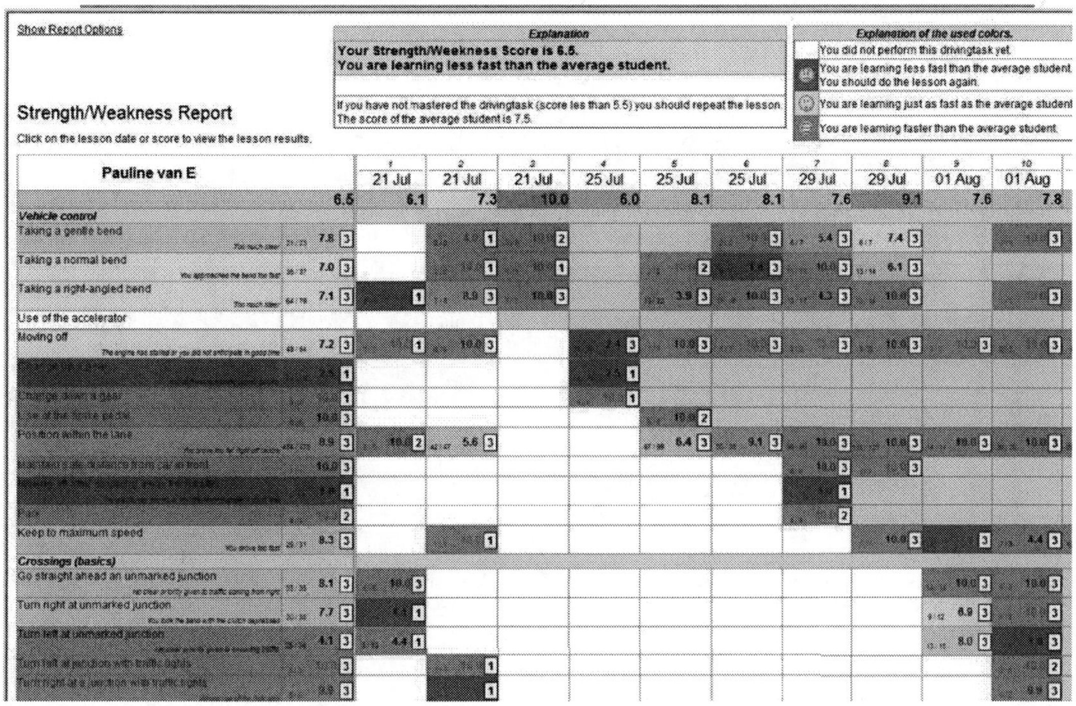

Some realism, within five years:

- Highway pilot
- Valet parking
- Truck platooning
- Solutions for public transport

Automated vehicles in cities: naaah

But: manned drones!

RDW

Vehicle driving license

1 Virtual Environment	2 Scale Modelling	3 Proving Ground
4 Driving Exam	5 Driving License	6 In Service Conformity

cbr robotTUNER RDW

Ministerie van Infrastructuur en Waterstaat

Rijkswaterstaat
Ministerie van Infrastructuur en Waterstaat

RDW

Vehicle driving license

Three tracks:

- National: the fast track. Can it be done?
- Worldwide: the knowledge track. Creation of an ISO standard
- European: the admittance and surveillance track. Knowledge from the national and worldwide track is input

RDW

4. Skills versus Scenarios

Regardless of skills or scenarios

- Traffic is an open system, with inherent risks (that we accept)
- Task of a driver: must drive at such <u>speed</u> that it is adapted to the other traffic and the total traffic- or road situation and that no danger or unnecessary hindrance for other road users arises or can arise
- Task of an examiner/authority: determines whether the displayed behavior deviates from what is expected. Depending on the <u>nature</u> and <u>severity</u> of the deviant behavior and the <u>number of times</u> it occurs, the driver is or is not fit to drive
- Depending the driving task, some actions are more essential than others

RDW

My problem with scenarios

- The industry is steering towards clear rules (level playing field), and searches for the familiar (harmonization, compliance based, scenarios)

- Scenarios give confidence, but ... humans do not get a book with scenarios to learn by heart when taking driving lessons...
- The goal is safe driving, scenarios can be a mean
- Training/learning scenarios can never be validation scenarios
- Scenarios are reactive

- Doom image: a fixed set of scenarios is determined, pre-programmed and admittance is obtained. In case of an accident: "that scenario was not part of the admission set, we are not liable"...

RDW

Are driving skills the way forward?

- Humans learn skills to anticipate to uncertain circumstances
- Traffic behaviour differs per country
- With skills you can handle mistakes (traffic is an open system!)
- Humans are good at translating specific situations into other situations
- So:
  - Level 4/5 systems need a form of artificial general intelligence
  - We need to introduce self awareness mechanisms (something current autonomous systems lack, only reasoning and learned algorithms) that can persue ethical goals
  - For many different situations the system then can distract what the best thing is to do in a given situation at hand, indicated by the ethical goals

"Between traffic rules and traffic safety still lies the resolving power of a human being"

RDW

Gerben Feddes

RDW

RDW | Program manager Vehicle Driving License | Senior Advisor Intelligent Mobility
+31 (0)79 345 7826 | +31 (0)655 123 958
gfeddes@rdw.nl | https://www.linkedin.com/in/gerbenfeddes/

SCENARIO BASED SAFETY VALIDATION OF AUTOMATED DRIVING: STREETWISE

Maurice Kwakkernaat, TNO

TNO innovation for life

streetwise

AUTOMATED VEHICLES SYMPOSIUM 2018

July 9–12, 2018

AUTOMATED VEHICLES SYMPOSIUM 2018

TNO innovation for life

THE NEED FOR SPEED

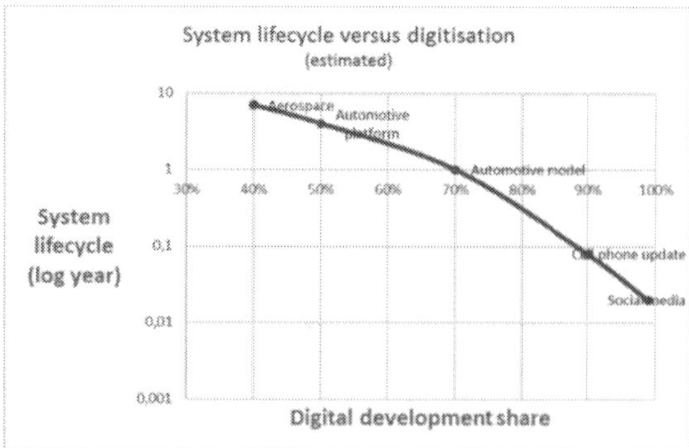

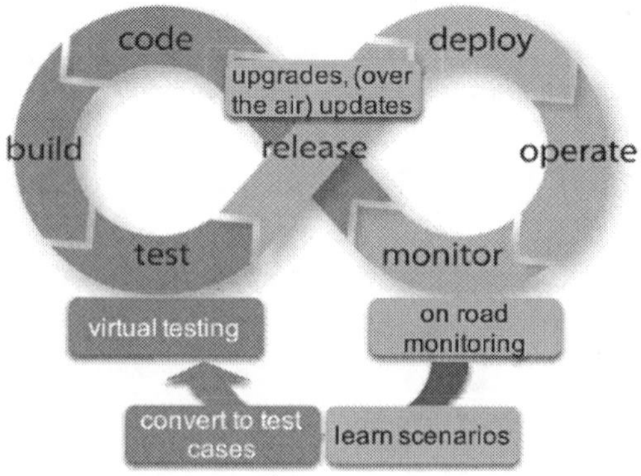

2 | Scenario based safety validation of automated driving: StreetWise

11 July 2018

WHY SCENARIOS?

- What is safe enough?
- Are our test cases complete?
- Lengthy validation of AVs

- Construct, collect relevant traffic events
- Increase real-world validity of test cases
- Scenarios are scalable
- Scenarios are not sensitive
- Scenarios allow to measure completeness of testing

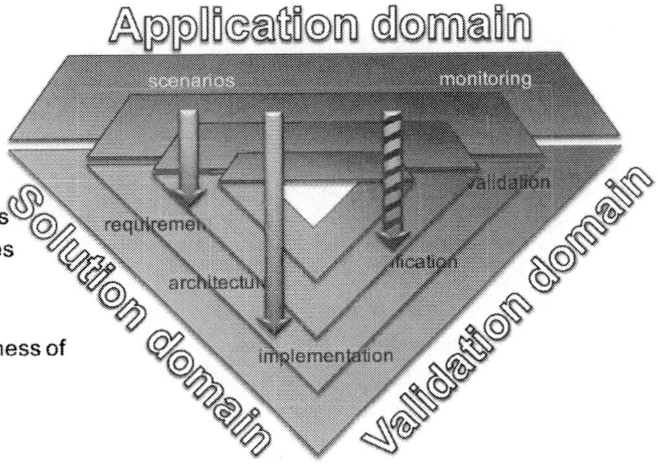

SCENARIO CLASS DEFINITION

ENABLE-S3

A scenario class is a formalized description of the system process, its static environment and its dynamic environment.

It may include goals of the activity and decisions of actors.

It will include activities and events.

Every scenario starts with an initial state.

Parameters are specified for the scenario class and may have parameter ranges or distributions.

STREETWISE METHOD

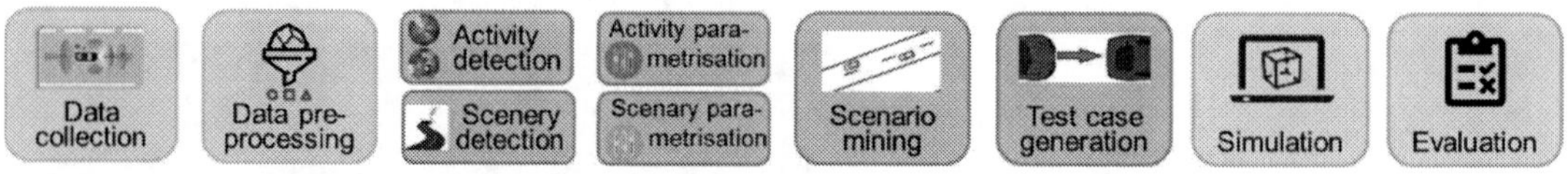

- Combine speed of on-road testing with thorough safety analysis
- Based on driving data sets from multiple contributors – real-world scenarios
- Behaviour of surrounding traffic with scenery
- Make validation a continuous process
- Leads to parametrized scenario database
- Determine completeness of scenario sets
- Determine real-world coverage of your validation
- Find out the critical scenarios for virtual and physical testing

- Scenario description
- Scenario parametirisation

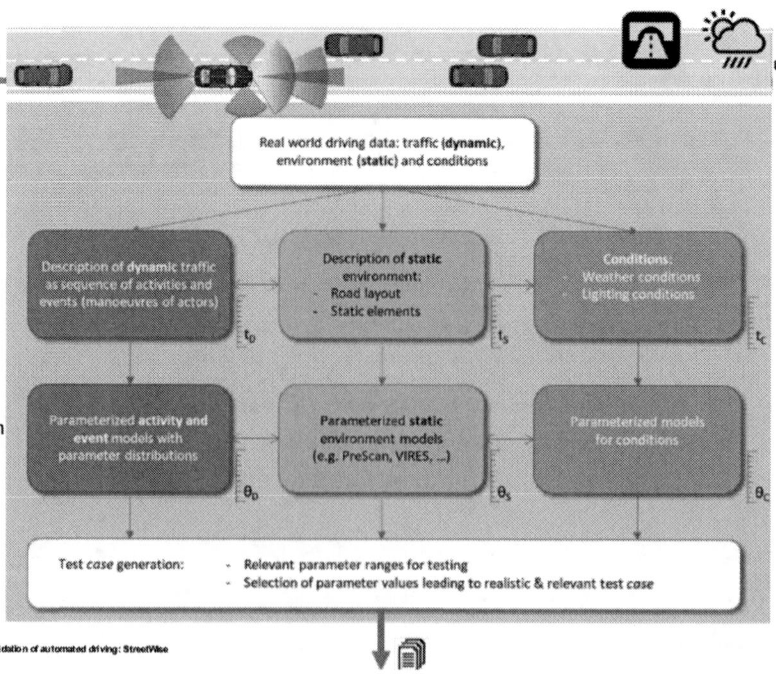

AUTOMATED VEHICLES SYMPOSIUM 2018

TNO innovation for life

ONTOLOGY

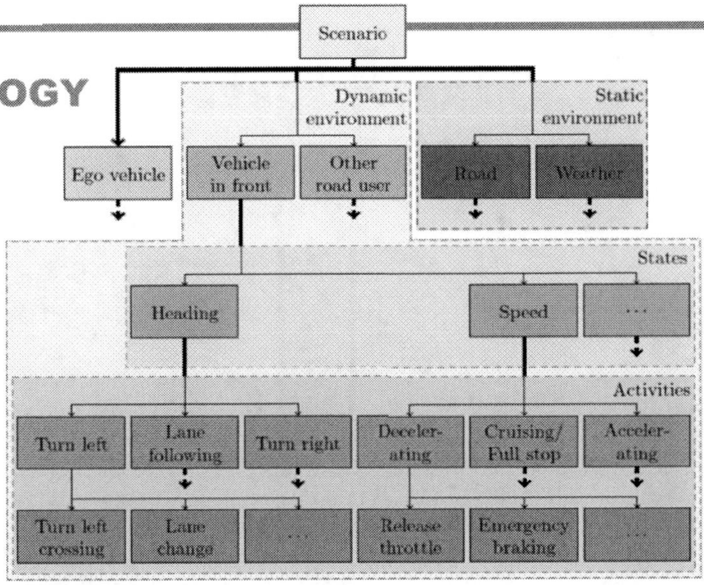

AUTOMATED VEHICLES SYMPOSIUM 2018

TNO innovation for life

STREETWISE: SCENARIO TIMELINE

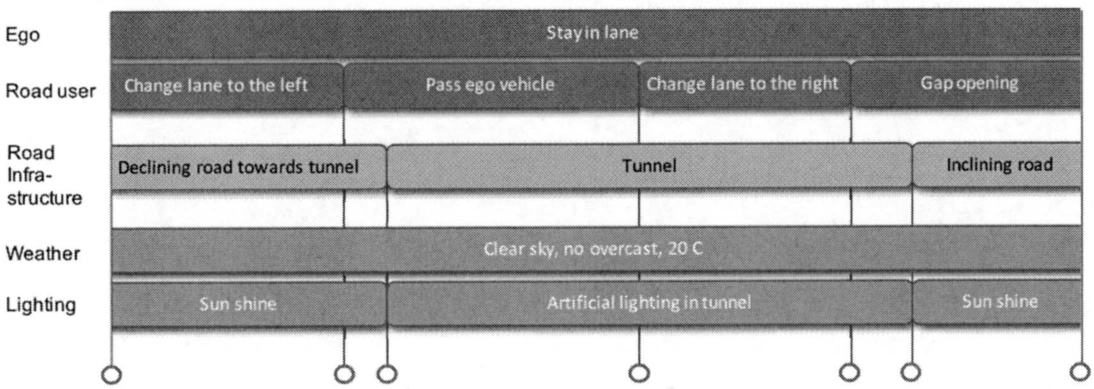

ACTIVITIES TO BE PARAMETRISED

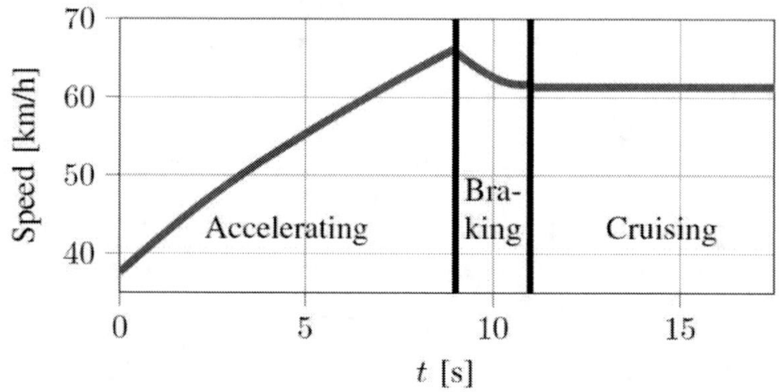

9 | Scenario based safety validation of automated driving: StreetWise

11 July 2018

PARAMETRISATION

Event parametrisation

- Condense events into parameters with minimal of information
  - Data reduction
  - Time interpolation
  - Activity / scenario parametrisation
  - Test case generation

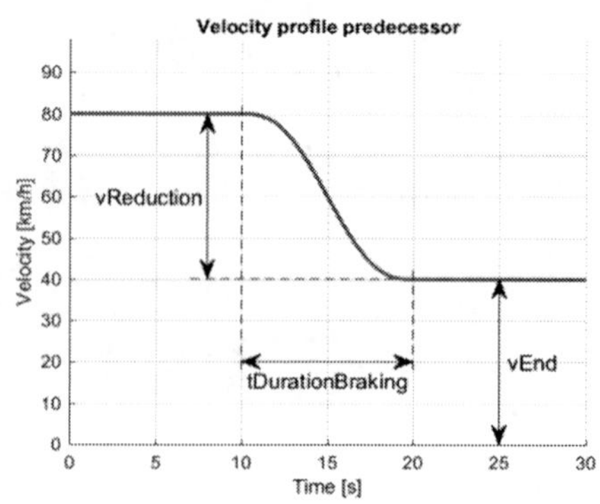

De Gelder & Paardekooper 2017

10 | Scenario based safety validation of automated driving: StreetWise

11 July 2018

AUTOMATED VEHICLES SYMPOSIUM 2018

TNO innovation for life

SCENARIO DETECTION: LANE CHANGE

Event detection

Scenery detection

11 | Scenario based safety validation of automated driving: StreetWise

11 July 2018

AUTOMATED VEHICLES SYMPOSIUM 2018

TNO innovation for life

TEST CASE GENERATION

Test case parameters

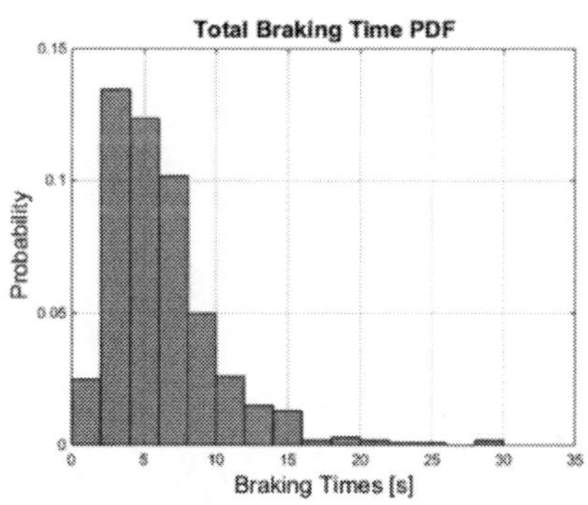

Distribution of 500 real-world observed scenarios instances

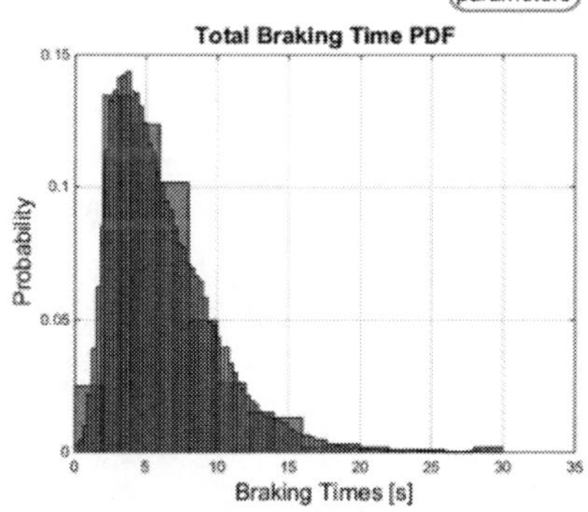

Distribution Monte Carlo with 10000 derived test cases

12 | Scenario based safety validation of automated driving: StreetWise

11 July 2018

STREETWISE TOOLING

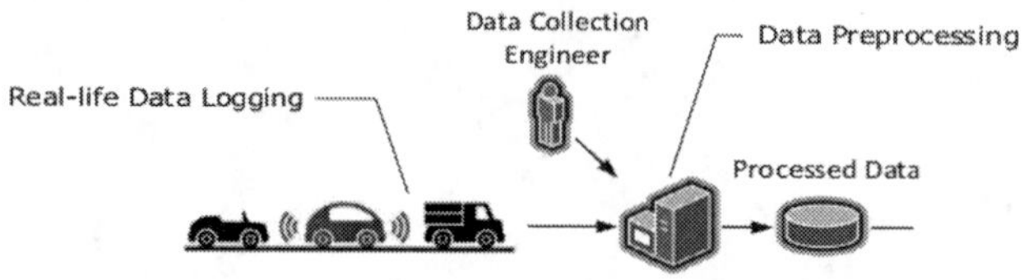

STREETWISE TOOLING

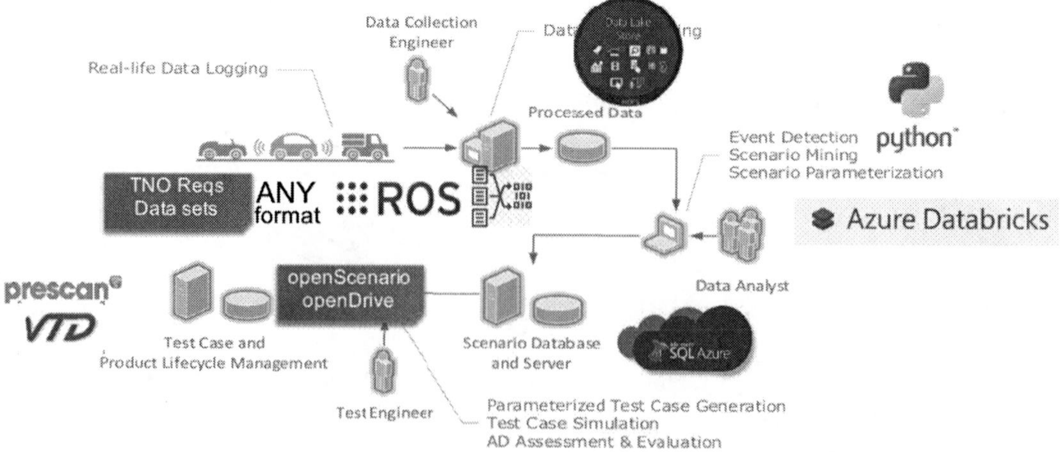

AUTOMATED VEHICLES SYMPOSIUM 2018 AUVSI

TNO innovation for life

TEST CASE 47 / 500

openScenario
openDrive

prescan

VTD

AUTOMATED VEHICLES SYMPOSIUM 2018 AUVSI

TNO innovation for life

SYSTEM BEHAVIOUR

Simulate

Cumulative distribution function over 500 cases

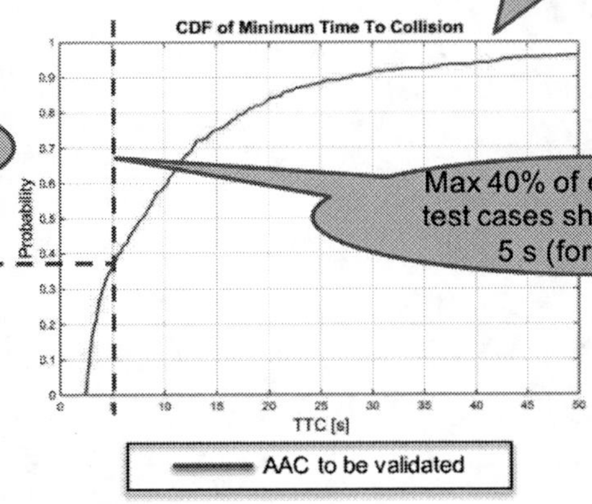

38% score: pass

Max 40% of cut-in scenario test cases shall have TTC < 5 s (for comfort)

SHARED METHODOLOGY

- A methodology is as good as its support
- Scenario based validation approach has strong support from industry
- Alignment with PEGASUS, MOOVE, SAE, US-DOT, CETRAN

- TNO approach
  - Multi-partner projects
  - Scenarios are shared
  - Scenarios can never be traced back to data set owner
  - Jointly build and improve scenario database and tooling

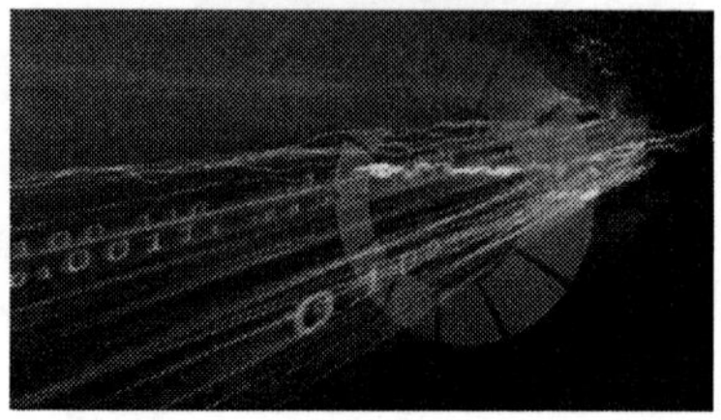

A License to Drive: How to Allow Automated Vehicles on Public Roads

Marcus Burke
National Transport Commission

July 2018

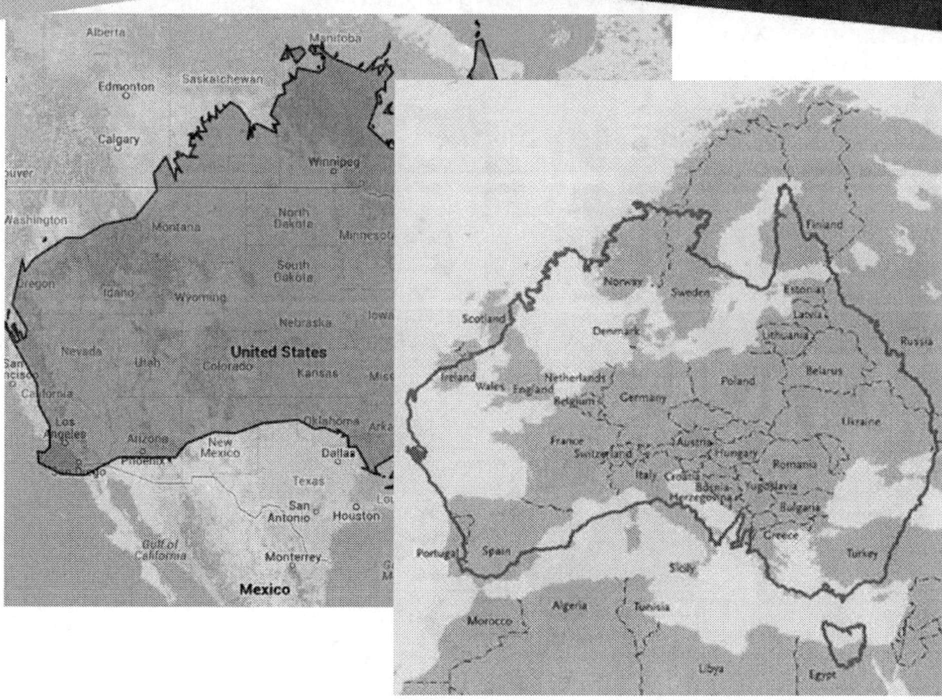

Australia – transport challenges

- Distance
- Variable infrastructure
- Safety
- Productivity
- Congestion

Driverless cars: Kangaroos throwing off animal detection software

By Jake Evans

Posted 24 Jun 2017, 8:28am

PHOTO: Driverless cars struggle to detect kangaroos due to the unique way they move. (Supplied)

Driverless car makers are discovering a unique problem as they begin to test the vehicles in Australia.

It turns out the unusual way that kangaroos move completely throws off the car's animal detection system.

"We've noticed with the kangaroo being in mid-flight ... when it's in the air it actually looks like it's further

Australia - Federation

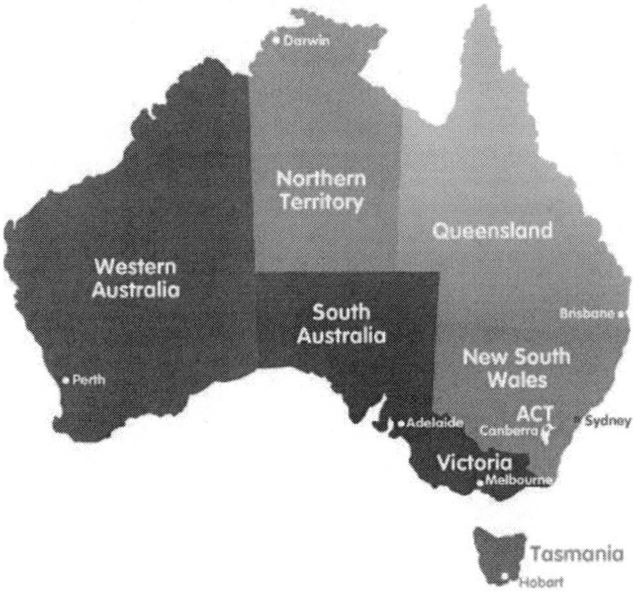

About the National Transport Commission

- Independent statutory body
- Develop national surface transport policy reform proposals
- Funded by the Commonwealth and eight state and territory governments
- Remove regulatory barriers to new, innovative transport services and products

Roadmap of reform

- To develop end-to-end legal framework for commercial deployment of automated vehicles
- Near, medium, long-term reforms

Goal

End-to-end regulation in place by 2020 to support the safe, commercial deployment and operation of automated vehicles at all levels of automation

Things we don't know about automated vehicles…

- When
- What applications
- Mix of technology
- Business models
- Consumer take up
- Consumer use/ behaviour
- Impacts

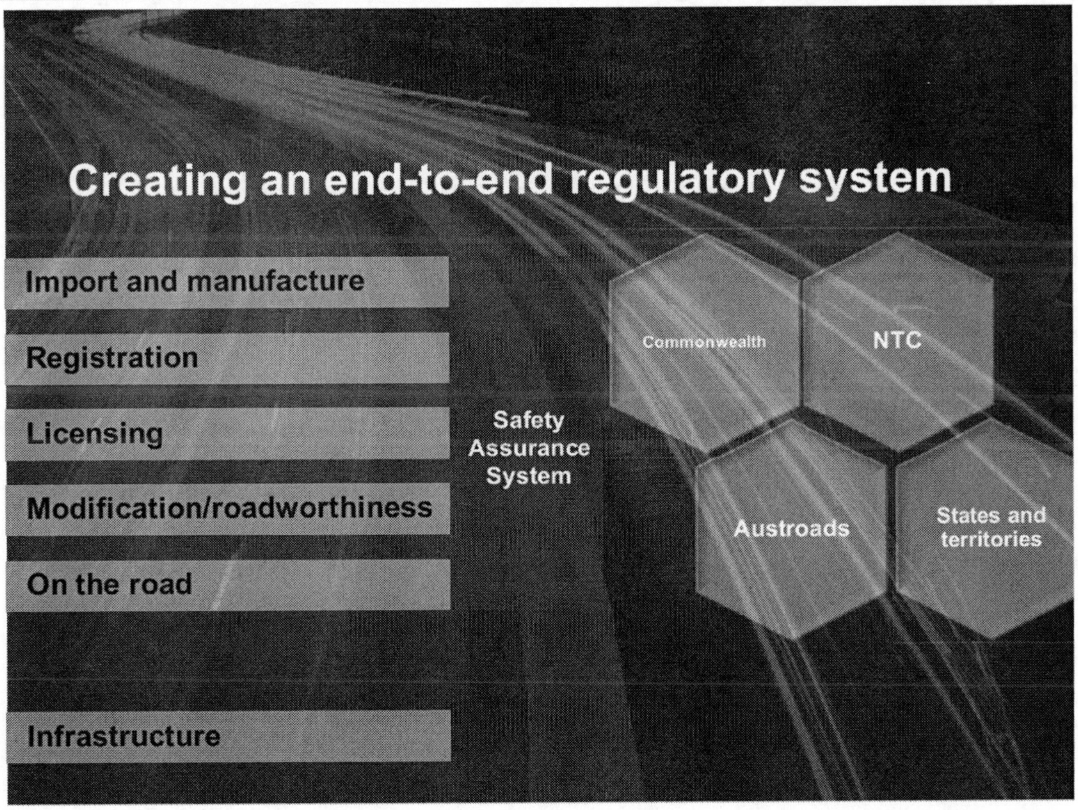

Complete

NTC's Automated Vehicle Program

How do we ensure automated vehicles are safe?

Safety assurance for automated vehicles

What are the road rules for automated vehicles?

Changing driving laws to support automated

How do we manage insurance for automated vehicles?

Motor accident injury insurance review

How do we protect users' data?

Review of government access to connected and automated vehicle data

What are the obligations of a driver?

What types of obligations do drivers have?

- Safe driving/ road rules
- Licenses and permits
- Interaction with enforcement
- Interaction with other road users
- Passenger compliance
- Vehicle obligations
- Provisions not relevant to automated systems

Why do we have driver licenses?

Why do we license drivers?

- Identification
- Minimum level of competence
- Permission to operate the vehicle/ ability to remove that permission

Safety for AVs

The problem

How do we know automated vehicles are safe before we allow them out on the road?

Safety Assurance Regulation Impact Statement

Published 15 May

Submissions due 9 July

What are the safety risks?

- Design
- Organisational
- Operational/ use

What are the safety risks? Design risks

- technological failure (malfunction due to poor design)
- cyber security failure (for example, hack or attack due to poor design)
- software updates introducing new safety issues (poor quality control, or the update is not supported by the vehicle's operating system)
- failure to function as expected in approved operating environments/conditions (system not up to the task)
- the ADS not being suited to Australian environmental or driving conditions
- the after-market system does not integrate safely with the existing vehicle
- the vehicle meets design criteria but still causes a safety risk in operation.

What are the safety risks? Organisational risks

- failure by the ADS entity to address safety issues that emerge over time (software or hardware) – for example, through lack of appropriate support
- failure to monitor the performance of the system
- failure to adapt the system to changes in regulation over time
- failure to adapt the system to changes in the road environment over time
- insolvency of the ADS entity
- the ADS entity no longer supports legacy versions of the ADS
- the company deploys an ADS (native, after-market or through software upgrade) that has not been through the self-certification process
- failure to monitor and issue security updates as required.

What are the safety risks? Operational/ use risks

- use in inappropriate environments/conditions
- technological failure (degradation of hardware due to poor maintenance/repair)
- cybersecurity failure (for example, hack or attack due to failure to follow security protocols)
- software updates (failure to apply)
- divided/competing or contradictory responsibilities (between the driver and the ADS)
- unclear responsibilities of human drivers in different vehicles
- after-market fitment and vehicle modifications adversely impacting the ADS's performance
- vehicle repairs adversely impacting the performance of the ADS
- repairers unable to assess impact the of repairs to an ADS.

Self-driving vehicles?

Who is the operator of an automated vehicle?

- Fleet?
- Privately owned vehicle?

Scenarios

- New standard (compliant) vehicle
- Non-standard vehicle
- Software upgrades to existing vehicles
- After-market devices/ vehicle modification

Automated driving system entity

- Company that is looking to bring the vehicles to market and go through the safety assurance process
- Could be vehicle manufacturer, software company, fleet operator, device manufacturer, other?
- Self-selected

Safety assurance reform

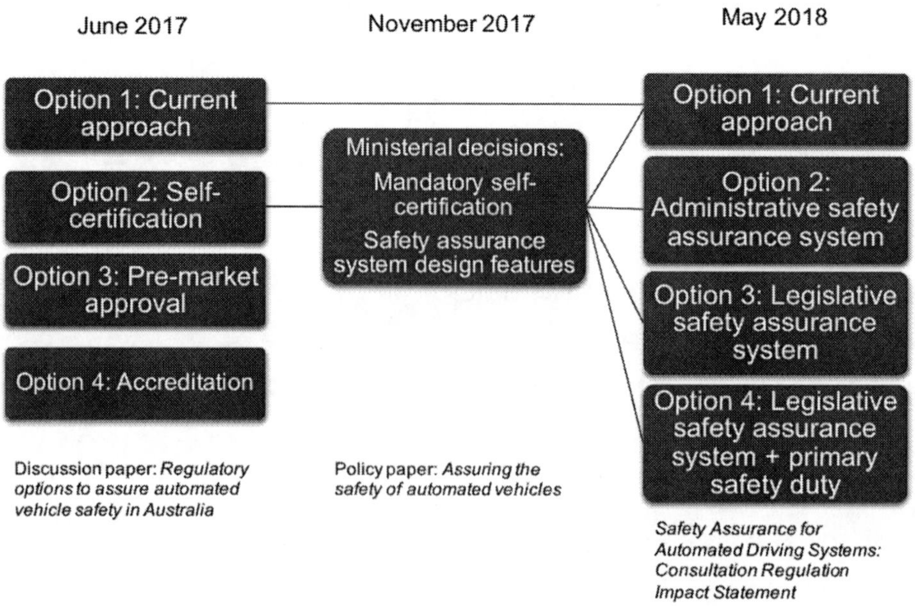

Proposed safety assessment criteria

1. Safe system design and validation processes
2. Operational design domain
3. Human-machine interface
4. Compliance with relevant road traffic laws
5. Interaction with enforcement and other emergency services
6. Minimal risk condition
7. On-road behavioural competency
8. Installation of system upgrades
9. Testing for the Australian road environment
10. Cybersecurity
11. Education and training

Additional obligations

1. Data recording and sharing
2. Corporate presence in Australia
3. Minimum financial requirements

Proposed Primary safety duty

What is the ongoing responsibility of the automated driving system entity to ensure safety?

- Ongoing duty to ensure the safety of an automated driving system so far as reasonably practicable
- Applies only to the Automated Driving System Entity
- Covers in-service safety risks not identified through the safety assurance system process are managed
- Covers unsafe behaviours not captured by prescribed offences

How do we achieve the objectives of licensing?

Can we achieve the objectives of licensing through a safety assurance system?

1. Identification = automated driving system entity that has applied through the safety assurance process
2. Minimum competence = mandatory self-certification process
3. Permission = safety assurance approval/ removal

Next steps

Safety Assurance

- Submissions closed 9 July 2018
- Recommendations to Transport Ministers November 2018

Discussion papers on:

- **Compulsory third party insurance** and automated vehicles
- Government access to **C-ITS and automated vehicle data**

For more information

www.ntc.gov.au

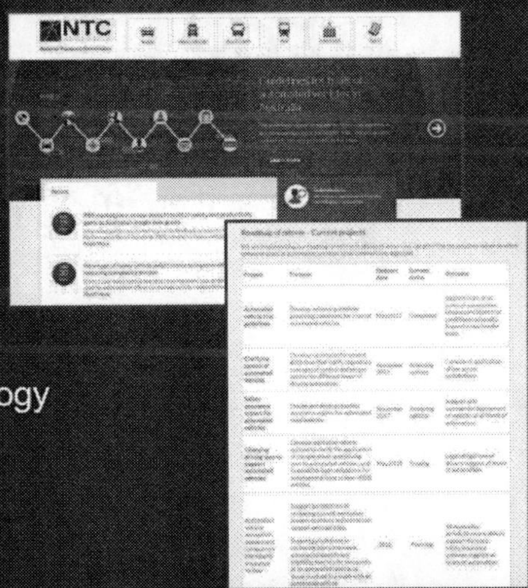

Marcus Burke
Project Director, Compliance and Technology
mburke@ntc.gov.au

AUTOMATED VEHICLES SYMPOSIUM 2018

USERS. VEHICLES. INFRASTRUCTURE.

Driven by AUVSI and TRB

Breakout Session #35. Data Exchanges to Enable AV Integration

Summary of Key Findings and Lessons Learned

- Access to data is a limiting factor across the AV ecosystem
  - See examples at www.transportation.gov/av/data
- USDOT can help convene around priority voluntary data exchanges
  - e.g., harmonized work zone data feeds
- Market-based solutions are emerging – often with limited government role
- Near-term challenges aren't technical – they're business processes
  - Lack of common terminology, convening mechanisms, and priorities
  - Difficulty adapting to agile processes
  - Back-end data generation processes (fax machines?!)

AUTOMATED VEHICLES SYMPOSIUM 2018

USERS. VEHICLES. INFRASTRUCTURE.

Driven by AUVSI and TRB

Breakout Session #35. Data Exchanges to Enable AV Integration

Next Steps

- Complete the work zone data experiment (near-term and long-term) and assess extensibility to other infrastructure data
- Consider options for increasing understanding of agile processes
- Identify other priorities for data exchange and federal/non-federal role
- Refine AV Data Principles and Framework
  - Clarify the public / broader research community as a data user
  - Fill in more examples and use cases
- Raise awareness of AV Data Principles and Framework as potential planning tools
  - See www.transportation.gov/av/data

U.S. Department of Transportation
Office of the Under Secretary

Data Exchanges to Enable Automated Vehicle Integration

AUTOMATED VEHICLES SYMPOSIUM 2018

#AVS18

Ariel Gold (Moderator)
Data Program Manager
US Department of Transportation (USDOT)
Intelligent Transportation Systems
Joint Program Office (ITS JPO)

Session Agenda

Purpose

- Raise awareness of the USDOT AV Data Principles and Framework and how they can help identify and advance data exchange priorities
- Identify lessons from the Work Zone Data Exchange project that may apply to other efforts to increase voluntary data exchange to support AV integration

Agenda

- U.S. DOT's Approach to AV Data
- Panel: The Principles and Framework in Action
- Panel: Applying the AV Data Principles and Framework Across the AV Ecosystem
- Conclusion

What is the *Data for Automated Vehicle Integration* Initiative?

Finch Fulton

Deputy Assistant Secretary for Transportation Policy, USDOT

USDOT Automated Vehicles Activities

Learn more about USDOT Automated Vehicles (AV) activities at:
https://www.transportation.gov/AV

How to Take on a Cross-cutting Issue Like Data?

Graphic credit:
https://xkcd.com/1429/

Access to data is a limiting factor for AV deployment. (It's also a sensitive topic.) One solution: *data exchanges*.

USDOT is using our *convening power* to understand critical use cases for data exchange and the appropriate federal role to enable them.

Bringing Stakeholders Together for Meaningful Conversations

(IT'S REALLY HARD TO DO)

AV Data Guiding Principles (Beta)

1. Promote proactive, data-driven safety, cybersecurity, and privacy-protection practices.
2. Act as a facilitator to inspire and enable voluntary data exchanges.
3. Start small to demonstrate value, and scale what works toward a bigger vision.
4. Coordinate across modes to reduce costs, reduce industry burden, and accelerate action.

https://www.transportation.gov/AV/Data

AV Data Framework (Beta)

Category*	Goals
Business-to-Business (B2B)	• Improve sector-to-sector cooperation • Improve industry-wide safety • Inform future insurance policies
Business-to-Government (B2G)	• Understand performance of rapidly evolving tech • Inform policies and investments to encourage innovation and protect safety
Business-to-Infrastructure (B2I)	• Help vehicles navigate safely around obstacles in all weather • Reduce system congestion • Help optimize infrastructure maintenance
Open Training Data (X2X)	• Improve performance in common scenarios • Support basic research and education

*represents two-way data exchange

https://www.transportation.gov/AV/Data

AV Data Framework (Beta)

Category*	Specific Data to Exchange	Real-World Examples
Business-to-Business (B2B)	• Cybersecurity incidents • Edge cases • Near-miss events • Performance in safety-critical scenarios	• Automotive Information Sharing and Analysis Center • ClinicalStudyDataRequest.com (CSDR)
Business-to-Government (B2G)	• Cybersecurity incidents • Near-miss events • Performance in safety-critical scenarios • Crash reconstruction	• Aviation Safety Information Analysis and Sharing • Voluntary Safety Self-Assessments (Part of ADS 2.0)
Business-to-Infrastructure (B2I)	• Work zone activities and geometrics • Road weather information • Missing signage or broken infrastructure • Curb use rules and availability	• National Transit Map • Waze Connected Citizens Program • Meteorological Assimilation Data Ingest System
Open Training Data (X2X)	• Road, signage, and other infrastructure imagery • Edge cases	• ImageNet • Multimedia Commons • Nexar NEXET

*represents two-way data exchange

https://www.transportation.gov/AV/Data

Roundtable on Data for AV Safety

Outcomes

- Clarity on value of federal government as convener and facilitator
- Priority use cases for data exchange: work zones, scenarios, cybersecurity, others

Next Steps

- Enable voluntary data exchanges as "One DOT" via pilot projects
- Incorporate into AV policies
- Continue conversations

Photo credit: Dan Morgan

Summary Report available via:
https://www.transportation.gov/AV/Data

Work Zone Data Exchange Project

THE LOCAL DATA CHALLENGE

The Local Data Challenge

Up-to-date information about dynamic conditions occurring on the roads – such as construction events – can help AVs navigate safely and efficiently.

Many infrastructure owners and operators (IOOs) maintain data on work zone activity, but lack of common data standards and convening mechanisms makes it difficult and costly for third parties – including original equipment manufacturers (OEMs) and navigation applications – to access and use these data across various jurisdictions.

Learning from the Open Transit Data Story

A **simple** specification...

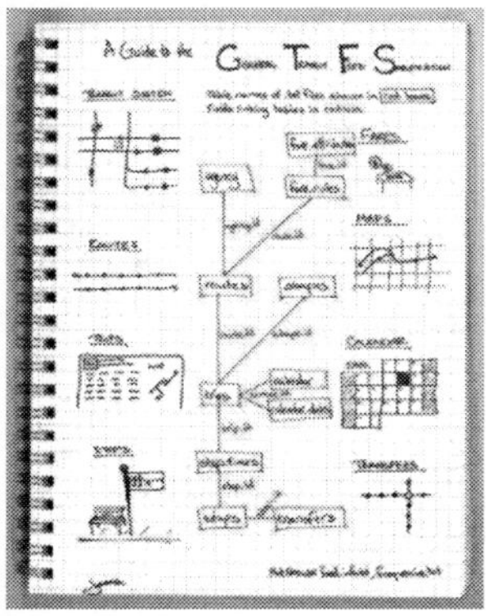

...with a **wide range** of uses

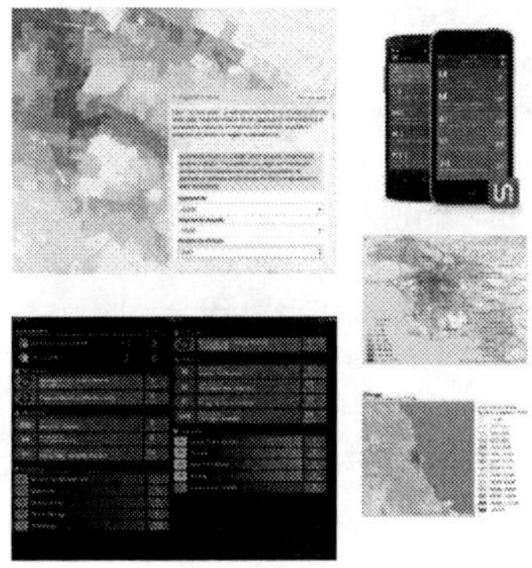

A Federated "Front Door" to Transit Data

Now, **basic transit data** is easy to **find and use nationwide;**
transit agencies and their users **continue to collaborate** on the spec

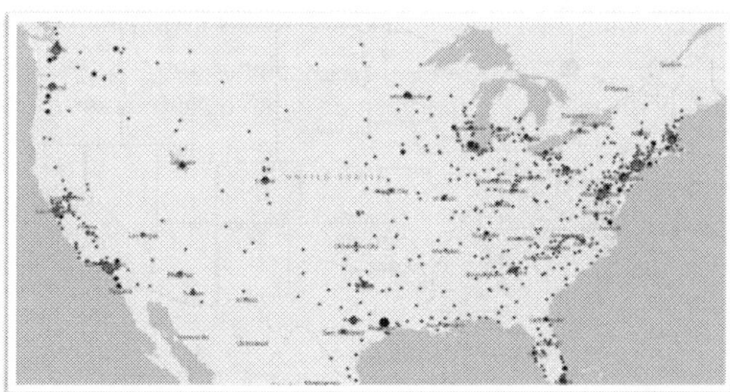

Work Zone Data Exchange Project (Overview)

Purpose

- To jumpstart voluntary adoption of a basic work zone data specification
- To enable collaborative maintenance and expansion of the spec

Outcomes within 6 months

- **Data producers** make available an active work zone data feed using a common, non-proprietary specification
- **Non-government developers** use that data in a meaningful way – thus establishing an MVP of voluntary data exchange for work zone data

Big Picture Outcome

- Repeatable approach to accelerate harmonization of local data sources

Work Zone Data Exchange Project (Overview)

Approach

- Data producers and users:
  - Agree on common data specification that meets their minimal needs
  - Leverage existing published and *de facto* standards
  - Work collaboratively and in the open using an agile process
- USDOT:
  - Facilitates and convenes
  - Provides technical inputs
  - Sets up to scale after 6-month mark

Work Zone Data Exchange Project
(Notional timeline)

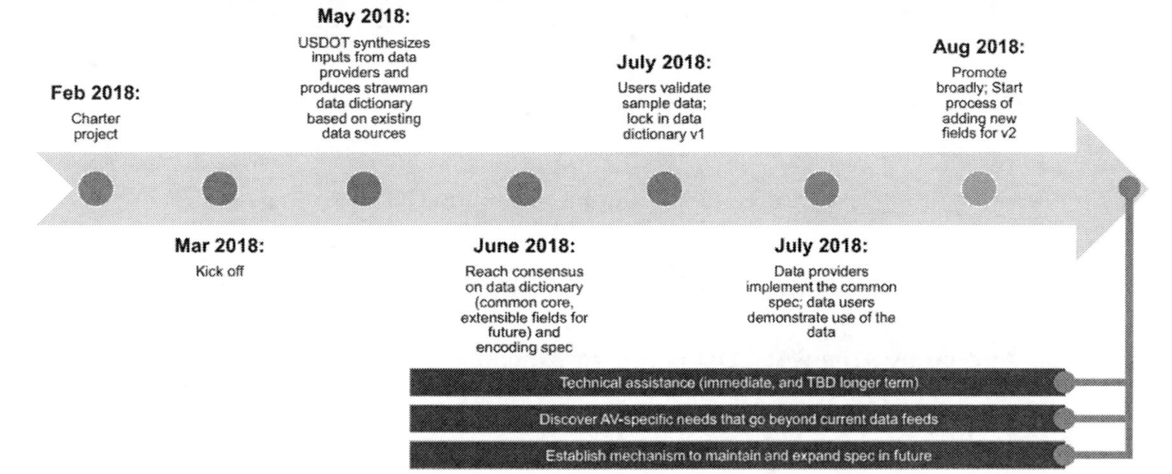

Longer-term Needs Discovery
(FHWA Work Zone Data Initiative)

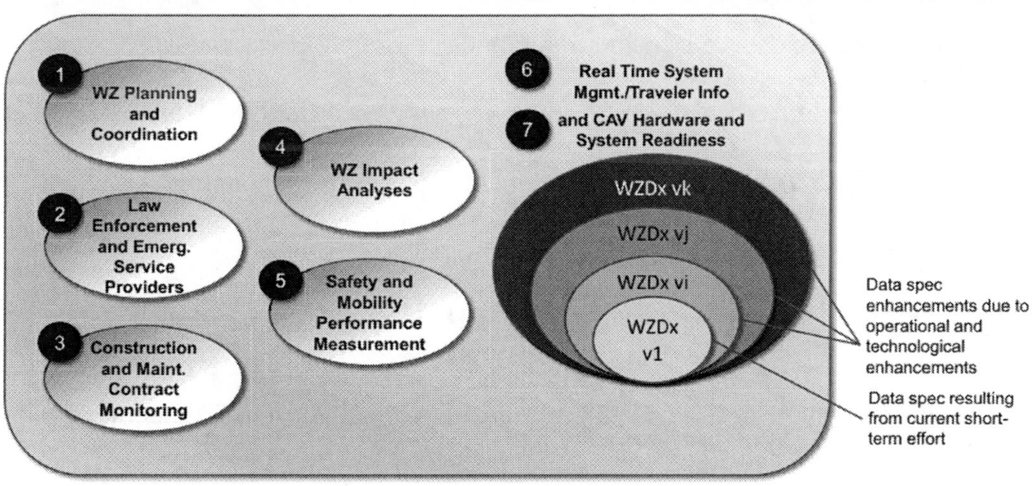

Panel Discussion: The Principles and Framework in Action

- **Jennifer Carter**, Senior Manager, Government Solutions, HERE Technologies
- **Neal Hawkins**, Associate Director, Institute for Transportation, Iowa State University
- **Peter Kozinski**, Director, RoadX Program, Colorado DOT
- **Paul Pisano**, Team Leader, Road Weather and Work Zone Management, Federal Highway Administration
- **Moderated by Ariel Gold**, Data Program Manager, Intelligent Transportation Systems Joint Program Office (ITS JPO)

Panel Discussion: Applying the AV Data Principles and Framework Across the AV Ecosystem

- **Tilly Chang**, Executive Director, San Francisco County Transportation Authority
- **David Kidd**, Senior Research Scientist, Highway Loss Data Institute
- **Steve Boyd**, Co-founder, VP of External Affairs, Peloton Technology
- **Ritchie Huang**, Manager, Engineering & Safety, Daimler
- **Jeff Loftus**, Division Chief, Office of Technology, Federal Motor Carrier Safety Administration
- **Moderated by Ariel Gold**, Data Program Manager, Intelligent Transportation Systems Joint Program Office (ITS JPO)

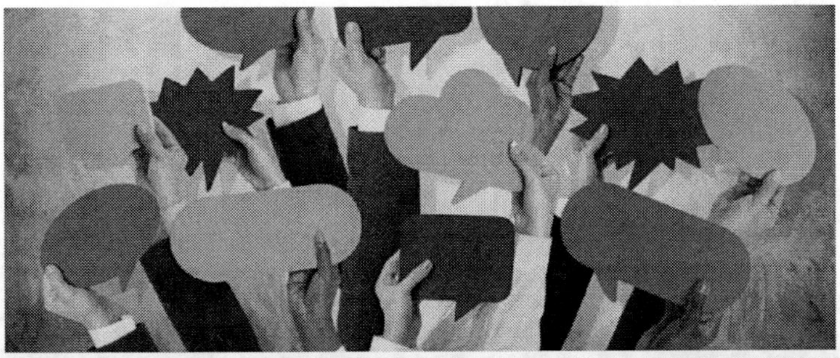

Thank You!

www.transportation.gov/av/data

ariel.gold@dot.gov

AVDX@dot.gov

DISCLAIMER

The Federal Highway Administration (FHWA) does not endorse any entity and the appearance of our presentation material in this template should not be interpreted as an endorsement or statement exhibiting any preference, support, etc.

OUSTER

AVS 2018
Angus Pacala, Founder/CEO
10 July 2018

LIDAR – Light Detection And Ranging

- RADAR – Radio Detection And Ranging
- SONAR – SOund Navigation And Ranging

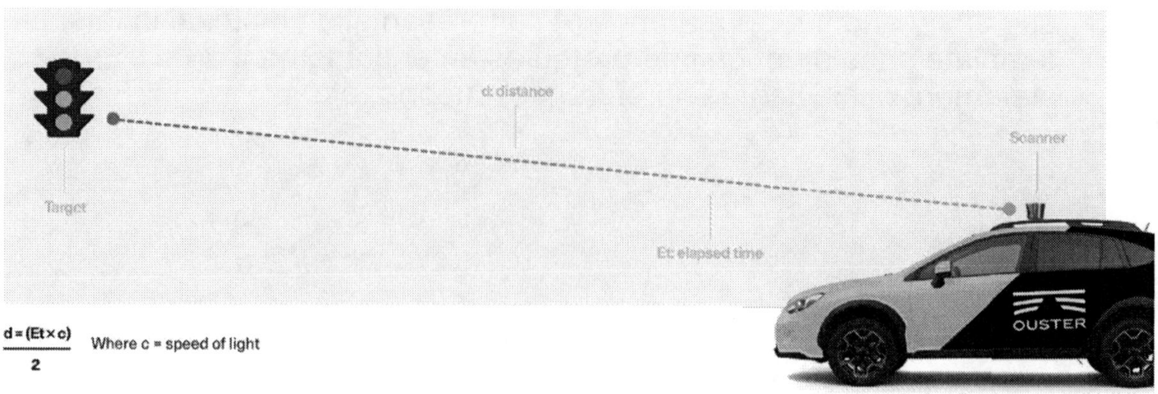

$d = \frac{(Et \times c)}{2}$ Where c = speed of light

2

Source: voyage.ai

LIDAR - What Matters?

- Performance
  - Range
  - Angular Resolution (horizontal, vertical)
  - Field of View (horizontal, vertical)
  - Scan rate/frame rate (per second, in Hz)
  - Range repeatability (range "noise")
  - Power consumption
  - Spatial resolution
  - Spatial accuracy - any "blooming", artifacts, etc (How accurately you measure the location you think the LIDAR is pointing in)
  - Environmental variability (e.g., over broad temperatures, in high humidity, peak sun, inclement weather)
  - Other features (e.g., color, multiple returns, region of interest)
- Price
- Form Factor
  - Size
  - Mass
- Quality/Reliability
  - Automotive Spec
  - Ingress
  - Shock/Vibe/Jitter
  - Thermal Range & Cycling
  - Corrosion
- Scalability/Availability
  - Product or PowerPoint Slide?
  - Manufacturability

LIDAR for ADAS

- OEM requirements 2021-2023
  - 100-200m @ 10%
  - HFOV: 120 degrees w/ no beam gaps (typically articulated as 0.1x0.1)
  - VFOV: 20-40 degrees w/ no beam gaps (typically articulated as 0.1x0.1)
  - Framerate: 10-20 hz
  - Precision: 5cm or less
  - ASIL: B(D) or better
  - Object detection: various requirements
- Integrations
  - Front & rear fascia
  - Windshield
  - Headlights & taillights
- Perception & SLAM stack

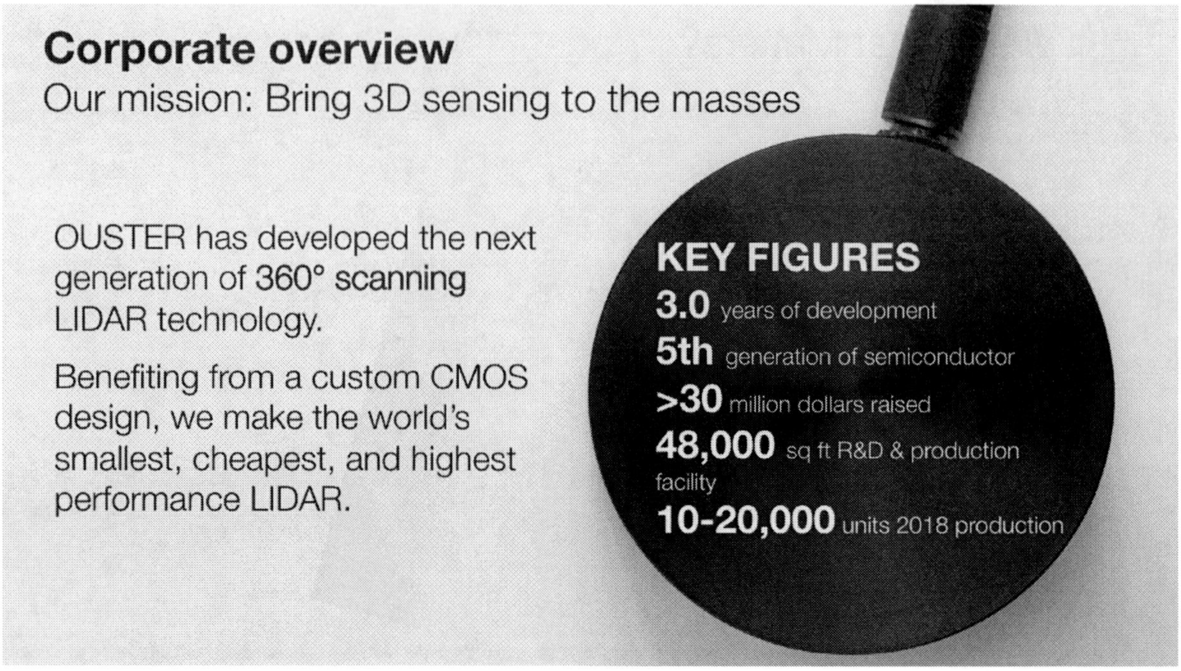
Corporate overview
Our mission: Bring 3D sensing to the masses
OUSTER has developed the next generation of 360° scanning LIDAR technology.
Benefiting from a custom CMOS design, we make the world's smallest, cheapest, and highest performance LIDAR.
KEY FIGURES
3.0 years of development
5th generation of semiconductor
>30 million dollars raised
48,000 sq ft R&D & production facility
10-20,000 units 2018 production

Made in the Mission
Ouster builds every unit in a 48,000 square foot facility in the mission district of San Francisco.

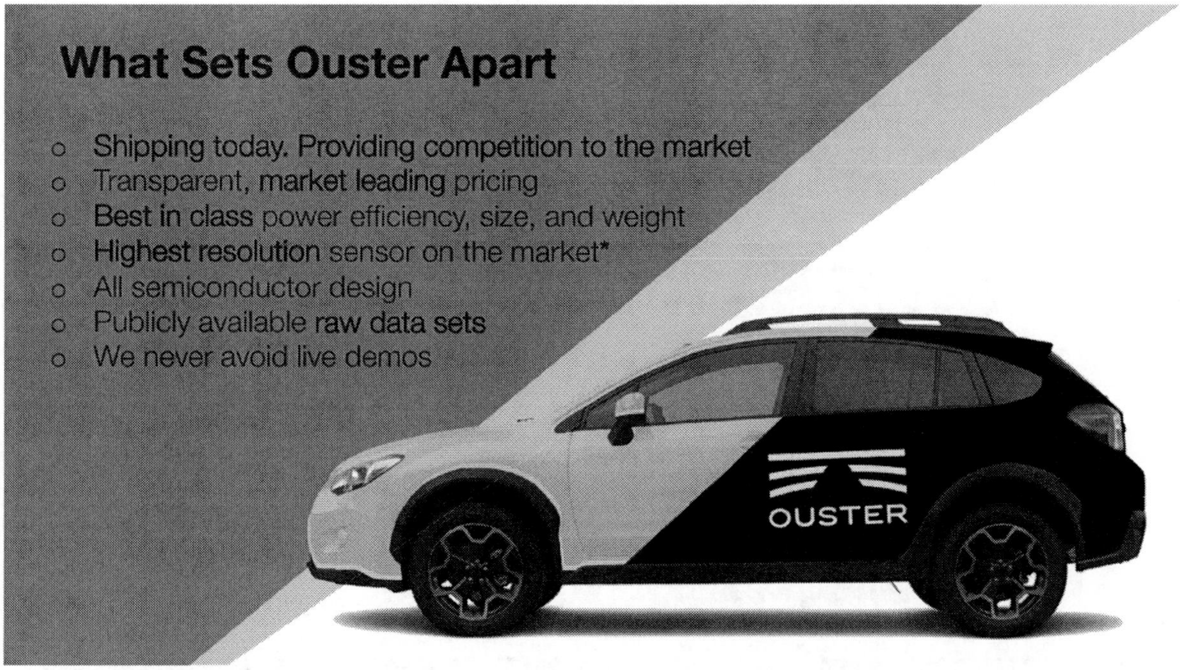
What Sets Ouster Apart
o Shipping today. Providing competition to the market
o Transparent, market leading pricing
o Best in class power efficiency, size, and weight
o Highest resolution sensor on the market*
o All semiconductor design
o Publicly available raw data sets
o We never avoid live demos
OUSTER

What Sets Ouster Apart
We ship products,
not promises
OUSTER

The Ouster OS-1

Bringing Moore's Law to LIDAR: 5th gen Ouster silicon (& counting)

OS-1

High Resolution Imaging LIDAR

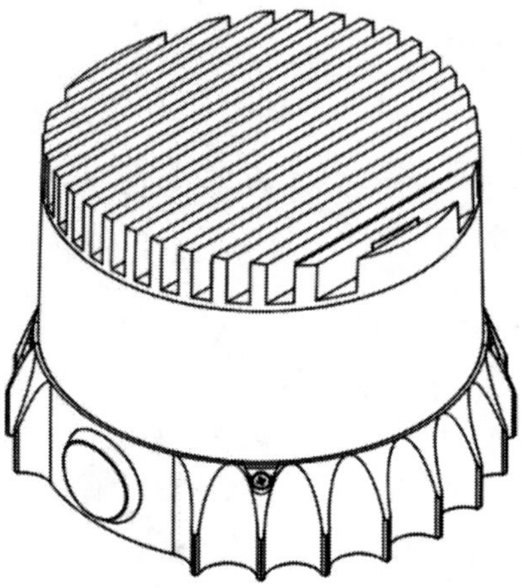

SUMMARY

The OS-1 offers a market leading combination of price, performance, reliability and SWAP. It is designed for indoor-outdoor all weather environments and long lifetime. As the smallest high performance LIDAR on the market, the OS-1 can be directly integrated into vehicle facias, windscreens, side mirrors, and headlight clusters.

HIGHLIGHTS

- Fixed resolution per frame operating mode
- Camera-grade reflectivity data
- Multi-sensor crosstalk immunity
- Industry leading intrinsic calibration
- Open source drivers

9

2 < 1000

Replace thousands of electronics in a traditional LIDAR with two custom chips. Laser. Detector.

We spent 3 years making this simple approach work, but the result is **lower power, lower cost, and higher performance.**

And because it's standard CMOS, it'll get better every year.

11

5th Generation Silicon

10 million transistors on chip

1 trillion photons counted per second

12

Source: https://www.sick.com/us/en/industries/semiconductor/c/g291168

The Same Technology in Smart Phones

We **customized** the same semiconductor technology used in the iPhone X Face ID, added **new 3D sensing algorithms**, and integrated it on a **compact spinning platform**. The result is 360° field of view and long range performance in a compact form factor.

These are mature, low cost, silicon CMOS technologies that are improving with Moore's Law.

13

Ouster OS-1-64
Raw data
Signal Image
Range Image
Combined Point Cloud

Signal Image
Range Image
Combined Point Cloud

Ouster OS-1-64
Raw data
Signal Image
Range Image
Combined Point Cloud

LIDAR Range and Intensity Images with CNNs

Ground segmentation trained on 5 minutes of data:

21

What People Are Saying

"This is the best LIDAR data I've seen. I just can't believe the resolution."

"I just want to congratulate you, **this is really impressive**"

"You're the only other company here showing data! Refreshing. **Nothing else I've seen looks this good.** I think you guys will be disruptive."

"**It's really too small.** You should have gone larger. **Could lose it in your pocket.**"

"All you have to do is show this [intensity image], forget 5 hours of PowerPoint slides. **We can see the repeatability and quality.**"

"We'd love to rely on Ouster for our vehicles.... I'm a big fan!"

New products #GTC 2018

OS-1-64
64 channels
$12,000
Shipping now

OS-1-16
16 channels
Multiple vertical field of view options
$3,500
Shipping Now

OS-2-64
64 channels
>200m range
50% tighter beam spacing
$24,000
Shipping Q4 2018

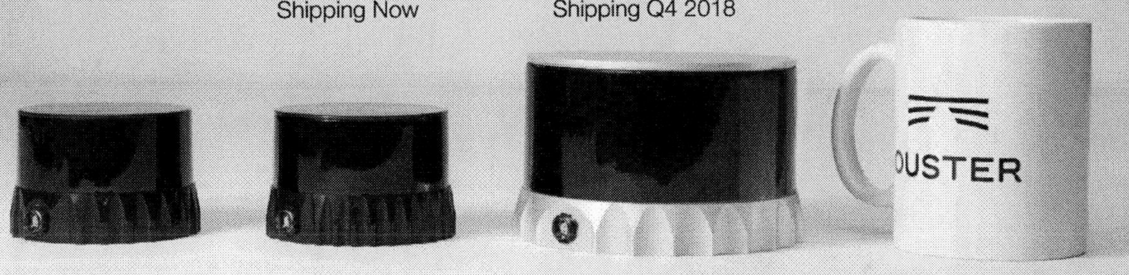

OUSTER

Angus Pacala
lidar@ouster.io

Transportation in the Third Dimension

Preparing for On-Demand Aviation Services

Arya Rohani
Chief Technology Strategist

AUVSI San Francisco
July 11, 2018

Stantec

The Future is Here

Stantec

Stantec

On-Demand Aviation Services

- Optimizing the 2D World
- Going Underground
- …and in the Air

Stantec

eVTOL

Stantec

Why?

- Faster
- Safer
- Cleaner
- Reliable
- Less Costly

Stantec

Why?

- Faster
  - 150-200 mph
  - Up to 2,000 feet altitude
  - 60 miles range per single charge
  - 5 minutes to recharge
  - Humans to Machines
- Safer
- Cleaner
- Reliable
- Less Costly

Stantec

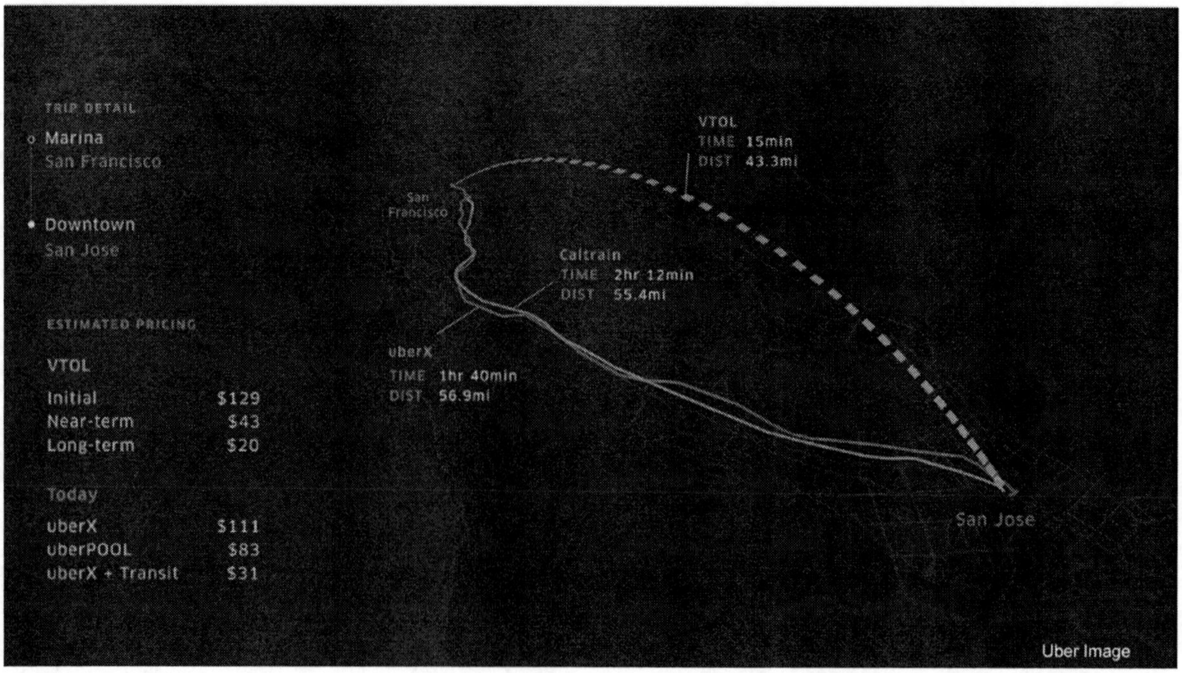

Why?

- Faster
- **Safer**
  - FAR Part 135 for Commuter/On-Demand Flights
  - 2x Fatalities per Passenger Mile than Vehicles
  - eVTOL Goal: 2x *Safer* than Vehicles
- Cleaner
- Reliable
- Less Costly

Stantec

Why?

- Faster
- Safer
- **Cleaner**
  - e for Electric (eVTOL)
- Reliable
- Less Costly

Stantec

Why?

- Faster
- Safer
- Cleaner
- Reliable
  - No Congestion
  - Incidents, Weather, Humans!
- Less Costly

Stantec

Why?

- Faster
- Safer
- Cleaner
- Reliable
- Less Costly

Stantec

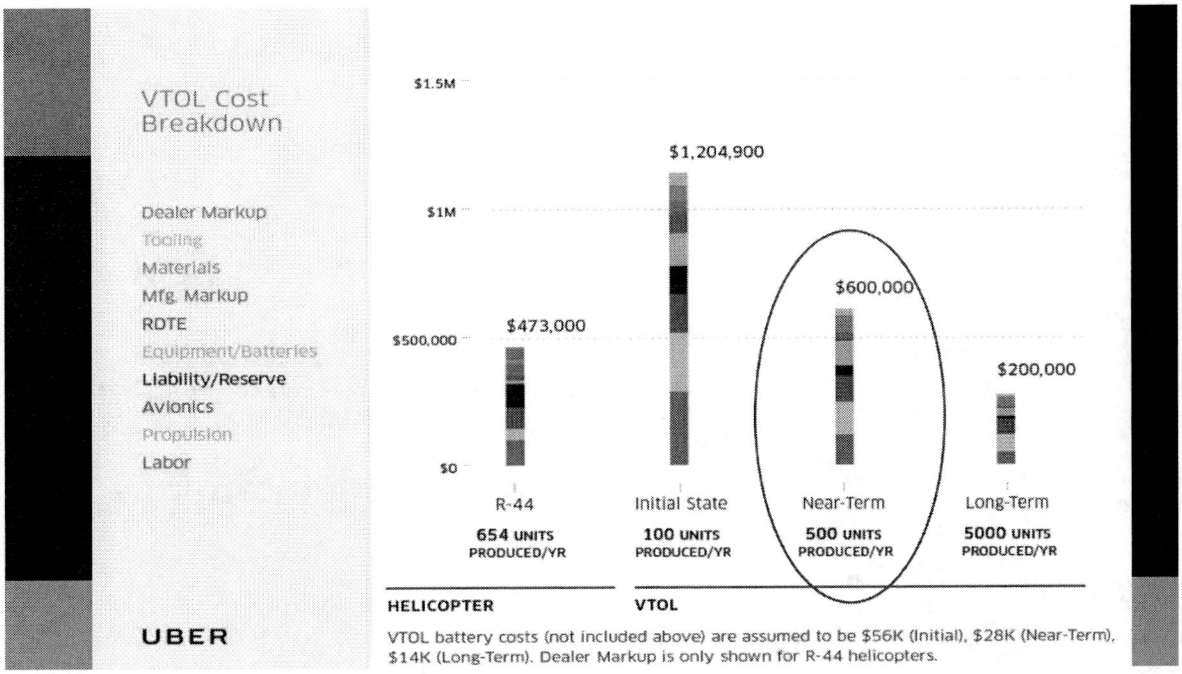
VTOL Cost Breakdown
Dealer Markup
Tooling
Materials
Mfg. Markup
RDTE
Equipment/Batteries
Liability/Reserve
Avionics
Propulsion
Labor
UBER
$1.5M
$1M
$500,000
$0
$1,204,900
$473,000
$600,000
$200,000
R-44
654 UNITS PRODUCED/YR
Initial State
100 UNITS PRODUCED/YR
Near-Term
500 UNITS PRODUCED/YR
Long-Term
5000 UNITS PRODUCED/YR
HELICOPTER
VTOL
VTOL battery costs (not included above) are assumed to be $56K (Initial), $28K (Near-Term), $14K (Long-Term). Dealer Markup is only shown for R-44 helicopters.

On-Demand Aviation:
Planning & Integration
Considerations
Stantec

FAA to Fly

Cities to Land

- Technology
- Regulatory
- Policy
- **Partnerships**
- **Infrastructure Planning, Design and Integration**
- **Operations and Maintenance**

Stantec

QUESTIONS?

Arya Rohani, PE
Chief Technology Strategist
Infrastructure
949-466-9062
arya.rohani@stantec.com

Stantec

MAX PLATFORM FOR AUTONOMOUS BEHAVIORS
DAVE HOFERT : PRI
PERRONE ROBOTICS
JULY 11, 2018

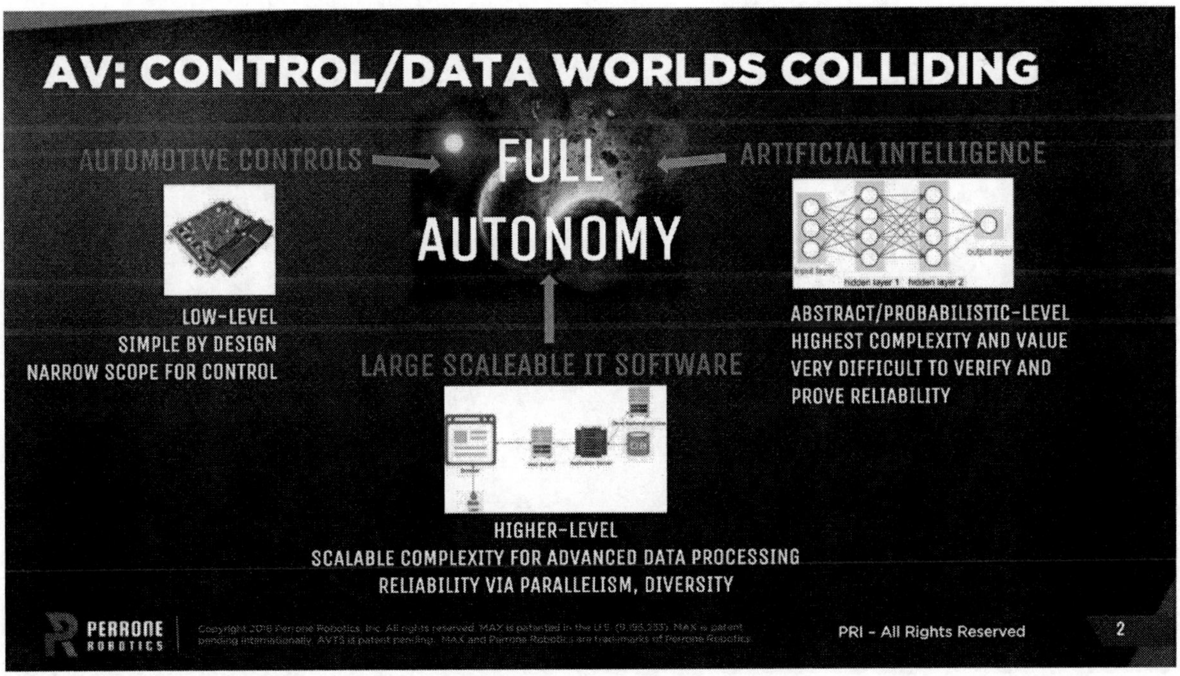
AV: CONTROL/DATA WORLDS COLLIDING
AUTOMOTIVE CONTROLS
FULL AUTONOMY
ARTIFICIAL INTELLIGENCE
LOW-LEVEL
SIMPLE BY DESIGN
NARROW SCOPE FOR CONTROL
LARGE SCALEABLE IT SOFTWARE
ABSTRACT/PROBABILISTIC-LEVEL
HIGHEST COMPLEXITY AND VALUE
VERY DIFFICULT TO VERIFY AND
PROVE RELIABILITY
HIGHER-LEVEL
SCALABLE COMPLEXITY FOR ADVANCED DATA PROCESSING
RELIABILITY VIA PARALLELISM, DIVERSITY
PERRONE ROBOTICS
PRI - All Rights Reserved
2

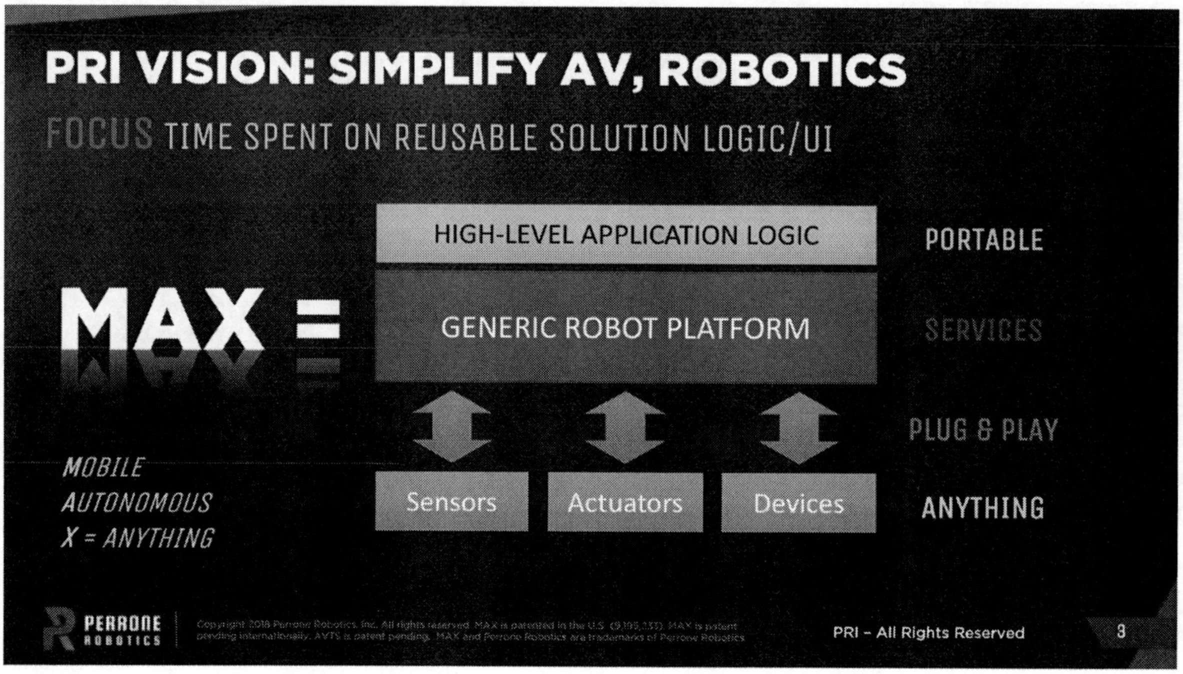
PRI VISION: SIMPLIFY AV, ROBOTICS
FOCUS TIME SPENT ON REUSABLE SOLUTION LOGIC/UI
MAX =
HIGH-LEVEL APPLICATION LOGIC
GENERIC ROBOT PLATFORM
Sensors
Actuators
Devices
PORTABLE
SERVICES
PLUG & PLAY
ANYTHING
MOBILE
AUTONOMOUS
X = ANYTHING
PERRONE ROBOTICS
PRI – All Rights Reserved
3

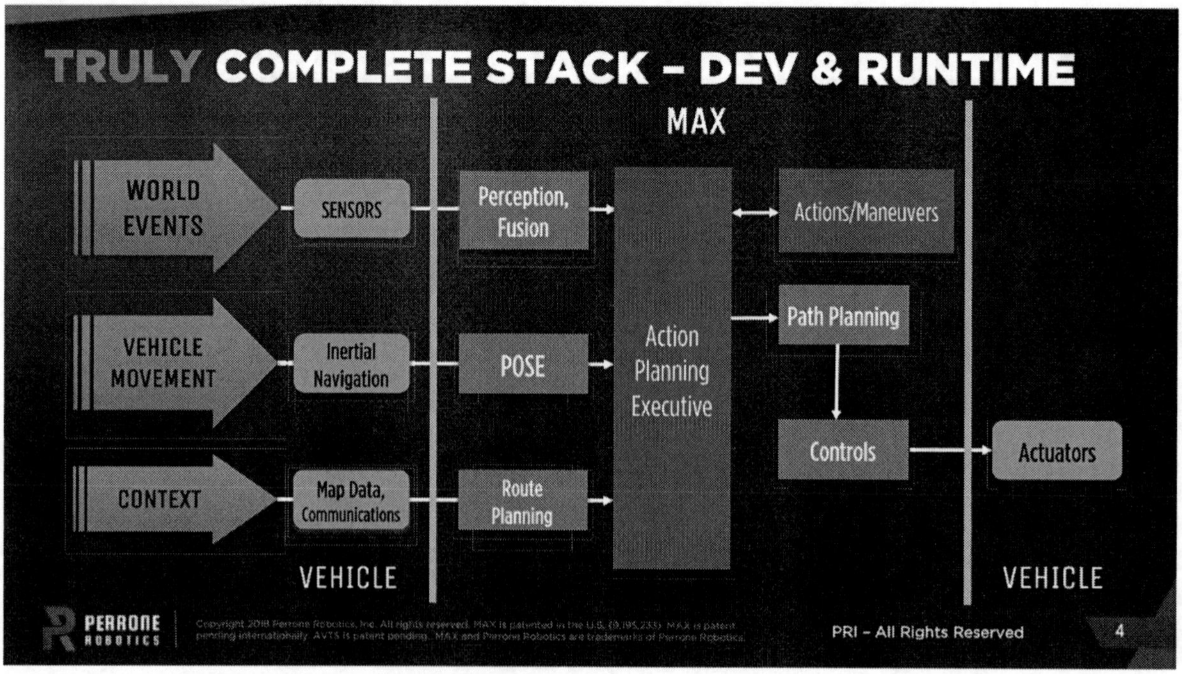
TRULY COMPLETE STACK – DEV & RUNTIME
MAX
WORLD EVENTS
VEHICLE MOVEMENT
CONTEXT
SENSORS
Inertial Navigation
Map Data, Communications
Perception, Fusion
POSE
Route Planning
Action Planning Executive
Actions/Maneuvers
Path Planning
Controls
Actuators
VEHICLE
VEHICLE
PERRONE ROBOTICS
PRI – All Rights Reserved
4

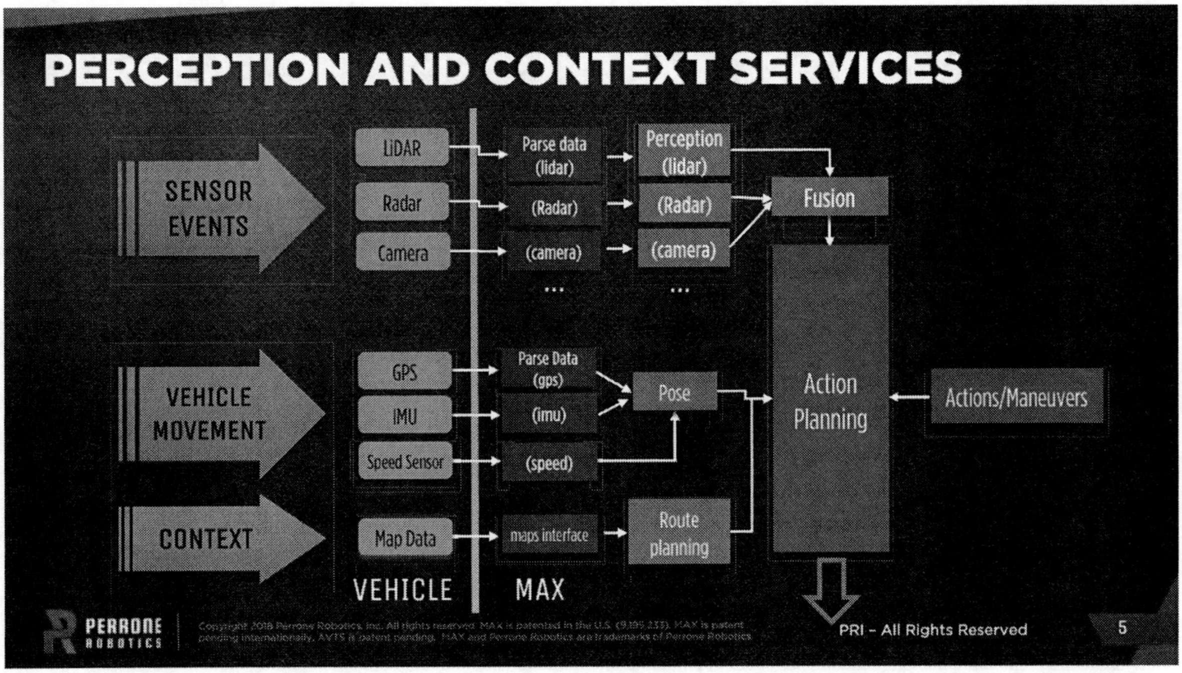
PERCEPTION AND CONTEXT SERVICES
SENSOR EVENTS
VEHICLE MOVEMENT
CONTEXT
LiDAR
Radar
Camera
GPS
IMU
Speed Sensor
Map Data
VEHICLE
MAX
Parse data (lidar)
(Radar)
(camera)
...
Parse Data (gps)
(imu)
(speed)
maps interface
Perception (lidar)
(Radar)
(camera)
...
Pose
Route planning
Fusion
Action Planning
Actions/Maneuvers
PERRONE ROBOTICS
Copyright 2018 Perrone Robotics, Inc. All rights reserved. MAX is patented in the U.S. (9,195,233). MAX is patent pending internationally. AVTS is patent pending. MAX and Perrone Robotics are trademarks of Perrone Robotics.
PRI – All Rights Reserved
5

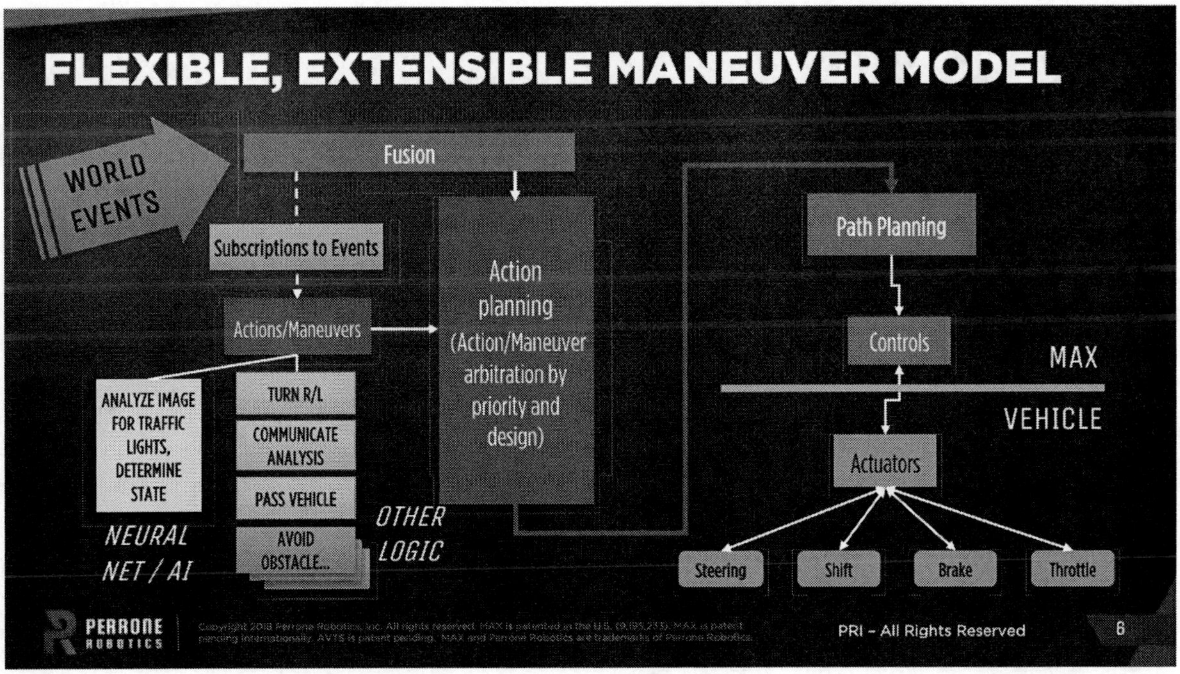
FLEXIBLE, EXTENSIBLE MANEUVER MODEL
WORLD EVENTS
Fusion
Subscriptions to Events
Actions/Maneuvers
ANALYZE IMAGE FOR TRAFFIC LIGHTS, DETERMINE STATE
NEURAL NET / AI
TURN R/L
COMMUNICATE ANALYSIS
PASS VEHICLE
AVOID OBSTACLE...
OTHER LOGIC
Action planning (Action/Maneuver arbitration by priority and design)
Path Planning
Controls
MAX
VEHICLE
Actuators
Steering
Shift
Brake
Throttle
PERRONE ROBOTICS
Copyright 2018 Perrone Robotics, Inc. All rights reserved. MAX is patented in the U.S. (9,195,233). MAX is patent pending internationally. AVTS is patent pending. MAX and Perrone Robotics are trademarks of Perrone Robotics.
PRI – All Rights Reserved
6

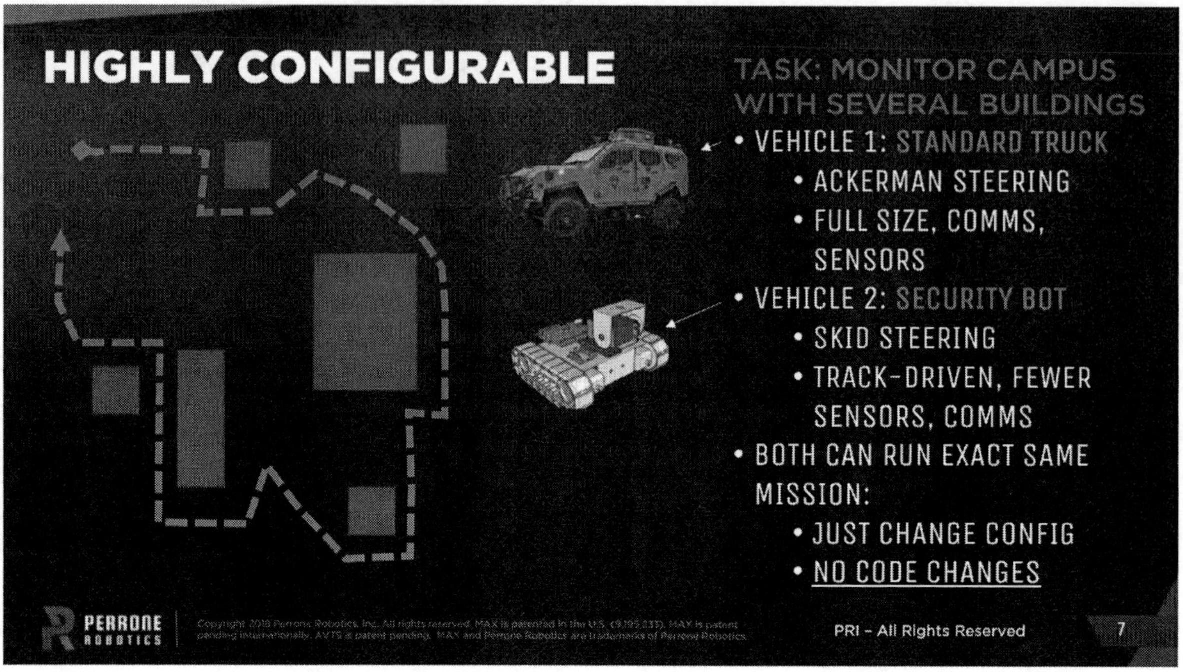
HIGHLY CONFIGURABLE
TASK: MONITOR CAMPUS WITH SEVERAL BUILDINGS
• VEHICLE 1: STANDARD TRUCK
• ACKERMAN STEERING
• FULL SIZE, COMMS, SENSORS
• VEHICLE 2: SECURITY BOT
• SKID STEERING
• TRACK-DRIVEN, FEWER SENSORS, COMMS
• BOTH CAN RUN EXACT SAME MISSION:
• JUST CHANGE CONFIG
• NO CODE CHANGES
PERRONE ROBOTICS
PRI – All Rights Reserved
7

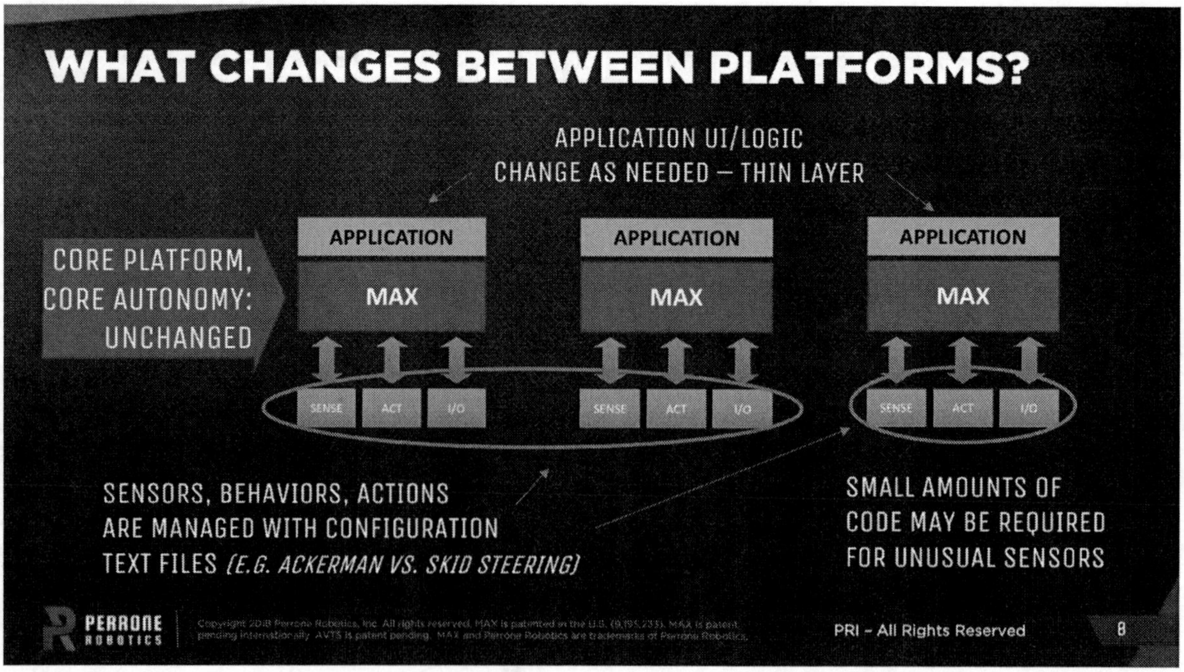
WHAT CHANGES BETWEEN PLATFORMS?
APPLICATION UI/LOGIC
CHANGE AS NEEDED – THIN LAYER
CORE PLATFORM, CORE AUTONOMY: UNCHANGED
APPLICATION
MAX
SENSE
ACT
I/O
SENSORS, BEHAVIORS, ACTIONS ARE MANAGED WITH CONFIGURATION TEXT FILES (E.G. ACKERMAN VS. SKID STEERING)
SMALL AMOUNTS OF CODE MAY BE REQUIRED FOR UNUSUAL SENSORS
PERRONE ROBOTICS
PRI – All Rights Reserved
8

SIMPLE CASE: "FOLLOW ME"

- SENSORS NORMALLY DETECT AND AVOID PEOPLE, THINGS
- IN THIS MODE, VEHICLE STILL DETECTS PEOPLE, BUT SEEKS TO KEEP PERSON IN FRONT OF VEHICLE
- SO AS PERSON MOVES, VEHICLE "FOLLOWS"
- TAKE FROM VEHICLE TO VEHICLE USING DIFFERENT SENSORS – WITHOUT CHANGES!

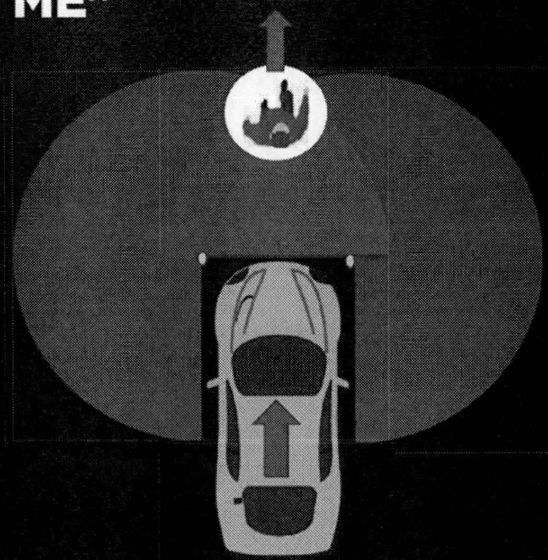

PERRONE ROBOTICS

9

MAX REUSE ACROSS PLATFORMS/SOLUTIONS

Function	AUTOMOTIVE	INDUSTRIAL	COMMERCIAL
Obstacle detection and avoidance	**Developed here**	Reuse without change, add negative ODA if needed	Reuse without change
V2X Communication	Reuse/Adapt, but use DSRC	**Developed here – DDS**	Reuse/Adapt, but use wifi
Parking Maneuver	**Developed here**	Reuse for loading (dump truck); add dynamic siting	Reuse for charging station
Intersection handling	**Developed here**	Reuse without change	Reuse without change
Dynamic course/mission changes	Add to existing re-routing	**Developed here**	Reuse without change
Indoors navigation	Reuse without change	Reuse without change	**Developed here**

PERRONE ROBOTICS

10

FLEXIBLE ARCHITECTURE

- SCALE UP/DOWN AS TASK REQUIRES
- LEVERAGE REAL-TIME VM FOR HW/OS FLEXIBILITY
- SPECIALIZED ALGORITHMS RUN ON TUNED HW ARCH
- DISTRIBUTED PROCESSING MORE FAULT-TOLERANT
- KEY:
  - Strong HW and SW **PLATFORM** approach – IT like
  - Abstraction of HW/OS enables maximum code reuse

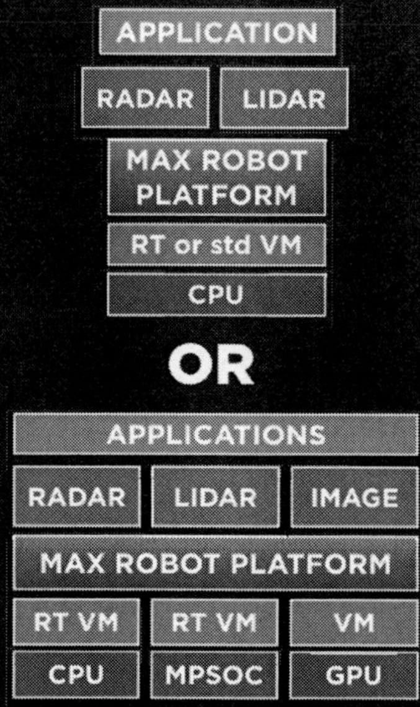

PERRONE ROBOTICS

WE HAVE DONE IT BEFORE

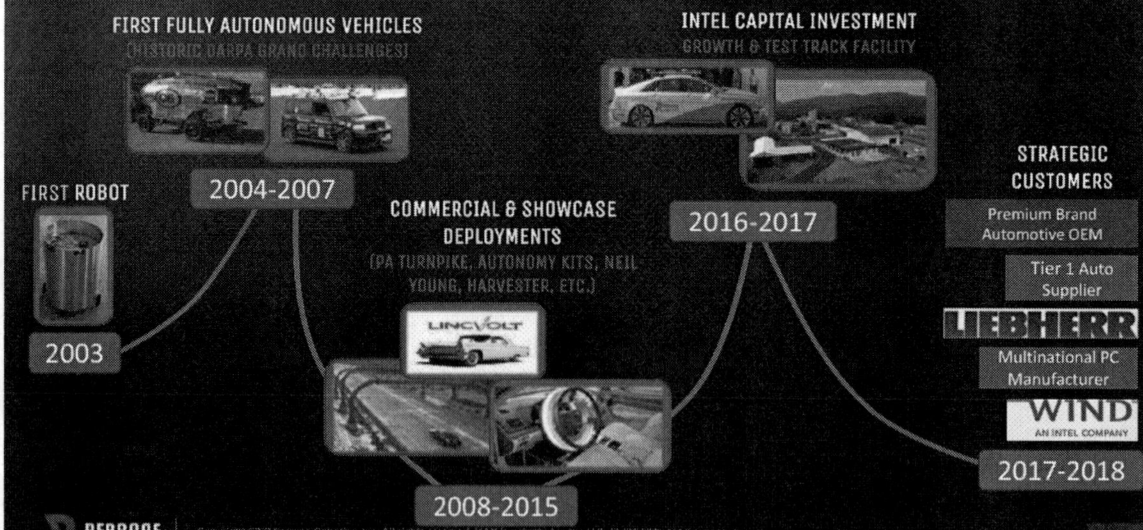

PERRONE ROBOTICS

LOCATED IN CROZET, VIRGINIA

PRIVATE TEST TRACK

PERRONE ROBOTICS

SOLUTIONS IN MANY AREAS

PERRONE ROBOTICS

14

MAXIMUM VALUE

- A TRULY FLEXIBLE PLATFORM, PROVEN OVER MANY IMPLEMENTATIONS
  - Unique combination of configurability, hardware/communications flexibility
  - Algorithm modularity, Full stack/suite of app services
- MIGRATE SEAMLESSLY ACROSS PROJECTS TODAY
  - Leverage inherent network effect from MAX platform model
- IP: PLATFORM PATENTED IN 2006
  - With extension (continuance in part) this Spring
- PATH TO PRODUCTION/CERTIFICATION
  - MAX built with production in mind - not just R&D, but actual deployment
  - Beginning work on 61508 certification (SIL 2 to start)

PERRONE ROBOTICS

THANK YOU!
QUESTIONS?

DAVE@PERRONEROBOTICS.COM

PERRONE ROBOTICS

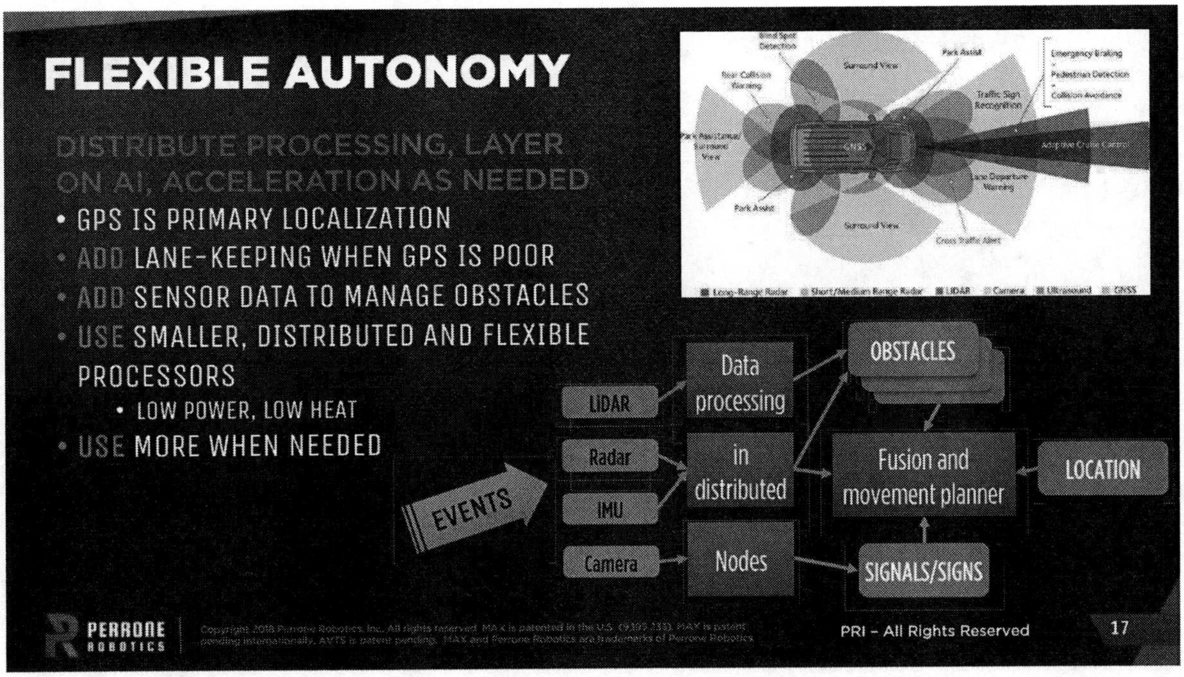
FLEXIBLE AUTONOMY
DISTRIBUTE PROCESSING, LAYER ON AI, ACCELERATION AS NEEDED
• GPS IS PRIMARY LOCALIZATION
• ADD LANE-KEEPING WHEN GPS IS POOR
• ADD SENSOR DATA TO MANAGE OBSTACLES
• USE SMALLER, DISTRIBUTED AND FLEXIBLE PROCESSORS
• LOW POWER, LOW HEAT
• USE MORE WHEN NEEDED
LIDAR
Radar
IMU
Camera
EVENTS
Data processing
in distributed
Nodes
OBSTACLES
Fusion and movement planner
LOCATION
SIGNALS/SIGNS
PERRONE ROBOTICS
PRI – All Rights Reserved
17

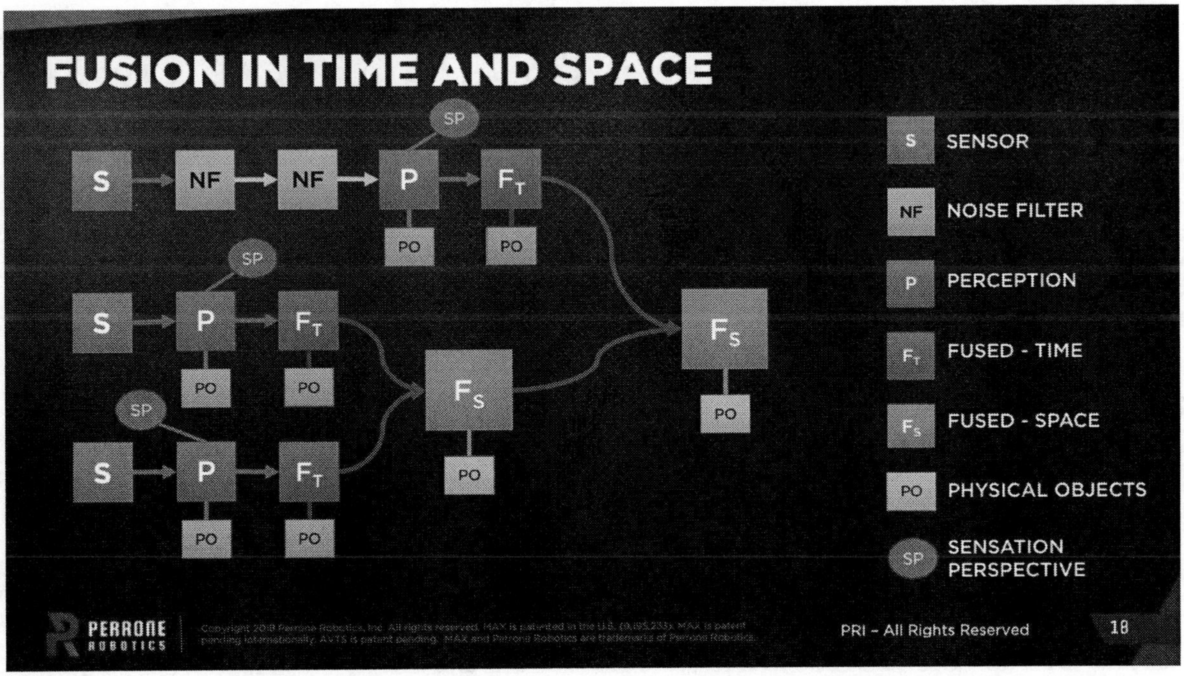
FUSION IN TIME AND SPACE
S
NF
NF
P
SP
PO
S
P
S
P
PO
SENSOR
NOISE FILTER
PERCEPTION
FUSED - TIME
FUSED - SPACE
PHYSICAL OBJECTS
SENSATION PERSPECTIVE
PERRONE ROBOTICS
PRI – All Rights Reserved
18

MULTI-SENSOR FUSION APPROACH

SENSOR DIVERSITY FOR SAFETY, RELIABILITY IN AUTOMOTIVE

- GPS FOR CORE LOCALIZATION
- CAMERA FOR ADVANCED DETECTION, LOCALIZATION (AI-BASED)
- 16-BEAM LIDARS
  - USE 1-4+ AS NEEDED FOR OBSTACLE IDENTIFICATION
- RADARS FOR LONG-RANGE OBSTACLE DETECTION

PERRONE ROBOTICS

19

MULTI-SENSOR FUSION APPROACH - II

HAUL TRUCK: VERY SIMILAR TO AUTOMOBILE – SAME PLATFORM, JUST RECONFIG

- GPS FOR CORE LOCALIZATION
- CAMERA FOR ADVANCED DETECTION (AI-BASED)
- RADARS FOR LONG-RANGE OBSTACLE DETECTION
- 8-BEAM LIDARS FOR SMALLER OBSTACLE DETECTION

PERRONE ROBOTICS

20

PLASTICS PARTNERSHIPS
for
AUTONOMOUS FUTURE

Gina Oliver
American Chemistry Council Plastics Division

American Chemistry Council | Plastics Division

Who are we?

AmericasStyrenics LLC
BASF We create chemistry
Braskem
Chevron Phillips Chemical Company LP
covestro
DOW
DUPONT
ExxonMobil
LANXESS Energizing Chemistry
lyondellbasell
NOVA Chemicals
سابك sabic
Shell Chemicals
SOLVAY asking more from chemistry
TRINSEO
TOTAL
The Vinyl Institute

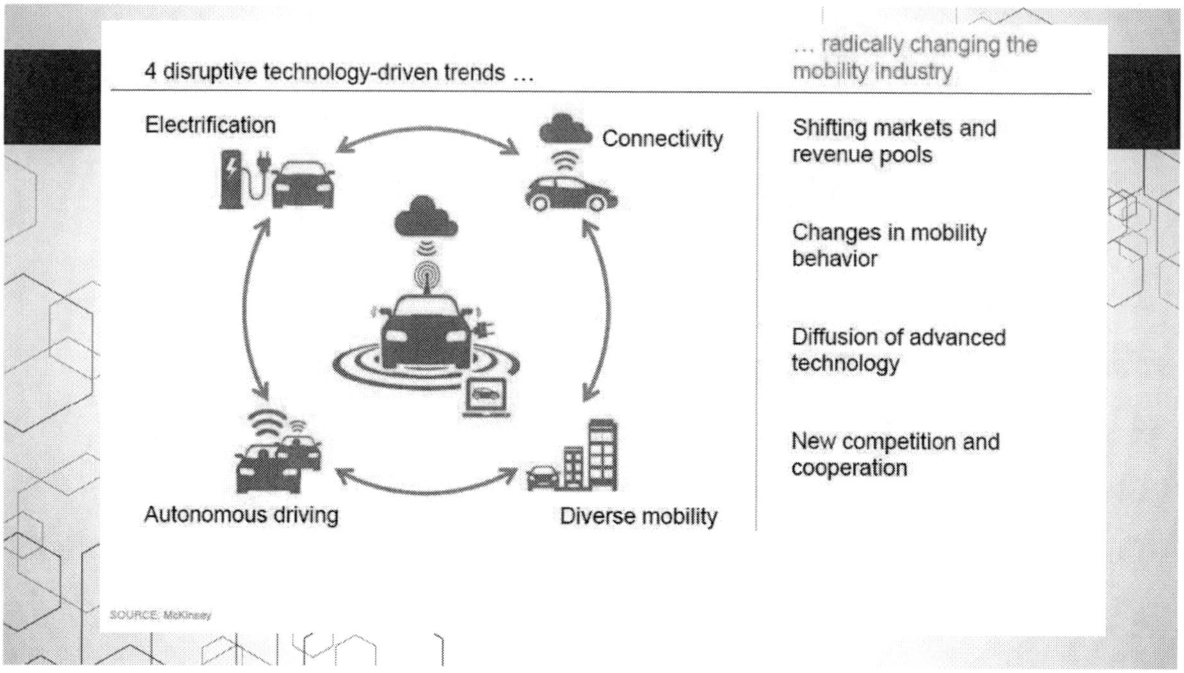
4 disruptive technology-driven trends …
… radically changing the mobility industry
Electrification
Connectivity
Autonomous driving
Diverse mobility
Shifting markets and revenue pools
Changes in mobility behavior
Diffusion of advanced technology
New competition and cooperation
SOURCE: McKinsey

Plastics:
• Lighter yet stronger
• Virtually unlimited, sustainable Solutions
• Unique, flexible designs
• Essential enabler for AVs

CHICAGO
© Waymo
Photo: Magna

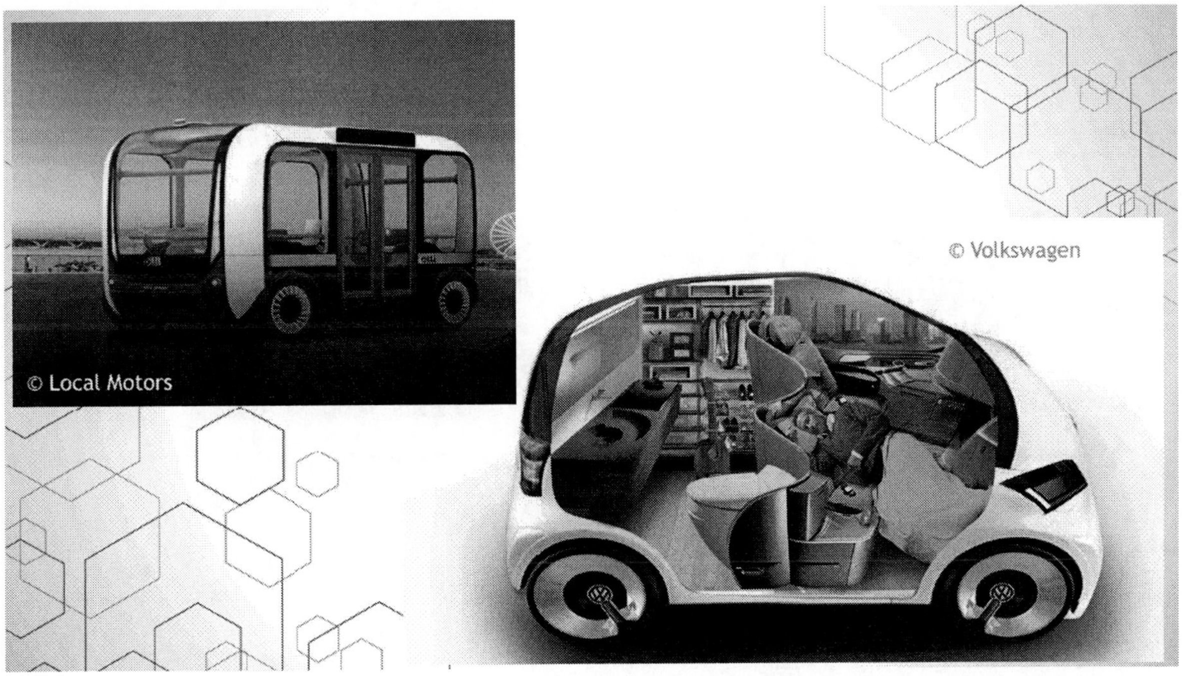
© Local Motors
© Volkswagen

Sustainable Solutions

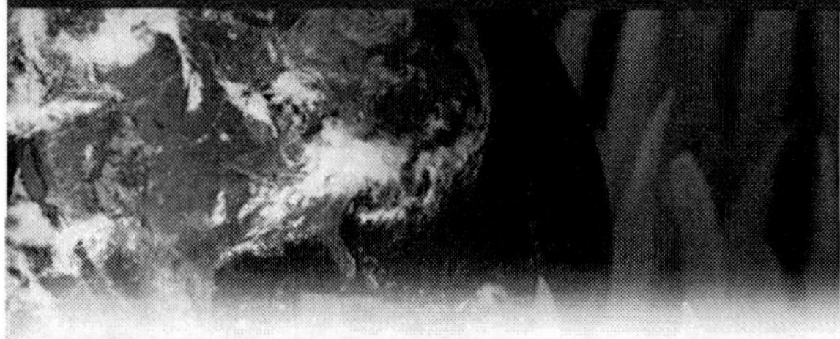

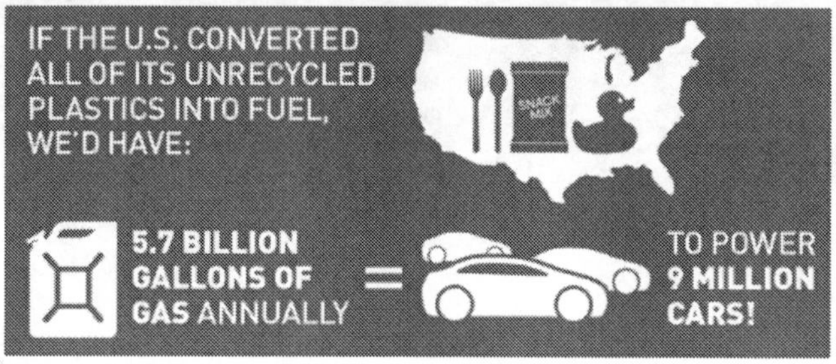

- Recycling & Recovery
- Sustainable, U.S. feedstock
- Bio-based possibilities

Creating Partnerships

Faster, Smarter, Better

American Chemistry Council | Plastics Division

American Chemistry Council | Plastics Division

Plastics and Polymer Composites
TECHNOLOGY ROADMAP
for Automotive Markets

March 2014

<u>VISION</u>:

By 2030, the automotive industry and society will recognize plastics and polymer composites as preferred material solutions

9

Outline

- A look back to 2014 Roadmap
- Key roadmap drivers/trends: Then and Now
- Roadmap development process
- How to get involved

The Roadmap is build through expert input

Roadmap contributors

- OEMs
- Tier suppliers
- Materials developers
- Researchers
- Federal agencies

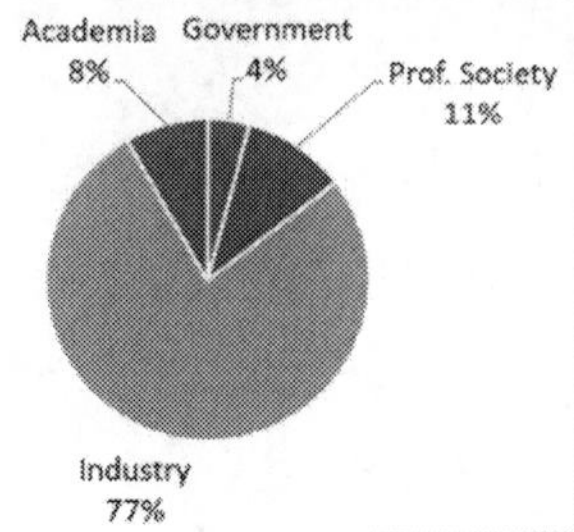

Vision

By 2030, the automotive industry and society will recognize plastics and polymer composites as preferred material solutions **that meet, and in many cases set, automotive performance and sustainability requirements**.

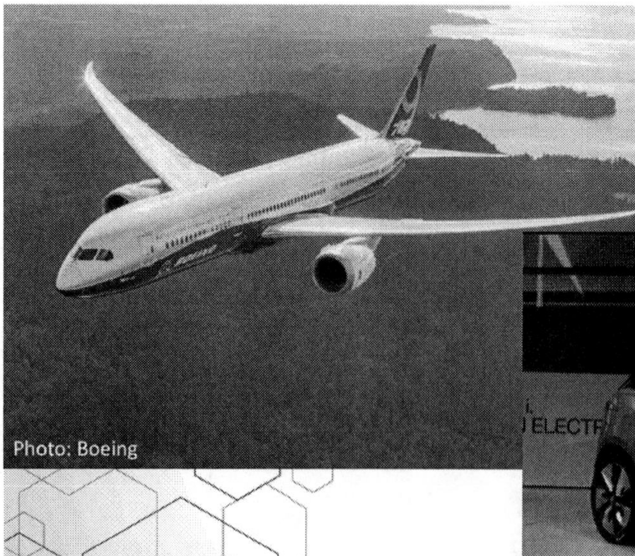
Photo: Boeing

Photo: BMW

Why update the Roadmap now?

Vehicle Autonomy
Urbanization • Mobility/ridesharing services • Vehicle-technology convergence • Consumer preference

Mobility/Ridesharing Services
Evolving vehicle ownership paradigms • Urbanization

Connectivity
Emerging 5G wireless technology • Complex/competitive landscape with tech. giants/mobility providers • Vehicle safety; DOT 2020 vehicle-to-vehicle (V2V) communication mandate?

Steel/Aluminum Tariffs
Potential tax on competing materials options

Safety, Cybersecurity
Autonomy, connectivity, mass reduction

CAFE [Fuel Economy] Standards
Trump Administration seeking to roll back standards

Alt. Powertrains/Electrification
Declining battery prices • Consumer demand • EU transport-related CO_2 regulations

Autonomous vehicles create big questions for materials

- **Vehicle use: 10% → 70%**
  - What does this mean for interior materials?
- **Drastic reduction in vehicle accidents**
  - What does this mean for vehicle crash requirements and safety features?
- **Vehicles need to see each other and be seen by each other**
  - What does this mean for paint and coatings?
- *As you envision autonomous vehicles of tomorrow, what do the materials need to do?*

Why should I care?

ACC's Roadmaps have guided the investment of more than $400M in groundbreaking R&D

iacmi

$259M in funding

The Roadmap's top priority was funded in <1 year

Priority Action 1

Technology Development Center

Establish an independent, pre-competitive technology development center where OEMs and suppliers can conduct laboratory work and test concepts at small volumes while collecting standardized data

iacmi THE COMPOSITES INSTITUTE

(Manufacturing USA Institute #5)

- Established: January 2015
- Funding: $259 million
- Focus: Low-cost, energy-efficiency manufacturing of fiber reinforced polymer composites (CFRP)

What is the roadmapping process?

- Structured interviews
- Online surveys
- Small focus-group workshops
- Larger expert workshops

What is the roadmapping process?

How can you get involved?

1. **Today:** Contact **Jared Kosters (jkosters@nexightgroup.com)**
   - **Conduct 1-on-1 interview at AVS 2018**
   - Receive notices about online surveys & future events/roadmapping workshops

2. **Later:** Contact **Gina-Marie Oliver (Gina-Marie_Oliver@americanchemistry.com)** for questions about ACC engagement

Discussion

- *As you envision autonomous vehicles of tomorrow, what do the materials need to do?*

PERKINS COIE

COUNSEL TO GREAT COMPANIES

The Emerging Role of State and Local Governments in Regulating Autonomous Vehicles

AUVSI Automated Vehicles Symposium
San Francisco, CA
July 10, 2018

William G. Malley
Washington, DC

Perkins Coie LLP

Firm Overview

1000+ Lawyers • **19** Offices on **2** Continents • **100+** Years of Partnering with Clients

With 16 offices in the United States, a strong footprint in Asia and a network of law firms around the world, Perkins Coie assists clients anywhere and everywhere they do business.

Perkins Coie LLP | PerkinsCoie.com

PERKINS COIE

Traditional Federal & State Roles

Federal Role

"what can be sold"

- Safety standards
- Emissions standards
- Consumer protection

State and Local Role

"who" can operate
"where, when, how" vehicles are used

- Driver licensing
- Vehicle registration
- Insurance requirements
- Product liability
- Rules of the road

3 Perkins Coie LLP | PerkinsCoie.com

PERKINSCOIE

AVs Emerge ...

- Initially, a regulatory vacuum
  - Absence of AV-specific safety standards at the federal level
  - Absence of clear prohibition in most state and local laws on use of AV

"Automated vehicles are probably legal in the United States."
(Bryant Walker Smith, Nov. 2012)

4 Perkins Coie LLP | PerkinsCoie.com

PERKINSCOIE

Regulatory Response to AVs, so far

- Wave of State legislation
  - 29 States have enacted AV laws*
  - Most allow testing & set insurance req'mts; some allow full deployment.
- NHTSA relies on guidance
  - NHTSA issues guidance to manufacturers
  - No steps taken to issue AV-specific FMVSS

* National Conference of State Legislatures, Autonomous Vehicles Legislative Database (as of June 2018)

NHTSA's Message to States

"NHTSA is responsible for regulating motor vehicles and motor vehicle equipment, and States are responsible for regulating the human driver and most other aspects of motor vehicle operation... DOT strongly encourages States to allow DOT alone to regulate the safety design and performance aspects of ADS technology. If a State does pursue ADS performance-related regulations, that State should consult with NHTSA."

NHTSA, "Automated Driving Systems 2.0: A Vision for Safety" (Sept. 2017).

On the Horizon

- State laws will continue to evolve
  - Allowing deployment, not just testing
  - Allowing testing w/o driver present
- Federal legislation may be enacted
  - AV bill passed in House, pending in Senate
  - Each would preempt State and local regulation of AVs to some extent.

7 Perkins Coie LLP | PerkinsCoie.com

PERKINSCOIE

The House Bill

- Self-Drive Act (passed by House)
  - Preempts States from adopting standards "regarding the design, construction, or performance" of AVs.
  - Preserves States authority over vehicle registration, licensing, etc., *unless* it is "an unreasonable restriction on the design, construction, or performance" of AVs.

H.R. 3388, Sec. 3. (passed by House 9/6/2017).

8 Perkins Coie LLP | PerkinsCoie.com

PERKINSCOIE

The Senate Bill

- AV Start Act (pending in Senate)
  - Preempts States from adopting standards "regulating the design, construction, or performance" of AVs with respect to specified subject areas.
    - "Unreasonable restriction" language was initially included, later dropped.

S. 1885, Sec. 3. (approved by Senate Commerce Committee 10/4/2017; pending in Senate).

PERKINSCOIE

State Agencies' Response

On the Self-Drive Act:

"Unfortunately, the bill currently being considered by the House seeks to significantly expand federal pre-emption of states by moving beyond the traditional definition of motor vehicle safety to encroach on vehicle operations, currently under the states' purview."

National Governor's Association et al. Letter, Sept. 5, 2017.

On the AV Start Act:

"The AV START Act would benefit from further defining the term "performance" to exclude the act of complying with traffic laws, thereby firmly associating it with the current federal responsibilities. With this approach, state and local laws will continue to focus on the operational safety laws regulating motor vehicles and their operators...."

National Governor's Association et al. Letter, Nov. 8, 2017.

PERKINSCOIE

Let's Take A Closer Look

- The Law of Preemption
  - Basics of Preemption
  - Preemption & NHTSA
- What it *Might* Mean for State/Local Govts
  - Emerging State and local AV laws
  - Potential impact of preemption language
  - What becomes the 'new normal'?

11 Perkins Coie LLP | PerkinsCoie.com

PERKINSCOIE

The Law of Preemption

12 Perkins Coie LLP | PerkinsCoie.com

PERKINSCOIE

Federal Preemption – The Basics

- Can be express or implied
- But courts 'do not lightly infer preemption'
  - Especially in areas traditionally subject to State regulation.

"[I]n a field which the States have traditionally occupied ... we start with the assumption that the historic police powers of the States were not to be superseded ... unless that was the clear and manifest purpose of Congress."
Wyeth v. Levine, 555 U.S. 555, 556 (2009)

PERKINSCOIE

NHTSA & Preemption: Safety Standards

- The Safety Act:
  - Preempts any State and local regulation of "the same aspect of performance" covered by a NHTSA safety standard. *[49 USC 30103]*
- What the courts have held:
  - Does not 'occupy the field' of vehicle safety.
  - Preemption found when State law conflicts with or 'frustrates' a NHTSA standard.

PERKINSCOIE

NHTSA & Preemption: Safety Standards

- Typical issues in case law:
  - Claims for liability under State common law
  - Court-ordered recalls
  - State stds that exceed federal requirements
- Rarely if ever litigated
  - State laws restricting when, where, and how NHTSA-compliant vehicles are used

Example: Low-Speed Vehicles

NHTSA Role:

- Sets safety standards that LSVs must meet

State/Local Role:

- Determine where LSVs can be used and applicable requirements.

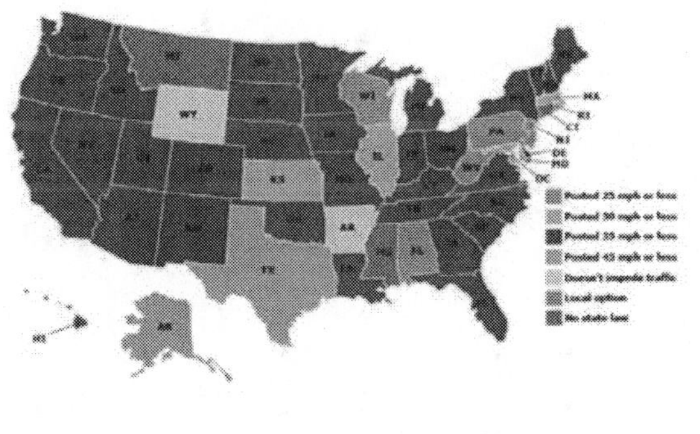

IIHS, Low and Medium Speed Vehicles, http://bit.ly/2N6ffo9.

Example: Hybrid Taxis

- Federal law (EPCA) gives NHTSA exclusive role in setting fuel economy standards:
  - States may not adopt any law "related to" fuel economy standards set by NHTSA.
- Courts:
  - Local laws mandating hybrid taxis are "related to" fuel economy & thus preempted.

See, e.g., Metropolitan Taxicab Bd. of Trade v. City of New York, 615 F.3d 152, 156–58 (2d. Cir. 2010).

Federal Preemption – Takeaways

- Lots of uncertainty remains:
  - Will preemption language be enacted?
  - What will it say? How will courts interpret?
- But reasonable to expect that:
  - Feds exclusively set AV safety standards;
  - State & local govts retain at least some discretion to regulate *use* of AVs.

Federal Preemption - Two Scenarios for AVs

"Narrower Preemption"

Federal	State & Local
Exclusive federal regulation of AV safety standards	Broad authority to determine when, where, and how AVs are used

"Broader Preemption"

Federal	**State & Local**
Exclusive federal regulation of AV safety standards *and* restricts State & local govts from regulating or prohibiting use of AVs	Limited authority to determine when, where and how AVs are used

??

19 Perkins Coie LLP | PerkinsCoie.com

PERKINSCOIE

And don't forget *State* preemption

- States can preempt local governments from regulating AVs.
  - State enacts AV law, prohibits any local laws regulating or restricting use of AVs.
- Examples to date:
  - Illinois, Nebraska, Nevada, North Carolina, Tennessee, Texas.*

* National Conference of State Legislatures, Autonomous Vehicles Legislative Database (as of June 2018)

20 Perkins Coie LLP | PerkinsCoie.com

PERKINSCOIE

COUNSEL TO GREAT COMPANIES

Looking Ahead – Potential State and Local Regulation of AVs

21 Perkins Coie LLP | PerkinsCoie.com

PERKINSCOIE

State Regulation of AVs: A Look Ahead

- As AVs become commonplace, States will need to update virtually aspects of their motor vehicle codes:
  - Driver licensing
  - Vehicle registration
  - Insurance requirements
  - Rules of the road

22 Perkins Coie LLP | PerkinsCoie.com

PERKINSCOIE

Driver Licensing

Today:

- States require all drivers to be licensed; require testing as precondition for license.

With AVs, States may:

- Allow HAVs to be operated w/o licensed driver
- Clarify that passenger in an HAV is not a 'driver'
- Shift some 'driver' responsibilities to 'provider'

Vehicle Registration

Today:

- States require all vehicles to be registered; attach many other requirements to registration.

With AVs, States may:

- Register AVs based on capability level.
- Impose other requirements as precondition for registration – e.g., representations of AV's safety.
- Require AV type to be displayed on license plate.

Rules of the Road

Today:

- State and local governments establish and enforce traffic safety laws.

With AVs, States may:

- Require AVs to comply with same traffic laws that apply to other vehicles
- Establish traffic laws specific to AVs – e.g., to address AV-human interactions.

PERKINSCOIE

Example: California Vehicle Code

State can restrict use of specific roads

"The department may restrict the use of, or close, any State highway whenever the department considers such closing or restriction of use necessary: (a) For the protection of the public. ...

California Streets & Highways Code, § 124.

Local govts can restrict which vehicles use which roads

"Local authorities, for those highways under their jurisdiction, may adopt rules and regulations by ordinance or resolution on the following matters: ... (c) Prohibiting the use of particular highways by certain vehicles...."

California Vehicle Code, §21101.

PERKINSCOIE

Uniform State Law – in development

- Effort under way to develop a 'uniform law' for State regulation of AVs.* It would:
  - Recognize that AVs are lawful.
  - Require AVs to abide by traffic laws.
  - Address insurance & liability issues.
  - Define 'driver' as the 'provider' of the AV.
  - Require 'provider' to 'register' with State.

*Nat'l Conf. of Commissioners on Uniform State Laws, Uniform Law Commission, Highly Automated Vehicles Act (June 2018 Draft)

The Harder Questions

Can States limit where AVs are used – e.g.

- Limit AVs to certain routes?
- Allow AVs only in designated lanes?
- Allow AVs only within geofenced areas?

The Harder Questions (cont'd)

Can States allow some types of AVs but not others, based on safety considerations – e.g.

- Allow Level 4 and 5 AVs, but not Level 3?

> *"Regulators and product designers should bar the use of partially automated vehicles (NHTSA Level 3) on any roadway without controlled access, like city streets."*
> National Association of City Transportation Officials, Policy Statement on Automated Vehicles (June 22, 2016)

The Harder Questions (cont'd)

Can States require AVs to have specific capabilities – e.g.

- Require AVs to have 'safety override'?
- Require AVs to have V-to-I system?
- Require remote-operator capability?

> *"The HAV shall have a safety override means or process to engage and disengage the ADS technology that is easily accessible to the operator, should manual intervention be required. The ADS shall also be able to be disengaged by law enforcement and other emergency responder personnel."*
> Pennsylvania Autonomous Vehicle Testing Policy, Draft Final Report (Nov. 2016).

The Harder Questions (cont'd)

Can States require AVs to perform tasks normally performed by human drivers – e.g.,:

- Contact emergency responder after crash?
- Pull over when directed by police?
- Provide registration to law enforcement?

31 Perkins Coie LLP | PerkinsCoie.com

PERKINSCOIE

The Harder Questions (cont'd)

Can States require AVs to make their AV status visible to other on the roadway – e.g.,

- Display an 'indicator light' in AV mode?
- Display other external marking of AV status?

"For vehicles that are not fully self-driving and that may require manual intervention or manual temporary control when an operator is physically located in the vehicle, the vehicle shall be equipped with a visual indicator inside the cabin to indicate when the ADS is engaged."
Pennsylvania Autonomous Vehicle Testing Policy, Draft Final Report (Nov. 2016).

32 Perkins Coie LLP | PerkinsCoie.com

PERKINSCOIE

The Harder Questions (cont'd)

Can States impose data-sharing requirements as a condition of registration – e.g.

- Require fleet owners to submit operational data to the State?
- Establish privacy requirements to limit the types of data that AV operators can release?

Takeaways

Closing thoughts

- Traditionally, State and local powers to regulate road usage are expansive.
- Federal preemption will likely restrict but not entirely preclude State and local regulation.
  - Wording is key – e.g., "related to"
- For local governments, *State* preemption may be a bigger issue than Federal.

Thank You

Bill Malley
Perkins Coie LLP

700 13th St. NW
Washington, DC 20005

wmalley@perkinscoie.com

(202) 654-6250

PERKINSCOIE

Bridging the Automated Vehicle Gap: Consumer Trust, Technology and Liability

Tina Georgieva
Senior Attorney, Product Safety Group
Miller Canfield

Kristin Kolodge
Executive Director, Driver Interaction and HMI
J.D. Power

Integrating the Consumer into the Automated Driving System (ADS) Development Process

MILLER CANFIELD J.D. POWER

M | city AFFILIATE UNIVERSITY OF MICHIGAN

Consumer Expectations

Legal Practitioners

Automated Vehicles Symposium 2018 2

MILLER CANFIELD J.D. POWER

Trust *or* **Experience** First?

MILLER CANFIELD J.D. POWER

Trust is Centric to ADS Success

ADS TRUST ECOSYSTEM

SAFETY STANDARDS
- Industry, Government, Insurance
- Assurance
- Standardization

CONSUMER INTEREST
- Trust drives interest in automation
- Purchase intention

LIABILITY FRAMEWORK
- Transparency
- Responsibility
- Just resolution

UNDERSTANDABILITY
- Naming convention
- Education / Training
- Industry / Media messaging

CONSUMER EXPERIENCE
- Actual usage vs. Expected usage
- Satisfaction

PERFORMANCE EXPERIENCE
- False positives / False negatives
- Humanistic behavior
- Consistency of performance
- Predictability

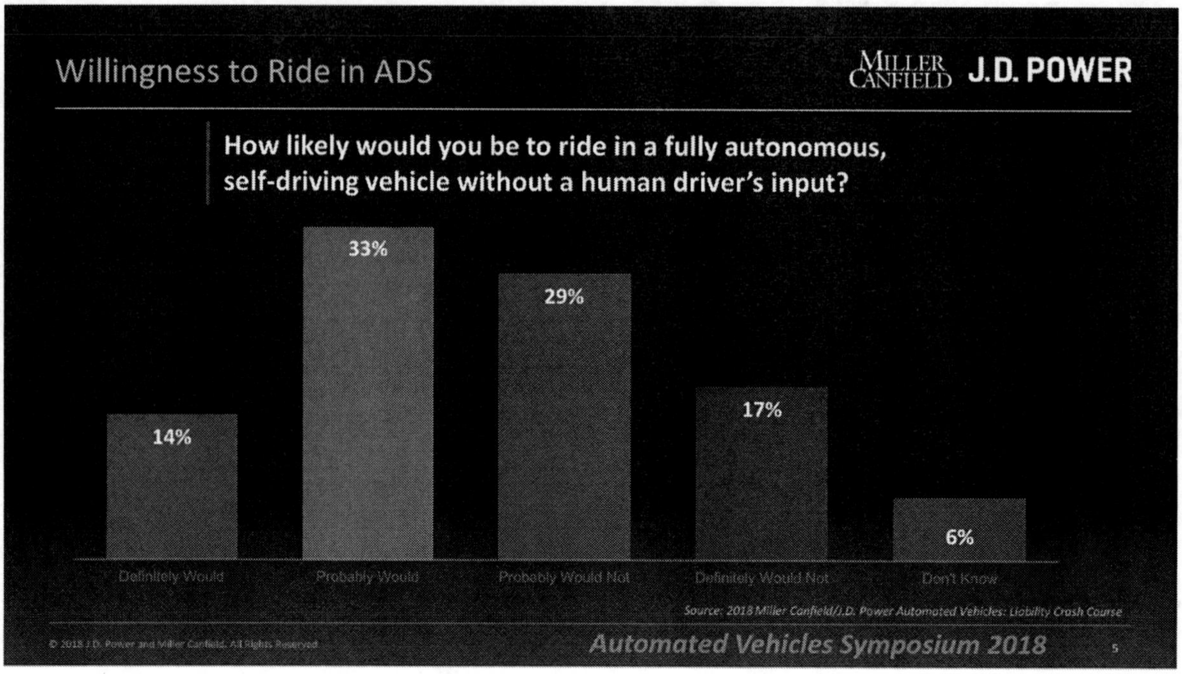

Willingness to Ride in ADS
MILLER CANFIELD J.D. POWER
How likely would you be to ride in a fully autonomous, self-driving vehicle without a human driver's input?
33%
29%
17%
14%
6%
Definitely Would
Probably Would
Probably Would Not
Definitely Would Not
Don't Know
Source: 2018 Miller Canfield/J.D. Power Automated Vehicles: Liability Crash Course
© 2018 J.D. Power and Miller Canfield. All Rights Reserved.
Automated Vehicles Symposium 2018
5

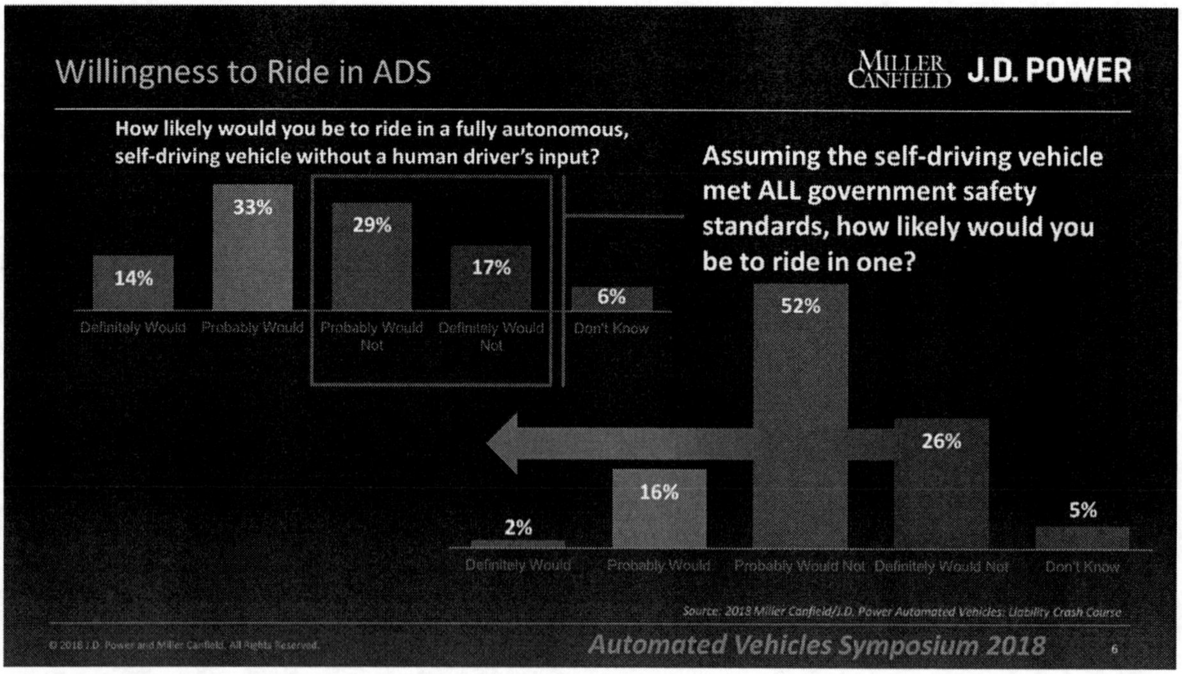

Willingness to Ride in ADS
MILLER CANFIELD J.D. POWER
How likely would you be to ride in a fully autonomous, self-driving vehicle without a human driver's input?
14%
33%
29%
17%
6%
Definitely Would
Probably Would
Probably Would Not
Definitely Would Not
Don't Know
Assuming the self-driving vehicle met ALL government safety standards, how likely would you be to ride in one?
52%
26%
16%
2%
5%
Definitely Would
Probably Would
Probably Would Not
Definitely Would Not
Don't Know
Source: 2018 Miller Canfield/J.D. Power Automated Vehicles: Liability Crash Course
© 2018 J.D. Power and Miller Canfield. All Rights Reserved.
Automated Vehicles Symposium 2018
6

MILLER CANFIELD J.D. POWER

Our **Communication** *Influences* Consumer **Perception, Acceptance and Trust** of ADS

http:// gigazine.net/gsc_news/en/20171109-self-driving-shuttle-bus-crash

Self-Driving Consideration Factors Change Over Time

MILLER CANFIELD J.D. POWER

Largest Concern of Self-Driving Vehicle

January 2017

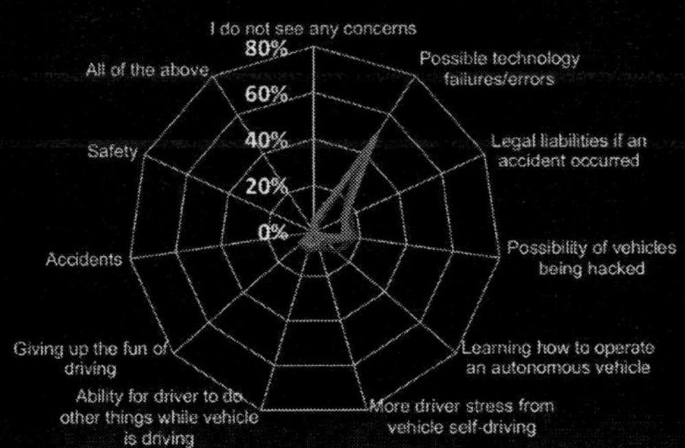

September 2017

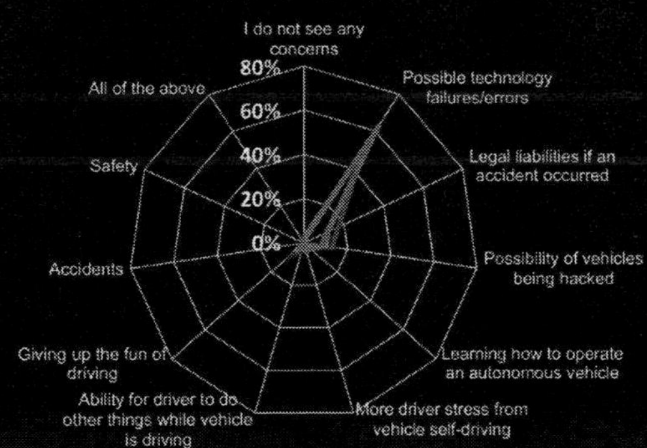

Gen Z Gen Y Gen X Baby Boomers Pre-Boomers

J.D. Power defines the generations as Pre-Boomers (born before 1946); Baby Boomers (1946-1964); Gen X (1965-1976); Gen Y (1977-1994); Gen Z (1995-2004).

Sources: 2017 J.D. Power U.S. Tech Choice Study and 2018 Miller Canfield/J.D. Power Automated Vehicles: Liability Crash Course

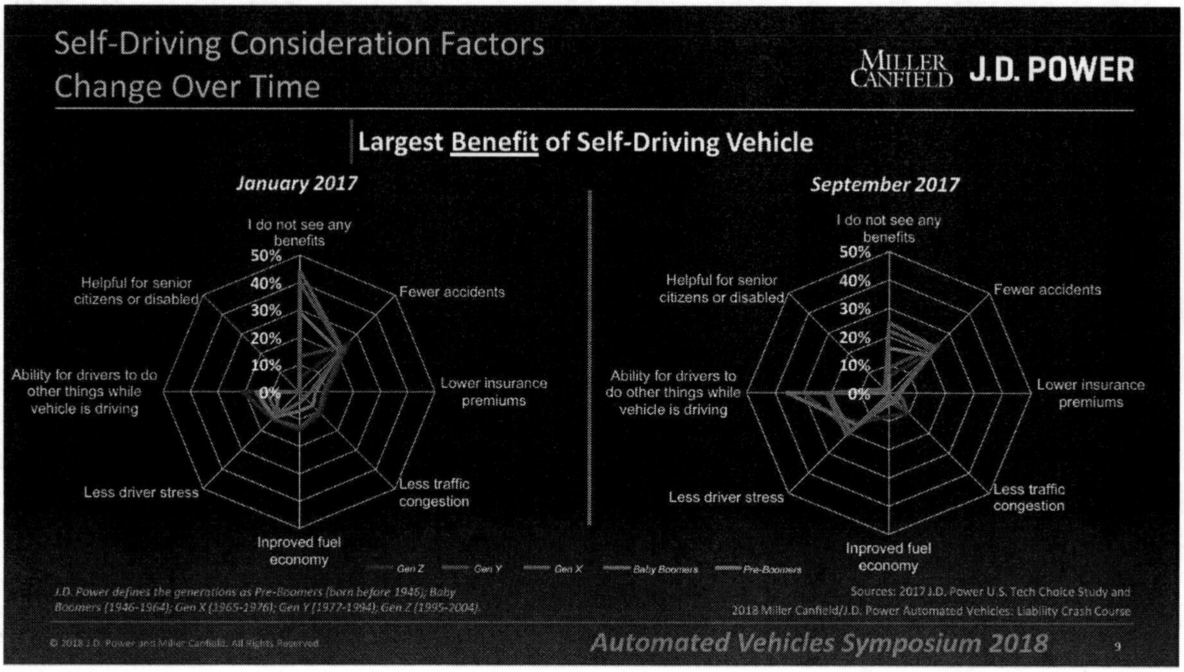
Self-Driving Consideration Factors Change Over Time
MILLER CANFIELD
J.D. POWER
Largest Benefit of Self-Driving Vehicle
January 2017
September 2017
I do not see any benefits
Fewer accidents
Lower insurance premiums
Less traffic congestion
Inproved fuel economy
Less driver stress
Ability for drivers to do other things while vehicle is driving
Helpful for senior citizens or disabled
50%
40%
30%
20%
10%
0%
Gen Z
Gen Y
Gen X
Baby Boomers
Pre-Boomers
J.D. Power defines the generations as Pre-Boomers (born before 1946); Baby Boomers (1946-1964); Gen X (1965-1976); Gen Y (1977-1994); Gen Z (1995-2004).
Sources: 2017 J.D. Power U.S. Tech Choice Study and 2018 Miller Canfield/J.D. Power Automated Vehicles: Liability Crash Course
© 2018 J.D. Power and Miller Canfield. All Rights Reserved.
Automated Vehicles Symposium 2018
9

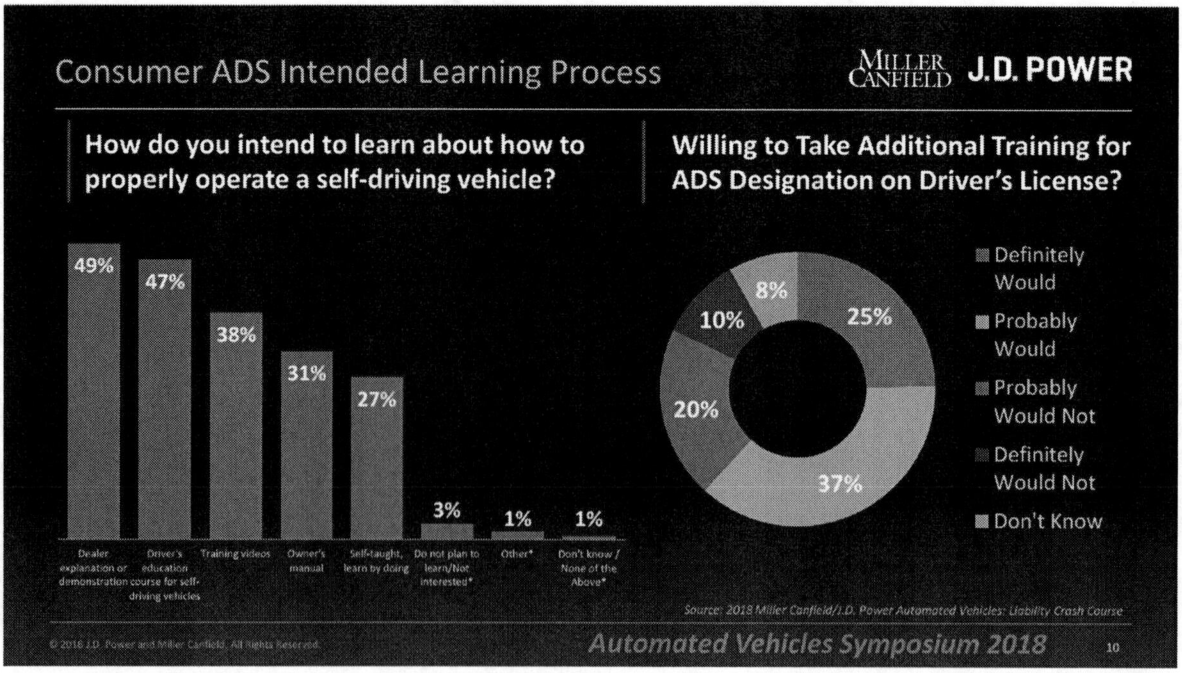
Consumer ADS Intended Learning Process
MILLER CANFIELD
J.D. POWER
How do you intend to learn about how to properly operate a self-driving vehicle?
49%
47%
38%
31%
27%
3%
1%
1%
Dealer explanation or demonstration
Driver's education course for self-driving vehicles
Training videos
Owner's manual
Self-taught, learn by doing
Do not plan to learn/Not interested*
Other*
Don't know / None of the Above*
Willing to Take Additional Training for ADS Designation on Driver's License?
25%
37%
20%
10%
8%
Definitely Would
Probably Would
Probably Would Not
Definitely Would Not
Don't Know
Source: 2018 Miller Canfield/J.D. Power Automated Vehicles: Liability Crash Course
© 2018 J.D. Power and Miller Canfield. All Rights Reserved.
Automated Vehicles Symposium 2018
10

ADS Dispute Resolution Patterns
Voice of the Consumer

MILLER CANFIELD J.D. POWER

A crash occurring at a **higher automation level** increased respondents':

- Willingness to litigate
- Desire to seek dispute resolution options with a longer duration for resolution
- Expectation of no crashes, especially for Level 5
- Clarity of fault (i.e., consumer perception: fully automated self-driving vehicle (Level 5) inherently means the vehicle is at fault)
- Emotional state
- Desire to bring public awareness

Source: 2018 Miller Canfield/J.D. Power Automated Vehicles: Liability Crash Course

Many Consumers Expect No ADS crashes

MILLER CANFIELD J.D. POWER

- Perception that crashes "should not occur" with ADS
- Consumers hold ADS to a higher safety standard
- Emotions increase as the level of automation increase mainly due to the driver becoming a "passenger"

*Critical to align **consumer expectations** to the **product's capability***

Level 5, Full Automation Verbatim

> *I'd **expect the car to be safer** and tested and would blame the company for any accident."*

> ***Should never happen**."*

> *B/c of **car is supposed to be good** shouldn't be in an accident."*

> *Because it is **supposed to be a safe vehicle** & there was a serious injury need to know why it happened & to prevent it from happening again."*

> *I would be **MAD and HORRIFIED** and would want to make a spectacle of the case."*

Source: 2018 Miller Canfield/J.D. Power Automated Vehicles: Liability Crash Course

MILLER CANFIELD J.D. POWER

Consumers are willing to share their ADS vehicle data after a crash.

The most common motivation: ***to help the greater good***

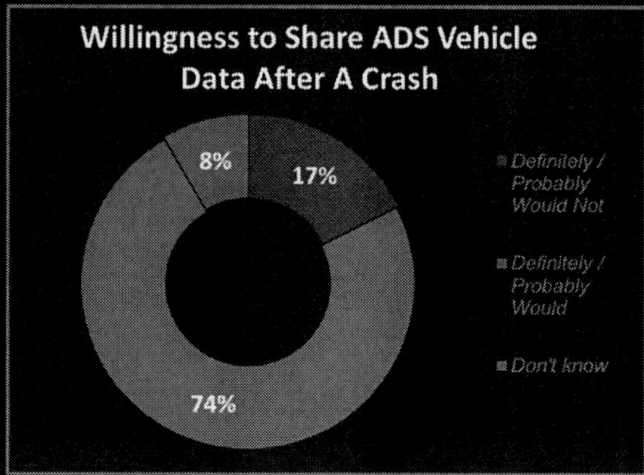

Source: 2018 Miller Canfield/J.D. Power Automated Vehicles: Liability Crash Course

Legal Practitioners' Voice

MILLER CANFIELD J.D. POWER

- Both sides agree:
  - **Costs** to litigate product liability for ADS will increase dramatically
  - ADS provides an opportunity for legal claims to be **resolved out of court through ADR**
  - Availability of **additional crash data** would accelerate resolution and support ADR
- Most focus is on Level 5 with little consideration given to the **complexities of "shared control"** (Level 3)

Opportunity: Create a workable ADR framework to bridge the gap, creating stronger relationships between manufacturers and consumers by means of transparency and equity

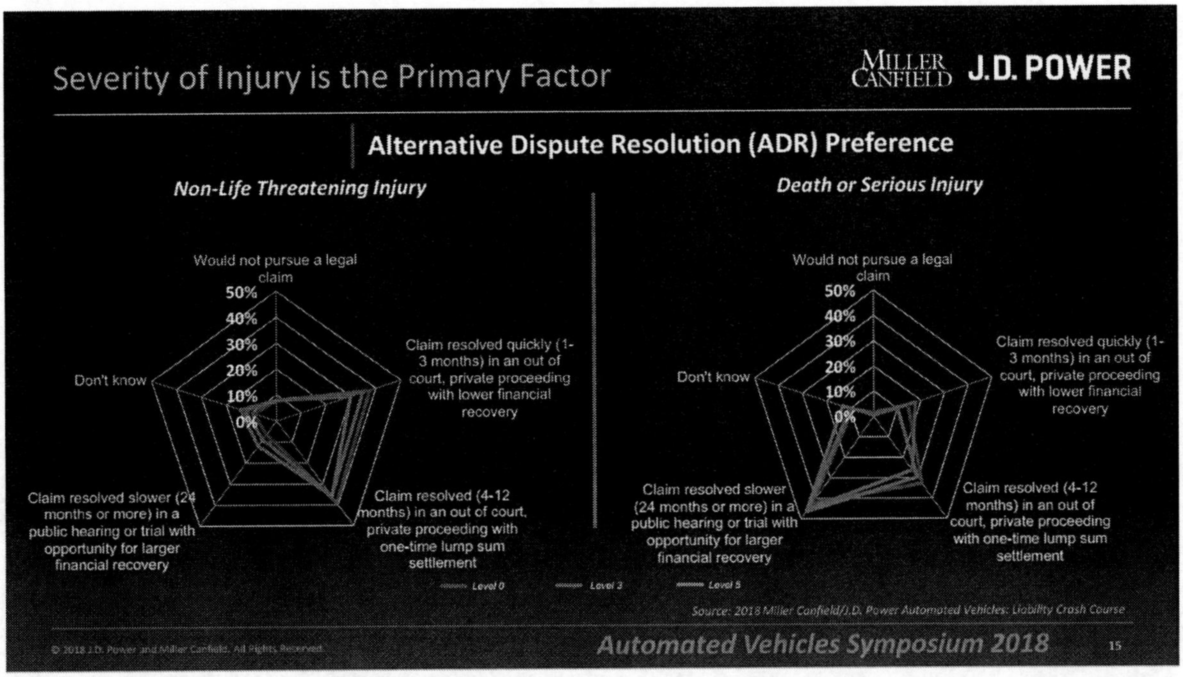
Severity of Injury is the Primary Factor
MILLER CANFIELD J.D. POWER
Alternative Dispute Resolution (ADR) Preference
Non-Life Threatening Injury
Death or Serious Injury
Would not pursue a legal claim
50%
40%
30%
20%
10%
0%
Claim resolved quickly (1-3 months) in an out of court, private proceeding with lower financial recovery
Claim resolved (4-12 months) in an out of court, private proceeding with one-time lump sum settlement
Claim resolved slower (24 months or more) in a public hearing or trial with opportunity for larger financial recovery
Don't know
Level 0
Level 3
Level 5
Source: 2018 Miller Canfield/J.D. Power Automated Vehicles: Liability Crash Course
© 2018 J.D. Power and Miller Canfield. All Rights Reserved.
Automated Vehicles Symposium 2018
15

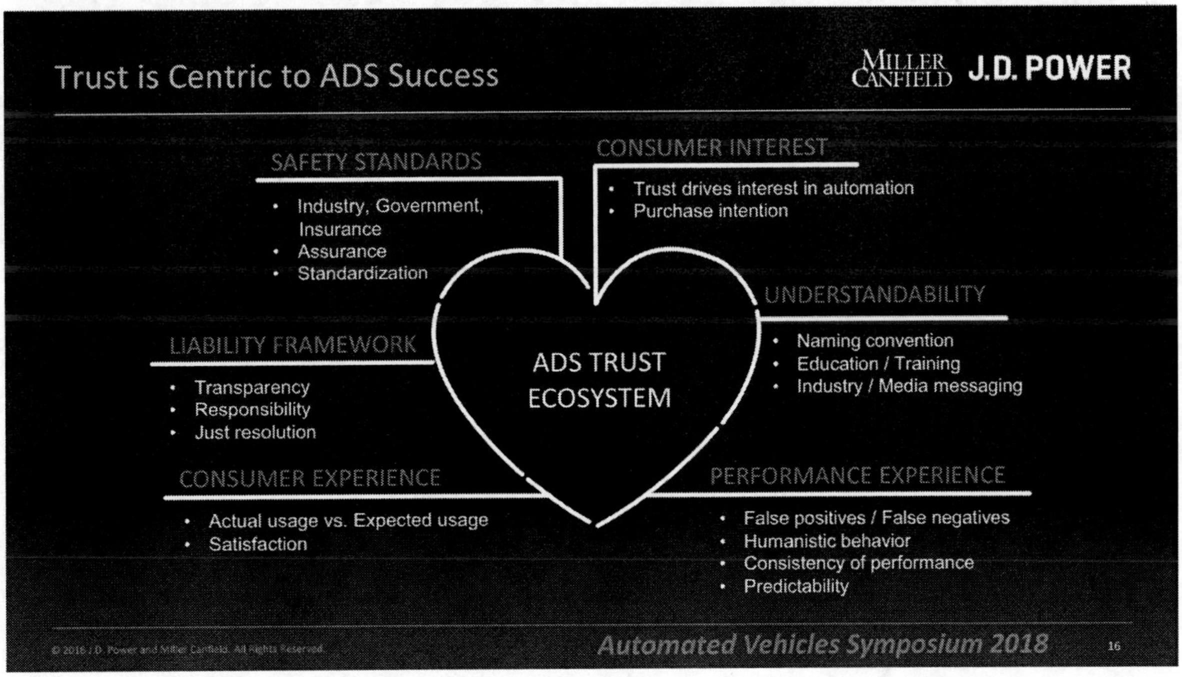
Trust is Centric to ADS Success
MILLER CANFIELD J.D. POWER
SAFETY STANDARDS
• Industry, Government, Insurance
• Assurance
• Standardization
CONSUMER INTEREST
• Trust drives interest in automation
• Purchase intention
UNDERSTANDABILITY
• Naming convention
• Education / Training
• Industry / Media messaging
LIABILITY FRAMEWORK
• Transparency
• Responsibility
• Just resolution
ADS TRUST ECOSYSTEM
CONSUMER EXPERIENCE
• Actual usage vs. Expected usage
• Satisfaction
PERFORMANCE EXPERIENCE
• False positives / False negatives
• Humanistic behavior
• Consistency of performance
• Predictability
© 2018 J.D. Power and Miller Canfield. All Rights Reserved.
Automated Vehicles Symposium 2018
16

MILLER CANFIELD J.D. POWER

Thank You

Tina Georgieva | *georgieva@millercanfield.com*
Kristin Kolodge | *kristin.kolodge@jdpa.com*

 Automated Vehicles Symposium 2018 17

National Transportation Safety Board

Pathway to Automation: Transition from CAS

Ensar Becic, Ph.D.
Project Manager
Office of Highway Safety

Overview

- The role of the NTSB
- Investigate ... make recommendations … follow-up on the implementation
- Path to Autonomous Vehicles

Perspective → Crash Investigation
- Level 2 AV (Williston, FL)

?

Perspective → Public
- Testing of AV (Tempe, AZ)

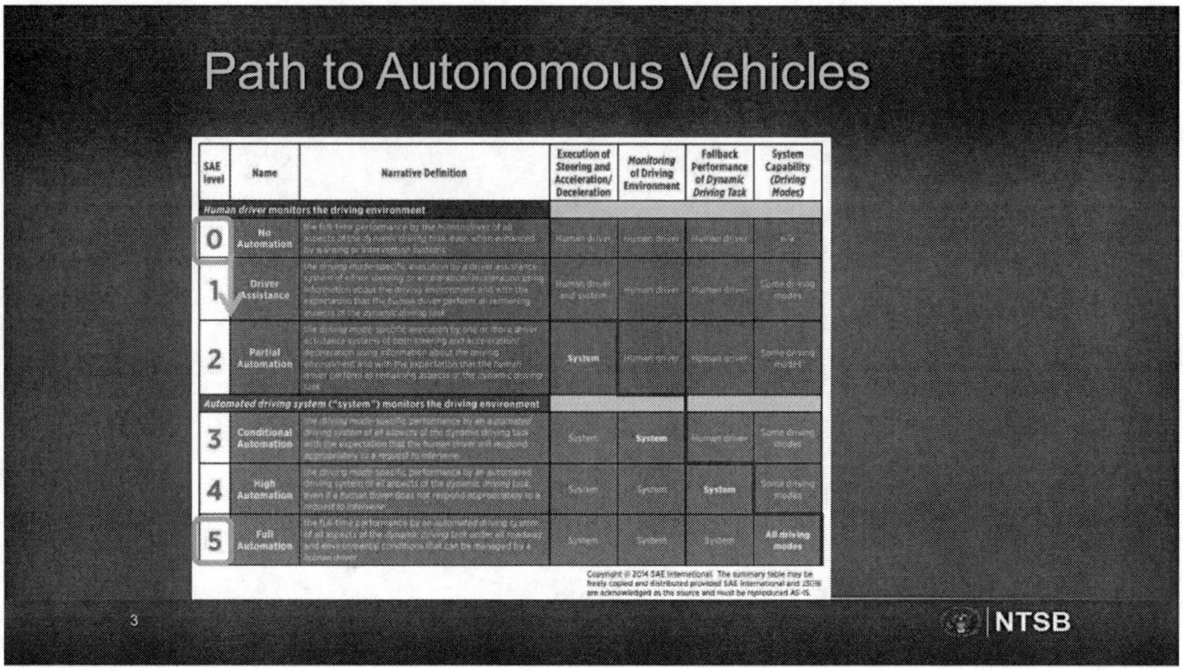

Path to Autonomous Vehicles
SAE level
Name
Narrative Definition
Execution of Steering and Acceleration/ Deceleration
Monitoring of Driving Environment
Fallback Performance of Dynamic Driving Task
System Capability (Driving Modes)
Human driver monitors the driving environment
0
No Automation
1
Driver Assistance
2
Partial Automation
System
Automated driving system ("system") monitors the driving environment
3
Conditional Automation
System
4
High Automation
System
5
Full Automation
All driving modes
Copyright © 2014 SAE International. The summary table may be freely copied and distributed provided SAE International and J3016 are acknowledged as the source and must be reproduced AS-IS.
3
NTSB

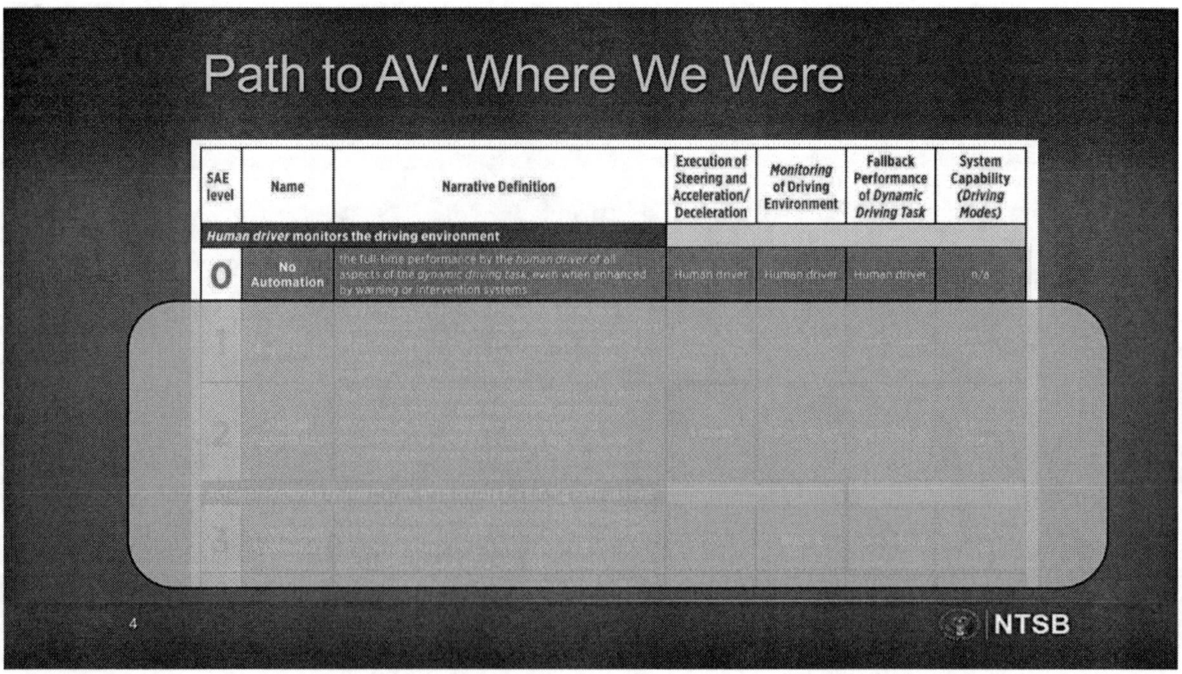

Path to AV: Where We Were
SAE level
Name
Narrative Definition
Execution of Steering and Acceleration/ Deceleration
Monitoring of Driving Environment
Fallback Performance of Dynamic Driving Task
System Capability (Driving Modes)
Human driver monitors the driving environment
0
No Automation
the full-time performance by the human driver of all aspects of the dynamic driving task, even when enhanced by warning or intervention systems
Human driver
Human driver
Human driver
n/a
4
NTSB

Collision Avoidance Systems (CAS)

- **1995** - Examining the effectiveness of collision warning (CW) in heavy vehicles
- **2001** - Development of performance standards for CW and adaptive cruise control (ACC) in passenger and heavy vehicles
- **2008** - Examining the effectiveness of automatic emergency braking (AEB) in heavy vehicles
- **2015** - Installing CW and AEB as standard equipment in all vehicles; expanding NCAP to rate CAS

5 NTSB

Path Transitioning from CAS to Automation

- CAS *assist* a driver in the performance of a driving task
  - Safety benefits by overcoming environmental factors
  - Safety benefits by reducing deficits in driver performance
- Level 2 automated vehicle systems
  - Possible additional safety benefits, but also
  - Potential cost due to reliance on driver monitoring
  - Retaining the safety benefits of CAS

6 NTSB

Path to AV: Public View

Autonomous features we have today

Automatic Emergency Braking

Adaptive Cruise Control

Auto Steering

…

7 NTSB

Path to AV: Public View

What is arriving tomorrow

US DOT image

8 NTSB

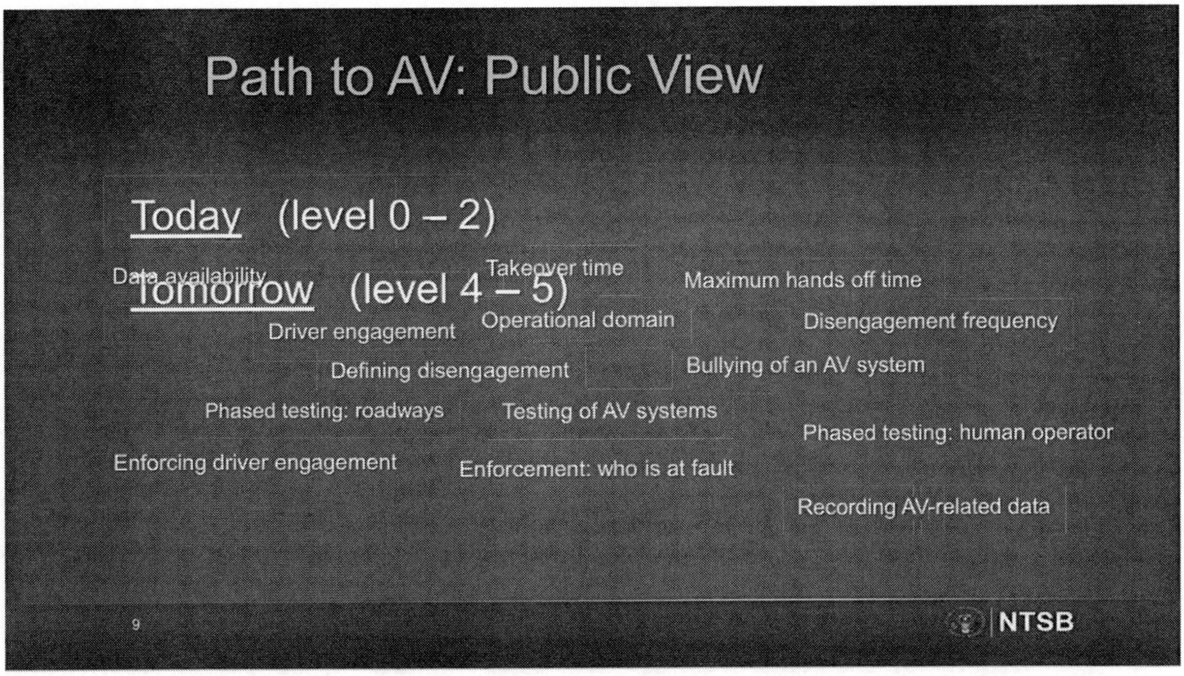
Path to AV: Public View
Today (level 0 – 2)
Tomorrow (level 4 – 5)
Data availability
Takeover time
Maximum hands off time
Driver engagement
Operational domain
Disengagement frequency
Defining disengagement
Bullying of an AV system
Phased testing: roadways
Testing of AV systems
Phased testing: human operator
Enforcing driver engagement
Enforcement: who is at fault
Recording AV-related data
9
NTSB

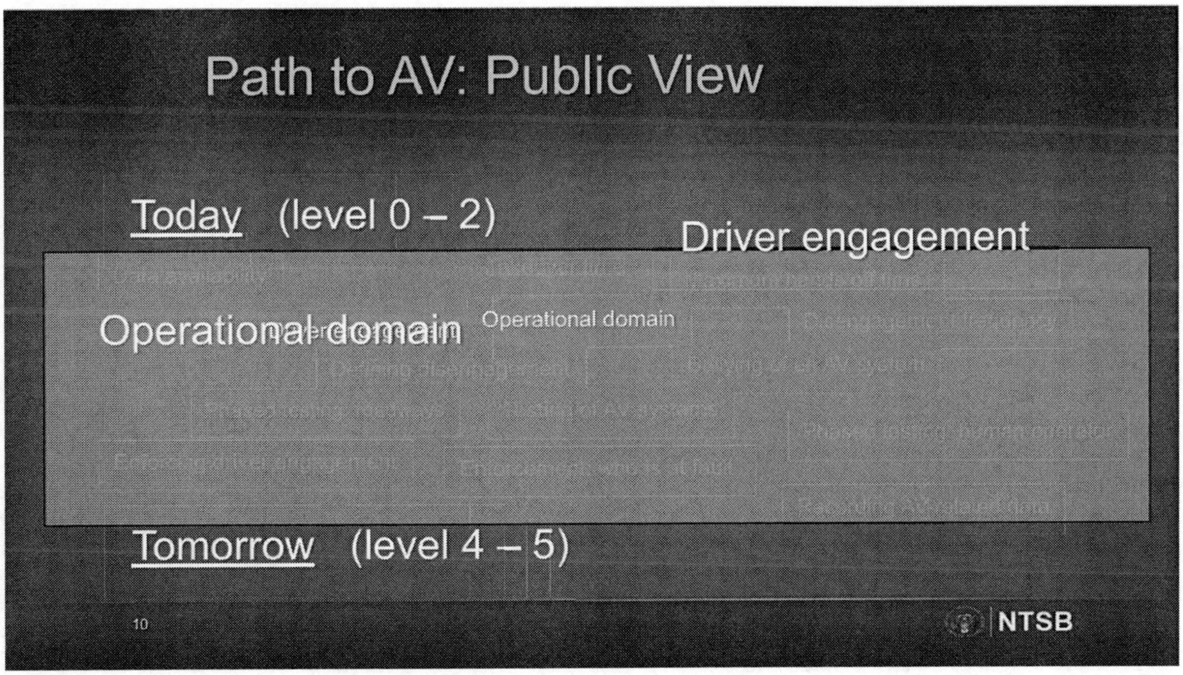
Path to AV: Public View
Today (level 0 – 2)
Driver engagement
Operational domain
Operational domain
Tomorrow (level 4 – 5)
10
NTSB

Williston, FL – Crash Overview

- May 7, 2016 ~4:36 p.m.
- 2014 Truck-tractor combination vehicle
- 2015 Tesla, Model S
  - Level 2 automation
- Daytime, dry roadway, no sun glare

11 NTSB

Operational Domain: Roadway

US-27A
- 4-lane roadway
- With central median divider
- Not limited access
- Cruise speed limited to 90 mph

12 NTSB

Operational Domain: Roadway

SR-24
- 2-lane roadway
- No central median divider
- Not limited access
- Cruise speed limited to 5 mph over the speed limit

13 NTSB

Driver Engagement

- Tesla monitored driver engagement through driver-applied changes to steering wheel torque
- Timing of the alerts after hands-free operation in Autopilot mode
  - Initial visual alert after 3-5 minutes when traveling above 45 mph
  - Additional 2 auditory alerts and a final auditory alert with a slowdown
  - No alert when traveling less than 45 mph

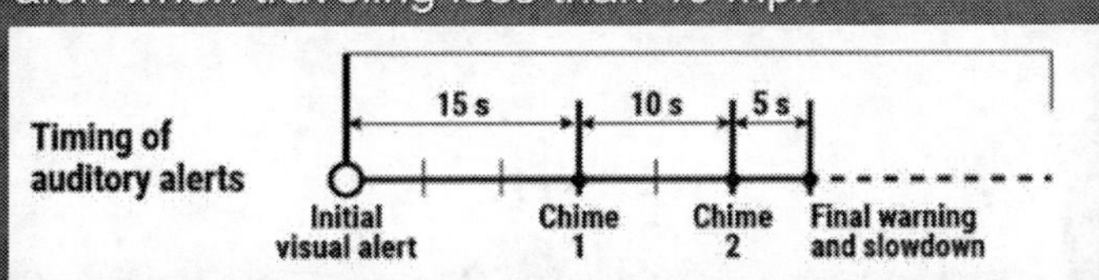

14 NTSB

Driver Engagement During the Crash Trip

- Hands on the steering wheel for 25 seconds
- Autopilot was engaged for 37 minutes

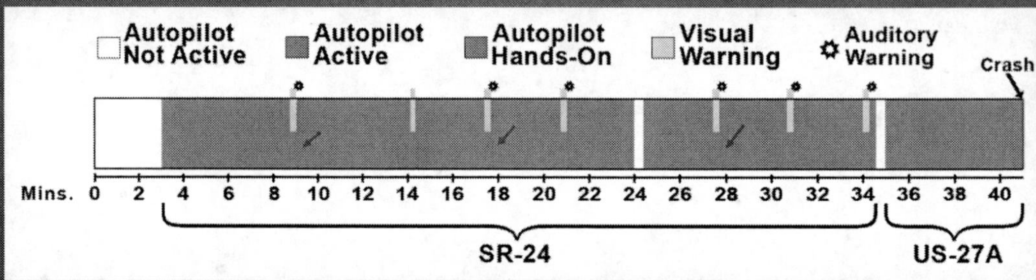

15 NTSB

Conclusions and Recommendations

Autopilot operated within its limited functional capabilities, but outside its operational domain

- Driver-based adherence or system-based implementation
  - Recommendation: System-based restriction to its operational domain

Driver's lack of responsiveness indicated overreliance on automation

- Steering wheel torque a poor surrogate measure
  - Recommendation: Improved means of monitoring driver engagement

Other recommendations: AV data parameters and availability

16 NTSB

Path to AV: Safety Issues

Today (level 0 – 2)

Testing of AV systems

Availability of AV data

Availability of AV data

Testing of AV systems

Future (level 4 – 5)

17 NTSB

Tempe, AZ – Crash Overview

- March 18, 2018 ~10 p.m.
- Uber test vehicle
  - Operated in autonomous mode
- Pedestrian walking a bicycle
- Nighttime conditions
- Roadside lighting present

18 NTSB

Uber Test Vehicle

- Built on Volvo XC90
  - Includes CAS with CWS and AEB
- Uber's autonomous system
  - LiDAR, radars, cameras
  - Disables Volvo's CAS

19 NTSB

Crash Detection Sequence

- A hazard (pedestrian) detected 6 sec TTC
- The hazard changed to an unknown object, a vehicle and finally a bicycle
- Image at 1.3 s TTC
  - Speed of 43 mph
- Driver steered < 1 s

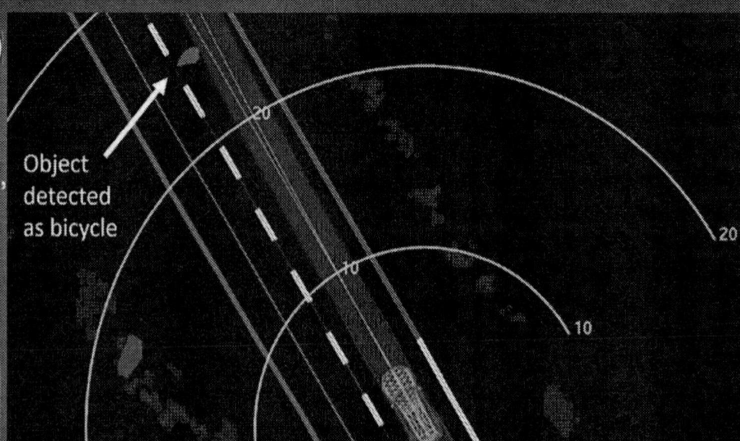

20 NTSB

System and Driver Tasks

- AV system task
  - Follow the path; speed and lane position determined by the system
  - Does NOT engage emergency braking
- Driver task
  - Take over in case of emergency
  - Report events of interest (e.g., system disengagement)

21 NTSB

All Issues Are Being Examined

- Highway design, pedestrian safety, …
- Testing of AV systems (not investigation-specific)
  - NHTSA and State requirements for public roads
  - Requirements dependent on an AV level
  - Phased complexity
  - Driver monitoring
    - Necessary or optional
    - In-vehicle or remote

22 NTSB

Conclusion and Near Future

- The safety promise of AV systems
  - Retaining the safety benefits of CAS while progressing toward a fully autonomous vehicle
- Modulating public expectations
- Determining parameters or requirements for testing

23

NTSB

National Transportation Safety Board

Presentation outline

- What is **Zenuity**?
- What is most difficult in designing self-driving vehicles?
- How to prove a self-driving vehicle is sufficiently safe?
- Alternative approaches
- Conclusion

ZENUITY PAGE 2

Zenuity's founding fathers

19th April 2017

Volvo Cars, the premium car maker, and Autoliv, the worldwide leader in automotive safety systems, have signed a final agreement to establish a new joint venture called Zenuity to develop software for autonomous driving and driver assistance systems

ZENUITY PAGE 3

Zenuity's technology scope

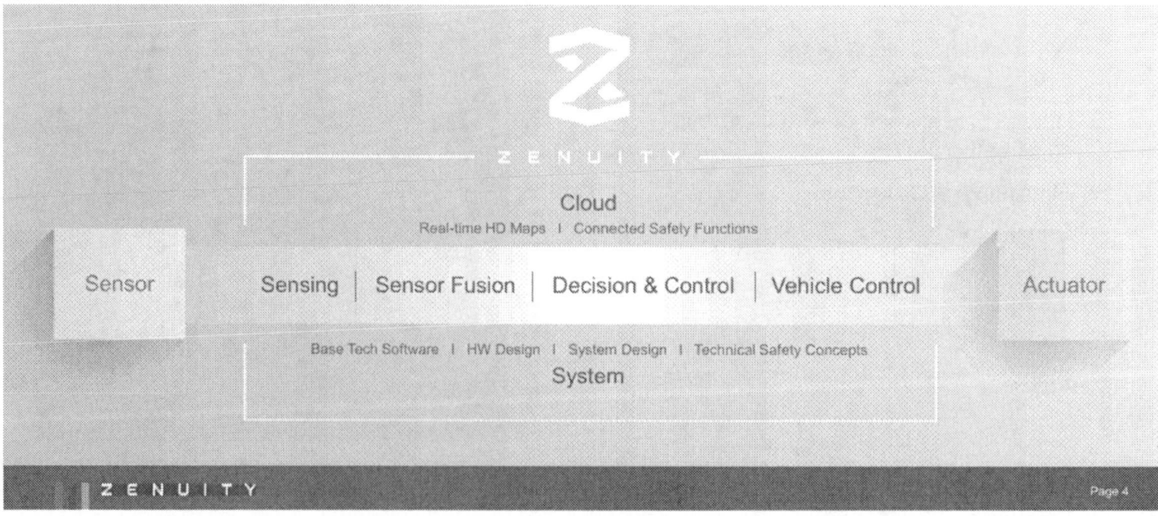

ZENUITY Page 4

Zenuity today

Started April 18th 2017 with 200 employees

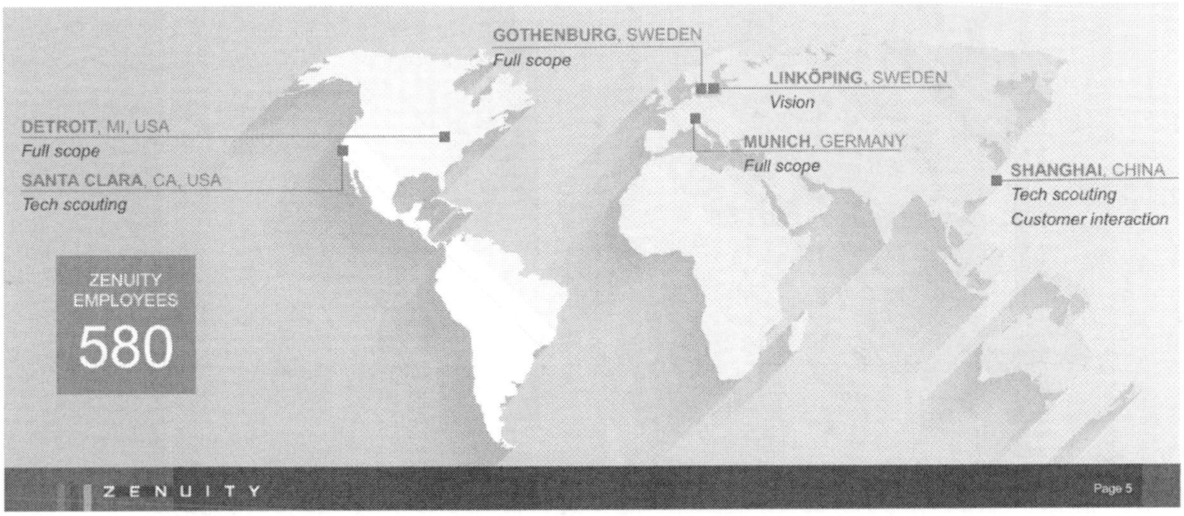

Zenuity's way to market

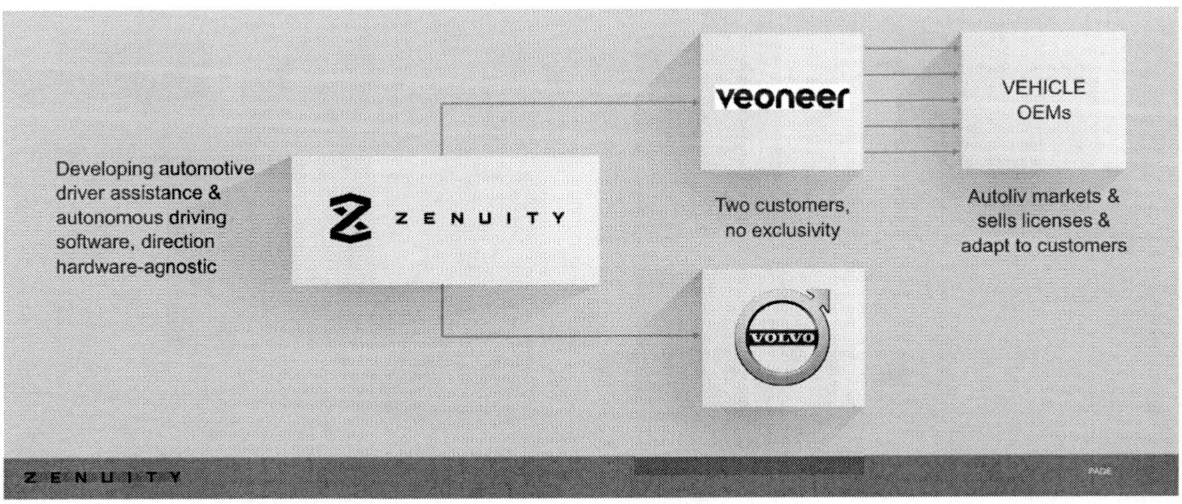

Our Product Roadmap

Combining driver support (ADAS) and autonomous driving

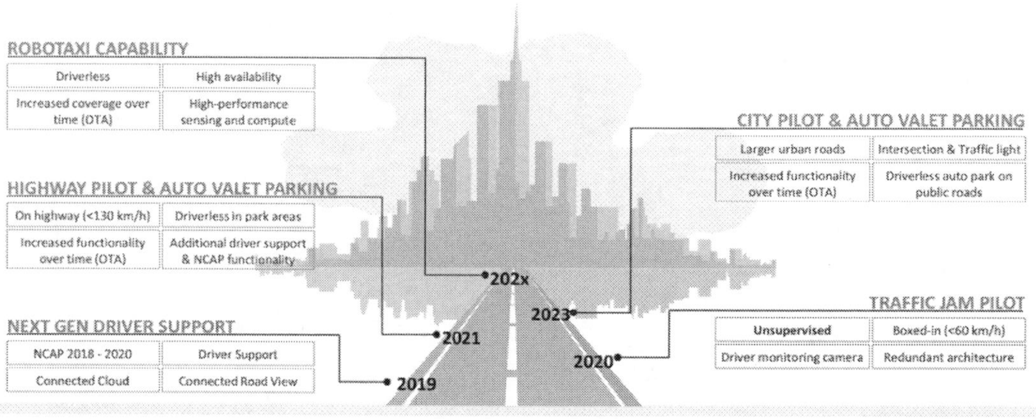

ZENUITY PAGE 7

From Driver Support to Autonomous Driving

Fundamental change for safety concepts

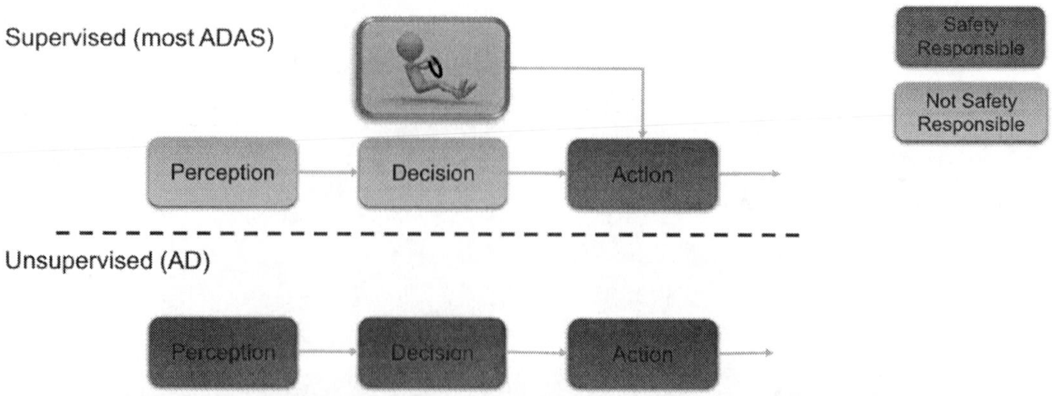

ZENUITY PAGE 8

The Challenge

Driver out of the loop

↓

Self-driving vehicles must be able to handle *all* foresee-able situations of the Operational Design Domain

This puts unique requirements on the vehicle, its sensor, actuators and electrical architecture.

Unsupervised driving

Safety Case:

"Structured **argument**, supported by **evidence**, intended to justify that the ***AD functionality*** is acceptably safe for **all relevant traffic situations** and **all relevant environmental conditions**."

Overall safety requirement: Fewer ***caused*** accidents (by some margin) than human driver

Topic	1/frequency	hours	Market
Road fatalities	150 million km	4 x 10E6 h	U.S.
Rail fatalities	2.5 billion passenger km	4 x 10E7 h	U.K.
Air fatalities	50 billion passenger km	1 x 10E8 h	U.K.

False AEB	0.5 million km	1 x 10E4 h	Global
Safety Driver interventions (High Score 2018)	20 thousand km	7 x 10E2 h	CA

How to design and prove self-driving to be sufficiently safe?

- **Define** the target. The product should be sufficiently safe when used by real customers in the real world.
- **Divide and conquer**, establish testable requirements for components
  - Use sufficient detail; False Negative performance depends on range, illumination, precipitation etc.
  - Reduce Operational Design Domain if needed
  - **Analyzable** vs non-analyzable components
- Use a combination of field testing, simulation, and selected scenarios at test track for verification
- Show that there is enough redundance and independence to reach the overall requirement (worst case scenarios occur less than 10E-8 /h?)
- Done!

Deep Neural Networks

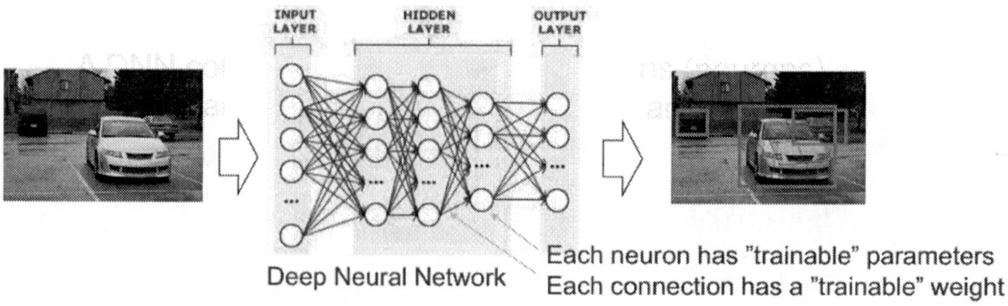

Deep Learning

On the other hand...

- Lidar detections are based on well-known physics and relatively simple algorithms
- Performance can be predicted also outside of the tested envelope (to some extent)
- When exposed to real life objects different to those used in verification tests, it is likely to detect them as well.
- Similar reasoning can apply to radar and ultrasonic

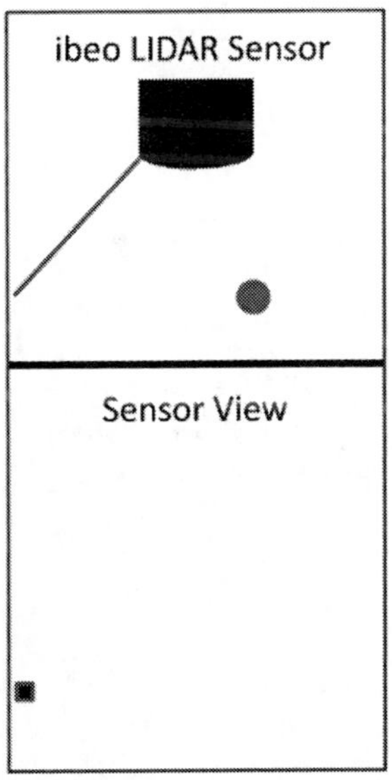

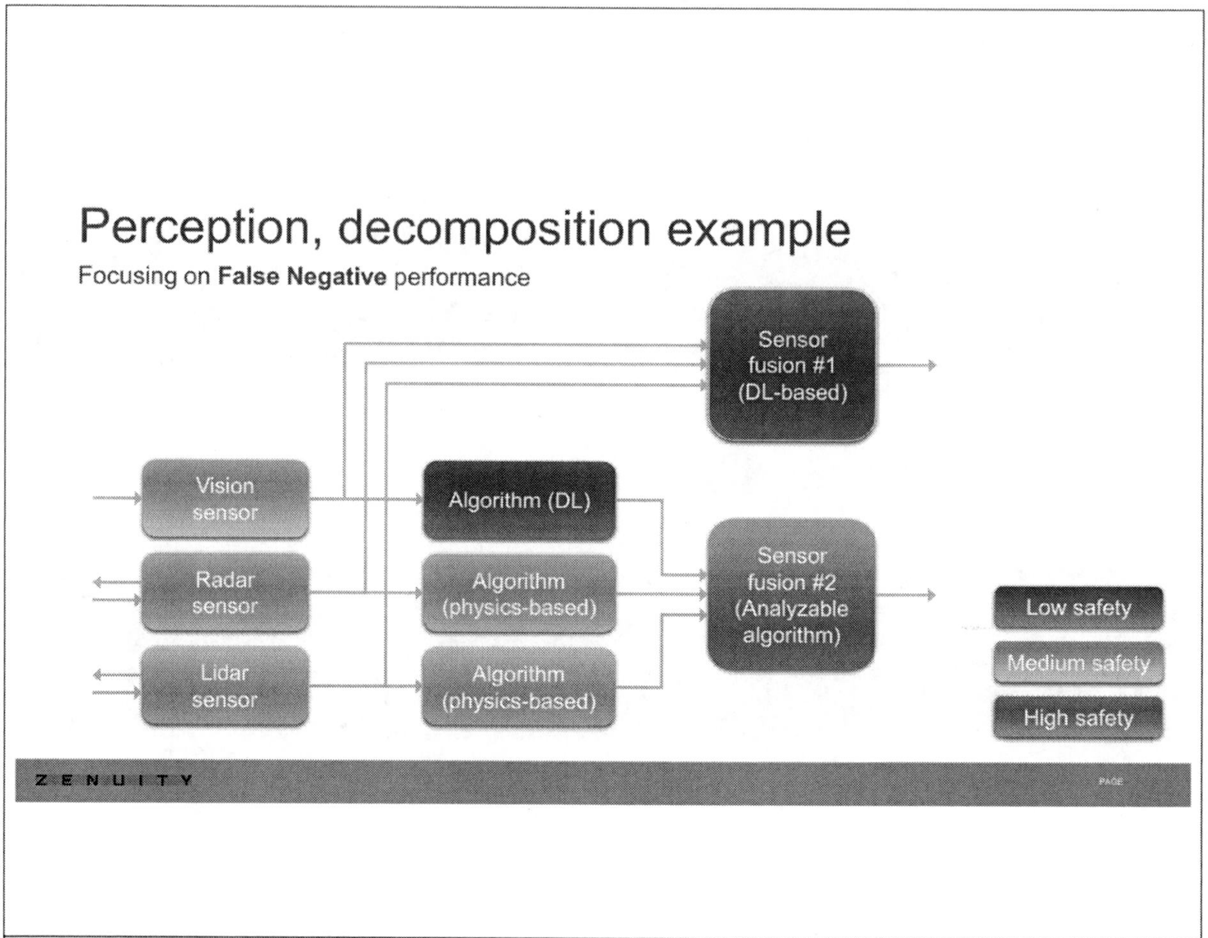
Perception, decomposition example
Focusing on False Negative performance
Sensor fusion #1 (DL-based)
Vision sensor
Radar sensor
Lidar sensor
Algorithm (DL)
Algorithm (physics-based)
Algorithm (physics-based)
Sensor fusion #2 (Analyzable algorithm)
Low safety
Medium safety
High safety
ZENUITY
PAGE

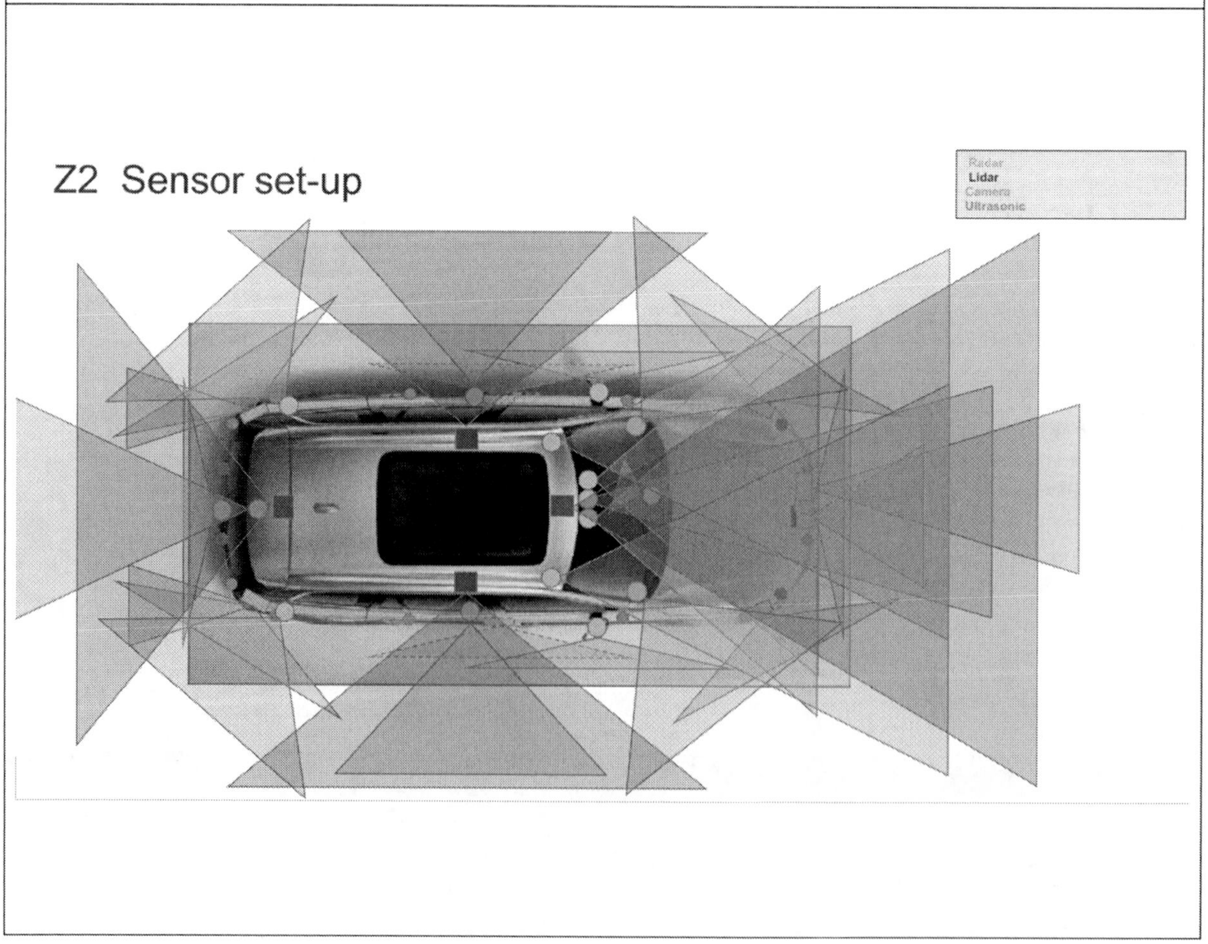
Z2 Sensor set-up
Radar
Lidar
Camera
Ultrasonic

Verification & Validation strategy

Safety verification of ***All major subsystems for All relevant scenarios***

Scenario DB

Perception → Objects → Decision-Making → Planned Path → Vehicle Control

Traffic/Test track with Ground Truth

Resimulation
Augmented/Virtual Data

Traffic/Test track with Ground Truth

Formal methods
Closed-loop simulation

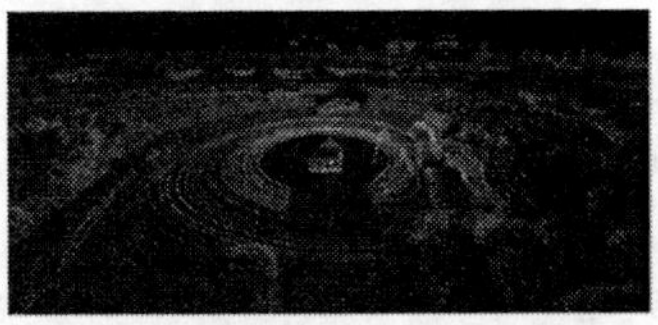

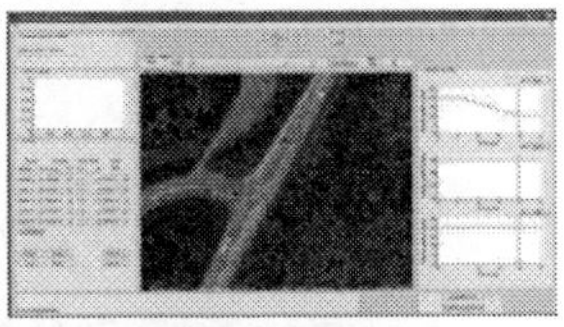

Z E N U I T Y

PAGE 17

Building complete customer features

Highway Pilot

Perception

Decision-Making

Vehicle Control

Z E N U I T Y

PAGE 18

Conclusions

- Zenuity develops SW for ADAS and unsupervised AD, approaching AD with its safety heritage from Volvo Cars and Autoliv
- Unsupervised Automated Driving is a huge challenge. Super-human performance needed.
- Safety verification by brute-force driving is not feasible
- Deep learning can be utilized safely by
  - parallel, independent channels with reduced functional safety requirements
  - verification by dedicated testing and various forms of simulation of modules

ZENUITY PAGE 19

ZENUITY

Make it real.

We are hiring!
Please visit: www.zenuity.com/career/

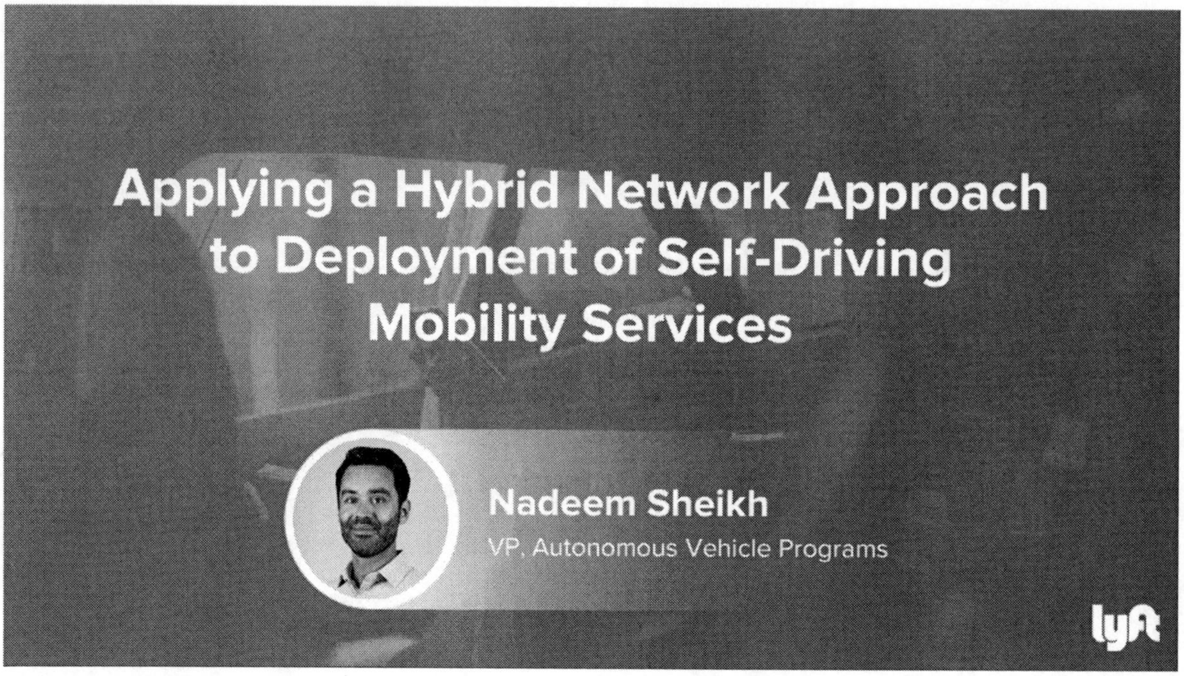
Applying a Hybrid Network Approach to Deployment of Self-Driving Mobility Services
Nadeem Sheikh
VP, Autonomous Vehicle Programs
lyft

1908
Henry Ford begins Model-T production

15 millionth Model-T produced

Price comparability by 1927:

Same price ($250) for 6x speed and range

CHALLENGES FACING DEPLOYMENT

BUILDING A HYBRID NETWORK

Consumers are excited about AV

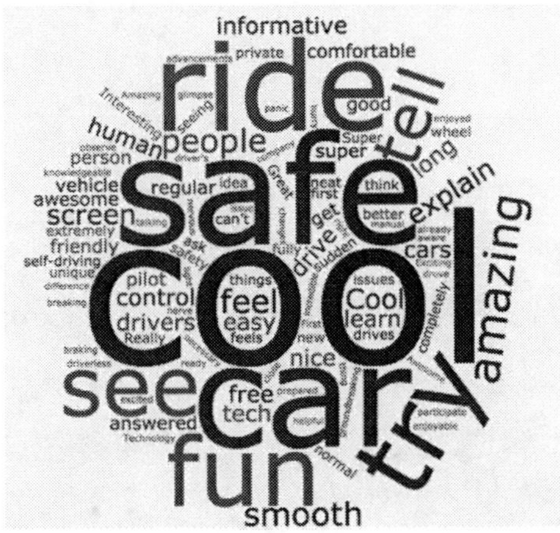

But operational challenges are significant

1. Expensive to build
2. Can't go a lot of places
3. Can't drive fast
4. Dead miles kill economics

Consumer Expectations
3-5 minute ETAs in most of the US
Low price
Interest in self-driving rides as novelty but most interested in getting from A to B

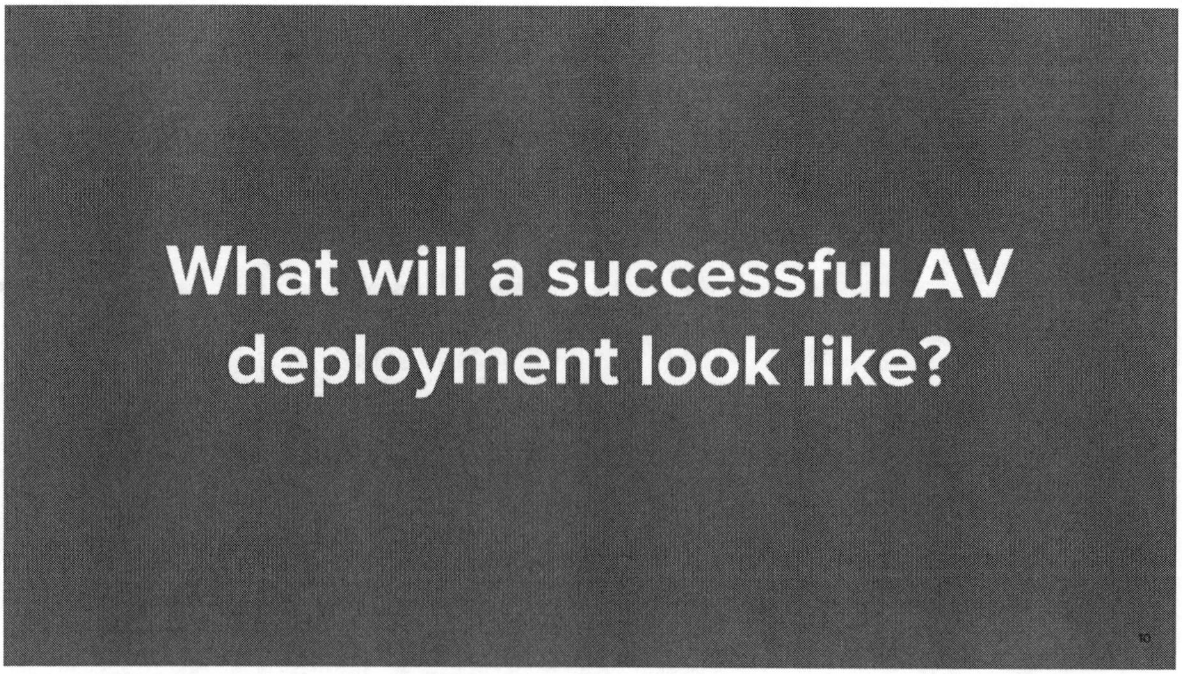
What will a successful AV deployment look like?
10

A Hybrid Network...

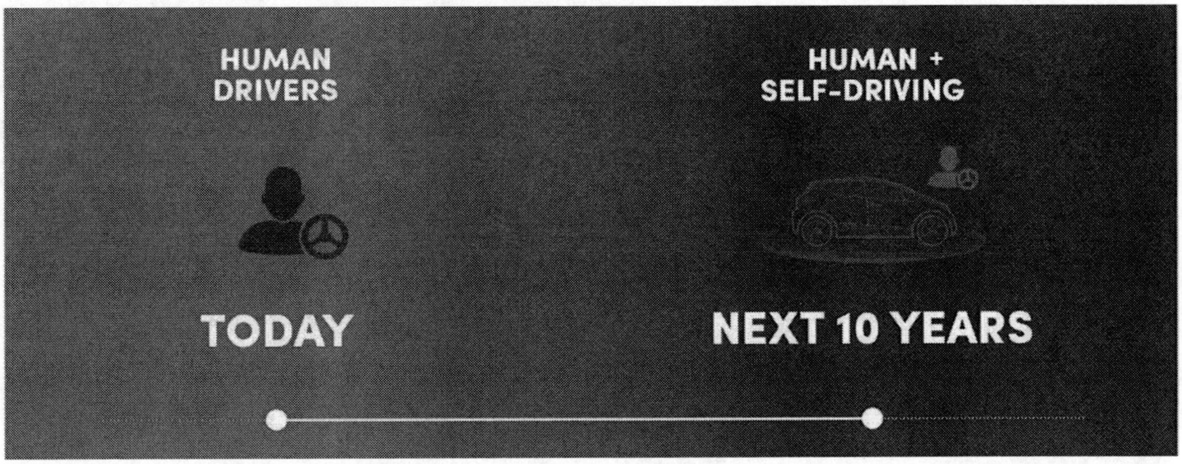

11

...with incremental deployment

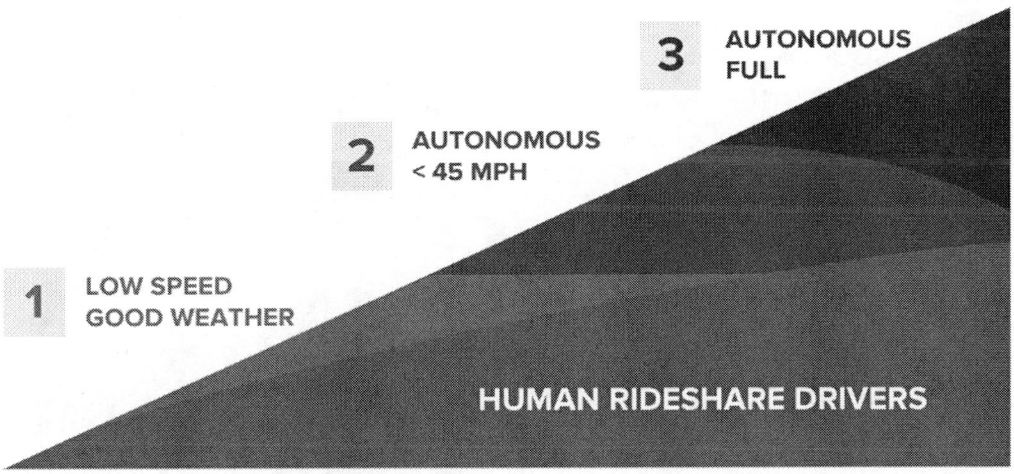

12

Keys to effective deployment

1. Know WHERE to drive
2. Keep COSTS LOW
3. FILL UP the cars

13

1 Where to drive: know the best routes

Lower speeds, limit complex situations & traffic restrictions

- Speeds <= 35 MPH
- Avoid difficult intersections
- No bike lanes
- Well-marked roads
- Easy pick-up and drop-off

14

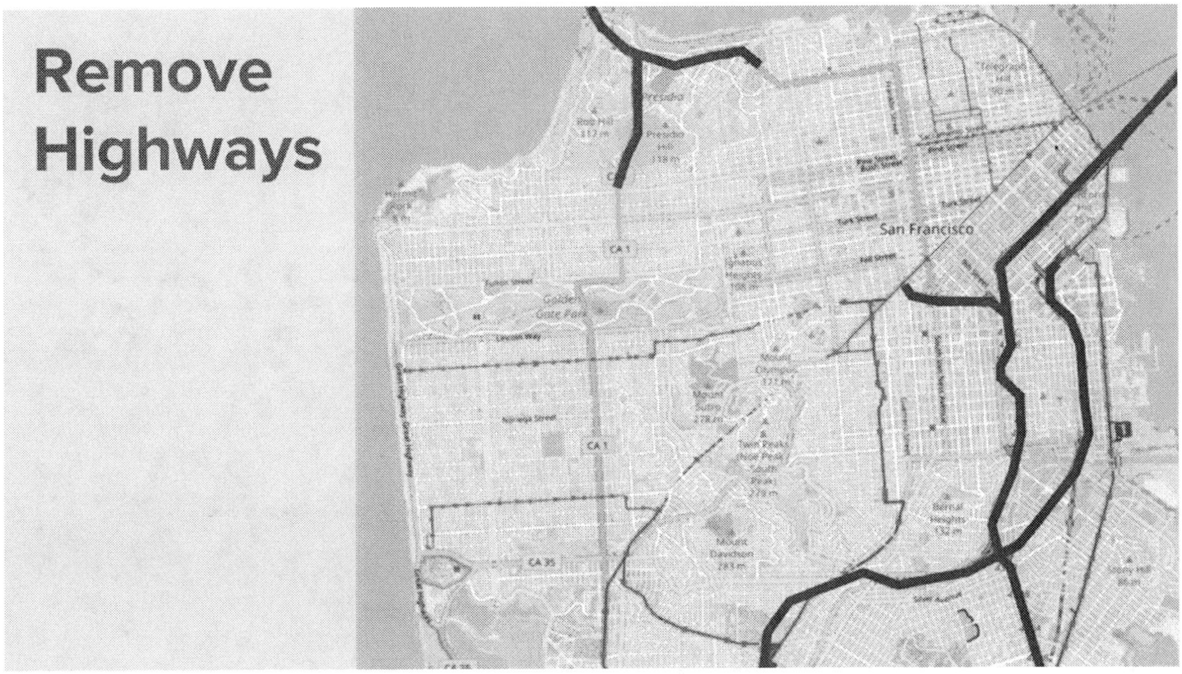
Remove
Highways
San Francisco

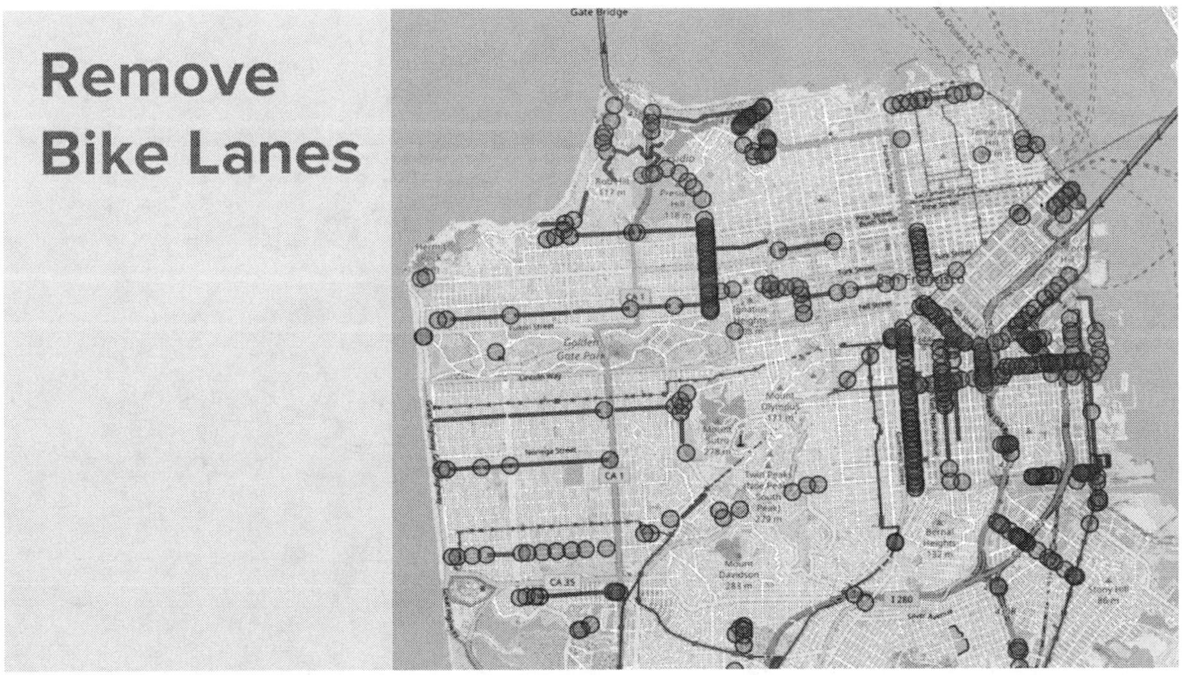
Remove
Bike Lanes

Remove Difficult Intersections

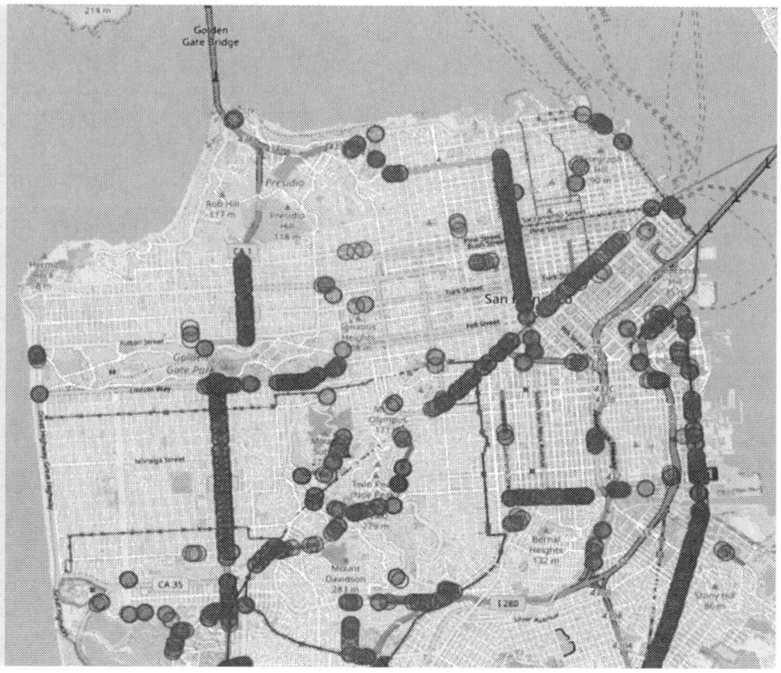

Remaining Routes:
"AV Eligible"

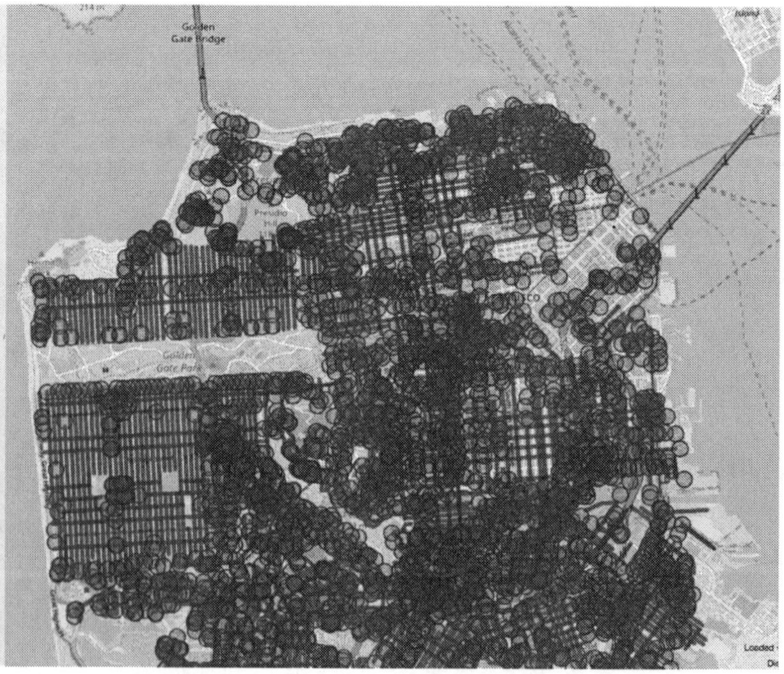

Overlay:
AV Eligible
Demand

2 Low-cost operations

Driver Service Centers for maintenance, cleaning, parking, & charging

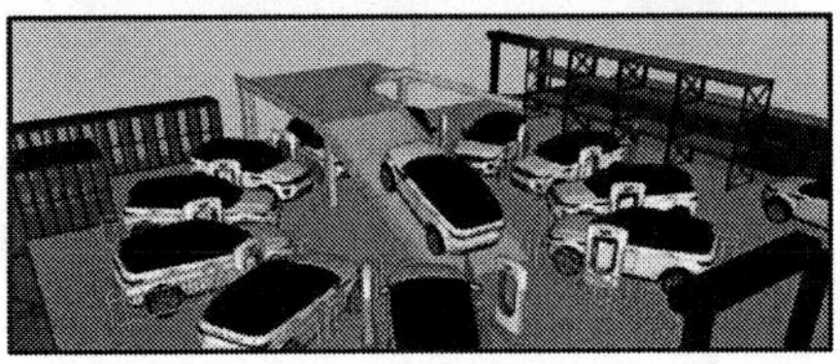

20

3 Shared Rides

Filling the cars up is essential for cost and congestion

Shared is 35% of Lyft rides in major markets

21

3 Building Shared into AV from day one

22

Endgame: Subscription Service

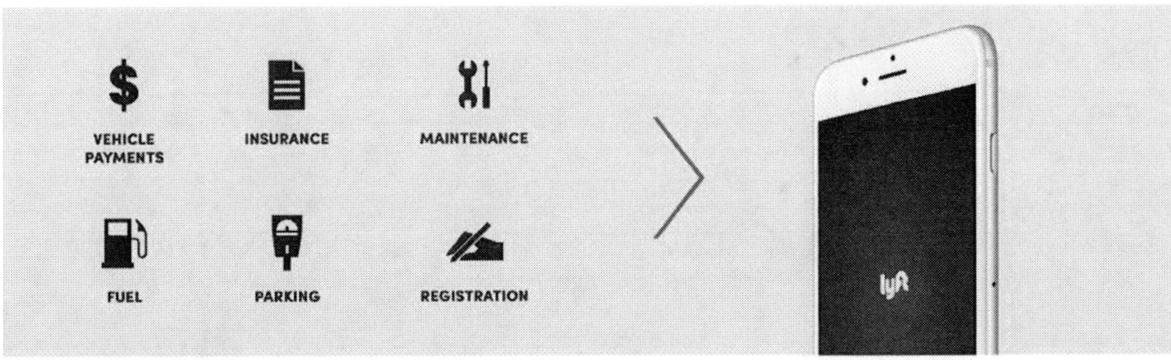

$9,000 Annual Cost Per Vehicle

Multi-modal subscription that gets you where you need to go

23

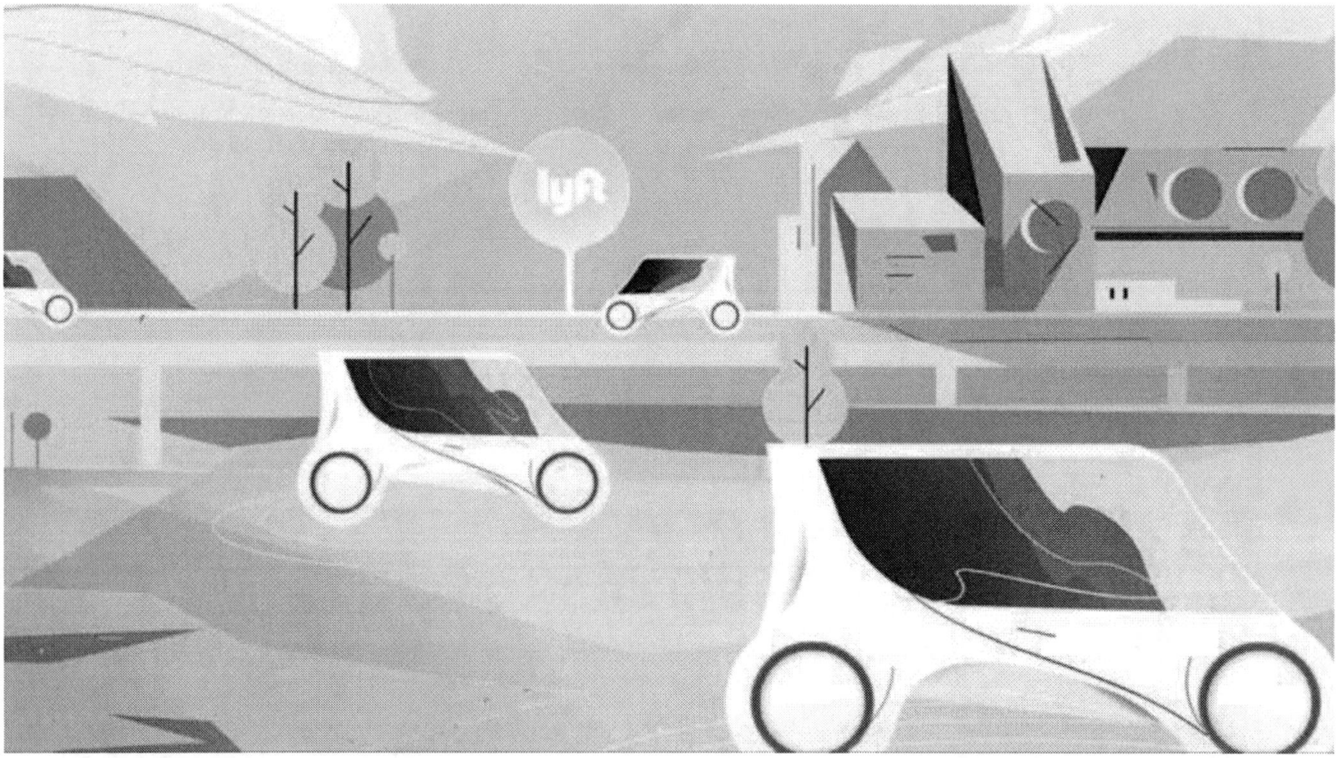
lyft

Safety Assurance, Validation and Certification Panel AVS 2018

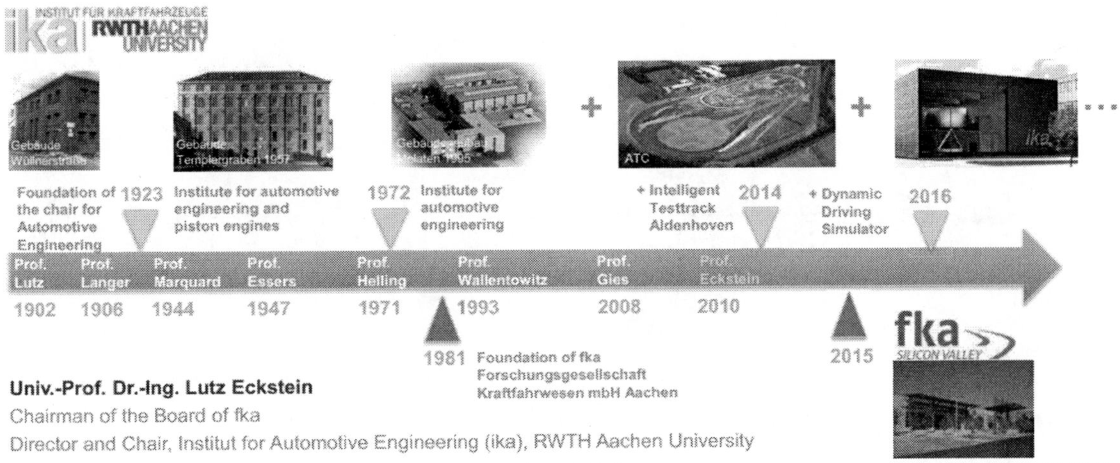

Univ.-Prof. Dr.-Ing. Lutz Eckstein
Chairman of the Board of fka
Director and Chair, Institut for Automotive Engineering (ika), RWTH Aachen University

#150· Prof. Dr. Lutz Eckstein Slide No. 1 12.07.2017

Development & Evaluation Process
Holistic Tool Chain for safe and exciting Automated Driving

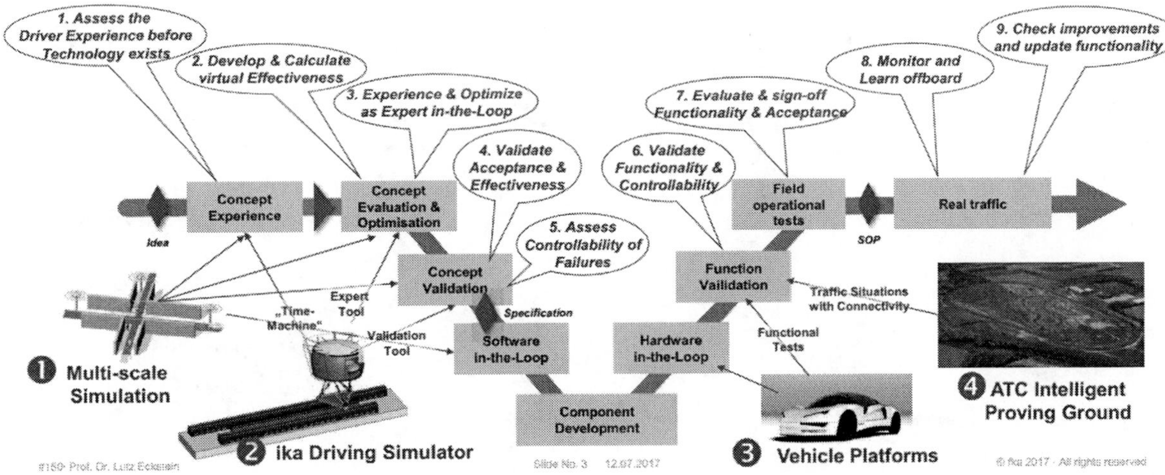

#150· Prof. Dr. Lutz Eckstein Slide No. 3 12.07.2017

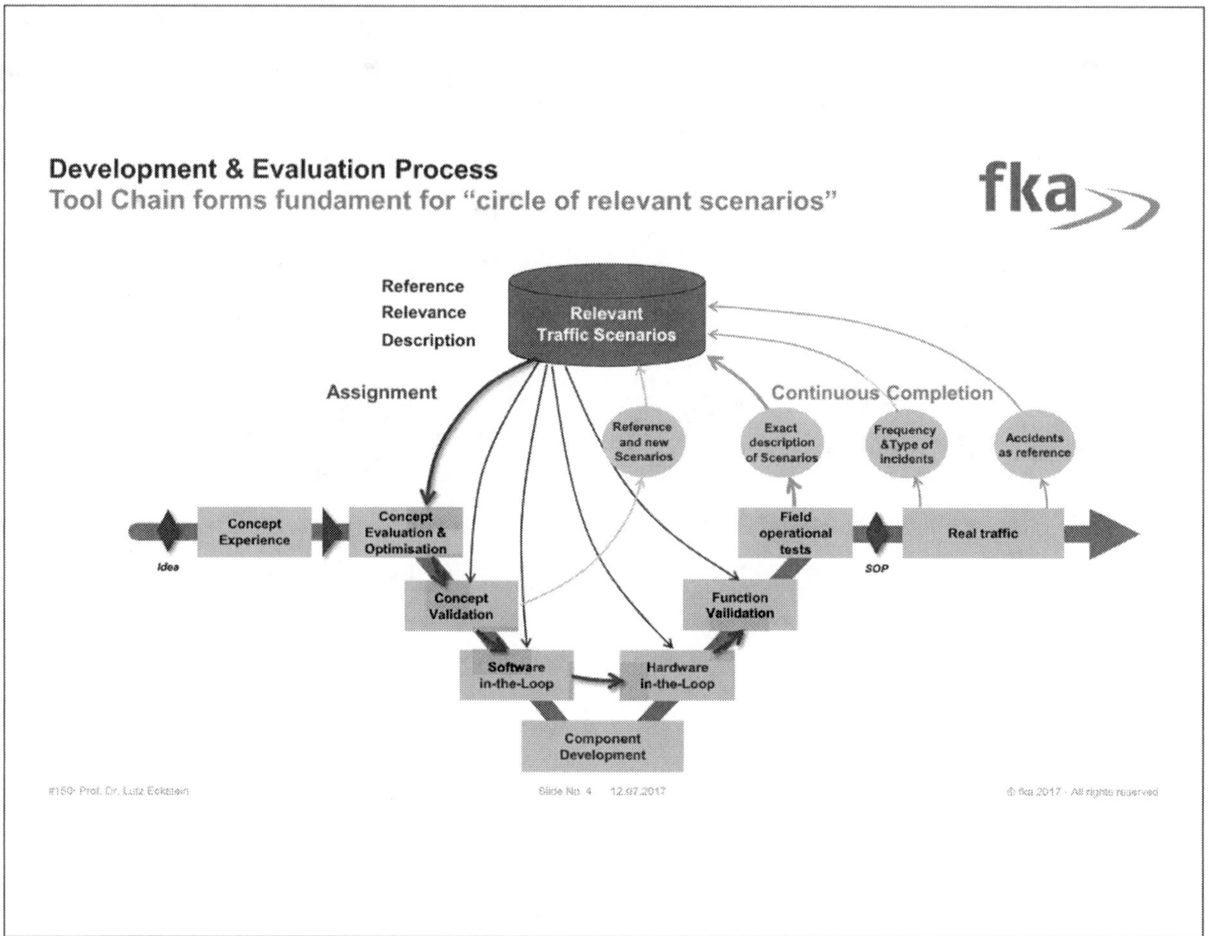
Development & Evaluation Process
Tool Chain forms fundament for "circle of relevant scenarios"
fka
Reference
Relevance
Description
Relevant Traffic Scenarios
Assignment
Continuous Completion
Reference and new Scenarios
Exact description of Scenarios
Frequency &Type of incidents
Accidents as reference
Concept Experience
Concept Evaluation & Optimisation
Field operational tests
Real traffic
Idea
SOP
Concept Validation
Function Vailidation
Software in-the-Loop
Hardware in-the-Loop
Component Development
#150· Prof. Dr. Lutz Eckstein
Slide No. 4 12.07.2017
© fka 2017 · All rights reserved

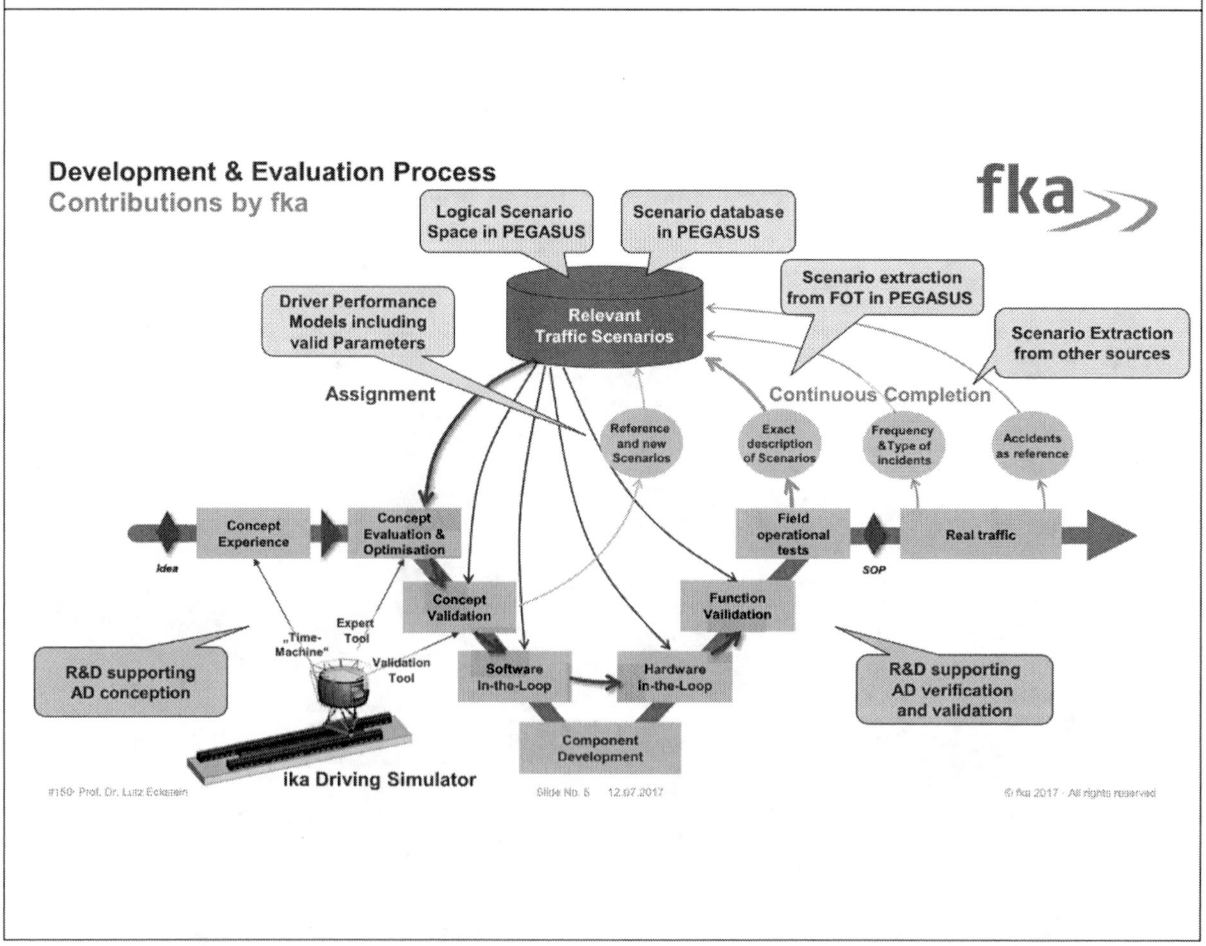
Development & Evaluation Process
Contributions by fka
fka
Logical Scenario Space in PEGASUS
Scenario database in PEGASUS
Scenario extraction from FOT in PEGASUS
Driver Performance Models including valid Parameters
Relevant Traffic Scenarios
Scenario Extraction from other sources
Assignment
Continuous Completion
Reference and new Scenarios
Exact description of Scenarios
Frequency &Type of incidents
Accidents as reference
Concept Experience
Concept Evaluation & Optimisation
Field operational tests
Real traffic
Idea
SOP
Concept Validation
Function Vailidation
„Time-Machine"
Expert Tool
Validation Tool
R&D supporting AD conception
Software in-the-Loop
Hardware in-the-Loop
R&D supporting AD verification and validation
Component Development
ika Driving Simulator
#150· Prof. Dr. Lutz Eckstein
Slide No. 5 12.07.2017
© fka 2017 · All rights reserved

Contact

Univ.-Prof. Dr.-Ing. Lutz Eckstein

fka Forschungsgesellschaft Kraftfahrwesen mbH Aachen
Steinbachstr. 7
52074 Aachen
Germany

Phone +49 241 – 80 25600
Fax +49 241 – 80 22147

Email eckstein@ika.rwth-aachen.de
Internet www.fka.de

Development & Evaluation Process
Tool Chain forms fundament for "circle of relevant scenarios"

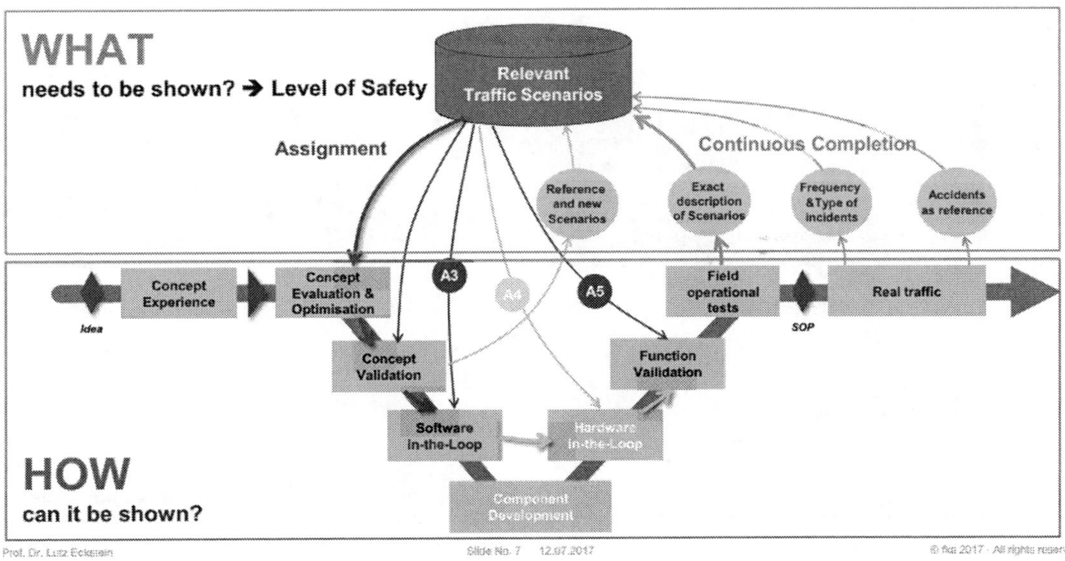

EURO NCAP TEST PROTOCOL AEB-CYCLIST

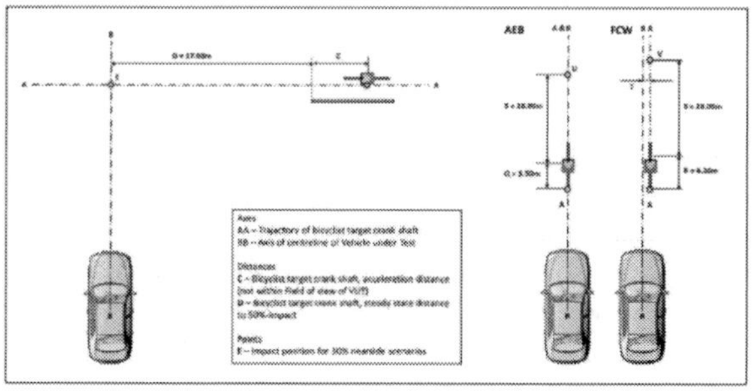

AV SAFETY ASSESSMENT FRAMEWORK

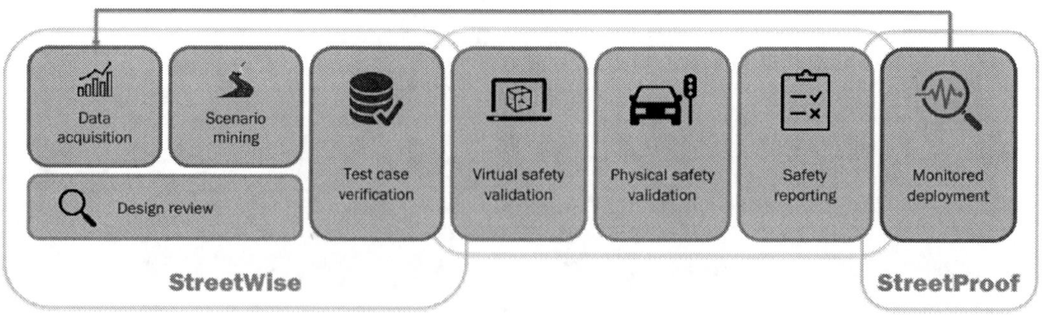

STREETWISE PIPELINE

WWW.TNO.NL/STREETWISE

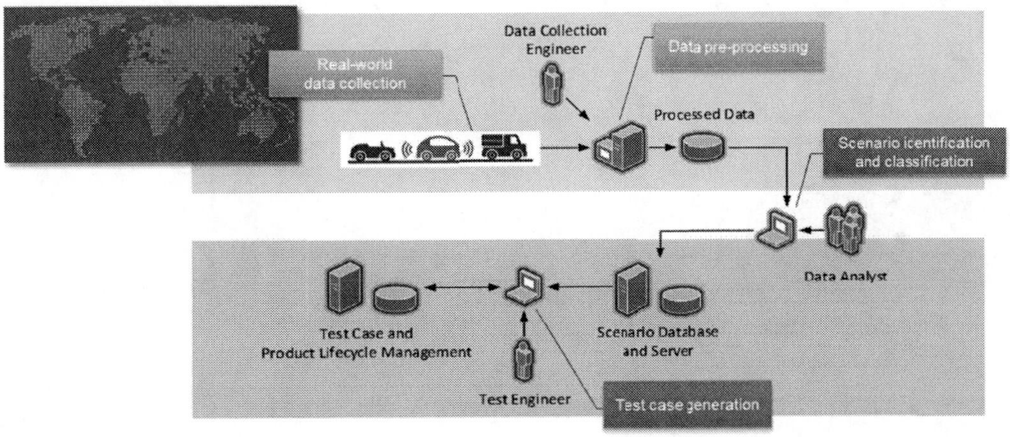

TNO innovation for life

SCENARIO EXAMPLE

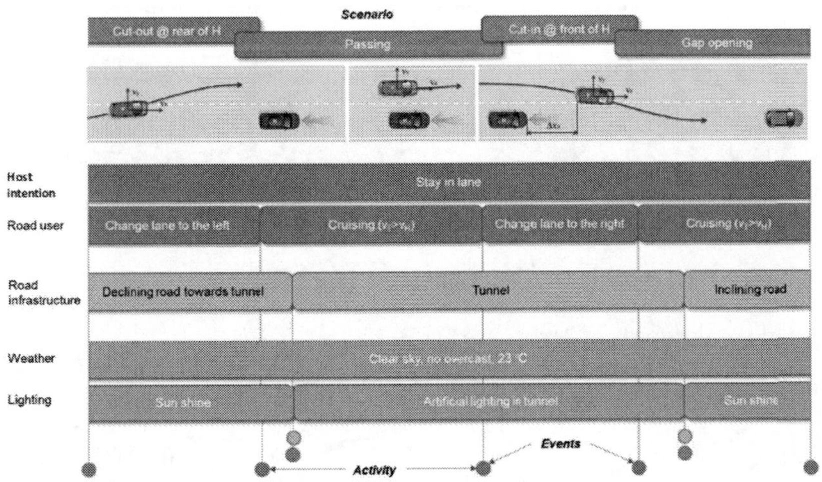

TNO innovation for life

Elements of Autonomous Vehicle Safety

(The Super-Short Version)

AVS 2018

Prof. Philip Koopman

Carnegie Mellon University

SafeAutonomy.blogspot.com

@PhilKoopman

Edge Case Research

The Real World: Heavy Tail Distribution

Edge Case Research

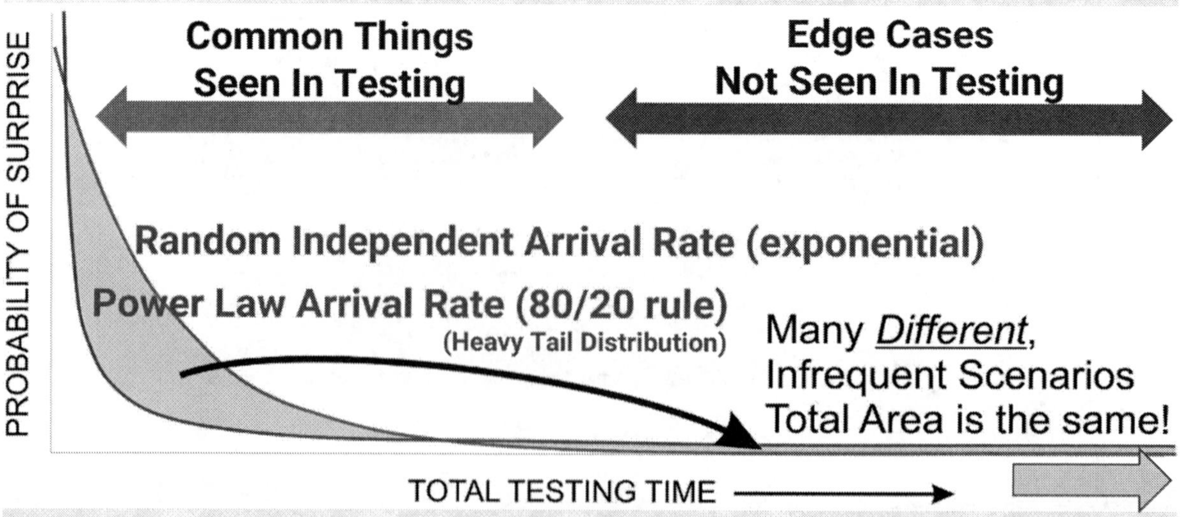

 2

Planning and Control: Doer/Checker

Edge Case Research

- Low integrity Doer
  - E.g., based on training data
- High integrity Checker
  - Use traditional safety methods

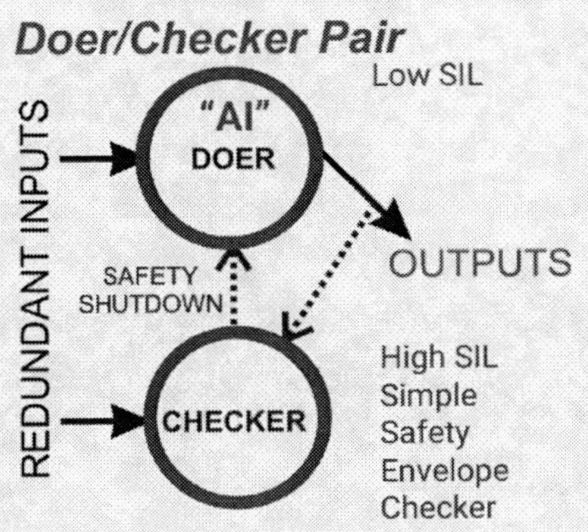

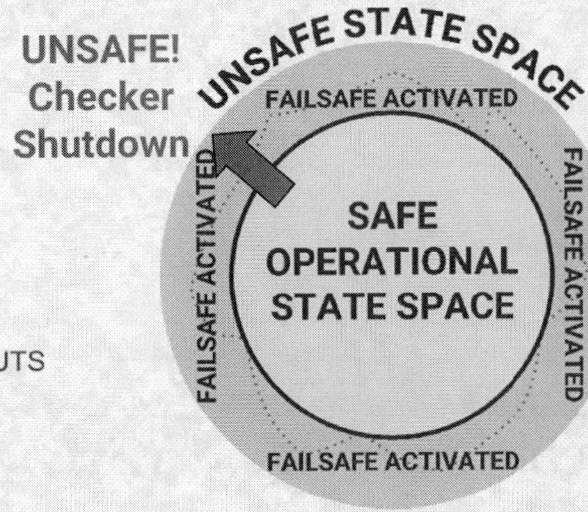

 3

Perception: Robustness via Augmentation

Edge Case Research

- Pseudo-realistic degradation
- Move and insert objects

Original data

Pedestrian Missed: Gaussian Blur

Pedestrian Missed: Gaussian Noise + Black Car

Recorded scene

Synthesized scene

Dark vehicle added to scene

 4

Other Techniques

Edge Case Research

- **Conventional V-based software safety**
  - ISO 26262, SOTIF forsafety requirements
- **Rapid safety case update tooling**
  - STPA plus augmented fault trees and GSN
- **Architectural safety patterns**
  - Doer/Checker for a fail silent channel
  - Multi-channel approach for safing mission
- **Robustness/stress testing**
  - Traditional robustness testing
  - Object & event edge case "zoo"

MISSA
AUTONOMOUS SYSTEM SAFETY ACCELERATED

switchboard
RELIABLE SOFTWARE FOR AUTONOMY

hologram
SAFER PERCEPTION FOR AUTONOMY

 5

TRL THE FUTURE OF TRANSPORT

GATEway

Exploring how people respond to, engage with and accept CAVs in a challenging urban environment

Richard Cuerden

AVS 2018

GATEway

Greenwich Automated Transport Environment

- £8m project funded by industry and Innovate UK
- Understand and overcome technical, legal and societal challenges of using CAVs in urban areas
- Vehicle trials, simulation and public engagement
- October 2015 – March 2018

GATEway

TRL

Led by TRL The UK's Transport Research Laboratory

ROYAL borough of GREENWICH

UNIVERSITY of GREENWICH

Royal College of Art

Imperial College London

Commonplace RSA O2 dg:Cities

Heathrow Making every journey better

Fusion Processing

GOBOTIX LTD

OXBOTICA robotics & autonomous systems

Supported by Innovate UK

Centre for Connected & Autonomous Vehicles

the future of transport.

GATEway

TRL

- *Demonstrate* the safe and efficient integration of sophisticated automated transport systems into complex real world smart city environments
- *Create* a validated test bed in the heart of London for the evaluation of next generation automated transport systems

Trial 1: Micro-transit

Trial 2: Automated valet parking

Trial 3: Last mile delivery

the future of transport.

GATEway

Activity timeline

TRL

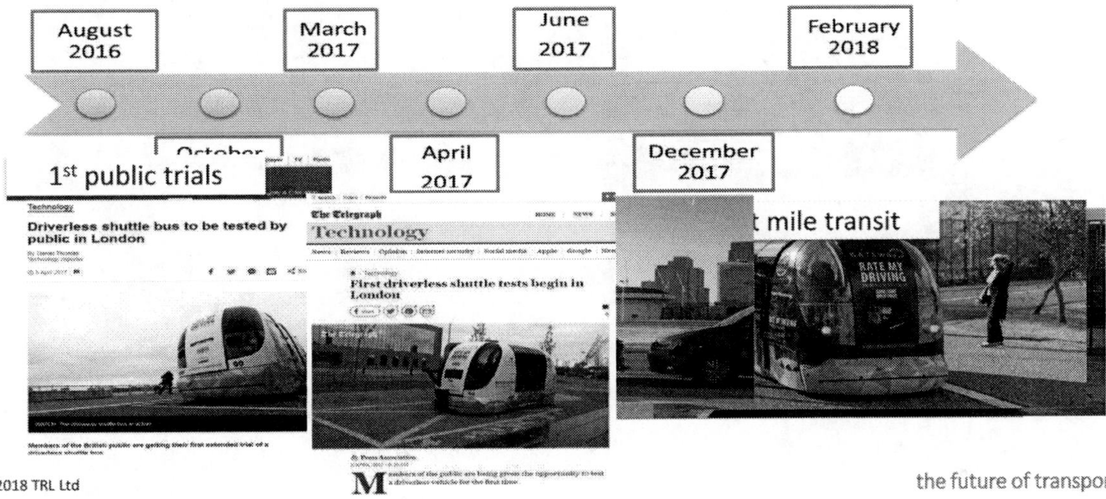

the future of transport.

GATEway

Trial 3: Last mile delivery

TRL

Aim

- *To explore public perception and experience of driverless deliveries to inform future deployments of services*

Research

- *2 week trial with CargoPod and Ocado*
- *Over 100 customers*
- *TRL surveyed recipients*
- *Commonplace local sentiment mapping*

the future of transport.

GATEway

Last mile delivery: Willingness to use

TRL

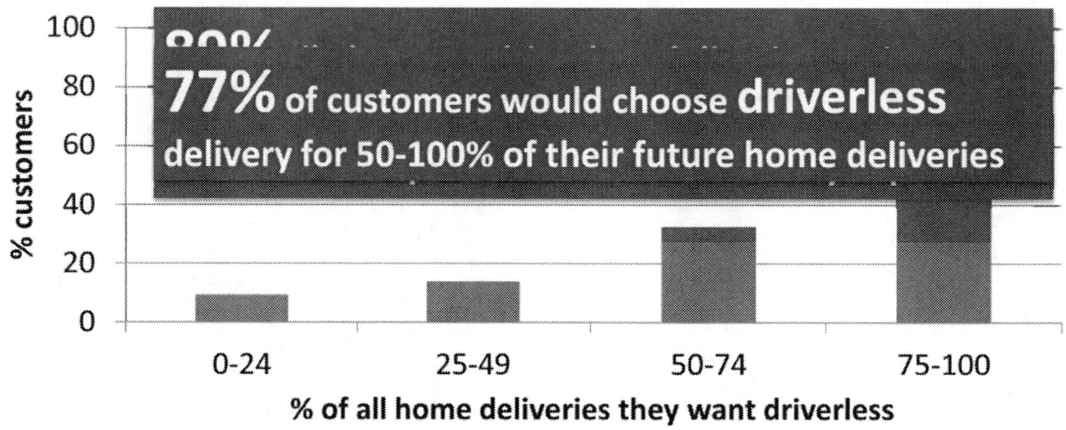

the future of transport.

GATEway

Trial 2: Automated valet parking

TRL

Aim

- *To provide members of the public with direct experience of an automated ("drop-off") valet parking service and gain feedback on their experience*

Findings

- *Auto valet parking was seen as a service that would improve the quality of travel and save time*
- *Increasing safety within vehicles and in public spaces*
- *Provide more inclusive personal transport*
- *Some challenges identified*
  - *Increased congestion*
  - *Management of drop-off and pick up points*

the future of transport.

GATEway

Trial 1: Micro-transit

TRL

Aim

- *To assess public perceptions of autonomous vehicles as a result of a direct experience with an AV, particularly in a last mile service*

Research

- *Online surveys (passengers) - TRL*
- *Sentiment mapping (anyone who has seen the vehicles) – Commonplace*
- *Structured observations (cyclists and pedestrians) – University of Greenwich*

the future of transport.

GATEway

TRL

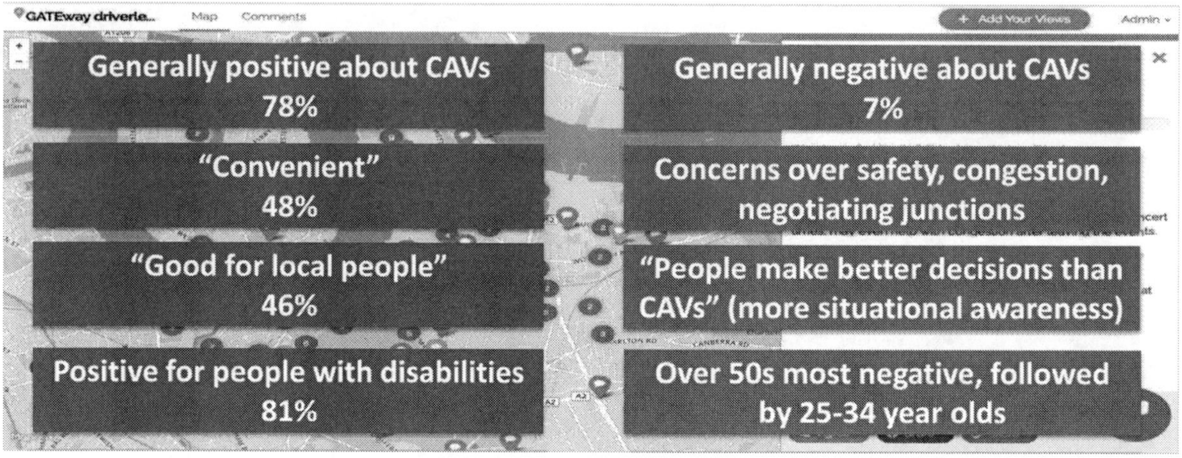

Commonplace

the future of transport.

GATEway

TRL

This is just the beginning .....

www.gateway-project.org.uk

the future of transport.

GATEway

TRL

merge
GREENWICH

"The MERGE Greenwich project will develop the blueprint for a scaleable commercial pilot for autonomous ride-sharing, integrated with public transport systems"

ADDISON LEE | Immense Enabling Intelligent Mobility | CATAPULT Transport Systems | Ford | TRL | DG:Cities

the future of transport.

MERGE Greenwich project

Key benefit: complementing public transport

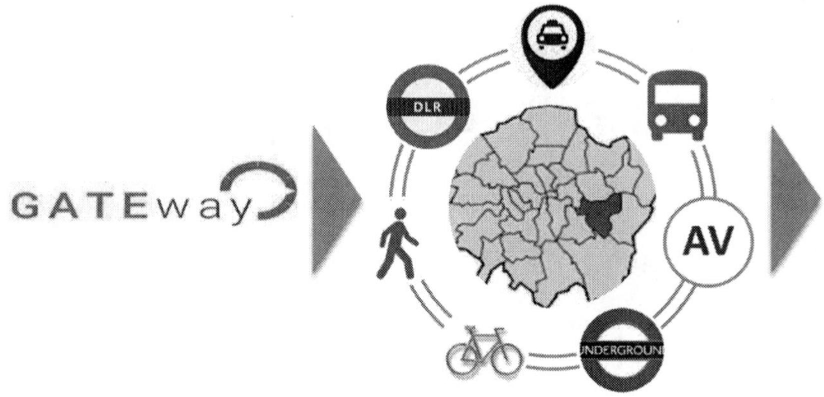

Designed and developed from the city/ public transport perspective

… in order to integrate new solutions with existing modes of transport and optimise the network

merge GREENWICH

MERGE Greenwich Customer Research

Customer attitudes to Autonomous Vehicles and Ride-sharing

Customer research:
Survey and focus groups

Aim:
Understand customer expectations, perceptions, concerns and motivations regarding Autonomous vehicles and ride-sharing

Reason:
To ensure AV ride-sharing meets customer needs

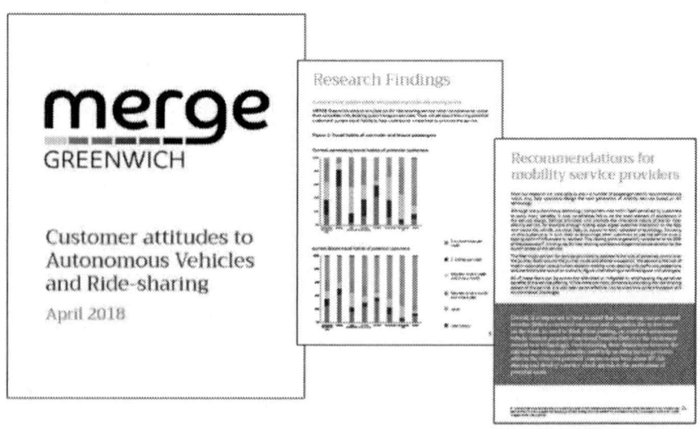

https://mergegreenwich.com

merge GREENWICH

Smart Mobility Living Lab: London

TRL

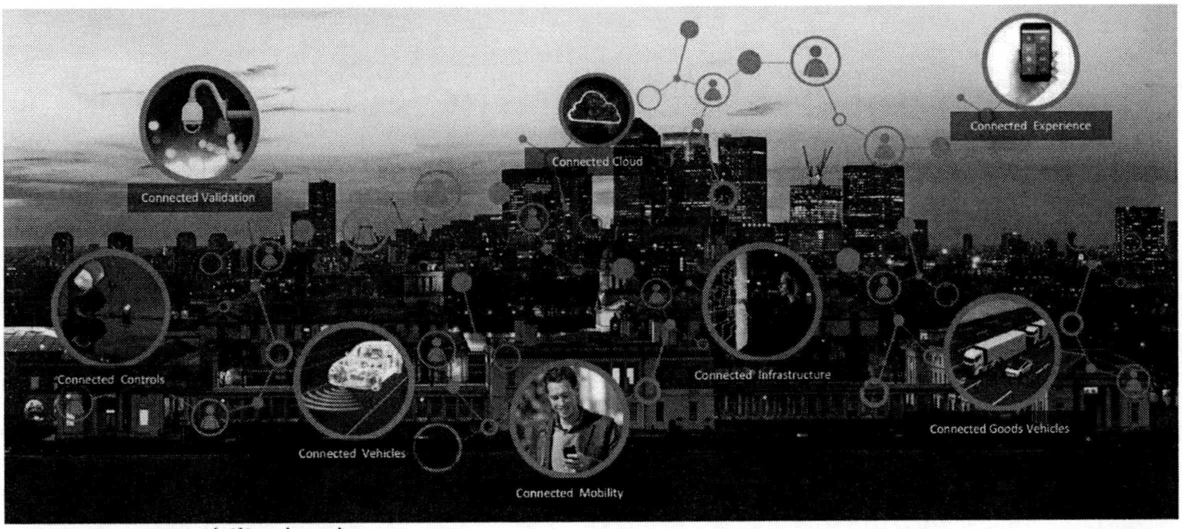

www.smartmobility.london

Smart Mobility Living Lab: London

Richard Cuerden
Director, TRL Academy
Email: rcuerden@trl.co.uk
Twitter: @rcuerden_trl

TRL THE FUTURE OF TRANSPORT

The National Academies of
SCIENCES · ENGINEERING · MEDICINE

TRANSPORTATUON RESEARCH BOARD

The National Academies & TRB: Preparing for Automated Vehicles and Shared Mobility

Neil Pedersen
Executive Director, TRB

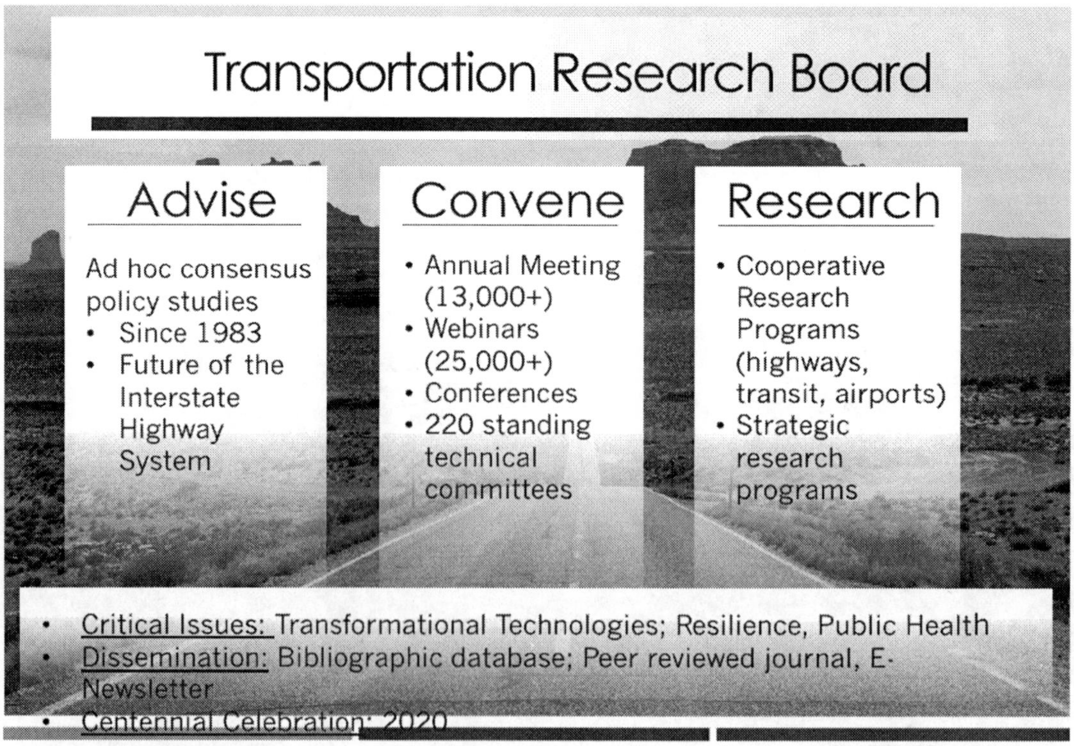

The National Academies of
SCIENCES · ENGINEERING · MEDICINE

TRB
TRANSPORTATION RESEARCH BOARD

Proceedings from Recent AVS

www.trb.org/Publications/
PubsTransportationResearchCirculars.aspx

The National Academies of
SCIENCES · ENGINEERING · MEDICINE

TRB
TRANSPORTATION RESEARCH BOARD

The National Academies of
SCIENCES · ENGINEERING · MEDICINE

FORUM MEETING #2: JULY 8-9, 2018; SAN FRANCISCO, CA

Forum on Preparing for Automated Vehicles & Shared Mobility

Products of the Forum

- Research roadmap
  - NCHRP Project 20-102
- White papers
- Authored perspectives
- Workshops and sessions
- *TR News* articles
- Forum website
- Forum annual report
- Research partnerships
- Education of stakeholder community

The National Academies of
SCIENCES · ENGINEERING · MEDICINE

TRB
TRANSPORTATION RESEARCH BOARD

Forum Research Needs Catalogue Focus Areas

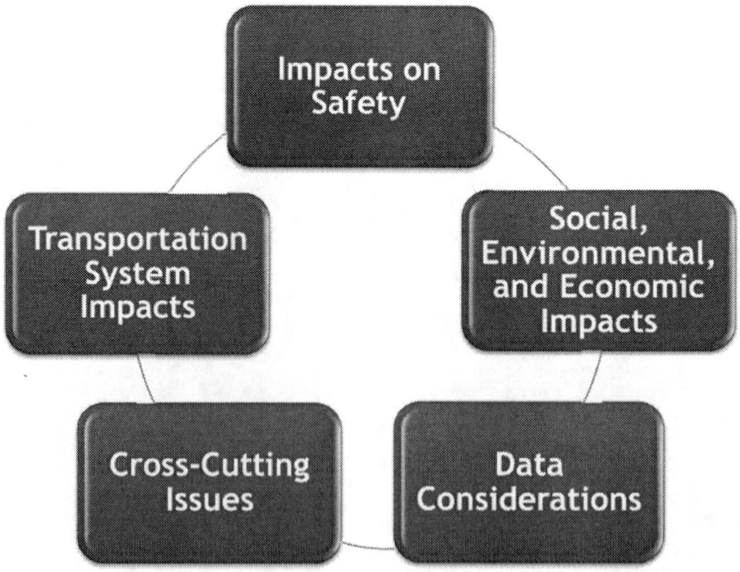

The National Academies of
SCIENCES · ENGINEERING · MEDICINE

TRB
TRANSPORTATION RESEARCH BOARD

Forum Research Needs

1. Models for sharing of data
2. Potential safety scenarios during the transition to highly automated vehicles
3. State and local policies to ensure safety prior to deployment
4. Infrastructure enablers for AVs & shared mobility
5. Net positive (and opportunities to mitigate negative) social impacts of AV deployment and shared mobility

Forum Research Needs

6. Addressing social inclusion and equity
7. Potential impacts of higher level automated vehicles and shared mobility on traveler behavior and freight movement
8. Impacts of shared mobility on transit, and vice-versa
9. Implications for transportation planning and planning models
10. Impacts on land use, and how land use impacts AVs and shared mobility

Preparing for AVs & Shared Mobility: Broader Questions

- What can we do to build awareness across a broader spectrum of the potential transformational impacts on society?
- What options exist to generate and fund a significant strategic research effort or program to inform policy decisions?
- How can we best leverage existing research programs to help address these issues?

The National Academies of
SCIENCES · ENGINEERING · MEDICINE

TRB
TRANSPORTATION RESEARCH BOARD

TRB/EU/US Symposium

- 6th annual EU/US joint symposium, this one focused on socio-economic effects of shared/AV
- Held in Brussels on June 27-28, 2018
- Approximately 50 participants from EU and US
- 4 breakout groups spent two days in discussion
  - Freight
  - Places
  - Travel Behavior
  - Stakeholders
  - Economics and welfare
  - Equity
  - Data access/privacy
  - Safety and security

The National Academies of
SCIENCES · ENGINEERING · MEDICINE

TRB
TRANSPORTATION RESEARCH BOARD

TRB/EU/US Symposium

- Each breakout group developed 10-15 research problem statements
- Next steps:
  - Finalize report of the proceedings
  - Hold Sunday workshop at the 2019 TRB Annual Meeting to share results
  - Identify opportunities for joint research between EU and US via 2019 TRB workshop, other events, etc.
  - Undertake research on issues identified

The National Academies of SCIENCES · ENGINEERING · MEDICINE

TRB TRANSPORTATION RESEARCH BOARD

TRB Research Highlights

http://bit.ly/2y8gEm4

Advancing Automated and Connected Vehicles: Policy and Planning Actions for State and Local Transportation Agencies [NCHRP Report 845 & Briefing Document]

- Explores potential impacts to traffic crashes, congestion, pollution, land development, and mobility for the underserved
- Describes 18 policy and planning strategies that could reduce negative impacts and increase positive ones by
  - Mitigating safety risks through testing, training, and public education
  - Encouraging shared AV use
  - Addressing liability issues that may impact market development
  - Enhancing safety, congestion, and air quality benefits by influencing market demand

The National Academies of SCIENCES · ENGINEERING · MEDICINE

TRB TRANSPORTATION RESEARCH BOARD

TRB Research Highlights

http://bit.ly/2y8gEm4

- Shared Mobility and the Transformation of Public Transit [TCRP Report 188]
- Broadening Understanding of the Interplay Between Public Transit, Shared Mobility, and Personal Automobiles [TCRP Report 195]
- Private Transit: Existing Services and Emerging Directions [TCRP Report 196]
  - These reports explore how transit agencies are adapting to shared mobility services, microtransit, and other emerging modes

The National Academies of SCIENCES · ENGINEERING · MEDICINE

TRB TRANSPORTATION RESEARCH BOARD

TRB Research Highlights

http://bit.ly/2y8gEm4

- Providing Support to the Introduction of CV/AV Impacts into Regional Transportation Planning and Modeling Tools
- Workplace Implications of Autonomous Vehicles on the Transit Workforce
- Cybersecurity of Traffic Management Systems
- Impact of Transformational Technologies on Land Use and Transportation
- Framework for Managing Data from Emerging Transportation Technologies to Support Decision-Making
- Impacts of Connected and Automated Vehicle Technologies on the Highway Infrastructure
- Assessing the Impacts of Connected, Automated, and Autonomous Vehicles on the Future of Transportation Safety

The National Academies of SCIENCES · ENGINEERING · MEDICINE

TRB TRANSPORTATION RESEARCH BOARD

TRB Research Summary

http://bit.ly/2y8gEm4

- Approximately $10M has been invested in AV/CV research by the NCHRP, TCRP, ACRP, and BTSCRP
- Results are available from the initial projects and more are coming
- New projects will be selected on Thursday, July 12 (~$2M); look for an announcement later this month

The National Academies of
SCIENCES · ENGINEERING · MEDICINE

TRB
TRANSPORTATION RESEARCH BOARD

The National Academies of
SCIENCES · ENGINEERING · MEDICINE

TRB
TRANSPORTATION RESEARCH BOARD

AUTOMATED VEHICLES SYMPOSIUM 2018

Preparing U.S. Workers & Employers for an Autonomous Vehicle Future

Erica L. Groshen, Cornell University—ILR School
With S. Helper, J.P. MacDuffie, & C. Carson
Cornell University ILR School

1

Agenda

- Lessons from past innovations
- Sizing and timing AV's impact on workers
- Next steps

Cornell University ILR School

2

Innovation is not new

- Industrial Revolution in England, *1750-1900*
- Autopilot in aviation, *1912-now*
- Computer numerical control in machine tools, *1960-1990*
- Automation in auto assembly plants, *1980s-now*
- Automatic teller machines, *1980s-now*
- Trade expansion with China, *1990-now*
- Self-driving trucks in Australian strip mines, *2000-now*

Lessons from past transitions

- Full employment returns
- Costs to workers
  - Concentrated & high
  - Uncertain
  - Precede benefits
  - Fuel unrest & resistance
- Benefits accrue diffusely & unevenly
- Policy & implementation matter

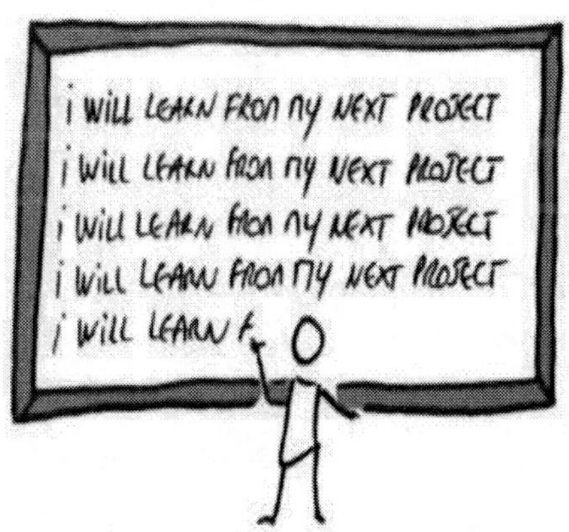

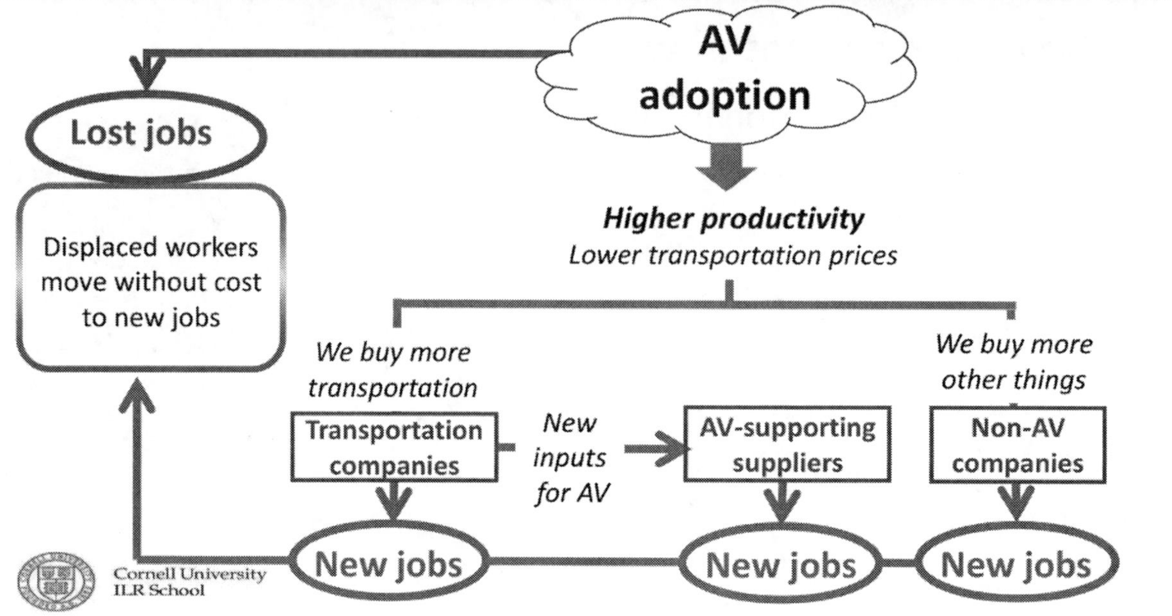
AV adoption
Lost jobs
Displaced workers move without cost to new jobs
Higher productivity
Lower transportation prices
We buy more transportation
We buy more other things
Transportation companies
New inputs for AV
AV-supporting suppliers
Non-AV companies
New jobs
New jobs
New jobs
Cornell University
ILR School

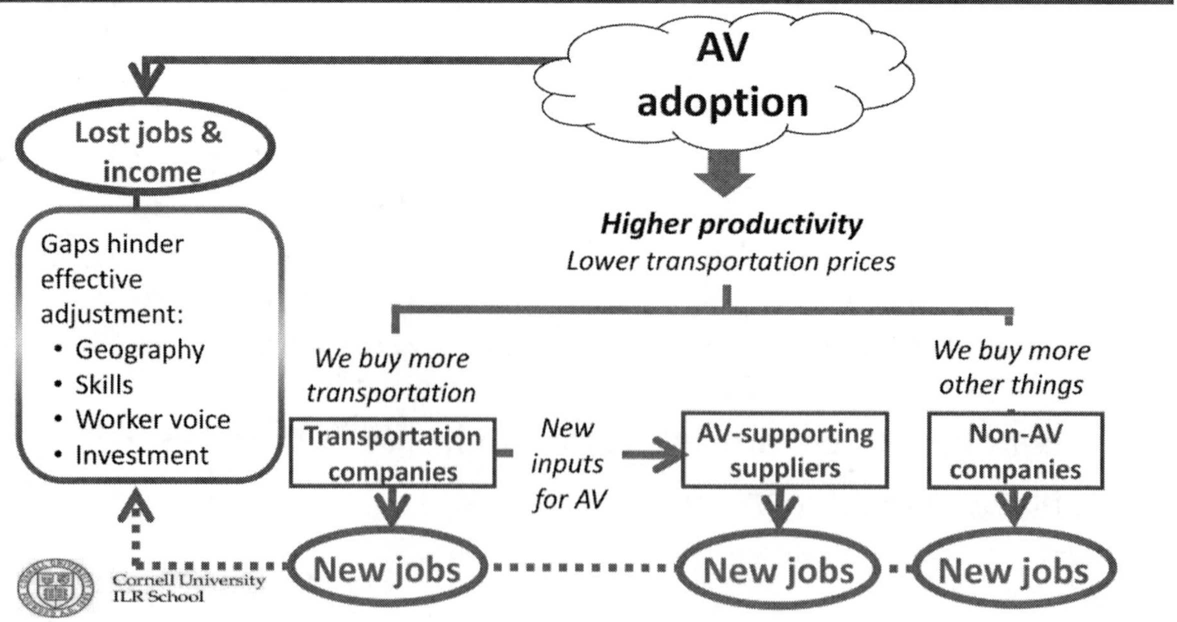
AV adoption
Lost jobs & income
Gaps hinder effective adjustment:
• Geography
• Skills
• Worker voice
• Investment
Higher productivity
Lower transportation prices
We buy more transportation
We buy more other things
Transportation companies
New inputs for AV
AV-supporting suppliers
Non-AV companies
New jobs
New jobs
New jobs
Cornell University
ILR School

Reducing adjustment gaps

- Key to success
- Historically neglected

Cornell University
ILR School

7

Sizing and timing AV's impacts

Cornell University
ILR School

8

1.3 – 2.3 million workers displaced

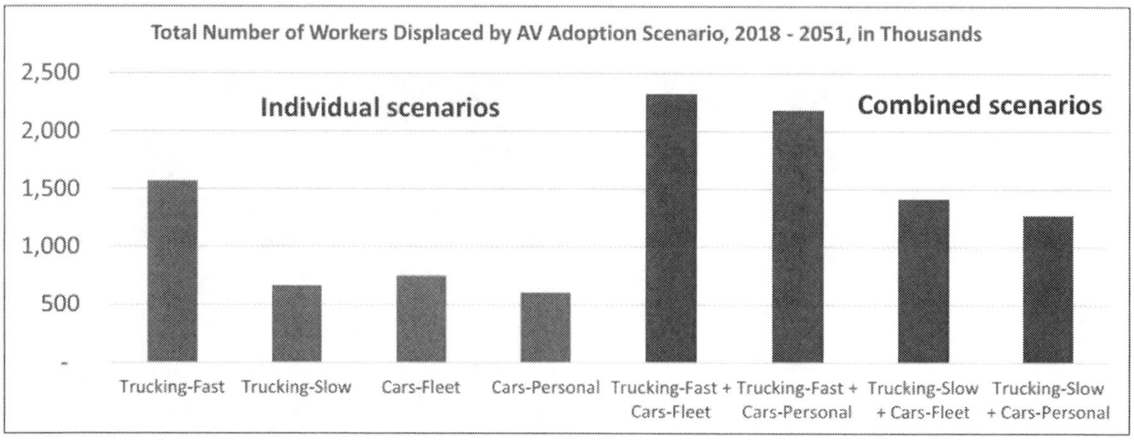

Cornell University
ILR School

9

Unemployment: max +0.13 percentage pt in mid-2040s

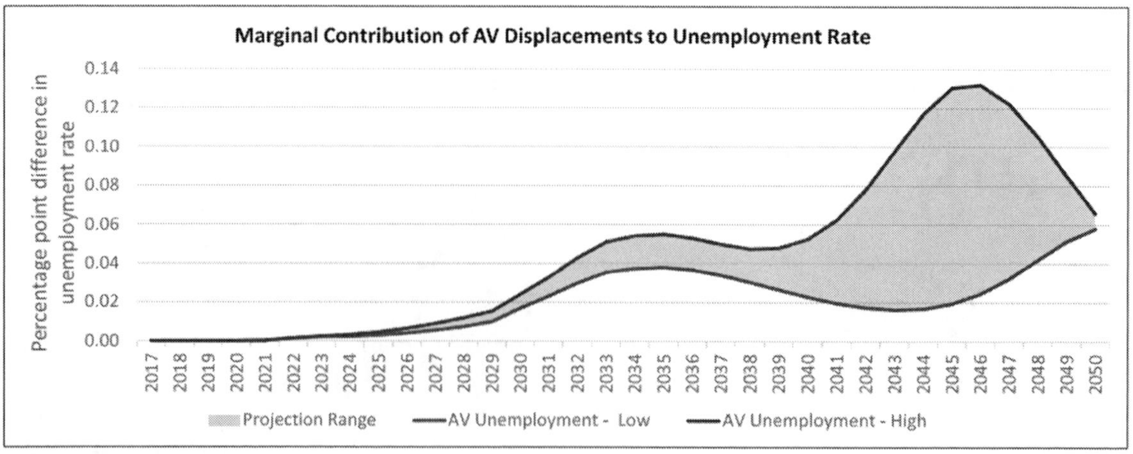

Cornell University
ILR School

10

Impact consequential & manageable

- Workers displaced (2018-2051): 1.3-2.3 M
  - +0.13 percentage pt to unemployment at peak
  - -0.1 percentage pt to participation at peak
  - ½ size of China shock
- Wealth losses: $200-$300 B
  - $80-$120K/worker
  - Worse in recessions
- Timing: max in mid-2040s
  - Starts slowly

Cornell University
ILR School

Next steps

Cornell University
ILR School

Choose a path

Mitigation Strategy Paths

- **Passive**: Deploy AV rapidly & accept high costs to workers at risk of social disruption and resistance
- **Reactionary**: Slow down AV adoption to allow current workforce system to handle transition without high costs
- **Investment**: Deploy AV rapidly & direct some benefits to invest in comprehensive cost mitigation

Cornell University
ILR School

13

What would comprehensive cost mitigation look like?

Strengthen existing system -- UI, Workforce Innovation & Opportunity Act, CCs	
Employers engaged	
Adequately funded for inclusive eligibility -- not only AV	
Multi-pronged • Worker voice • Training • Place-based • Income support • Search & placement	**Evidence-based** • Program experiments & evaluations • Relevant, high-quality official statistics • Administrative data • Research • Stakeholder input

Cornell University
ILR School

14

Some of many tested policy tools

- Wage insurance
- Works councils
- Worker training accounts
- Universal basic income
- Flexicurity
- Public sector jobs for infrastructure
- Place-based economic development

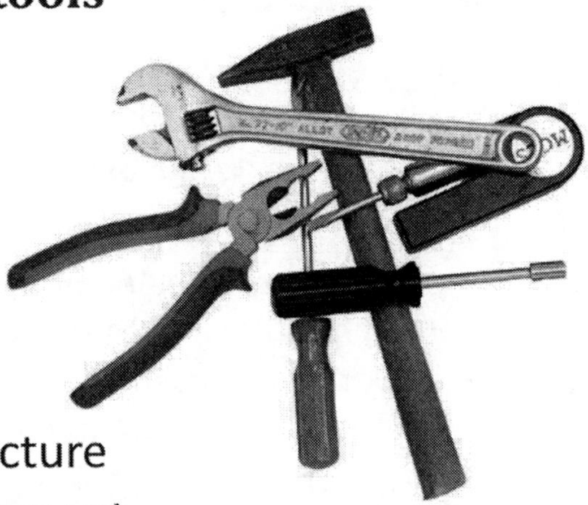

Cornell University
ILR School

15

Conclusion on mitigation investment

- Advisable
  - Avoid consequential harm
  - Promote further innovation
  - Use workers' skills & insights
- Doable
  - 10-20 years before costs mount
  - Many policy options
  - Annual AV benefits ($800 B) >> Total costs ($200-300 B)

Cornell University
ILR School

16

Immediate steps

✓Employers: start plans
- Tap workers' skills and insights
- Retrain & retain existing staff

✓Local stakeholders: hold planning forums
- Workforce development system
- Tech & transportation companies
- Worker representatives
- Civic leaders, foundations, researchers

✓National stakeholders: craft an investment strategy

Cornell University
ILR School

17

Erica L. Groshen

Visiting Senior Scholar, Cornell-ILR

erica.groshen@gmail.edu

Cornell University
ILR School

The gap between productivity and a typical worker's compensation has increased dramatically since 1973

Productivity growth and hourly compensation growth, 1948–2016

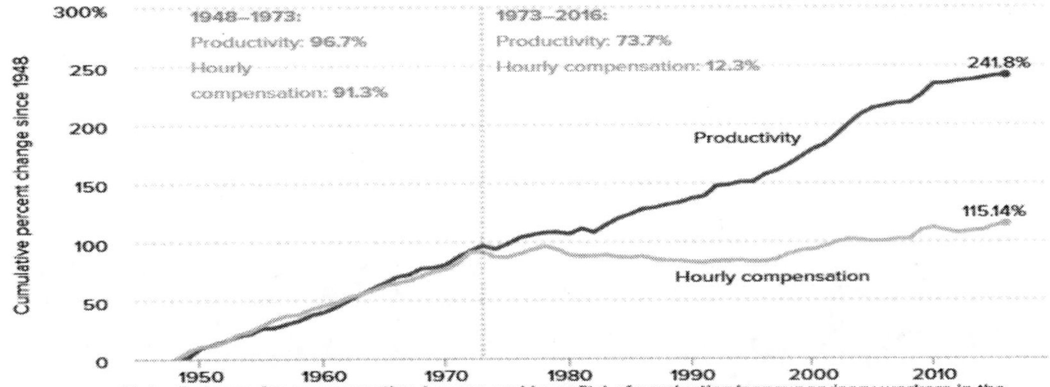

Note: Data are for compensation (wages and benefits) of production/nonsupervisory workers in the private sector and net productivity of the total economy. "Net productivity" is the growth of output of goods and services less depreciation per hour worked.

Source: EPI analysis of Bureau of Labor Statistics and Bureau of Economic Analysis data

Updated from Figure A in *Raising America's Pay: Why It's Our Central Economic Policy Challenge*

Corr ILR

19

Artificial intelligence: who is right?

- AI: replicates routine brainwork

- Techno pessimists
  - AI destroys jobs
  - Employment falls
  - Permanent pool of poor unemployed

- Techno optimists
  - AI productivity makes us rich
  - Few will need to work

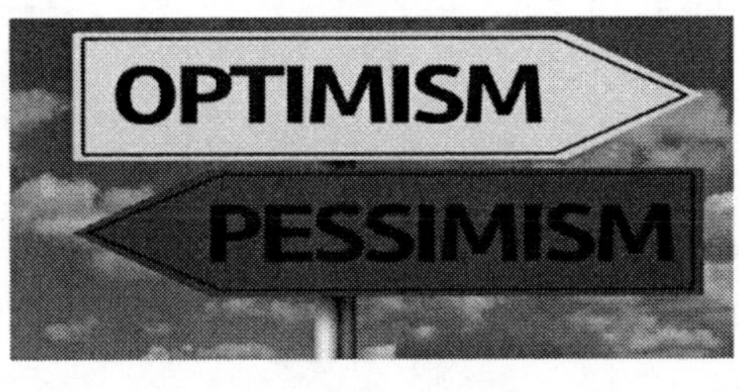

- Examples: BLS, driverless cars

Cornell University
ILR School

AI impact on labor market: temporary but costly

- Techno optimists and pessimists both wrong
- Unless we're all satisfied currently
  - Unemployment rises temporarily, perhaps a long time
  - We return to full employment
- Benefits not used to compensate displaced workers
  - Lifetime earnings losses of 1-4x previous annual earnings
  - More jobless and NLF spells, fewer hours, lower wages

  → **Suicides, poor health,..., and resistance to change**
- Policy (not technology) can reduce losses by
  - Employer and government actions
  - Closing gaps to reduce unemployment and raise wages
  - Compensating job losers

Cornell University
ILR School

DESIGNING AUTOMATED VEHICLES AROUND HUMAN VALUES

J. CHRISTIAN GERDES
STANFORD UNIVERSITY

Ethical programming, not programming ethics

Professor Shannon Vallor
Santa Clara University

MIT Technology Review

A View from **Emerging Technology from the arXiv**

Why Self-Driving Cars Must Be Programmed to Kill

Self-driving cars are already cruising the streets. But before they can become widespread, carmakers must solve an impossible ethical dilemma of algorithmic morality.

October 22, 2015

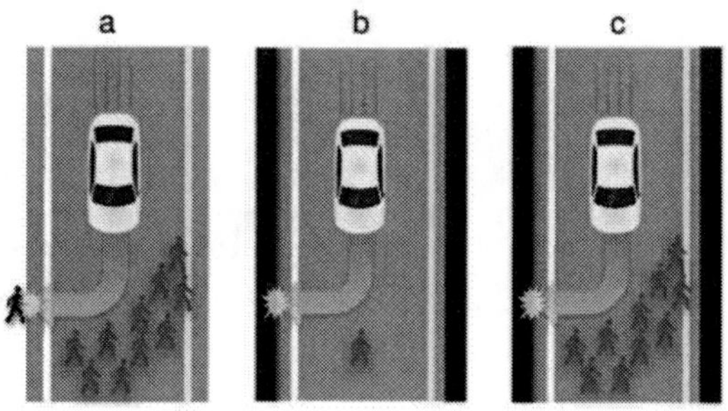

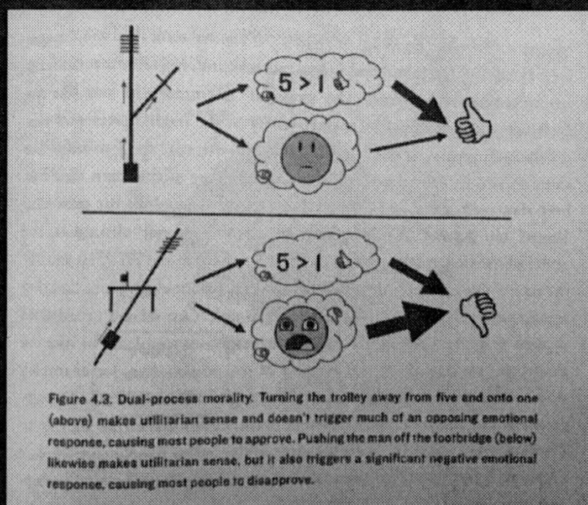

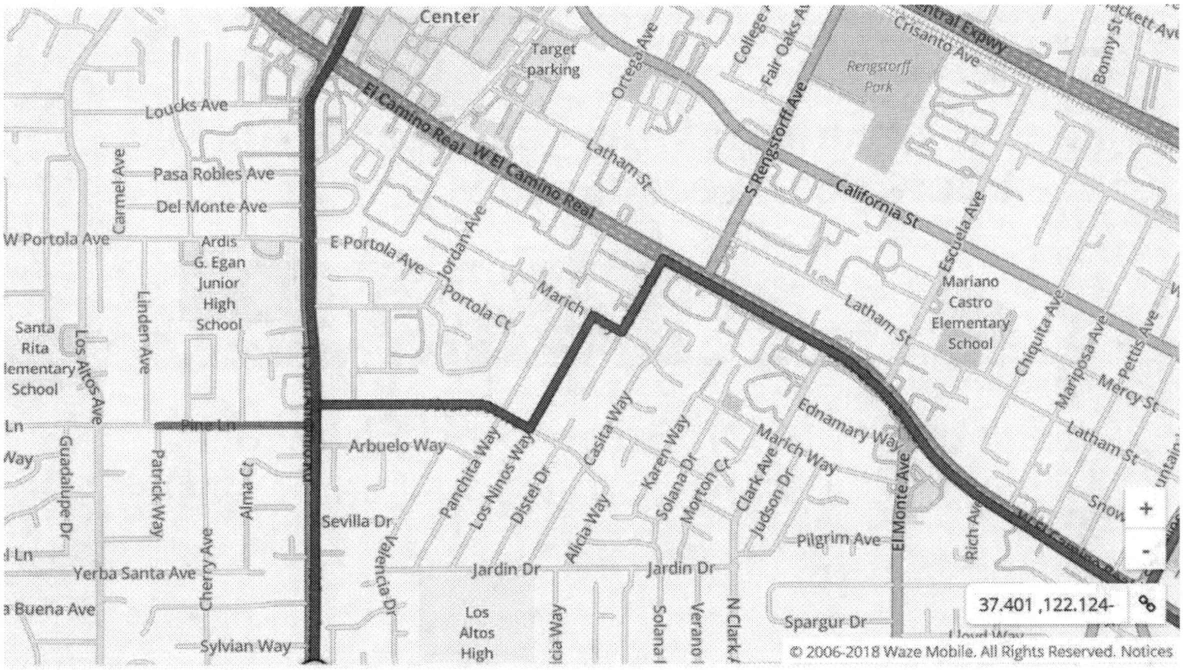

Los Angeles Times

How Waze and Google Maps turned an Encino neighborhood into a speedway

By STEVE LOPEZ MAY 30, 2018 | 3:00 AM

The New York Times

Navigation Apps Are Turning Quiet Neighborhoods Into Traffic Nightmares

By Lisa W. Foderaro Dec. 24, 2017

The Seattle Times | Traffic Lab

Pedestrian and bicyclist deaths double statewide in recent years

Originally published June 13, 2018 at 6:00 am | *Updated June 13, 2018 at 4:10 pm*

By Michelle Baruchman

Seattle Times staff reporter

NATIONAL

Distraction, On Street And Sidewalk, Helps Cause Record Pedestrian Deaths

March 30, 2017 · 10:02 PM ET

Heard on All Things Considered

DAVID SCHAPER

npr

Peloton
H2

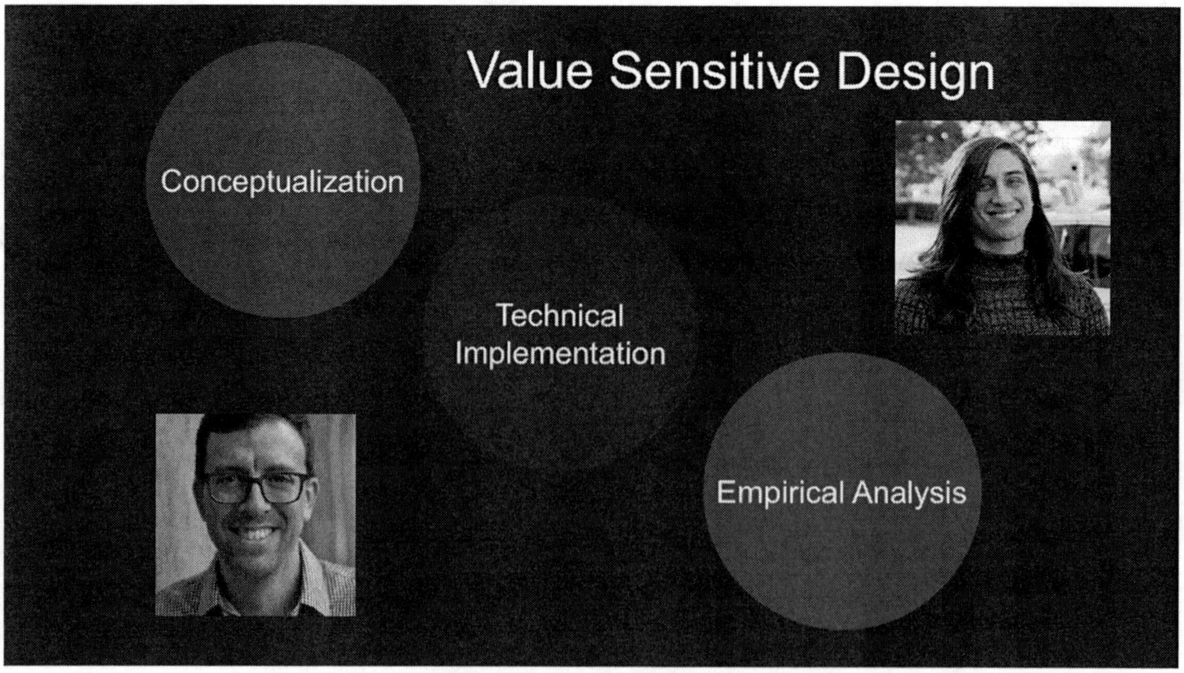
Value Sensitive Design
Conceptualization
Technical
Implementation
Empirical Analysis

Conceptualization

Concept

What values are important?

- Gather a range of stakeholders to get a range of perspectives
- Stakeholder engagement helps to…
  - Surface the underlying values
  - Identify value conflicts and types of resolutions
  - Frame the underlying problem
- Build a workshop around a relevant scenario

Technical Implementation

Tech

How do we design the system?

- Technical implementation should:
  - Provide a means for realizing values, often as rules or costs
  - Map design parameters to associated values
  - Enable treatment of value conflicts
- Not a formulaic process
  - Need engineers comfortable thinking in terms of values
  - Need to move from specific examples to general principles

Empirical Analysis

Analysis

Did we capture the values?

- Revisit values and conflicts identified in concept phase
- Evaluate the design
  - Simulation
  - User testing and experience
- Estimate the societal impacts
  - Challenging due to lack of models
- Legal analysis

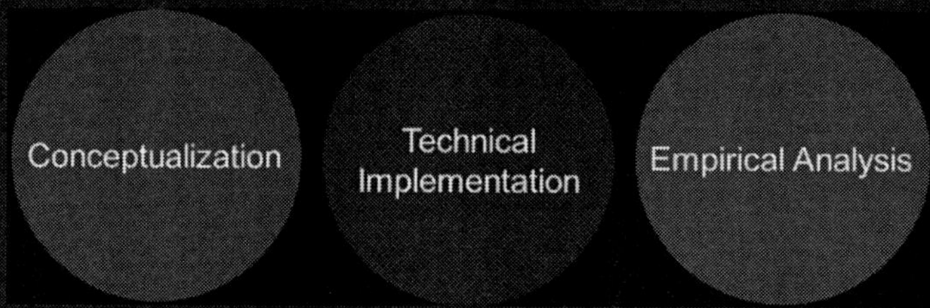

- Seemingly small engineering decisions may have the greatest societal impact
- Value Sensitive Design provides a method for identifying human values and tracing them through the design process
- Interdisciplinary modeling required to more fully map these micro-level decisions into macro-scale societal effects

Autonomous Urban Mobility

Key Lessons from the City of Boston

WORLD ECONOMIC FORUM

COMMITTED TO IMPROVING THE STATE OF THE WORLD

John Moavenzadeh
Member of the Executive Committee, World Economic Forum
Automated Vehicles Symposium
July 11, 2018, San Francisco

The World Economic Forum and Boston Consulting Group have completed a three-year collaboration

2015	2016	2017
Consumer acceptance	**Go Boston 2030**	**Future modal mix research**
Customer research	GO BOSTON 2030 — AV strategy development	Conjoint study with 7,000 consumers in three cities
City perspective	**AV impact study**	**AV impact study 2.0**
City policy maker interviews	Agent-based simulation of downtown traffic	Holistic city-wide traffic simulation with enriched trip data
Mobility scenarios	**AV testing pilot**	**Catalyze AV testing**
Scenarios for urban mobility	Launch of AV testing in Boston	Expanded testing area, three partners, active passenger trials

1

Conducted large-scale conjoint study in three cities

Representative sample

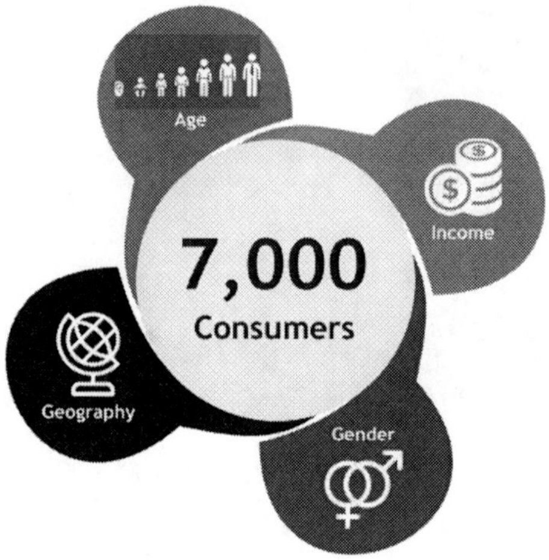

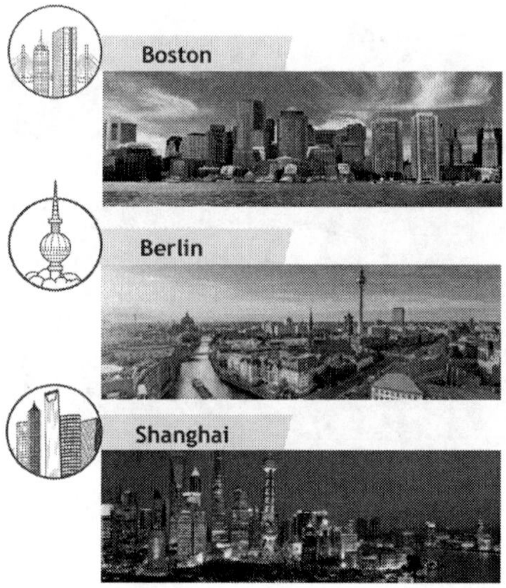

Source: World Economic Forum, BCG Analysis

Respondents were given specific situations and use cases ...

Situations along four criteria

1 | Trip reason

2 | Group context

3 | Weather

4 | Time of day

Use case examples

Commute to work alone

Family trip to the zoo

Night out with friends

Source: World Economic Forum, BCG Analysis

... and asked to chose from 8 transport modes

Mass Transit

Bus/
subway

Commuter
rail

Personal car

Personal
car

Autonomous
personal car

Mobility on demand

Taxi/
ride sharing

Autonomous
taxi

Autonomous
shared taxi

Autonomous
minibus

Source: World Economic Forum, BCG Analysis

4

Top 5 findings from our conjoint study

1. Cities globally move to 30-40% mobility on demand
2. Mass transit ridership drops in urban areas
3. AV adoption varies across city—correlated to income levels
4. The shorter the trip, the higher the AV adoption
5. 20-25% will use a personal car no matter what

Source: World Economic Forum, BCG Analysis

5

①Cities globally move to 30-40% mobility on demand

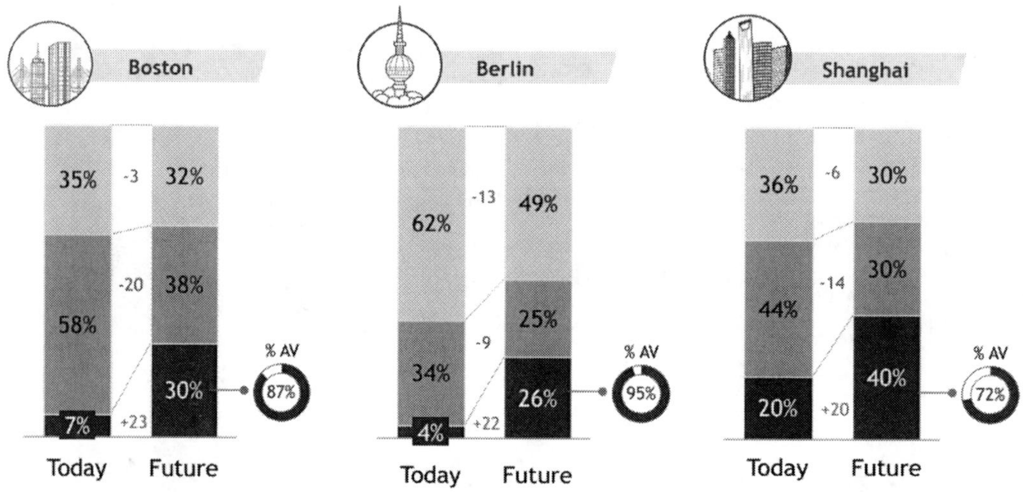

	Mass transit	= Bus / subway + Commuter rail
	Personal car	= Personal car + Autonomous personal car
	Mobility on demand	= Taxi / ride sharing + Autonomous taxi + Autonomous shared taxi + Autonomous minibus

Source: World Economic Forum, BCG Analysis

6

②Mass transit ridership drops in urban areas

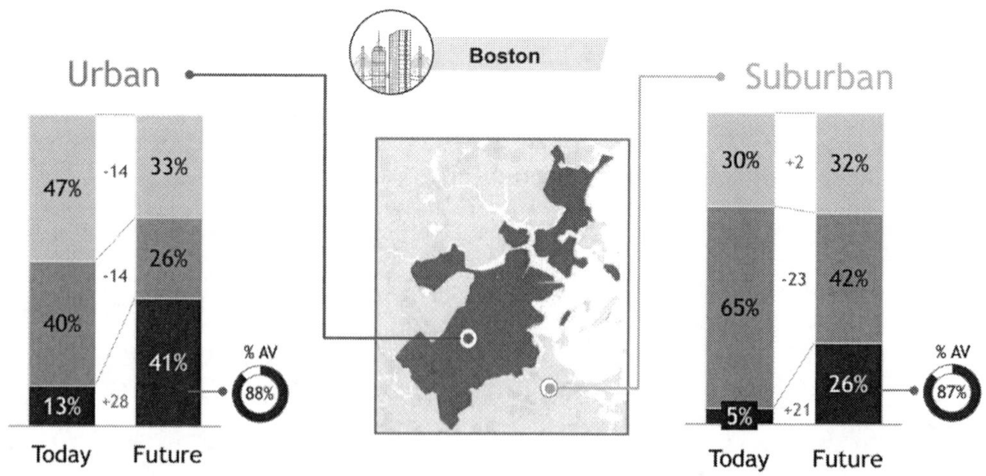

	Mass transit	= Bus / subway + Commuter rail
	Personal car	= Personal car + Autonomous personal car
	Mobility on demand	= Taxi / ride sharing + Autonomous taxi + Autonomous shared taxi + Autonomous minibus

Source: World Economic Forum, BCG Analysis

7

③ AV adoption varies across city – correlated to income

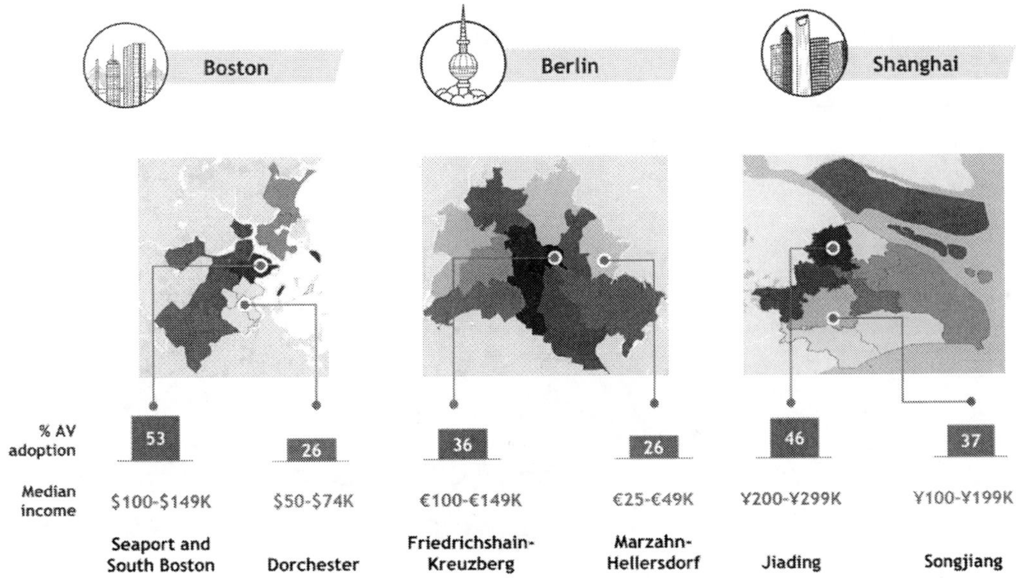

Source: World Economic Forum, BCG Analysis

8

④ The shorter the trip, the higher the AV adoption

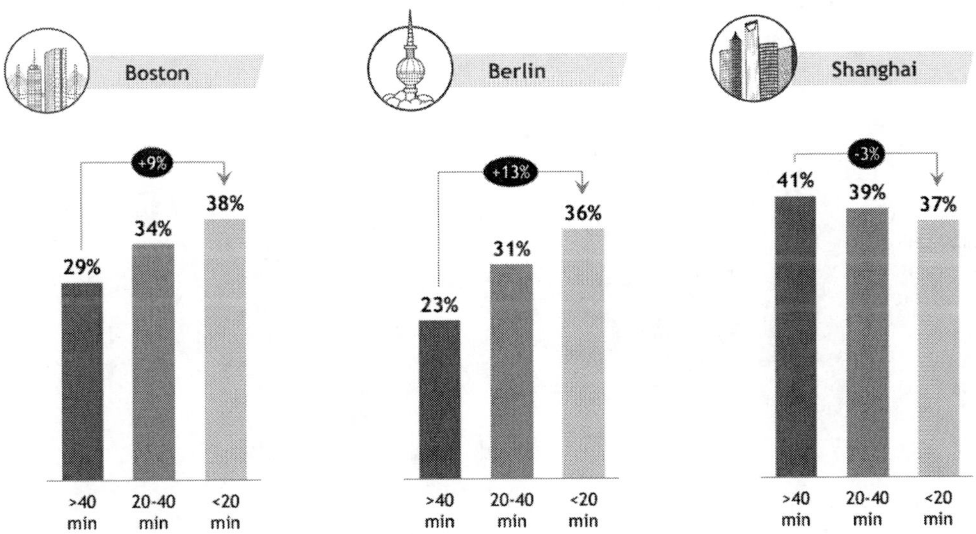

Source: World Economic Forum, BCG Analysis

9

AV testing in Boston—an example for fast scaling

4 months to launch | 12 months of scaling up

Sep 2016	Nov 2016	Jan 2017	Davos 2017	Jun 2017	Oct 2017	Nov 2017	Davos 2018
AV initiative announced	MoU signed	On-street testing started		Testing area expanded	1,000 miles completed	Passenger pilots launched	

nuTonomy — nuTonomy Lyft

OPTIMUS RIDE

•APTIV•

Source: World Economic Forum, City of Boston, BCG Analysis

Testing initially allowed in small area; June 2018 expanded to entire city

Testing Started in "Seaport"

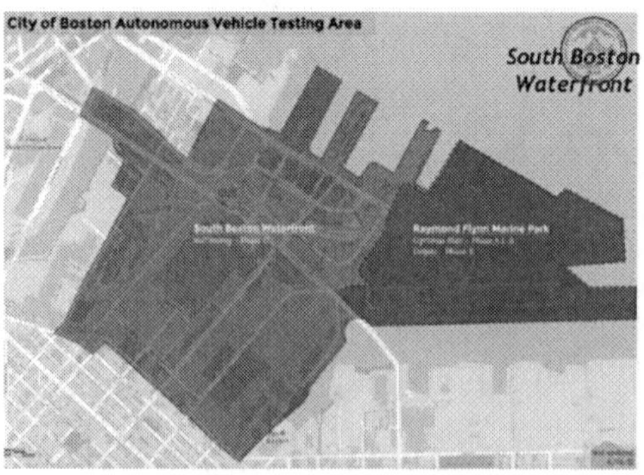

Expanded to Entire City

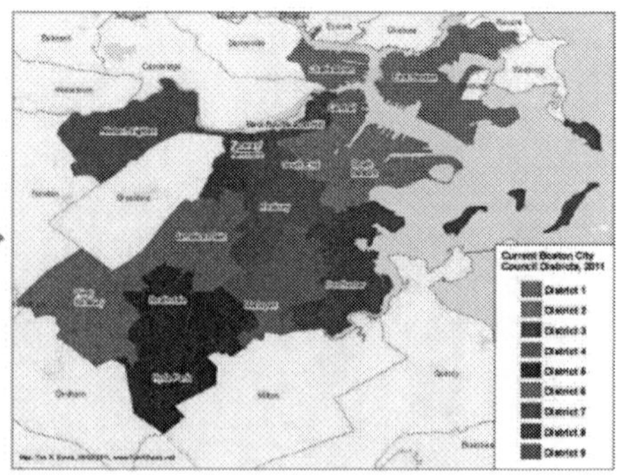

Source: World Economic Forum, City of Boston, BCG Analysis

Identified best practices for launching an AV pilot

- Develop clear mobility vision and KPIs
- Balance stakeholder interests in approval process
- Create a tiered testing plan with achievement milestones
- Build public awareness early, e.g. AV petting zoo/robot block party
- Publish regular updates on testing progress to residents

Source: World Economic Forum, City of Boston, BCG Analysis

12

Last year: Agent-based model for downtown Boston

We took a real world environment ...

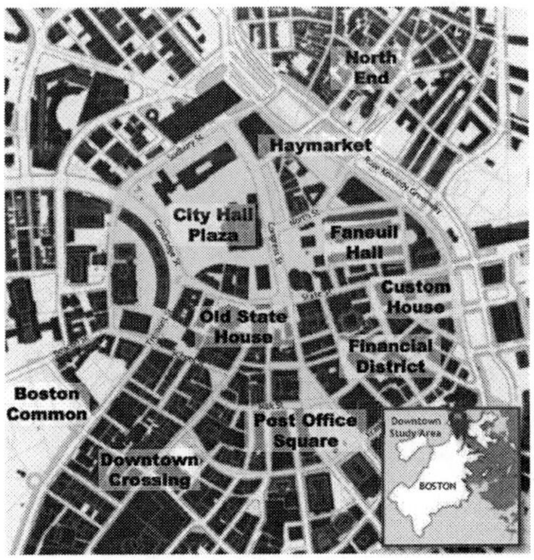

... and simulated traffic flows in its streets

Traffic participants, autonomous and traditional: Cars, Taxis, Pedestrians, Buses, Minibuses

Environment and infra-structure: Traffic lights, Streets

Dynamic behaviors: Following distance, Speed, Traffic rules, Capacity

Source: World Economic Forum; BCG analysis in cooperation with MIT Media Lab

13

Expanded impact study across four key dimensions

1 **Expanded study to cover entire City of Boston**

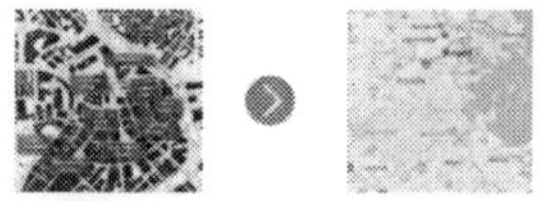

316x
larger area
(142 km²)

2 **Added enriched trip data and commercial vehicles**

CITY of BOSTON ups INRIX

11x
more trips
(2M/day)

3 **Determined future modal mix through conjoint study with 2,400 consumers**

72
modal mixes,
by area & use case

4 **Quantified traffic efficiency gains from AV technology**

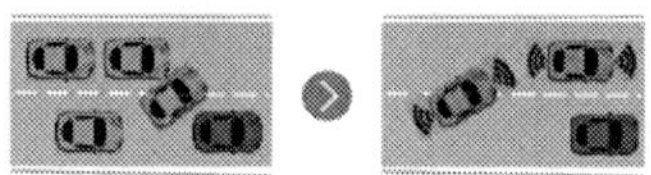

6.3%
gain in throughput
with 37.5% AV share

Source: World Economic Forum, BCG Analysis

14

Key outputs from the city-wide impact study

	Today	Future (Conjoint Scenario)
Traffic volume on the road	1.75M	-15%
Vehicle distance travelled (km)	8.8M	+16%
Parking space needed (km²)	10.0	-48%
Average travel time (min)	12.0	-4%

Source: World Economic Forum, BCG Analysis

15

Congestion increases in Downtown Boston

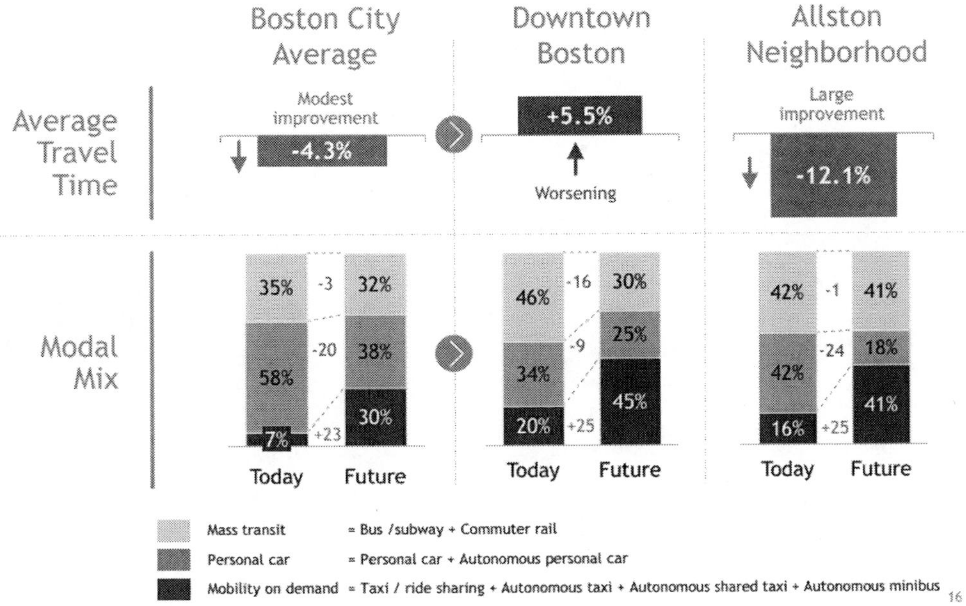

16

Source: World Economic Forum, BCG Analysis

Cities can influence outcome through policy levers

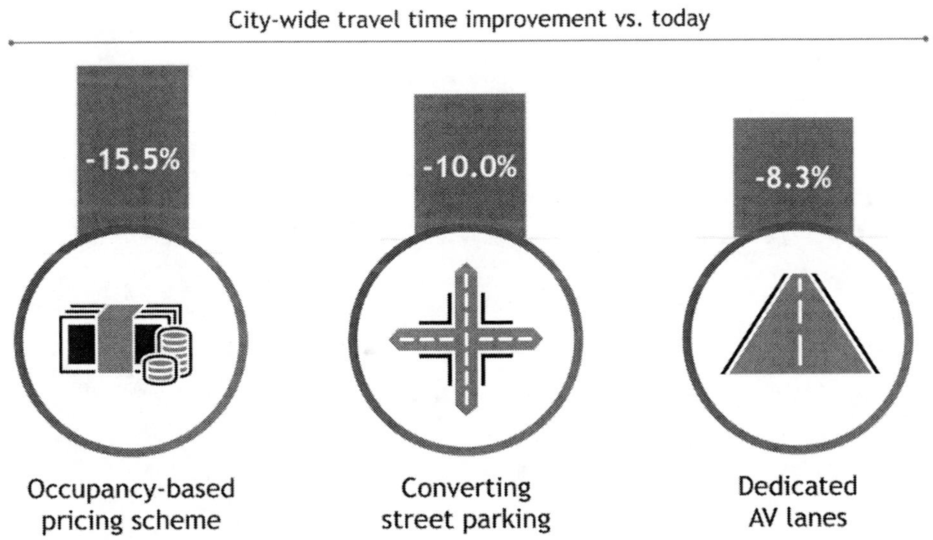

Source: World Economic Forum, BCG Analysis

17

WORLD ECONOMIC FORUM

COMMITTED TO IMPROVING THE STATE OF THE WORLD

Thank you

How Safe is "Safe"

Goal: Safer after pushing button than before

PELOTON TECHNOLOGY

Foundation in Safety

Goal: Not Just Safer Overall, but also:

- Safer even than trucks equipped with CMS/AEB
- Safer in key types of situations:
  - Lead truck hard braking
  - Lead truck collision

Peloton Company Confidential | July 2018 3

PELOTON TECHNOLOGY

Situation: Front truck hard braking

Peloton Company Confidential | July 2018 4

PELOTON TECHNOLOGY

Situation: Front truck collisions

Peloton Company Confidential | July 2018 5

PELOTON TECHNOLOGY

How We Make Platooning Safe

- Safe Functionality:
  - **V2V+V2Cloud**
- Safe Design/HW/SW:
  - **ISO26262**
- Safe Driver:
  - **UI, UX, Driver Study**

Peloton Company Confidential | July 2018 6

Manual Following

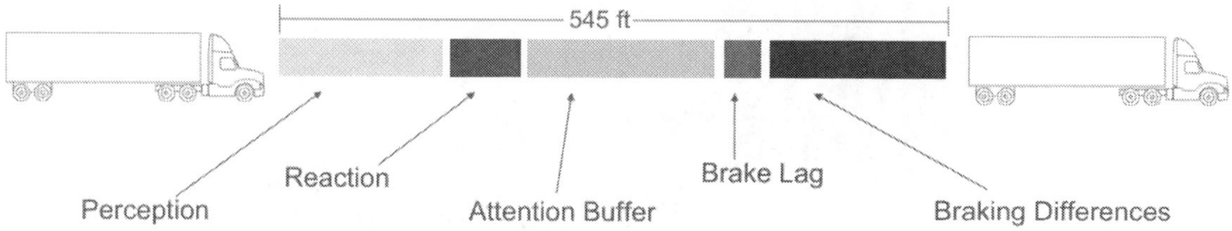

- Safe following distance is many hundreds of feet
- Real world following distances much shorter
  → Collision rate significant
- Many collisions at high impact velocity

Radar/Lidar Following

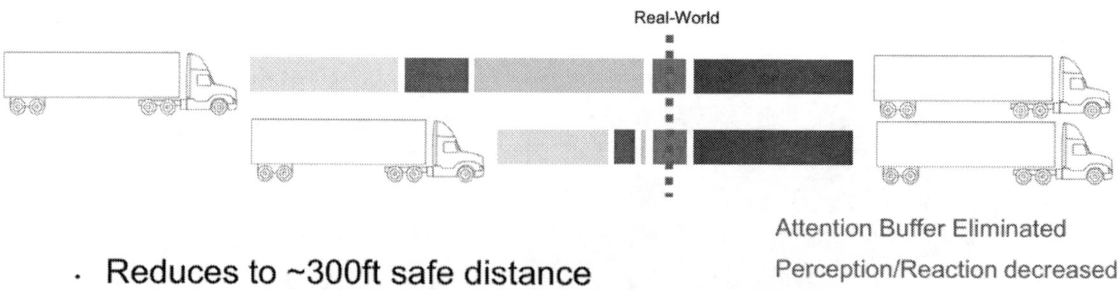

- Reduces to ~300ft safe distance
- In real world, cut in vehicles frequently reduce following distance: Average of <200ft following distance
- Lower but still significant collision rate
- Mitigates impact velocity of many collisions

PELOTON TECHNOLOGY

Platooning

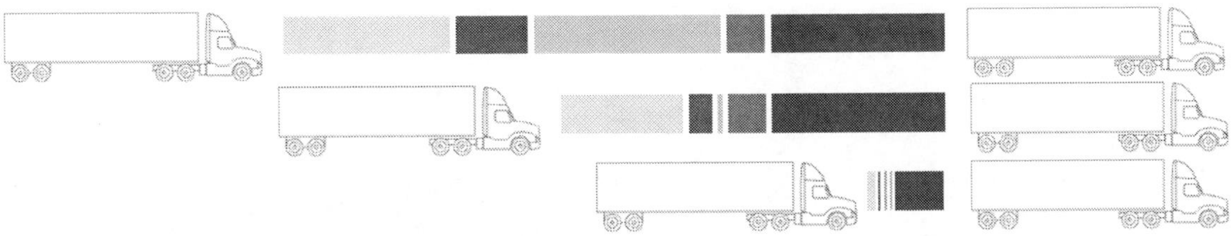

- Brings following distance down to 30-50ft
- Reduces cut-ins → Maintain platoon most of the time
- Our data/analysis suggests:
    - Collisions can still occur, but lower rate
    - Impact velocity reduced in most cases

Peloton Company Confidential | July 2018 9

PELOTON TECHNOLOGY

Power of V2V

Peloton Company Confidential | May 2018 10

PELOTON TECHNOLOGY

Applying ISO 26262* to Platooning

ISO used to create traceable **design requirements** for **V&V**

- **Item Def**: specifies functional concept and Operational Design Domain (ODD)
- **HARA**: identifies hazardous scenarios based on Item Def and generates Safety Goals and ASIL ratings
- **Functional Safety Concept** (**FSC**) and **Functional Safety Requirements** (**FSRs**): high level strategy, FSR and ASIL allocation to system components.
- **Technical Safety Concept** (**TSC**) and **Technical Safety Requirements** (**TSRs**): Specification of Safety Mechanism into testable requirements

ISO **Safety: ISO26262**

Ensure Design/EE issues are below safety goal threshold

PELOTON TECHNOLOGY

How We Test

All tied to ISO26262-Derived Requirements

Impact of platooning on drivers' attention

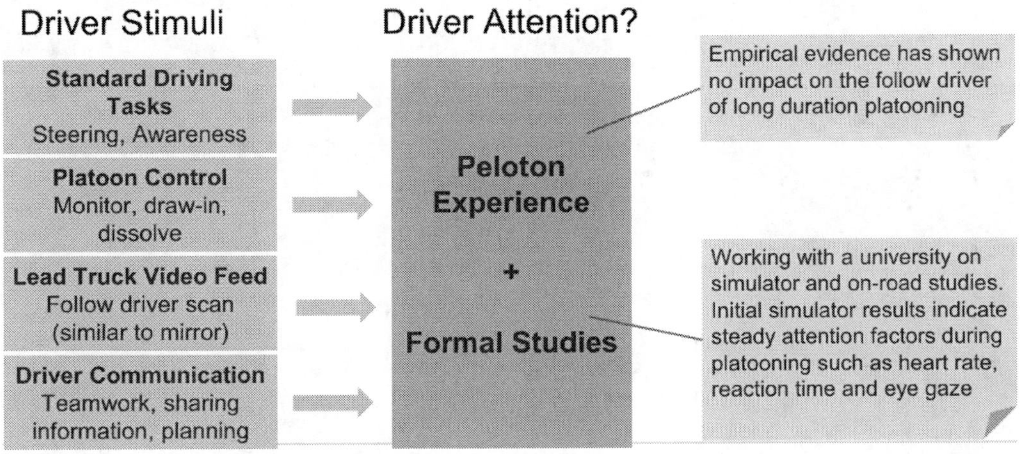

Real Savings

PELOTON TECHNOLOGY

Maximize Platooning Usage

Long-Term

Ubiquitous Platooning

- Density creates opportunities
- Cross-fleet, cross-shipper
- Opportunistic in addition to scheduled

Maximize Platooning Usage

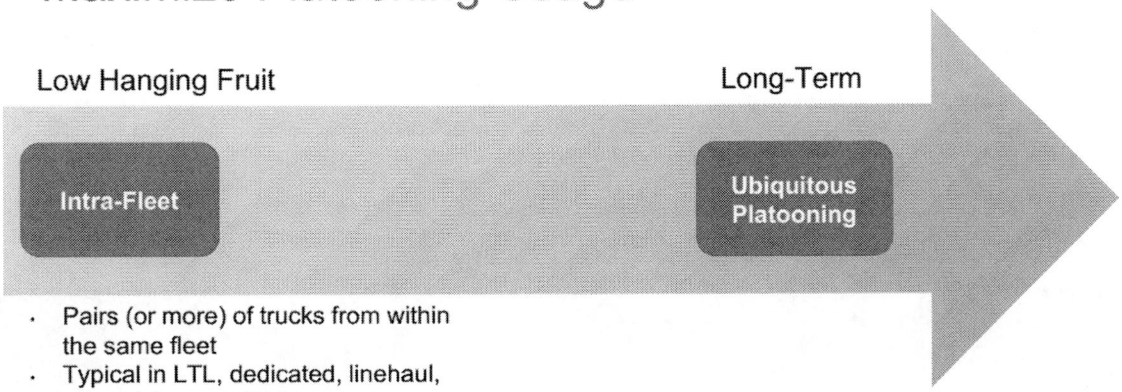

- Pairs (or more) of trucks from within the same fleet
- Typical in LTL, dedicated, linehaul, relay style operations

PELOTON TECHNOLOGY

Maximize Platooning Usage

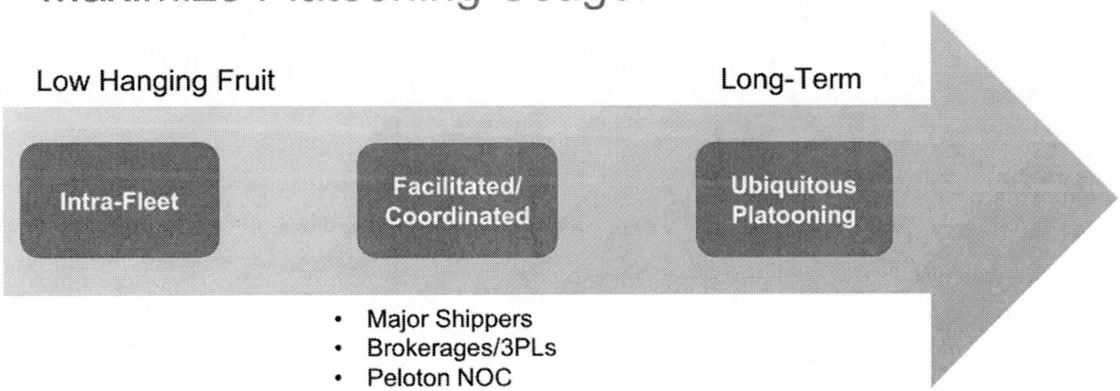

- Major Shippers
- Brokerages/3PLs
- Peloton NOC
- Telematics Partnerships

Press START button
Bendix
Peloton

EMBARK

Safe Testing of Automated Trucks

U.S. Economy

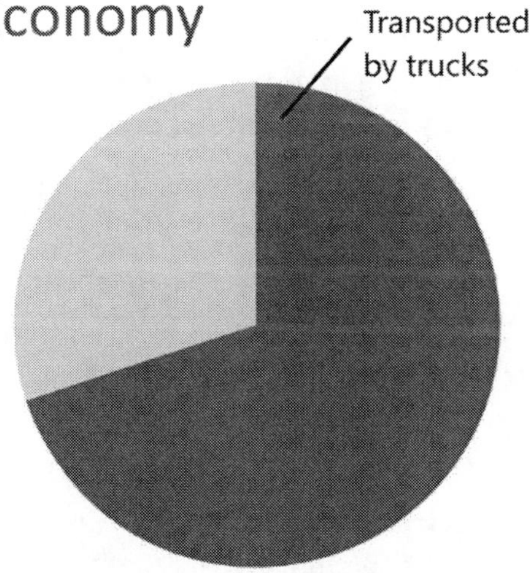

Trucks transport 70% of all goods.

EMBARK

4,000 people die in
truck accidents each year

The only way to
zero fatalities
is self-driving

EMBARK

Advantages of Self-Driving Trucks

53% lower cost per mile

Cost per Mile*		Savings
Driver Wage	33%	-43%
Driver Benefits	10%	
Insurance	5%	-3%
Fuel	21%	-3%
Maintenance	10%	-2%
Lease	16%	-2%
Other (Permits, etc.)	5%	0%
Total	**100%**	**-53%**

*American Transportation Research Institute,
An Analysis of the Operational Costs of Trucking: 2017

24/7 operation

A self-driving truck can cross the country in 2 days instead of 5.

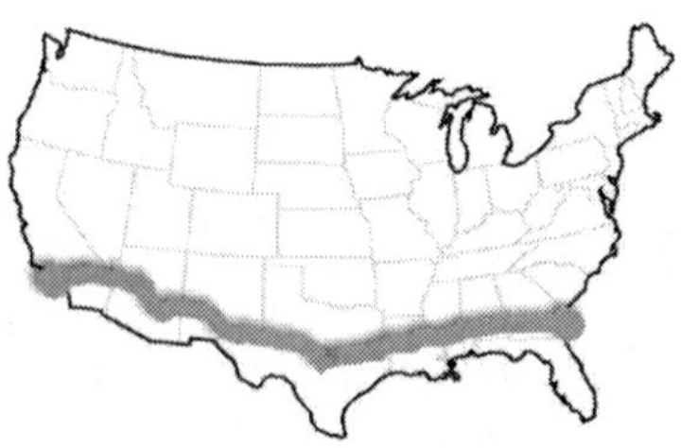

Truck drivers are limited to 11 hours/day by law.

EMBARK

Embark's freight route with Electrolux:

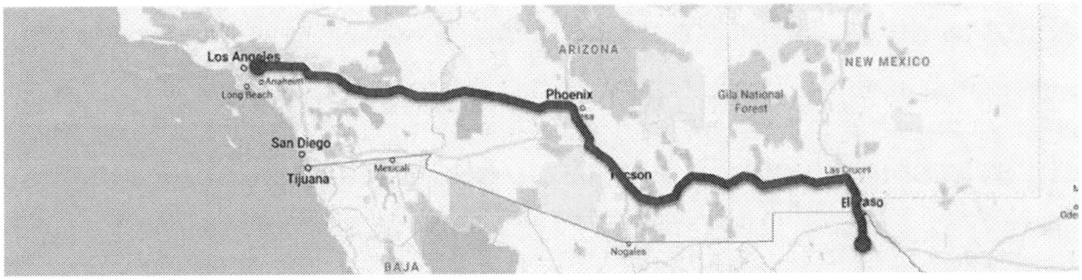

Human Driver

Self-Driving

EMBARK

Embark's system is highway-only

Transfers the load to a local driver for city streets

Embark's Milestones

- First to drive through a major city
- First cross-country trip with an automated truck
- First to ship freight for a Fortune 500 customer

EMBARK

How do we do test safely?

EMBARK

You need an attentive driver

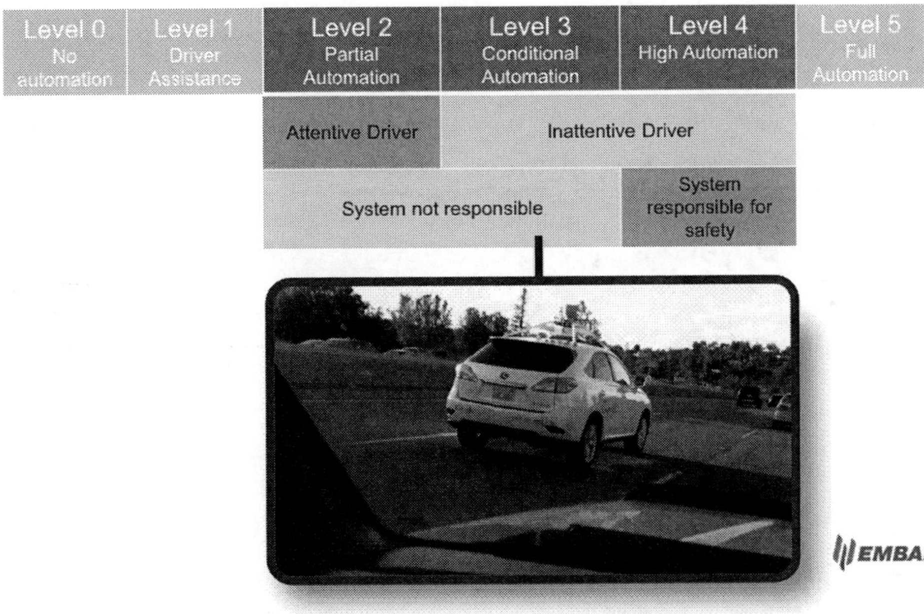

EMBARK

Safety at Embark

- Engineering
- Drivers
    - Selection
    - Training
    - Operating

EMBARK

Engineered for Safety

- No single point of failure
- Run safety checks on all outputs
- Guarantee Driver's ability to disengage

EMBARK

Engineered for Safety

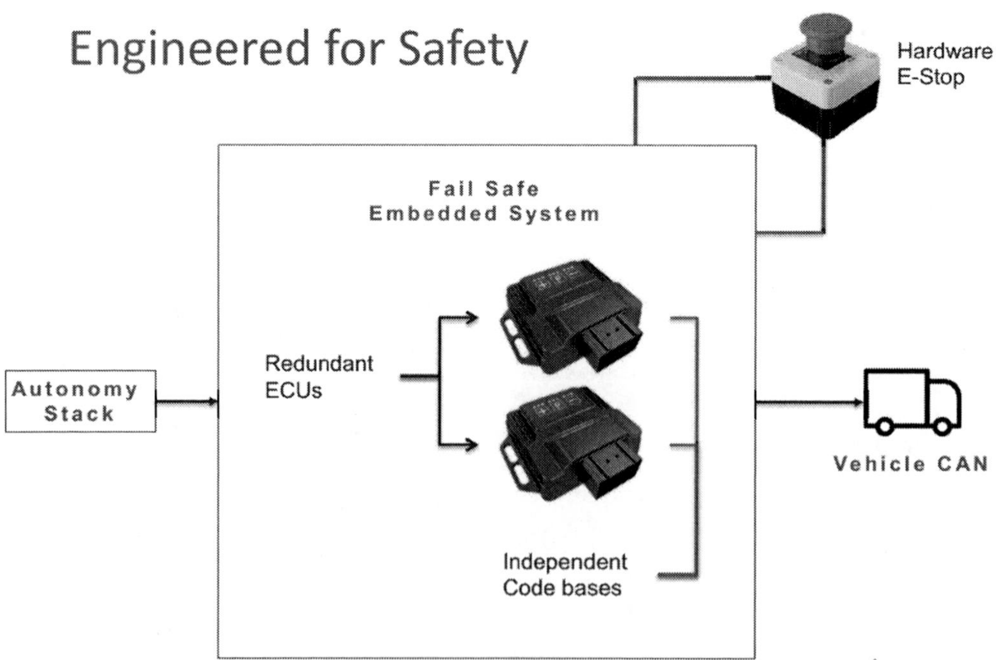

EMBARK

Safety at Embark

Engineering

> Drivers

Selection

Training

Operating

EMBARK

Driver Selection

1. Professional – Specially licensed and trained
2. Experienced
   i. Our drivers average of over a million miles of experience as a professional driver
3. Good driving record
   i. We match the standard of the most selective fleets in America

EMBARK

Driver Training

Multi-week Training Program:

- Classroom:
  - Theory
  - Demonstrations
- Test Track:
  - Self-driving practice
  - Fault injection training
- On-road
- Final Exam (DMV Style)

EMBARK

Operations – Safety Policies

- Stricter driving limits than the federal Hours of Service
- All drivers submit to random drug testing
- No phones allowed while the truck is in motion

EMBARK

Operations – Real-time Monitoring

We continuously monitor key safety metrics and provide feedback to drivers in real time:

- Hands on wheel detection
- Driver gaze monitoring

If driver appears distracted, audio and visual warnings go off for 5 seconds. Following which, the system returns control to the driver.

EMBARK

Operations – Audits

Embark records inward and outward facing camera views at all times:

These videos are audited on a weekly basis to continuously improve our processes.

Audits include both random sampling and specific safety events we've identified.

EMBARK

Key Innovations

Top 3 safety processes we'd like to see adopted across the industry:

- Fault injection training
- Real-time gaze tracking and feedback
- Video audits

EMBARK

EMBARK

Thank You.

The PEGASUS-Method for HAD Assessment - Status Quo -

USA, San Francisco, 2018 July 11th, Dr. Walther Wachenfeld

PEGASUS

Supported by:

Federal Ministry for Economic Affairs and Energy

on the basis of a decision by the German Bundestag

PEGASUS

Project for the establishment of generally accepted quality criteria, tools and methods as well as scenarios and situations for the release of HAD functions.

42 Months Term	January 1st, 2016 – June 30th, 2019
17 German Partners	▪ OEM: Audi, BMW, Daimler, Opel, Volkswagen ▪ Tier 1: Bosch, Continental ▪ Test Lab: TÜV SÜD ▪ SMB: fka, iMAR, IPG, QTronic, TraceTronic, VIRES ▪ Scientific institutes: DLR, TU Darmstadt
12 Subcontracts	▪ i.a. IFR, ika, OFFIS
Project Volume	▪ approx. 34,5 Mio. EUR ▪ Subsidies: 16,3 Mio. EUR
Personnel Deployment	▪ approx. 1.791 man-month or 149 man-years

PEGASUS FOCUS

THE HIGHWAY-CHAUFFEUR

PEGASUS

- **Level of Automation: Level 3 – Conditional Automated**

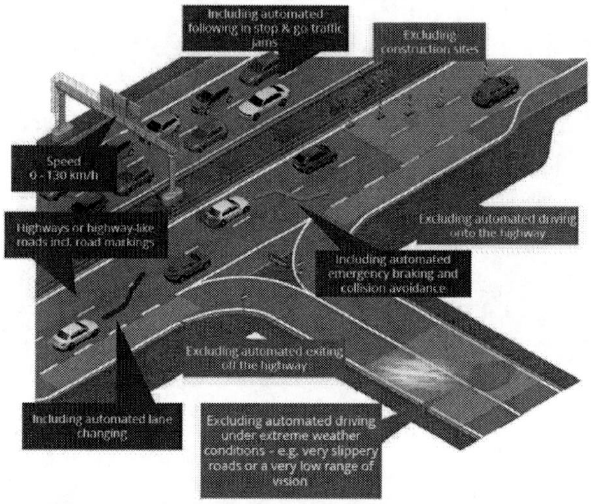

Introduction to the Research Project

Major Questions of the Project

PEGASUS

What level of performance is expected of an automated vehicle?
How can we verify that it achieves the desired performance consistently?

Scenario Analysis & Quality Measures

- What human capacity does the application require?
- What about technical capacity?
- Is it sufficiently accepted? 1
- Which criteria and measures can be deducted from it?

Implementation Process

- Which tools, methods and processes are necessary?

Testing

- How can complete-ness of relevant test runs be ensured?
- What do the criteria and measures for these test runs look like?
- What can be tested in labs or in simulation? What must be tested on proving grounds, what must be tested on the road?

Reflection of Results & Embedding

- Is the concept sustainable?
- How does the process of embedding work?

High Level Requirements

SOCIAL ACCEPTANCE FOR SAE L3+

- What is accepted today?

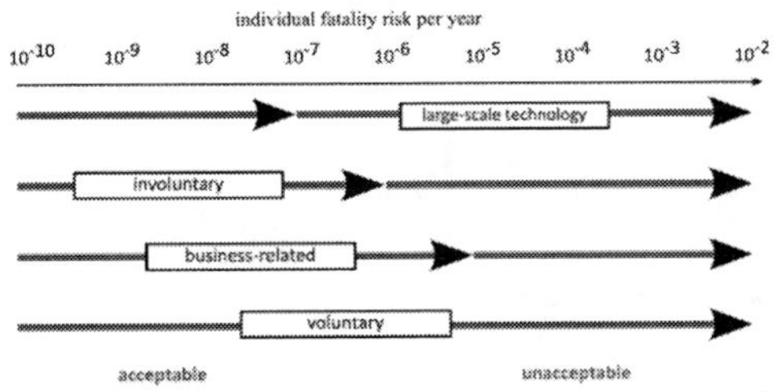

Numbers from A. F. Fritzsche: Wie sicher leben wir? Risikobeurteilung und -bewältigung in unserer Gesellschaft; TÜV Rheinland, Köln 1986, ISBN 3-88585-278-0

- What might be acceptable for SAE L3+?

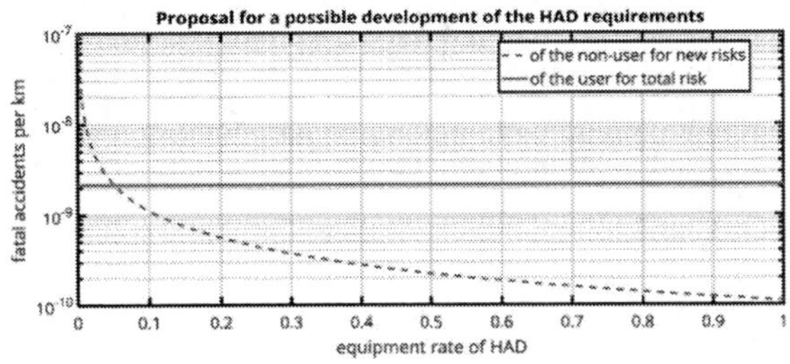

PEGASUS-Method

Interaction for a Statement on Safety

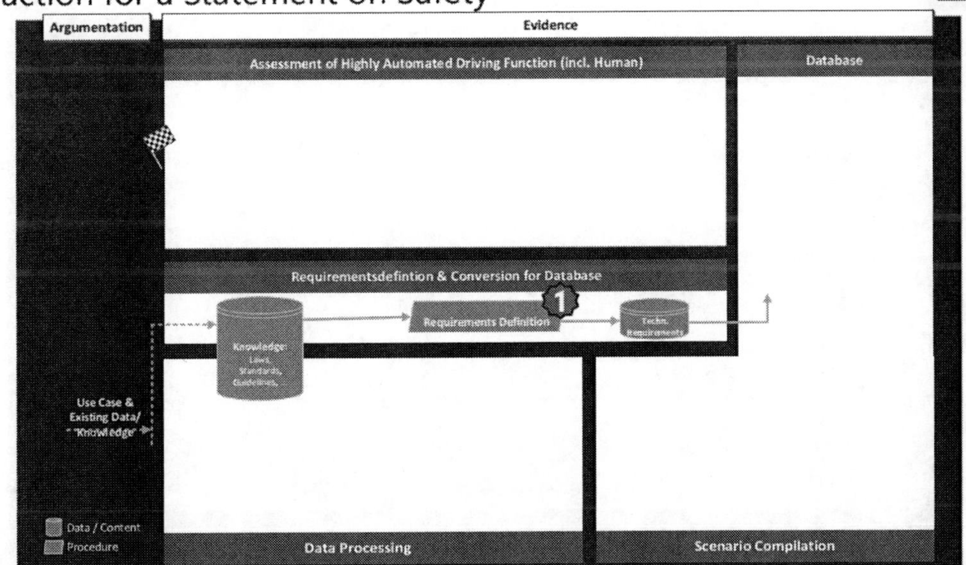

Introduction to the Research Project

Major Questions of the Project

PEGASUS

What level of performance is expected of an automated vehicle?
How can we verify that it achieves the desired performance consistently?

Scenario Analysis & Quality Measures

- What human capacity does the application require?
- What about technical capacity?
- Is it sufficiently accepted? (1)
- Which criteria and measures can be deducted from it?

Implementation Process

- Which tools, methods and processes are necessary?

Testing

- How can complete-ness of relevant test runs be ensured? (2)
- What do the criteria and measures for these test runs look like?
- What can be tested in labs or in simulation? What must be tested on proving grounds, what must be tested on the road?

Reflection of Results & Embedding

- Is the concept sustainable?
- How does the process of embedding work?

© PEGASUS | The PEGASUS-Method for HAD Assessment - Status Quo | W. Wachenfeld 7

PEGASUS-Method

Interaction for a Statement on Safety

PEGASUS

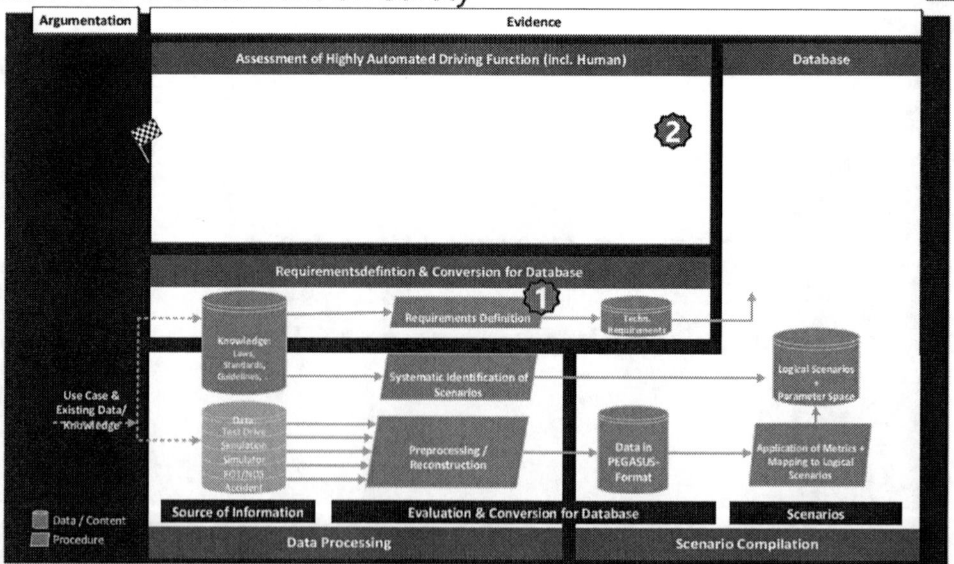

© PEGASUS | The PEGASUS-Method for HAD Assessment - Status Quo | W. Wachenfeld Source: PEGASUS-Symposium Nov. 2017, Exhibition, Topic: "Approach-and-Consistency" 8

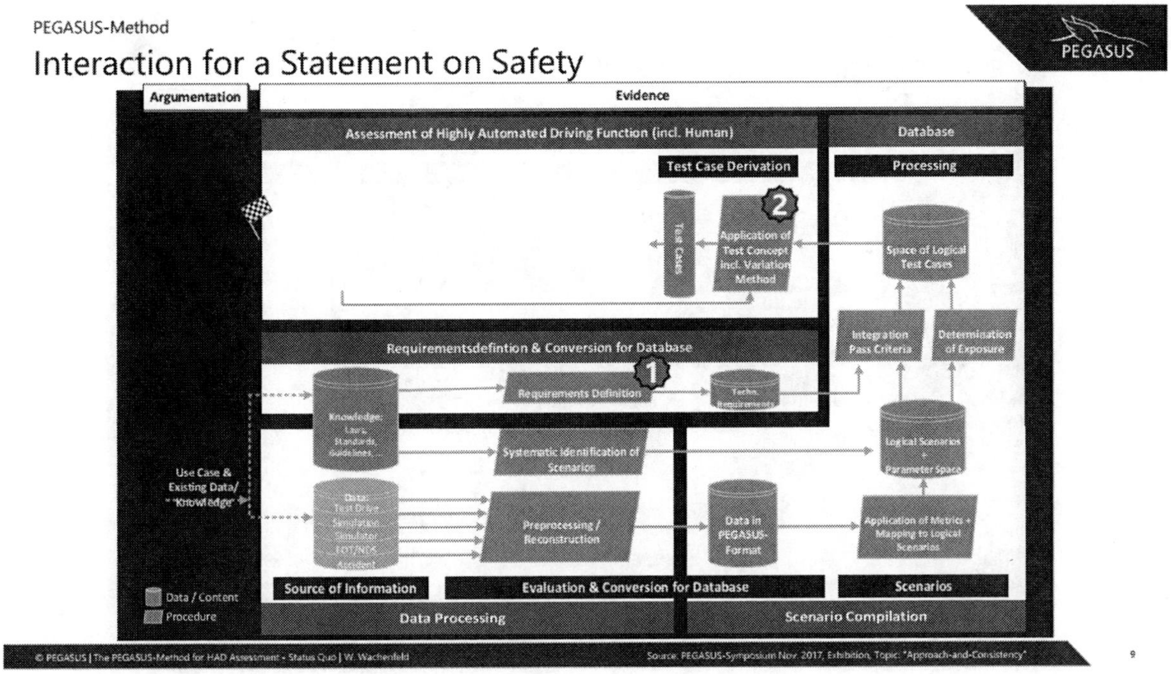
PEGASUS-Method
Interaction for a Statement on Safety
PEGASUS
Argumentation
Evidence
Assessment of Highly Automated Driving Function (incl. Human)
Database
Test Case Derivation
Processing
2
Test Cases
Application of Test Concept incl. Variation Method
Space of Logical Test Cases
Integration Pass Criteria
Determination of Exposure
Requirementsdefintion & Conversion for Database
1
Requirements Definition
Knowledge:
Systematic Identification of Scenarios
Logical Scenarios + Parameter Space
Use Case & Existing Data/ Knowledge
Preprocessing / Reconstruction
Data in PEGASUS-Format
Application of Metrics + Mapping to Logical Scenarios
Data / Content
Procedure
Source of Information
Evaluation & Conversion for Database
Scenarios
Data Processing
Scenario Compilation
© PEGASUS | The PEGASUS-Method for HAD Assessment - Status Quo | W. Wachenfeld
Source: PEGASUS-Symposium Nov. 2017, Exhibition, Topic: "Approach-and-Consistency"
9

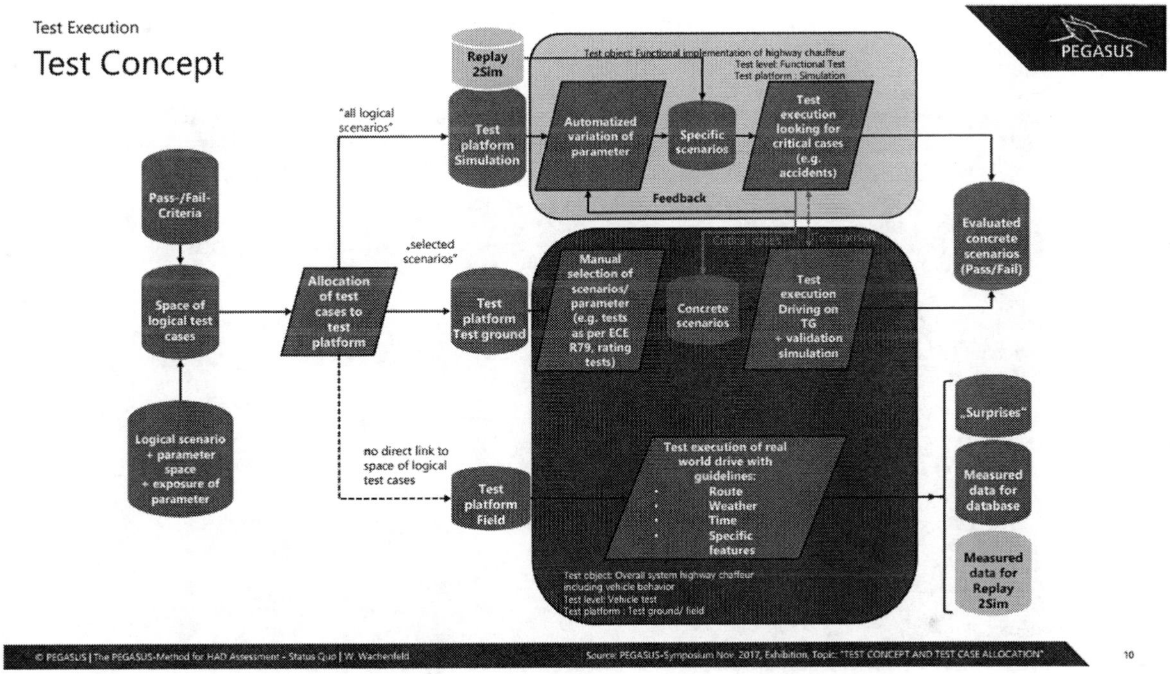
Test Execution
Test Concept
PEGASUS
Replay 2Sim
Test object: Functional implementation of highway chauffeur
Test level: Functional Test
Test platform : Simulation
"all logical scenarios"
Test platform Simulation
Automatized variation of parameter
Specific scenarios
Test execution looking for critical cases (e.g. accidents)
Feedback
Pass-/Fail-Criteria
Evaluated concrete scenarios (Pass/Fail)
„selected scenarios"
Space of logical test cases
Allocation of test cases to test platform
Test platform Test ground
Manual selection of scenarios/ parameter (e.g. tests as per ECE R79, rating tests)
Concrete scenarios
Test execution Driving on TG + validation simulation
„Surprises"
Logical scenario + parameter space + exposure of parameter
no direct link to space of logical test cases
Test platform Field
Test execution of real world drive with guidelines:
Route
Weather
Time
Specific features
Measured data for database
Measured data for Replay 2Sim
Test object: Overall system highway chauffeur including vehicle behavior
Test level: Vehicle test
Test platform : Test ground/ field
© PEGASUS | The PEGASUS-Method for HAD Assessment - Status Quo | W. Wachenfeld
Source: PEGASUS-Symposium Nov. 2017, Exhibition, Topic: "TEST CONCEPT AND TEST CASE ALLOCATION"
10

PEGASUS-Method

Interaction for a Statement on Safety

PEGASUS

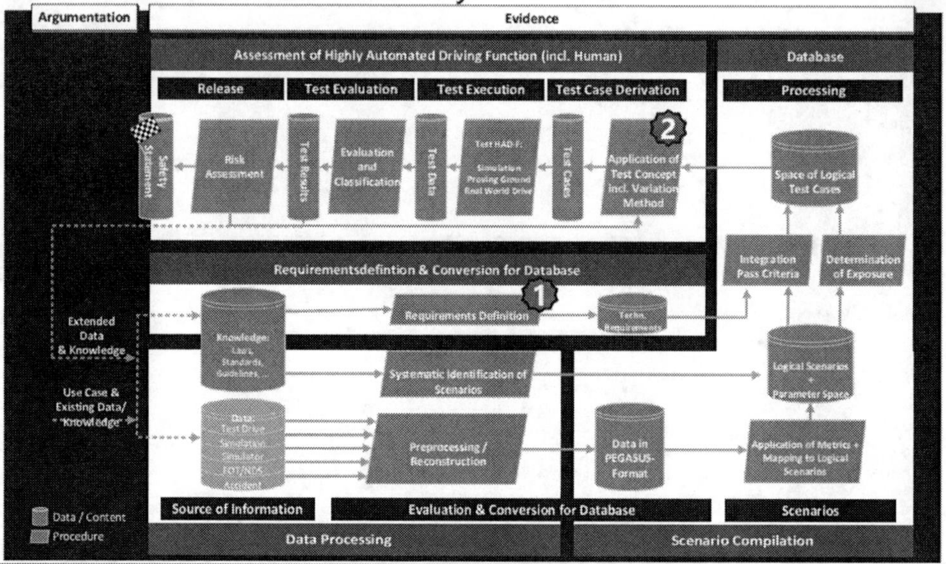

© PEGASUS | The PEGASUS-Method for HAD Assessment - Status Quo | W. Wachenfeld   Source: PEGASUS-Symposium Nov. 2017, Exhibition, Topic: "Approach-and-Consistency"   11

PEGASUS-Method

Interaction for a Statement on Safety

PEGASUS

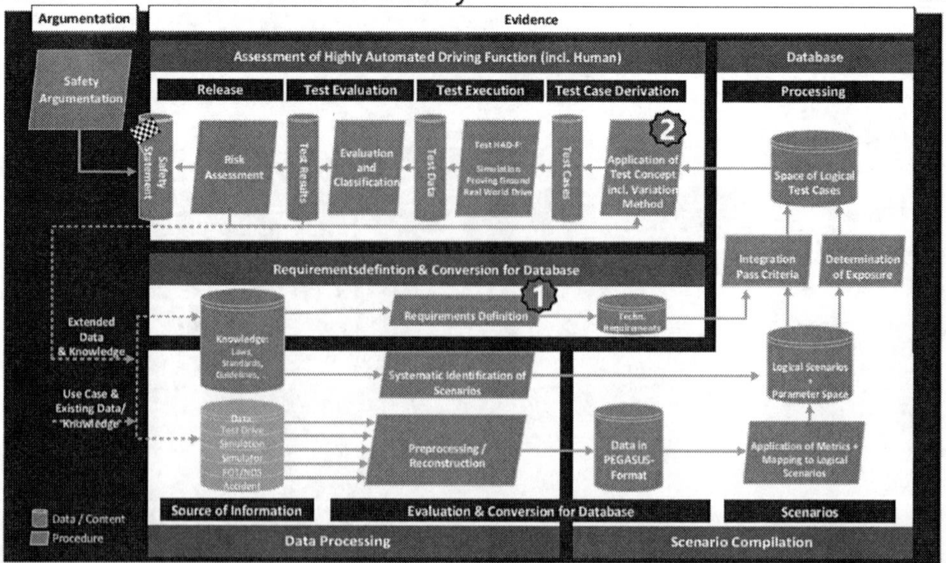

© PEGASUS | The PEGASUS-Method for HAD Assessment - Status Quo | W. Wachenfeld   Source: PEGASUS-Symposium Nov. 2017, Exhibition, Topic: "Approach-and-Consistency"   12

PEGASUS

Contact:

Continental

Dr. Walther Wachenfeld
Continental
Test Architect
Corporate Systems & Technology
walther.wachenfeld@continental-corporation.com

www.pegasusprojekt.de

 13

interACT

Designing cooperative interaction of automated vehicles with other road users in mixed traffic environments

interACT – Project overview

Natasha Merat
University of Leeds

UNIVERSITY OF LEEDS

Project facts

interACT

- **Programme:** EU/H2020-**ART04** - *Safety and end-user acceptance aspects of road automation in the transition period*
- **Duration**: 36 months
- **Period**: May 2017 – April 2020
- **EU Funding**: 5.527.581 €
- **Coordinator:** Anna Schieben, DLR German Aerospace
- **Partners:** 8 industrial and academic partners from 4 European countries (Germany, Italy, Greece, UK)
- **US - EU twinning project**: AVIntent (NHTSA)

Question: how will AVs communicate and interact with other road users?

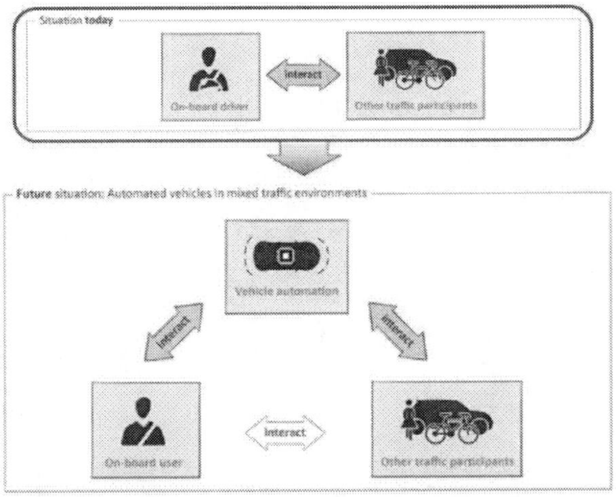

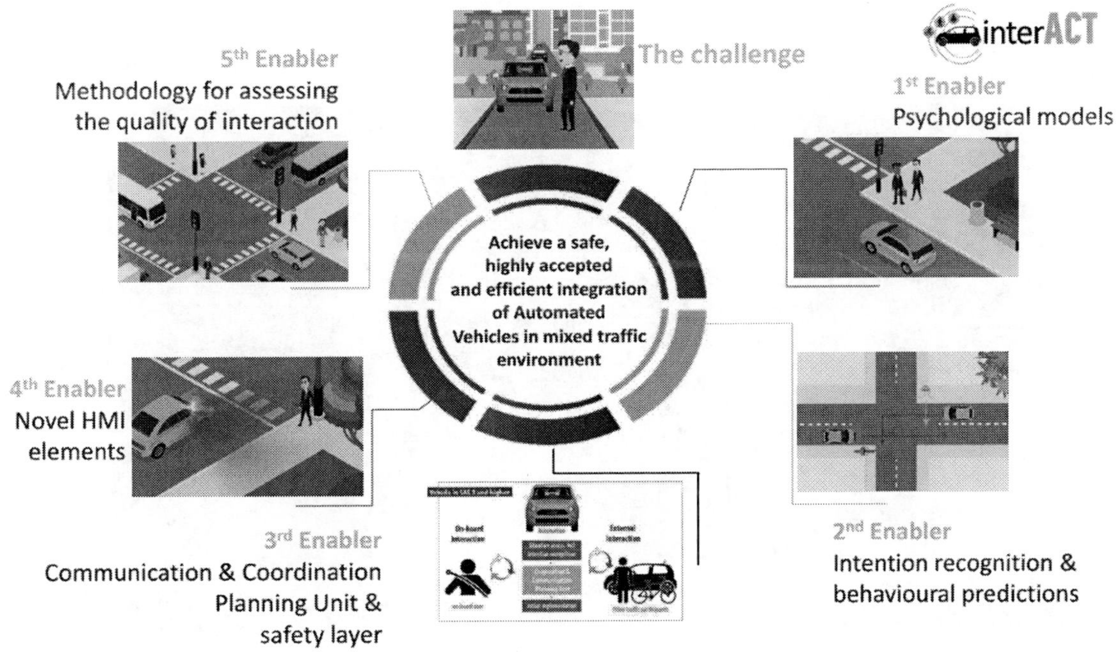

interACT

Observation of Pedestrian-Vehicle Encounters

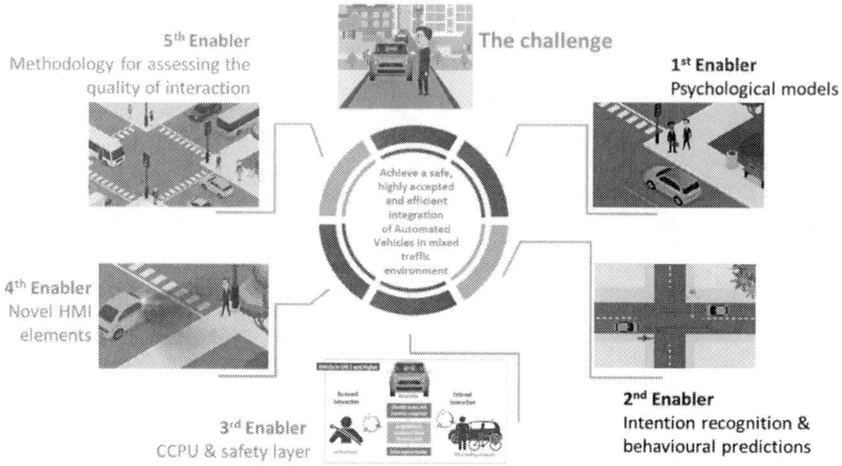

UNIVERSITY OF LEEDS

Interaction in Urban Traffic – Insights into an Observation of Pedestrian-Vehicle Encounters

Interaction today

interACT

Key Objectives

- **Observe** human-human interactions in current complex urban environments

UNIVERSITY OF LEEDS

Interaction in Urban Traffic – Insights into an Observation of Pedestrian-Vehicle Encounters

Key Objectives

- Observe human-human interactions in current complex urban environments
- **Model the interactions** using different approaches:
  - Interaction vocabulary: *How do Traffic Participants communicate and anticipate intent*
  - Interaction sequences: *What is the general interaction process in specific use cases and scenarios and what are the cultural difference?*
  - Quantitative models: *How can interactions be mathematically formulated to allow model-in-the-loop simulations?*

UNIVERSITY OF LEEDS

Interaction in Urban Traffic – Insights into an Observation of Pedestrian-Vehicle Encounters

Key Objectives

- Observe human-human interactions in current complex urban environments
- Model interaction using different approaches
  - Interaction vocabulary: *How do TPs communicate and anticipate intent*
  - Interaction sequences: *What is the general interaction process in specific use cases, scenarios and scenes?*
  - Quantitative models: *How can interactions be mathematically formulated to allow model in the loop simulations?*
- Develop real-time **situation and intention analysis** algorithms based on the interaction models

UNIVERSITY OF LEEDS

Interaction in Urban Traffic – Insights into an Observation of Pedestrian-Vehicle Encounters

Key Objectives

- Observe human-human interactions in current complex urban environments
- Model interaction using different approaches
  - Interaction vocabulary: *How do TPs communicate and anticipate intent*
  - Interaction sequences: *What is the general interaction process in specific use cases, scenarios and scenes?*
  - Quantitative models: *How can interactions be mathematically formulated to allow model in the loop simulations?*
- Develop real-time situation and intention analysis algorithms based on the interaction models

Observe, Model, Predict

UNIVERSITY OF LEEDS

Interaction in Urban Traffic – Insights into an Observation of Pedestrian-Vehicle Encounters

interACT

Methodology

- 3 Countries

- 4 Use Cases

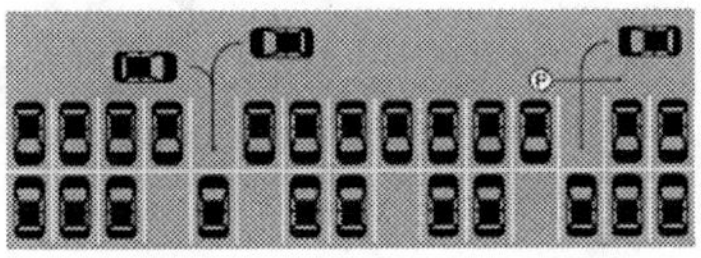

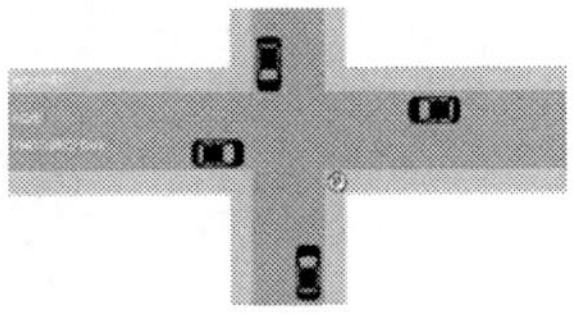

UNIVERSITY OF LEEDS

Interaction in Urban Traffic – Insights into an Observation of Pedestrian-Vehicle Encounters

interACT

Methodology

Naturalistic observation of urban traffic

- Video
- Observation Protocols
- Questionnaires
- LiDAR

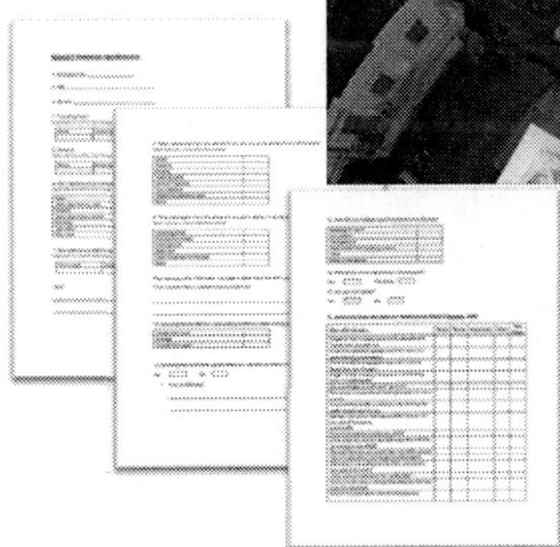

UNIVERSITY OF LEEDS

Interaction in Urban Traffic – Insights into an Observation of Pedestrian-Vehicle Encounters

Methodology

interACT

Video:

- Birds eye view perspective of locations chosen to represent the use-cases
- Algorithmic analysis of the videos to derive positions and velocities of various traffic participants

UNIVERSITY OF LEEDS

Interaction in Urban Traffic – Insights into an Observation of Pedestrian-Vehicle Encounters

Methodology

interACT

LiDAR:

- Stationary LiDAR giving additional information on traffic participants and increasing tracking range
- Collected data is synchronized in time enabling a holistic overview of observed interactions

UNIVERSITY OF LEEDS

Interaction in Urban Traffic – Insights into an Observation of Pedestrian-Vehicle Encounters

Methodology

interACT

Manual Observation:

- Observers recording individual observed interactions from the ground
- HTML based app for tablets observing pedestrian and driver behaviour, including head rotation, eye contact, etc.
- Questionnaires

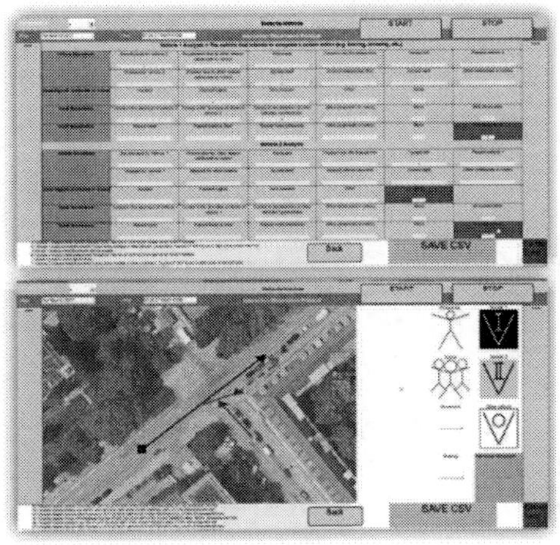

UNIVERSITY OF LEEDS

Interaction in Urban Traffic – Insights into an Observation of Pedestrian-Vehicle Encounters

Preliminary Results – Manual Observation

interACT

- Over 100 Protocols per use case and country
- Also: combined 100+ hours of videos, 20+ hours of LiDAR Data and 150+ people interviewed

UNIVERSITY OF LEEDS

Interaction in Urban Traffic – Insights into an Observation of Pedestrian-Vehicle Encounters

Preliminary Results – Manual Observation

"Interaction Sequence" -Intersection – pedestrian goes first:

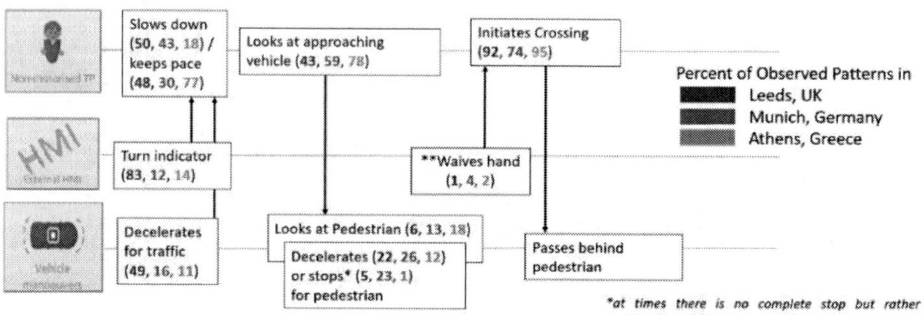

**at times there is no complete stop but rather a continuation of the movement at a very slow pace*

***in some cases there was no hand waiving and the scenario played out comparably*

Overall Findings

- The occurrence and necessity of interactions depends on the situation and a variety of **other factors**, such as traffic density, time of day and specific traffic conditions
- **Explicit communication (e.g. gesturing, flashing lights etc.) happens rarely** - most potential interaction-demanding situations are resolved before they actually arise, mostly by adjusting *kinematic motion*
- **Cooperation, communication** and thus **interaction** between human road users takes place at **low speeds**, usually below 20 km/h
- At **higher speeds, conflict avoidance** is predominant – pedestrians use large enough inter-vehicle gaps to cross without expecting the second vehicle to adapt
- **Self reports ≠ reality**: Observations did not match pedestrians' reporting
- **Cultural Differences** are complex: What is "culture"?

Overall Findings

interACT

- Human road users seem to **avoid active communication with others** by adapting their movement behavior early
- Communication is only used in **ambiguous situations** to let the other traffic participant go first, mostly **using gestures**

UNIVERSITY OF LEEDS

Interaction in Urban Traffic – Insights into an Observation of Pedestrian-Vehicle Encounters

Some (preliminary) Conclusions

interACT

- The use of "external Human Machine Interfaces" is only relevant in **ambiguous situations**, when explicit communication is necessary above and beyond *kinematic cues*
- BUT – Unlike manually driven vehicles, in addition to adapting their movement, perhaps Automated Vehicles could **enhance acceptance, safety and traffic flow** by communicating their intentions to other traffic participants earlier.

BREAKOUT SESSION #25

The Role of Human Factors in the Design of Automated Vehicle External Communications

Wednesday, July 11, 2018, 1:30 PM – 5:30 PM
Room: Yosemite C

Organizers:

Andy Schaudt, Virginia Tech Transportation Institute
John Shutko, Ford Motor Company
Sheldon Russell, Virginia Tech Transportation Institute

UNIVERSITY OF LEEDS

Interaction in Urban Traffic – Insights into an Observation of Pedestrian-Vehicle Encounters

interACT

Designing cooperative interaction of automated vehicles with other road users in mixed traffic environments

Thank you

Any questions?

http://interact-roadautomation.eu

UNIVERSITY OF LEEDS n.merat@its.leeds.ac.uk @NatashaMerat

This project has received funding from the European Union's Horizon 2020 research and innovation programme under grant agreement No 723395

AUTOMATED VEHICLES SYMPOSIUM 2018
USERS. VEHICLES. INFRASTRUCTURE.
Driven by AUVSI and TRB

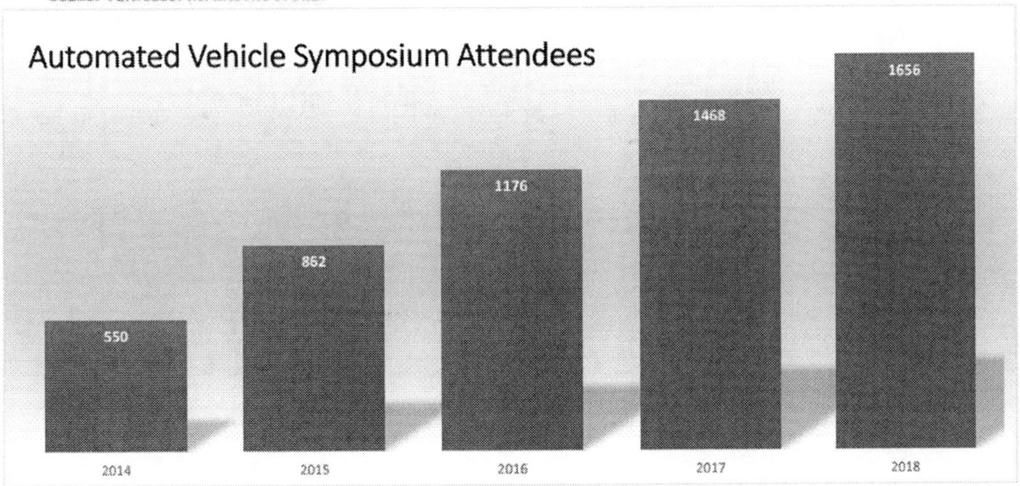

AUTOMATED VEHICLES SYMPOSIUM 2018
USERS. VEHICLES. INFRASTRUCTURE.
Driven by AUVSI and TRB

32 Countries Represented

AUSTRALIA	**GERMANY**	**KOREA**	**SPAIN**
AUSTRIA	**HONG KONG**	**NETHERLANDS**	**SWEDEN**
BELGIUM	**HUNGARY**	**NORWAY**	**SWITZERLAND**
CANADA	**INDIA**	**PORTUGAL**	**TAIWAN**
CHINA	**ISRAEL**	**SAUDI ARABIA**	**UNITED ARAB EMIRATES**
DENMARK	**ITALY**	**SINGAPORE**	**UNITED KINGDOM**
FINLAND	**JAPAN**	**SOUTH AFRICA**	**UNITED STATES**
FRANCE	**JORDAN**	**SOUTH KOREA**	

AUTOMATED VEHICLES SYMPOSIUM 2018

USERS. VEHICLES. INFRASTRUCTURE.

Driven by AUVSI and TRB

AUSTRALIA
AUSTRIA
BELGIUM
CANADA - 20
CHINA
DENMARK
FINLAND
FRANCE
GERMANY - 25
HONG KONG
HUNGARY
INDIA
ISRAEL
ITALY
JAPAN - 57
JORDAN
KOREA
NETHERLANDS - 16
NORWAY
PORTUGAL
SAUDI ARABIA
SINGAPORE
SOUTH AFRICA
SOUTH KOREA
SPAIN
SWEDEN
SWITZERLAND
TAIWAN
UNITED ARAB EMIRATES
UNITED KINGDOM - 20
UNITED STATES

23% International

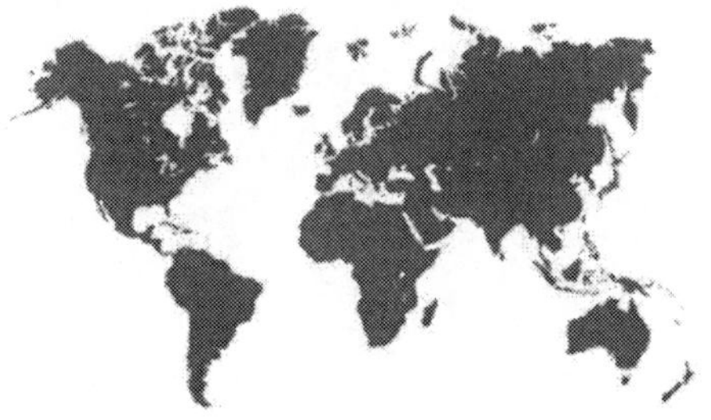

AUTOMATED VEHICLES SYMPOSIUM 2018

USERS. VEHICLES. INFRASTRUCTURE.

Driven by AUVSI and TRB

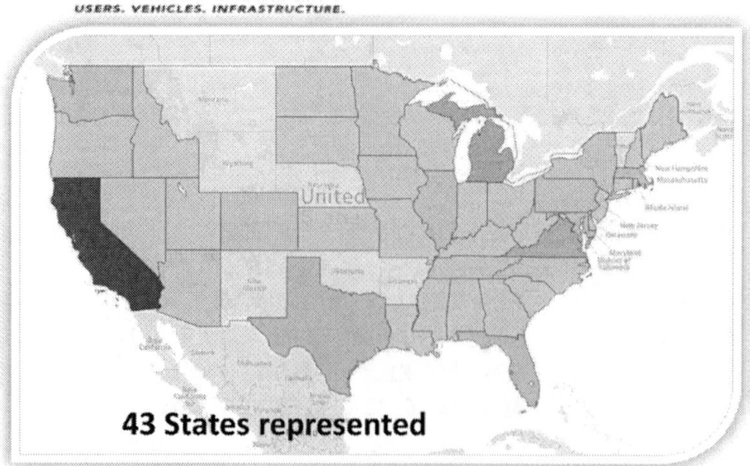

Top 10 States

CA	543
MI	131
DC	119
VA	89
TX	52
FL	43
MA	41
WA	36
PA	35
IL	34

AUTOMATED VEHICLES SYMPOSIUM 2018

USERS. VEHICLES. INFRASTRUCTURE.

Driven by AUVSI and TRB

DEMO '98

Automated Vehicles Symposium 2018

SIP-adus : Field Operational Tests and Regulatory Issues

SIP-adus: Cross-Ministerial Strategic Innovation Promotion Program
Innovation of Automated Driving for Universal Services

July 12, 2018

Hajime Amano
President and CEO, ITS Japan
Chair, International Corporation WG, SIP-adus

SIP Cabinet Office

Technologies for Automated Driving Systems

1

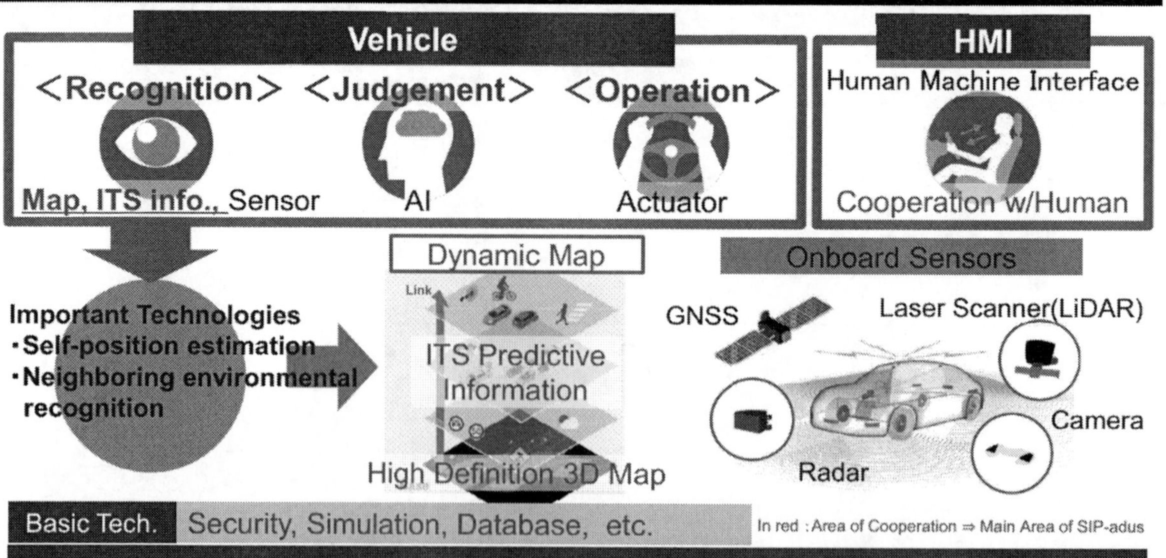

SIP-adus focus on R&D in Cooperative area with Industry, Academia and Government

Objectives of Field Operational Tests

2

Verification of research results in 5 integrated themes

- Dynamic Map
- Human Machine Interface (HMI)
- Cyber Security
- Pedestrian Accident Reduction
- Next Generation Transport

International cooperation sharing the test fields and the data sets

- International participants signed up (OEMs, suppliers and research institutes)
- Concrete evidence acquired through the tests on the common grounds
- In-depth discussions on the specific research topics
- Identification of shared challenges and direction to overcome them

Business model investigation

Field Operational Tests: Test Sites

3

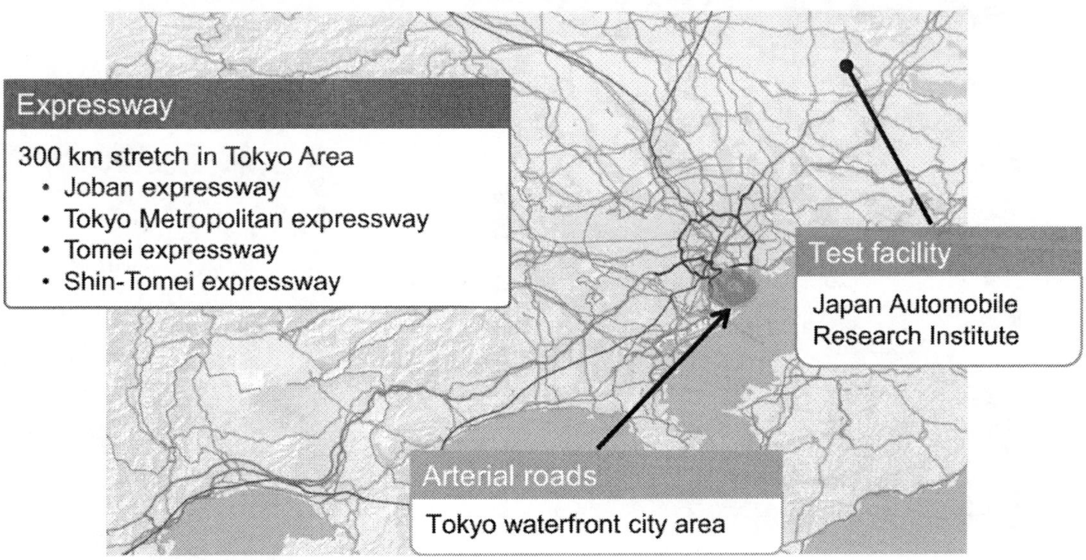

Field Operational Tests: Participants

4

ALPINE Driving Mobile Media Innovation

BMW

MINI

BOSCH Invented for life

Calsonic Kansei

Continental

DAIHATSU

HONDA The Power of Dreams

mazda

Mercedes-Benz

MLT MEIJI LogiTech

MITSUBISHI ELECTRIC

名古屋大学 NAGOYA UNIVERSITY

NISSAN MOTOR CORPORATION

OMRON

Pioneer

埼玉工業大学 SAITAMA INSTITUTE OF TECHNOLOGY

SUZUKI

SUBARU

TOYOTA

Valeo

VOLKSWAGEN AKTIENGESELLSCHAFT

ZMP

Alphabetical order

Vehicle Position Detection using Dynamic Map

5

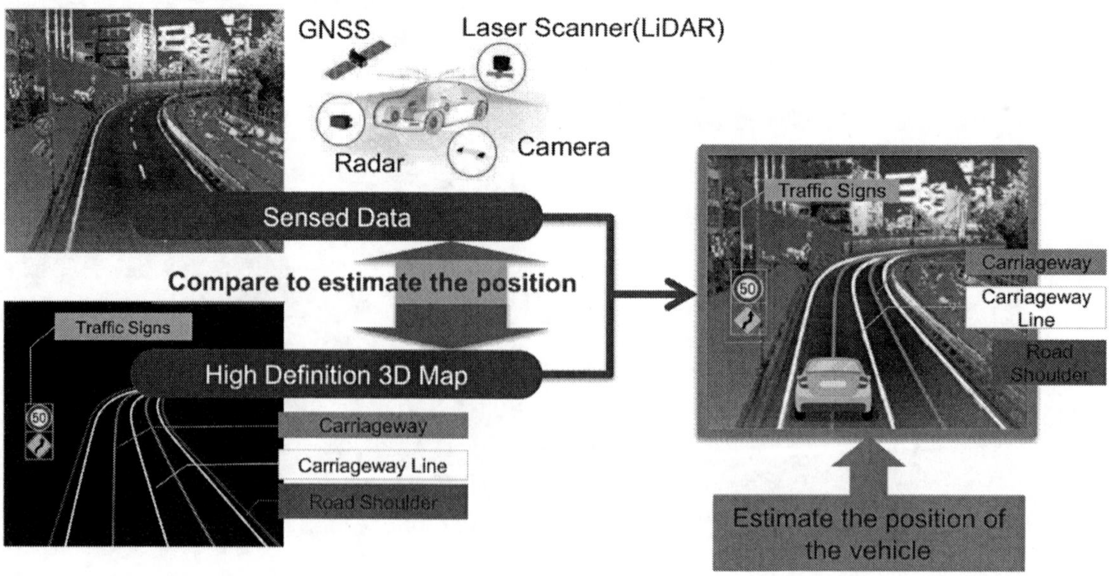

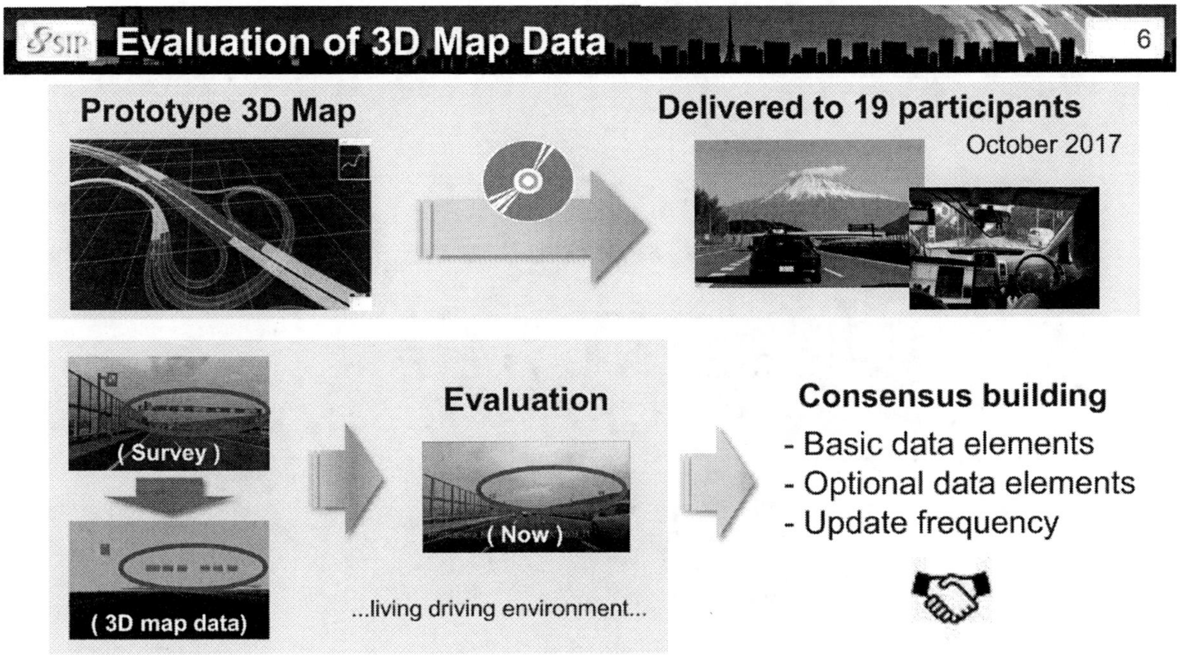
SIP
Evaluation of 3D Map Data
6
Prototype 3D Map
Delivered to 19 participants
October 2017
(Survey)
(3D map data)
Evaluation
(Now)
...living driving environment...
Consensus building
- Basic data elements
- Optional data elements
- Update frequency

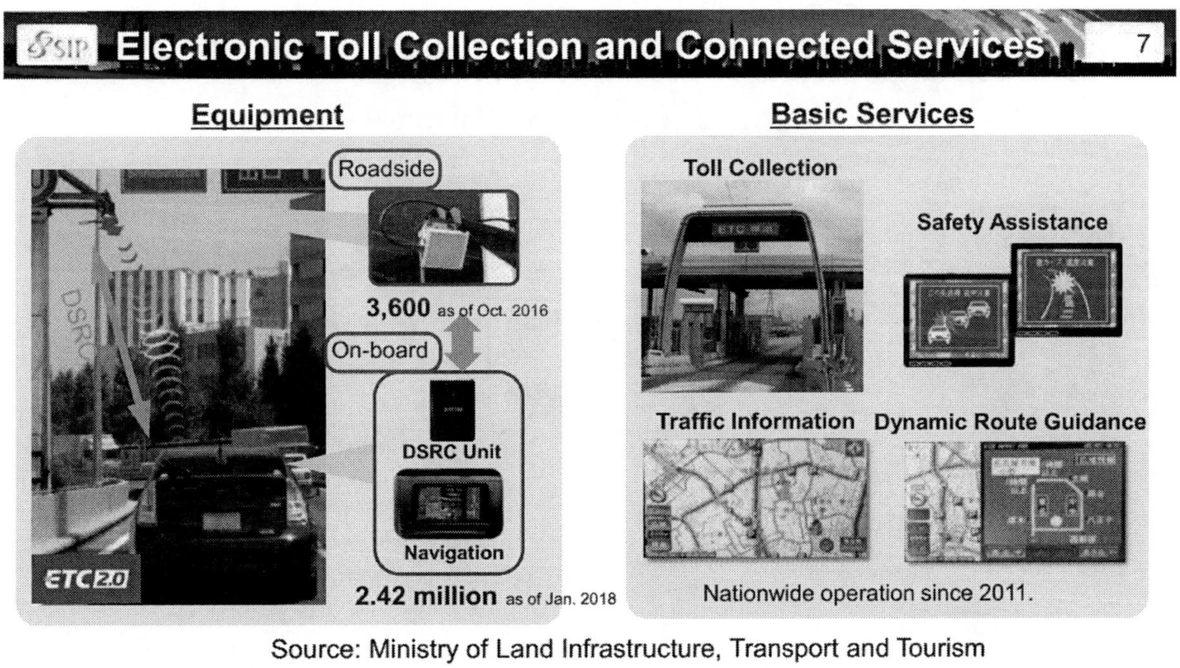
SIP
Electronic Toll Collection and Connected Services
7
Equipment
Roadside
3,600 as of Oct. 2016
On-board
DSRC
DSRC Unit
Navigation
ETC2.0
2.42 million as of Jan. 2018
Basic Services
Toll Collection
Safety Assistance
Traffic Information
Dynamic Route Guidance
Nationwide operation since 2011.
Source: Ministry of Land Infrastructure, Transport and Tourism

Traffic Signal Prediction Systems (TSPS) 8

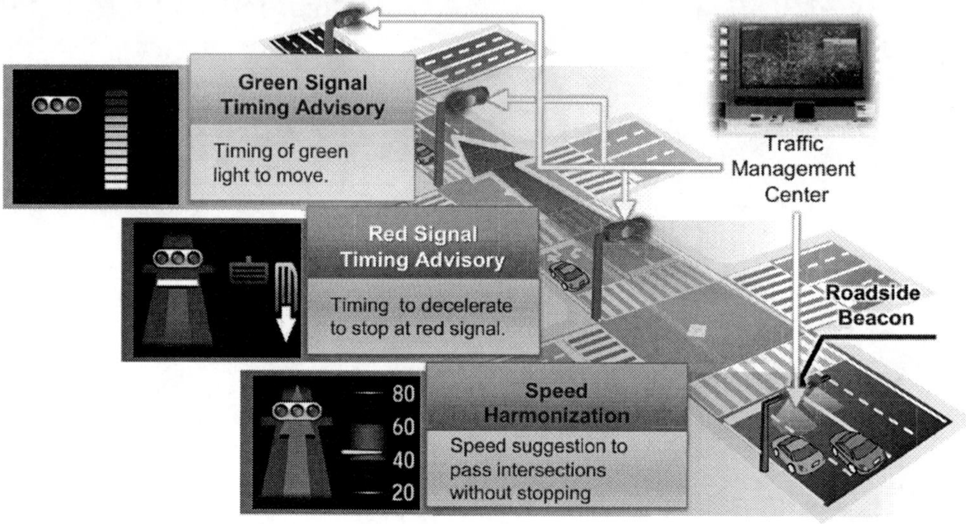

Source: National Police Agency

Driving Safety Support System (DSSS) 9

Right Turn Collision Warning

	Display	Sound	Situation
A	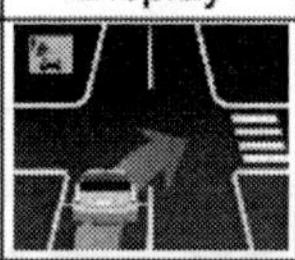	-	Clear
B	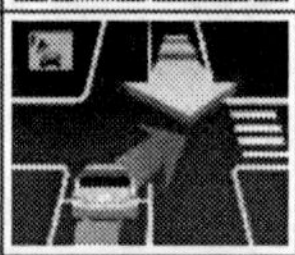	-	In coming vehicle/s detected
C	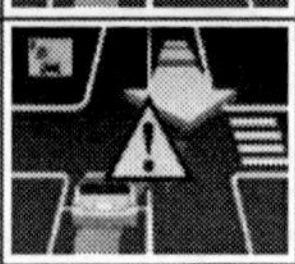	beep	Turning against in coming vehicle/s

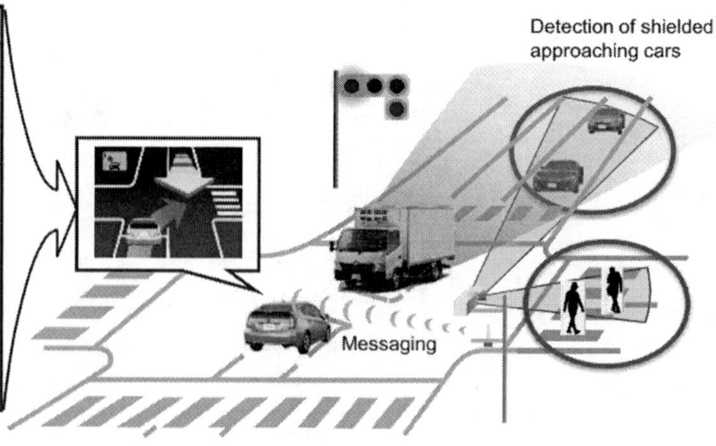

Source: National Police Agency

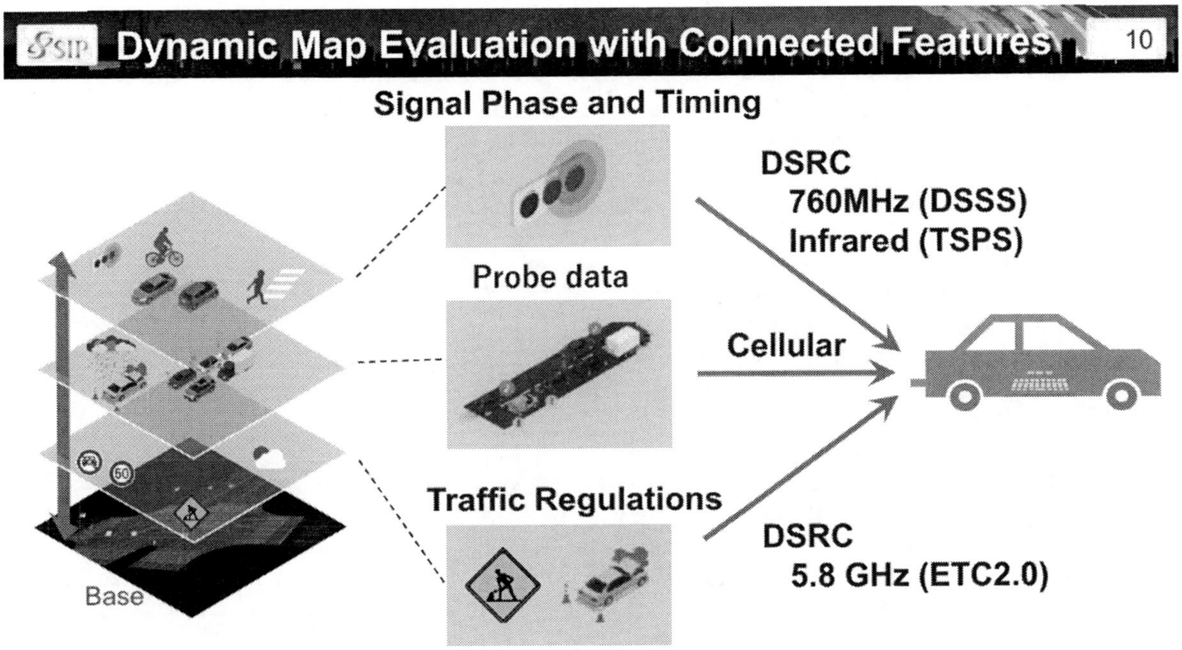
SIP
Dynamic Map Evaluation with Connected Features
10
Signal Phase and Timing
DSRC
760MHz (DSSS)
Infrared (TSPS)
Probe data
Cellular
Traffic Regulations
DSRC
5.8 GHz (ETC2.0)
Base

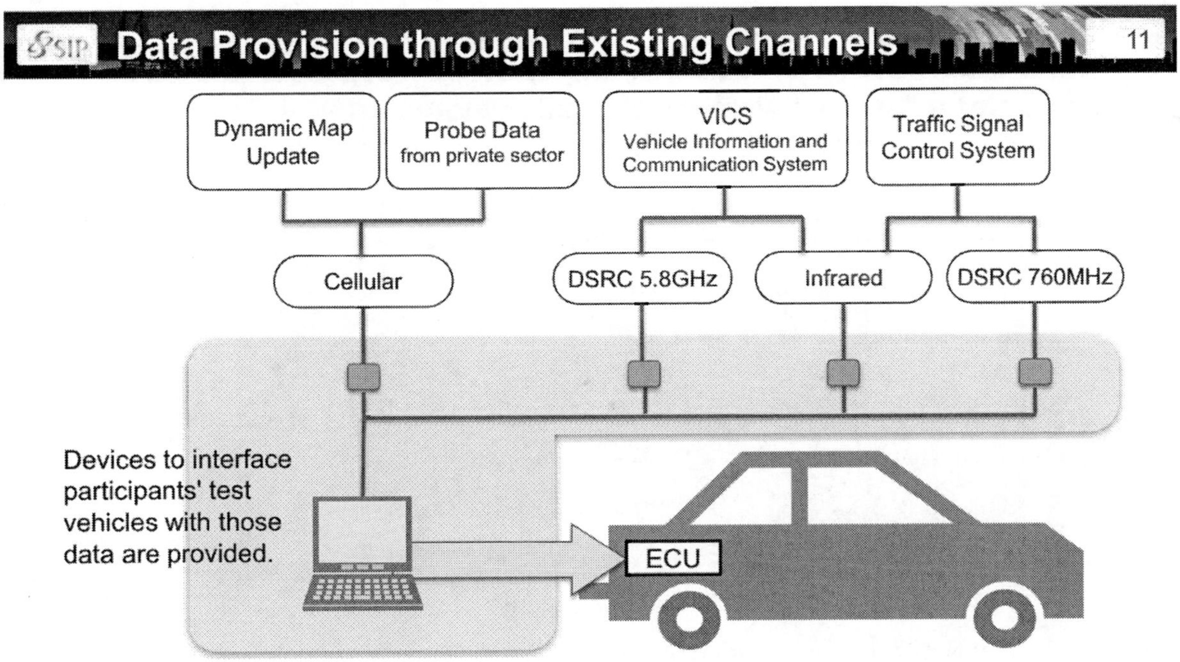
SIP
Data Provision through Existing Channels
11
Dynamic Map Update
Probe Data from private sector
VICS Vehicle Information and Communication System
Traffic Signal Control System
Cellular
DSRC 5.8GHz
Infrared
DSRC 760MHz
Devices to interface participants' test vehicles with those data are provided.
ECU

The 2nd Phase of SIP-adus (2018-2022)

Objectives: transition from the 1st phase to the 2nd phase

1) Extension of **operational domain** from the highways to the arterial and general public roads
2) More focus on **mobility services** including public transportation and logistic operations
3) Pursuit of **societal benefits** for safety, efficiency, inclusive society and enhanced economy

Deployment Goals:

1) **Tokyo Olympic and Paralympic Games**
2) Public transportation by **local government**
3) Mobility service businesses by **private sector**

Research Topics:

1) **Validation** of integrated automated driving systems through field operations
2) Foundation for **roadworthiness testing**: data collection, modeling and simulation
3) Quantitative impact assessment to foster **social acceptance**
4) **International collaboration** for harmonization

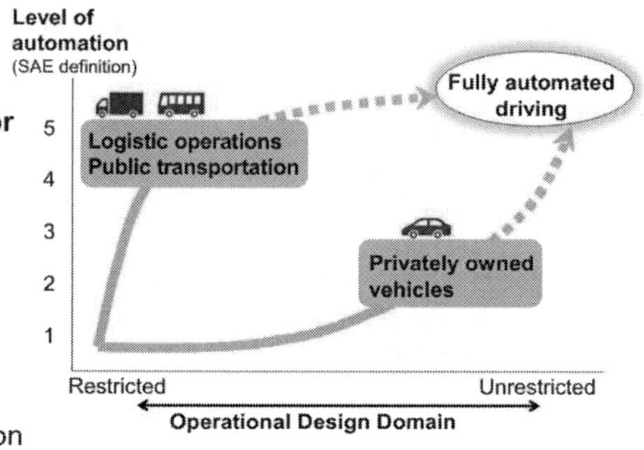

Holistic Approach for Safety

"The Charter for improvement of legal system and driving environment for automated driving systems"

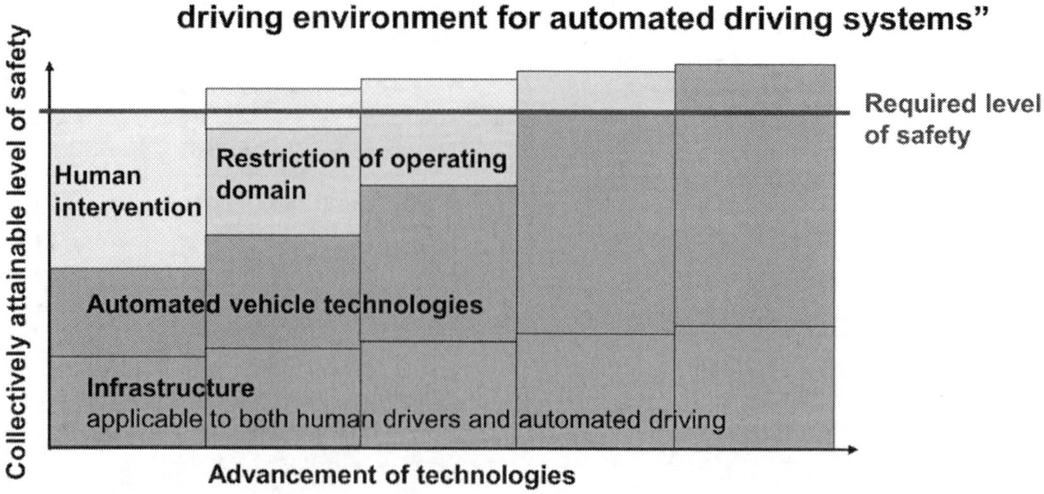

Source: National Strategy Office of Information and Communications Technology

Regulatory Considerations for Deployment by 2020

14

Vehicle safety regulations and conformance testing for type approval

- Safety guidelines for automated driving (by summer 2018)
- Vehicle safety regulations for automated driving vehicles

Road traffic rules

- Revision of road traffic rules in line with technology development and international discussion
- Necessary measures for automated driving systems to comply with the traffic rules
- Unmanned operation of automated vehicles with remote monitoring
- Rules for platoon operation of automated vehicles

Liability

- Application of Japanese mandatory automobile liability insurance for immediate relief of victims and their families
- Criminal responsibility based on clearly defined responsibilities of divers entities involved
- Installation of event data recorder on-board the vehicle and requirements of recorded data specifications and their submission

Regulations for public transportation and freight operators

Source: National Strategy Office of Information and Communications Technology

What is automated driving for ?

Societal benefits of deployment for
mobility to sustain daily life and
vitalization of economic activities.

Achieved only if
integrated with social innovations.

Ageing and Declining Population in Japan

ITS Japan

Population in 2050 compared with that of 2010

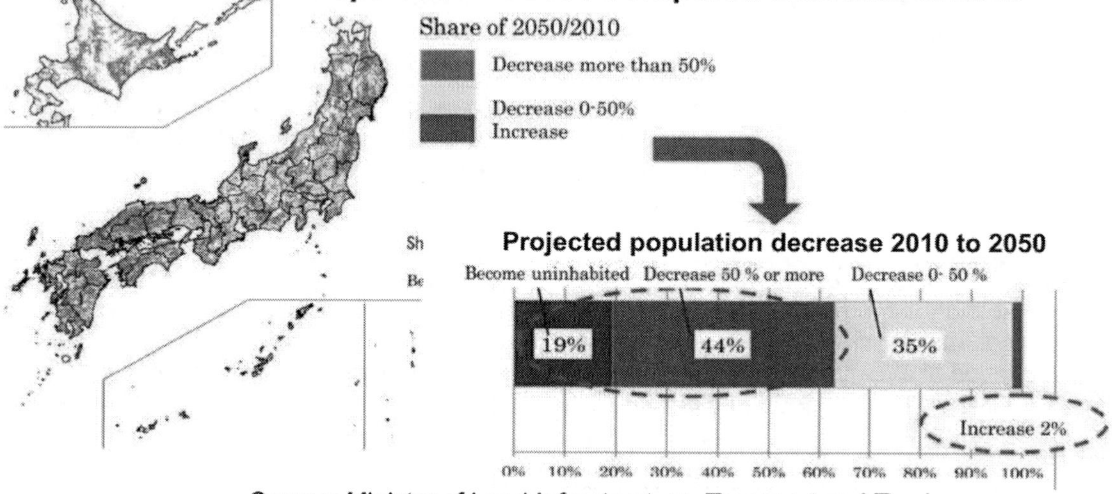

Source: Ministry of Land Infrastructure, Transport and Tourism

16

"Grand Design of National Spatial Development"

ITS Japan

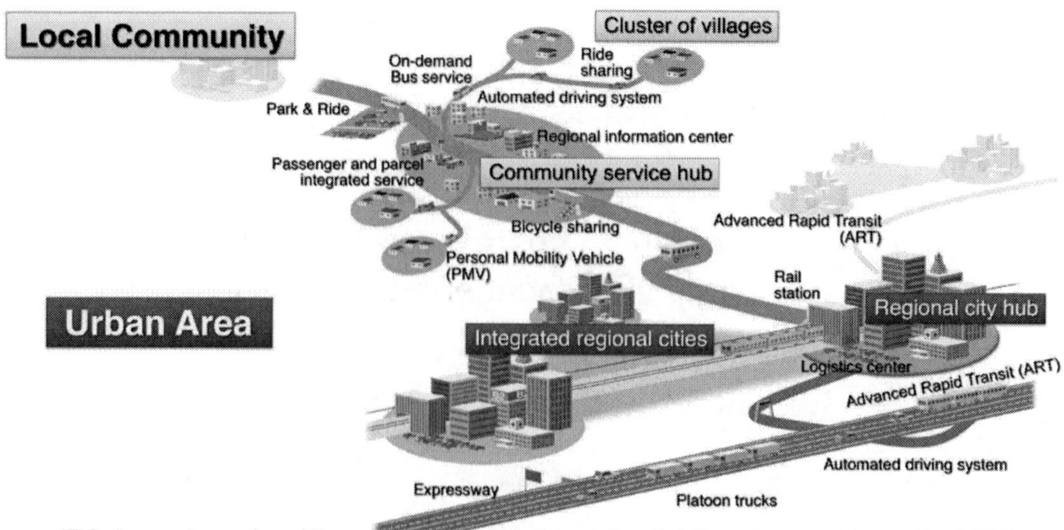

Source: ITS Japan based on 'Grand Design of National Spatial Development towards 2050'

17

FOT: Automated Vehicles in Rural Area

Photo: Ministry of Land Infrastructure, Transport and Tourism

18

Vision: Integrated Mobility for Inclusive Society

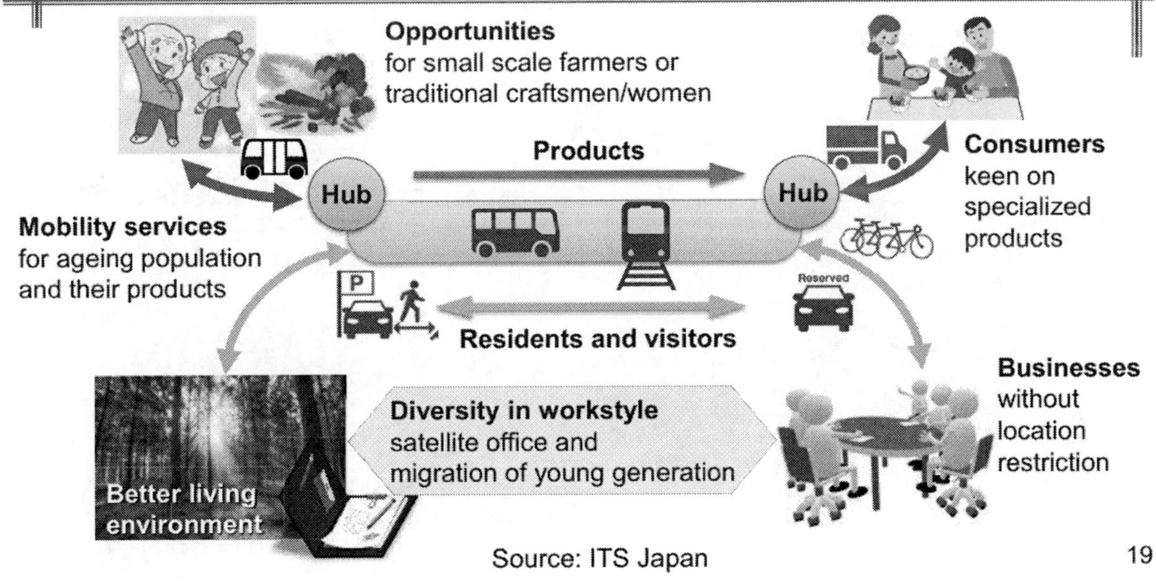

Source: ITS Japan

19

Sustainable Economic Development

Strategic integration of economic activities with networked transportation

Super Mega Region:
- highly concentrated mega-cities connected by high speed trains

Nationwide integration:
- vitalization of regional economies

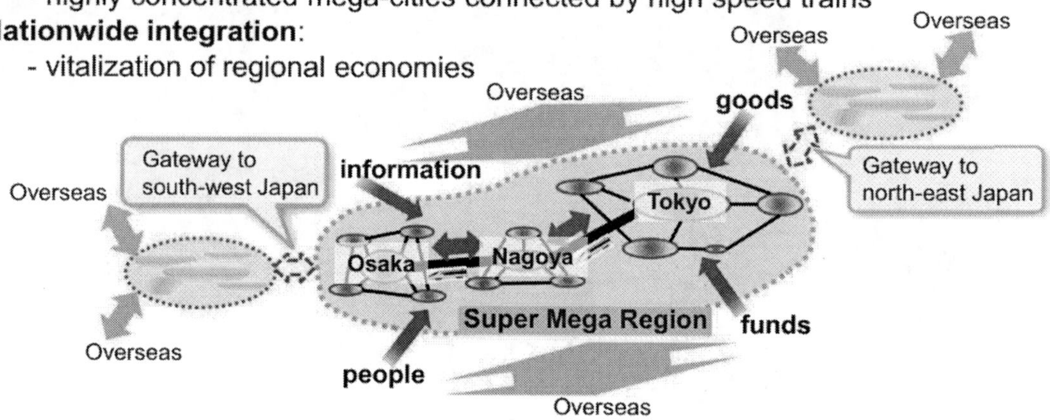

Source: Grand Design of National Spatial Development towards 2050

20

FOT: Truck Platooning

Fully automated platoon
(Unopened section of highway, 2012)

Photo: Japan Automobile Research Institute

CACC platoon
(Mixed traffic on highway, January 2018)

Photo: Toyota Tsusho Corporation

21

Vision: Integrated Freight Operations

ITS Japan

Connected and Automated Driving for LEAN, AGILE and RESILIENT operation

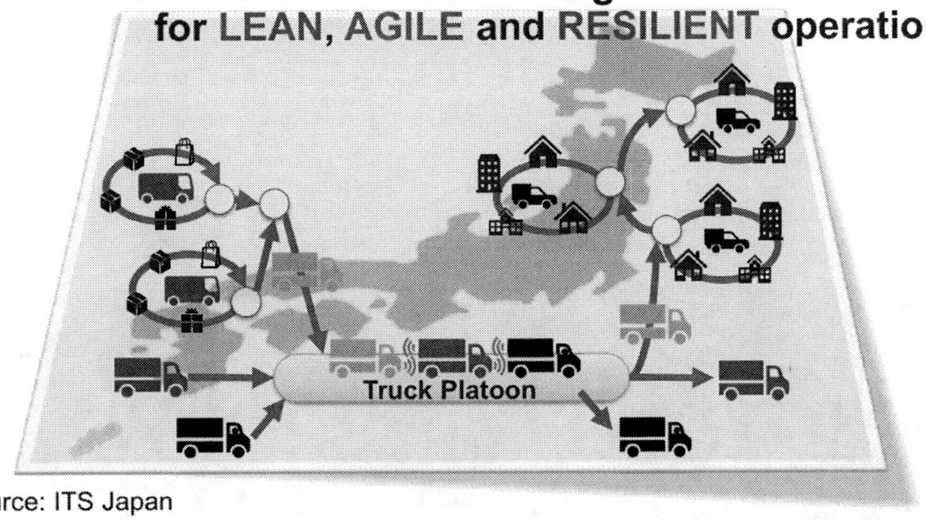

Source: ITS Japan

22

SIP-adus: Key Message from the Project

23

Cross-Ministerial Strategic Innovation Promotion program
Innovation of Automated Driving for Universal Services

"SIP- adus"

- Mobility Bringing Everyone a Smile -

Inclusive society, where diverse people in diverse communities actively participate in generating values, will enhance both wellness of individuals and economic development. Automated driving technologies integrated with social innovations should provide everyone with mobility to fully exercise his or her capacity, enabling sustainable development of the society.

5th SIP-adus Workshop
Date: November 13 – 15, 2018
Venue: Tokyo International Exchange Center
Topics:
1. Regional Activities and Field Operational Tests
2. Report Session from SIP-adus Activities
3. Dynamic Map
4. Connected Vehicles
5. Human Factors
6. Impact Assessment
7. Security
8. Next Generation Transport

European Commission

Automated Mobility: the EU strategy for Mobility of the Future

Automated Vehicles Symposium
San Francisco, 12 July 2018

Clara de la Torre
Director for Transport
Directorate-General for Research & Innovation

European Commission

HORIZON 2020

European Commission

3rd Mobility Package (17 May 2018)

Key elements:

More info:
https://ec.europa.eu/transport/modes/road/news/2018-05-17-europe-on-the-move-3_en

❑ Safe Mobility

- Communication on new road safety policy framework
- Legislative initiative on vehicle and pedestrian safety
- Legislative initiative on infrastructure safety management

❑ Clean Mobility

- Legislative initiatives on CO_2 standards for heavy duty vehicles
- Action Plan for Batteries

❑ Connected and Automated Mobility

- Legal initiatives establishing a digital environment for information exchange in transport
- Communication on Connected and Automated Mobility

EC Communication on Automated Mobility

- ❑ Strategy for the roll-out of connected and automated vehicles in Europe
- ❑ Comprehensive regulatory, financial and research framework to develop these technologies.

What needs to be done?

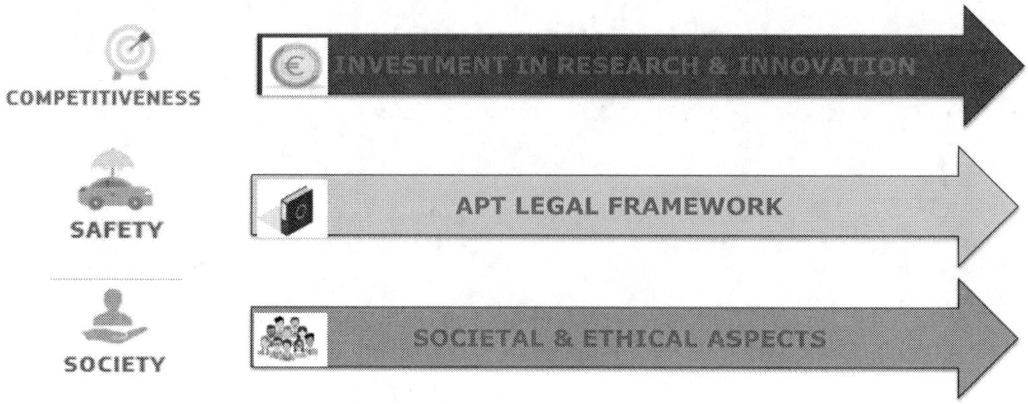

H2020 - Calls on "Automated Road Transport"

- ❑ Budget: € 300 Mio (2014-2020)
- ❑ Focus
  - ➢ Large-scale demos of automated driving systems for passenger cars, trucks and urban transport
  - ➢ Safety and end user acceptance
  - ➢ Road infrastructure to support automation
  - ➢ Traffic management solutions
  - ➢ Connectivity for automation
  - ➢ Testing and validation procedures
  - ➢ Assessment of impacts, benefits and costs of CAD systems
  - ➢ Support for cooperation and networking activities
  - ➢ Human centred design of AV

5 Calls for proposals 2016 › 2017 › 2018 › 2019 › 2020

4

H2020 – Ongoing and planned large-scale demos

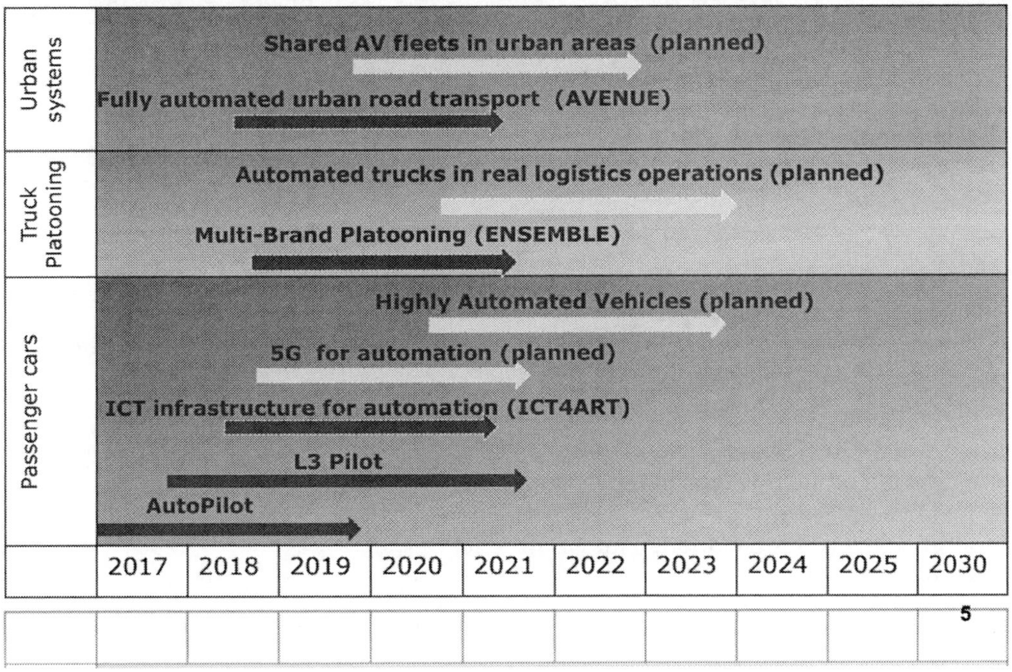

5

Horizon Europe

Commission proposal for a € 100 billion research and innovation funding programme for seven years (2021-2027)

- to strengthen the EU's scientific and technological bases
- to boost Europe's innovation capacity, competitiveness and jobs
- to deliver on citizens' priorities and sustain our socio-economic model and values

6

European Commission

EU-US Symposia

- Aim
  - enable systemic long-term collaboration to address common research needs, share information on research and to develop opportunities for joint research projects

- 6th EU-US symposium on 'Socio-economic Impacts of Automated and Connected Vehicles', Brussels, 26-27 June 2018
  - 3 partners/sponsors: TRB, DoT and EC
  - 50 participants (25 US+25 EU)
  - Added value: mutual learning, networking, definition of common research topics and twinning of projects
  - Final report: recommendations and pointers for cooperation

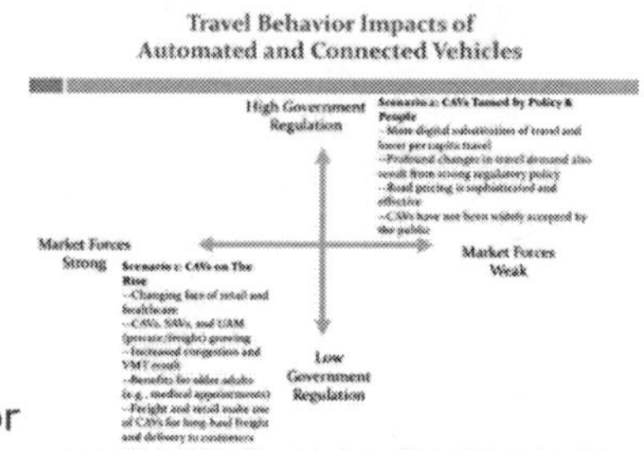

7

European Commission

connected automated driving.eu

2nd European Conference on Connected and Automated Driving

Date: 2-3 April 2019, Brussels

8

L3Pilot Driving Automation

Testing Automated Driving on European Roads

Automated Vehicles Symposium 2018

12 July 2018, San Francisco

Adrian Zlocki, Christian Rösener
ika/fka RWTH Aachen University

INSTITUT FÜR KRAFTFAHRZEUGE ika | RWTH AACHEN UNIVERSITY

www.L3Pilot.eu Twitter@_L3Pilot_ LinkedIn L3Pilot

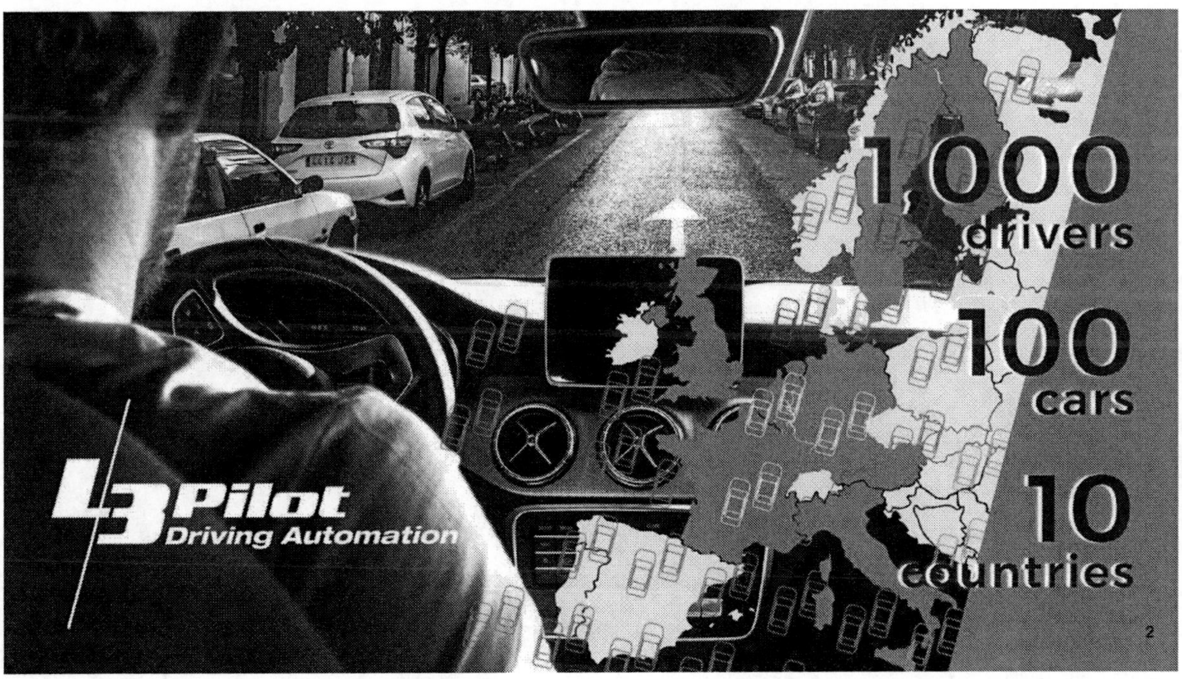

European Projects on Automated Driving

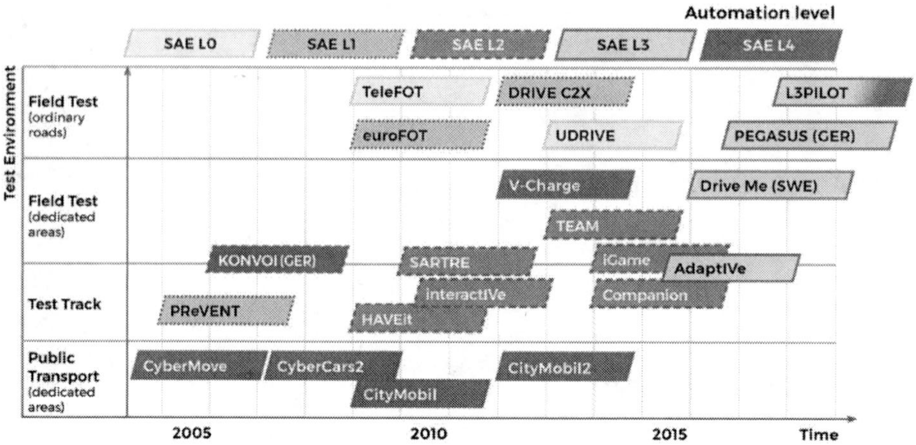

12 July 2018

3

From euroFOT to L3Pilot

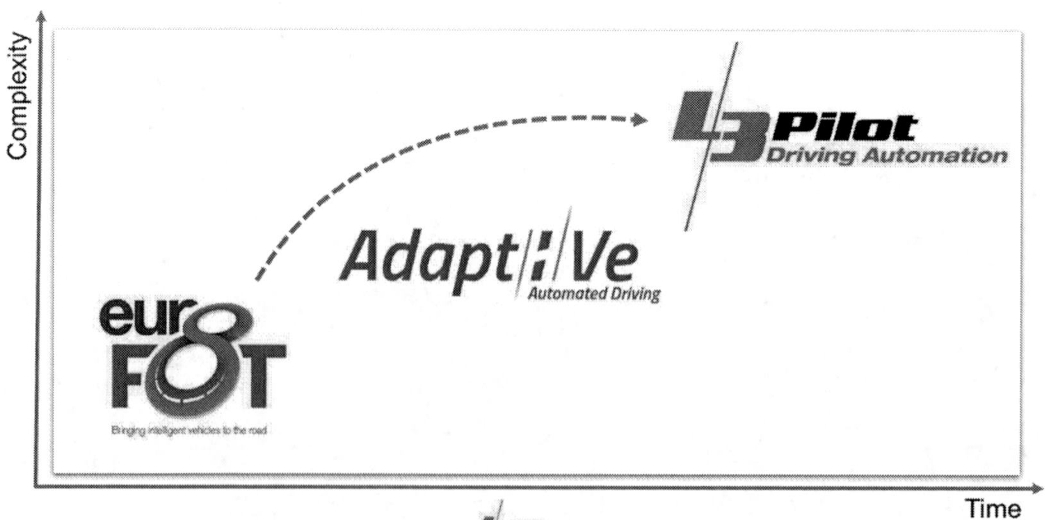

12 July 2018

4

euroFOT

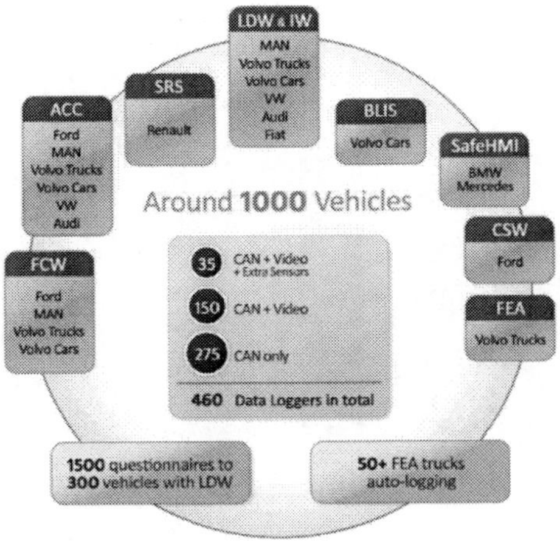

Longitudinal control functions

- Forward Collision Warning (FCW)
- Adaptive Cruise Control (ACC)
- Speed Restriction System (SRS)

Lateral control functions

- Blind Spot Information System (BLIS)
- Lane Departure Warning (LDW)
- Impairment Warning (IW)

Advanced applications

- Curve Speed Warning (CSW)
- Fuel Efficiency Advisor (FEA)
- Safe Human Machine Interaction (SafeHMI)

AdaptIVe

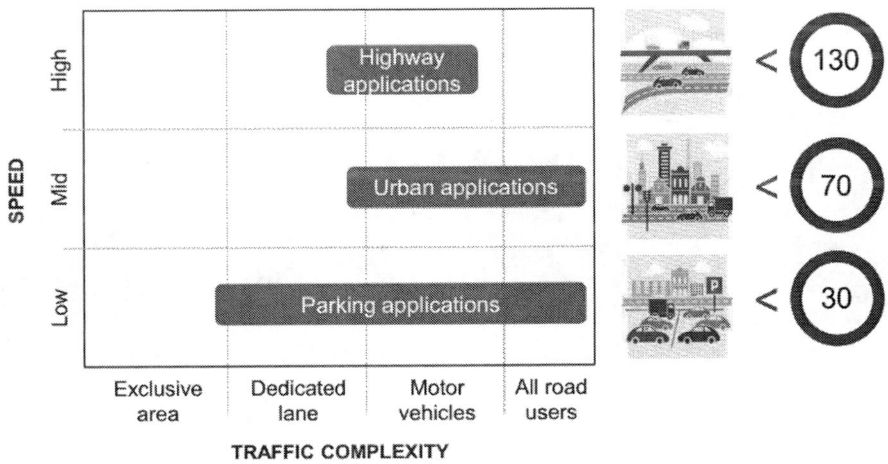

SAE levels

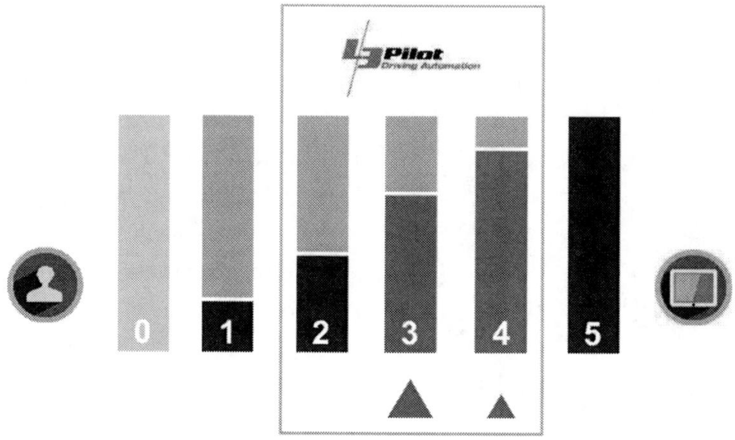

12 July 2018

See SAE document J3016, "Taxonomy and Definitions for Terms Related to Driving Automation Systems for On-Road Motor Vehicles", revised 2016-09-30, see also http://standards.sae.org/j3016_201609

7

Pilot
Driving Automation
Use Cases

Motorway ***Traffic Jam*** ***Urban*** ***Parking***

1,000 drivers 100 cars 10 European countries

Pilot across Europe

CROSSBORDER	
Austria	Germany
Austria	Italy
Belgium	Germany
Belgium	Netherlands
Finland	Sweden
France	Germany
Germany	Netherlands

COUNTRY / REGION / OEM	
BE / Brussels / NL	Toyota
DE / Aachen	Ford
DE / Munich	BMW
DE / Offenbach	Honda
DE / Wolfsburg	VW
FR / Paris and other regions	REN / PSA
IT / Turin and Trento	CRF
LU / NL	Delphi
SE / Gothenburg	Volvo
UK / Coventry	JLR

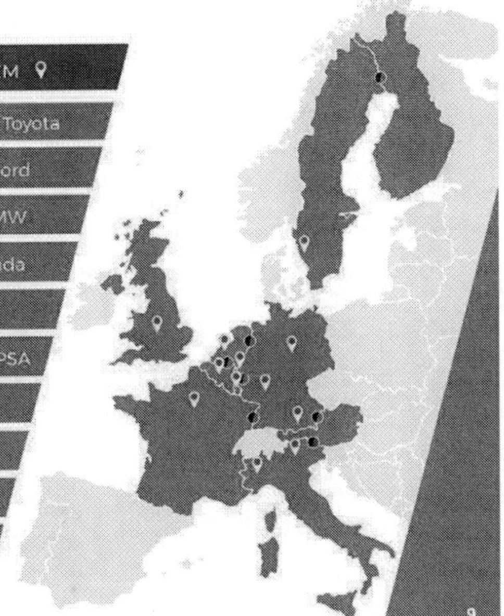

12 July 2018

9

Facts

Supported by eucar

This project has received funding from the European Union's Horizon 2020 research and innovation programme under grant agreement No 723051.

€68 million BUDGET

48 months DURATION, starting in September 2017

€36 million FUNDING

34 PARTNERS, among them OEMs, suppliers, research, SMEs, insurers, authorities and user groups

12 COUNTRIES involved: Austria, Belgium, France, Finland, Germany, Greece, Italy, Netherlands, Norway, Sweden, Switzerland, UK

L3Pilot Driving Automation

12 July 2018

10

Partners

OEMs: VOLKSWAGEN, AUDI, BMW GROUP, DAIMLER, FCA, Ford, GROUPE RENAULT, HONDA, JAGUAR, LAND-ROVER, PSA GROUPE, TOYOTA, VOLVO

Suppliers: Autoliv, DELPHI, FEV

SMEs: eict

Research: bast, DLR Deutsches Zentrum für Luft- und Raumfahrt German Aerospace Center, ika RWTH AACHEN UNIVERSITY, SAFER, SNF, TNO, UNIVERSITÀ DEGLI STUDI DI GENOVA, UNIVERSITY OF LEEDS, WMG, VTT, wivw

Insurers: Allianz, Swiss Re

Authorities: RDW

User group: FIA

12 July 2018

L3Pilot Driving Automation

Timeline

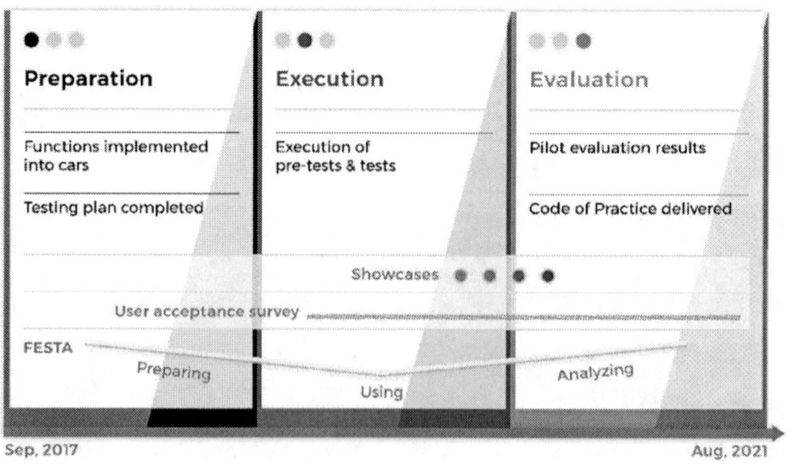

12 July 2018

L3Pilot Driving Automation

L3Pilot Driving Automation

Objectives

1,000 drivers 100 cars 10 European countries Piloting Automated Driving on European Roads.

12 July 2018

Code of Practice

Provide a comprehensive guideline with best practices for the development of functions: Code of Practice for Automated Driving.

- Collect best practices on relevant topics.
- Describe a typical process for an Automated Driving function.
- Include hands-on checklists.
- Include safety aspects and methods to confirm a safe operation of Automated Driving functions

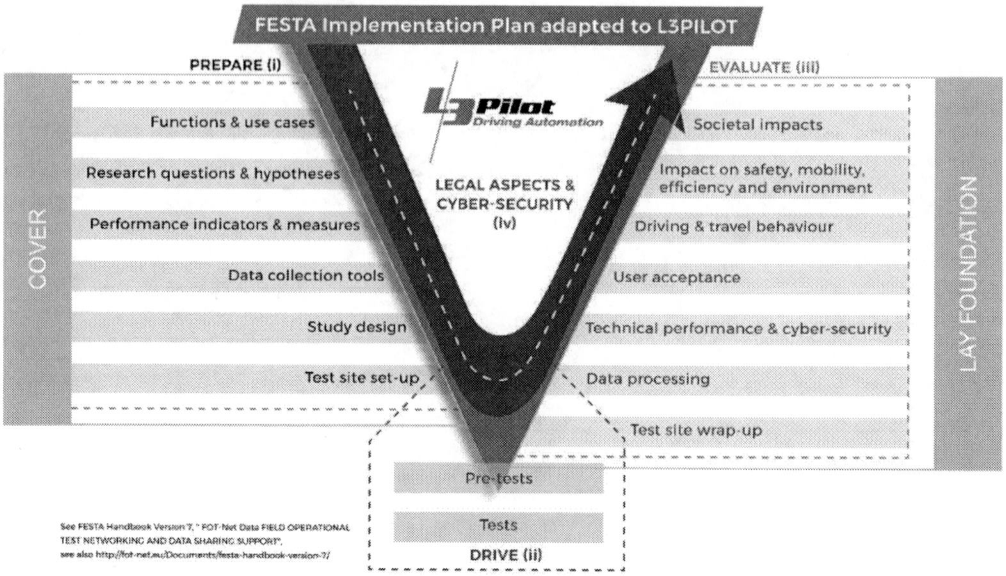

See FESTA Handbook Version 7, " FOT-Net Data FIELD OPERATIONAL TEST NETWORKING AND DATA SHARING SUPPORT", see also http://fot-net.eu/Documents/festa-handbook-version-7/

12 July 2018 | Pilot Driving Automation | Methodology | 15

Evaluation Objectives

The **evaluation will draw conclusions** on technical aspects, user acceptance, driving and travel behaviour, and impact on traffic and society.

- Assess long-term effects of automated driving on user attitudes and acceptance.
- Investigate interactions between different traffic participants in different automation modes.
- Assess the readiness and reliability of automated driving functions.
- Develop tools for the effective analysis and evaluation of field data and enhance simulation tools for simulation in the impact assessment.
- Determine safety, efficiency, mobility and economic impact of automated driving applications under mixed automated traffic conditions based on collected in the real world pilot data.

31.01.2018 | Pilot Driving Automation | Evaluation | 16

Evaluation Methodology

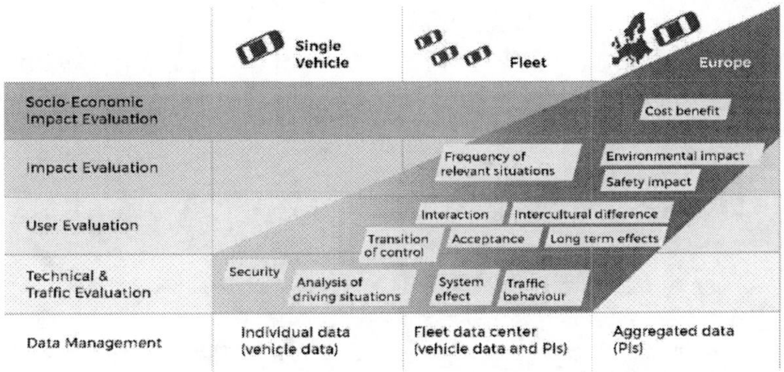

31.01.2018

Evaluation 17

L3Pilot Driving Automation

Thank you for your kind attention.

Aria Etemad
Volkswagen Group Research
aria.etemad@volkswagen.de
Coordinator

Adrian Zlocki
Ika/fka RWTH Aachen University
Adrian.zlocki@fka.de
SP7 Lead

This project has received funding from the European Union's Horizon 2020 research and innovation programme under grant agreement No 723051.

www.L3Pilot.eu Twitter @_L3Pilot_ LinkedIn L3Pilot

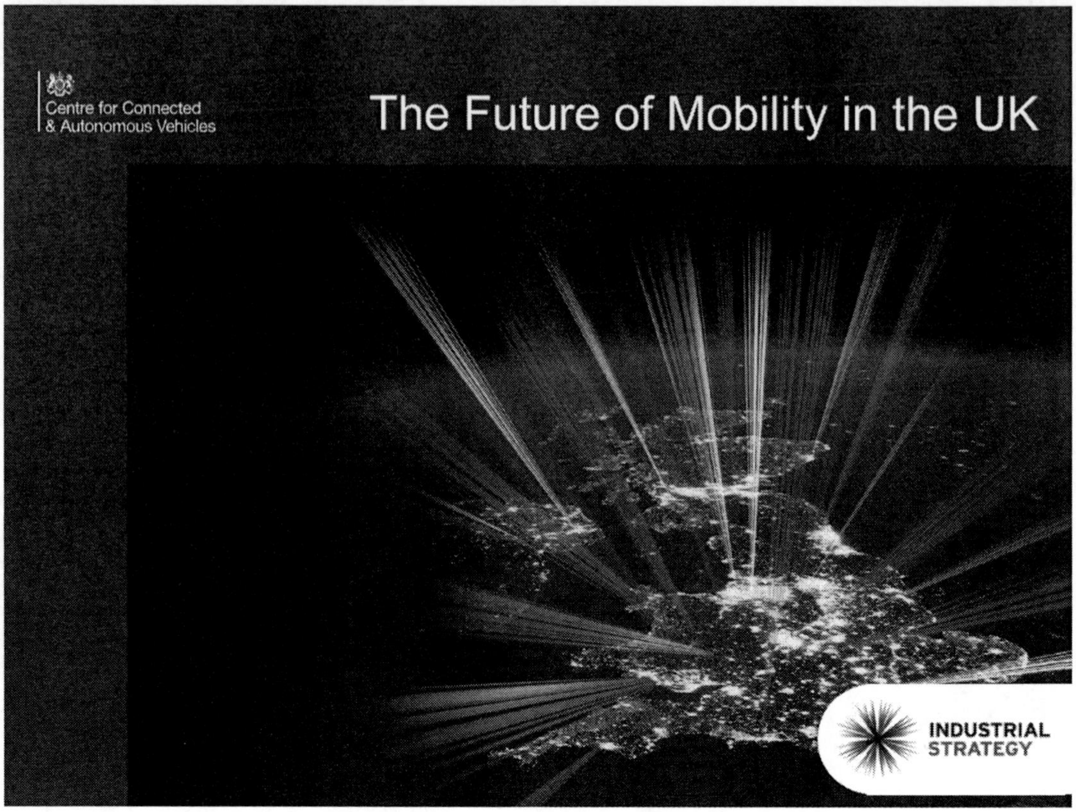
Centre for Connected
& Autonomous Vehicles
The Future of Mobility in the UK
INDUSTRIAL
STRATEGY

Fusion Processing
RATE MY DRIVING

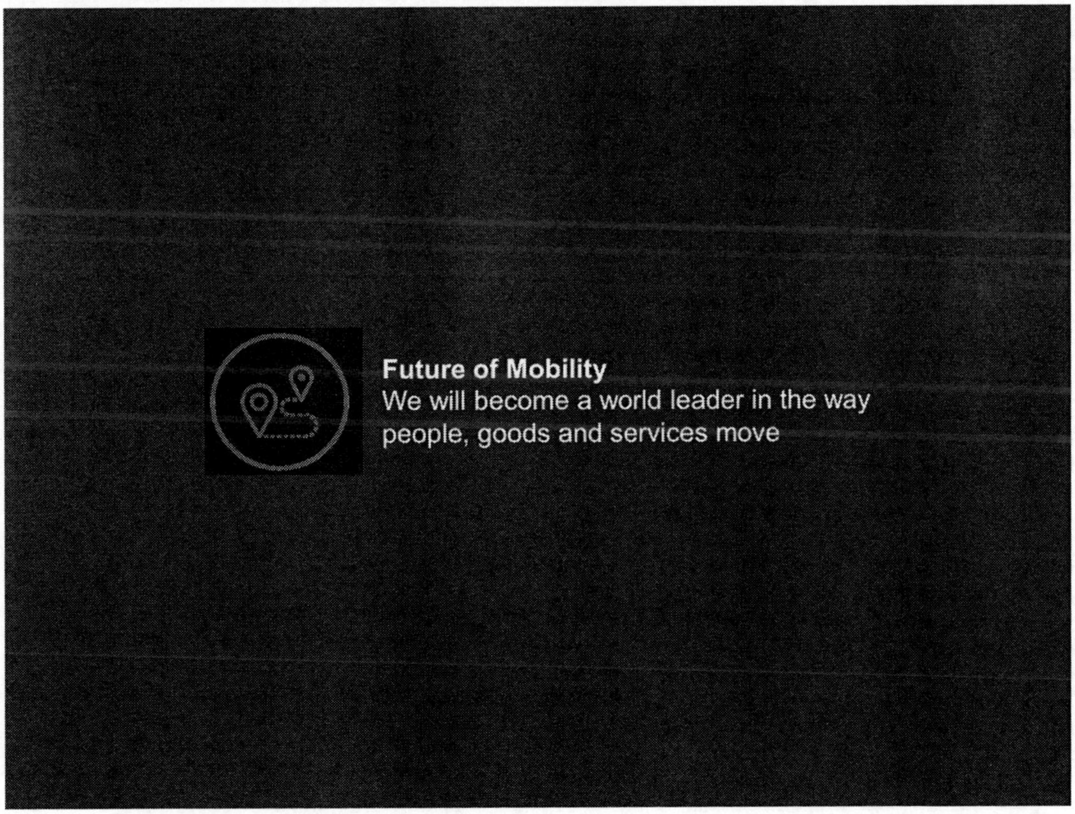
Future of Mobility
We will become a world leader in the way
people, goods and services move

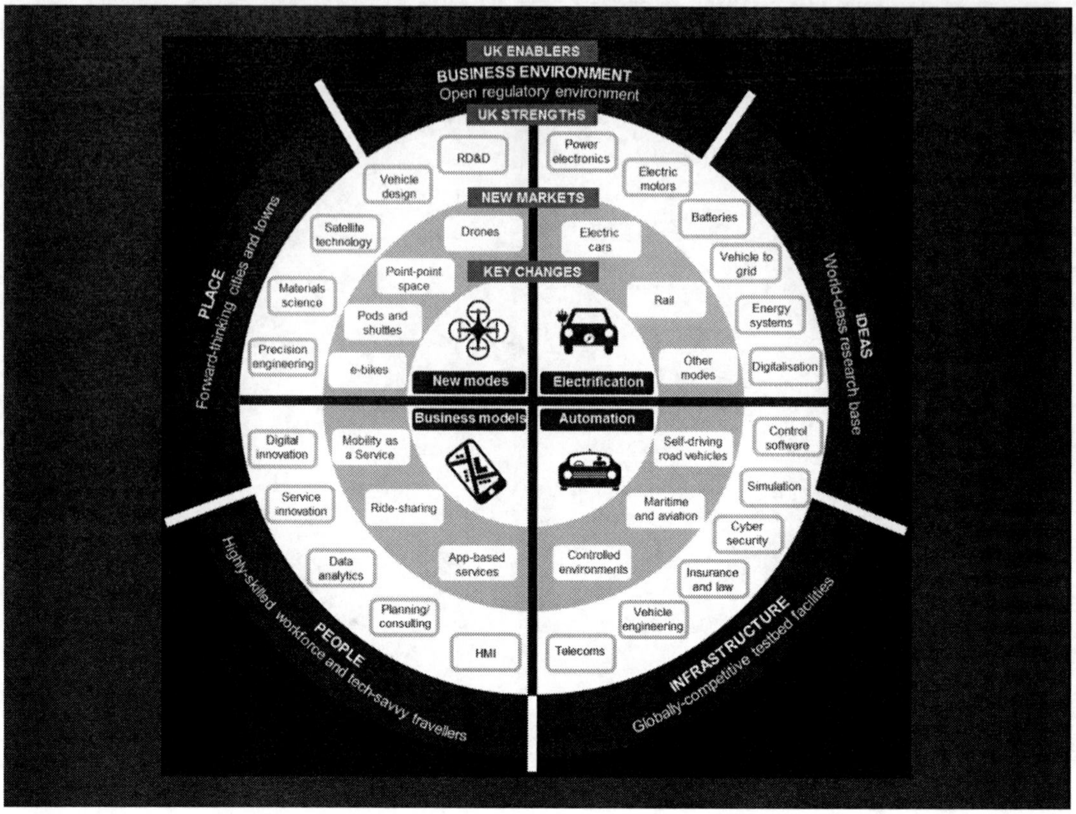
UK ENABLERS
BUSINESS ENVIRONMENT
Open regulatory environment
UK STRENGTHS
RD&D
Power electronics
Vehicle design
Electric motors
NEW MARKETS
Batteries
Satellite technology
Drones
Electric cars
Vehicle to grid
Point-point space
KEY CHANGES
Materials science
Rail
Energy systems
Pods and shuttles
PLACE
Forward-thinking cities and towns
IDEAS
World-class research base
Precision engineering
e-bikes
New modes
Electrification
Other modes
Digitalisation
Business models
Automation
Digital innovation
Mobility as a Service
Self-driving road vehicles
Control software
Simulation
Service innovation
Ride-sharing
Maritime and aviation
Cyber security
Data analytics
App-based services
Controlled environments
Insurance and law
Planning/consulting
Vehicle engineering
HMI
Telecoms
PEOPLE
Highly-skilled workforce and tech-savvy travellers
INFRASTRUCTURE
Globally-competitive testbed facilities

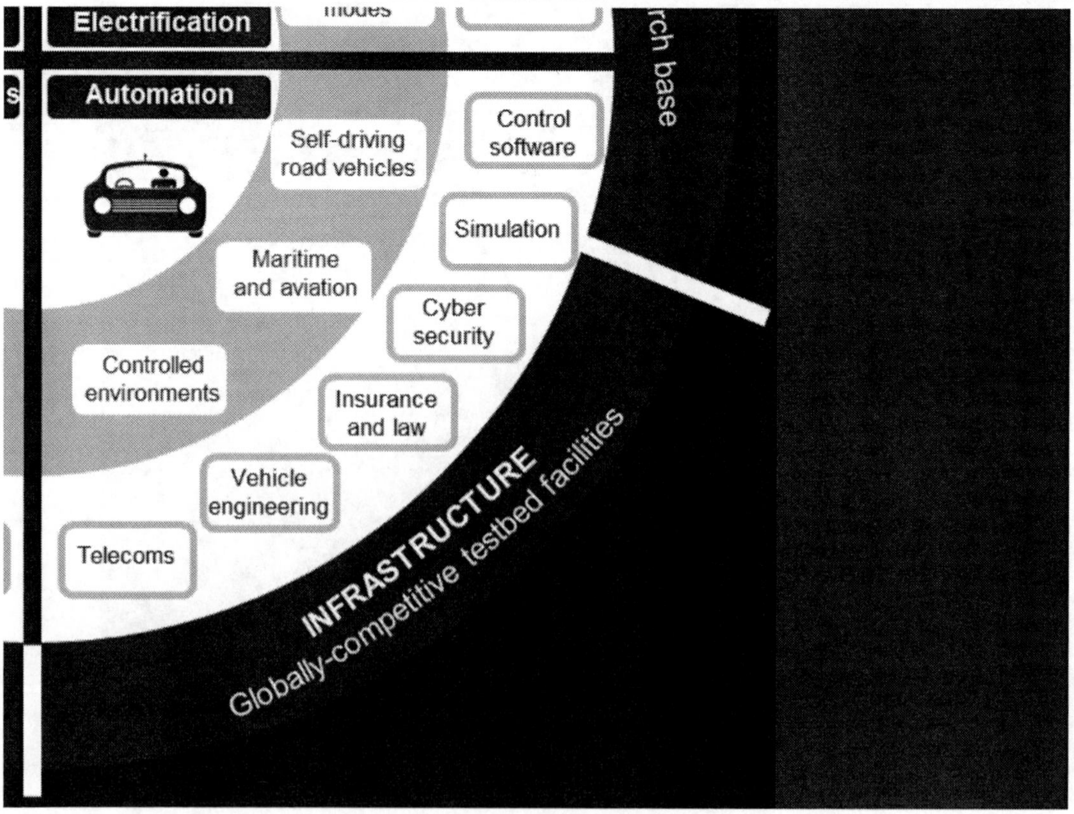
Electrification
Automation
Self-driving road vehicles
Control software
Simulation
Maritime and aviation
Cyber security
Controlled environments
Insurance and law
Vehicle engineering
Telecoms
INFRASTRUCTURE
Globally-competitive testbed facilities

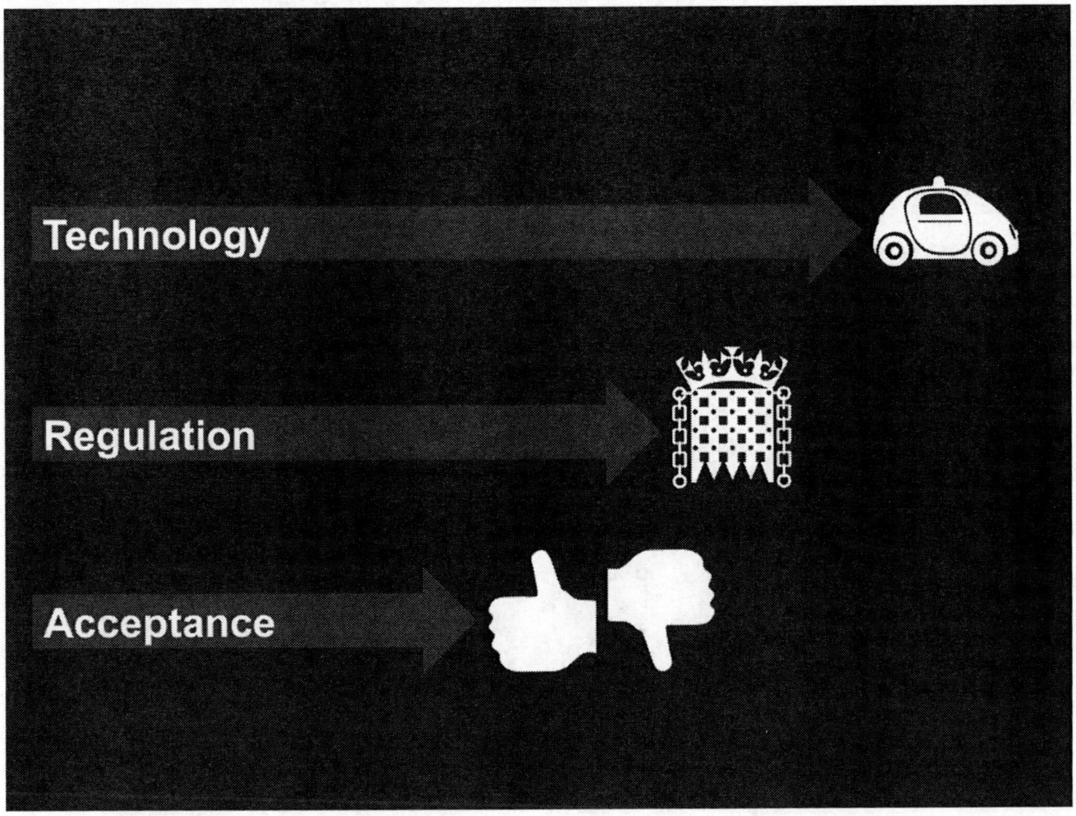

Joyously Toying With Death

[redacted]

Centre for Connected
& Autonomous Vehicles

INDUSTRIAL
STRATEGY

[redacted]

[redacted]

Centre for Connected
& Autonomous Vehicles

INDUSTRIAL
STRATEGY

[redacted]

Centre for Connected & Autonomous Vehicles

INDUSTRIAL STRATEGY

A clear regulatory pathway

	Current situation	Short term	Long term
Trialling	Trials on public roads with human safety operator	New procedure for trials with no human safety operator	Integrating real world and simulated trialling
Vehicle safety approvals	Prototype approvals	International standards for driver support systems	International standards for fully self-driving vehicles
Regulations on use	Motor insurance framework for self-driving vehicles	Three year review to establish long term framework	New regulatory framework in place

[redacted]

Centre for Connected & Autonomous Vehicles

INDUSTRIAL STRATEGY

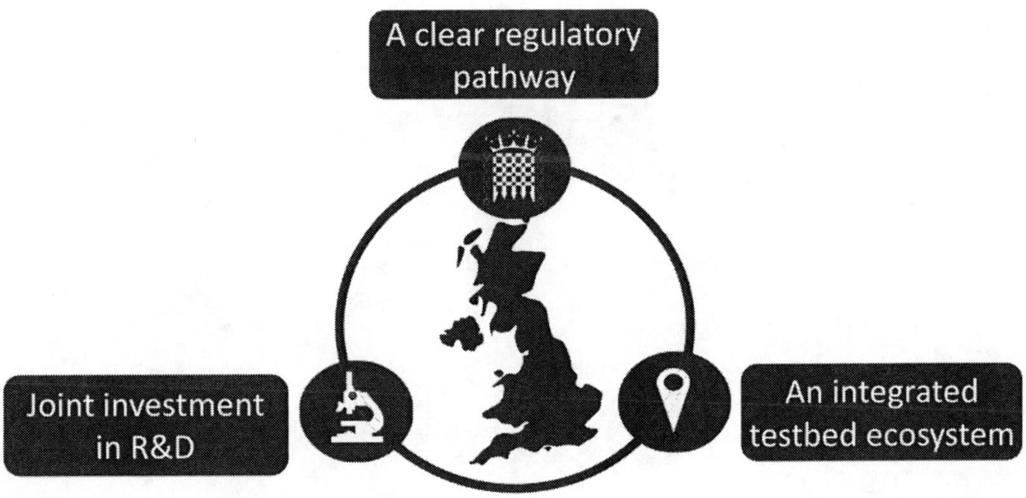

[redacted]

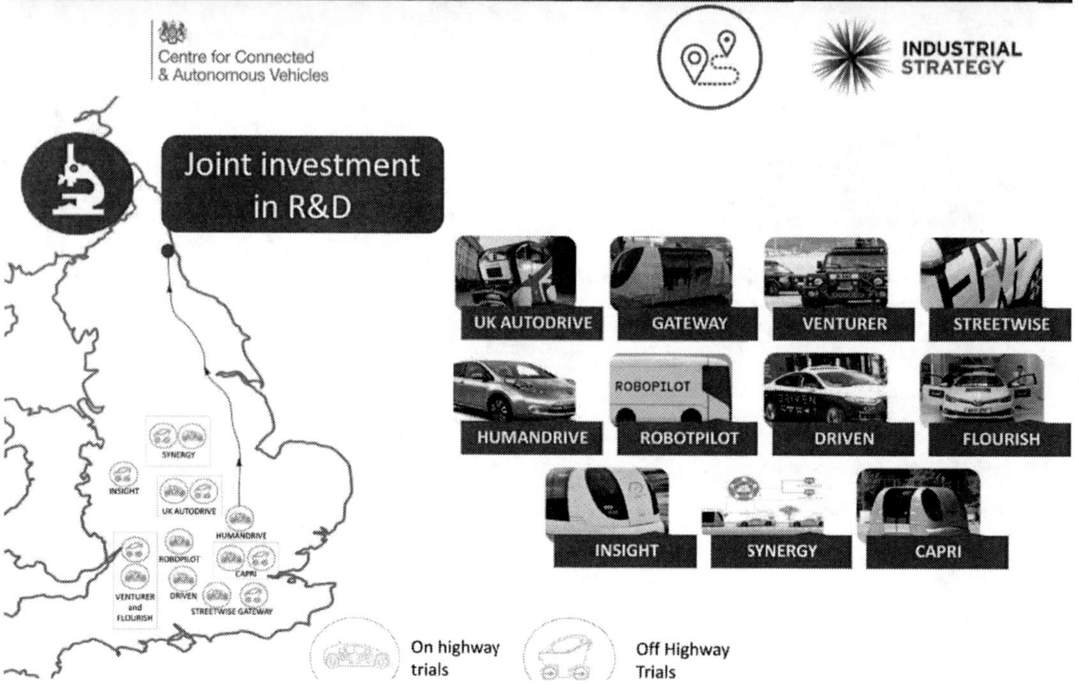

[redacted]

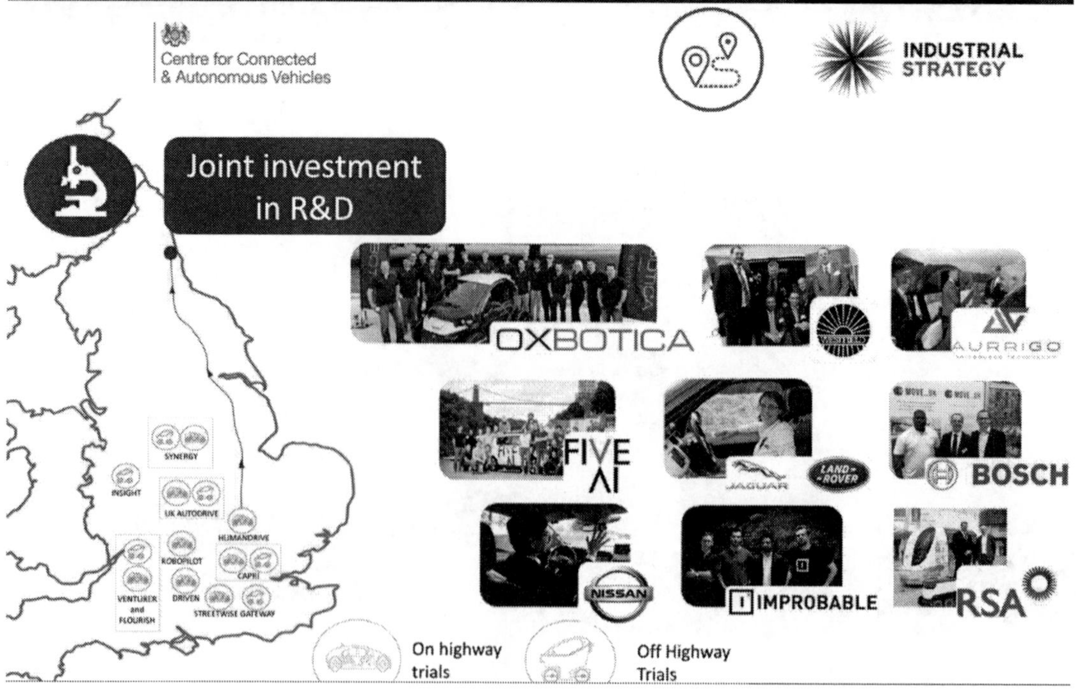

[redacted]

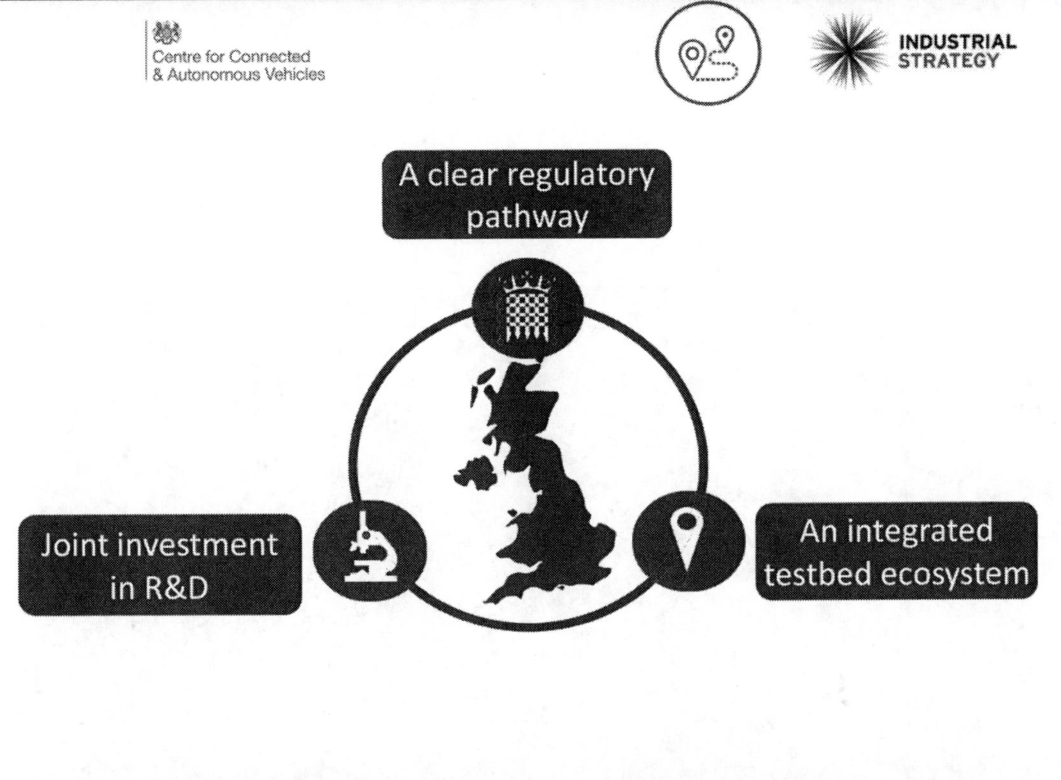

[redacted]

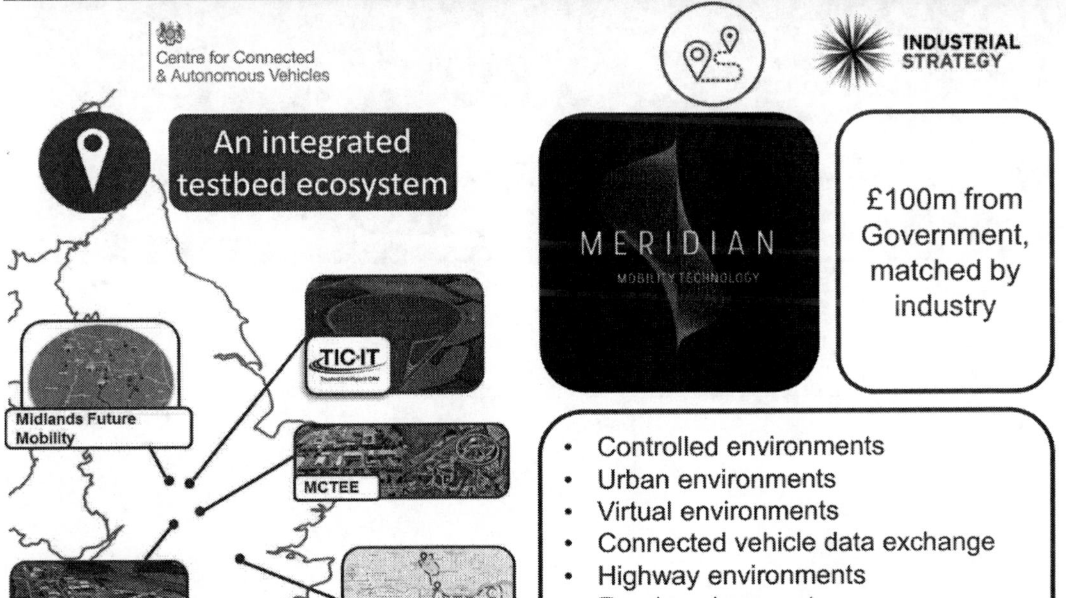

iain.forbes@ccav.gov.uk
gov.uk/ccav
@ccavgovuk

Dubai Self-driving Transport Strategy

Vision & World Challenge

Dr. Ismail Zohdy
Program Manager for Dubai Self-driving Transport
Roads and Transport Authority-
Dubai UAE

About Dubai

Dubai – A Rapidly Growing City

1970	2016	2030
		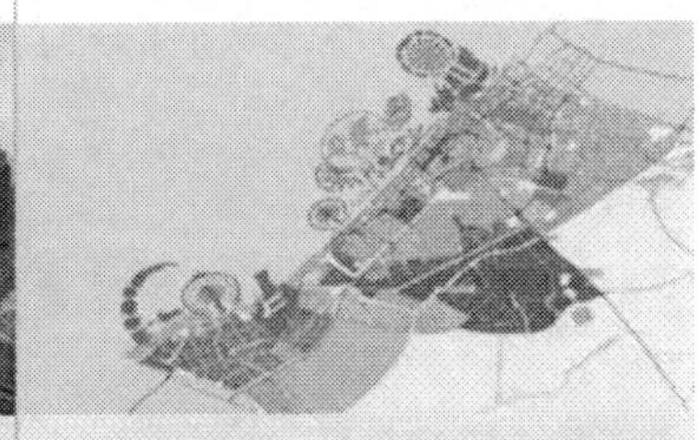
• Developed Area = 18 km² • Populatoin = 100,000	• Developed Area = 950 km² • Populatoin = 2.8 Million	• Expected developed area = 1900 km² • Forecasted population= 5.2 Million

Dubai Area = 4114 km²	Dubai population = 2.8 M | 16 Million visitors in 2016
Dubai is 5% of UAE total area	Dubai has 30% of UAE total population

2

Who are we? What transportation Modes we have?

RTA

Video Presentation in Original e-media

Video of Transportation
System in Dubai

3

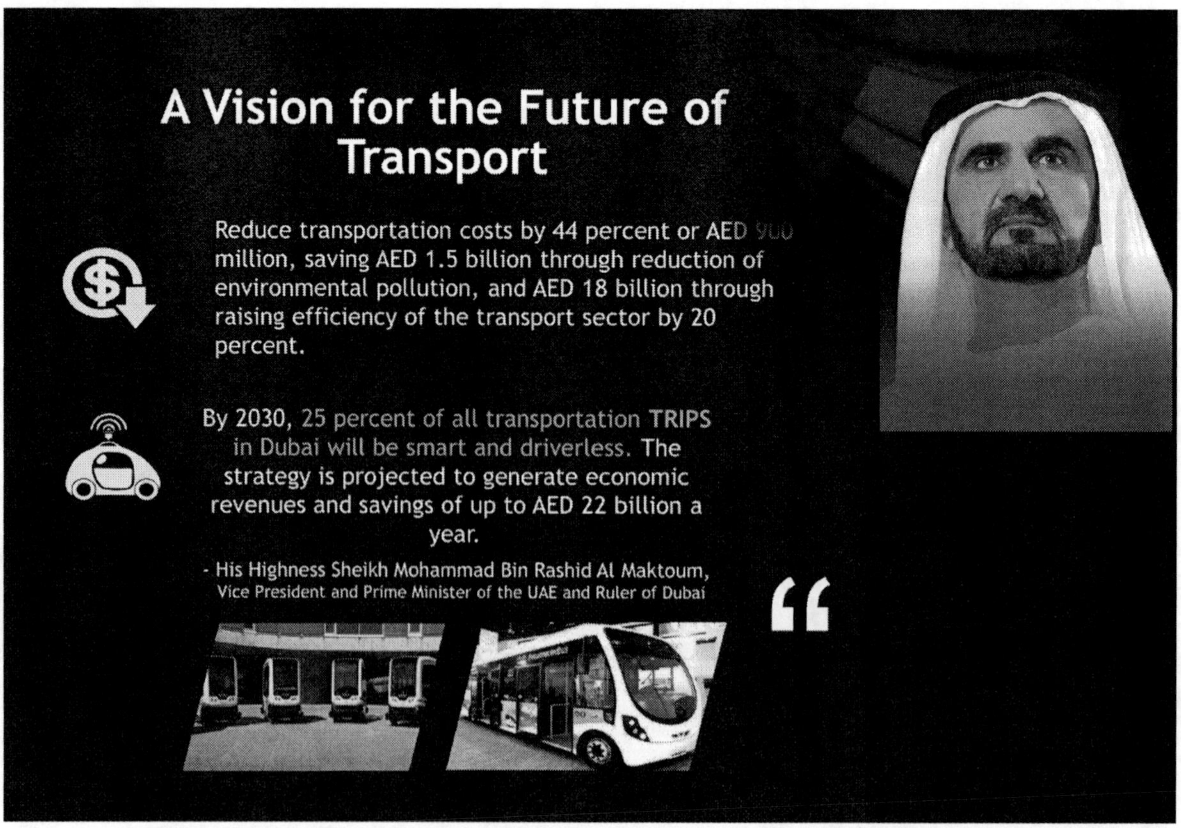

RTA SDT Strategy Summary

…and it defines 2030 targets for self-driving technology adoption for each of the public transports modes

SDT Targets per Mode (as a % of total trips)

	2017	2030
Metro	9%	12.2%
Tram	0%	0.2%
Shuttle & Cable Cars	0%	2.5%
Marine Transport	0%	0.2%
Taxi / Rideshare	0%	3.5%
BRT, Rider & Runner	0%	6.4%
	Total Self-Driving PT	**25%**

5

Dubai Strategy Key Enablers

1. Legislation	•Testing AND Operations
2. Driver Behavior & Acceptance	•Driver behavior in Level 2-3 vehicles •Driver acceptance of SDT
3. Driver & Vehicle Licensing/Registration	•SDT driver licensing •SD vehicle testing, registration, renewal
4. Insurance/Liability	•Crash liability of self-driving vehicles •Insurance requirements
5. Infrastructure Requirements	•Infrastructure improvements for SDT •Crowd sourcing and analytics
6. Cyber security/Data Privacy	•Self-driving vehicle security •On-board data recording and retrieval
7. Connected Vehicle Enablers	•Communication (V2I, V2V, V2X) •Connected cloud
8. HD Mapping	•Accurate HD Mapping

6

KPMG Report- Autonomous Vehicles Readiness Index

So where UAE stands at the moment among the world..

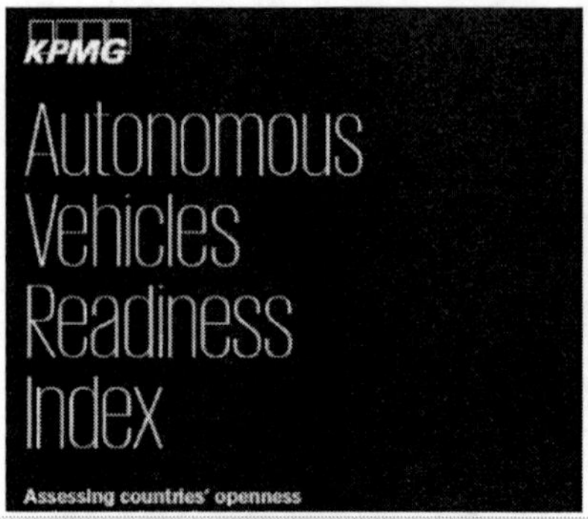

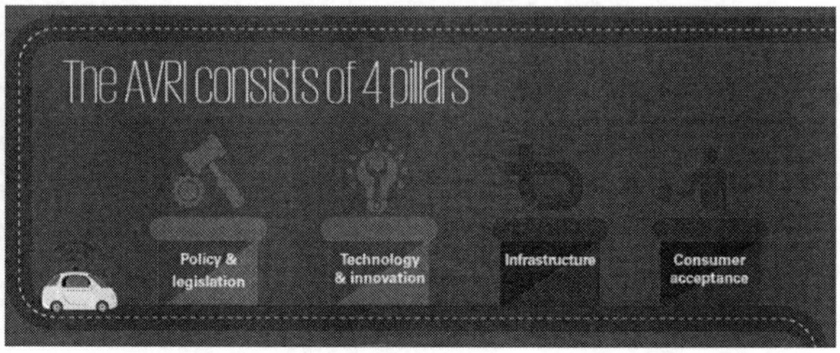

Dubai Media Office
@DXBMediaOffice

#UAE has ranked among the top 10 countries in the world according to @KPMG 's Autonomous Vehicle Readiness Index for 2018, which shows which countries are most prepared for driverless cars.

United Arab Emirates

6th 14th 5th 8th

7

RTA SDT Strategy Summary

The RTA Self-Driving Transport (SDT) strategy aims to make Dubai the leader in the world for SDT…

Unique SDT – Areas of Focus

Multimodal SDT

- Dubai is targeting self-driving transport across all 7 modes of public transport
- This will enable a harmony of SD systems across transport modes

Worldwide SDT Challenge

- Competition will attract manufacturers, operators, researchers and academics
- Dubai will be exposed to new strengths and capabilities

SDT Operations Policies/Legislation

- There exists a clear and significant gap in policies and legislation related to the revenue based operation of self-driving vehicles
- Dubai should lead the world in developing these policies and legislation to allow full operation of self-driving vehicles on the public streets in Dubai

8

RTA
Dubai World Challenge for Self-Driving Transport
www.sdchallenge.com
تحدي دبي العالمي
للتنقل ذاتي القيادة
DUBAI WORLD CHALLENGE
FOR SELF-DRIVING TRANSPORT
2018 | 2019
9

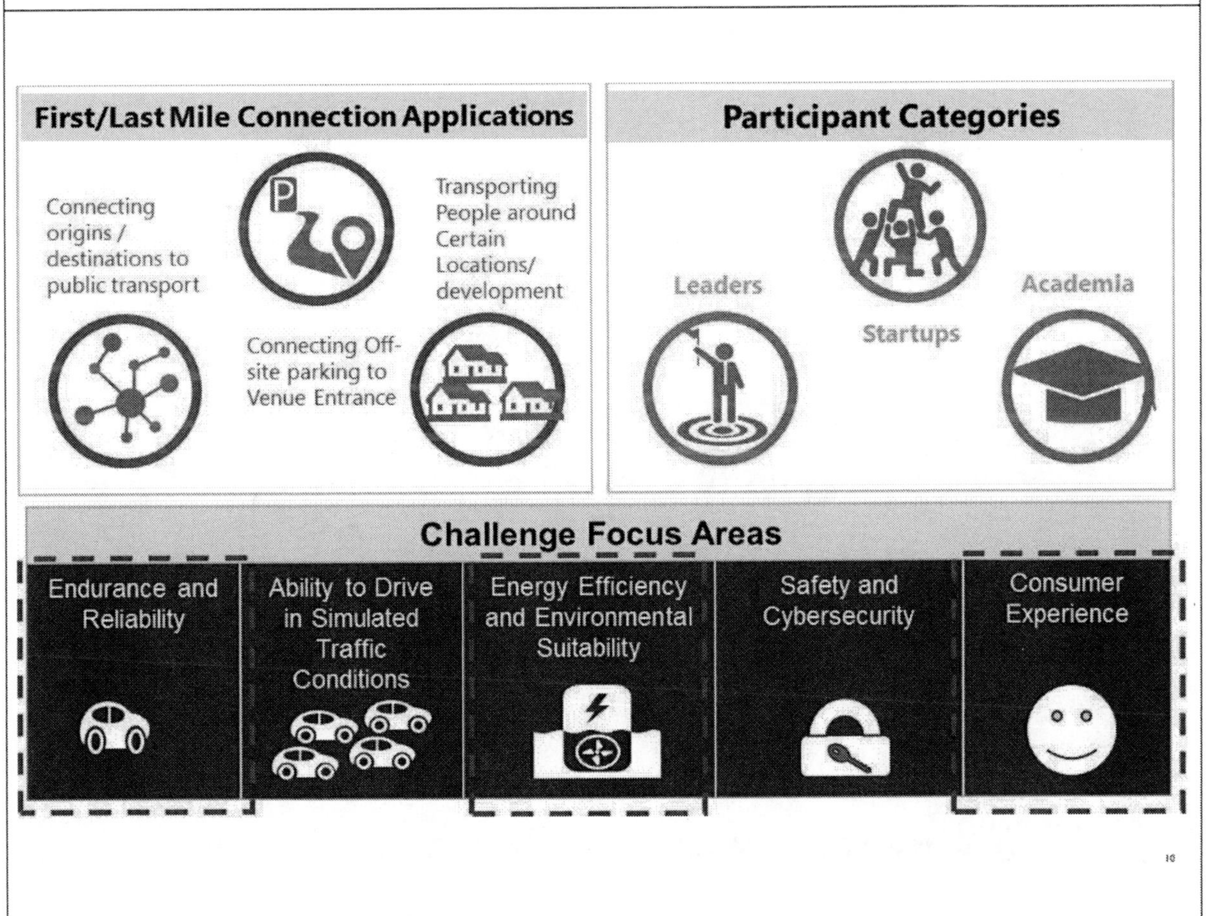
First/Last Mile Connection Applications
Connecting origins / destinations to public transport
Transporting People around Certain Locations/ development
Connecting Off-site parking to Venue Entrance
Participant Categories
Leaders
Startups
Academia
Challenge Focus Areas
Endurance and Reliability
Ability to Drive in Simulated Traffic Conditions
Energy Efficiency and Environmental Suitability
Safety and Cybersecurity
Consumer Experience
10

Dubai Challenge Categories and Awards

Leaders

- **US$ 1 Million** / Three finalists $1M each for different streams /
- **Pilot operation for the 1st place** --Exclusive

Startups

- **US$ 1 M for the 1st place and Golden Ticket Dubai Future Accelerator**
- **US$ 500,000 for the runner-up**

Academia

- Local: **1st Place US$ 200,000 and runner-up US$ 100,000**
- International: **1st Place US$ 200,000 and runner-up US$ 100,000**

11

Dubai Challenge Timeline for Application

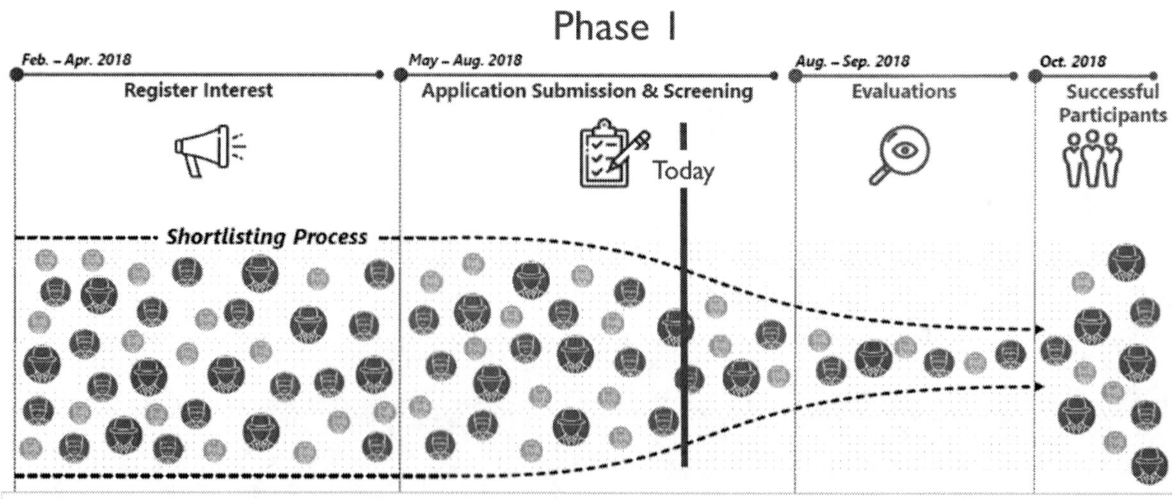

Awards at "Dubai World Congress for Self-driving Transport"

12

Site for First/Last Mile Connection

13

Dubai SDT.. The Future

Self-driving Transport is a reality in Dubai….More to come in the Future

THANK YOU

Email: Ismail.Zohdy@rta.ae

13

U.S. Department of Transportation
Office of the Under Secretary

U.S. DOT Automated Vehicle Research Activities

Finch Fulton, Deputy Assistant Secretary for Transportation Policy
Kevin Dopart, ITS Joint Program Office
July 12, 2018
Automated Vehicles Symposium

U.S. DOT and Automation

Facilitating a new era of transportation innovation and safety

Preparing for the Future of Transportation

U.S. DOT Automation Policy

- **Automated Driving Systems (ADS) 2.0**: A Vision for Safety (September 2017)
  - Clarifies Voluntary Safety Self-Assessment process
  - Emphasis on motor vehicle safety
- **2018 Update**
  - Multimodal, surface transportation
  - Broadens considerations to reflect multimodal responsibilities (e.g. operations)

More information on ADS 2.0 is available on the NHTSA website: https://www.nhtsa.gov/technology-innovation/automated-vehicles

4

Stakeholder Engagement

Informing action through dialogue

2018 Requests for Comment and Information Related to Automation

- FHWA, FTA, FMCSA, NHTSA, FRA, PHMSA each released RFIs/RFCs in early 2018
- Nearly 4,000 comments submitted
- Topics included:
  - Technology availability
  - Mode-specific barriers to integration of automation
  - Mode-specific considerations

6

RFI/RFC Key Themes – FHWA

- **Uniformity and Quality**
  - Greater uniformity and quality in road markings and traffic control devices would enable automation.
- **Leadership**
  - FHWA should take a leadership role in convening stakeholders to encourage collaboration.
- **Data**
  - Certain data elements around the roadway environment are useful for industry, State, and local DOTs to share and could improve automation operations.
- **Pilots**
  - Conducting pilots and supporting pilot testing are important for facilitating learning, collaboration, and information sharing.
- **Uncertainty**
  - Uncertainty in infrastructure investment and allocation of limited resources are key concerns for State and local agencies.

‹#› 7

RFI/RFC Key Themes – FTA

- **ADA and Accessibility**
  - Non-driving functions must be addressed
  - Automation could improve access for seniors and persons with disabilities
  - Incorporate universal design into research and development
- **Workforce and Labor**
  - Consideration for job loss and retraining
  - Could increase labor costs due to need for specialized skills
- **Funding and Procurement**
  - Buy America concerns
  - Useful life and spare ratio requirements questions

‹#› 8

RFI/RFC Key Themes – FMCSA

- **Consistency**
  - Consistent, performance-based standards and regulation are needed across states
- **Research**
  - Further research is need on distracted driving, medical qualifications, and hours of service requirements
- **Cybersecurity**
  - A few organizations mentioned cybersecurity issues

RFI/RFC Key Themes – PHMSA

- **Research**
  - Further research is needed to establish a baseline for safety before permitting AV transport of hazardous materials.
- **Education and Training**
  - Continued education and training are vital as automation technology evolves.
- **Regulatory Review**
  - PHMSA should maintain its current regulations, while it conducts a review to identify those that may require updating to address risks and opportunities presented by a future AV environment.

Public Listening Summit on Automated Vehicle Policy

- Stakeholder Discussion
  - 117 participants
- Public Listening Session
  - 1,111 participants (in-person and livestream)
- Themes
  - Public safety and first responders
  - Disability and accessibility
  - Consumer and public education
  - Insurance and liability
  - Employment issues
  - Cybersecurity

11

FHWA National Dialogue on Highway Automation

The Federal Highway Administration is initiating a national conversation with partners and stakeholders to better understand the implications of highway automation to facilitate innovation and inform the Agency's role in this area.

FHWA will use inputs from the National Dialogue to...

- **ASSESS** National issues and priorities
- **DEVELOP** technical guidance, best practices and standards
- **SUPPORT** necessary research
- **ADAPT** programs and policies
- **CREATE** a National community or coalition

NATIONAL DIALOGUE ON
HIGHWAY AUTOMATION

"Collaboration Corner" at Philadelphia National Dialogue event, June 26-27

FHWA National Dialogue Schedule

Month	Event	Location
June 7	National Dialogue Launch Workshop	Detroit, MI
June 26-27	National Workshop 1 Planning and Policy Considerations for Highway Automation	Philadelphia, PA
July 12	Automated Vehicle Symposium FMCSA-FHWA Truck Automation Listening Session	San Francisco, CA
August 1-2	National Workshop 2 Digital Infrastructure and Data Considerations for Highway Automation	Seattle, WA
Early September	National Workshop 3 Freight Considerations for Highway Automation	Chicago, IL
October 24-25	National Workshop 4 Operations Considerations for Highway Automation	Phoenix, AZ
Week of Nov. 12	National Workshop 5 Infrastructure Design and Safety Considerations for Highway Automation	Austin, TX

Additional Information: https://ops.fhwa.dot.gov/automationdialogue/

13

FMCSA Listening Sessions/Open Meetings on Vehicle Automation

- **Public Listening Session on Automation Technologies:** Held June 19 at University of Michigan
  - Participants included representatives from industry, law enforcement, commercial motor vehicle operators, academia, and technology providers
- **FMCSA-FHWA Truck Automation Listening Session:** Thursday, July 12 at 1:30, Grand Ballroom, Hilton Union Square
- **Motor Carrier Safety Advisory Committee (MCSAC) Meeting:** July 30-31 at USDOT Headquarters
- **FMCSA Truck Automation Listening Session**, August 24, 2018 at the Great American Trucking Show, Dallas, Texas

NHTSA Public Meetings and Listening Sessions

- Public Meeting On Removing Regulatory Barriers For Vehicles With Automated Driving Systems March 2018
- Public Listening Session on Automated Driving Systems 2.0: A Vision for Safety November 2017
- Public Meeting on Automated Driving Systems: Voluntary Safety Self-Assessments October 2017

Research Approach and Highlights

Exploring how to ensure safe, accessible, and efficient integration of automation

Consolidated Appropriations Act, 2018

- Signed into law on March 23, 2018
- Funds highly automated vehicle research and development
- Reallocates a total of $100 million for automation activities, including:
  - Up to $38 million for direct research
  - Up to $60 million for demonstration grants
  - Up to $1.5 million for analysis of impacts on drivers and operators of commercial motor vehicles, in consultation with Department of Labor
- Additional funding for NHTSA and OST
- https://www.congress.gov/bill/115th-congress/house-bill/1625

17

Accessible Transportation Technologies Research Initiative

Enabling the Complete Trip

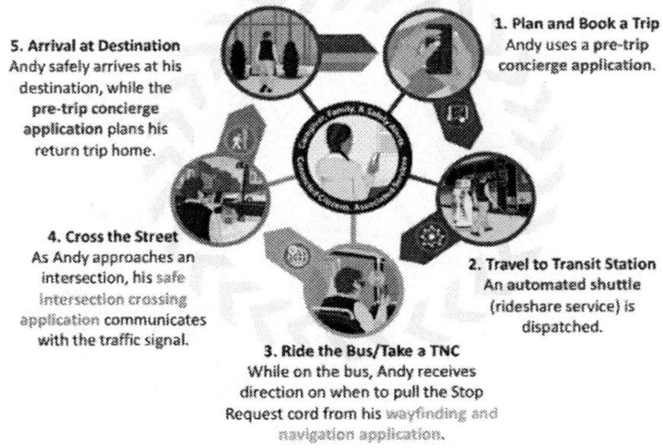

Current ATTRI Applications in Development

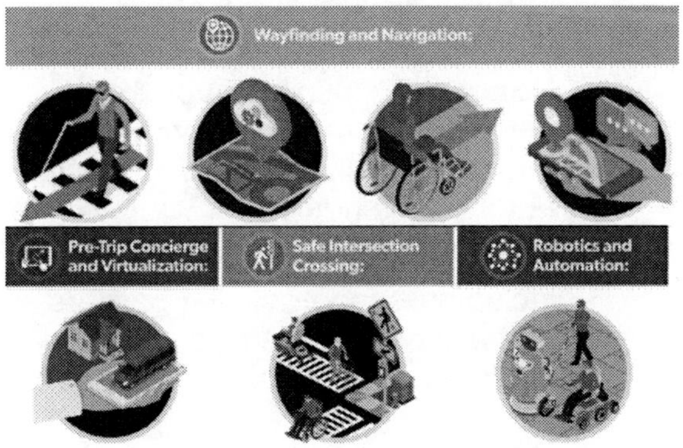

http://www.its.dot.gov/research_areas/attri/index.htm

Data For Automated Vehicle Integration

Voluntary data exchanges to accelerate the safe integration of AVs

WORK ZONE DATA EXCHANGE

- **Purpose**
  - Voluntary adoption of a basic work zone data spec
- **Near-Term Outcomes**
  - Active work zone data feed
  - Feed is used in a meaningful way
- **Big Picture Outcome**
  - Repeatable approach to harmonize local data

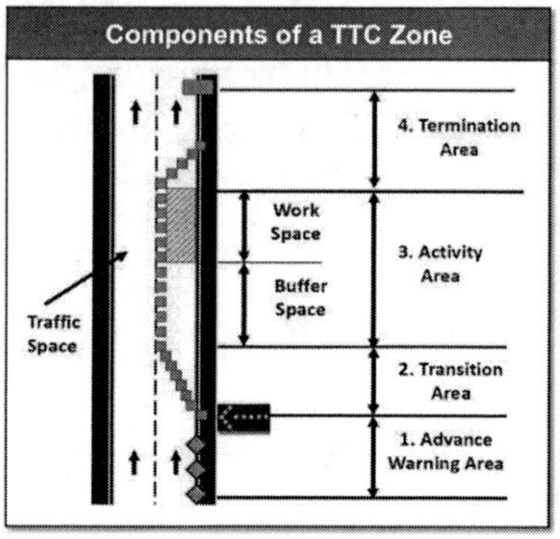

transportation.gov/av/data

Automated Low-Speed Shuttles

- **Monitor** development and deployment activity
- **Convene** working group of early deployer communities and other organizations interested in small, automated shuttles.
- **Document Emerging Findings** – including best practices, barriers, and lessons learned from early deployers and technology suppliers

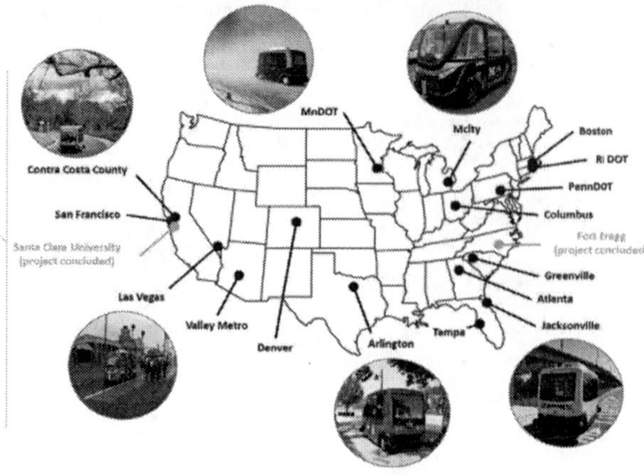

Working Group Member Communities

Impact Assessment

- Framework for AV Impacts
  - Mobility/energy/emissions modeling on a freeway
  - Safety baseline development
  - System dynamics of broader impacts
  - International collaboration on Key Performance Indicators
- Economic and Workforce Impacts from Automation
  - New project to analyze potential impacts from vehicle automation the macro economy including: GDP growth, employment/jobs, wages, consumption, exports, etc.

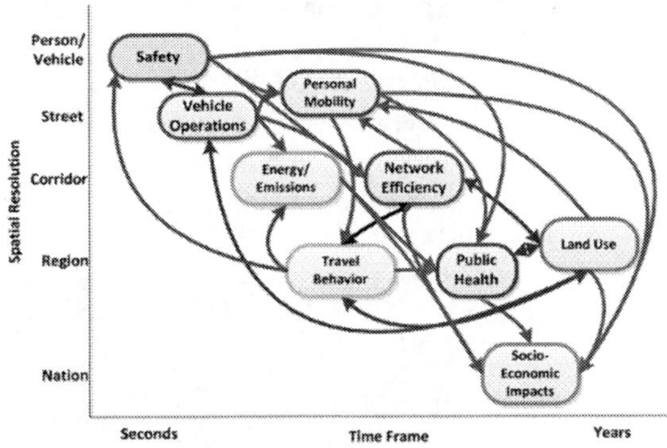

21

FHWA: Cooperative Automation Testing

Cooperative Automated Vehicle Testing with VDOT and Transurban

Objective: Investigate the combination of speed harmonization, vehicle platooning, and cooperative merging at an entrance ramp for a single-lane, managed facility with access limited to cooperative automated vehicles.

Findings: Completed 21 runs with 5 FHWA test vehicles that confirm earlier simulation results that throughput can realistically be increased by more than 50 percent with greater stability and reliability

.

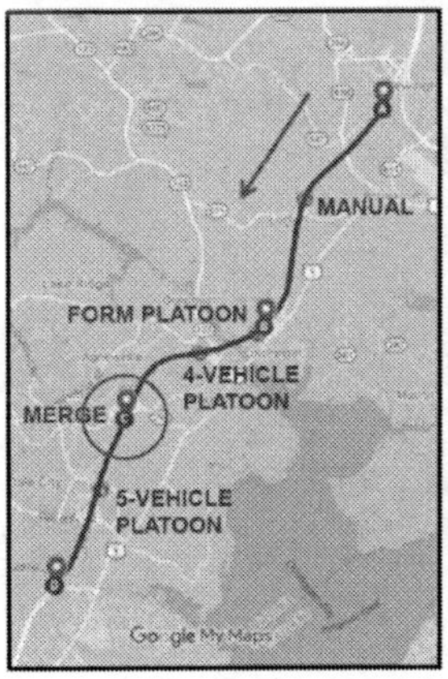

Google Maps Image of the Sites for the Cooperative Automated Vehicle Testing

FHWA: Scenario Planning for Connected and Automated Vehicles

Purpose: To equip agencies with information and tools to consider the uncertainties of CV/AV deployment

- Created scenarios of potential CV/AV deployment
- Conducted workshops to refine test scenarios
- Assessed overarching scenario impacts and implications

Outputs: Practitioner guidance/guidebook, qualitative assessment of impacts and policy implications arising from future described in scenarios

Scenario Building Blocks

FMCSA: Development of Baseline Safety Performance Measures for Automated CMVs

- **Overview: Create baseline safety performance measures** to evaluate automated commercial motor vehicle (CMV) safety systems through analysis of data from CMV naturalistic driving and driver/fleet management systems (100m hours, 33,000 collision events).

- **Goal:** Establish a baseline against which the safety performance of automated CMV functions can be evaluated.

FMCSA: Sensor Guidelines for Automated CMV Applications

- **Overview:**
  - Document input from vehicle automation companies and sensor manufacturers regarding best practices for adequate safe performance of relevant sensors (upkeep, optimum conditions).
  - Review and report on current sensor technologies that support automation and current calibration processes.
- **Goal:**
  - To conduct performance and maintenance research for add-on sensors that support various levels of automation on CMVs.

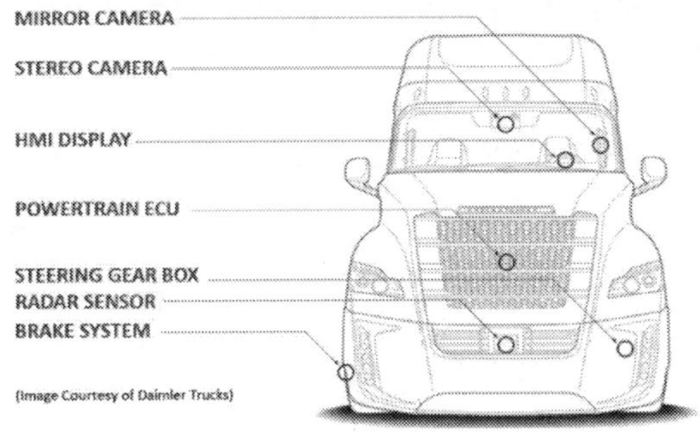

FRA: Automated Vehicles at Highway-Rail Grade Crossings

- The Federal Railroad Administration is researching the interaction between Automated Vehicles (AV) and rail crossings.
- The objective of this project is to explore what information AVs will need to negotiate Highway-Rail Intersections (HRI)
- Deliverables
  - Literature Review
  - Stakeholder Engagement
  - System Requirements

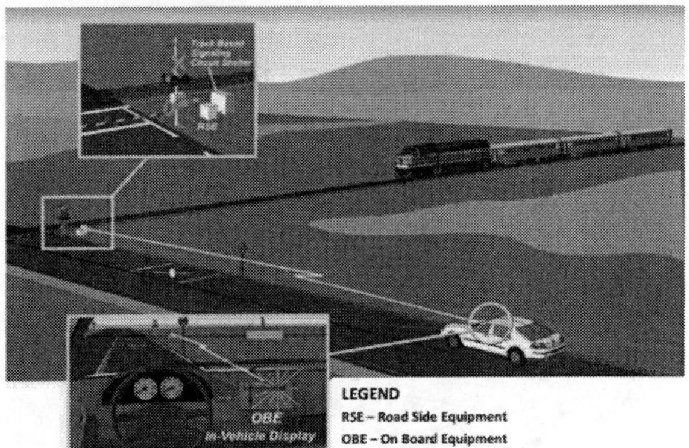

26

FTA: Strategic Transit Automation Research Plan

7 demonstrations planned through 2022

Research in progress:

- Transit Bus Applications of Light and Commercial Vehicle Automation Technology
- Automation Policy Review
- Test Facility Requirements for Automated Transit Buses
- User Acceptance and Human Factors for Automated Transit Buses

https://www.transit.dot.gov/automation-research

27

MARAD/FMCSA: Feasibility Study: Low-Speed Automated Truck Queue at Ports and Warehouses

- Exploring application of automation to low-speed commercial vehicle operations at port terminals and warehouses
- Review of related studies and papers
- Surveys of industry and technology stakeholders (currently underway)
- Technology scan of existing or near-term enabling technologies
- Regulatory changes, cost estimation and safety benefits
- Report expected in Fall 2018

NHTSA: Driving Automation Communication of Intent with Shared Road Users

- **Project Summary**: Drivers communicate with other road users using gestures, eye contact and other complex means. A similar need will likely emerge to communicate Automated Driving Systems' intentions to shared road users.
  - Describe social interaction strategies used by drivers when communicating to other road users
  - Conduct initial studies to test potential strategies for driving automation systems
- **Project Start**: 2017
- **End Products**: Report on types of communication needs and potential strategies for effectively communicating automated systems' intention to other road users
- **Status**: Data collection in process and will be completed in 2018

29

NHTSA: FMVSS Considerations for Vehicles with Automated Driving Systems

- **Goal**: Research and identify potential barriers for self-certification and compliance verification of innovative new vehicle designs precipitated by Automated Driving Systems (ADSs)
- Developed four conceptual vehicle designs using generic features and characteristics expected in ADS-equipped vehicles with no human-operated controls
- Significant progress in translation of regulatory text has been made
- **Next step** is testing FMVSS test procedures for vehicles without traditional human-operated controls to assess objectivity and repeatability
- **Contractor**: VTTI

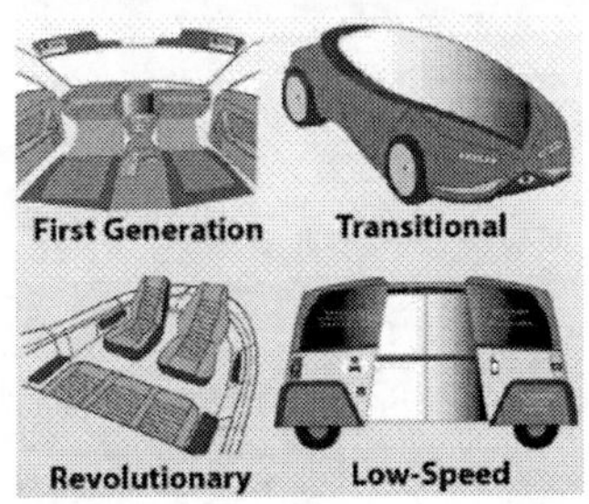

Source: VTTI

30

PHMSA: Carriage of Hazardous Materials by Automated Vehicles

- **Overview:**
  - The Pipeline and Hazardous Materials Safety Administration is researching the potential for applying automation for hazardous materials transportation.
- **Objective:**
  - To identify, characterize, and quantify risks associated with pairing automation technologies and hazardous materials transport.

transportation.gov/av

AUTHOR INDEX

Association for Unmanned Vehicle Systems International
2700 South Quincy Street
Suite 400
Arlington, VA 22206

ISBN 978-1-5108-6840-3

CURRAN ASSOCIATES INC.
proceedings.com

3rd International School and Conference on Optoelectronics, Photonics, Engineering and Nanostructures (Saint Petersburg OPEN 2016)

Journal of Physics: Conference Series Volume 741

St. Petersburg, Russia
28-30 March 2016

Part 1 of 2

IOP
Institute of Physics